Engineering Materials

This series provides topical information on innovative, structural and functional materials and composites with applications in optical, electrical, mechanical, civil, aeronautical, medical, bio- and nano-engineering. The individual volumes are complete, comprehensive monographs covering the structure, properties, manufacturing process and applications of these materials. This multidisciplinary series is devoted to professionals, students and all those interested in the latest developments in the Materials Science field, that look for a carefully selected collection of high quality review articles on their respective field of expertise.

Indexed at Compendex (2021) and Scopus (2022)

Qing Xie
Editor

Electrical Materials

Performance Improvement, Recent Advances and Engineering Applications

Springer

Editor
Qing Xie
Department of Electric Power Engineering
North China Electric Power University
Baoding, China

With Contributions by
Peng Wang
School of Energy, Power and Mechanical Engineering
North China Electric Power University
Baoding, China

Jun Xie
Department of Electrical Engineering
North China Electric Power University
Baoding, China

Yiyi Zhang
School of Electrical Engineering
Guanxi University China
Guangxi, China

ISSN 1612-1317 ISSN 1868-1212 (electronic)
Engineering Materials
ISBN 978-981-99-9049-8 ISBN 978-981-99-9050-4 (eBook)
https://doi.org/10.1007/978-981-99-9050-4

This Springer imprint is published by the registered company Springer Nature Singapore Pte Ltd.
The registered company address is: 152 Beach Road, #21-01/04 Gateway East, Singapore 189721, Singapore

Paper in this product is recyclable.

Contents

Study on the Construction Method of Epoxy Resin Crosslinking Model

Kexin Fu, Qing Xie, Jingxuan Song, and Guowei Xia

1 Introduction

The application of molecular dynamics (MD) simulation method in the field of polymers has developed rapidly and has become a mature research tool [1]. Due to the complex structure of epoxy cured material, which is different from linear polymers, it is a three-dimensional network structure formed by the reaction of epoxy oligomers with curing agents [2]. Therefore, how to obtain an accurate molecular model of epoxy resin crosslinking network is a key problem in the application of MD simulation method in epoxy resin research.

In order to obtain an accurate epoxy resin crosslinking network model, the development of its construction method has gone through two main stages: one-step crosslinking method and iterative crosslinking method. The one-step crosslinking method is to crosslink the reactive sites within a certain cut-off distance to form an epoxy resin crosslinking network [3, 4]. It is characterized by a small number of model molecules and low network conversion rate or crosslinking density. The iterative crosslinking method, in which a system equilibrium is performed after the formation of a crosslinked bond, followed by a continuation of the crosslinking reaction until the end of the crosslinking, is characterized by a high network conversion rate or crosslink density [5–7]. Compared with the one-step crosslinking method, the iterative crosslinking method is more in line with the actual polymer model and has received more attention from scholars. Wu and Xu [5] used "active" molecular fragments of Bisphenol A diglycidyl ether (DGEBA) and IPD for modelling. When the truncation distance was 4–10 Å and the distance between the two reaction sites was minimum, a crosslinking bond was formed between the two reaction sites. After the formation of the crosslinking bond, energy minimization and MD simulation are performed to eliminate the unreasonable structure and interaction caused by the

K. Fu · Q. Xie (✉) · J. Song · G. Xia
Department of Electrical Engineering, North China Electric Power University, Baoding, China
e-mail: xq_ncepu@126.com

Q. Xie (ed.), *Electrical Materials*, Engineering Materials,
https://doi.org/10.1007/978-981-99-9050-4_1

formation of new construction. Then, the above process is repeated until the conversion is within the experimental value. This method can achieve a high conversion rate, but the calculation time is long. Varshney et al. [6] and Wu et al. [5] used "active" molecular fragments for modelling. Within a certain truncation distance, all potential non-crosslinked reaction pairs were identified and crosslinked bonds were formed. Next, a multi-step relaxation process is used to relax the new topology generated by the introduction of new bonds, angles, dihedral angles, etc. Then the bonding and relaxation processes are repeated until the crosslinking reaction is completed. This method can effectively improve the computational efficiency.

To further improve computational efficiency and avoid manual operation errors, more and more researchers are using computer modelling with the help of scripting programs. Xin et al. [8] wrote an epoxy resin crosslinking reaction program in Perl language based on the iterative crosslinking method and compared it with experimental data to verify the validity of the model. The algorithmic flow of the crosslinking procedure is shown in Fig. 1. Liu [9] wrote an epoxy resin crosslinking reaction program in Perl language, refined the iterative process and investigated the structure–property relationship of epoxy resin.

In recent years, the iterative crosslinking method based on the use of scripting procedures for epoxy crosslinking calculations has been widely used, but the constructed epoxy resin crosslinking systems are all epoxy-amine curing systems. However, the epoxy resin systems for high voltage insulation used in electrical insulation are epoxy-anhydride curing systems with different cross-linking mechanisms. The rapid acquisition of an accurate and realistic epoxy-anhydride crosslinking model is the basis for the study of existing resins and the development of new resins.

2 Design Principle of Epoxy Resin Crosslinking Model Based on Iterative Crosslinking Method

2.1 Reaction Mechanism of Anhydride Curing Epoxy Resin

The main reactions of anhydride with epoxy resin in the absence of accelerator are as follows, as shown in Fig. 2: (a) The hydroxyl group in the epoxy resin first opens the anhydride ring to form a carboxylic acid with an ester chain; (b) The carboxylic acid reacts with the epoxy group in the epoxy resin to form a secondary hydroxyl group; (c) The generated hydroxyl group reacts with another anhydride group, while reaction (d) takes place simultaneously; (e) The generated secondary hydroxyl group reacts with another epoxy group, ultimately forming a three-dimensional crosslinked network of epoxy resin.

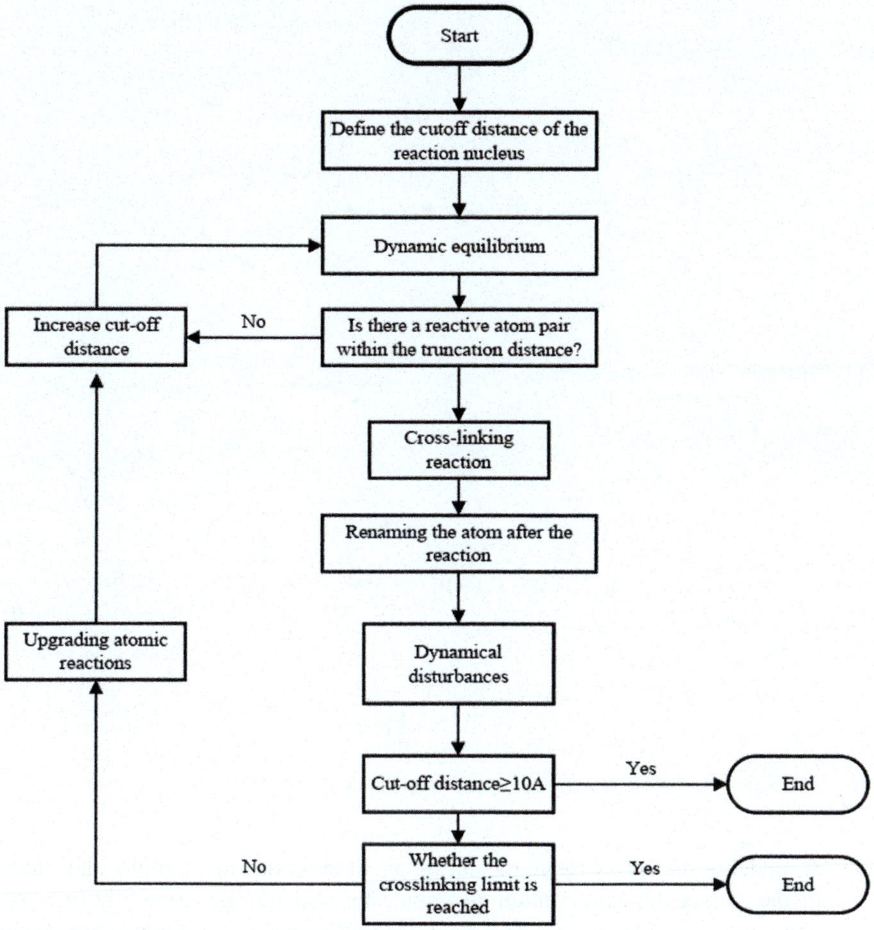

Fig. 1 Flowchart of a perl-based scripting program [10]

2.2 *Design Principles of Epoxy Resin Crosslinking Model*

The mechanism of the epoxy resin anhydride curing reaction shows that the crosslinking reaction is the alternating ring-opening polymerization of curing agent and epoxy compound [2]. It has also been shown that the homopolymerization of epoxy groups can also occur in the presence of catalysts or excessive epoxy at high temperatures [10]. The curing temperature of epoxy resin for high voltage insulation is below 473.15 K, and the epoxy resin is not excessive. The curing reaction mechanism of epoxy anhydride does not include homopolymerization. Therefore, in this paper, the construction of an epoxy-anhydride crosslinking system was implemented using a Perl-based scripting program based on the reaction mechanism of anhydride-cured epoxy resins (Fig. 2). First, the pre-crosslinking model is processed:

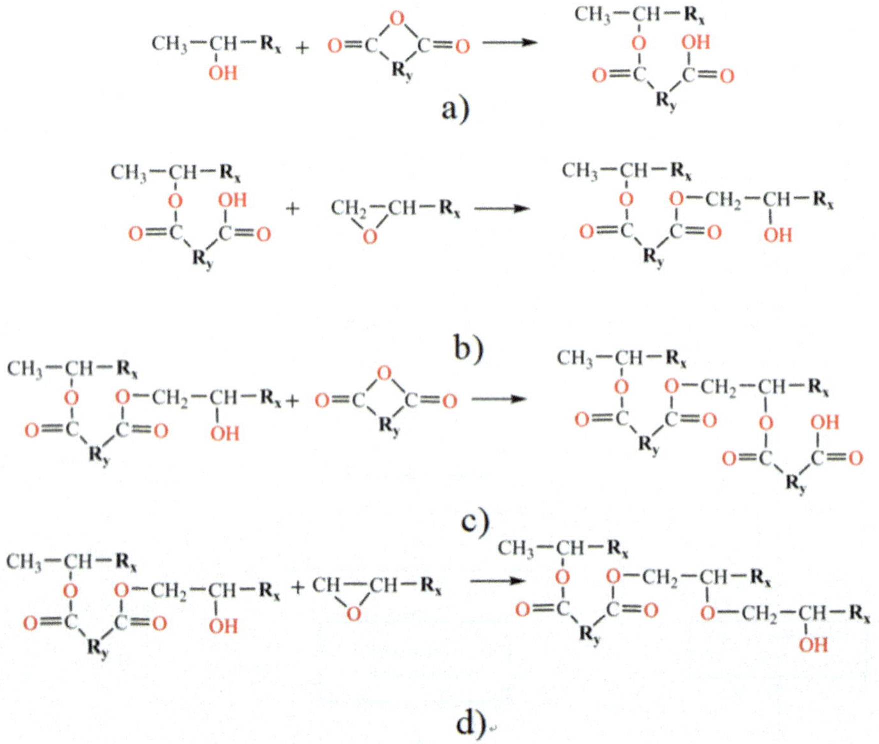

Fig. 2 Reaction mechanism diagram of epoxy resin cured by anhydride

(1) A model of the epoxy resin, the curing agent monomer and a molecular model of the carboxylic acid-containing monoester structure generated by the ring-opening reaction of the epoxy resin with the curing agent (as shown in Fig. 2a and hereinafter referred to as the primary crosslinked structure) were constructed. Labelling the epoxy resin with the C atom in CH_2 in the epoxy group and one C atom in the anhydride group of the primary crosslinked structure as reaction atom R_1 and the O atom in –OH in the carboxyl group of the primary crosslinked structure as reaction atom R_2.
(2) The molecular model of the monomer and primary cross-linked structure constructed in (1) is used to construct an amorphous model of the epoxy-anhydride system at a specific value of the molar ratio of the epoxy group to the anhydride group, which is ready for use after equilibration in MD. Then, in order to simplify the script, the following design was made in the script program while ensuring the accuracy of the epoxy-anhydride crosslinking model: labelling the O atom in the carboxyl and hydroxyl groups generated by the reaction as the reaction atom R_2, with R_1 and R_2 being the reaction atom pair; Suppose that each reactive group has the same activity; The reaction is diffusion controlled and

each reaction is carried out simultaneously. Finally, the automatic crosslinking program flow chart is shown in Fig. 3.

The flowchart is described as follows:

(1) For the amorphous model of the epoxy-anhydride system with labelled reaction atoms, first set the cross-linking temperature *T*, the target crosslinking degree *TCD*, and the range of cut-off radius for the formation of crosslinked bonds.
(2) The radius of the current stage is R_n, the maximum truncation radius is R_{max}, and the distance between the reaction atom pairs is calculated as d. If $d < R_n$, the atomic pair generates a new crosslinked bond according to the reaction principle, followed by a 50 ps NPT molecular dynamics equilibrium to ensure the rationality of the new structure; If $d > R_n$ then proceed to step (3);

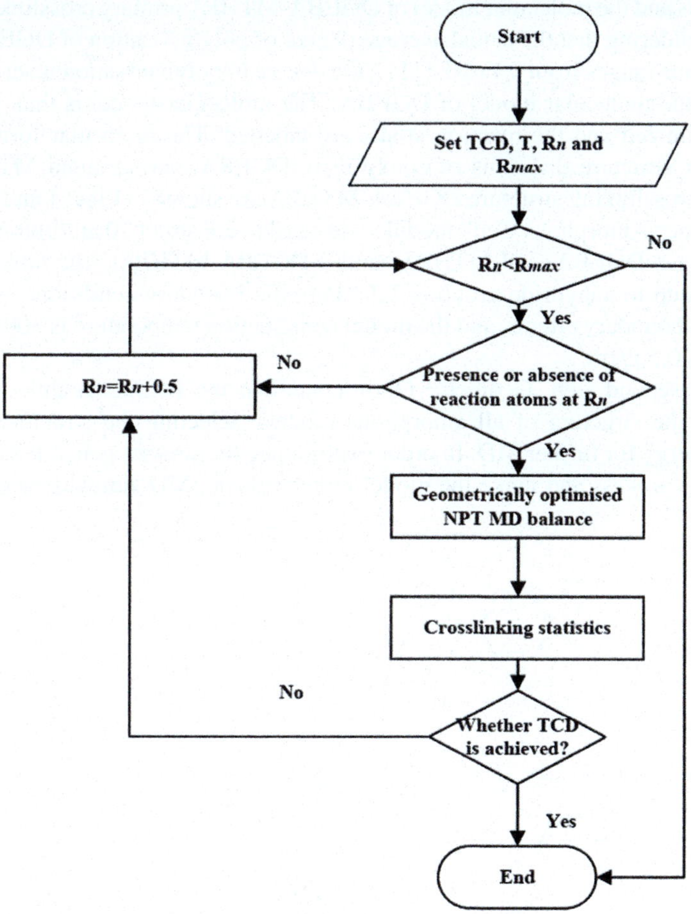

Fig. 3 Flow chart of automatic crosslinking reaction program

(3) Set $R_n = R_n + 0.5$ Å and repeat step (2). Until R_n is greater than R_{max} or the target crosslinking degree is reached, the crosslinking process is completed.

3 Epoxy-Anhydride Crosslinking Model Construction

The crosslinking network model of DGEBA/methyl-tetrahydrophthalic anhydride (MTHPA) was constructed, which can be decomposed into two main stages: (1) Constructing an amorphous model; (2) Based on the mechanism of anhydride curing reaction, the formation of DGEBA/MTHPA crosslinking network was realized by using the automatic crosslinking program written in Perl language.

In the first step, the core module "Materials Visualizer" in "Materials Studio 7.0" software is used to construct the molecular models of DGEBA and MTHPA monomers and the molecular model of DGEBA-MTHPA primary crosslinking structure. Considering that the actual average degree of polymerisation of DGEBA in the experiments ranges from 0.1 to 0.2 [11], the degree of polymerisation is set to 0 when building the molecular model of DGEBA. The molecular model is then geometrically optimized and the reaction atoms are labelled. The molecular formulas and molecular structure diagrams of epoxy resin DGEBA, curing agent MTHPA and primary crosslinking structure DGEBA-MTHPA are shown in Figs. 4 and 5. Subsequently, the "Amorphous Cell" module was used to construct 10 amorphous models containing 8 DGEBAs, 18 MTHPAs and 2 DGEBA-MTHPAs (the molar ratio of epoxy group to anhydride group is 1:1). Periodic boundary conditions are used to eliminate boundary effects, and the model construction temperature is 300 K and the density is 0.6 g/cm^3.

In the second step, Geometry Optimization in the Forcite module is used to optimize the structure of all amorphous models, selecting the structure with the lowest energy for further MD. In order to eliminate the stresses generated during the modelling process and make the model more realistic, MD simulations of NVT at

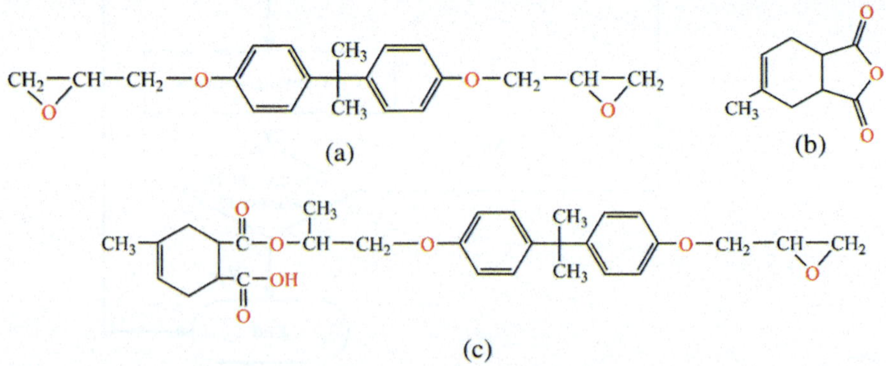

Fig. 4 Molecular formula of epoxy resin, curing agent and primary crosslinking structure: **a** DGEBA, **b** MTHPA, **c** DGEBA-MTHPA Primary crosslinked structure

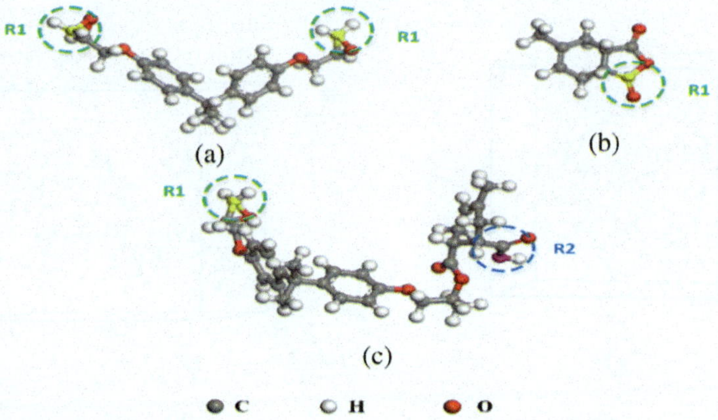

Fig. 5 Molecular model of epoxy resin, curing agent and primary crosslinked structure: **a** DGEBA, **b** MTHPA, **c** DGEBA-MTHPA primary crosslinked structure

100 ps and NPT at 200 ps are performed sequentially, with the temperature of 300 K and the pressure of 1 atmosphere. The COMPASS force field is selected for dynamic process, Andersen [12] and Berendsen [13] methods are used for temperature and pressure control respectively, and the van der Waals and electrostatic interaction calculation methods are Atom-based and Ewald methods respectively.

The third step is to embed the general program script of epoxy resin/anhydride curing agent automatic crosslinking reaction in "Materials Studio 7.0", which is compiled according to the anhydride curing reaction mechanism and the design principle of epoxy-anhydride system crosslinking method. Set the target cross-linking degree (taking 89% crosslinking degree as an example), set the temperature at 300 K, and set the truncation radius and maximum truncation radius at 3.5 Å and 7.5 Å respectively. The epoxy-anhydride crosslinking model at the target crosslinking degree can be obtained by running the crosslinking program. The molecular structures of the DGEBA/MTHPA amorphous and crosslinked models are shown in Fig. 6. Taking the crosslinking degree of 89% as an example, the basic parameters of the system before and after crosslinking are shown in Table 1.

In the fourth step, after optimizing the structure of the obtained epoxy-anhydride crosslinking model, MD simulation of NVT and NPT ensemble was carried out successively at 600 K, and the simulation parameters are consistent with the previous paper. Finally, the 600 K crosslinked-anhydride model is "annealed" to cool it down: The crosslinked models are simulated at 600 K for 100 ps NVT and 200 ps NPT respectively and are heated up to 600 K and then cooled down by "annealing" from 600 to 300 K, with a temperature interval of 20 K and a kinetic equilibrium time of 100 ps at each temperature. Then the crosslinked structure models of epoxy-anhydride at different temperatures are output for the subsequent calculation of structure and performance parameters.

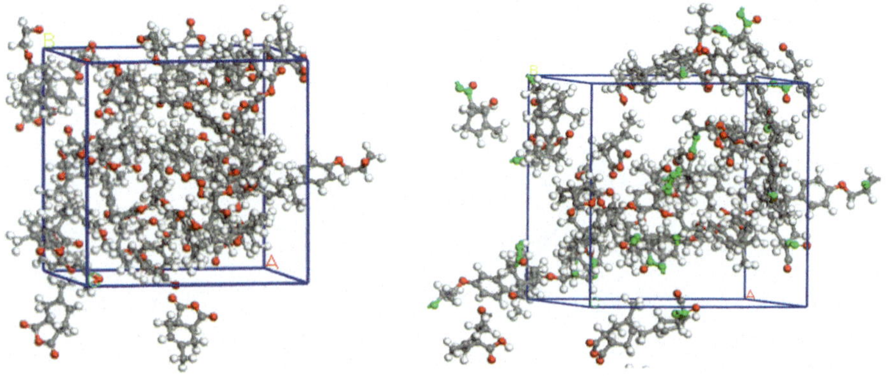

(a) (b)

Fig. 6 DGEBA/MTHPA molecular model: **a** Amorphous model, **b** crosslinked model

Table 1 Basic parameters of the DGEBA/MTHPA(DM) system before and after crosslinking

System	Number of monomer molecules			Amorphous models		Crosslinked model (89%)		System shrinkage
				Volume	Density	Volume	Density	
DM	DGEBA	MTHPA	DGEBA-MTHPA	Å^3	g/cm^3	Å^3	g/cm^3	%
	8	18	2	9782.61	1.143	9693.98	1.153	0.91%

4 Epoxy-Anhydride Crosslinking Model Applications

4.1 *Preferred Epoxy Resin Monomer for High Voltage Insulation*

There are many types of epoxy resins and curing agents, and the curing products of different epoxy resins and curing agents have complex structures and vary greatly in thermodynamic properties. Therefore, the selection and matching of epoxy resin and curing agent base materials, as well as the correlation between the microstructure of epoxy materials and their macroscopic properties, are of great research significance. In this section, DGEBA, dicyclopentadienyl alicyclic epoxy resin (DCPDE), MTHPA and maleic anhydride (MA) anhydride curing agents are selected as monomers to construct crosslinking structure models of four epoxy resin systems: DGEBA/MTHPA, DGEBA/MA, DCPDE/MTHPA and DCPDE/MA. The effect of the crosslinking degree on the microstructure and thermodynamic properties of epoxy has been investigated by MD simulations. Multiple parameters of the four epoxy crosslinking systems are compared, and the correlation between the microstructure and thermodynamic properties of the epoxy resin substrates is

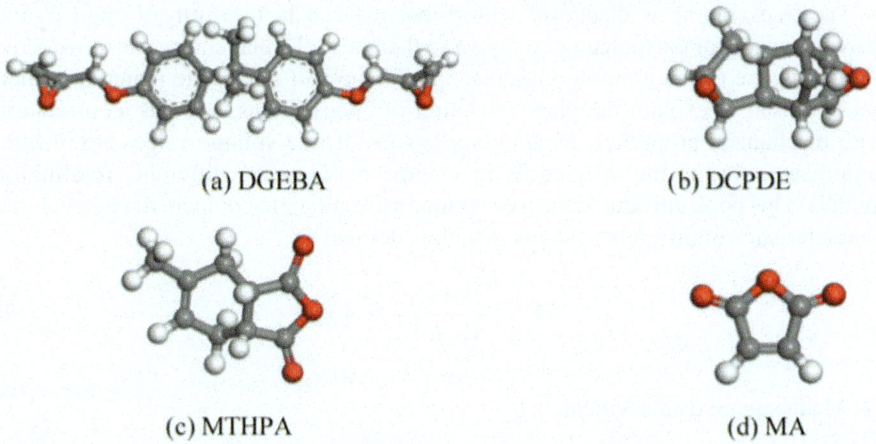

Fig. 7 Molecular structure of epoxy resins and curing agents

constructed. The molecular structures of the epoxy resins DGEBA, DCPDE and the anhydride curing agents MTHPA and MA are given in Fig. 7 (Table 2).

The microstructure of a material is intrinsically important in influencing its performance. Analyzing structural parameters and predicting macroscopic performance is crucial for the optimization of epoxy resins and curing agents. Therefore, in this section, relevant parameters such as free volume, mean square displacement, glass transition temperature, coefficient of thermal expansion and modulus are calculated for the four epoxy resin systems.

The basic calculations used to characterize the structural and performance parameters are based on the following.

(1) Free volume ratio

According to the free volume theory, the volume V_T of liquid and solid substances consists of two parts: one is the volume V_0 occupied by molecules, and the other is the unoccupied volume V_{f}, namely the free volume. That is:

$$V_T = V_0 + V_f \tag{1}$$

Table 2 Basic parameters of the 4-group crosslinking model

Models	Volume (Å^3)	Density (g/cm^3)	Number of atoms
DGEBA/MTHPA	47,985.63	1.17	4726
DGEBA/MA	35,213.85	1.27	2137
DCPDE/MA	22,575.40	1.33	3410
DCPDE/MTHPA	34,853.13	1.18	3360

The free volume is dispersed within the material in the form of "pores" and provides space for the molecules to move so that the molecular chains can move. Free volume is one of the important microscopic features of polymeric materials which is often used to explain the glass transition of epoxy resins, and has a correlation with mechanical properties. Direct comparisons of free volume values are of little significance due to the differences in volume between the different crosslinking models. This book introduces the free volume ratio (FFV) to compare the free volume characteristics of different systems, which is defined as:

$$FFV = \frac{V_f}{V_0 + V_f} \times 100\% \tag{2}$$

(2) Mean square displacement

During the MD process, the position of the molecule in the model is constantly changing and the mean square displacement (MSD) of the molecule characterizes the ability of the molecule to move in the model. The mean square displacement of a molecule is defined as:

$$MSD = \left\langle |R(t) - R(0)|^2 \right\rangle \tag{3}$$

where $R(t)$ and $R(0)$ respectively represent the displacement vector of any atom i in the system at time t and at the initial moment, respectively. For a system with N atoms, the mean square displacement can be expressed as:

$$MSD = \sum_{i=0}^{N-1} \left[|\boldsymbol{R}_i(t) - \boldsymbol{R}_i(0)|^2 \right] \tag{4}$$

(3) Glass transition temperature

Glass transition is a transition that can occur in both amorphous polymers and amorphous regions of crystalline polymers. According to the free volume theory, the glass transition temperature (T_g) is a transition temperature at which the polymer chain segments transition from freezing to moving, which macroscopically shows that the material changes from glass state to rubber state and loses its high stiffness. Therefore, T_g is the highest temperature of epoxy resin as an insulating support material. In general, an increase in temperature will lead to a decrease in the density of the material, and the curve of polymer density with temperature will show an inflection point at the T_g temperature value. Therefore, the information such as density and temperature of the model under each crosslinking degree during the "annealing" cooling process can be extracted and the T_g of epoxy resin can be predicted by the temperature corresponding to the inflection point of the density curve with temperature. The specific method is shown in Fig. 8, where linear fitting is performed on the

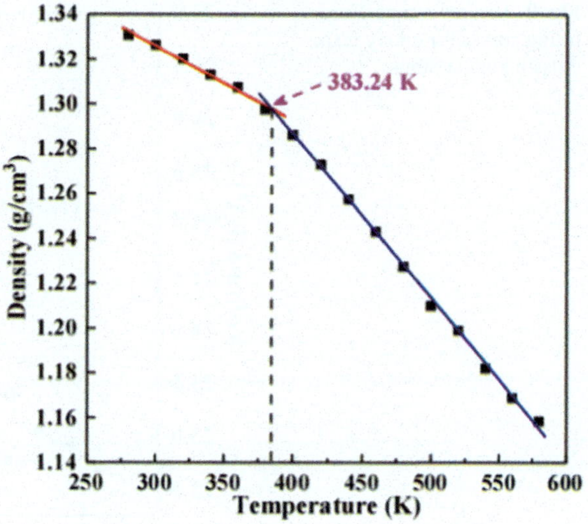

Fig. 8 Density-temperature fitting curves for epoxy resin crosslinking systems

density values on both sides of the inflection point, and the intersection of the two lines is T_g.

(4) Coefficient of thermal expansion

The Coefficient of thermal expansion (CTE) is one of the main physical properties of a material and can be used to calculate the internal stress due to thermal expansion, and is an important indicator of the thermal stability of a material. Its calculation formula is:

$$CTE = \frac{1}{V_0}\left(\frac{\partial V}{\partial T}\right)_P \tag{5}$$

where V_0 is the volume of the initial system, here taken as the volume of the system at 300 K, and P is taken as one standard atmospheric pressure. Epoxy resins correspond to different states (glassy and rubbery) before and after the glass transition and their CTE can change abruptly. A linear fit to the volume-temperature data before and after the glass transition, respectively, gives the rate of change of volume with temperature in both states. As shown in Fig. 9.

(5) *Modulus*

MD simulations enable the static normal strain method to be used to calculate the mechanical properties of crosslinked systems, that is, to analyze the stress and strain of a three-dimensional epoxy system with slight deformation, and then obtain its mechanical properties related parameters such as bulk modulus, shear modulus, Young's modulus, Poisson's ratio, etc. The process can be described as follows: a small strain is applied to the system, causing the model to undergo uniaxial tensile

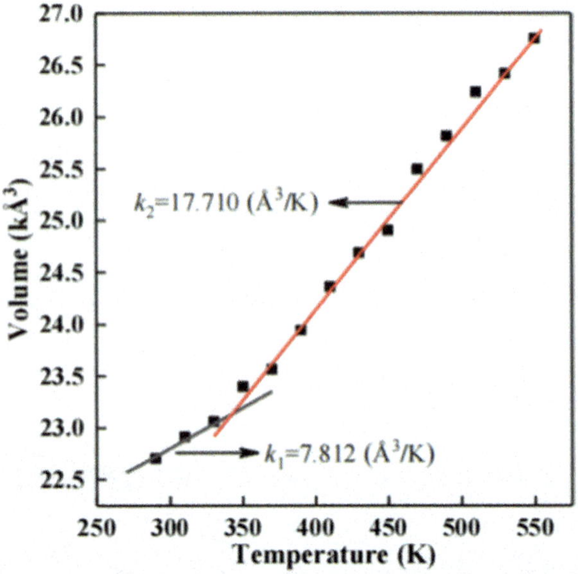

Fig. 9 Volume-temperature fitting curves for epoxy resin crosslinking systems

and compressive deformation along the x, y, and z axes, respectively, i.e., shear deformations in the xy, xz, and yz planes, respectively, and the stiffness matrix C_{ij} can be obtained from the model's response to the strain, which can be described as follows for isotropic materials.

$$C_{ij} = \begin{bmatrix} \lambda + 2\mu & \lambda & \lambda & 0\ 0\ 0 \\ \lambda & \lambda + 2\mu & \lambda & 0\ 0\ 0 \\ \lambda & \lambda & \lambda + 2\mu & 0\ 0\ 0 \\ 0 & 0 & 0 & \mu\ 0\ 0 \\ 0 & 0 & 0 & 0\ \mu\ 0 \\ 0 & 0 & 0 & 0\ 0\ \mu \end{bmatrix} \tag{6}$$

where λ and μ are elastic constants which can be derived from the relevant elements in the stiffness matrix.

$$\begin{cases} \lambda = \frac{1}{6}(C_{12} + C_{13} + C_{21} + C_{23} + C_{31} + C_{32}) \\ \mu = \frac{1}{3}(C_{44} + C_{55} + C_{66}) \end{cases} \tag{7}$$

The bulk modulus K, Young's modulus E, shear modulus G and other mechanical properties of epoxy system can be obtained from elastic constants λ and μ. The calculation methods are as follows:

$$\begin{cases} E = \mu \dfrac{3\lambda + 2\mu}{\lambda + \mu} \\ G = \mu \\ K = \lambda + \dfrac{2}{3}\mu \end{cases} \tag{8}$$

The results of the structural properties of the four groups of epoxy resin systems involved in the calculations are shown in Fig. 10.

The thermodynamic properties of the four groups of epoxy crosslinking models are different. The T_g of the four systems with 90% crosslinking is DGEBA/MA > DCPDE/MTHPA > DCPDE/MA > DGEBA/MTHPA in descending order; the CTE in the glassy state is DCPDE/MA < DGEBA/MA < DCPDE/MTHPA < DGEBA/MTHPA. The DGEBA/MTHPA system has low modulus and the worst mechanical properties; The DCPDE/MTHPA system is the most resistant to shear stresses and the DCPDE/MA system is the best in terms of tensile and compressive stresses. In addition, there is a certain correlation between the microscopic and thermodynamic properties of the epoxy resin base material. There is a positive correlation between FFV and the chain segment kinematic properties of epoxy resins; generally lower FFV and weaker chain segment kinematic models have better thermodynamic properties.

4.2 *Study of the Effect of Polymerisation on the Structural Properties of Epoxy Resins*

DGEBA is usually a mixture of homologues with different molecular sizes (chain length, molecular weight and polymerization degree). However, most of the current simulation studies have focused on the case where the polymerization degree of DGEBA is 0 and the effect of polymerization degree on the performance of crosslinked epoxy resins has been rarely studied. Previous experimental studies have confirmed that different molecular weights do have an effect on the properties of the cured product. Therefre, it is particularly important to investigate the effects of different polymerization degrees on the structure and properties of epoxy resins at the simulation level.

In this section, DGEBA, a bisphenol A epoxy resin with a polymerization degree of 1 and 2, is used as the base material for the epoxy resin, and MTHPA is used as the curing agent (the molecular structure of the monomer is shown in Fig. 11). The epoxy resins are mixed in different proportions to study the effect of different polymerization levels on the performance of the crosslinked epoxy resin system.

To investigate the effect of polymerization on the performance of crosslinked epoxy resin systems, four types of ratios of 8:2, 6:4, 5:5 and 3:7 are determined for DGEBA (n = 1) and DGEBA (n = 2). Combining the actual epoxy resin with the reaction mass ratio of the curing agent, the molar ratio of the epoxy resin mixed

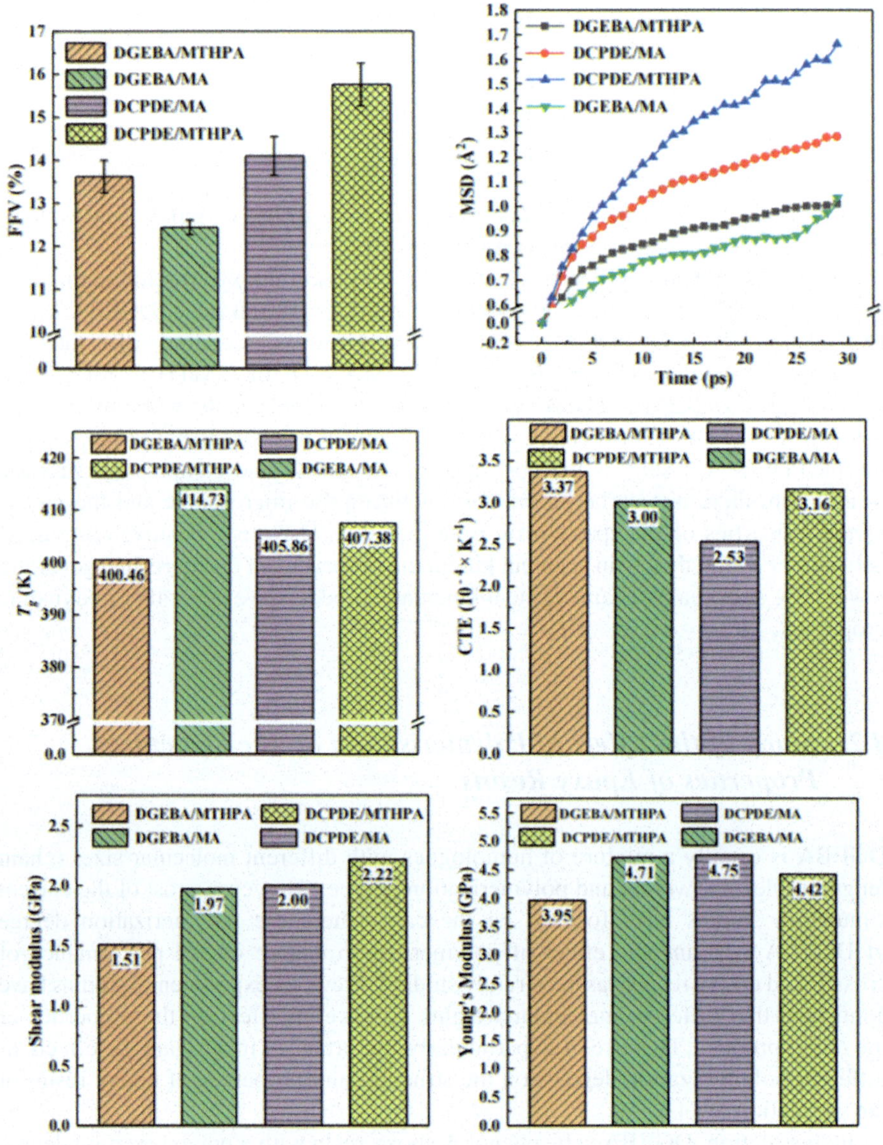

Fig. 10 Properties of four epoxy-crosslinked systems: *FFV*, *MSD*, T_g, *CTE* and modulus

with the anhydride curing agent in this section is epoxy resin: curing agent = 1:2. Therefore, the total number of DGEBA molecules is selected as 30 and MTHPA as 60. And the following molecular models are constructed respectively: (1) DGEBA (n = 1):DGEBA (n = 2):MTHPA = 24:6:60; (2) DGEBA (n = 1):DGEBA (n = 2):MTHPA = 18:12:60; (3) DGEBA (n = 1):DGEBA (n = 2):MTHPA = 15:15:60; (4) DGEBA (n = 1):DGEBA (n = 2):MTHPA = 9:21:60.

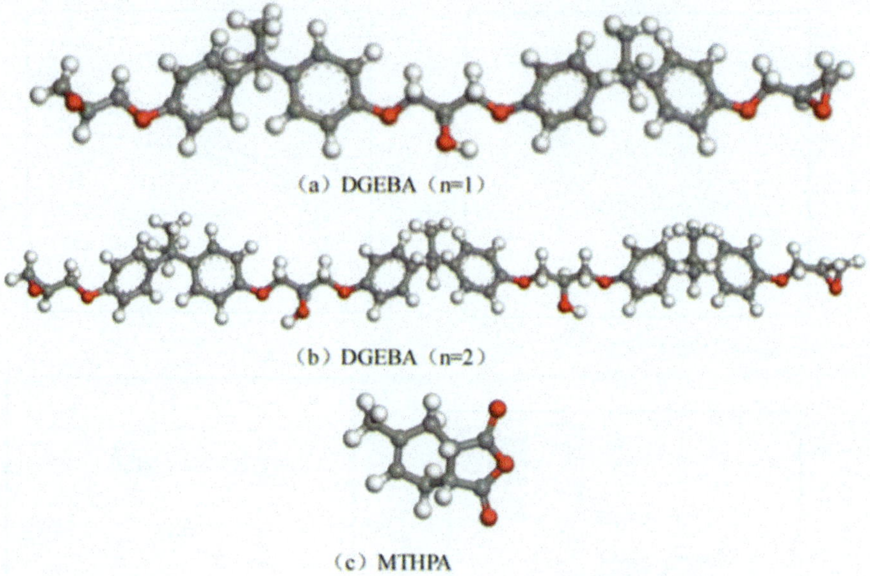

Fig. 11 Molecular structures of monomers of DGEBA and MTHPA with different degrees of polymerization

FFV, *MSD*, T_g, *CTE* and modulus are calculated for different polymerization proportioning models respectively.

Figure 12 reveals the effect of different polymerization degrees on the thermodynamic properties of crosslinked epoxy resins. It is found that as the average molecular weight increased, the T_g of the system first increased and when it exceeded a certain value, T_g decreased to the initial level. The change in average molecular weight has a significant improvement on the glassy *CTE* of the system, while the effect on the rubbery *CTE* is not significant. The increase of average molecular weight will have a more significant improvement on the mechanical properties of the system. When the ratio is 6:4 in the polymerization degree model, the increase in each elastic modulus is the largest.

4.3 Study on the Effect of Capping Agents on the Structural Properties of Epoxy Resins

In this section, DGEBA (0 degree of polymerization) is selected as the epoxy resin matrix. MTHPA is the curing agent and ethyl salicylate (ES) is the capping agent. The molecular structural formulae and primary crosslinking structures are shown in Figs. 13, 14 and 15.

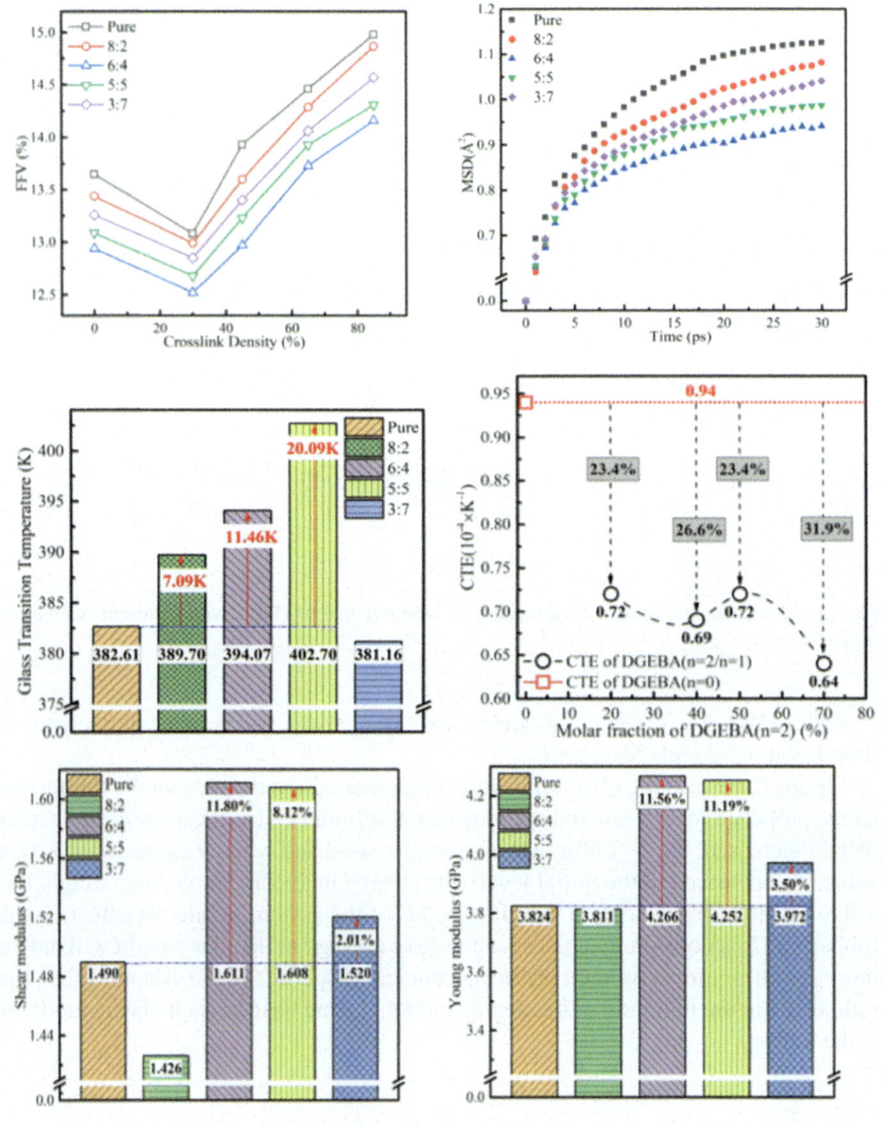

Fig. 12 *FFV*, *MSD*, T_g, *CTE* and Modulus of epoxy resin crosslinking systems with different polymerization degrees

In this section, four cases of capping agent ES with 2, 4, 6 and 8% mass fraction of epoxy resin are selected. The components of the system are shown in Table 3.

The cohesion energy density (CED), hydrogen bonding number, *FFV*, *MSD*, T_g, *CTE* and modulus are calculated for different capping agent proportioning models respectively (Fig. 16).

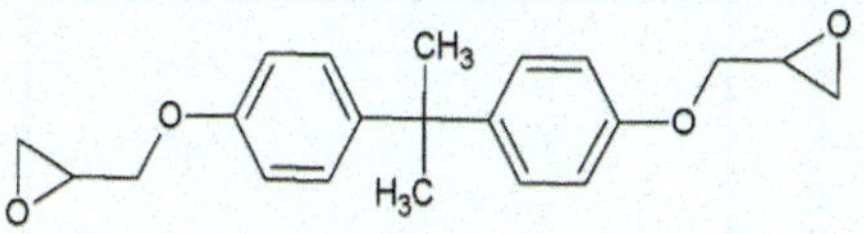

(a) DGEBA constitutional formula

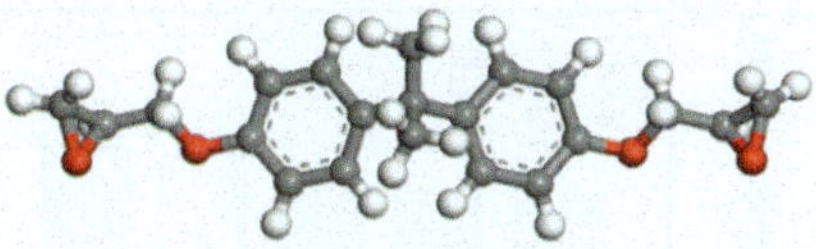

(b) DGEBA monomer molecule

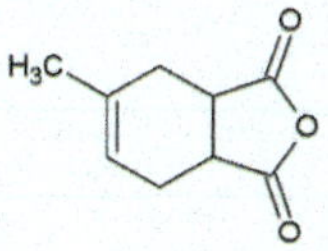

(c) MTHPA structural formula

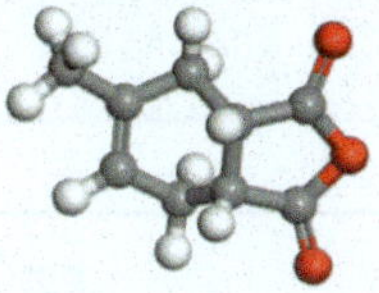

(d) MTHPA monomer molecule

Fig. 13 Molecular structure formula of epoxy resin and curing agent

Fig. 14 Molecular structure formula and molecular model of capping agent

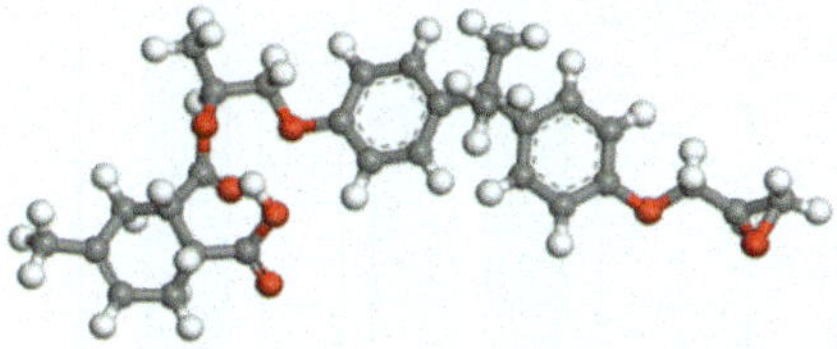

(a) DGEBA-MTHPA primary crosslinking molecular model

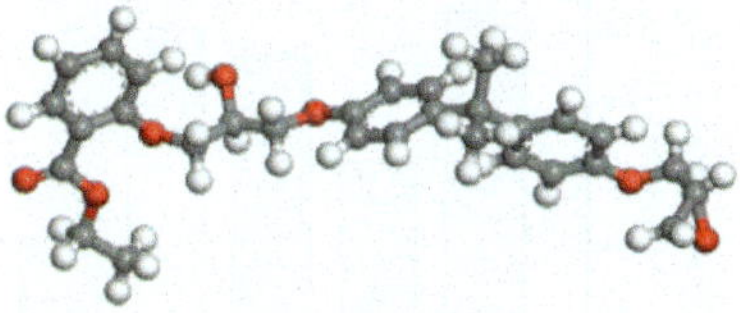

(b) DGEBA-ES primary crosslinked molecular model

Fig. 15 Primary crosslinked molecular model

Table 3 Amorphous model components for different mass fraction systems

ES content (wt%)	Epoxy resin/strip	Curing agent/strip	One cross-linked molecule/strip	
	DGEBA	MTHPA	DGEBA-MTHPA	DGEBA-ES
2	61	136	10	4
4	56	131	10	9
6	50	125	10	15
8	44	119	10	21

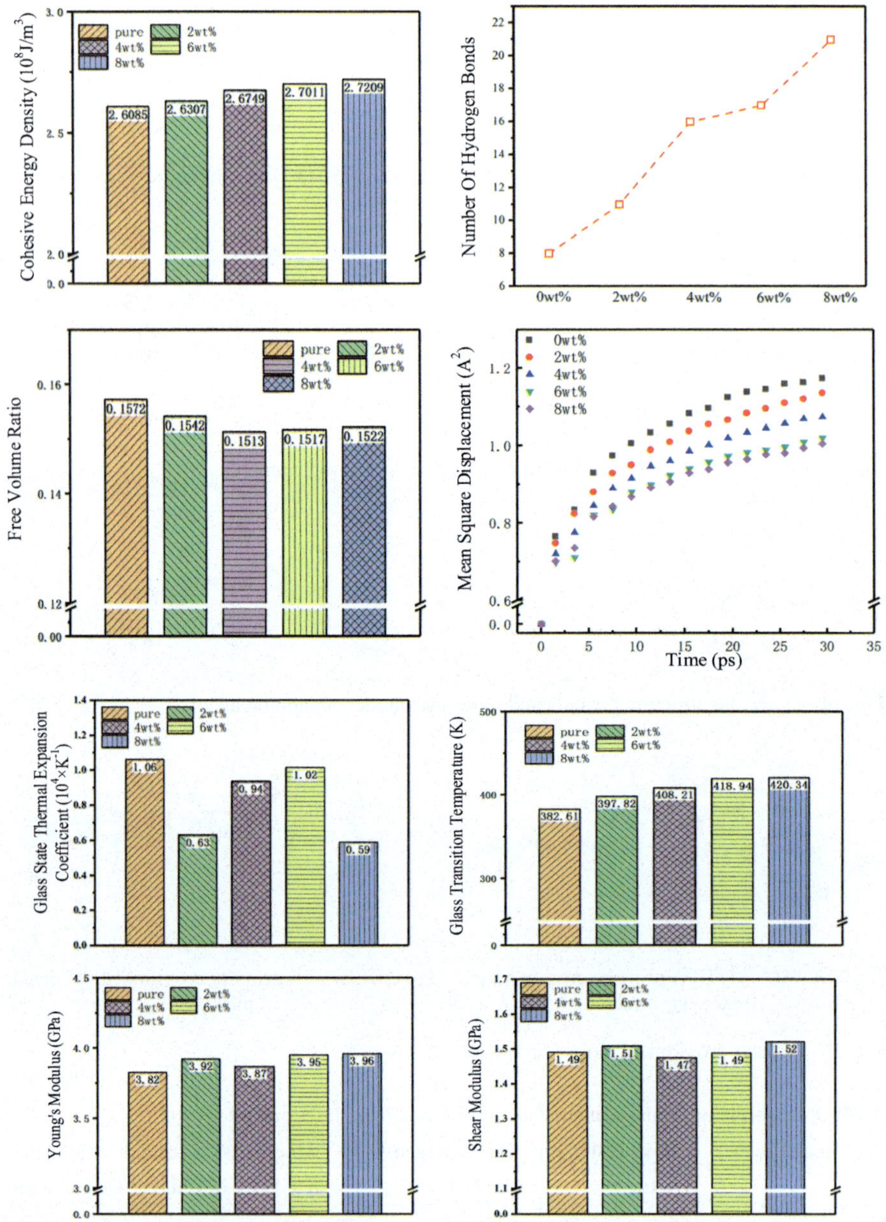

Fig. 16 CED, hydrogen bonding number, *FFV*, *MSD*, T_g, *CTE* and Modulus of different capping agent epoxy resin crosslinking systems

The results indicate that:

(1) With the increase of end-capping agent content, CED and FFV show an upward trend, which is also related to the benzene ring and double bond contained in ES. The order of FFV is pure > 2 wt% > 8 wt% > 6 wt% > 4 wt%, and the order of MSD is 8 wt% < 6 wt% < 4 wt% < 2 wt% < pure, which is due to the synergistic effect of the ES polar group with the methyl group. The number of hydrogen bonds increased with the increase of capping agent content, and the final number remain unchanged.
(2) With the increase of the content of the capping agent, the T_g of the epoxy resin is obviously increased, and the increase is gradually reduced. The thermal expansion coefficients of the ES systems with capping agents are all lower than those of the pure systems; and the trend is to increase and then decrease.
(3) After adding the capping agent ES, the Young's modulus of each system increases, and the size relationship is the same as the mean square displacement of each system. The bulk modulus and shear modulus of each system do not increase or decrease significantly with the increase of capping agent content.

4.4 Study of the Effect of Matrix Fluorination on the Properties of Epoxy Resins

In this section, bisphenol A epoxy resin DGEBA is used as epoxy substrate and MTHPA is used as an anhydride curing agent. Four monomers, diglycidyl trifluoromethylaniline (DGTFA), dodecafluoroheptanol (FTOH), 4 (-1,1,1,3,3,3-hexafluoro-2-hydroxy-2-propyl) phenol (HFIP) and hexafluorobisphenol A (BPAF), are selected as fluorine-modified reagents to fluorinate the epoxy resin matrix. The molecular formulae of the epoxy resin, curing agent and fluorinated monomer are shown in Fig. 17.

Firstly, the effect of the fluorination method on the structure and properties of the epoxy crosslinking system is investigated. The molecular models of the four groups of fluorinated resins with different matrix fluorination methods and the number of molecules contained in the blank group are shown in Tables 4.

CED, T_g, thermal conductivity and modulus are calculated for the models with different fluorination methods respectively.

(1) CED calculation analysis

Calculations of the cohesive energy density of different fluorinated epoxy systems are carried out at room temperature and the results are shown in Table 5.

The table shows that the epoxy system with the physical fluorination method has the lowest cohesive energy density of 113.26 J cm^{-3}, which indicates that the interaction force of the molecular chain segments of this epoxy system is weak. Among the fluorine-containing systems of the three chemical fluorination methods, the CED of the system using the median fluorinated fluorine-containing monomer is relatively high. Further comparison shows that the BPAF/DGEBA epoxy system

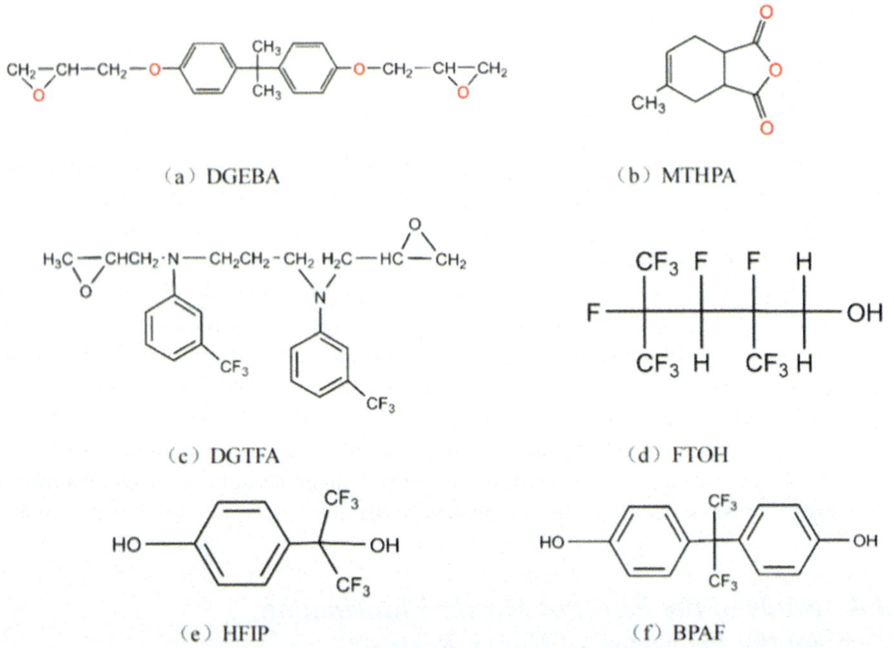

Fig. 17 Molecular structure formula of epoxy resin, curing agent and fluorinated monomer

Table 4 Components of epoxy resin and hardener in the amorphous model

Fluorinated epoxy systems	Number of molecular strips of epoxy substrate		Substrate fluorination		
	DGEBA	MTHPA	Fluorinated monomers	Number of molecules	Fluoridation method
Pure	50	100	–	–	Unfluorinated
DGTEA/DGEBA	50	100	DGTFA	10	Physical co-mingling
FTOH/DGEBA	50	100	FTOH	10	End-site fluorination
HFIP/DGEBA	50	100	HEIP	10	Median fluoride
BPAF/DGEBA	50	100	BPAF	10	Median fluoride

Table 5 Cohesive energy density of resin systems with different fluorination methods

Epoxy systems	Fluoridation method	CED/J cm^3
Pure	Unfluorinated	130.03
DGTEA/DGEBA	Physical co-mingling	113.26
FTOH/DGEBA	End-site fluorination	118.94
HFIP/DGEBA	Median fluoride	135.34
BPAF/DGEBA	Median fluoride	141.69

has the highest cohesion energy density of 141.69 J cm^{-3}, which indicates that the intermolecular forces in the chain segments of this epoxy system are relatively strong and the interaction forces in the crosslinked network are high.

The analysis shows that the physical fluorination is a simple blending of the fluorinated monomer with the epoxy matrix, with no chemical bonding between the fluorinated monomer and the DGEBA molecule. Therefore, the intermolecular interaction force is weak and the CED of the system is relatively low, resulting in poor compatibility of lend phase and uneven dispersion of the two phases, thus compromising the overall performance of the fluorinated epoxy system. After chemical fluorination, the fluorine-containing segments are grafted into the epoxy resin network system by chemical reaction between the fluorination reagent and the epoxy group on the epoxy matrix. The obtained fluorine-containing epoxy system has stronger intermolecular force, more stable product structure, and relatively high CED parameters.

(2) Synergistic reversal barrier analysis

The essence of molecular segment flexibility is the ease of single bond rotation, which can be expressed by the cooperative torsion barrier of single bond with representative position in molecular segment during rotation. The larger the synergistic torsional potential at key locations for monoliths with approximate molecular structure, the less flexible their corresponding chain segments are.

From the CED calculations, it is clear that the CED values of BPAF/DGEBA and HFIP/DGEBA fluorinated epoxy resin systems are similar. This section therefore further calculates the torsional energy barriers for the chain segments of these two fluorinated monomer-containing molecules. The conformational behavior of the molecular structure is used to compare the information on the local structure and the flexibility of the molecular chain segments in the crosslinked network of the epoxy resin. The conformational analysis of the molecular chain segments of the two fluorinated monomers, HFIP and BPAF, is shown in Fig. 18. Figure 19 gives the calculation results of the co-twisting energy barriers at key bonding positions for HFIP and BPAF monomer molecules. The figure shows that the torsional energy barrier at the two hydroxyl positions of BPAF is 1131.36 kcal/mol higher than the energy barrier of the HFIP monomer molecule. This indicates that BPAF molecular chain segments require more energy to rotate around a single bond and that their molecular chains are more rigid. The analysis suggests that the two fluorinated monomers with similar structures at both ends, the BPAF molecule contains two benzene rings, which bring about a greater spatial resistance, resulting in a less free molecular conformation. Therefore the torsional energy barrier of the BPAF monomer molecule is calculated to be higher and it can be assumed that the BPAF/DGEBA epoxy system is more rigid.

(3) Performance calculation analysis

The results for T_g, thermal conductivity and modulus are shown in Fig. 20.

The analysis of the T_g calculation results shows that after fluorination of the epoxy resin, the introduction of fluorine elements and CF_3 groups will effectively improve the heat resistance of the epoxy resin, while the T_g enhancement of the

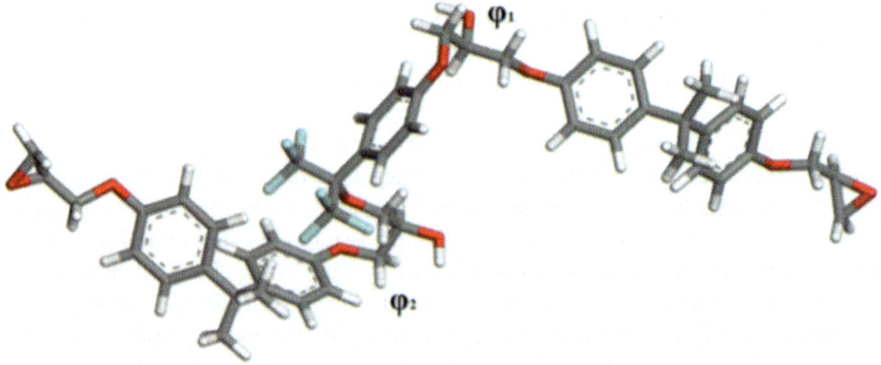

(a) Schematic diagram of HFIP key positions

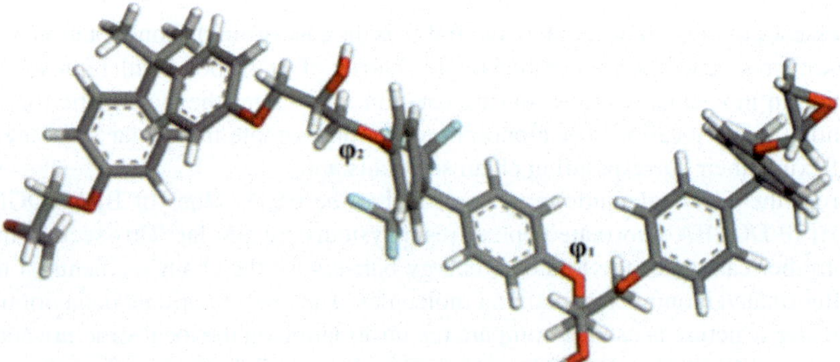

(b) Schematic diagram of BPAF key positions

Fig. 18 Schematic diagram of the bonding of HFIP and BPAF

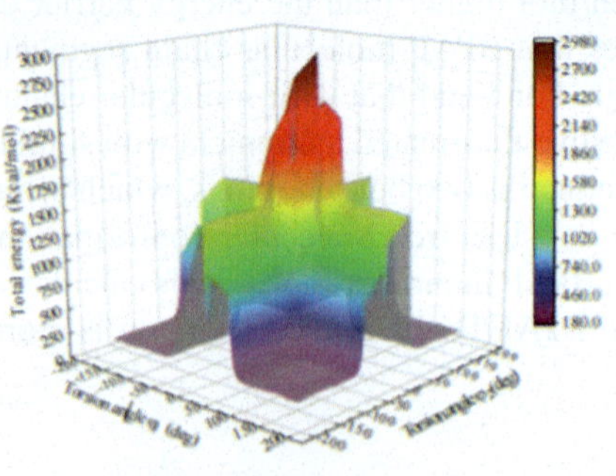

(a) HFIP Collaborative Twisting Energy Barrier

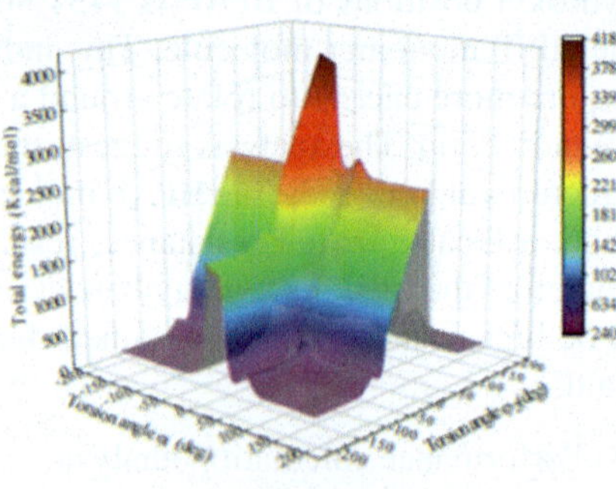

(b) BPAF Collaborative Twisting Energy Barrier

Fig. 19 Calculated results of the synergistic torsional energy barrier

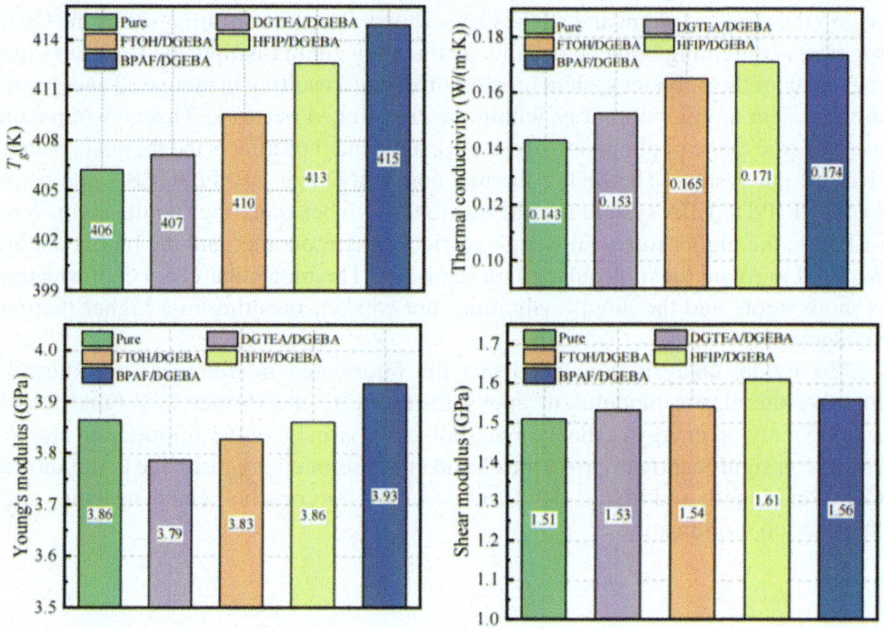

Fig. 20 T_g, thermal conductivity and modulus of epoxy systems with different fluorination methods

system by fluorinating the resin by physical blending is relatively low, and the T_g enhancement by chemical grafting fluorination is better. The highest T_g enhancement for the DGEBA/MTHPA epoxy system is achieved by the neutral fluorination of BPAF. Combined with the calculation results of micro parameters, it can be found that for the analysis of crosslinking network structure parameters, the system with relatively high polymer CED has relatively strong intermolecular interaction forces and a higher glass transition temperature. From the perspective of the fluorinated monomer configuration, the higher the molecular chain segment co-twisting energy barrier, the more rigid the molecular chain segment and the higher the T_g of the corresponding fluorinated system. Therefore, the glass transition temperature of the BPAF/DGEBA fluorinated resin system is the highest.

The analysis of the thermal conductivity calculations concluded that the epoxy resin material mainly relies on internal phonons for thermal conductivity. When the heat generated by the external environment is transferred to the epoxy molecular chain segments, it will cause the atoms to vibrate, thus enabling the transfer of heat. In the case of fluorinated resins with physical fluorination, the interaction between the fluorinated monomer and DGEBA is weaker, the molecular chain segments are more mobile, the atoms vibrate more vigorously when heated, the phonon excitation number increases and the average phonon free range decreases, so the thermal conductivity of the system is relatively low. The use of dodecafluoroheptanol chemical graft fluorination increases the number of branched molecular chain segments within the epoxy system, and studies have shown that the conformation of the chain segments containing branched chains on the main chain disrupts the symmetry and regularity of the polymer system to a certain extent, resulting in increased energy and more violent atomic vibrations within the crosslinked network. This constrains the average free range of phonons to a certain extent and thus affects the thermal conductivity of the system. The higher thermal conductivity of BPAF/DGEBA compared to the HFIP/DGEBA system may be due to the two benzene rings on the main chain of BPAF, the higher torsional energy barrier of the monomer and the higher energy required to rotate the molecular chain segments. The molecular chain conformation is more stable and the atomic vibrations are weaker, resulting in a higher thermal conductivity.

The modal analysis concluded that the fluorinated modifier can significantly improve the elastic modulus of epoxy resin due to the shorter C–F bond length and higher bond energy in the fluorocarbon chain structure after fluorination modification can significantly improve the elastic modulus of epoxy resin due to the shorter C–F bond length and higher bond energy in the fluorocarbon chain structure after fluorination modification.

References

1. Sun, Y.G., Guo, Y.F., Yang, H.: A molecular dynamics study of crosslinked epoxy networks: construction of atomistic models. Mol. Simul. **46**(2), 121–127 (2020)
2. Chen, P., Liu, S.P., Wang, D.Z., et al.: Epoxy Resin and Its Applications. Chemical Industry Press (2011)
3. Yarovsky, I., Evans, E.: Computer simulation of structure and properties of crosslinked polymers: application to epoxy resins. Polymer **43**(3), 963–969 (2002)
4. Gou, J., Minaie, B., Wang, B., et al.: Computational and experimental study of interfacial bonding of single-walled nanotube reinforced composites. Comput. Mater. Sci. **31**(3), 225–236 (2004)
5. Wu, C., Xu, W.: Atomistic molecular modelling of crosslinked epoxy resin. Polymer **47**(16), 6004–6009 (2006)
6. Varshney, V., Patnaik, S.S., Roy, A.K., et al.: A molecular dynamics study of epoxy-based networks: cross-linking procedure and prediction of molecular and material properties. Macromolecules **41**(18), 6837–6842 (2008)
7. Yang, H., Yu, K., Mu, X.M., et al.: A molecular dynamics study of bond exchange reactions in covalent adaptable networks. Soft Matter **11**(31), 6305–6317 (2015)
8. Xin, D.R., Han, Q.: Investigation of moisture diffusion in cross-linked epoxy moulding compound by molecular dynamics simulation. Mol. Simul. **39**(4), 322–329 (2013)
9. Liu, W.S.: Structure and Properties of Crosslinked Epoxy Resin TGDDM: A Molecular Dynamics Simulation. North University of China (2016)
10. Komarov, P.V., Chiu, Y.T., Chen, S.M., et al.: Highly cross-linked epoxy resins: an atomistic molecular dynamics simulation combined with a mapping/reverse mapping procedure. Macromolecules **40**(22), 8104–8113 (2007)

11. Wang, D.Z.: Production and Application of Epoxy Resin. Chemical Industry Press, Beijing (2001)
12. Andersen, C.H.: Molecular dynamics simulations at constant pressure and/or temperature. J. Chem. Phys. **4**(72), 2384–2393 (1980)
13. Berendsen, H.J., Postma, J., Van Gunsteren, et al.: Molecular dynamics with coupling to an external bath. J. Chem. Phys. **8**(81), 3684–3690 (1984)

Study of Model Construction Methods for Epoxy Resin Composites

Qijun Duan, Qing Xie, Yuyao Zhong, and Jun Xie

1 Model Construction and Performance Study of SiO_2/EP Composites

The addition of nanomaterials to the epoxy resin (EP) matrix can significantly improve the comprehensive performance of the material. In recent years, many scholars have conducted numerous studies on nano-modification to improve the thermal property, mechanical property, electrical property and other aspects of EP. These studies mainly focus on the effects of filler type, particle size, shape and dosage on the performance of EP composites. SiO_2 is the most commonly used additive filler in EP insulating materials, but there is no clear conclusion on the mechanism of improving polymer performance. Moreover, molecular simulation plays an important role in revealing the mechanism of interaction between nanoparticles and matrix. Nevertheless, there are few molecular dynamics studies researching on the effect of nano-SiO_2 size on EP. In this section, DGEBA, MTHPA and nano-SiO_2 are used as EP matrix, curing agent and filler respectively. Different molecular models of nano-SiO_2 EP composites are constructed based on molecular dynamics. And the effects of filler particle size, morphology, content and surface modification on the microstructure and macroscopic properties of nanocomposite epoxy resins are studied.

Q. Duan · Q. Xie (✉) · Y. Zhong · J. Xie
Department of Electrical Engineering, North China Electric Power University, Baoding, China
e-mail: xq_ncepu@126.com

Q. Xie (ed.), *Electrical Materials*, Engineering Materials,
https://doi.org/10.1007/978-981-99-9050-4_2

1.1 *Study on the Effect of SiO_2 Particle Size on the Structure and Properties of Its Epoxy Composites*

1. Construction of SiO_2 monomer molecules

Molecular models of SiO_2 crystals can be imported from MS material library. In this section, the typical cell model of SiO_2 quartz crystal is selected as shown in Fig. 1. This cell model is used to construct molecular models of spherical SiO_2 with radius of 6.5 Å, 7.6 Å, 8.8 Å and 10 Å respectively, which are oxidized according to the actual process, as shown in Fig. 2.

Nanofillers are introduced according to the crosslinking model construction method mentioned above. Each model is filled with only one nano-SiO_2 particle to simulate a well-dispersed nanocomposite epoxy system, with the filling mass fraction of SiO_2 in each system being maintained to be 6.5%. The model components of SiO_2 nanocomposite systems with various particle sizes are given in Table 1.

After running the crosslinking procedure, molecular models with crosslinking degrees of 10, 34, 67 and 96% were selected from each composite system. Among these selections, molecular models with four crosslinking degrees of 8.8 Å particle size composite system are displayed in Fig. 3, and the crosslinking points are marked with spherical atoms. It can be seen from the figure that in addition to the crosslinking of DGEBA and MTHPA, the hydroxyl groups on the surface of SiO_2 are also involved in the curing reaction of EP.

The variation trends of *FFV*, T_g and modulus of epoxy composites with different particle sizes according to the crosslinking density are then analyzed.

(1) FFV analysis

Figure 4 presents the variation of FFV with crosslinking degree of each system. It can be seen from the figure that the addition of nano-SiO_2 results in a significant decrease in FFV compared to pure epoxy, which is due to the fact that the nanoparticles occupy some of the small pores and a considerable portion of the space in the epoxy resin, resulting in a smaller range of segment motion and a decrease in FFV. When the crosslinking degree is lower ($\leq$34%), the FFV does not change obviously

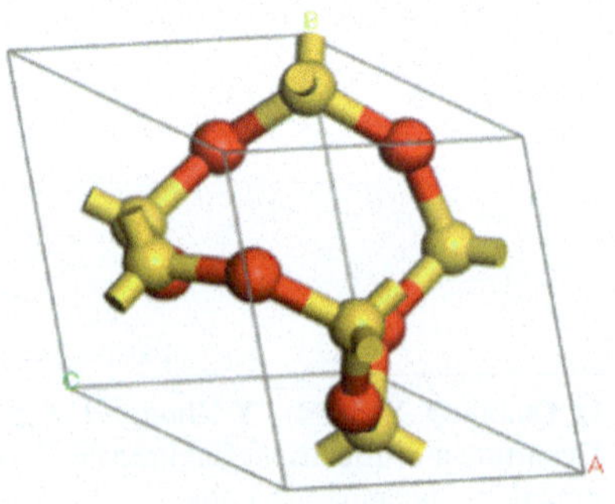

Fig. 1 Molecular model of SiO_2 cell

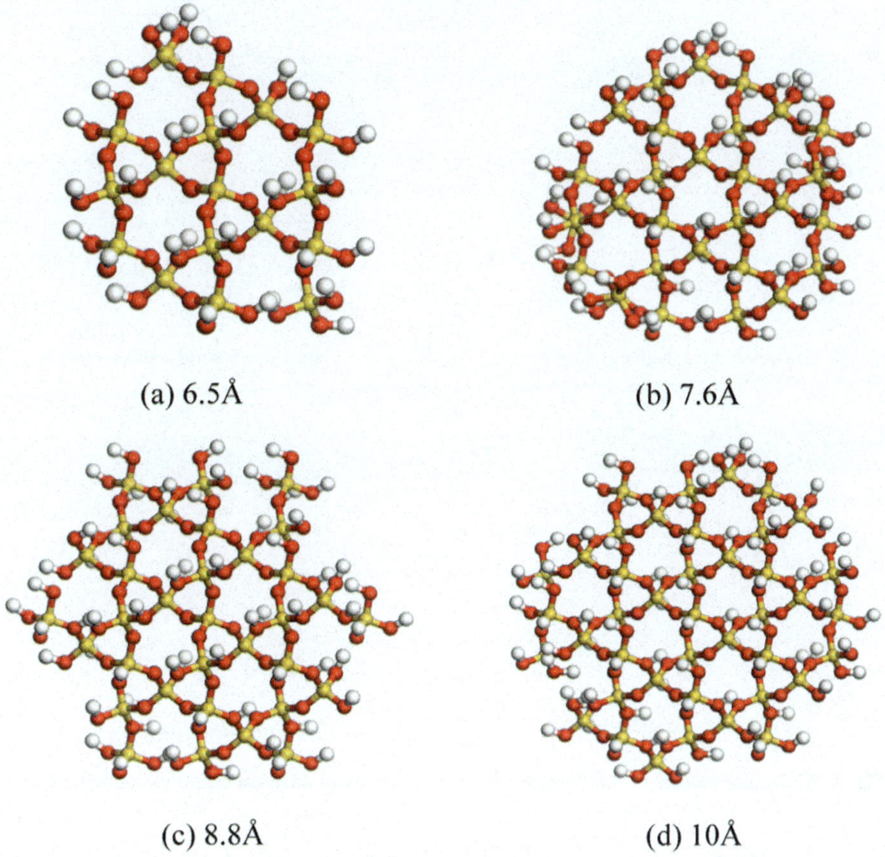

Fig. 2 Molecular models of SiO_2 with different particle sizes

Table 1 Model components of SiO_2 nanocomposite systems with different particle sizes

SiO_2 (Å)	DGEBA	MTHPA	DGEBA-MTHPA	Atomic number	SiO_2 mass fraction/ %
6.5	40	90	10	4838	6.5
7.6	65	146	16	7823	6.5
8.8	88	199	23	10,708	6.5
10	124	288	40	15,830	6.5

with the particle size. And with the increase of crosslinking degree, FFV decreased with decreasing particle size. This may be due to the fact that SiO_2 is bonded to the system molecules when a more complete cross-linked network is formed, and the interaction force between small particle size nano-SiO_2 and the matrix is stronger. According to previous studies by scholars, the mechanical properties of polymers

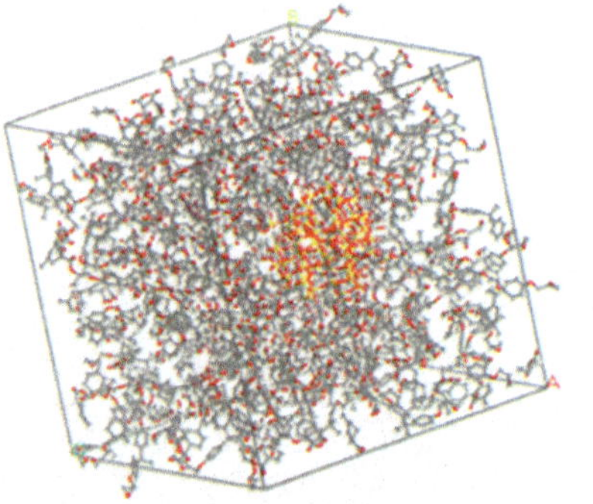

(a) Crosslinking degrees of 10%

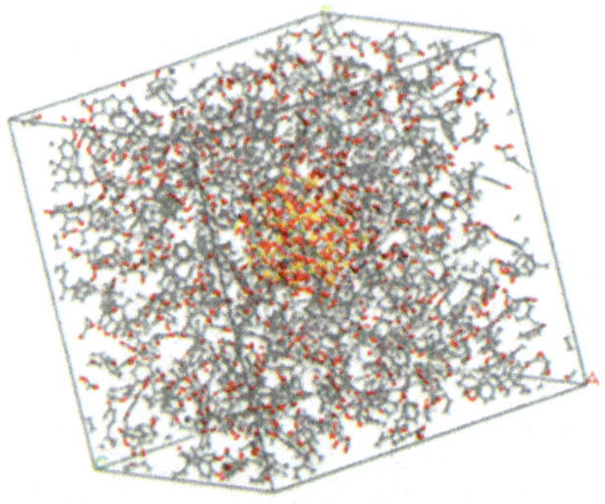

(b) Crosslinking degrees of 34%

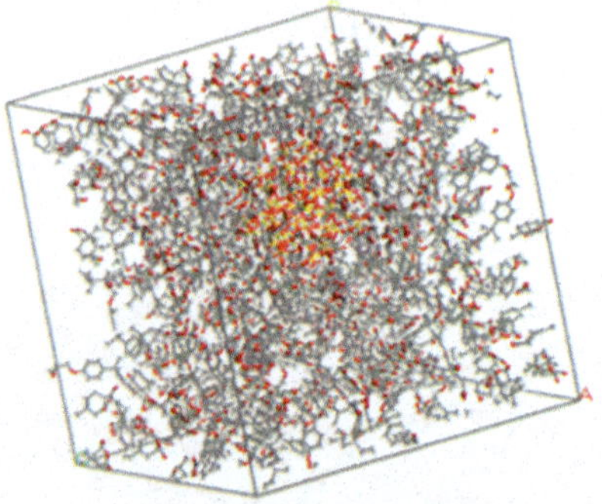

(c) Crosslinking degrees of 67%

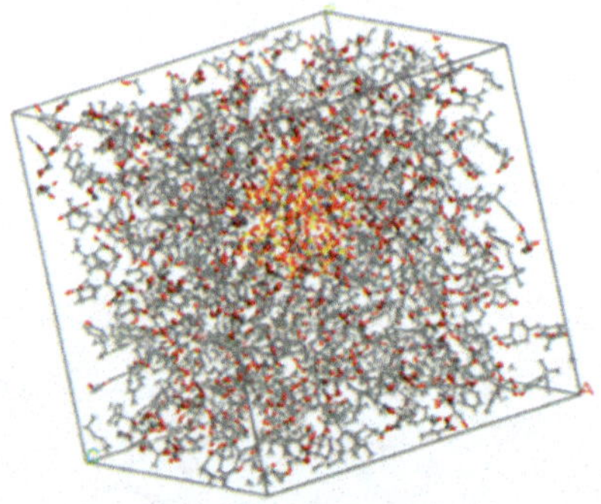

(d) Crosslinking degrees of 96%

Fig. 3 Molecular models of 8.8 Å nanocomposite epoxy with different crosslinking degrees

have a certain dependence on the free volume. The smaller the free volume is, the better the properties are. It can be predicted that the mechanical properties of the system with filler are better than those of the pure epoxy system. Furthermore, when the crosslinking degree is higher, the smaller the particle size is, the better the mechanical properties are.

(2) T_g analysis

As shown in Fig. 5, T_g of the pure epoxy system is higher than that of the filler-containing system at the beginning of crosslinking. While the opposite result is shown as the crosslinking degree increased. Because when the system molecules completed a certain stacking, the filler is bonded to the resin matrix and curing agent, and the filler forms an interface layer with the matrix, which increases the T_g. From the filler particle size, the variation trend of T_g with particle size is not obvious at low crosslinking degree. However, when the crosslinking degree is ≥67%, the T_g of the composites basically increases with the decrease of the filler particle size. Moreover, when the crosslinking degree reached 96%, the T_g of the 6.5 Å system is particularly large. This shows that when the filler and polymer are bonded to a certain extent, the interfacial bonding force between the filler with small particle size and the matrix

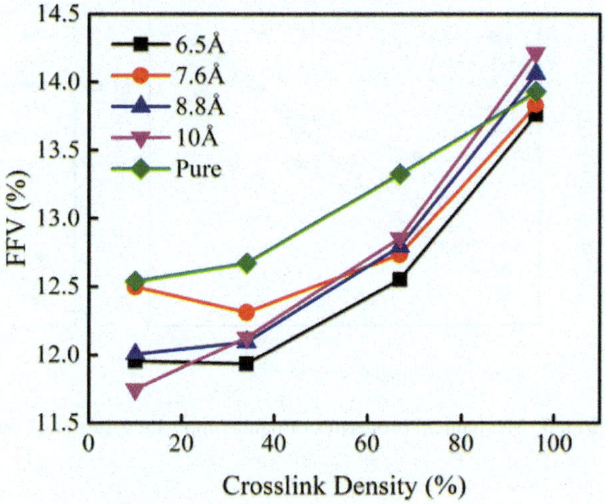

Fig. 4 Variation of FFV for each system

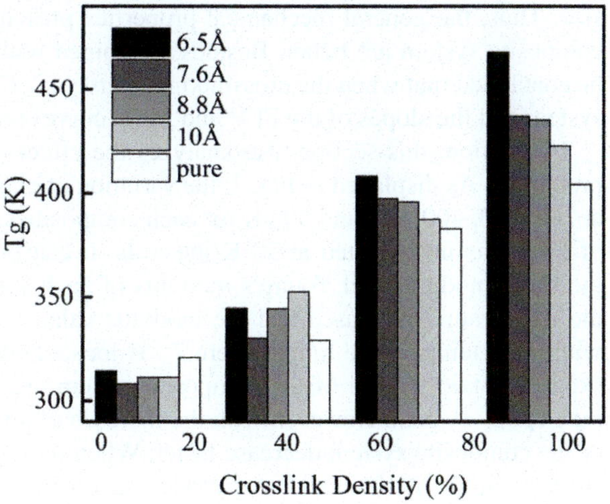

Fig. 5 T_g value of pure epoxy and filler containing system

is stronger, which is more conducive to the formation of a stable and high-rigidity system.

(3) Analysis of the modulus

It can be seen from Fig. 6 that the elastic modulus value of each system increases with the increase of crosslinking degree. Meanwhile, comprehensive static mechanical properties of the filler system were significantly superior to those of the pure epoxy system. Moreover, when the crosslinking degree is ≥67%, the smaller the particle size of SiO_2, the higher each modulus value of the composite system. This is owing to the fact that, on the one hand, due to the large modulus of nano-SiO_2, adding SiO_2 can remarkably upgrade the mechanical properties of matrix. On the other hand, when

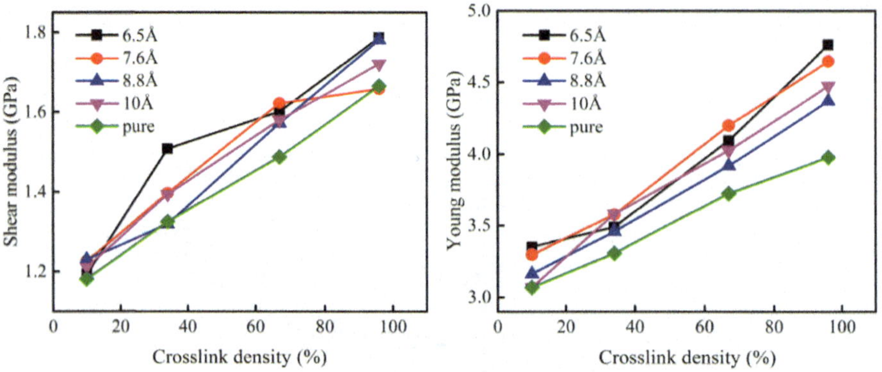

Fig. 6 Mechanical properties of nanocomposite system and pure epoxy system

the crosslinking reaches a certain degree, the bonding force between the nano-SiO_2 and matrix gets strengthened and enlarges with the reduction of the filler particle size. Thus, the general mechanical properties presented by the small particle size composite system are better. Besides, combined with the previous analysis, it can be concluded that when the crosslinking degree is $\geq$67%, the modulus values of the system and the slopes of the FFV and MSD curves are negatively correlated.

In addition, this section also analyzes the effect of temperature on mechanical properties. As displayed in Fig. 7, the variation of elastic modulus with temperature at 300, 350, 400, 450 and 500 K for each composite system at a crosslinking degree of 67% was investigated at 50 K intervals. It can be noticed from the figure that the shear modulus and Young's modulus of each system reduced continuously as the temperature increased and the modulus value of each system decreased most within the temperature range where T_g is located. When the temperature is lower (<350 K), two moduli of the composite system are both higher than those of the pure epoxy system. However, with the increase of temperature, the modulus values of the composite system decrease faster. When the temperature exceeds 400 K, the shear modulus of 10 and 8.8 Å particle size composite system is inferior to that of pure epoxy system. In addition, the reduction in modulus is also severe for the 6.5 and 7.6 Å particle size systems. Nevertheless, their modulus value is still higher than that of the pure epoxy system at this temperature owing to their obvious bonding enhancement effect. When the temperature reaches 500 K, the overall modulus value of the composite system is lower than that of the pure epoxy system.

In general, the modulus value of the filler system is more sensitive to temperature shifts at the crosslinking degree of 67%. In particular, the 6.5 Å composite system experienced the most remarkable falling range in each modulus value with increasing temperature after the glass transition. It suggests that the unique nano small size effect of the system after the glass transition might gradually collapse at high temperature, and the combination between filler and matrix is sharply weakened. For the whole temperature range from 300 to 500 K, 10 Å system and pure epoxy system exhibited better heat resistance and their CTE are also relatively small under this condition.

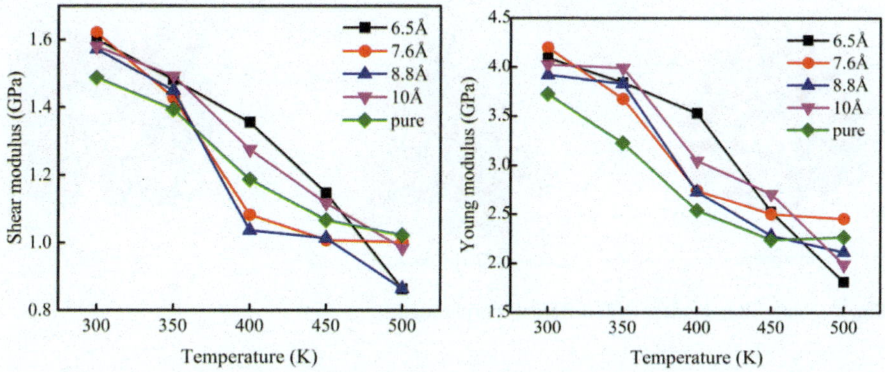

Fig. 7 Modulus of elasticity at different temperatures for each system with 67% crosslinking degree

However, when $T \leq 400$ K, the mechanical modulus value of the system with filler is generally higher than that of the pure epoxy system. Furthermore, every mechanical modulus value of the 6.5 Å system were consistently maintained within a high level. In accordance with the actual working condition, the insulation material is generally no longer used after undergoing glass transformation. For this reason, researching on the modulus value and heat resistance for each system before the temperature reaches T_g is of more engineering significance. On the whole, it can be concluded that the 6.5 and 10 Å system possess the best heat resistance.

1.2 Study About the Influence of SiO_2 Morphology and Content on the Structure and Properties of Its Epoxy Composites

In this section, three morphological Nano-SiO_2 filler models are constructed, and EP/SiO_2 cross-linking models with different morphologies and filling amounts were taken as the research objects, and the cross-linking models were simulated by MD.

Nano-SiO_2 is a familiar inorganic filler in EP composites, and its model construction can be divided into in three steps as described below.

(1) Firstly, import SiO_2 crystal structure as shown in Fig. 1.
(2) Secondly, construct filler particles with different morphologies. SiO_2 unit cells can be shaped into Nano-SiO_2 particle models with different shapes by using the Build Nanostructure function of MS software. This paper selects Nano-SiO_2 models with three shapes of sphere, tetrahedron and cube to simulate fillers with different morphologies [1]. To ensure the comparability of the calculation results, the packing model is constructed based on the principle of consistent number of Si atoms, with the packing for all three morphologies possessing 47 Si atoms.

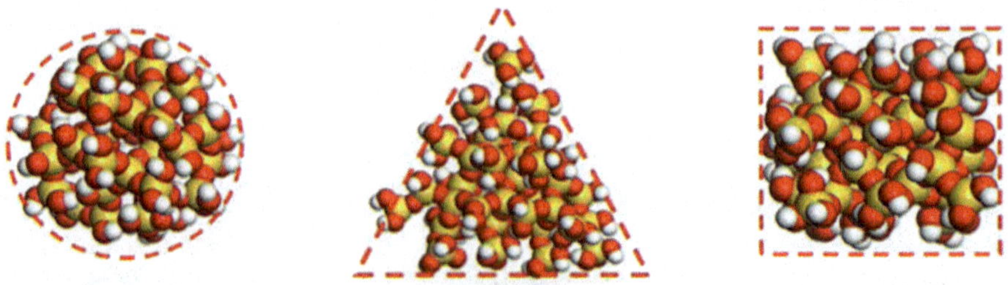

(a) Sphere (b) Tetrahedron (c) Cube

Fig. 8 Nano-SiO_2 models with different morphologies

(3) Thirdly, handle the oxidation on filler surface. In the formation of SiO_2 from the cell structure to filler particles, the Si–O bond on its surface is partially broken, which is a surface defect easily oxidized by air. Therefore, in this paper, hydroxyl groups are attached to the unsaturated Si atoms on the surface of the Nano-SiO_2 model to simulate the actual oxidation process. Figure 8 presents the model of Nano-SiO_2 filler after geometric optimization. Due to the various morphologies, the surface and dimensional parameters of the packing models are different. Additionally, the specific parameters are given in Table 2.

The effects of filler morphology and filling amount on the microstructure and thermodynamic properties of EP nanocomposites are taken as the research target. In this section, EP/Nano-SiO_2 composites with filling mass fraction of 10 wt% containing three filler shapes, namely sphere, tetrahedron and cube, and composite models with 5 wt%, 10 wt% and 15 wt% filled spherical fillers are modeled respectively. In order to simulate the idealized dispersion state of the filler, each model contained a Nano-SiO_2 filler, adjusting the filler mass fraction by changing the amount of the substrate. The detailed components of the model are listed in Table 3.

Subsequently, the interfacial binding energy, FFV, MSD and thermomechanical properties of the EP/Nano-SiO_2 model are calculated.

(1) Interfacial binding energy

The process of doping and compatibility of fillers into the substrate is completed relying on the physical interaction between substances. From this, the compatibility

Table 2 Dimensional parameters and atomic number of nano-SiO_2 with different morphologies

Filler profile	Dimensional parameters	Number of Si atoms	Number of O atoms	Surface light base
Sphere	Diameter 15 Å	47	126	64
Tetrahedron	Side length 23 Å	47	133	76
Cube	Side length 12 Å	47	129	67

Table 3 Component composition of different EP/nano-SiO_2 models

Models	Filler profile	Filling mass fraction (%)	Number of molecules		
			Nano-SiO_2	DGEBA	MTHPA
EP/5 wt% spherical SiO_2	Spherical	5	1	96	192
EP/10 wt% spherical SiO_2		10		45	90
EP/15 wt% spherical SiO_2		15		29	58
EP/10 wt% cubic SiO_2	Cube	10		47	94
EP/10 wt% tetrahedral SiO_2	Tetrahedron	10		47	94

between filler and substrate will directly affect whether the filler can be successfully doped into the substrate. Furthermore, the inorganic–organic bonding interface is established after the crosslinking of the composite system, and the strength of the interfacial binding effect will impact the modification effect of nano-doping. Binding energy between filler and substrate is the energy required to separate substrate and filler to "infinity", which is an essential parameter to evaluate the tightness of the bonding between composite substrate and filler. The binding energy can be calculated by the following equation [2]:

$$E_{Binding} = E_{EP} + E_{SiO_2} - E_{Total} \tag{1}$$

where $E_{Binding}$ is the binding energy of SiO_2 filler and EP substrate in composite model. E_{Total} is the total intermolecular interaction energy of the composite model. E_{EP} is the interaction energy between the substrate molecules and E_{SiO_2} is the interaction energy of the filler.

In this section, the binding energy of SiO_2 filler and EP substrate in the composite model without crosslinking is calculated. The binding energy of filler-substrate in different models is summarized in Table 4. It can be noticed that it is the van der Waals force (VDW) that plays a dominant role in the model proposed in this paper, which is consistent with the conclusion of the literature [3]. The interfacial binding energy of SiO_2 filler with DGEBA molecules given in the literature [2] is close to 0. Consequently, the positive binding energy of SiO_2 filler with EP substrate (DGEBA/MTHPA) is mainly caused by the anhydride curing agent, which indicates that SiO_2 is suitable for the filler of DGEBA/MTHPA system in terms of compatibility. According to Table 4, the binding energy of SiO_2 to the substrate exhibits little change as the mass fraction of filler increases. While compared with the three shapes of filler, spherical SiO_2 offers the strongest binding energy and the greatest compatibility with the substrate. For this reason, the spherical SiO_2 is chosen for the analysis of the effect of filler amount on the structural and thermodynamic properties of the composites.

Table 4 Binding energy of filler-substrate for different models

Models	Combination of energy	Van der Waals interaction components	Static energy fraction
EP/5 wt% spherical SiO	1706.83	1829.18	−122.35
EP/10 wt% spherical SiO_2	1822.09	1943.49	−121.39
EP/15 wt% spherical SiO_2	1817.00	1918.54	−101.53
EP/10 wt% cubic SiO_2	1433.10	1542.97	−109.87
EP/10 wt% tetrahedral SiO_2	1537.89	1666.10	−128.21

(2) FFV analysis

The introduction of nanofillers will induce the variation of free volume in the materials. Accordingly, in this section, FFV is also used to characterize the free volume characteristics of the composite model, the influence of Nano-SiO_2 filler on the free volume characteristics of the composite model at room temperature is analyzed. The FFV calculations of EP/Nano-SiO_2 models for different systems are presented in Fig. 9.

There are multifaceted impact mechanisms by which fillers affect the free volume properties of polymer composites. For one thing, doping filler will introduce inorganic–organic interface layer, in which the filler and substrate are bonded together primarily by VDW interaction. And the presence of "pores" leads to an enlargement of free volume. For another, the filler will occupy a certain amount of free volume space and increase the total volume of the composite. Thus, the variation trend of FFV depends on the dominant force of the two factors. As revealed in Fig. 9, Nano-SiO_2 filler can reduce the FFV of EP composites, and both the morphology

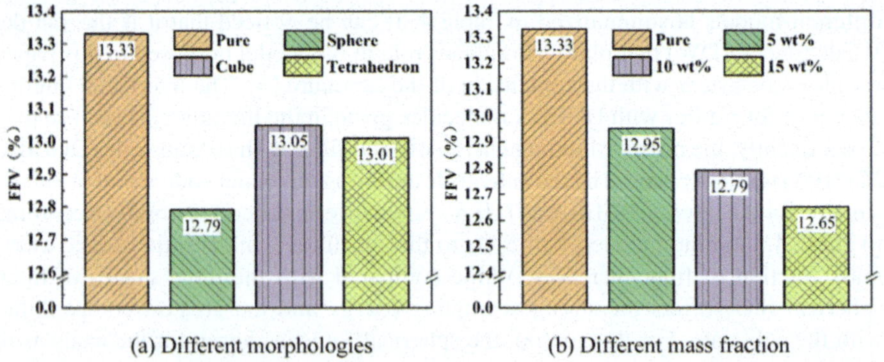

Fig. 9 FFV of EP/nano-SiO_2 models for different systems

and addition amount of filler will affect the free volume properties of the composites. Compared with the fillers of different morphologies, the FFV of the composite models with three types of fillers decreased to some extent, and the spherical SiO_2 shows the greatest influence on the FFV of material. Specifically, the model FFV of the three fillers are ranked as sphere < tetrahedron < cube. This sorting has an opposite magnitude relationship with the model filler-substrate binding energy and also confirms the conclusion that there is more free volume in the interface region. Including that the mass fractions of the three fillers are the same and their capacity to occupy the free volume is equivalent, while the lower the binding energy, the weaker the binding effect. As a result, the interfacial layer is relatively loose and the free volume is larger. Besides, the FFV of the model gradually decreases with increasing mass fraction of filler. The same pattern can be also observed in the nano-SiO_2 and PVA/PVP blending system [4] together with micron SiO_2/EP composites [5, 6]. It can be summarized that the occupying effect of Nano-SiO_2 filler is greater than the introducing effect of interfacial layer. Therefore, the free volume fraction gradually declines with the increasing filling amount.

(3) MSD analysis

In order to characterize the Nano-SiO_2 filler on the chain segment motion properties of the composite model, the MSD values of all atoms in the model at room temperature are calculated by MD simulation in this section. The calculated MSD results for EP/Nano-SiO_2 models of different systems are shown in Fig. 10. It can be found that nano-SiO_2 filler doping can significantly reduce the MSD of molecular chain segments and limit the chain segment motion of material molecules. In addition, the model MSD of the three fillers are ordered as sphere < cube < tetrahedron and the MSD decrease gradually as the filler mass fraction increase.

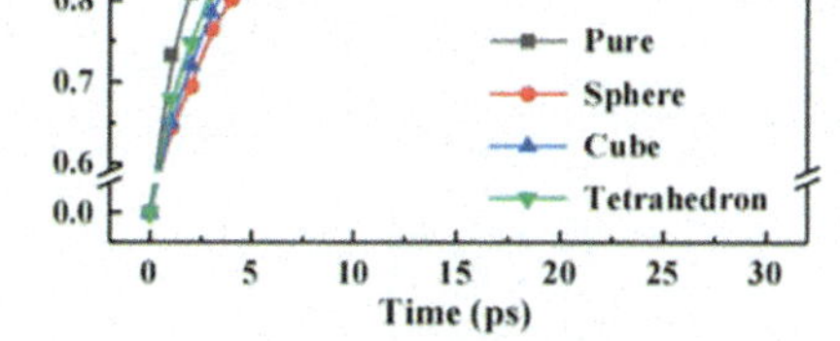

(a) Different morphologies

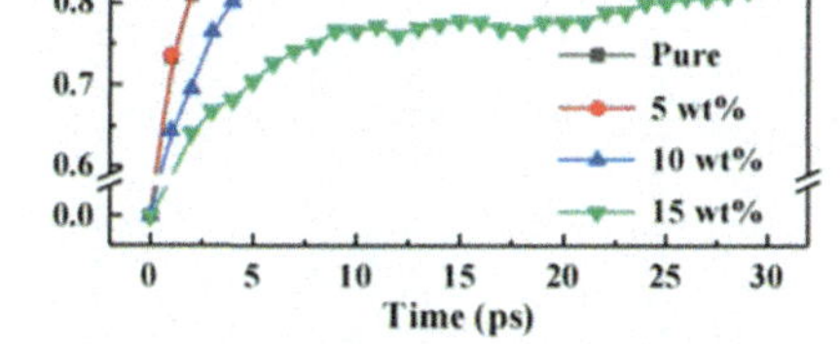

(b) Different filling amount

Fig. 10 MSD of EP/nano-SiO_2 models for different systems

The free volume can provide space for the molecular chain segments of the material to move [7]. Therefore, for homogeneous materials, the free volume of the material is the primary factor that affects the motion characteristics of the chain segments. According to Fig. 10, the FFV of the material decreased obviously with the increase of filler mass fraction. Furthermore, the FFV and MSD of the doped spherical SiO_2 model are both evidently less than those of the other two systems. Nevertheless, the MSD of the model doped with tetrahedral SiO_2 is larger than that of the cube system, suggesting that there are other factors influencing the chain segment motion of molecules inside the material.

(4) Analysis of glass transition temperature

By linear fitting of the density-temperature curves during the annealing process of the model, the glass transition temperature of EP/Nano-SiO_2 composites with different filler morphologies and filling amounts can be obtained. And the results are displayed in Fig. 11. As can be seen from Fig. 11a, the SiO_2 filler doping can increase the T_g of epoxy by more than 20 K (the T_g of pure epoxy substrate is 400.46 K). The lifting effects of the three fillers were equivalent, and in comparison, the spherical SiO_2 filler had the most remarkable effect. According to Fig. 11b, the T_g of epoxy composites increased gradually with the rising mass fraction of filler. When the mass fraction of SiO_2 is 15 wt%, the T_g of the material reached 430.35 K, indicating that the filling amount has a much higher degree of influence on the T_g of material compared to the morphology of filler.

The effect of Nano-SiO_2 filler on the T_g of EP composites is a macroscopic manifestation of the change in its internal microstructure. It is generally believed that the process of glass transition occurs with some kind of chain segment movement within the material. As shown in Figs. 10 and 11, the MSD of the material presented a consistent variation pattern with T_g, because the T_g of material is strongly influenced by the movement of molecular chain segments within the material. A reduction in MSD indicates that the molecular motion of material is restricted. Moreover, with

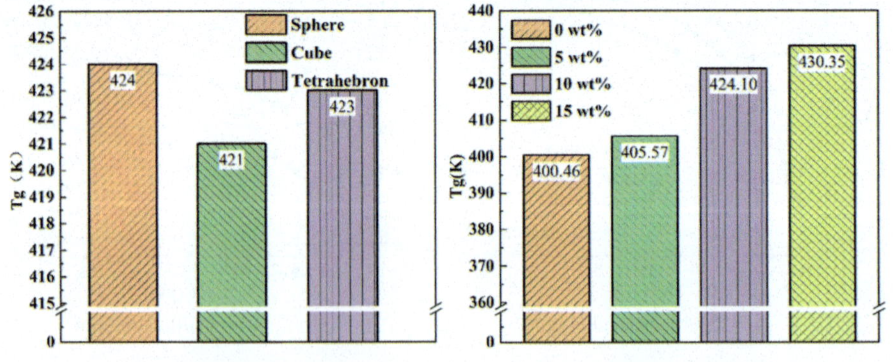

(a) Different morphologies (b) Different filling amounts

Fig. 11 T_g of EP/nano-SiO_2 models for different systems

the increasing temperature, the degree of molecular motion is small and the glass transition is unlikely to occur, leading to a higher T_g.

(5) CTE analysis

The calculated results for CTE of EP/Nano-SiO_2 composites with different filler morphologies and filling amounts are exhibited in Figs. 12 and 13. According to Fig. 12, the thermal expansion characteristics of EP composites in glassy state and rubbery state remain remarkable are distinct. Additionally, the CTE of glassy state is smaller than that of rubbery state, which is consistent with the findings for the epoxy substrate. Compared with the pure epoxy substrate, the Nano-SiO_2 filler doping has a certain enhancement on the thermal expansion characteristics of the composites, but the gap of the filler modification effect among the three morphologies is obvious. For glassy epoxy materials, spherical SiO_2 doping works best, reducing the material CTE by about 19%. For epoxy materials in rubbery state, cubic SiO_2 doping has the best effect, reducing the material CTE by approximately 18%. However, the improvement effect of tetrahedral filler is not obvious in either glassy or rubbery state.

The filling amount of SiO_2 has considerable influence on the CTE of epoxy composites. As revealed in Fig. 13, the CTE of the material reduced gradually as the mass fraction of SiO_2 in the composite system raised. Additionally, the calculated results suggested that the CTE was close to a linear relationship with the mass fraction of SiO_2. This is attributed to the fact that the SiO_2 filler doped into the substrate and occupied part of the free volume, reducing the contribution of the free volume to the CTE by thermal expansion. Also, the CTE of the filler itself is far less than that of the organic substrate. Compared to the substrate with the equivalent volume, the CTE of composites is degraded by the reduction of organic molecules in composites. As the filler mass fraction raised, more free volume of the system was occupied and less organic matter content was contained. For this reason, the CTE of composites brought down with the increasing mass fraction of SiO_2 filler.

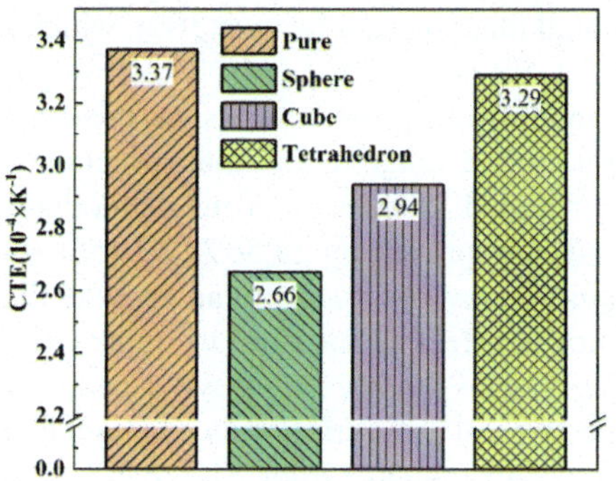

(a) Glassy state

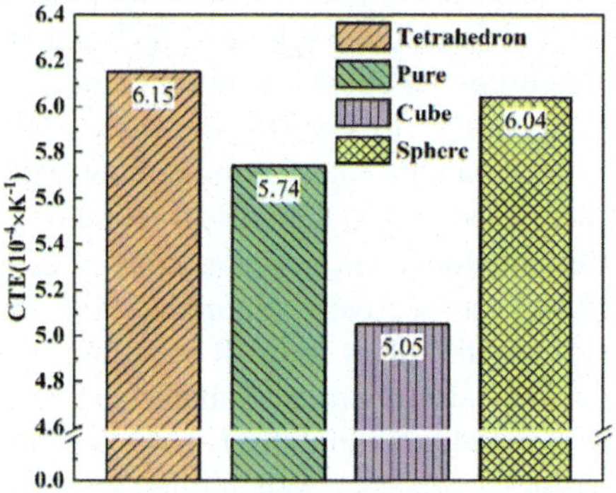

(b) Rubbery state

Fig. 12 CTE of EP/nano-SiO_2 composites with different filler morphologies

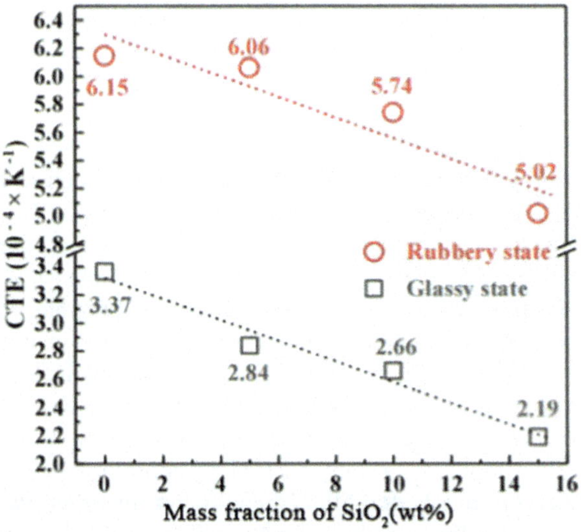

Fig. 13 CTE of composites with different nano-SiO_2 mass fractions

(6) Mechanical property analysis

The Young's modulus, shear modulus and bulk modulus of EP/Nano-SiO_2 composites with different filler morphologies (all with 10 wt% mass fraction) and filling amounts (spherical) at room temperature (300 K) are calculated by the static constant strain method. And the results are illustrated in Figs. 14 and 15. It can be noted that the three shapes of fillers improved the elastic modulus of materials to varying degrees. In order to compare the modification effects of fillers with different morphologies more accurately, the lifting rates for three elastic moduli of the material after doping are calculated. From Fig. 14, it can be concluded that the Young's modulus of the material was dramatically boosted by the three fillers. Nonetheless, the three fillers had unstable enhancing effects on the shear modulus and bulk modulus of the material. In contrast, the spherical SiO_2 enhanced all three moduli by over 10%, which is superior to the cubic and tetrahedral fillers.

According to Fig. 15, the three elastic moduli of epoxy composites presented an approximate linear enlargement within a certain range as the mass fraction of filler raised. When the mass fraction of filler reached 15 wt%, the Young's modulus, shear modulus and bulk modulus of composites improved by 24.56%, 32.45% and 23.24%, respectively, compared with the pure epoxy substrate. It can be seen that doping nanofillers into EP substrate is a proven effective measure to improve the mechanical properties of materials. And the reasons mainly lie in two aspects. As an inorganic material, SiO_2 has excellent mechanical properties of its own and high hardness properties. When filled into the epoxy substrate as a filler, SiO_2 can play an excellent supporting role in the material, making it less susceptible to be deformed by external force [8]. Besides, the SiO_2 filler occupies a certain free volume space in the substrate, which falls the movement of chain segments in model and likewise

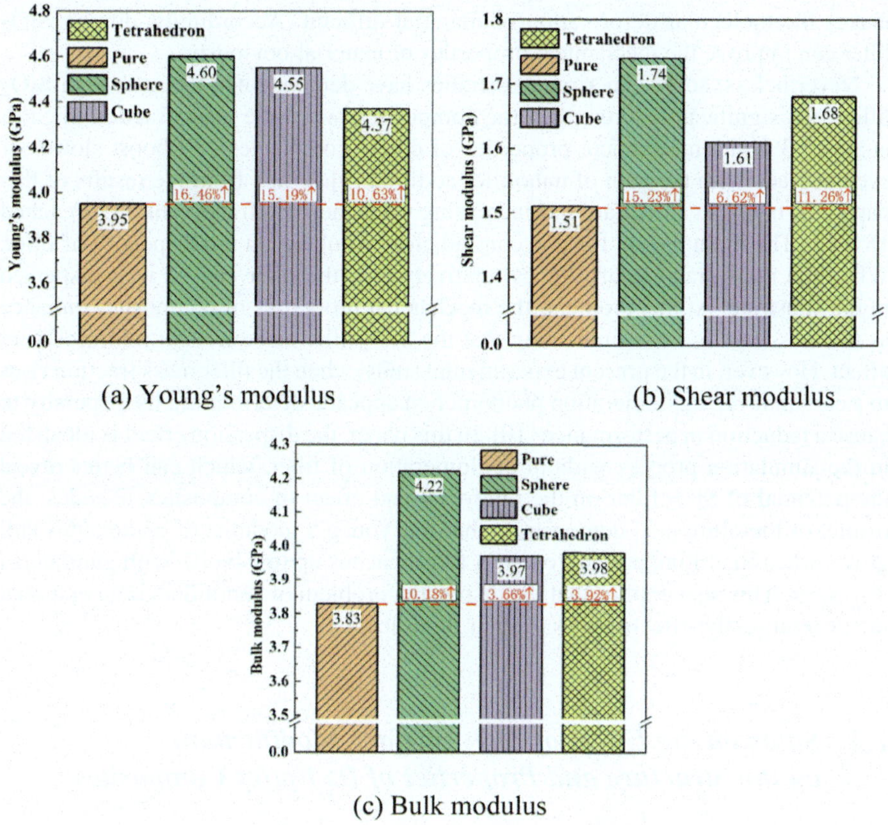

Fig. 14 Mechanical properties of composites with different filler morphologies

Fig. 15 Variation of mechanical properties of EP/Nano-SiO_2 model with SiO_2 mass fraction

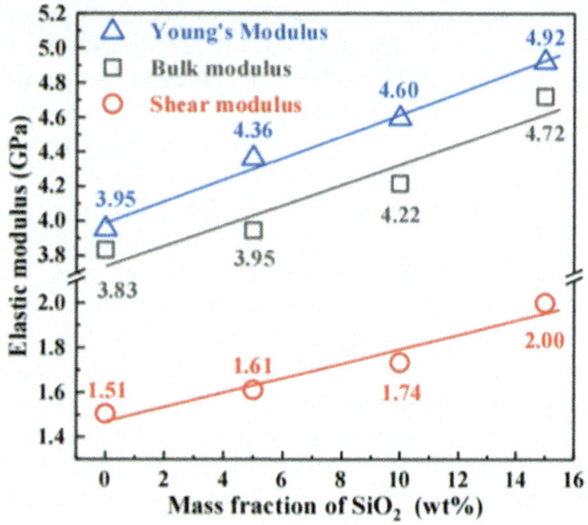

makes the molecular deformation of material difficult. Accordingly, doping SiO_2 filler can improve the mechanical properties of material obviously.

Nevertheless, some experimental studies have demonstrated that although SiO_2 fillers can significantly strengthen the composites, when the mass fraction of SiO_2 is above 3%, the mechanical properties of epoxy composites will boost slower or even degrade with the gain of nanofiller addition [9]. While from the results of this paper, the modulus values are still increasing when the mass fraction of SiO_2 reached 15 wt%. The main reason for this contradictory result lay in the dispersity of filler. When the mass fraction of filler is relatively low, the filler can be well dispersed in the substrate. At this moment, the mechanical properties of composites enhance as the mass fraction rise, indicating that the SiO_2 filler has excellent reinforcement effect. However, in the present experimental study, when the filler mass fraction rises to a certain level, agglomeration phenomenon appears, deteriorating its dispersity to cause a reduction in performance [10]. In this paper, the dispersion effect is idealized in the simulation process without agglomeration of filler, which can better reveal the potential of SiO_2 filler on the reinforcement effect of composites. Besides, the results of literature [11] demonstrate that the Young's modulus of composites still grows when the filling amount of SiO_2 filler reaches up to 14vol% with guaranteed dispersity. This also implies that the dispersion problem of nanofillers is an essential factor limiting the effectiveness of their modification.

1.3 Study on the Effect of SiO_2 Surface Modification on the Structure and Properties of Its Epoxy Composites

Nano-SiO_2 is extremely prone to agglomeration during the doping process since it has great surface energy and high surface activity [12], resulting in an enlargement of the filler particle size and loss of numerous properties as a nanomaterial, which leads to poor modification [13, 14]. Additionally, the surface of inorganic nanofillers is easily oxidized with a huge number of hydrophilic hydroxyl groups, contributing to a poor affinity between the filler and EP substrate. These problems restrict the filling amount and modification effect of nanofillers in EP composites. Studies have revealed that the dispersity of filler in the substrate can be improved by grafting the SiO_2 filler surface with silane coupling agent. However, current experimental methods pay more attention to the compatibility amelioration of filler with substrate by grafting on the surface of the coupling agent. While the influence of coupling agent itself on the microstructure and properties of material has rarely been reported. Furthermore, selecting the appropriate coupling agent for a particular composite system could be a lengthy and highly cost exploration process. To this end, three silane coupling agents are selected to graft Nano-SiO_2 filler in this section, with a Nano-SiO_2 surface modification model being constructed. The effects of various silane coupling agents and grafting rates on the microstructure and thermodynamic properties of EP composites are compared and analyzed based on MD simulations.

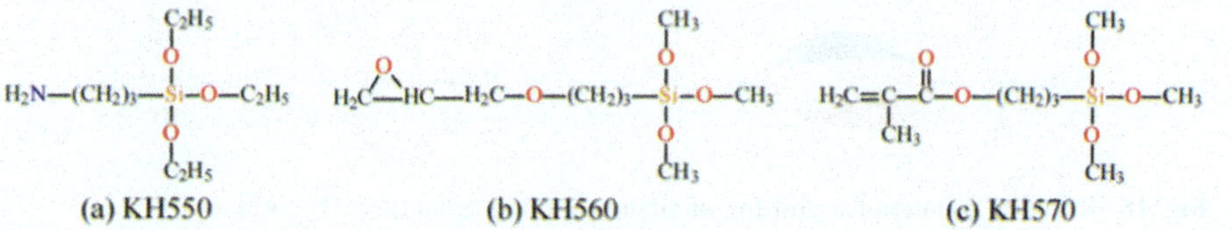

Fig. 16 Molecular structure formula of 3 common silane coupling agents

It is expected to provide some guidance for the selection of silane coupling agents in the production of Nano-SiO_2/EP composites.

Silane coupling agent is an organic substance with silyl groups, which is commonly used for surface treatment of micro and nano inorganic fillers. The commonly used coupling agent is generally trisilyl-based, and its molecular structure formula can be expressed as R_1–$Si(OR_2)_3$, where –R_1 is an organic group and –$Si(OR_2)_3$ is a silyl group. In this chapter, we select three silane coupling agents, 3-aminopropyltriethoxysilane (APTES, KH550), 3-glycidyl ether propyltrimethoxysilane (GOTMS, KH560), and 3-methylpropyltrimethoxysilane (MPTS, KH570) to simulate the surface modification of Nano-SiO_2 with a particle size of 15 Å. The structural formula is shown in Fig. 16.

After the hydrolysis of silane coupling agent, the formation of –OH groups on the Si atomic surface are required for grafting onto the SiO_2 surface. The mechanism of hydrolysis is that –H of water molecule replaces –R_2 in the coupling agent molecule, forming –OH group [15]. As shown in Fig. 17, the hydrolysis of a silane coupling agent results in the formation of three –OH groups on Si atoms.

In this section we used three coupling agents which have different functional groups, KH550 with an amino group, KH560 with an epoxy group, and KH570 with a methacrylate oxide group. The amino group will be involved in the crosslinking reaction of the epoxy resin, which needs to be pretreated in order to simplify the model crosslinking mechanism. For the amino groups of KH550, a DGEBA molecule is attached to each amino group by referring to the reaction mechanism of amino groups with epoxy groups in the literature [16]. In some conditions, the coupling agent can undergo a shrinkage reaction with the SiO_2 surface, forming covalent bonds and thus grafting the coupling agent onto the SiO_2 surface [17]. Figure 18 shows the reaction mechanism. Ideally, all the three –OH groups on the Si atom of the silane coupling agent will participate in the reaction. A total of five surface modification models of

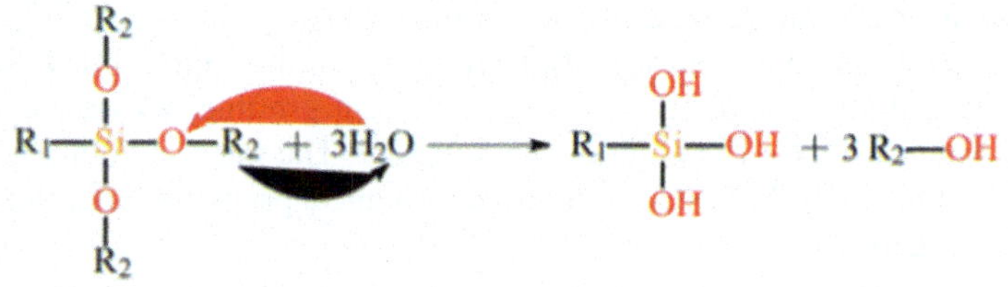

Fig. 17 Hydrolysis flux of silane coupling agent

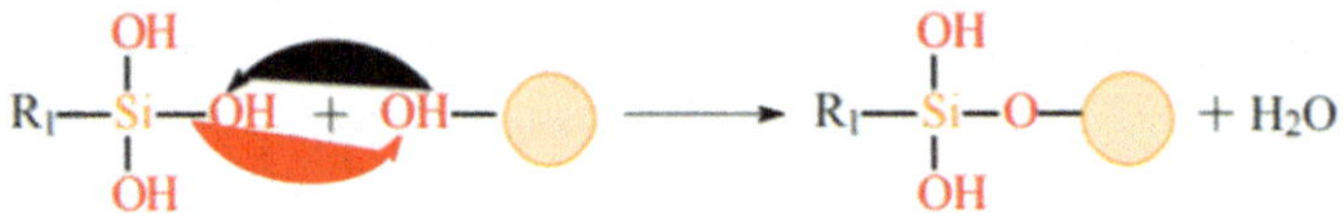

Fig. 18 Reaction equation for grafting of silane coupling agent on SiO_2 surface

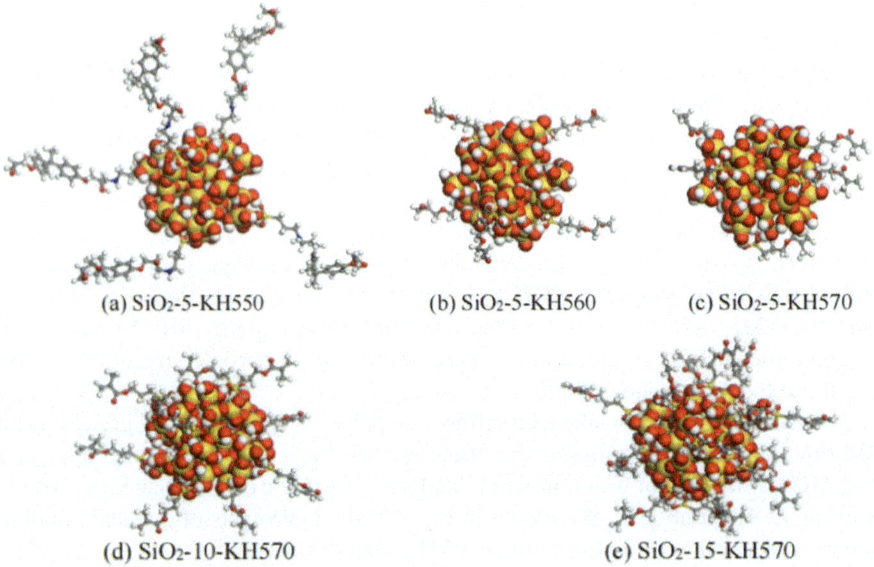

Fig. 19 Surface modification model of nano-SiO_2

Nano-SiO_2 are constructed in this section, as shown in Fig. 19. The naming rule is "SiO_2—number of graft—type of coupling agent", and the surface modification model of Nano-SiO_2 can be divided into two categories:

(1) Different coupling agent types: five treated KH550, KH560 and KH570 molecules are grafted onto the surface of spherical SiO_2 with a particle size of 15 Å, respectively, as shown in Fig. 19a, b, c; (2) Different grafting rates: 10 and 15 KH570 molecules are grafted on the surface of spherical SiO_2 with a particle size of 15 Å, respectively. According to Table 3, there are 64 –OH groups on the surface of spherical SiO_2 with a particle size of 15 Å. The grafting rates of the models shown in Fig. 19c, d, and e can respectively be calculated as 23%, 47%, and 70%.

After constructing the surface modification model, the EP/modified Nano-SiO_2 models with different coupling agents and different grafting rates are constructed using the Amorphous Cell module of MS software. The interfacial binding energy, diffusion coefficient, FFV, MSD and thermomechanical properties are calculated and analyzed, respectively.

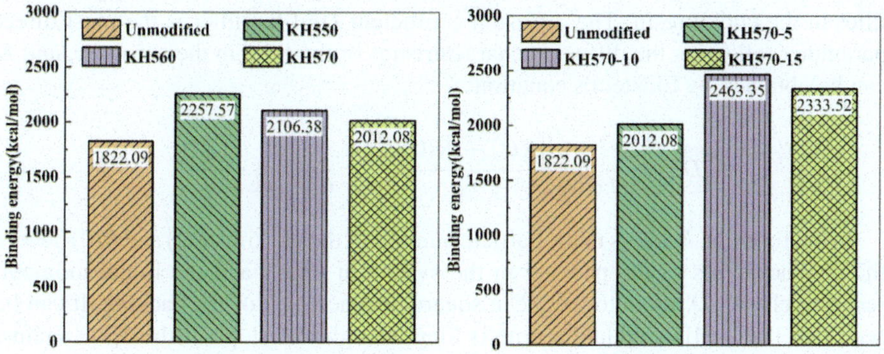

(a) Different coupling agents (b) Different grafting rates

Fig. 20 Bonding performance of modified SiO_2 with epoxy resin substrates

(1) Interface binding energy

In order to characterize the effect of Nano-SiO_2 surface coupling agent modification on the compatibility of filler and substrate, we calculate the binding energy between modified Nano-SiO_2 and epoxy resin substrate in this section, as shown in Fig. 20. It can be seen that the modification of SiO_2 surface by silane coupling agent can significantly improve the binding energy of SiO_2 filler and epoxy resin substrate. Among the three coupling agents, KH550 coupling agent modified system has the highest binding energy and KH570 coupling agent is relatively the lowest. The improvement of the bonding energy between filler and substrate indicates that the coupling agent improves the interfacial compatibility of filler and substrate. It enhances the bonding performance of both [18]. The reasons for this are twofold: one is that the coupling agent can reduce the hydrophilic hydroxyl groups distributed on the SiO_2 surface, thus weakening the hydrophilicity of the filler surface. The second is that the silane coupling agent is a lipophilic organic substance, and its grafting on the surface of the filler can improve the lipophilicity of the filler.

According to Fig. 20b, the binding energy of the filler to the substrate increases and then decreases as the amount of coupling agent increases. It indicates that there is an optimal amount of silane coupling agent to enhance the dispersion of filler, and too much amount will lead to a decrease in binding energy. We also found from the literature [19] that excessive amounts of coupling agents reduce the affinity of the filler with the organics. The reason for this may be that too much coupling agent wrapped around the filler surface increases the distance between the filler and the substrate, and it result in weaker Van der Waals interactions and lower binding energy [20].

(2) Diffusion characteristics analysis

To further analyze the compatibility of the modified Nano-SiO_2 filler with the substrate, in this section we analyzed the diffusion characteristics of the modified filler in the epoxy resin. The diffusion coefficient D of the filler is the most direct parameter reflecting the diffusion characteristics of the filler in the substrate, and D can be obtained by Einstein's equation:

$$D = \frac{1}{6N} \lim_{t \to \infty} \frac{\langle |R(t) - R(0)|^2 \rangle}{6t} = \frac{1}{6N} \lim_{t \to \infty} \frac{MSD}{6t} \tag{2}$$

In the formula, N refers to the particle number of the packing, $R(t)$ and $R(0)$ denote the displacements of the packing in the system at time t and the initial moment, respectively, MSD refers to the mean square displacement of the packing. It can be seen that if the MD simulation time is long enough, the slope of the MSD against time t can approximately characterize the diffusion properties of the filler and the larger the slope of the MSD, the larger the value of the diffusion coefficient.

The MSD of Nano-SiO_2 filler in EP/SiO_2 in composites before and after modification is shown in Fig. 21. The coupling agent can significantly enhance the diffusion behavior of the filler in the substrate. Compared to the three coupling agents, the diffusion behavior of the modified fillers was KH550 > KH570 > KH560 in order of strength to weakness. As the grafting rate of coupling agent increases, the diffusion coefficient of filler first increases and then decreases.

(3) FFV analysis

We gave the FFV of the EP/modified SiO_2 composite model in Fig. 22. All three silane coupling agents resulted in an increase in FFV values of the composites by about 1%, and the type of coupling agent have little effect on the free volume properties. The FFV of the composites tends to increase slightly with the increase of the coupling agent grafting rate. It shows that the coupling agent increases the free volume fraction of the composites.

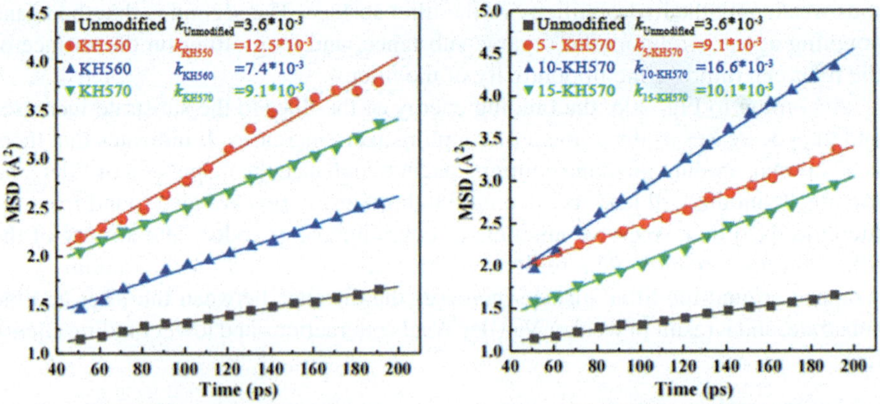

(a) Different coupling agents (b) Different grafting rates

Fig. 21 MSD versus time for nano-SiO_2 filler

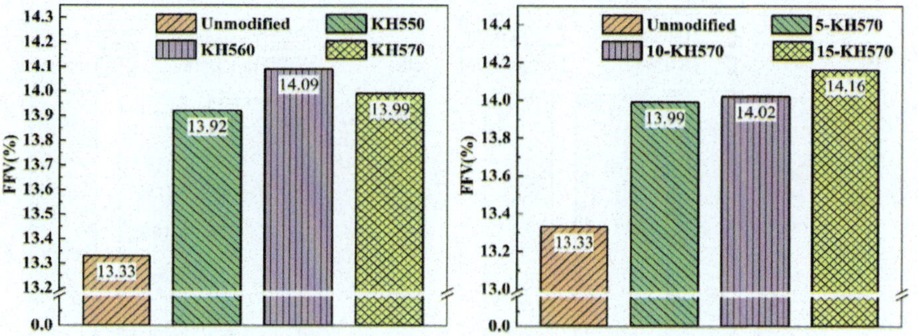

(a) Different coupling agents (b) Different grafting rates

Fig. 22 The FFV of EP/modified SiO_2 composite model

(4) MSD analysis

At normal temperature, Fig. 23 show the results of the all-atom MSD calculations for different system modification models. It can be seen that the MSD of the EP/ modified SiO_2 composite model is greater than that of the unmodified system. The MSD magnitudes of the three coupling agent-modified composite models are in the order of KH550 < KH560 < KH570. With the increase of coupling agent grafting rate, the MSD gradually increases. In other words, the coupling agent leads to enhanced chain segment motion in the composite system. In our analysis, the increase in MSD leads to a larger space of motion of the molecules in the model due to the increase in the free volume of the system by the coupling agent. On the other hand, according to Fig. 21, the coupling agent enhances the diffusion characteristics of the filler, resulting in an increase in the MSD of the filler. So the MSD of the composite system rises (Fig. 23).

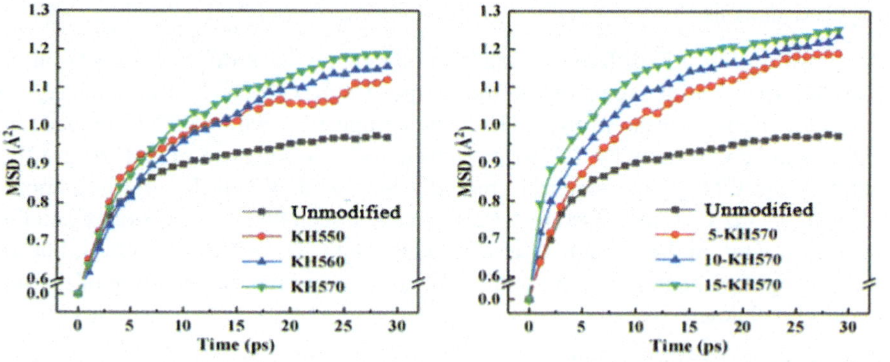

(a) Different coupling agents (b) Different grafting rates

Fig. 23 The MSD of EP/modified SiO_2 composite model

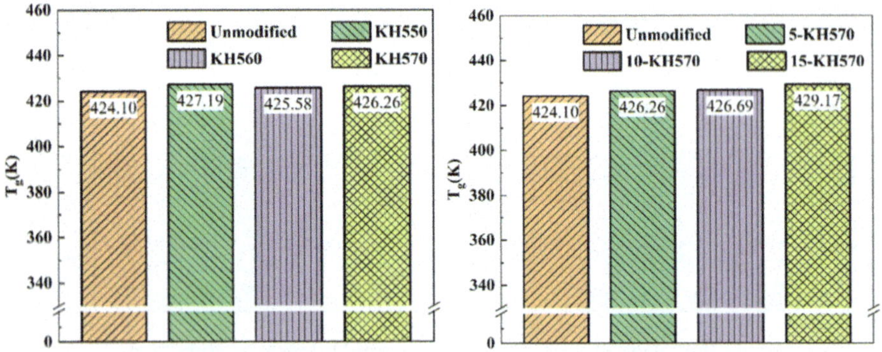

(a) Different coupling agents (b) Different grafting rates

Fig. 24 The T_g of EP/modified SiO_2 composite model

(5) Glass transition temperature

By linear fitting of the density-temperature curves during model annealing, we can obtain the T_g of EP/modified Nano-SiO_2 composites with different coupling agent types and grafting rates as shown in Fig. 25. When the filler is in the ideal dispersion state, the surface modification of SiO_2 filler by silane coupling agent can improve the T_g of the composite to some extent. Compared to the three coupling agents, the KH550 modified model is the most effective, with an increase in T_g of 3.09 K. The T_g of the composite increased with the increase in the amount of coupling agent grafted. When the grafting amount of KH570 coupling agent reached 15, the T_g increased by 5.07 K. However, when the grafting amount is 15, the grafting rate on the surface of SiO_2 filler has reached 70%, but the T_g is only elevated by 1.2%. It can be seen that the silane coupling agent has limited ability to enhance the T_g of the composite with good dispersion of the filler.

(6) CTE analysis

As shown in Fig. 25, different silane coupling agents have different effects on the thermal expansion characteristics of the composites. KH550, KH560 coupling agents lead to a slight increase in the CTE of the composites, while KH570 reduces the CTE of the composites by 7.14%. However, with the increase of KH570 grafting amount, the CTE of the material gradually increased. When the grafting amount reached 15, the CTE increased by 9.02%. It can be seen that the selection of suitable coupling agent modified SiO_2 filler can reduce the CTE of EP/Nano-SiO_2, but too much coupling agent will increase the CTE and lead to the deterioration of thermal expansion characteristics.

(7) Mechanical property analysis

In this section, we calculate the Young's modulus, shear modulus and bulk modulus of the EP/modified Nano-SiO_2 system with different coupling agents at 300 K to

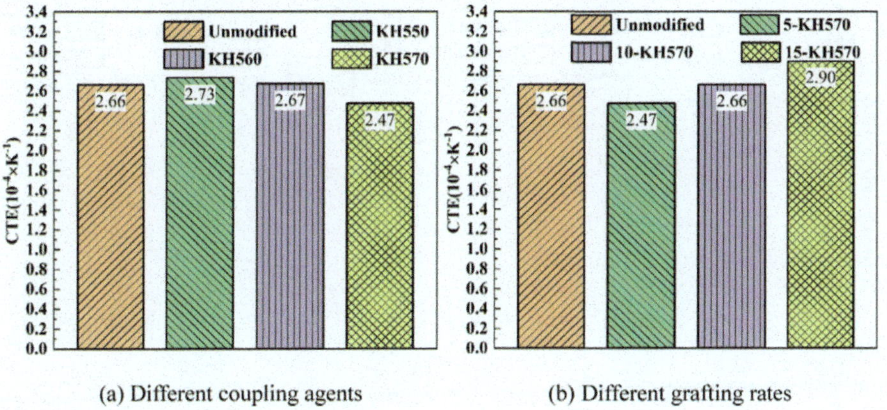

Fig. 25 The CTE of EP/modified SiO_2 composite model

characterize the mechanical properties of the composites. The results are shown in Fig. 26.

Compared with the unmodified composite model, all three moduli of the KH550 modified models show a significant decrease and the mechanical properties deteriorated. Both coupling agents KH560 and KH570 affect the mechanical properties of the composites to a comparable extent, but this effect is also slightly lower than that of the unmodified system. In Fig. 27 we give the effect of the amount of coupling agent on the mechanical properties of EP/modified SiO_2 composites. With the increase of coupling agent, the Young's modulus, shear modulus and bulk modulus of the composite show an overall fluctuating downward trend. When the grafting amount reached 15, the three modulus reduction rates are 6.74%, 13.22%, and 9.72%, respectively.

2 Model Construction and Performance Study of POSS/Epoxy Resin Composites

2.1 Study on the Effect of Nano-filler POSS Doping on the Properties of Epoxy Composites

Nanomodification of insulating materials is a new way to enhance the performance of epoxy nanocomposites. The doping of nanofillers can combine the excellent properties of fillers such as high toughness and high thermal conductivity with epoxy resin. The doping of nano-fillers can also enhance the thermal, electrical and mechanical properties of epoxy composites. In this section, two kinds of nano-POSS filler models were constructed, and two kinds of nano-POSS fillers with different mass fractions were doped into the epoxy resin substrate. According to MD simulation,

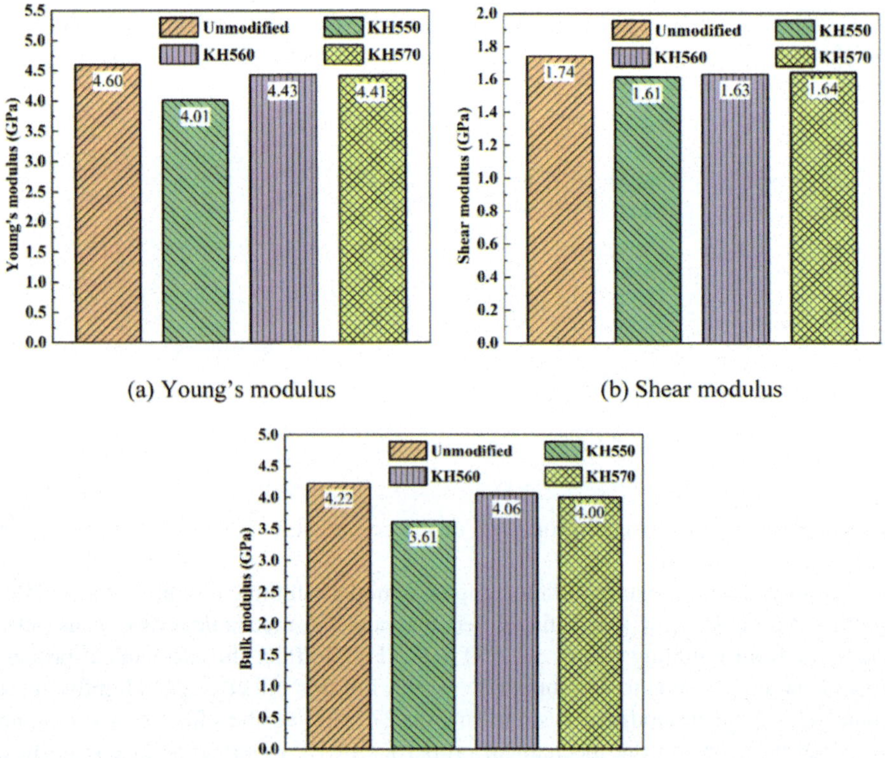

(a) Young's modulus

(b) Shear modulus

(c) Bulk modulus

Fig. 26 Mechanical properties of EP/modified SiO_2 composite model

we analyze the effects of the type and filling amount of nanofiller on the microstructure and thermal and mechanical properties of the double crosslinked epoxy resin composites.

Polyhedral oligomeric silsesquioxane (POSS) is an organic–inorganic hybrid material with the molecular formula $(RSiO_{1.5})_8$. The structure is shown in Fig. 28. It is connected by Si–O bonds to form a unique cage-like skeleton. Different organic groups R/X attach outside the backbone, such as methyl, phenyl, hydroxyl, etc. The particle size of POSS ranges from 1 to 3 nm and seven R organic groups are generally not active. The presence of seven R organic groups improves the compatibility of the POSS nano-structure with the polymer matrix. The X-active group can be chemically bonded to the polymer molecule, acting as a "rivet". Because of its low dielectric constant, good thermal stability and high mechanical strength, nano-POSS has great value in the field of thermosetting resin modification.

Fig. 27 Variation of model mechanical properties of EP/modified SiO_2 composites with the amount of coupling agent

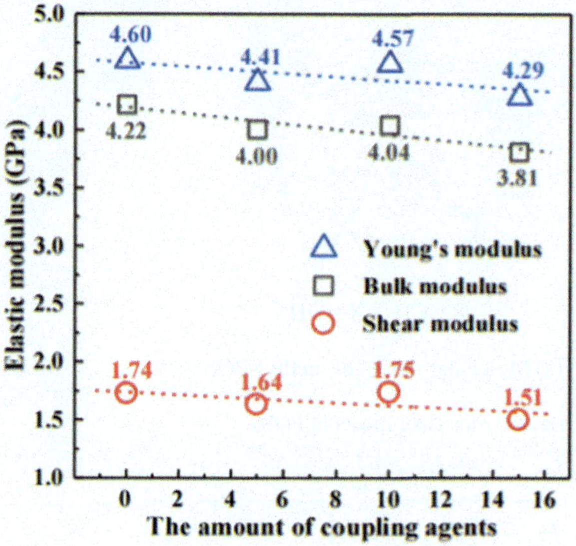

Fig. 28 Molecular structure formula of POSS

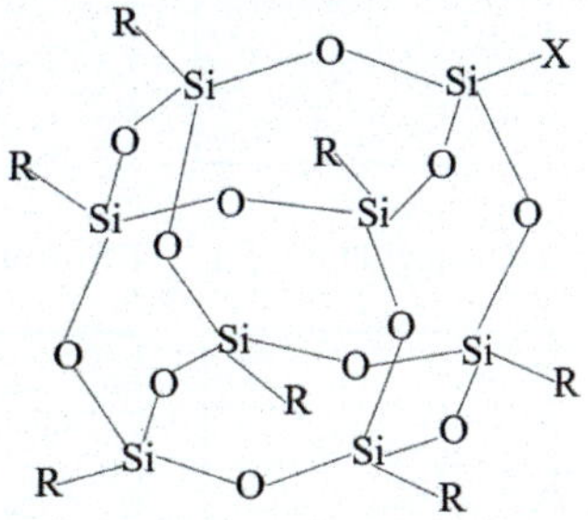

According to the molecular structure formula of POSS, two different nanofiller models are constructed in MS software: heptamethyl caged sesquisiloxane (methyl-POSS) and heptaphenyl caged sesquisiloxane (phenyl-POSS). And we optimize the structure, as shown in Fig. 29.

Based on the composite model construction method above, we constructed EP/POSS composites with filler (methyl-POSS) mass fractions of 5.1%, 9.6%, 15.1%, and 19.9% and filler (phenyl-POSS) of 6%, 11.4%, 16.2%, and 20.4%, respectively, as shown in Fig. 30.

Subsequently, this article analyze the FFV, MSD, radial distribution (RDF) and thermomechanical properties of the EP/POSS composite model.

(1) FFV analysis

As shown in Fig. 31, the nano-POSS filler can reduce the FFV of the epoxy composite system, and both the type and content of the filler affect the free volume percentage of the material. With the increase of nanofiller POSS content, the FFV of the system

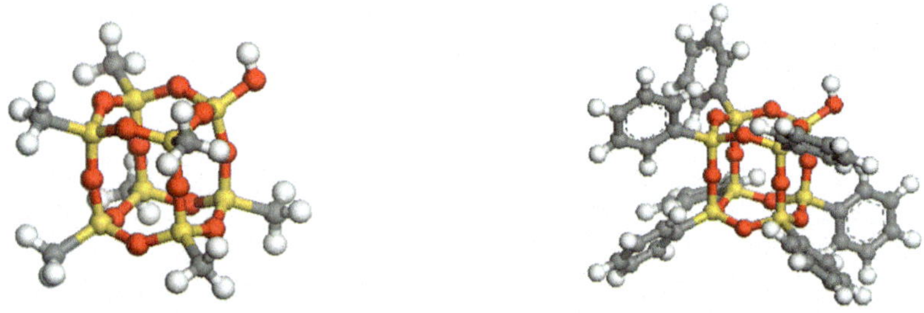

$R=-CH_3$, X=-OH

$R=-C_6H_5$, X=-OH

(a) Molecular model of methyl-POSS

(b) Molecular model of phenyl-POSS

Fig. 29 Molecular model of POSS

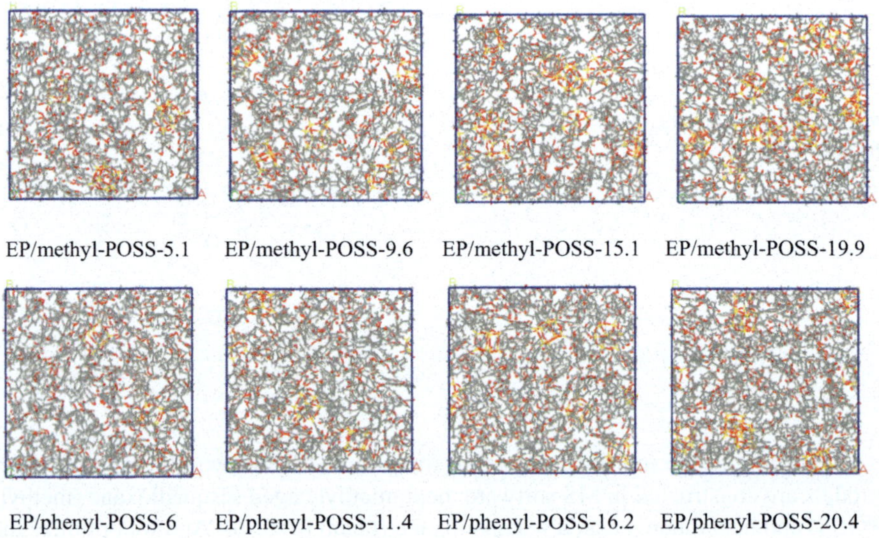

Fig. 30 Molecular modeling of nano-composite epoxy materials

show a law of first decreasing and then increasing. The magnitude of FFV at a filler mass fraction of 20% is approximately equal to the FFV of the double crosslinked epoxy system. It can be seen that the bonding interaction between POSS and epoxy molecular chain segments is greater than its own filling effect, when the content of POSS filler is around 0–10%. While the content of POSS filler is around 10–20%, the filling effect of nano-POSS itself is greater than its bonding with epoxy molecular chain segments. Meanwhile, the FFV of EP/phenyl-POSS system is smaller than that of EP/methyl-POSS system, which may be due to the better compatibility of the phenyl group of nano phenyl-POSS with the epoxy substrate.

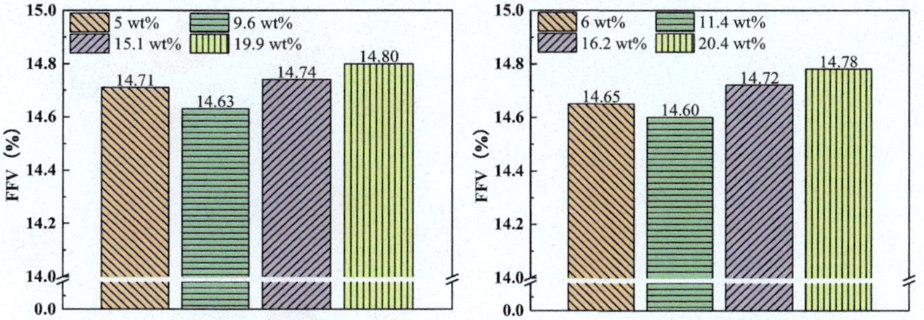

(a) FFV of EP/methyl-POSS system (b) FFV of EP/phenyl-POSS system

Fig. 31 FFV for different EP/POSS systems

(2) MSD analysis

As seen in Fig. 32, nano-POSS filler doping can restrict the motion of epoxy molecular chain segments, thus reducing the MSD of the epoxy system. With the increase of the nanofiller POSS content, the MSD value of the system show a pattern of decreasing first and then becoming larger. This trend is positively correlated with the results of the free volume share in the previous section. Therefore, for different EP/POSS systems, the free volume of the material is the main factor affecting the ability of the molecular chain segments to move.

(3) Radial distribution function (RDF) analysis

Radial Distribution Function (RDF) can reflect the characteristics of the material microstructure [21], whose value $g(r)$ reflects the magnitude of the relative probability of finding another atom or group of atoms at a distance r to the reference particle. The equation is:

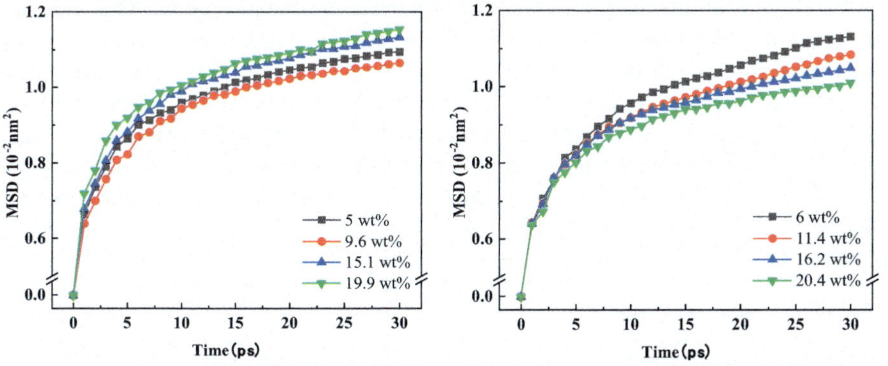

(a) MSD of EP/methyl-POSS system (b) MSD of EP/phenyl-POSS system

Fig. 32 MSD for different EP/POSS systems

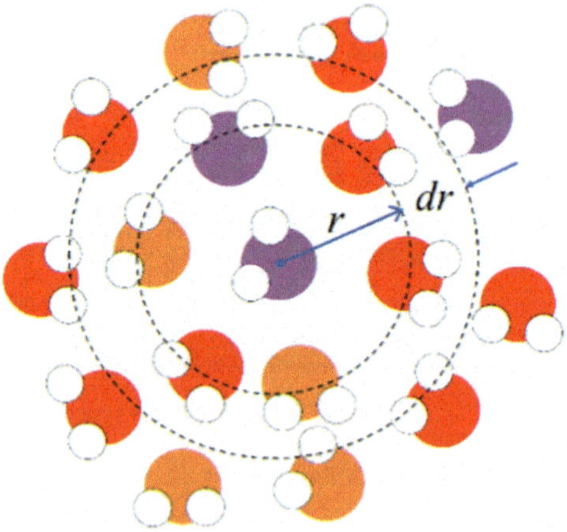

Fig. 33 Schematic diagram of RDF

$$g_{\mathrm{A-B}}(r) = \frac{n_B \cdot V}{4\pi r^2 dr \cdot N_B} \tag{3}$$

In the above equation: n_B is the number of B atoms at a distance r around the A atom. N_B is the total number of B atoms. V is the volume of the whole system. The schematic diagram is shown in Fig. 33.

By analyzing the radial distribution function of nanofillers in the epoxy system, we can obtain the influence of POSS on the stacking of epoxy molecular segments. Figure 34 shows the images of the radial distribution function obtained based on the nano-POSS in the epoxy system. In terms of the general trend, the RDF images of the different epoxy systems varied almost equally, achieving significant peaks at $r = 1.1$, 1.35, and 1.65 Å, respectively. The absence of any spikes and the gradual convergence of $g(r)$ to 1 at $r > 8$ Å are considered as evidence of the amorphous character of the epoxy system. As the POSS mass fraction increases, the peak $g(r)$ becomes progressively smaller. This suggests that the presence of nano-POSS reduces the probability of finding another particle within the same distance range. This indicates an increase in the distance between the chain segments of the epoxy molecule. This may be due to two reasons: first, the cage structure of nano-POSS plays a supporting role for the epoxy molecular chain segments, which increases the distance between molecular chain segments and is not conducive to the tight bonding of the crosslinked network. Second, when the mass fraction of nano-POSS gradually increases, it is easy to form agglomerates due to its own surface properties, which further increases the distance between epoxy molecular chain segments. This effect becomes stronger as the POSS mass fraction continues to increase.

(4) Glass transition temperature

In this section, we used the density-temperature fitting method for the study of T_g of EP/POSS system. The calculation results of T_g of different EP/POSS systems are

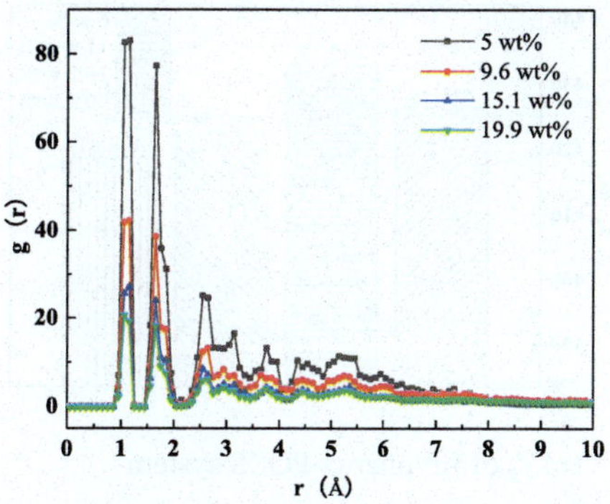

(a) RDF based on nano-methyl-POSS

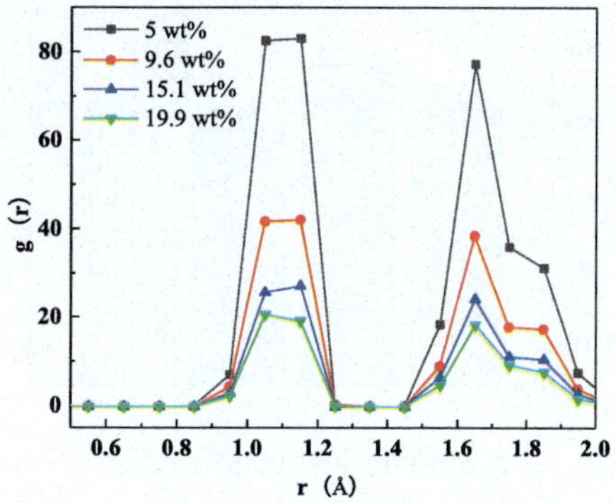

(b) Partial enlarged view of (a)

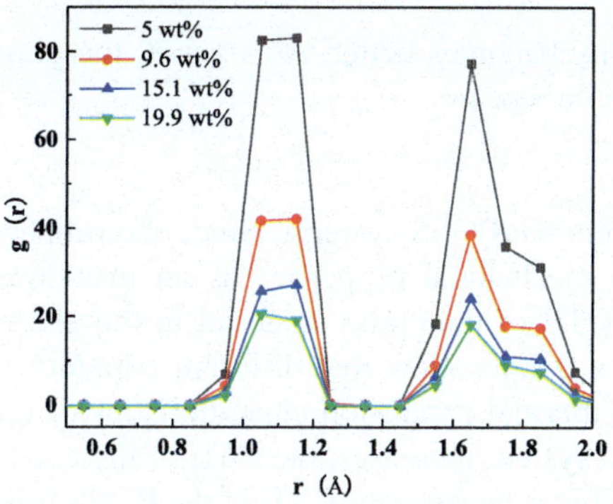

(c) RDF based on nano-phenyl-POSS

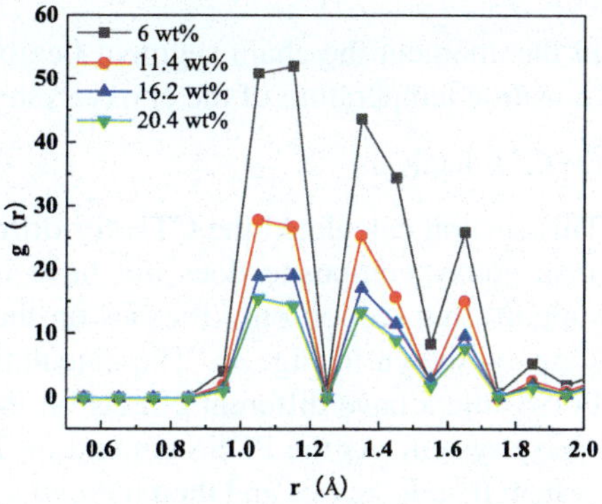

(d) Partial enlarged view of (c)

Fig. 34 RDF images of different EP/POSS systems

shown in Fig. 35. The calculation results show that: when the mass fraction of nano-POSS is low (about 0–10 wt%), it is uniformly distributed in the epoxy system. And it connected to the molecular chain segments in the form of chemical bonds. This connection acts as a strong mechanical interlocking effect limiting the movement of molecular chain segments and increasing the rigidity of molecular chain segments. At the same time, the addition of a small amount of nano-POSS has little effect on the build-up of the crosslinked network of the epoxy system. And the intermolecular chain segment forces are not significantly changed. As a result, the glass transition temperature of the epoxy system increases. When the POSS content continues to increase (about more than 10 wt%), it tends to agglomerate in the epoxy system. This leads to an increase in the distance between the epoxy molecular chain segments, the structure becomes loose and it reduce the forces between the molecular chain segments. At the same time, the bonding between nano-POSS and epoxy molecular chain segments is weakened, and the molecular chain segment motility is enhanced.

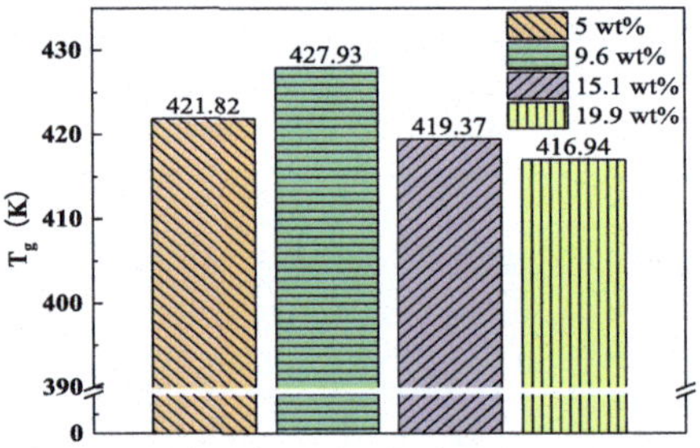

(a) T_g of EP/methyl-POSS system

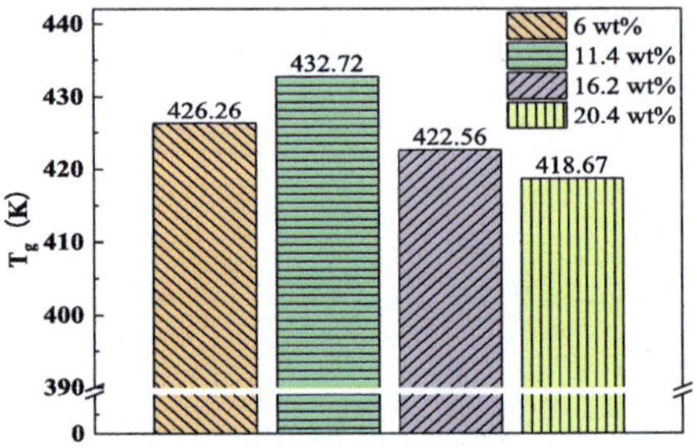

(b) T_g of EP/phenyl-POSS system

Fig. 35 T_g for different EP/POSS systems

In this moment the chain segment flexibility becomes better. As a result, the glass transition temperature of the epoxy system decreases.

(5) CTE analysis

This section calculated the CTE for different EP/POSS systems. Since the rubbery state epoxy composite does not have its mechanical properties as an insulating support, this section only focuses on the CTE of the epoxy material in the glassy state, as shown in Fig. 36. The calculation results show that different contents of POSS fillers have different effects on the thermal expansion characteristics of the epoxy system. As the POSS content of the system increases, the CTE of the epoxy system first decreases and then increases. This is because the CTE of the POSS filler itself is much smaller than that of the epoxy substrate. The introduction of a small amount of filler reduces the contribution of volume expansion by heat to the CTE. And as the mass fraction of filler increases, it hinders the accumulation of epoxy molecular chain segments. The crosslinked network structure becomes loose, which leads to poor thermal expansion characteristics. It can be seen that the selection of appropriate amount of nano-filler POSS can reduce the CTE of epoxy system.

(6) MSD analysis

The mechanical property parameters such as Young's modulus, shear modulus and bulk modulus of EP/POSS composites with different filler contents at room temperature (300 K) were calculated by the static normal strain method. The results are shown in Fig. 37.

From the figure, it can be seen that the mechanical properties of the system containing nanofillers are significantly better than those of the double crosslinked epoxy resin. And the performance enhancement effect is more significant when the nano-POSS content is 10%. Among them, Young's modulus increased by 4.20%, shear modulus increased by 6.02% and bulk modulus increased by 16.83%. On the one hand, the nano-POSS itself has excellent mechanical properties and a large elastic modulus. When it is introduced into the epoxy base material in appropriate amount as filler, it can play a good supporting role in the material, so that it is not

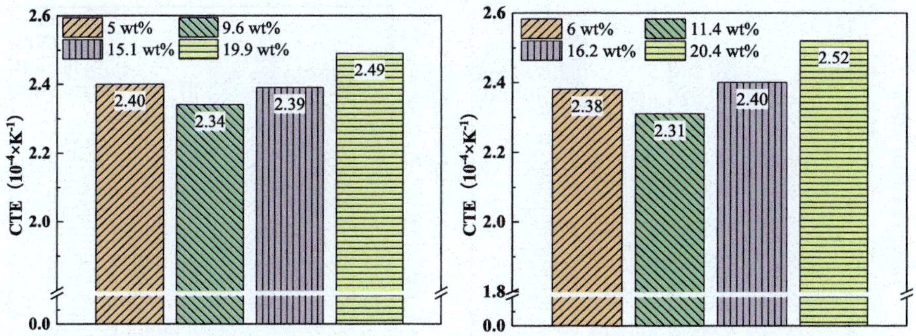

(a) Glassy CTE for EP/methyl-POSS system (b) Glassy CTE for EP/phenyl-POSS system

Fig. 36 Glass state CTE for different EP/POSS systems

easy to deformation due to external force. On the other hand, as the nano-POSS content continues to increase, its own emergence of agglomeration will affect the distance between epoxy molecular chain segments to a certain extent. This agglomeration enhances the motion of the molecular chain segments, which can lead to easy deformation of the material. Therefore, the epoxy system doped with appropriate amount of POSS nanoparticles presented better mechanical properties. Meanwhile, the mechanical properties of the EP/phenyl-POSS system are better than those of the EP/methyl-POSS system, which may be due to the greater rigidity of the R group (phenyl) of the nano phenyl-POSS.

2.2 *Study on the Effect of POSS Surface Modification on the Performance of Epoxy Composites*

In this section, we select three different functional groups to functionalize the X reactive group of heptaphenyl caged sesquisiloxane. These are phenol-functionalized heptaphenyl caged sesquisiloxane (P-phenyl-POSS), amine-functionalized heptaphenyl caged sesquisiloxane (A-phenyl-POSS) and fluorinated heptaphenyl caged sesquisiloxane (F-phenyl-POSS), respectively, as shown in Fig. 38.

Based on the above composite model construction method, three groups of EP/PPOSS composites with 10% filler mass fraction are constructed as EP/P-phenyl-POSS, EP/A-phenyl-POSS and EP/F-phenyl-POSS.

For the above model, we calculated and analyzed the FFV, MSD and thermomechanical properties respectively.

(1) Free volume characteristics

In order to determine the impacts of different surface modification methods on the free volume of EP/POSS systems, we calculate the FFV of different EP/POSS systems at

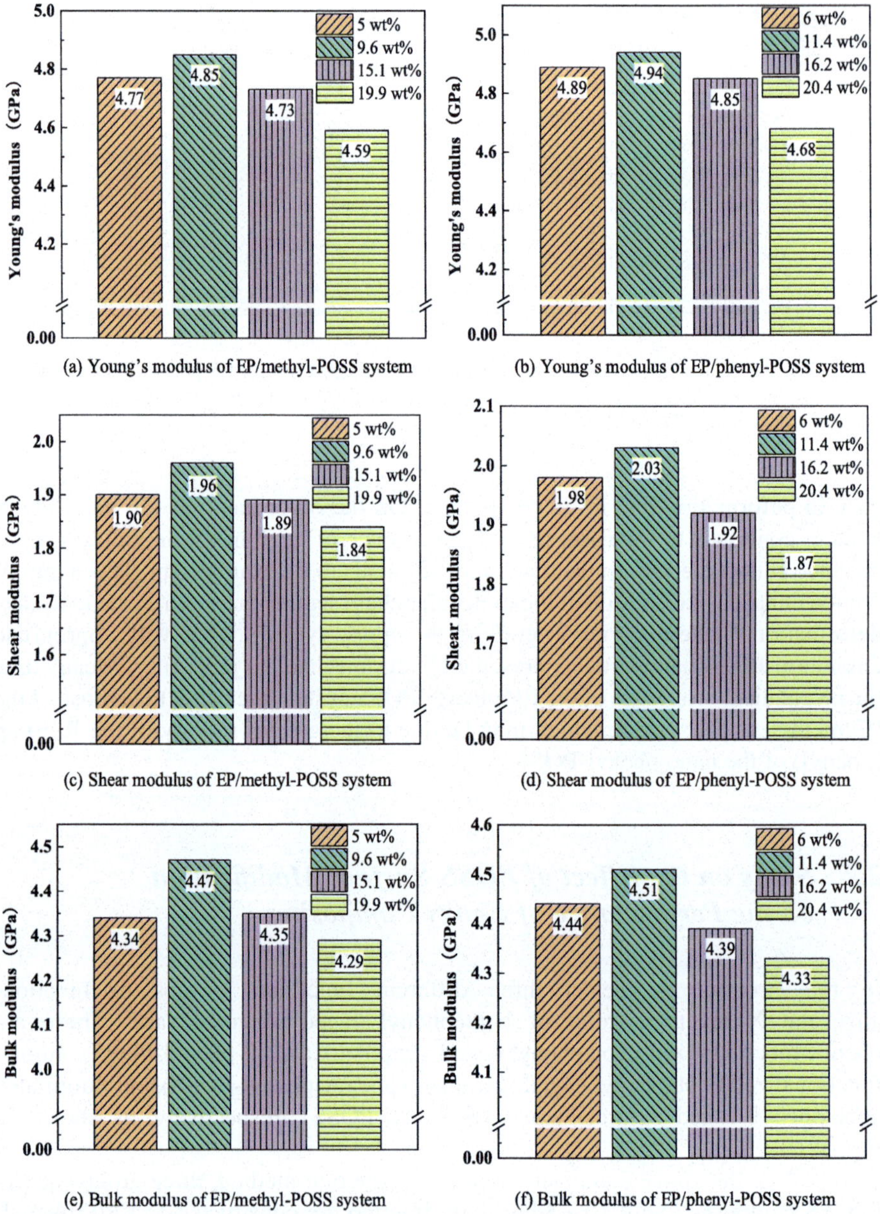

(a) Young's modulus of EP/methyl-POSS system

(b) Young's modulus of EP/phenyl-POSS system

(c) Shear modulus of EP/methyl-POSS system

(d) Shear modulus of EP/phenyl-POSS system

(e) Bulk modulus of EP/methyl-POSS system

(f) Bulk modulus of EP/phenyl-POSS system

Fig. 37 Mechanical properties of different EP/POSS systems

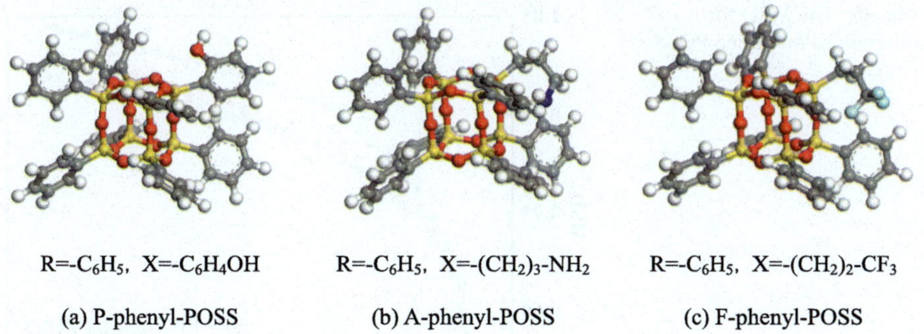

Fig. 38 Molecular model of modified POSS

room temperature (300 K), and the results are depicted in Fig. 39. The three surface modifications have a small effect on the free volume percentage of the epoxy system. Compared to the EP/phenyl-POSS system, the FFV of EP/P-phenyl-POSS system and EP/A-phenyl-POSS is slightly reduced and the FFV of EP/F-phenyl-POSS was slightly enlarged. It can be attributed that the nano P-phenyl-POSS and A-phenyl-POSS connect the epoxy molecular chain segments in a chemically bonded manner, while their compatibility with the epoxy substrate is better, consequently reducing the free volume. In contrast, nano F-phenyl-POSS mainly depends on van der Waals forces to bond with the epoxy substrate, and the interaction forces are weaker, thus increasing the free volume.

(2) Molecular chain segment motility

In order to characterize the effect of various surface modification techniques on the molecular chain segment motion properties of the EP/POSS system, we utilize molecular dynamics simulations to calculate the MSD values of different systems at room temperature (300 K). The results are shown in Fig. 40. It can be seen that the MSD

Fig. 39 FFV for different EP/modified POSS systems

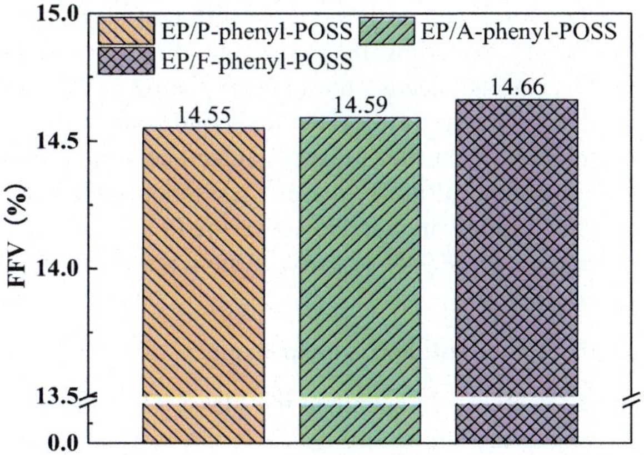

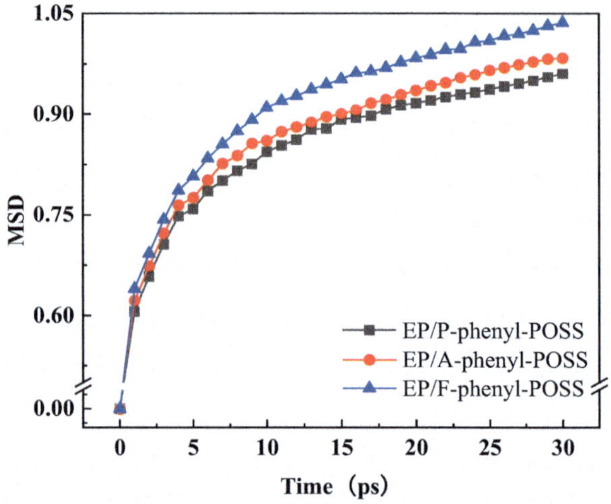

Fig. 40 The MSD for different EP/modified POSS systems

sort of the three surface-modified EP/PPOSS systems are ranked as EP/P-phenyl-POSS < EP/A-phenyl-POSS < EP/F-phenyl-POSS, which indicates that nano-filler surface modification can reduce the MSD of epoxy molecular chain segments and restrict the chain segment movement of the material molecules. Meanwhile, these findings are in accordance with the previous analysis of the free volume. It is once more demonstrated that there is a positive correlation between the molecular chain segment motility and the free volume of the epoxy system.

(3) Glass transition temperature

We can obtain the T_g of the EP/modified POSS system with different surface modification methods via linear fitting of the density-temperature curves during the annealing of the epoxy system. The result is shown in Fig. 41. It can be seen that when the mass fraction of filler is about 9%, the functionalized modification of nano POSS filler surface functional groups can improve the T_g of the composites to a certain extent. Compared with the three different surface modifications, the EP/P-phenyl-POSS system exhibits the greatest modification effect, resulting in a 6.4 K increases in T_g. This is because the nano P-phenyl-POSS reacts with the epoxy molecular chain segment to form a chemical bond connection. The interfacial interaction force is larger, which makes the cross-linked network structure tightly bound and confines the movement of molecular chain segments and thus the T_g of the epoxy system is correspondingly higher. However, the POSS surface modification has limited ability to enhance the T_g of the epoxy system under the doping with appropriate amount of nanofillers.

(4) Coefficient of thermal expansion

In this section, we calculate the glassy CTE of the EP/modified POSS system and the results are shown in Fig. 42. It can be concluded that different surface modifications have different impacts on the thermal expansion properties of the EP/modified

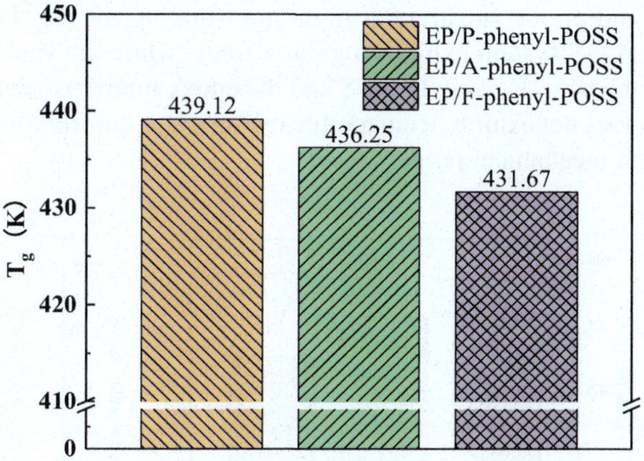

Fig. 41 The T_g for different EP/modified POSS systems

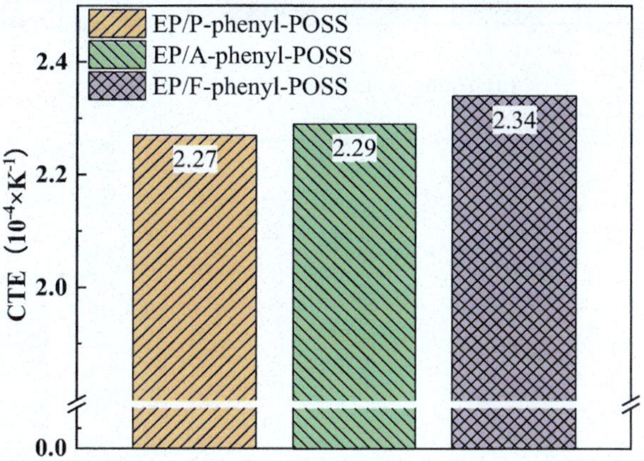

Fig. 42 The CTE for different EP/modified POSS systems

POSS system. In comparison to the EP/phenyl-POSS system, the CTE of EP/P-phenyl-POSS system and EP/A-phenyl-POSS is slightly reduced and the CTE of EP/F-phenyl-POSS is slightly enlarged. This could be attributed to the small thermal expansion coefficient of nano phenyl-POSS itself and the strong interaction between P-phenyl-POSS, A-phenyl-POSS and epoxy molecular chain segments.

(5) Static mechanical properties

In this section, we calculate the Young's modulus, shear modulus and bulk modulus of different EP/modified POSS systems at 300 K to characterize the mechanical properties of the composites, and the results are as follow in Fig. 43. Compared with the unmodified composite system, the mechanical properties of EP/P-phenyl-POSS system and EP/A-phenyl-POSS show somewhat enhancement, while the mechanical properties of EP/F-phenyl-POSS system are not significantly improved. This is due to there are strong interaction between nano P-phenyl-POSS and A-phenyl-POSS and epoxy substrate at high cross-linking degree. Furthermore, the EP/P-phenyl-POSS system exhibits a greater force, while the van der Waals interaction between the nano-F-phenyl-POSS and the epoxy matrix is weaker. Consequently, the system does not exhibit a remarkable resistance to deformation, and no significant improved in mechanical properties.

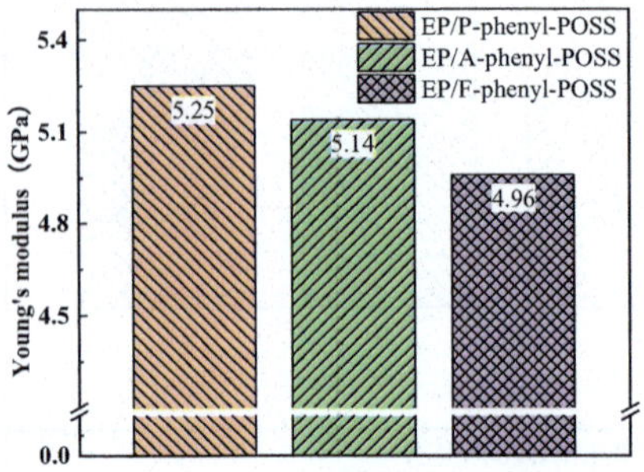

(a) Young's modulus

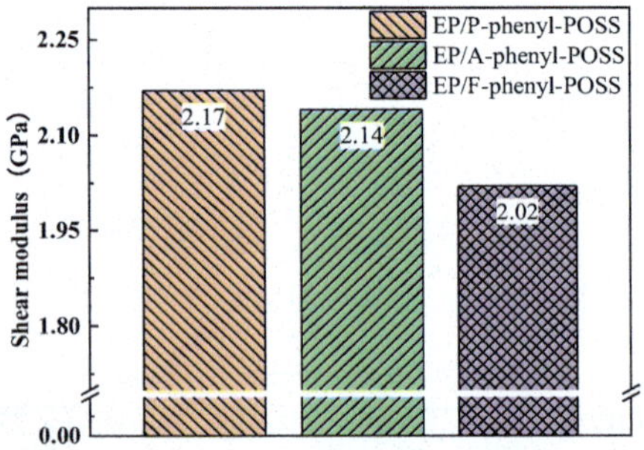

(b) Shear modulus

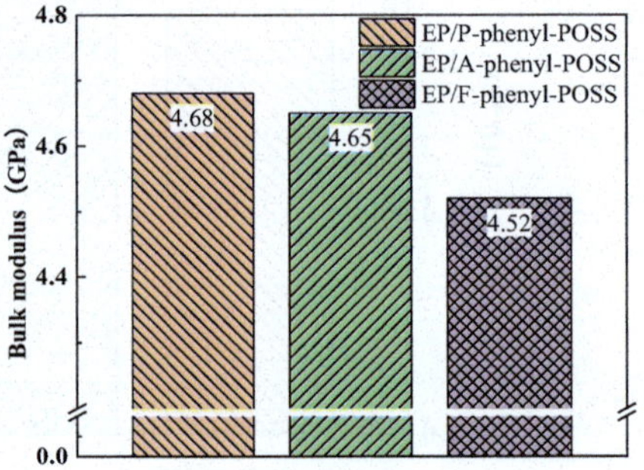

(c) Bulk modulus

Fig. 43 Mechanical properties of different EP/modified POSS systems

3 Model Construction and Performance Study of Graphene/Epoxy Composites

3.1 Research on the Effect of Graphene Doping on the Structure and Performance of Fluorine Resin Composites

The rapid growth of ultra-high voltage technology is driving the demand for higher performance of high-voltage epoxy insulation material for electrical equipment. The physicochemical properties of graphene are excellent and its derivatives can effectively be used as fillers to prepare better performance composites. This chapter takes BPAF/DGEBA fluorine resin as the fluorine epoxy matrix, and respectively chooses graphene (GR), graphene oxide (GO), and fluorinated graphene oxide (GF) as the fillers of the composites to construct models of various graphene/epoxy cross-linked network. Through molecular dynamics (MD) simulation, this research explores the influence law of modification method and doping amount of the graphene on microstructure and macro-performance of the fluorine epoxy resin.

Graphene is a two-dimensional carbon nanomaterial consisting of carbon atoms connected by sp2 hybrid orbitals, and it is a common inorganic filler in epoxy resin composites. There are 3 steps to construct the models of graphene with different modification methods.

(1) Import the crystal structure of graphene. Import the unit cell structure of graphite from the crystal base of Materials Studio, the coordination number of carbon atoms in the graphene model is 3 and the included angle of the carbon–carbon bond is 120°, as shown in Fig. 44.
(2) Construct the monomer model of graphene. After selecting the unit cell, we use the "Symmetry" of the Build module in the MS software and select Supercell in the menu bar to further construct a single-layer graphene supercell. This

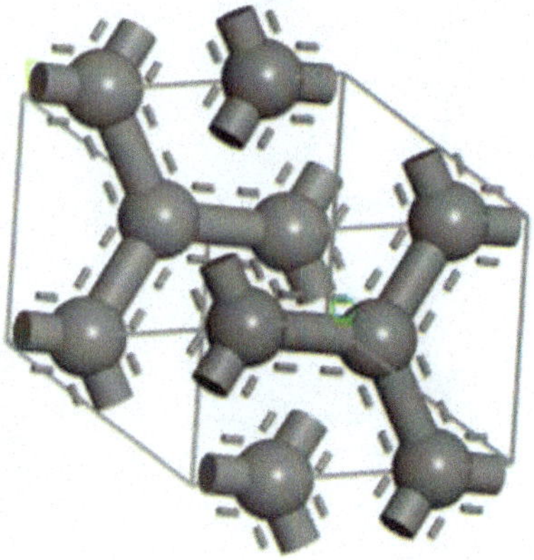

Fig. 44 Molecular model of graphite unit cell

method conducts geometry optimization of graphene monomer molecule and optimize the molecular configuration of it to the lowest energy configuration. The monomer molecular model of graphene constructed is shown in Fig. 46a.

(3) Construct the monomer model of modified graphene. Based on the "Lerf-Klinowski" theory, we construct the monomer molecular model of GO, and the simplest molecular formula of GO is $C_{10}O_1(OH)_1(COOH)_{0.5}$. The epoxy group and carboxyl group of GO model are randomly attached to the surface, while the carboxyl group is distributed at the edge of graphene oxide sheets. In this paper, we construct a GO sheet containing 82 carbon atoms and 22 oxygen atoms in order to represent the graphene oxide layer, and add hydrogen atoms to prevent the generation of unsaturated edge. The model of GO is shown in Fig. 45b. We further create fluorination modification of graphene oxide, and fluorine atoms are added to the edge of the graphene oxide according to a certain proportion of the total number of atoms to construct the model of modified graphene filler with a fluorinated surface. The model of GF is shown in Fig. 45c, and Fig. 45d–f shows different functional groups on the modified graphene surface.

The model of the GO/EP cross-linked network based on the method of constructing a composites model is shown in Fig. 46. Table 5 shows the compositions of each model. The subsequent articles calculate and analyze the interface binding energy, FFV, MSD, and thermomechanical properties of each model.

(1) The interface binding energy

Table 6 shows the binding energy of four different composites. The results indicate that the filler of the GF/F-EP system has maximum binding energy with the epoxy

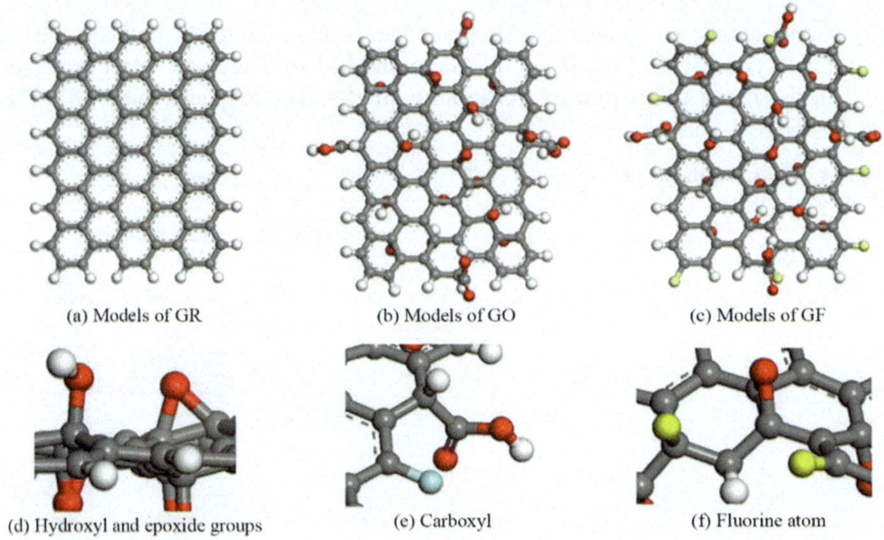

Fig. 45 Models of different modified graphene monomer molecules

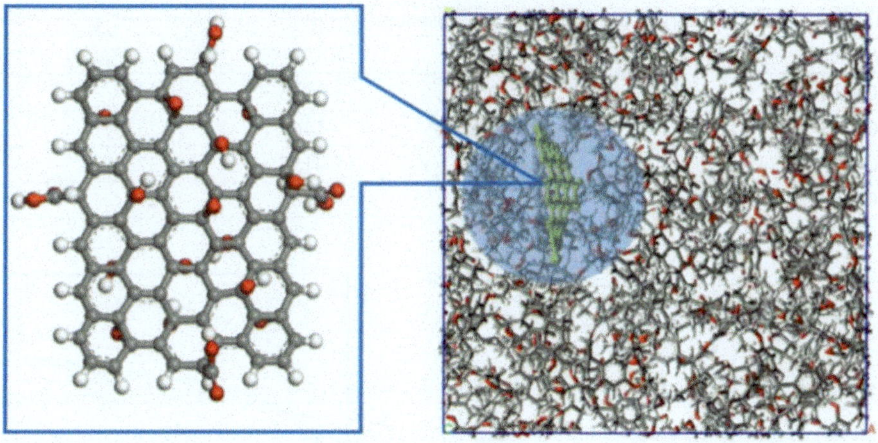

Fig. 46 Model of GO/EP cross-linked network

Table 5 Compositions of GO/EP epoxy nanocomposite system

Systems	Mass fraction of filler (wt%)	Number of molecules (number of articles)			
		Fillers	DGEBA	MTHPA	BPAF
GR/EP	2.8	1	50	100	0
GO/EP	3.9	1	50	100	0
GO/F-EP	3.2	1	50	100	25
GF/F-EP	3.5	1	50	100	25
GF/F-EP	6.7	2	50	100	25
GF/F-EP	9.8	3	50	100	25
GF/F-EP	12.6	4	50	100	25

matrix, while that of the GR/EP system is comparatively low, among the four epoxy composites systems. Our analysis demonstrates that modified graphene possesses various functional groups, such as the carboxyl group and hydroxyl group, which can react with the epoxy group to generate strong bonding effects, thus reinforce the combination between filler and epoxy matrix and improve the binding energy of nanofiller and epoxy matrix significantly. It also explicates that the surface modification of filler improves the interfacial compatibility between filler and base material and thus improve the interaction energy between the two.

(2) The FFV and MSD

Figure 47 shows the calculation results of FFV and MSD. The results indicate that nanofiller can notably reduce the free volume of the epoxy composite system and limit the motion ability of molecular chain segments effectively simultaneously. And a positive correlation appears between the FFV and MSD of the epoxy system.

Table 6 Binding energy of different epoxy composites

System	E_{resin}	E_{fiber}	E_{total}	$E_{interface}$
GR/EP	−1962	−2795	−3876	881
GO/EP	−1962	−769	−1467	1263
Cube	−4761	−769	−4157	1373
GF/F-EP	−4761	−805	−3766	1799

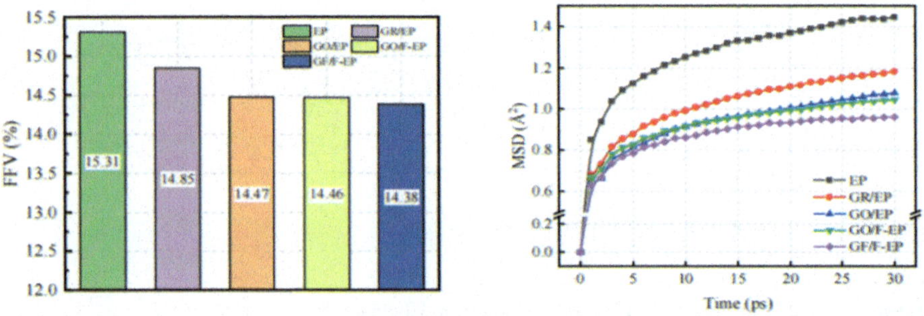

(a) FFV (b) MSD

Fig. 47 The FFV and MSD of different composites systems

Compared with the microscopic parameter of the cross-linked network of different systems, we can find that both FFV and MSD of GF/F-EP Epoxy composites system are the best.

The introduction of modified graphene will cause alterations in the free volume of the materials. On the one hand, the oxidation functional groups on the surface of the modified graphene can participate in the cross-linking and curing reaction with the epoxy system. They will generate bonding interaction with the epoxy group and create an inorganic–organic interface layer between the nanofiller and the epoxy molecules and then give rise to a strong mechanical interlocked effect and limit the movement of molecular chain segments, thus reducing the FFV of composites. On the other hand, part of the space volume of the epoxy system is occupied by the nanofiller, leading to an increase in the distance between molecular chain segments and impeding the stacking of molecular chain segments. This consequently promotes the free volume. Consequently, the leading factors of the above two elements determine the changing trend of the free volume of the epoxy system. Fluorinated graphene oxide has more functional groups when the mass fraction of filler is the same and has more sites for reacting with epoxy matrix. Moreover, higher levels of fluorine will generate hydrogen bonding, which enhances the intermolecular force in the system and effectively constrains the movement of molecular chain segments. GF at a low

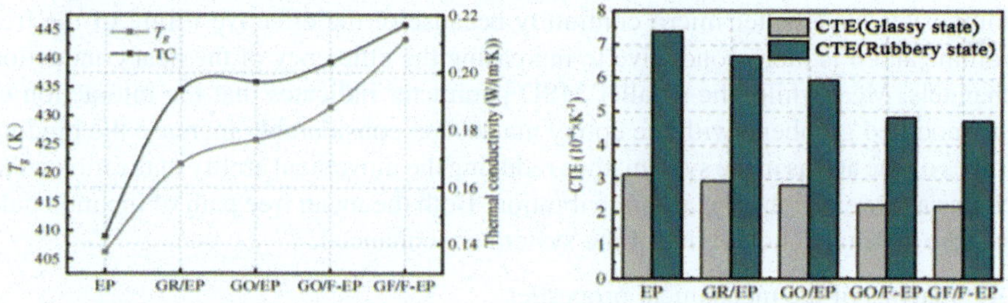

Fig. 48 The thermal properties of different composites systems

mass fraction can also fill the free volume of the cross-linked network better, thus reducing the FFV and MSD of composites.

(3) Analysis of the thermal properties

This section calculates the T_g, CTE, and thermal conductivity of different epoxy composites. And the results of the calculation are shown in Fig. 48.

The results demonstrate that the GF/F-EP system has the most ideal comprehensive thermal properties. Its T_g is 443 K, the thermal conductivity is 0.217 W/(m K), and its coefficient of thermal expansion is also improved considerably. For the glass transition temperature and coefficient of thermal expansion of the composites after doping modified graphene, both thermal properties of the composites are enhanced remarkably, compared with the DGEBA/MTHPA epoxy system. This suggests that doping modified graphene can boost the heat resistance of epoxy system.

The analysis concludes that graphene and modified graphene have better heat resistance than epoxy molecules, so the heat resistance of composites after doping modified graphene is noticeable increased. Following modification of graphene, a variety of functional groups on the surface of the modified graphene filler can react with the epoxy matrix, making a better dispersion and interfacial compatibility of GF in the epoxy resin. Combined with the calculation of FFV and MSD, we can find that a relatively dense cross-linked network and low molecular chain segments motion ability can promote the heat resistance of system effectively. Therefore, the T_g of the GF/F-EP fluorine resin system is relatively high and its thermal expansion performance is better.

It can be found that the epoxy composites doped with GF exhibit the highest thermal conductivity, compared with other different systems. It is likely that doping graphene or modified graphene can enhance the performance of the epoxy system effectively due to its impressive characteristics, such as ultrahigh carrier mobility of 15,000 $cm^2/(V\ s)$, excellent thermal conductivity of 5300 W/(m K) and large specific surface area. While the free volume ratio of the fluorinated modified graphene oxide/fluorine resin system is relatively small, and the cross-linked network is denser. The composites have better micro continuity because of the effective filling of the free volume and it is more conducive to increasing the efficiency of the heat conduction channels. Meanwhile, the smaller MSD parameter indicates that the interaction of the modified graphene with the epoxy matrix can considerably increase the binding forces on the atoms in the system, thus reducing the movement ability of the molecular chains and weakening the atomic vibration. Both the mean free path of the molecule and the thermal conductivity of the system are enhanced.

(4) Analysis of the mechanical properties

This section calculates the Young's modulus, shear modulus, and bulk modulus of different modified graphene doped composites at 300 K, and the results are shown in Fig. 49.

As illustrated in Fig. 49 that the mechanical properties of the modified graphene doped system are far more superior compared to that of the undoped epoxy resin system, and the mechanical properties improvement of the GF doped fluorine resin

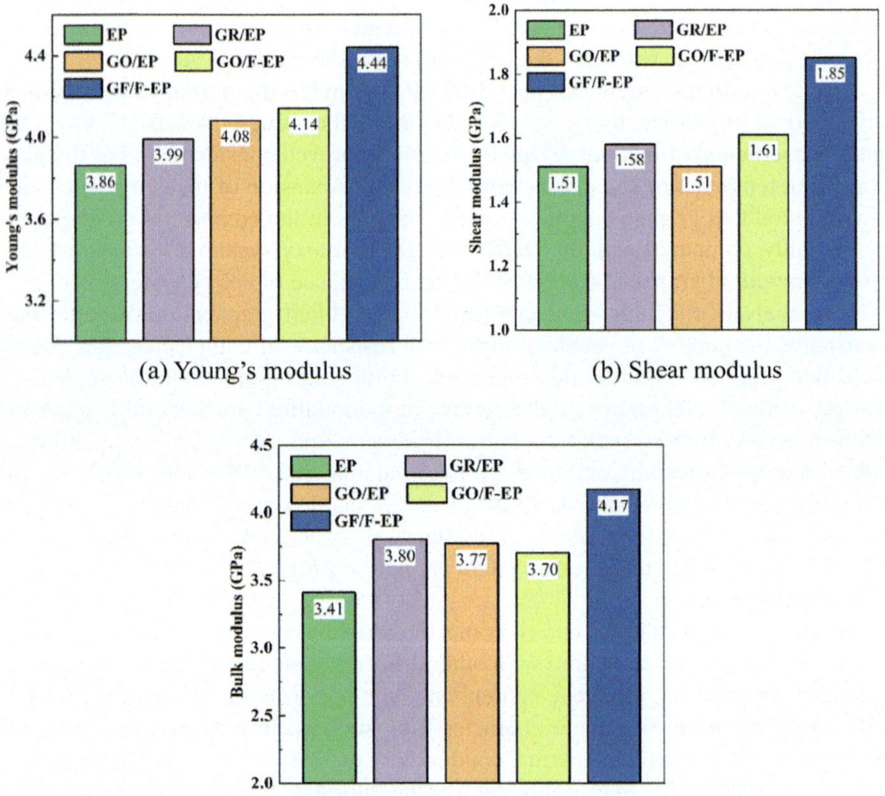

(a) Young's modulus (b) Shear modulus

(c) Bulk modulus

Fig. 49 The elastic modulus of different composites systems

is more significant. The GF/F-EP system has a 15.03%, 22.52% and 22.29% increase in Young's modulus, shear modulus and bulk modulus respectively, compared with the DGEBA/MTHPA system.

The analysis concluded that the modified graphene itself has excellent mechanical properties and a large elastic modulus. After introducing into the epoxy base material as a reinforcing filler, the modified graphene can support the material and fill the free volume space of the system. Therefore, it can form a dense cross-linked network and make the composites difficult to deform due to external force. Nevertheless, the molecules of GF have functional groups which can provide reaction sites. The interaction forces (encompassing Van der Waals force and hydrogen bonding) between modified graphene and the molecular chains of the epoxy matrix have been upgraded substantially by bonding the modified graphene and the epoxy group. And this also makes the combination between the epoxy matrix and the graphene closer. When the system has been exposed to external forces, this type of strong intermolecular force can meaningfully transfer the stress to the modified graphene filler, and further improve the mechanical properties of the composites. For this reason, doping modified graphene can enhance the mechanical properties of epoxy composites well.

3.2 *Research on the Effect of Orderly Filling of Modified Graphene on the Structure and Performance of Fluorine Resin Composites*

Because of the special molecular configuration, it is easy to generate a strong intermolecular interaction force between the different layers of graphene, which eventually leads to agglomeration in the composites and severely limits the application of graphene in high-voltage epoxy composites. This chapter chooses fluorinated graphene oxide/fluorine resin composites as the object. And we guide the ordered distribution of the internal structure of the epoxy composites by regulating the distribution state and dispersity of GF in the epoxy composites. Based on the BPAF/DGEBA fluorine epoxy base material, we add the GF filler with a mass fraction of 9wt% and then construct four epoxy composites models with fluorinating graphene oxide sheets spacing are 3 Å, 6 Å, 9 Å and 12 Å respectively and orderly filling by four fillers. And the influence law of the distribution state of modified graphene on the microstructure and thermal properties of the epoxy composites is studied.

In this section, we construct four models of the ordered GF molecules with sheets spacing of 3 Å, 6 Å, 9 Å, 12 Å and one disorderly distributed GF model respectively based on the "Construction" function of the "Amorphous Cell Calculation" module in the MS. And we fix the position coordinates of the GF molecules in the model. The fillers models are shown in Fig. 50.

The molecular models of the orderly distributed GF/F-EP composites are also constructed in this section, using the DGEBA/MTHPA as epoxy base materials. The mass fraction of the BPAF is 20 wt% and the mass fraction of the modified graphene

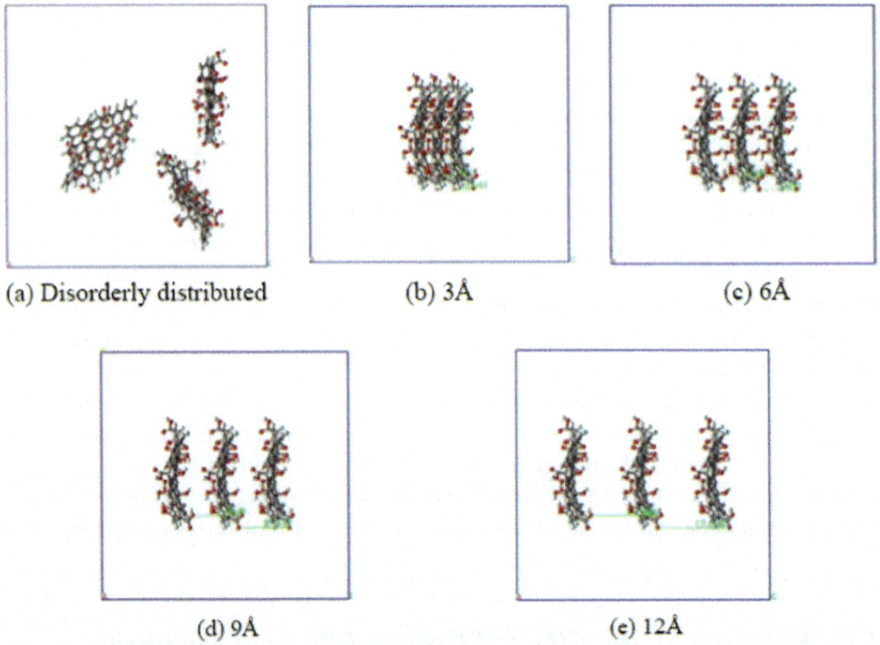

Fig. 50 The models of GF molecules with different distribution state

is about 9 wt%. The specific components of each system are shown in Table 7, and the composite material model is shown in Fig. 51.

An analysis of density distribution, FFV, MSD, and thermomechanical properties of the composites model have been conducted in subsequent articles.

(1) Density distribution

In this section, we studied the effects of the distribution state of graphene on the distribution of matrix molecules in the composites by calculating the relative concentration distribution curves of the modified graphene/epoxy composites models in the z direction (the direction perpendicular to the surface of graphene). By observing the

Table 7 The compositions of the EP/modified GF orderly filled systems

System	Sheets spacing of GF (Å)	Number of molecular (strip)			
		GF	DCEBA	MTHPA	BPAF
0-GF/F-EP	Random	3	50	100	25
3-GF/F-EP	3	3	50	100	25
6-GF/F-EP	6	3	50	100	25
9-GF/F-EP	9	3	50	100	25
12-GF/F-EP	12	3	50	100	25

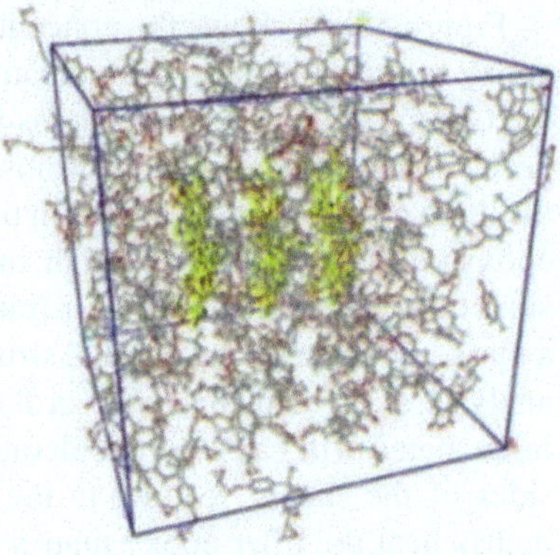

Fig. 51 The model of orderly filled GF/F-EP cross-linked network

density change of the composites, we can investigate the effect of the ordered distribution of modified graphene on the configuration of epoxy molecules. The epoxy matrix along the z-axis is divided into several small areas to calculate the density distribution. The results are shown in Fig. 52. Figure 52a shows the concentration distribution of the system of the GF disorderly distributed epoxy composites. It can be seen from the figure that the density distribution is uneven and has large fluctuations for the modified graphene disorderly filled epoxy composites.

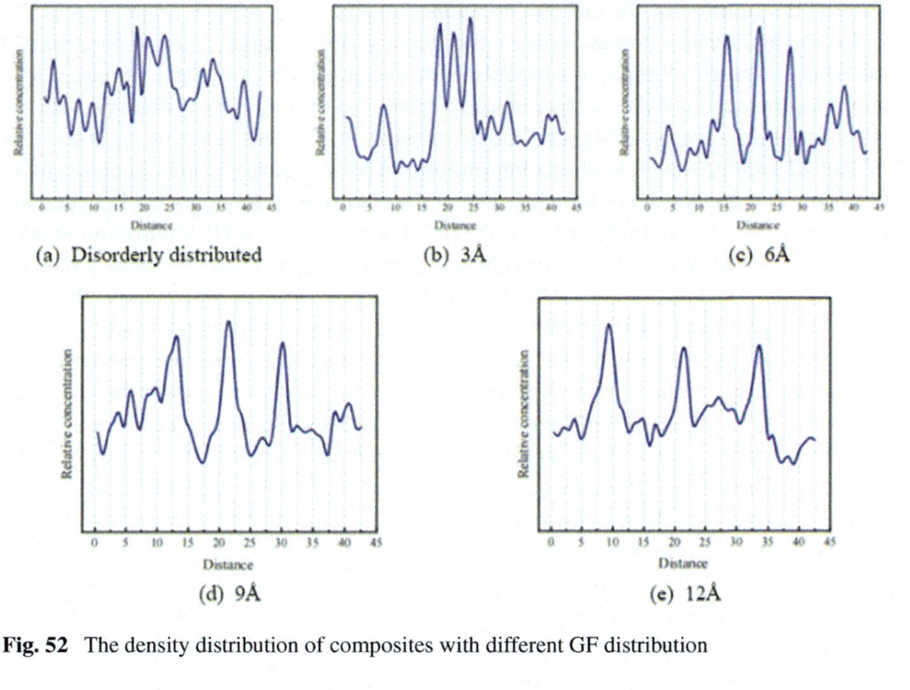

Fig. 52 The density distribution of composites with different GF distribution

Figure 52b–e shows the concentration distribution of the GF ordered distributed epoxy composites system, indicating that the concentration distribution curves of the GF/epoxy composites systems with different sheet spacing all appear with three peak values, and the peak positions are the z-direction coordinates corresponding to the three graphene insert layers in the system. The density of the system near the GF/epoxy interface is relatively high, and the density value sharply decreases along both sides of the interface, which illustrates that the ordered distribution of the fillers exerts a marked impact on the internal structural distribution of the composites system. The analysis suggested that after orderly filling the modified graphene, the graphene intercalated will cause a considerable amount of epoxy resin to accumulate on both sides of the sheets, leading to the adsorption zone in the composites. The epoxy matrix near the filler-epoxy matrix interface presents the distribution of "adsorption type". This also reveals that the layers spacing control proposed in this paper can acquire the epoxy composites with ordered structure. The formation of the adsorption interface layer is attributed to the strong interaction between the modified graphene and the epoxy resin where near the GF surface, which causes the graphene surface to adsorb more epoxy resin polymers, and consequently presents a peak in the density distribution curve at this location. Furthermore, the bonding interaction generated among the active groups such as carboxyl and hydroxyl groups will further enhance the interaction between the filler and the epoxy matrix, resulting in a remarkable surge in the adsorption capacity of the modified graphene to the epoxy resin, and a directional stacking of specific chain segments. These can more efficiently guide the ordered distribution of the GF/epoxy composite structure.

Following further comparison of the density distribution in the z-direction of the four GF/epoxy composites models with varying layer spacings, we can notice that the peak value of the composites adsorption layer gradually decreases with the increase of the layer spacing, and the decreasing in the density curve on both sides of the adsorption layer tends to be smooth. And the curve values become more uniform. The analysis concluded that with the increase of the distance between the GF sheets, the adsorption strength of the modified graphene on the matrix molecules diminishes steadily, and the interaction between the sheets is also weakened, thereby resulting in a decrease in the peak value of the adsorption layer gradually. Whereas the multilayer modified graphene enables the matrix molecules in the system to form multiple adsorption layers. When the GF layer spacing is too small, the graphene piles up and the dispersion is poor. Only the matrix molecules surrounding the graphene are able to occur "absorption", while the epoxy matrix molecules over a farther range from the graphene surface are largely unaffected by the adsorption of the graphene. Currently, the internal structural distribution of the composites is not uniform enough. As the dispersion of the graphene increases, the contact surface between the GF and the epoxy matrix broadens further, thus extending the area within the composites that are affected by adsorption. When the graphene in the composites is completely dispersed, the "adsorption" area within the composite is the maximized, and at this moment the orderly filling of the modified graphene has the pronounced effect on the structural distribution of the epoxy composites. However, excessive layer spacing will weaken the interaction forces between the modified graphene sheets, but diminish the adsorption of the GF in the matrix material.

To conclude, the orderly filling of the modified graphene will lead to the matrix molecules forming "adsorption" on both sides of the graphene, thus resulting in a more uniform distribution of the structure of the matrix material. An appropriate layer spacing will maximize the influence of ordered fillers on the adsorption force and adsorption range of the epoxy system. Therefore, it can be inferred that the orderly filling of the fillers can well guide the structural distribution of the epoxy system and can significantly contribute to the controllable interface of the base material.

(1) Analysis of the FFV

In this section, we calculate the free volume proportion of the GF/F-EP system with different layer spacing, and the results are shown in Fig. 53. As the layer spacing of the GF filler increases, the FFV of the composites system tends to decline before rising. Compared to the GF/F-EP system with randomly distributed fillers, the FFV of the composites is substantially reduced when the layer spacing of the modified graphene sheets is in the range of 3–9 Å. When the layer spacing is further increased, the FFV of the system increases slightly.

The analysis revealed that the randomly distributed modified graphene is not homogeneously distributed throughout the system and shows a tendency to form agglomeration. Due to the strong interaction forces, the adsorption occurs between the modified graphene sheets, which causes agglomeration, increases the free volume of the system and has a larger FFV. While the modified graphene with appropriate layer spacing will fill the epoxy system uniformly and introduce an inorganic–organic interface layer, with adsorption occurring on both sides of the GF sheets. In the interface layer, the filler and base material are mainly combined through Van der Waals interaction. The regularly distributed modified graphene layer will effectively direct the organized disposition of epoxy molecules and supplement the free volume better, thus reducing the proportion of the free volume of the system. When the layer spacing is excessively wide, the molecular distribution inside the model of

Fig. 53 The FFV of the composites with different GF distribution state

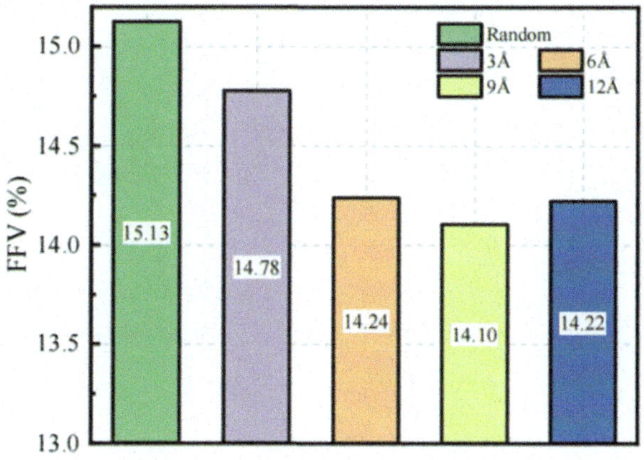

the modified graphene filled composites is too loose and the intermolecular force is relatively weak. More holes are generated, thus increasing the FFV.

(2) Analysis of the MSD

We calculate the MSD of the GF/F-EP systems with different filler distribution states at 300 K and the results are shown in Fig. 54. The MSD size of the fluorine epoxy composite systems with different GF distribution states is 9 Å < 6 Å < 12 Å < 3 Å < Random. This result appears in line with the calculation results of the FFV of the systems.

The analysis suggested that the adsorption layer can be formed in the system via the interaction force after orderly filling by the modified graphene. The oxygen-containing functional groups modified on the GF layer can bond with the epoxy matrix and further direct the distribution of the matrix chain segments, thus limiting the movement ability of the molecular chain segments usefully. A reasonable regulation of layer spacing can limit the segment motion of the material molecules effectively and reduce the MSD of the epoxy molecular chain segments.

(3) Analysis of the thermal properties

The influence law of the distribution state of the modified graphene on the T_g, CTE and thermal conductivity of the orderly filled GF/F-EP composites is shown in Fig. 55.

Comparing the T_g and CTE of the GF/F-EP systems with different filler distribution states, we can find that when the layer spacing of the GF is below 6 Å and above 9 Å, the orderly filling of the GF will lead to a certain degradation of the thermal properties of the GF/F-EP composites system. When regulating the layer spacing of the GF in the range of 6–9 Å, the orderly filling of the GF will increase the glass transition temperature of the fluorine epoxy composites significantly and decrease the CTE of the system.

Fig. 54 The MSD of the composites with different GF distribution state

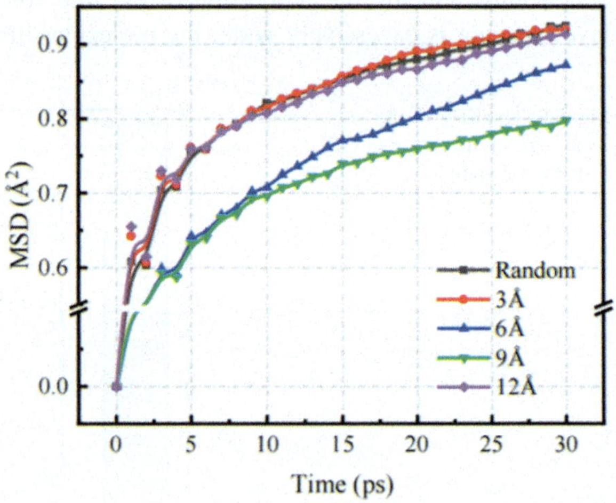

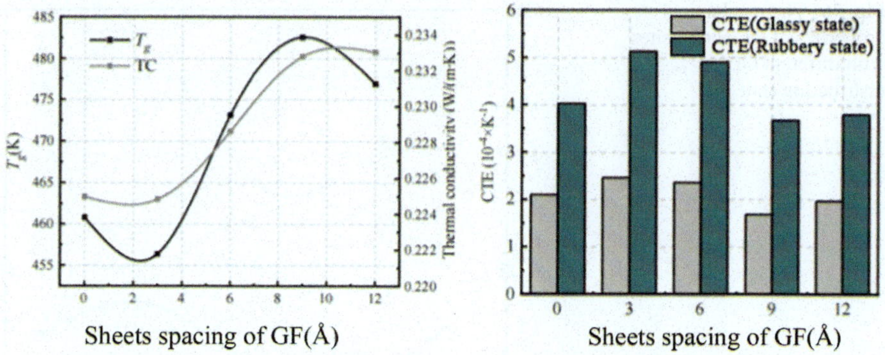

Fig. 55 The thermal properties of the composites with different GF distribution state

Analysis results suggested that when the layer spacing is too close, the modified graphene is not uniformly dispersed in the epoxy composites, thus occurring agglomeration and accumulation. However, excessive layer spacing will weaken the interaction forces between the graphene and between the graphene and matrix. In both cases, the advantages of structure regulation cannot be fully realized, resulting in the reduction of thermal properties.

A comparison of the Thermal Conductivity (TC) data of the composites system with different GF distribution states shows that the orderly filled modified graphene can significantly promote the thermal conductivity of the fluorine epoxy system. When the GF layer spacing is set to 9 Å, the TC of the system is increased by 3.47% in comparison to that of the disorderly distributed composites. The analysis shows that the graphene itself has typical anisotropic thermal conductivity. For the filler with anisotropic thermal conductivity, the common blending method only leads to a random distribution of the thermal conductive fillers within the matrix, making it difficult to take full use of their advantage of high thermal conductivity, and subsequently resulting in relatively low thermal conductivity values of the composites. After orderly filling the modified graphene, the uniformly dispersed oriented graphene sheet builds a bridging role for the BPAF/DGEBA composites on both sides of the sheets. This will make a significant reduction of the interface thermal resistance during heat conduction, forming an effective thermal conductive network. At the same time, the interconnected network structure between the graphene and matrix will be generated relying on strong interaction forces and thus generates a uniformly distributed adsorption layer. As the temperature rises, the heat flow is dispersed more rapidly along the uniform "adsorption" network structure, thus significantly improving the thermal conductivity of the epoxy composites. Nevertheless, if the layer spacing of the modified graphene is too close, the adsorption layer will be concentrated in the middle of the matrix and cannot form a uniformly dispersed heat conduction channel. While if the layer spacing is too large, the bridging effect will be invalidated, the interaction force in the system will be weakened, the structure of

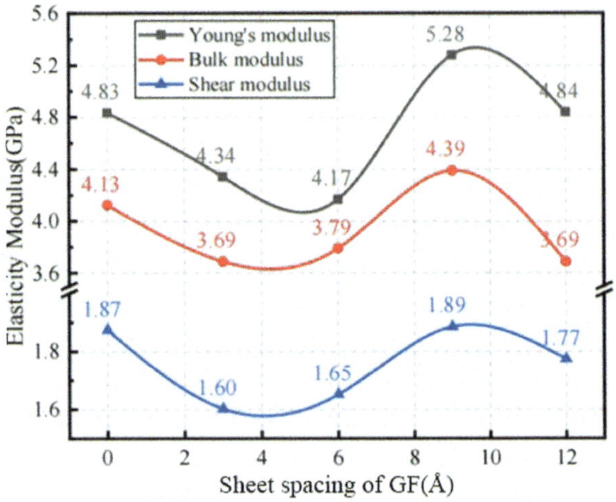

Fig. 56 The mechanical properties of the composites with different GF distribution state

the system will not be dense enough, and the thermal conductivity of the system will be remarkably weakened.

In conclusion, after reasonably controlling the layer spacing of the modified graphene, the orderly filling of the nanofillers can effectively promote the homogeneity of the epoxy composites structure, and substantially improve the thermal properties of the GF/F-EP composites.

(4) Analysis of the mechanical properties

The Young's modulus, shear modulus and bulk modulus of the GF/F-EP system are shown in Fig. 56. In comparison to the GF randomly distributed composites system, there is a significant enhancement of the mechanical properties in the GF/F-EP system with a layer spacing of 6–9 Å.

The analysis concluded that when the layer spacing is small, the distribution of the composite structures is uneven, and the characteristics of the high elasticity modulus of the modified graphene cannot be fully utilized. When the layer spacing is too large, the Van der Waals interaction force between the GF and epoxy matrix is weak, resulting in the weak resistance to deformation of the system and the decline of its mechanical properties. After regulating to a suitable layer spacing, the modified graphene can fully contact with the matrix material and form a uniformly dispersed adsorption layer. The external stresses imposed on the matrix material can be effectively buffered in the adsorption layer and transferred to the modified graphene sheets with excellent mechanical properties. The interaction force between the GF filler and the epoxy matrix also restricts the movement of the molecular chains within the system, resulting in a denser cross-linked network. The functional groups adhered to the modified graphene surface allow the epoxy molecular chains to be well anchored to the reinforcing filler, which also disperses the stress on the epoxy matrix. Therefore, the orderly distributed GF/epoxy structure will improve the mechanical properties of the composites.

References

1. Tang, C., Zhang, S., Li, X., et al.: Molecular dynamics simulations of the effect of shape and size of SiO_2 nanoparticle dopants on insulation paper cellulose. AIP Adv. **6**(12), 125106 (2016)
2. Li, F., Lu, Z., Yang, Z., et al.: Surface interaction energy simulation of ceramic materials with epoxy resin. Polimery **60**(07/08), 468–471 (2015)
3. Xiaoxing, Z., Xiaoyu, C., Song, X., et al.: Molecular dynamics simulation of thermal-mechanical properties of modified SiO_2 reinforced epoxy resin. High Volt. Eng. **44**(3), 740–749 (2018)
4. Wei, Q., Zhang, Y., Wang, Y., et al.: A molecular dynamic simulation method to elucidate the interaction mechanism of nano-SiO_2 in polymer blends. J. Mater. Sci. **52**(21), 12889–12901 (2017)
5. Marzocca, A.J., Somoza, A., Goyanes, S.N., et al.: Characterization of free volume in particulate-filled epoxy resin by means of dynamic mechanical analysis and positron annihilation lifetime spectroscopy. Polym. Int. **51**(11), 1277–1284 (2002)
6. Yaping, Z., Rongchang, N.: The study on structure characters and properties of PHMS/VAC/NMA composite emulsion. Polym. Mater. Sci. Eng. **05**, 148–150 (2002)
7. Liu, L., Xie, Q., Liang, S., et al.: Molecular dynamics simulation of the influence of molecular number on the properties of cross-linked epoxy resin system. High Volt. Appar. **54**(05), 80–86 (2018)
8. Nikje, M.M.A., Garmarudi, A.B., Tehrani, Z.M., et al.: Thermal and mechanical evaluation of epoxy resin composites by synthesis of amine-based coupling agent-nano silica complex. Polym.-Plast. Technol. Eng. **50**(6), 646–650 (2011)
9. Xueqing, L., Jiyan, L., Huyi, Z., et al.: Study of mechanical properties of rice husk silica/epoxy resin nanocomposites. China Plast. Ind. **11**, 18–20 (2009)
10. Wu, Z., Wang, M., Wang, Z.: The gas phase SiO_2/epoxy nanocomposites with enhanced mechanical and thermal properties. High Perform. Polym. **27**(4), 469–475 (2014)
11. Johnsen, B.B., Kinloch, A.J., Mohammed, R.D., et al.: Toughening mechanisms of nanoparticle-modified epoxy polymers. Polymer **48**(2), 530–541 (2007)
12. Kashfipour, M.A., Mehra, N., Zhu, J.: A review on the role of interface in mechanical, thermal, and electrical properties of polymer composites. Adv. Compos. Hybrid Mater. **1**(3), 415–439 (2018)
13. Li, H., Wang, C., Li, L., et al.: Effects of silane coupling agent treatment on properties of SiO_2/epoxy nanocomposites. Insul. Mater. (02), 1–8 (2017)
14. Xu, M., Cao, X., Yu, B.: Study on properties of nano-SiO_2/epoxy resin composite—electrical and mechanical properties. Polym. Mater. Sci. Eng. (01), 153–155 (2005)
15. Gao, Z., Jiang, X., Guo, K.: Study of the hydrolysis of 3-aminopropyltriethoxysilane (KH550) and the surface modification of silica. J. Beijing Univ. Chem. Technol. (Nat. Sci.) (02), 7–12 (2012)
16. Han, Z., Zou, L., Xin, Z., et al.: Molecular dynamics simulation of vital physical properties of epoxy/carbon nanotube composite coatings on DC GIL insulators. Trans. China Electrotech. Soc. (20), 4692–4703 (2018)
17. Xie, Q., Liang, S., Liu, B., et al.: Structure, microparameters and properties of crosslinked DGEBA/MTHPA: a molecular dynamics simulation. AIP Adv. **8**(7), 75332 (2018)
18. Yang, Q., Chen, X., Lan, F., et al.: Molecular dynamics study of the interfacial properties of polyethylene and ethylene-vinyl acetate copolymer composites. High Volt. Eng. (11), 3626–3633 (2016)
19. Li, J., Wang, B.: Study on the surface modification process of quartz sand with silane coupling agent KH570. Non-Metall. Mines (02), 26–29 (2013)
20. Zhao, Y., Qi, X., Ma, J., et al.: Interface of polyimide-silica grafted with different silane coupling agents: molecular dynamic simulation. J. Appl. Polym. Sci. **135**(4), 45725 (2018)
21. Liu, W., Wang, J., Miao, R., et al.: Molecular dynamics simulations of the cross-linked epoxy resin TGDDM/DDS. Polym. Mater. Sci. Eng. **32**(5), 109–114 (2016)

Modulation of Surface Properties of Epoxy Resin by Plasma Modification

Jiyuan Yan, Qing Xie, Yanze Song, and Yixiao Zhang

1 Introduction

The formation and development of surface flashover of insulating materials has the interwovenness connection with the surface conductivity and surface charge movement characteristics. And these electrical parameters are associated with the surface morphology and chemical composition of the insulating materials [1]. If a method can be found to selectively modulate the physicochemical and electrical properties of insulating materials, it is expected to further enhance the surface flashover performance of insulating materials and help reveal the enhancement mechanism. Studies have shown that plasma modification methods can change the surface properties of insulating materials and improve surface flashover performance [2, 3]. Moreover, the method is easy to operate, diverse in methods, and easy to control experimental parameters, providing an experimental instrument for the modulation of epoxy resin surface properties.

Plasma modification technology refers to the use of multiple effects generated when high-energy plasma comes in contact with the material surface to achieve changes in the physical and chemical properties of surface. With the characteristics of high efficiency, simple operation and low equipment cost, it has become a popular modification technology for insulating materials [4, 5]. In addition, the premise of plasma modification technology is how to generate plasma. According to whether the heavy particles and electrons in the plasma are in thermodynamic equilibrium or not, the plasma is divided into thermal plasma and non-thermal plasma. Since the former has almost the same temperature of heavy particles and electrons, the overall temperature reaching nearly 20,000 K, while the latter has much higher electron temperature than that of the heavy particles and also has the characteristics of high

J. Yan · Q. Xie (✉) · Y. Song · Y. Zhang
Department of Electrical Engineering, North China Electric Power University, Baoding, China
e-mail: xq_ncepu@126.com

Q. Xie (ed.), *Electrical Materials*, Engineering Materials,
https://doi.org/10.1007/978-981-99-9050-4_3

energy and high activity. Therefore, non-thermal plasma is often used to achieve surface modification of materials [6, 7].

Plasma can be generated in the form of gas discharge [8, 9] because it is composed of positive and negative ions in gas. Currently, the main plasma discharge methods involving in the modification of insulating materials are Dielectric Barrier Discharge (DBD) and Atmospheric-Pressure Plasma Jet (APPJ). DBD refers to the insertion of quartz glass, polytetrafluoroethylene (PTFE) and other insulating materials between discharge electrodes. In this way, the infinite growth of inter-electrode currents can be effectively inhibited, avoiding the discharge transitioning to arc discharge [10]. APPJ refers to introducting working gas into the tubular electrode at a certain rate. When the gas is discharged, the plasma generated is blown out of the discharge region, thereby separating the discharge area from modified area [11].

The selection of precursors in plasma modification determines which effect is used to modify the material surface. The precursors refer to the substances involved in plasma discharge. Due to the extremely high energy generated during the plasma discharge, the chemical bonds of the precursors are broken and rearranged. According to the properties of precursors and the effects on the material, the effects of plasma on the material surface can be divided into plasma etching, plasma grafting and plasma polymerization [12]. For plasma modification of insulating materials, the plasma discharge form and precursors are usually selected appropriately. And plasma etching, grafting and polymerization are used to regulate the surface physicochemical properties of insulating materials, thus improving the electrical performance.

Plasma etching effect refers to the breaking and reorganizing of the chemical bonds on the surface of the material under non-reactive atmosphere conditions (such as inert gas Ar or He) by high-energy plasmas, resulting in changes in the original morphology of material surface [13]. Since the plasma acts on the material at a depth of only a few hundred nanometers, plasma etching effect can change the microstructure of material surface without damaging the internal physicochemical structure of the material [14]. Studies demonstrate that the surface morphology of insulating materials can affect their electrical properties. Therefore, many scholars take advantage of the plasma etching effect to improve the surface flashover performance of insulating materials. With the help of a radio-frequency plasma jet device, Luan et al. used an inert gas Ar as carrier gas and mixed it with a certain proportion of oxygen as precursor, thereby completing the surface etching of polystyrene, PMMA and polyvinyl alcohol, and the relationship between oxygen concentration and etching depth was investigated [15]. Dai et al. developed a set of separated maskless etching devices based on the atmospheric pressure He/O_2 plasma jets and micro-electro-mechanical system nozzles. By regulating the distance between nozzles and insulating materials, oxygen concentration and etching time, this device can etch different micropore characteristics in the insulating materials [16]. Seok et al. adopted DBD method to complete plasma etching of SiC and modulated the etching depth of SiC surface by varying the proportion of NF_3 gas in He and altering the power supply parameters [17]. Shao Tao's team, at the Institute of Electrical Engineering of the Chinese Academy of Sciences, used microsecond pulsed power to perform APPJ etching on PMMA and investigated the mechanism of plasma interaction with

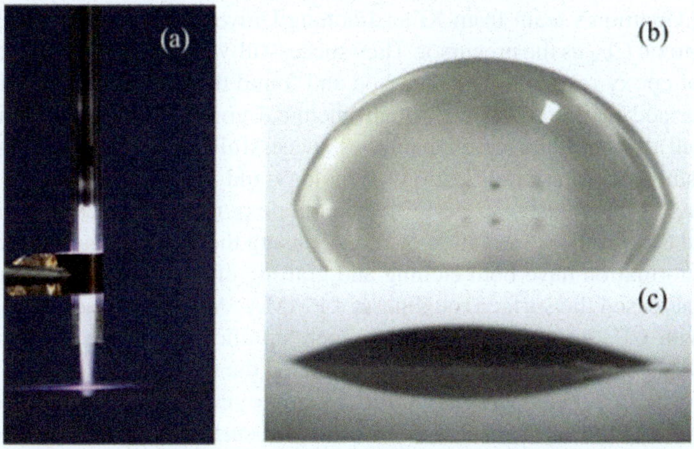

Fig. 1 Effect of plasma modification on surface hydrophilicity of insulating materials [18]

PMMA surface. It was found that APPJ etching can not only change the surface roughness of PMMA, but also enhance the hydrophilicity of PMMA in the plasma, which in turn improves the surface flashing performance of PMMA in vacuum [18]. The team used a similar approach to perform plasma etching on epoxy resin, finding that increasing the surface roughness of epoxy resin can effectively improve their surface flash performance. In addition, the surface chemical components of epoxy resin are also changed during plasma etching, which affects the trap energy level distribution and surface charge dissipation rate [19] (Fig. 1).

Studies above show that the adoption of plasma etching not only changes the surface micromorphology of insulating materials, but also introduces other groups with the participation of non-inert gases. Therefore, the plasma modification technology tends to utilize multiple plasma effects, and determine which effect dominates through the selection of plasma discharge form and precursor. The plasma grafting effect refers to the precursor contains functional groups or elements that need to be introduced onto the material surface. And the energy generated during plasma discharge used to break the chemical bonds between the precursor and the material surface at the same time. In this way, the functional groups in the precursor become free and a large number of reactive sites appear on the surface of material, and they will re-bond when coming into contact [20]. From the perspective of chemical composition, functional groups or elements in the precursors are grafted onto the surface of material. In the introduction of method for regulating the surface properties of epoxy resin, scholars propose the surface fluorination technique based on fluorine-nitrogen gas mixture, which introduces fluorine, the most electronegative element, into the material surface. However, fluorine gas itself is a highly toxic gas and has the drawbacks of low reaction effect and expensive equipment. Therefore, the plasma fluorination of insulating materials by using the plasma grafting effect has become a new direction of fluorination research.

Zhang Guanjun's team from Xi'an Jiaotong University used inert gas He mixed with non-toxic CF_4 as the precursor. They successfully implemented the surface fluorination of epoxy resin via APPJ method and found the chemical bond was formed on the surface of epoxy resin through chemical composition characterization. It was proved that the fluorine element was successfully grafted onto the surface of epoxy resin by plasma modification technology and plasma fluorination improved the vacuum surface flashover performance of epoxy resin [21]. Zhang Cheng et al. completed the plasma fluorination of PMMA using the DBD method and found that plasma fluorination have both etching and grafting effects, that is, the plasma fluorination increased the surface roughness of PMMA. And the surface of PMMA was grafted with CF_x groups additionally, which significantly improve the hydrophobicity of PMMA [22]. Zhan Zhenyu et al. studied the surface flashover performance of epoxy resin in C_4F_7N/CO_2 gas mixture by employing plasma fluorination method. They found that plasma fluorination increased the surface roughness of epoxy resin, leading to an increase in the creepage distance between electrodes. The presence of fluorine-containing groups accelerates the dissipation rate of negative charge, which ultimately enhances the surface flashover performance of epoxy resin [23]. Wan Fangchao et al. used Ar/CF_4 plasma fluorination to change the microscopic surface morphology of cross-linked polystyrene. The surface conductivity and surface charge capture ability of material were improved, finally enhancing the surface flashover performance [24, 25]. Xie Qing's team from North China Electric Power University investigated the timeliness of physicochemical properties of epoxy resin surface after plasma fluorination. They found that the epoxy resin after plasma fluorination had a tendency to return to its pre-modification physicochemical properties after a period of placement [26]. In addition, many scholars have compared the conventional surface fluorination technology with plasma technology. Peyroux et al. studied the effect of both surface fluorination and plasma etching on the hydrophobicity of insulating materials [27]. Zhang Cheng et al. thoroughly investigated the different effects of surface fluorination and plasma etching on the surface properties of epoxy resin/alumina basin insulators. It was found that the surface charge dissipation rate of the plasma etched sample will change considerably after a period of time, which suggested that plasma etching has a certain timeliness, while that of surface fluorination is not obvious [28, 29] (Fig. 2).

Plasma fluorination can not only modify the insulation surface, but also change the surface structure of micro-nano fillers, thus enhancing the mechanical and electrical properties of substrate. Matsubara et al. improved the hydrophobic properties of polystyrene powder by plasma fluorination [30]. Duan Qijun et al. carried out plasma fluorination on BTO nanofibers and incorporated the fillers into epoxy resin before and after modification respectively. Plasma fluorination of the filler was found to further enhance the surface flashover performance of epoxy resin composites [31]. Ruan Haoou et al. fluorinated Al_2O_3 and TiO_2 nanofillers respectively to study the effect of fluorination on the surface charge dissipation rate and surface flashover performance of epoxy resin matrix [32, 33]. Zhan Zhenyu et al. similarly improved the surface flashover performance of epoxy resin matrix by plasma fluorination of micron AlN [34], studied mixed the plasma fluorinated SiC nanofillers with epoxy

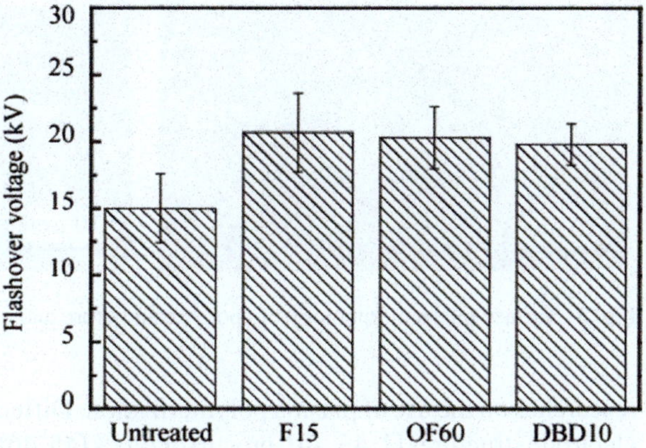

Fig. 2 Comparison of plasma modification and surface fluorination on surface flashover voltage [29]

resin to form a new coating and the effect of the coating on the physicochemical and electrical properties of epoxy resin surface [35].

The plasma breaks the chemical bonds of both the material surface and precursor, and recombines the free groups in the precursor with the active sites on the material surface. So is it possible to use plasma to directly break and recombine precursors to generate new substances and deposit them on the material surface? Many scholars have conducted in-depth research on this and concluded that the method takes advantage of the polymerization effect of plasma. Plasma polymerization usually refers to the method of forming coatings by polymerizing monomers in plasma environment. In the research of insulating materials, the plasma polymerization effect is usually used to cover the film on the surface of the material, directly changing the physicochemical properties of gas–solid interface, which in turn modulates the surface electrical properties of insulating materials [6]. Armenise et al. successfully deposited films containing carboxylic acid groups on polyurethane using APPJ with helium, acrylic acid and ethylene as precursors [36]. Kakiuchi et al. similarly used APPJ to successfully prepare SiOC films at atmospheric pressure with hexamethyldisilane and oxygen as precursors [37]. Using $(CH_3CH_2O)_3SiCH_2CH_2(CF_2)_7CF_3$, $(CH_3O)_3SiCH_2CH_2CF_3$ and $F_5ArSi(OCH_2CH_3)_3$ as precursors, Yim et al. deposited hydrophobic layers on the surface of ultrahigh molecular weight polyethylene (UHMWPE) by plasma polymerization [38]. The team of Xie Qing at North China Electric Power University used tetraethyl orthosilicate (TEOS) as precursor. SiO_x films were deposited on the epoxy resin surface using APPJ, and it was found that the SiO_x films covered the original defects on the epoxy resin surface. And a large number of shallow traps were introduced, which greatly increased the surface charge dissipation rate. This inhibits the accumulation of charge on the epoxy resin surface, which in turn enhances the surface flashover performance [39–41]. Additionally, the team successfully deposited ZrO_2 films on the epoxy resin surface taking zirconium nitrate as precursor, effectively improving the surface charge dissipation rate and surface flashover performance of epoxy resin [42]. Shao Tao's team at the Institute of Electrical Engineering, Chinese Academy of Sciences, has conducted extensive

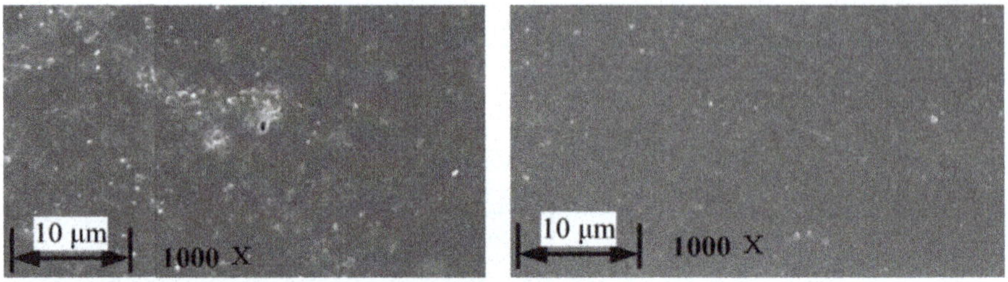

Fig. 3 Surface micro morphology of epoxy resin before and after plasma SiO_x deposition [39]

researches on the use of plasma polymerization. Different effects of plasma discharge electrode structures [1, 43, 44], precursor types [45, 46] and concentrations [47, 48] on the surface physicochemical and electrical properties of various modified substances [49–52] were deeply studied, providing substantial theories and experience for the development of plasma modification (Fig. 3).

The above mentioned studies fully illustrate that plasma modification can effectively alter the physicochemical and electrical properties of insulating materials, as well as engineer their surface characteristics by modifying process parameters. Nonetheless, most experts merely compare the comparison of surface properties between modified and unmodified plasma samples. Little attention has been paid to the different modulation mechanisms on surface properties of insulating materials by different plasma modification techniques [53–55], leading to restricted modulation range on the surface properties of those materials. This chapter will introduce sub-atmospheric pressure glow discharge modification, plasma fluorination modification and plasma SiO_x deposition modification methods. The modulation effects and mechanisms of various modification methods on the surface micromorphology, chemical components and electrical properties of epoxy resin are thoroughly illustrated. Subsequent to this, a comparison between the advantages and disadvantages of various means will be provided to present theoretical and experimental basis for the plasma modification technology of insulating materials.

2 Plasma Modification Platform for Epoxy Resins

Although plasma modification can regulate the surface micromorphology and chemical composition of epoxy resins, thus contributing to the surface charge distribution and polydisperse properties and ultimately improving the flashing performance along the surface. Different plasma modification methods use different modification mechanisms and have different modification effects, so it is necessary to build specific modification platform for each of them, and unify the measurement so as to compare the effects of effects of the different modification methods on the surface micromorphology, chemical composition and electrical properties of epoxy resins.

This section outlines the construction methods of the three modification platforms, namely sub-atmospheric glow discharge, plasma dielectric barrier discharge and atmospheric pressure plasma jet, respectively. It then describes the specific steps and experimental parameters of the three methods, namely sub-atmospheric glow, plasma fluorination and plasma SiO_x deposition. The measurement methods for surface micromorphology, chemical components and electrical properties were evaluated using methods previously outlined and will not be repeated here.

2.1 Plasma Modification Platform for Epoxy Resins

2.1.1 Sub-atmospheric Glow Discharge Modified Platform

The group deployed a sub-atmospheric pressure glow discharge low temperature plasma surface treatment system (HPD-280) from Nanjing Suman Plasma Technology Limited for material modification of epoxy resins (as Fig. 4). The system is capable of performing various surface modification such as surface polymerization, surface limb joining, surface catalysis, metal nitriding and insulating material modification. The driving power source for the system is a high frequency high voltage power supply embedded into the cavity. During the experiments, the material modification can be regulated in a stepwise fashion by adjusting parameters such as discharge time, discharge power, discharge spacing, gas atmosphere and discharge air pressure.

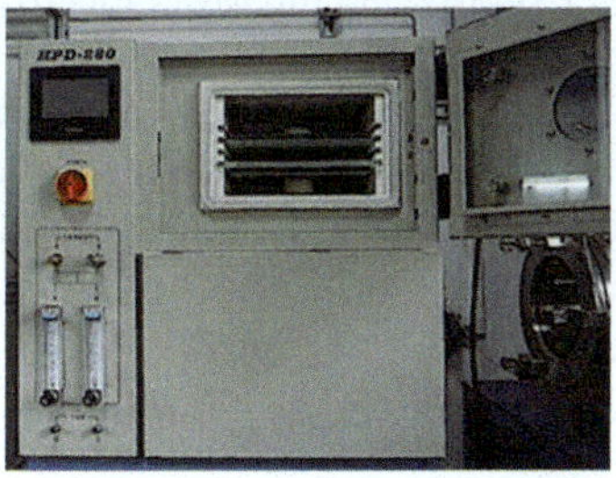

(a) Physical view of the glow discharge modified cavity

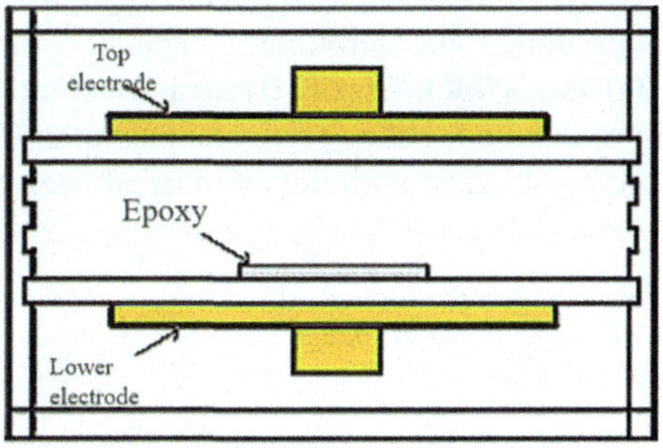

(b) Schematic diagram of the glow discharge modified cavity

Fig. 4 Glow discharge modified cavity and parameter setting interface

2.1.2 Plasma Dielectric Barrier Discharge Modification Platform

The group constructed a plasma dielectric barrier discharge platform, featuring both a circuit part and a gas circuit part. The circuit part comprises a plasma power supply, a dielectric barrier discharge reactor, a grounding capacitor, an oscilloscope and a voltage probe. The gas circuit section consists of gas cylinders, mass flow meters, precursors and their supporting devices. Each piece of equipment is discussed in details below.

(a) Circuit:

Nanjing Suman Plasma Technology Co., Ltd. offers Model CTP-2000K, as seen in Fig. 5. Plasma is formed by applying a high-energy excitation to a gaseous substance, thereby inducing ionization of the internal molecules of the gas, resulting in the presence of both positive and negative ions and neutral substances, forming an overall externally electrically neutral plasma. The group employs a high-frequency, high-voltage power supply to excite the gas for plasma generation. The power supply has a center frequency of 50 kHz and a maximum voltage of 30 kV, it also displays its output voltage and current for output power estimation, and has an internal power supply detection port for monitoring the output voltage waveform with an oscilloscope and voltage probe.

Dielectric barrier discharge reactor: The reactor, also manufactured by Nanjing Suman Plasma Technology Co., Ltd., consists of a metal flat plate electrode, an adjustable support, a quartz reactor and a quartz glass cover sheet, as illustrated in Fig. 6. The flat metal electrodes have a diameter of 100 mm and are held in a fixed distance between the two electrodes. During the operation, the upper electrode is connected to the high-frequency high-voltage power supply and the lower electrode is grounded via a grounding capacitor. The quartz reactor has an outer diameter of 200 mm, a thickness of 10 mm and a circular recess of 100 mm in diameter and 8 mm in thickness for the epoxy resin sample; There are a concentric cylinder on the left and right sides, with the cylindrical axes coinciding which serves as a pathway for

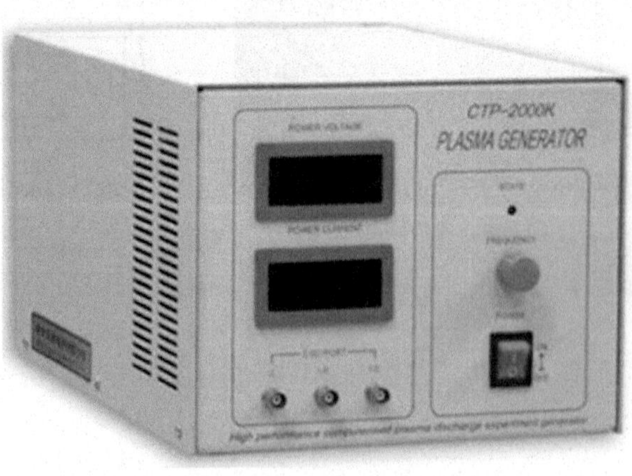

Fig. 5 Actual photograph of high frequency high voltage power supply

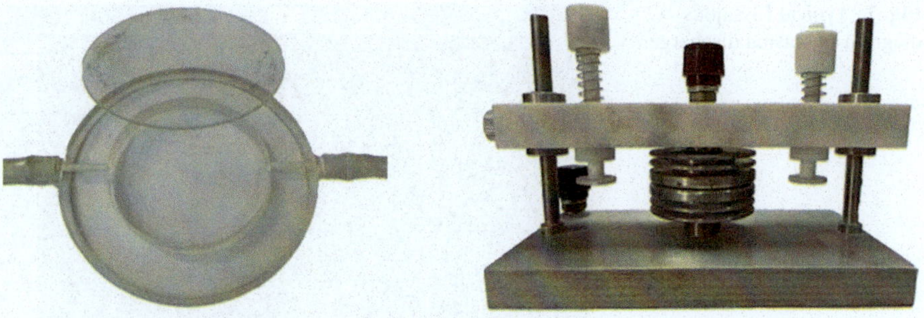

(a) Quartz reactor and glass sheet (b) DBD reactor

Fig. 6 Actual photograph of DBD reactor

various gas circuits. The quartz glass, 120 mm in diameter and 2 mm thick, is sealed on the quartz reactor to form a confined reaction space in the internal recesses.

Grounding capacitance: by connecting a 0.1 μF grounding capacitor to the flat electrode on the grounding side, the plasma discharge power can be calculated by monitoring the voltage waveforms on both sides of the capacitor during the plasma discharge and by combining this with the voltage waveform of the power supply using the Lissajous graph method. The method is based on the following principle: let the ground capacitor be C_1, the voltage across the capacitor during discharge be U_1, the corresponding supply voltage be U_2, and let the discharge current be I, the corresponding charge be Q, and the time of one discharge cycle be T. Then we have [23]:

$$I = \mathrm{d}Q/\mathrm{d}t = \mathrm{d}(C_1U_1)/\mathrm{d}t = C_1\mathrm{d}U_1/\mathrm{d}t \tag{1}$$

The discharge power P is:

$$P = \frac{1}{T}\int_0^T U_2I\mathrm{d}t = \frac{C_1}{T}\int_0^T U_2\frac{\mathrm{d}U_1}{\mathrm{d}t}\mathrm{d}t = fC_1\oint U_2\mathrm{d}U_1 \tag{2}$$

As evidenced above, the discharge power can be can be determined by ascertaining the discharge frequency, the value of the earth capacitance, the supply voltage and the earth capacitance voltage waveform.

The oscilloscope and voltage probe: oscilloscope model Tektronix DPO 2002B. For the purpose of computing the discharge power with the Lissajous graph method, an essential step is to deduce the integration of the supply voltage and ground capacitance voltage. As depicted in Fig. 7, when the supply voltage value serves as the horizontal coordinate and the ground capacitance voltage as the vertical coordinate at the identical point in time, the points thereafter in a fixed interval of time form a typical Lissajous diagram. The data thus derived is imported into Origin data

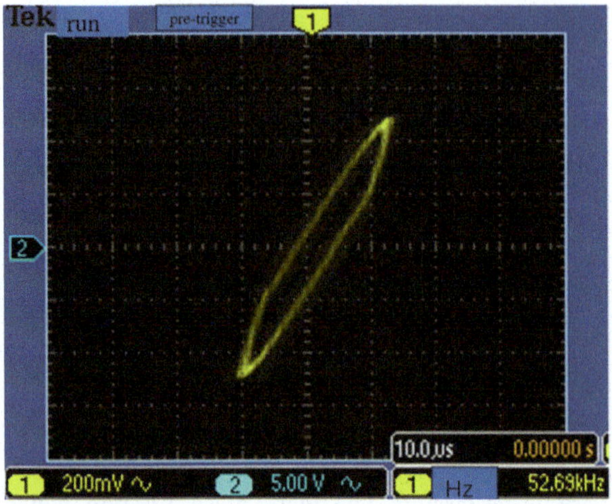

Fig. 7 Typical Lissajous diagram of plasma discharge

processing software, and the discharge power can be computed by the formula $P = f \times C_1 \times A$. The graph displays a supply frequency of f = 53 kHz and a ground capacitance of 0.1 μF, which gives a approximately discharge power of 35 W.

(b) Air circuit:

Gas cylinders: argon (Ar) and carbon tetrafluoride gas (CF_4), both with a purity of 99.99%. Despite the fact that a wide range of gases such as air can be used as a plasma source, they introduce a large number of functional groups such as hydroxyl groups when used to treat materials, which can impede the treatment outcome. Hence, the group used argon as the plasma source because it is an inert gas, generating plasma with minimal impurities and with comparatively little energy input. It can be used independently to etch the surface of the material or in tandem with the precursor as a carrier gas. Since direct plasma discharge with CF_4 is difficult to achieve and the discharge phenomenon is quite unbalanced and intensive, the group opted CF_4/Ar as the plasma source to achieve plasma fluorination of epoxy resin. On the one hand, Ar is easy to generate plasma discharge and can catalyse the ionisation of CF_4, on the other hand, Ar is an inert gas and does not react with CF_4 after ionisation and does not leave other impurities on the material surface.

Mass flow meter: it composed of a mass flow controller and a flow indicator, its physical diagram is shown in Fig. 8. The mass flow controller models D07-26C and D07-19B, and the flow indicator model D08-2F, both from Beijing Seven Star Huachuang Flowmeter Co., Ltd. The mass flow meter can be used to precisely regulate the flow rate and ratio of each gas, while maintaining accuracy. Estimating the type of gas and the approximate flow rate required for the experiment enables the selection of the necessary scope of mass flow controller, thus necessitating a dedicated mass flow meter for each way of gas. Specifically, the D07-26C mass flow controller designed to manage the Ar flow rate, with a standard range of 4 standard liters per minute (slm), and the D07-19B mass flow controller for CF_4 display with a

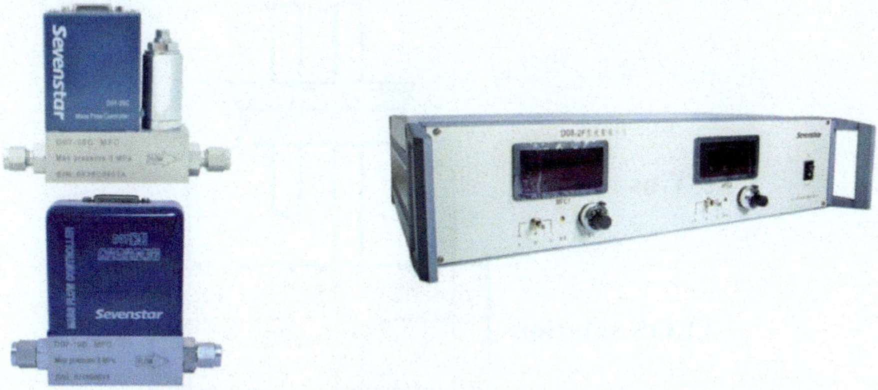

(a) Mass flow controller (b) Flow indicator

Fig. 8 Actual photograph of mass flowmeter

standard range of 400 sccm. The flow indicator simultaneously shows the flow rate of the two gases, enabling precise control in aspect of each gas ratio.

Precursors and their supporting devices: Precursors are substances containing the functional groups or elements necessary for the modified material in plasma modification. They are often used in conjunction with a carrier gas, such as ethyl orthosilicate (TEOS) which is used in the study to achieve plasma SiO_x deposition on the surface of epoxy resin using Argon (Ar). TEOS, which has a molecular formula of $Si(OC_2H_5)n(OH)_{4-n}$, is a colorless liquid at room temperature and is difficult to mix directly with gas. In order to tackle this issue, the group the group designed a supporting device, depicted in Fig. 9, which allowed for the TEOS to be conveniently mixed with Ar. The system was composed of a gas flask with a double-hole stopper and two glass tubes, one short and one long, inserted into each of the holes. The long tube submerged in the TEOS at one end and connected to the Ar gas line via a mass flow meter at the other end. Meanwhile, the short tube not submerged in the liquid and connected to the reactor at the other end. In order to make the TEOS more volatile, the bottom of the flask was outfitted with a water bath heater, which was heated to 70 °C during the experiment.

The plasma dielectric barrier discharge modification platform is shown in Fig. 10. Its circuit section is connected as follows: the high frequency high voltage power supply is connected to the high voltage terminal of the reactor, the low voltage side of the reactor is coupled to the oscilloscope's Channel 1 through a grounding capacitor, while the power monitoring port of the high frequency high voltage power supply is connected straight to the oscilloscope's Channel 2. As for the gas section, Ar and CF_4 are supplied to one end of the reactor via a mass flow meter, Ar, in turn, is fed to the long tube of the gas wash bottle through its mass flow meter the short tube of the gas wash bottle is connected to the reactor, the other end act as a gas collection device for the exhaust gas released post-experiment.

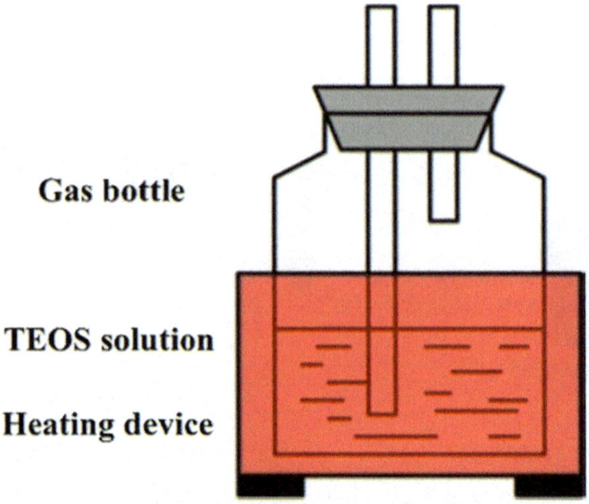

Fig. 9 Schematic diagram of precursors and supporting devices

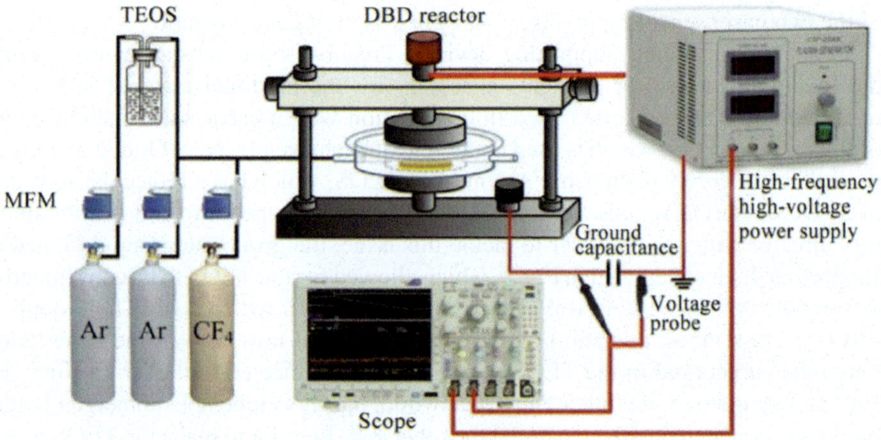

Fig. 10 Schematic diagram of DBD modification platform

2.1.3 Atmospheric Pressure Plasma Jet Modification Platform

As with the plasma dielectric barrier discharge (DBD) platform, the atmospheric pressure plasma jet platform is also composed of an electrical section and a gas section, with the exception of the discharge electrodes, the rest of the equipment is identical. The main difference between the two platforms is the structure of the discharge electrodes. While the discharge area and the treatment area coincide in the DBD platform, with the epoxy resin sample placed between the two flat electrodes and the quartz glass as a barrier medium, for the jet platform the separation of the

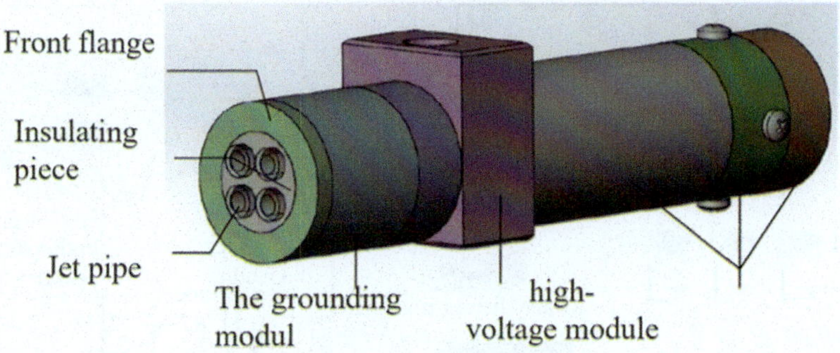

Fig. 11 Structure diagram of jet electrode

discharge area and the treatment area requires, a redesign of the discharge electrodes. The plasma jet electrode used by the group is still made by Nanjing Suman Plasma Co., Ltd. and can be seen schematically in Fig. 11, with the following structure.

The jet electrode consists of a circuit module, a gas circuit module and insulation and fixing components. The circuit module comprises two metal cylinders electrodes with a diameter of 40 mm and a thickness of 20 mm as arranged in parallel with a distance of 50 mm, the upper electrode is connected to the high-frequency high-voltage power supply and the lower electrode is grounded via a grounding capacitor. Both electrodes have four small 2×2 holes with a diameter of 8 mm for the quartz glass tube. The gas path module consists of four quartz glass tubes with an external diameter of 8 mm, an internal diameter of 4 mm and a length of 180 mm. One end is connected to the gas path of the plasma jet platform and the other end passes through the two cylindrical electrodes via a small hole. The rest of the jet electrode is filled with PTFE to act as an insulating support for the entire electrode.

The atmospheric pressure plasma jet modification platform is depicted in Fig. 12. The circuit section is connected as follows: the high frequency high voltage power supply is wired to the high voltage side of the jet electrode, the low voltage side of the jet electrode is linked through a grounding capacitor, both sides of the grounding capacitor are attached to channel 1 of the oscilloscope through a voltage probe, and the power monitoring port of the high frequency high voltage power supply is directly coupled to channel 2 of the oscilloscope. The gas section is connected as follows: one way Ar and one way CF_4 are connected to one end of the jet tube via a mass flow meter, the other way Ar is linked to the long tube of the gas wash bottle via a mass flow meter, and the short tube of the gas wash bottle is connected to the jet tube. As the small cross-sectional area of the plasma plume produced by a single jet tube is relatively small, it is difficult to cover the entire sample area. In order to improve the uniformity of the plasma jet treatment, the group added a lifting platform and a rotating platform, on which the epoxy resin sample placed. The rotating platform is situated on the lifting platform, with the centre of the lifting platform on the axis of the jet electrode. In this way, the epoxy resin sample is completely exposed to the plasma plume, while the contact area between the plasma plume and the epoxy resin

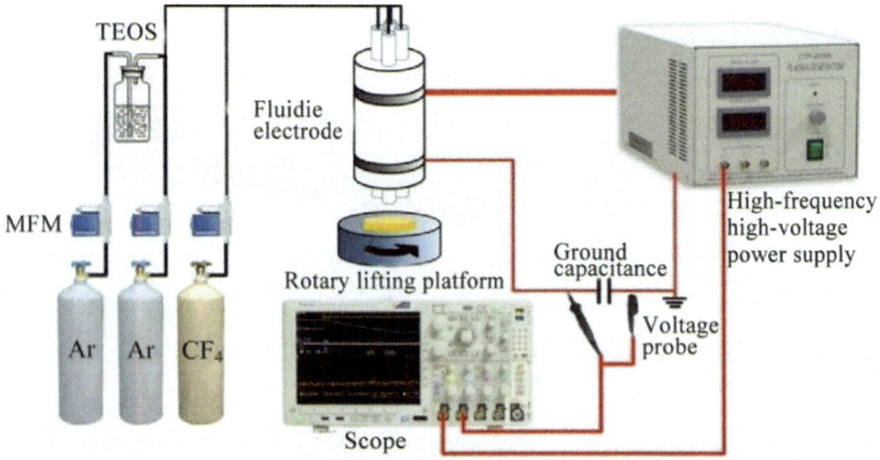

Fig. 12 Schematic diagram of APPJ modification platform

sample is constantly changed by the rotational rotation, thereby leading to a more even plasma jet treatment from an overall perspective.

2.2 *Plasma Modification Methods*

2.2.1 Sub-atmospheric Glow Discharge Modification

The concrete steps for the modification of epoxy resins by sub-atmospheric pressure glow discharge are:

(1) The pre-prepared epoxy resin samples are subjected to an ultrasonic cleaner and ultrasonically cleaned with anhydrous ethanol and deionized water for 10 min respectively, and then put into a vacuum oven to air dry for 24 h.
(2) Place the epoxy resin into the sub-atmospheric pressure glow discharge low temperature plasma surface treatment system, adjust the parameters of discharge spacing, discharge power and discharge time in the operation interface, ensuring the vacuum level within the range of 2000 Pa.
(3) After the selection of the working mode, the evacuation process is initiated and once the vacuum chamber reaches within the specified discharge vacuum, the vacuum pump is switched off and the discharge is taken place.
(4) A stable glow discharge is obtained at a voltage of 2.8 kV and a current amplitude of 7 A without significant current pulses. In addition, the discharge was stable and uniform over the entire surface of the electrode. The discharge emitted a lavender glow without any discernible discharge filaments.
(5) Following the end of the discharge, the pressure relief process is triggered and the modified sample is removed.

2.2.2 Plasma Fluorination

A plasma DBD platform is employed to realize the plasma fluorination modification of epoxy resin. The experimental steps are as follows:

(1) Put the prepared epoxy resin sample into the ultrasonic cleaning machine, undergoing cleaning increments with absolute ethanol and deionized water for 10 min respectively. Then put it into the vacuum oven to air dry for 24 h.
(2) Take out the dried sample and place it in the center of the glass reactor encompassed with quartz glass. And adjust the discharge electrode be intimately adhered to the upper side of the quartz glass sheet and the lower side of the glass reactor, with the electrode spacing at 10 mm.
(3) Open the pneumatic valves of Ar and CF_4, with the mass flowmeter setting of 5 slm for Ar and 0.5slm for CF_4, giving the total flow rate a $Ar:CF_4 = 10:1$ ratio [55]. Pass the mixed gas into the reactor for 1 min to achieve a complete depletion of air.
(4) Turn on the oscilloscope and set it to X–Y display mode, turn on the high-frequency high-voltage power supply and set the supply voltage to 5 kV, continuously adjust the supply frequency and monitor the current value on the power supply display. When the current value rises sharply and a large amount of orange filamentary discharge occurs in the reactor, these indicate that Ar/CF_4 has been ionized. At this time, the supply frequency is observed to be about 53 kHz through the oscilloscope. Simultaneously commence recording treatment time at the same time.
(5) Set the plasma fluorination treatment time of each sample to 1 min, 2 min, 3 min, 4 min, 5 min, and lable them as DBD-F1, DBD-F2, DBD-F3, DBD-F4, DBD-F5 respectively. After the treatment was completed, the power supply voltage was gradually reduced to 0, and it is found that the discharge then gradually disappears. Since the former plasma discharge produced such a high temperature in the reactor and sample, a short period of a mixed-gas ventilation is demanded for cooling. After the reactor is cooled down, lift the flat electric board and take out the sample, and the plasma DBD fluorination experiment is completed.

The plasma fluorination modification of epoxy resin can also be carried out by using the APPJ modification platform, albeit with quite different experimental steps. The specific steps are as follows:

(1) Following the cleaning and drying steps of epoxy resin sample, similar to that of plasma DBD modification experiment.
(2) Take out the dried sample and place it in the center of the rotating platform. Adjust the lifting platform so that the sample is 40 mm away from the bottom of the jet pipe to ensure that the sample will not contact the plasma plume at the beginning of the generation of the plasma plume.
(3) Open the pneumatic valves of Ar and CF_4, use the mass flowmeter to set the flow rate of Ar to 6 slm and CF_4 to 0.5 slm. At this time, with the sound of mixed gas discernible when ejecting from the jet pipe.

(4) Turn on the oscilloscope and set it to X–Y display mode, and activate the high-frequency high-voltage power supply, adjusting the supply voltage to 5 kV. Monitor the current value displayed on the power supply adjust the supply frequency. When the current value rises sharply and orange plasma plume continuously ejected from the jet pipe, this indicate that Ar/CF_4 has been ionized. At this time, turn on the rotating platform to make the sample rotate at a uniform speed, and lift the sample to 15 mm away from the bottom of the jet pipe. Ensuring that it's adequately exposed to plasma plume, while avoiding instability or the disappearance of plasma plume due to close proximity to the jet pipe. When the sample reaches the specified position, and the supply frequency is observed to be about 53 kHz through the oscilloscope, start recording treatment time.
(5) Set the plasma fluorination treatment time of each sample to 1 min, 2 min, 3 min, 4 min, 5 min, and lable them as APPJ-F1, APPJ-F2, APPJ-F3, APPJ-F4, APPJ-F5 respectively. After the treatment was completed, the power supply voltage was gradually reduced to 0, and it is found that the plasma plume then gradually disappears. At this time, the temperature of the sample is too high due to the plasma discharge, and it needs to be cooled before being taken out from the rotating platform. The APPJ fluorination experiment is completed.

2.2.3 Plasma SiO_x Deposition Modification

Because the precursor of plasma SiO_x deposition is different from plasma fluorination, the corresponding experimental steps are also different. Achieve the plasma SiO_x deposition of epoxy resin by using plasma DBD. The specific steps are as follows:

(1) Following the cleaning and drying steps of epoxy resin sample, similar to those of plasma fluorination experiment, which will not be repeated here.
(2) The method of placing the sample into the reactor is the same as that of plasma fluorination experiment, which will not be repeated here.
(3) Set the temperature of water bath to 70 °C. After heating for 5 min, open the two-way Ar, and set mass flowmeter to 5 standard liters per minute (slm) of Ar going directly into the reactor and 0.5 slm Ar going into the gas-washing bottle containing TEOS, yielding an Ar:Ar/TEOS ratio of 10:1. When there is a small amount of bubbles in the gas-washing bottle, this indicates gas mixed with TEOS and Ar has been blown out of the gas-washing bottle. Allow the mixed gas to continue flowing into the reactor for 1 min to deaerate the reactor.
(4) Also turn on the oscilloscope set it to X–Y display mode and turn on the high-frequency high-voltage power supply. Unlike plasma fluorination, the existence of CF_4 will increase the threshold of plasma discharge, but TEOS has little effect on plasma discharge. Therefore, set the supply voltage to only 4.5 kV during the plasma SiO_x deposition, and continuously adjust the supply frequency and monitor the current value on the power supply display. When lilac filamentary discharge occurs in the reactor, it indicates that Ar/TEOS has been ionized.

Oscilloscope readings show that the frequency is around 52 kHz at this point. Start recording treatment time.

(5) Additionally, set the plasma SiO_x deposition treatment time of each sample to 1 min, 2 min, 3 min, 4 min, 5 min, and lable them as DBD-Si1, DBD-Si2, DBD-Si3, DBD-Si4, DBD-Si5 respectively. Follow the same sample cooling step utilized for the plasma fluorination experiment. The plasma DBD SiO_x deposition experiment is completed.

The plasma SiO_x deposition modification of epoxy resin can also be achieved by the APPJ modification platform, but the experimental steps are quite different. The specific steps are as follows:

(1) Following the cleaning and drying steps of epoxy resin sample, similar to those of plasma fluorination experiment.
(2) Same as in a plasma fluorination experiment, the sample is placed in its designated position.
(3) Set the temperature of water bath to 70 °C. After have heating for 5 min, open the two-way Ar, and set the flow rate of the Ar directly flowing into the reactor to 6slm through the mass flowmeter, and the flow rate of the Ar connected to the gas-washing bottle containing TEOS to 0.6 slm. When a small amount of bubbles is visible in the gas-washing bottle, it indicates that the gas mixed with TEOS and Ar has been blown out of the gas-washing bottle. At this time, the sound of mixed gas ejecting from the jet pipe can be heard.
(4) Also turn on the oscilloscope and set it to X–Y display mode, turn on the high-frequency high-voltage power supply, set the supply voltage to 4.5 kV, continuously adjust the supply frequency and monitor the current value on the power supply display. When the current value rises sharply and lilac plasma plume continuously ejected from the jet pipe, it indicate that Ar/TEOS has been ionized. At this time, start the rotating platform to move at an even speed, and lift the sample to 15 mm away from the bottom of the jet pipe. This measure allows the epoxy resin sample to be fully contacted with the plasma plume in order to achieve modification without bringing it too close to the jet pipe, which will result in instability or even disappearance of the plasma plume. Upon reaching the desired position, the supply frequency is observed to be about 52 kHz through the oscilloscope, and begin recording treatment time.
(5) Set the plasma SiO_x deposition treatment time of each sample to 1 min, 2 min, 3 min, 4 min, 5 min, and lable them as APPJ-Si1, APPJ-Si2, APPJ-Si3, APPJ-Si4, APPJ-Si5 respectively. Subsequently, the voltage was gradually decreased to 0, resulting in a gradual weakening of the plasma plume. As the temperature of the samples had raised significantly due to the plasma discharge, they needs to be cooled down before being taken out from the rotating platform. The plasma SiO_x deposition experiment is completed.

3 Modulation of Plasma Modification on Epoxy Resin Surface Micromorphology

Studies have shown that altering the surface morphology of insulating material on micro and nano scales has an substantial effect on electrical properties of material. However, most existing studies only focus one kind of plasma modification technique. Even when accounting for factors such as processing time, concentration of precursors and other parameters, it is difficult to control surface morphology at multiple scales, and there is inadequate research on the varying effects of varying effects of different plasma modification devices on insulating material surface morphology. Hence, there is a lack of clarity regarding how surface morphology affects the electrical properties of insulating materials. To address this gap, other plasma-modification techniques must be explored and incorporated.

This chapter investigates how sub-atmospheric glow discharge, plasma fluorination and plasma SiO_x deposition modify the micromorphology of epoxy resin surface. Furthermore, it explores the effect rules of plasma treatment under different conditions on epoxy resin micromorphology.

3.1 Modulation of Sub-atmospheric Glow Discharge on Epoxy Resin Surface Micromorphology

The surface of epoxy resin was subjected to modification for 1, 2, 3, 5 and 10 min at the power of 320 W and the distance of 30 cm. Figure 13 exhibits an optical 3-D microscope image of the material surface magnified 500x, as presented below. Figure 13a shows the surface of epoxy resin without glow discharge, which can be observed that the smoothness of material surface is excellent before modification. A few scratches and some large particles are randomly distributed on the surface. This condition provided a strong reference of the resin's intrinsic refractive index. Figure 13b, c and d depict the surface morphology of epoxy resin after glow discharging for 1, 3 and 5 min. After modification, it uniformly distributed granular matter forms on the surface of epoxy resin, and as the time increases, the distribution of granular matter become more uniform. The 3-D ultra-depth field microscope was used for a deep synthesis study of the height information at the same position on the surface. As shown in Fig. 14, with increasing treatment time, the original irregularly particles gradually increased in size, and small particles merged to form large bumps. The distribution of bumps possesses a comparably regular pattern. Meanwhile, the corresponding height difference increases with the prolongation of the treatment.

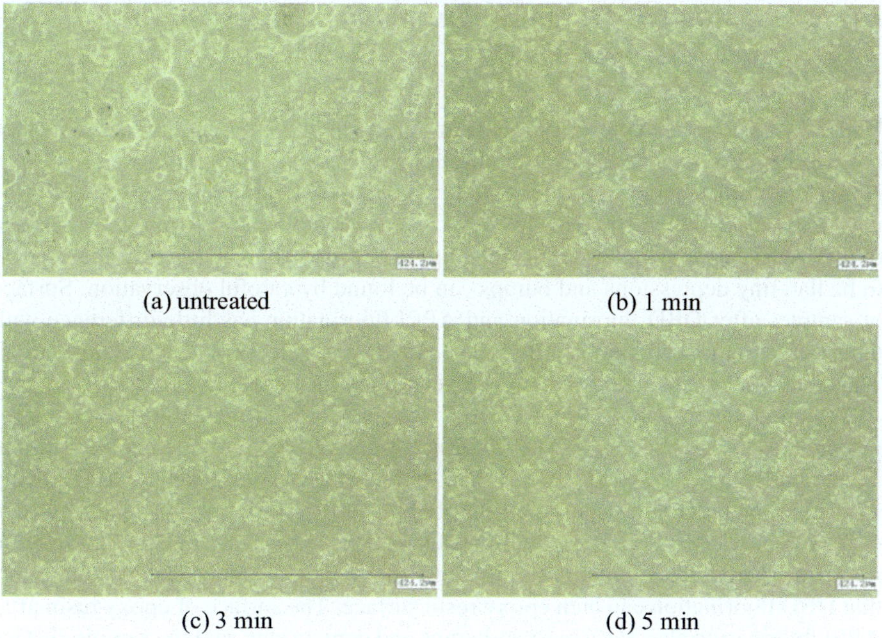

Fig. 13 Surface morphology at different modification time

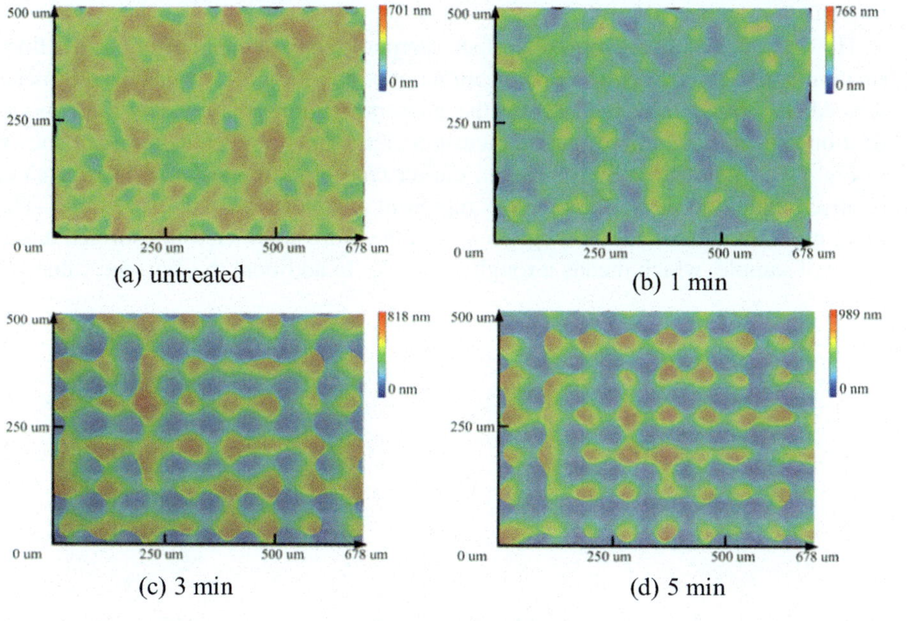

Fig. 14 Surface 3-D height maps at different modification times

3.2 *Modulation of Plasma Fluorination on Epoxy Resin Surface Micromorphology*

Under the same plasma modification means and different treatment time, the surface morphology of epoxy resin has little difference. Therefore, sample treated for 3 min is treated as representation for all various plasma modification. Figure 15 shows the unprocessed, DBD fluorinated, APPJ fluorinated epoxy resin. Visually, the surface of unprocessed epoxy resin sample is light gray. Besides, though the surface appears to be flat, tiny depressions and bumps can be found by careful observation. Surface of samples after DBD fluorination and APPJ fluorination has little difference with unprocessed sample, making it difficult to validate the success of plasma fluorination. Thus, microscopic characterization is essential for further study.

Scanning electron microscopy (SEM) was used to investigate the surface micromorphology of unprocessed, DBD fluorinated and APPJ fluorinated epoxy resin at a magnification of 100 μm (Fig. 16). The untreated epoxy resin sample as relatively flat with a small amount of particles distributing on itself. In contrast, a large amount of bumps and depressions exist on the surface of epoxy resin after DBD fluorination, rendering the overall sample is significantly rough in appearance. This result shows that DBD fluorination can etch epoxy resin surface. The surface of epoxy resin after APPJ fluorination also has a mass of bumps and depressions, and the sample surface is also quite rough, though the sample is less rough than the DBD fluorinated one, demonstrating a comparatively weaker etching effect of APPJ fluorination than that of DBD fluorination.

However, SEM observation of surface morphology before and after plasma fluorination is not considered to be an accurate enough metric. A surface profilometer is needed to quantitatively measure the micromorphology before and after modification. The surface micro profile measurement results of each sample are shown in Fig. 17. As indicated by the figure, darker red colors represent areas of greater relative height and darker blue colors represent areas of lower relative height. The more pronounced the color change in figure, the more significant the surface undulation of sample, which means a rougher sample. In addition, the extreme values of

(a) Untreated (b) DBD fluorination (c) APPJ fluorination

Fig. 15 Actual photograph of epoxy resin samples before and after plasma fluorination

(a) Untreated (b) DBD fluorination (c) APPJ fluorination

Fig. 16 SEM figure of epoxy resin before and after plasma fluorination

relative height of each sample are given on the right side of the figure, allowing a direct comparison of the roughness of different samples in color. It is observed that a large number of particles distributed across its surface and a few large bumps and depressions. The number of particles on epoxy resin surface after DBD fluorination is larger and the bumps become rougher. Compared to the untreated epoxy resin, the overall height difference of sample is bigger, indicating the sample's surface has roughened due to DBD fluorination. More particles on the surface of the bumps indicates that DBD fluorination can etch epoxy resin and preferentially etch the elevated area. The surface of epoxy resin after APPJ fluorination is similar to that of DBD fluorinated sample, retaining rougher bumps. But the overall height difference of epoxy resin after APPJ fluorination is between the DBD fluorinated and untreated samples, which proves that APPJ fluorination can also etch epoxy resin though with a lesser effect than that of DBD fluorination.

Internationally, the arithmetic mean of the absolute value of the surface profile deviation is commonly used to quantitatively characterize the surface roughness of material. It is named as contour arithmetic mean deviation and denoted as "Ra" wherein higher values correspond to rougher surfaces [56]. In order to calculate the effect of plasma fluorination modification on epoxy resin surface morphology at different discharge electrodes and treatment time, the surface roughness value before and after plasma fluorination can be worked out by the surface micro profile shown in Fig. 17. The calculation result is shown in Fig. 18. It is observed that the result is in line with the SEM pictures and surface micro profile pictures. As the surface roughness value (Ra) increase after plasma fluorination, and Ra has a positive correlation to the treatment time. Besides, the improvement of surface roughness modified by DBD fluorination is greater than that via APPJ fluorination. The untreated epoxy resin had a Ra value of 426 nm. which is remarkably increased to 1055 nm via DBD-F5, the maximum value among all the plasma modification of all plasma modification, while Ra of APPJ fluorination reaches only achieved 889 nm at most.

Based on the experimental results and analysis above, it is evident that plasma fluorination can etch the surface of epoxy resin and then increase the surface roughness of it. The discharge power of DBD electrode is higher than that of APPJ electrode, making its fluorination yield better results in terms of etching performance and a greater variance in the surface roughness value. Furthermore, it is possible to adjust

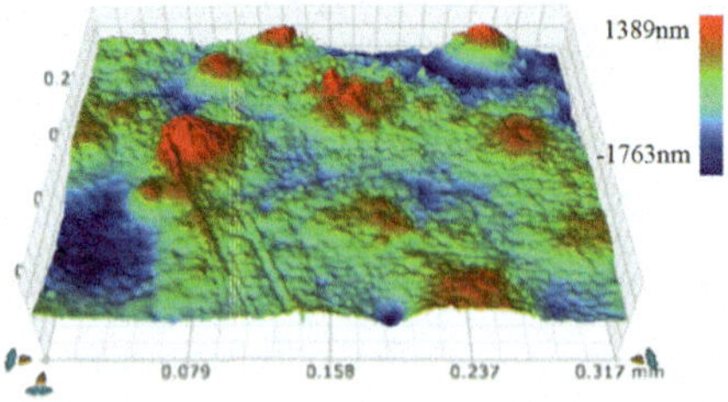

(a) Untreated

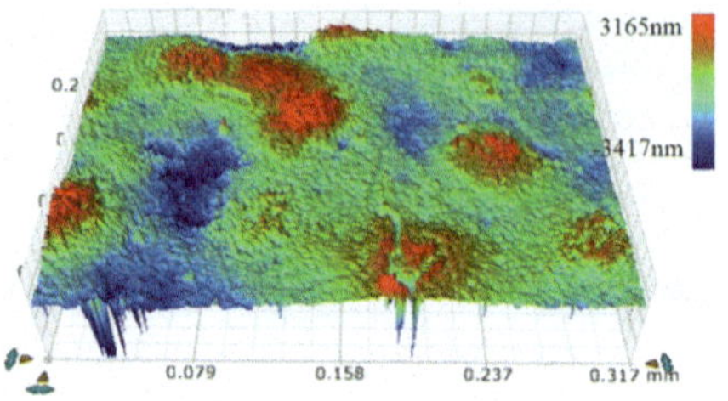

(b) DBD fluorination

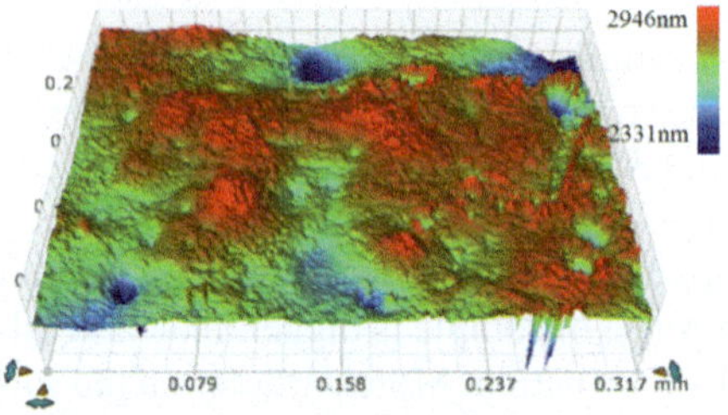

(c) APPJ fluorination

Fig. 17 Surface micro profile of epoxy resin before and after plasma fluorination

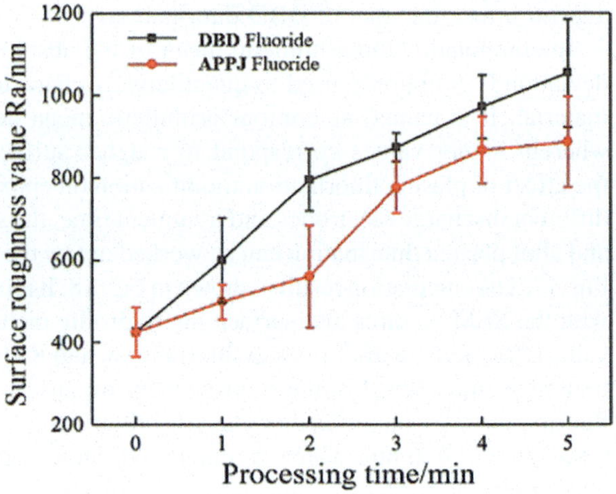

Fig. 18 Surface roughness of epoxy resin before and after plasma fluorination

the surface roughness value Ra of epoxy resin to a range of 400–1000 nm. By varying the structure of plasma discharge electrode and treatment time.

3.3 Modulation of Plasma SiO_x Deposition on Epoxy Resin Surface Micromorphology

The same research concept regarding modulation of plasma SiO_x deposition on epoxy resin surface micromorphology has been implemented, and the sample treated by plasma SiO_x deposition for 3 min is used as a representative. Untreated, DBD-SiO_x deposited and APPJ-SiO_x deposited epoxy resin is shown in Fig. 19. Unlike plasma fluorination, upon irradiation by natural light at a specific angle, on the surface of plasma SiO_x deposited epoxy resin will reflect light waves of multiple colors. This phenomenon is called film interference. The phenomenon occurs due to the existence of thin films, close to visible light wavelengths, on the surface of plasma SiO_x deposited epoxy resin. Thereby, when the light passes through such thin film, light waves will be reflected respectively from its upper and lower surfaces, subsequently leading to their interference, and formation of a new light wave. Consequently, some hues in natural light are intensified, while the intensity of other hues is reduced, and finally, multiple hues of light waves can be perceived by the human eyes. In essence, this distinguishing effect of film interference indicates that plasma SiO_x deposition deposits nanoscale films on epoxy resin surface. Moreover, it was also found that the film deposited by DBD-SiO_x is irregular in shape, while the film deposited by APPJ-SiO_x is approximately circular. Then wiping the film with absolute alcohol resulted in no disappearance or deformation, which indicates that the formed film is dense and solid.

The surface micromorphology of epoxy resin after plasma SiO_x deposition is further investigated and the SEM figure of samples is shown in Fig. 20. It was found that particles on surface of epoxy resin were obviously covered by film after plasma SiO_x deposition, but bright or dark spots are observed on the surface of epoxy resin after DBD-SiO_x deposition. The thickness of film deposited on surface is not uniform, which leads to the difference of intensity of reflected light. So different bright and dark areas will show in SEM picture. In comparison, the color distribution of SEM picture of APPJ-SiO_x deposition is more uniform than that of DBD-SiO_x deposition, which indicates the film formed by APPJ-SiO_x deposition is more homogeneous.

(a) Untreated (b) DBD SiO_x deposition (c) APPJ SiO_x deposition

Fig. 19 Actual photograph of epoxy resin samples before and after plasma SiO_x deposition

(a) Untreated (b) DBD SiO_x deposition (c) APPJ SiO_x deposition

Fig. 20 SEM figure of epoxy resin before and after plasma SiO_x deposition

The measurement scale of SEM is extended to 1 μm and 200 nm, and the obtained SEM figure of plasma SiO_x deposition is shown in Fig. 21. Upon inspection of the SEM image on 1 μm magnification, numerous particles can be observed to have deposited on the surface of epoxy resin after plasma SiO_x deposition. However, it is evident that the particle distribution of APPJ-SiO_x deposition yields a more uniform coverage, signifying a more homogenous film formation. Analyzing SEM picture at 200 nm magnification, reveals that the film generated by plasma SiO_x deposition is coralloid, which is aggregated by a large number of nanoscale particles.

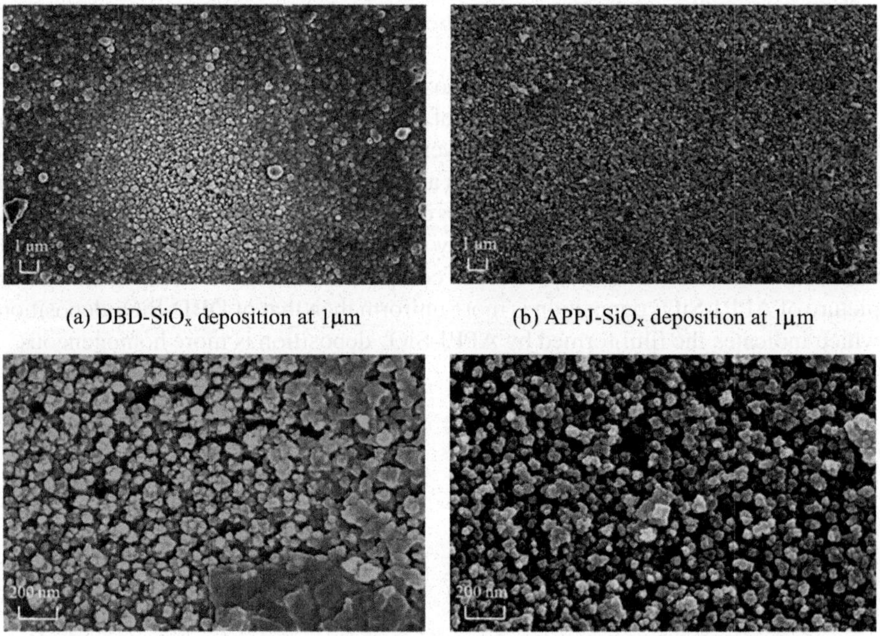

(a) DBD-SiO_x deposition at 1μm (b) APPJ-SiO_x deposition at 1μm

(c) DBD-SiO_x deposition at 200nm (d) APPJ-SiO_x deposition at 200nm

Fig. 21 SEM images of SiO_x deposition samples with high magnification

It evident from both visual inspection and SEM imaging that the plasma SiO_x deposition method can deposit film on surface of epoxy resin. But it is difficult to discern whether the surface of epoxy resin become rougher after modification. Therefore, surface micro profile of epoxy resin after plasma SiO_x deposition is needed to be measured, as shown in Fig. 22. The figure shows that plasma SiO_x deposition can notably decrease the overall relative height difference of epoxy resin surface. It implies that film generated by plasma SiO_x deposition covers the original defects of untreated epoxy resin surface, thus diminishing its roughness to a certain extent. Nevertheless, many large bumps appear on the surface of epoxy resin after DBD-SiO_x deposition, indicating that the film generated by DBD-SiO_x deposition has an uneven thickness. To the contrary, the surface of epoxy resin after APPJ-SiO_x deposition exhibited more flatness, hinting that APPJ-SiO_x yields films with a more uniform thickness.

To quantify characterize variation of epoxy resin surface roughness before and after plasma SiO_x deposition, surface roughness value (Ra) of epoxy resin before and after modification by plasma SiO_x deposition at different electrodes and treatment time was also calculated. The results are shown in Fig. 23. Like that of untreated sample, the Ra of epoxy resin after DBD-SiO_x deposition fluctuates varied between 375 and 579 nm with increasing treatment time. Ra of sample after APPJ-SiO_x deposition decrease monotonically with the increasing treatment time, and it is obviously smaller than that of untreated epoxy resin. Of these samples, Ra of APPJ-Si5 reaches 171 nm, which is the smallest value of plasma modified sample. These analysis above

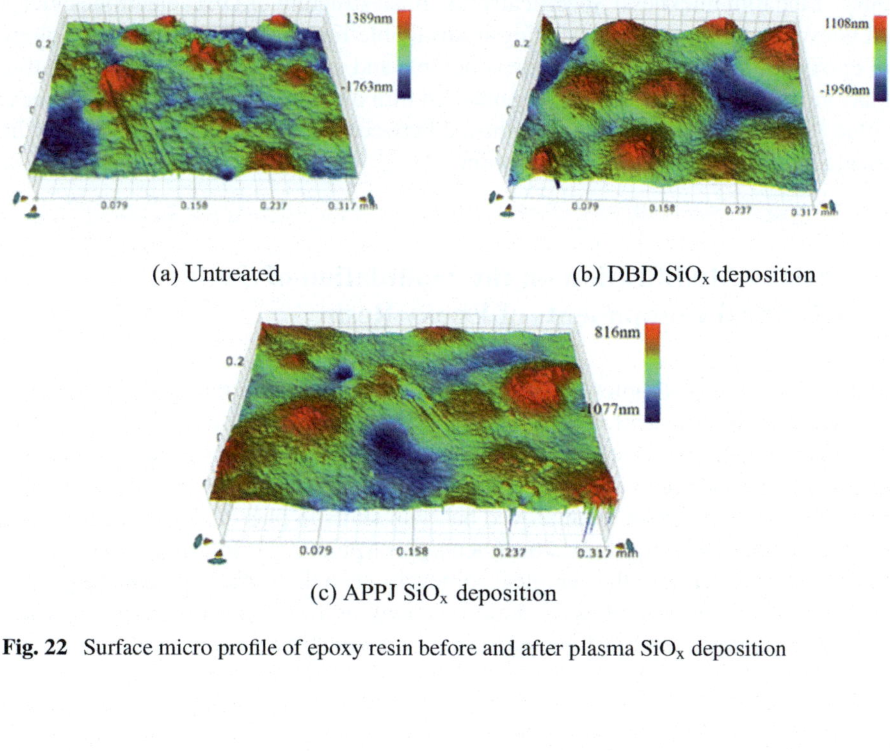

(a) Untreated

(b) DBD SiO_x deposition

(c) APPJ SiO_x deposition

Fig. 22 Surface micro profile of epoxy resin before and after plasma SiO_x deposition

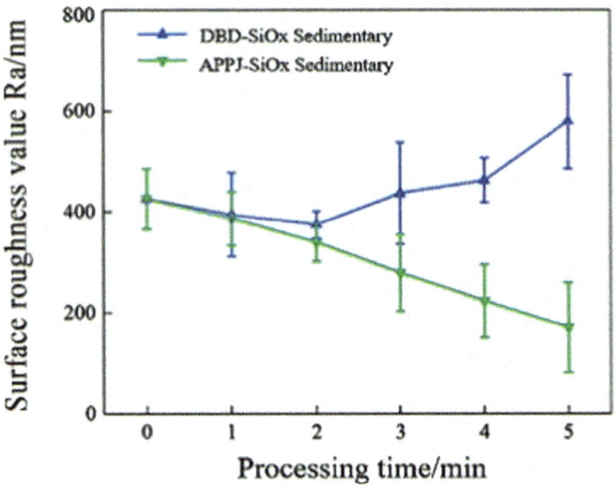

Fig. 23 Surface roughness of epoxy resin before and after plasma SiO_x deposition

imply that APPJ-SiO_x deposition method effectively decrease surface roughness of epoxy resin by depositing uniform film to cover original defects of epoxy resin. For DBD-SiO_x deposited samples, the inconstancy of the surface roughness of epoxy resin results from the irregular thickness of film.

By combining the experimental results and analysis above, it can be concluded that plasma SiO_x deposition can decrease surface roughness by depositing film on the surface of epoxy resin, though the amount of reduction depends on whether uniformity of the film thickness. Specifically, the fixed structure of DBD electrodes yields an unevenly deposited film. Making it unsuitable for regulating micromorphology of epoxy resin. The APPJ electrode is flexible and can be combined with a rotating platform to make the thickness of generated film more even. The surface roughness value (Ra) of epoxy resin can be adjusted between 100 and 400 nm by varying the treatment time of APPJ-SiO_x deposition.

4 Plasma Modification on the Modulation of Surface Chemical Components of Epoxy Resin

Studies have revealed noteworthy differences in the distribution of main chemical elements on the surface of insulating materials following the use of different precursors, thereby altering the electrical properties. Similarly, it has been determined that changes to the surface functional groups, and thus electrical properties, are attainable when the same precursor is employed but with shifts in plasma discharge form and treatment times. Notably, these adjustments play a pivotal role in better understanding of the interplay between the chemical composition on the surface of insulating materials and electrical properties. In addition, studying the chemical composition and functional group changes of epoxy resin before and after plasma modification can clarify the reaction form between plasma and epoxy resin surface, reveal the mechanism of plasma's influence on the physicochemical properties of epoxy resin, and realize the regulation of the chemical composition of epoxy resin surface.

This section begins by presenting the discharge electrode form, treatment time, the control effects of subatmospheric pressure glow discharge, as well as plasma fluorination and plasma SiO_x deposition on the surface chemical composition of epoxy resin. It then analyzes the change pattern and reaction mechanism of the plasma treatment under different conditions on the chemical composition of the epoxy resin surface.

4.1 Modulation of Chemical Composition on the Surface of Epoxy Resin by Sub-atmospheric Glow Discharge

Attenuated total reflection-Fourier transform spectroscopy was employed to investigate the effect of glow discharge plasma on the surface composition of materials from both qualitative and quantitative aspects. Figure 24 reveals the typical molecular formula of epoxy resin. It is observable from the molecular formula that there are a large number of organic functional groups on the surface of the material: benzene ring, $–CH_x$, –CH(O)CH–, etc.

According to the composition of functional groups and chemical bonds of epoxy resin, Table 1 shows the typical infrared spectrum absorption peaks of epoxy resin. Together with the infrared spectrum, it is used to examine the influence of sub-atmospheric pressure glow discharge on the chemical structure of epoxy resin in combination.

The infrared spectra of epoxy resin before and after glow discharge modification are portrayed in Fig. 25. The FTIR spectrum of EP includes C–H absorption peaks at 3033–2871, 1180 and 827 cm^{-1}, CH_2 and CH_3 absorption peaks at 1460, C=O characteristic peaks at 1730 cm^{-1} and C=C absorption peaks at 1607 and 1506 cm^{-1}. Glow discharge modification did not change the surface absorption peak of epoxy resin.

To further analyze the effect of plasma modification on the chemical composition of the material surface, XPS technology was employed to determine the change of material elements and the composition ratio. Querying the XPS binding energy comparison table shows the binding energy of the chemical bond of each functional group is obtained, as depicted in Table 2.

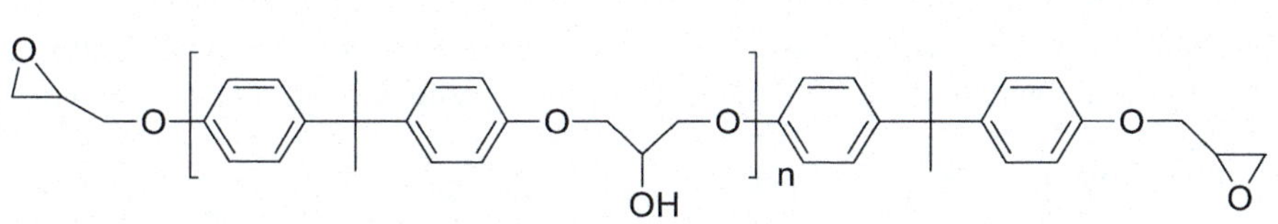

Fig. 24 Molecular formula of epoxy resin

Table 1 Typical infrared absorption peaks of epoxy resin

Absorption peak (cm^{-1})	Molecular structure
3033–2871	C–H
1730	C=O
1607	C=C
1506	C=C
1460	CH_2, CH_3
1180	C–H
827	C–H

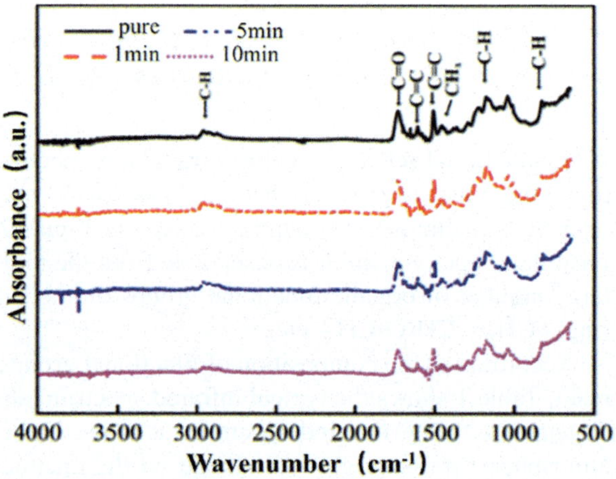

Fig. 25 FTIR spectral lines under different modification times

Table 2 Binding energy of chemical bonds of each functional group

Functional group chemical bond	Binding energy (eV)
C–C/C–H	284.7
C–O	286.3
C=O	288.8

Figure 26a and b are the XPS analysis spectra of the untreated and the samples treated for 3 min, respectively. It can be observed from the figure that the surface of untreated epoxy resin mainly contains an O 1s peak, C 1s peak, soft Si 2s peak, Si 2p peak and Al 2p peak. The appearance of the Al 2p peak is primarily attributed to the filler alumina in EP. the appearance of Si 2s and Si 2p is primarily attributed to the bisphenol A-based raw materials. After glow discharge treatment, the peak value of the O 1s peak increased significantly, and the peak value of the C 1s peak decreased slightly.

Figure 27 shows the C 1s peak spectrum on the epoxy resin surface before and after glow discharge treatment. As shown in Fig. 27a, C 1s of untreated EP can

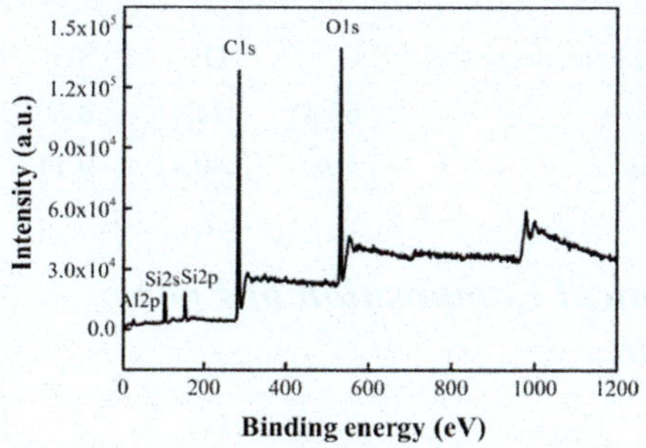

(a) Unprocessed

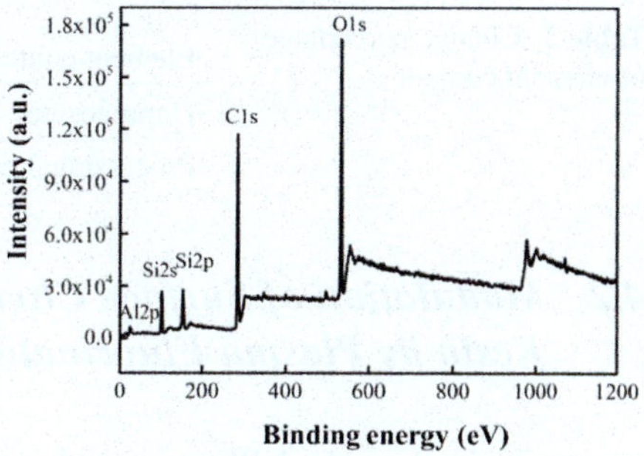

(b) Processing 3 min

Fig. 26 XPS spectra of material surface before and after modification

be decomposed into two peaks, C–C/C–H and C–O. After glow discharge treatment, the peak height and peak width of C 1s in the high binding energy part of EP were increased, indicating that glow discharge changed the composition and proportion of carbon-containing groups on the surface of epoxy resin. It can be seen that after glow discharge modification, the C–C/C–H content of the epoxy resin surface decreases, the C–O content increases, along with the integration C=O oxygen-containing groups.

The XPS PEAK software was used to analyze the percentages of the sample elements before and after treatment, as shown in Table 3. It was found that the content of carbon elements decreased from 63.41 to 50.07% after 3 min of treatment, while the oxygen increased from 24.41 to 40.41%.

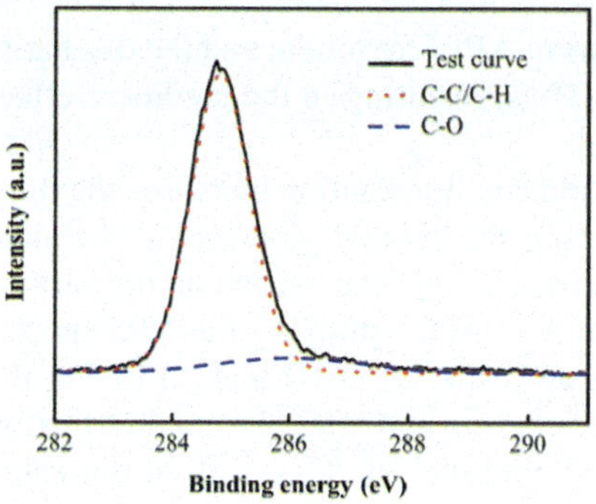

(a) Unprocessed C 1s split peak line

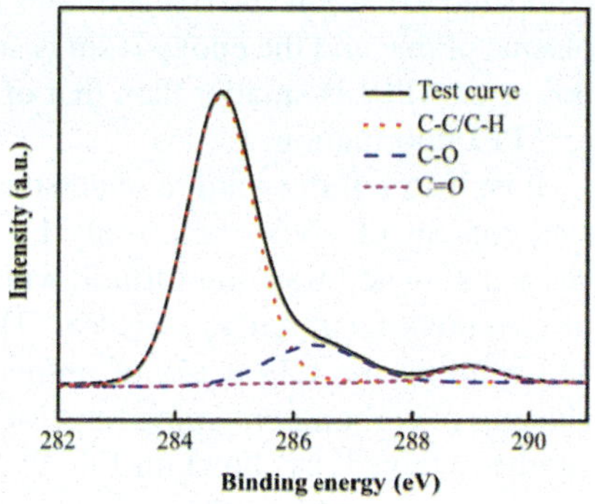

(b) C 1s split peak line after processing

Fig. 27 XPS peak splitting lines on the surface of epoxy resin before and after modification

Table 3 Change percentage of element content

Element content percentage (%)	C	O	Si
Unprocessed	63.41	24.41	8.36
Processing 3 min	50.07	40.41	9.14

4.2 *Modulation of Surface Chemical Composition of Epoxy Resin by Plasma Fluorination*

Figure 28 presents the XPS spectra of epoxy resin surfaces before and after fluorination of all DBD and APPJ. Observing the XPS spectrum, it was found that compared with the untreated epoxy resin sample, the C1s peak, O1s peak and F1s peak changed significantly after plasma fluorination, especially the F1s peak of DBD fluorination increased significantly. Based on the XPS spectrum, the relative content of the main elements can be calculated, as shown in Table 4. The epoxy resin sample is a three-dimensional network structure formed by carbon chain crosslinking filled with a considerable amount of micron alumina. This structure leads to the untreated epoxy resin sample being dominated by carbon elements and containing some oxygen elements and negligible fluorine elements. The content of fluorine element on the surface of the epoxy resin sample fluorinated by DBD increased obviously. It was observed that the fluorine content increased continuously with the increase of treatment time, up to 24.31%. The increase in fluorine content shows that plasma fluorination can increase fluorine elements' content on the epoxy resin surface and the treatment effect increases monotonically with the treatment time. Similarly, the fluorine elements' content on the epoxy resin's surface changed little after APPJ fluorination with the highest proportion of fluorine is only 5.14%, showing a negligible impact compared to carbon and oxygen. This can be attributed the contact area between the plasma plume and the epoxy resin is small during APPJ treatment and the discharge power of APPJ is smaller than that of DBD, Thus, resulting in the mediocre effect of APPJ fluorination.

It isn't straightforward to understand how plasma fluorination increases the fluorine content of epoxy resin only by comparing the relative content of chemical elements. In addition, the form in which fluorine element introduced on the surface of the epoxy resin remains unclear. Therefore, by further studying the XPS spectra and splitting the C1s peaks of untreated epoxy resin, DBD-F3 and APPJ-F3, the changes of carbon-containing functional groups before and after plasma fluorination can be obtained, as shown in Fig. 29. It was found that the C1s peak of untreated epoxy resin can be further subdivided into three peaks. In contrast, the C1s peak after plasma fluorination can be divided into more peaks, up to five peaks of DBD fluorination, and the peaks of each peak with distinct characteristics.

To quantitatively study the types and changes of carbon-containing functional groups before and after plasma fluorination, the peak values of each peak can be compared with the binding energy of specific functional groups in the XPS binding energy comparison table. At the same time, the ratio of the peak area of each spectrum

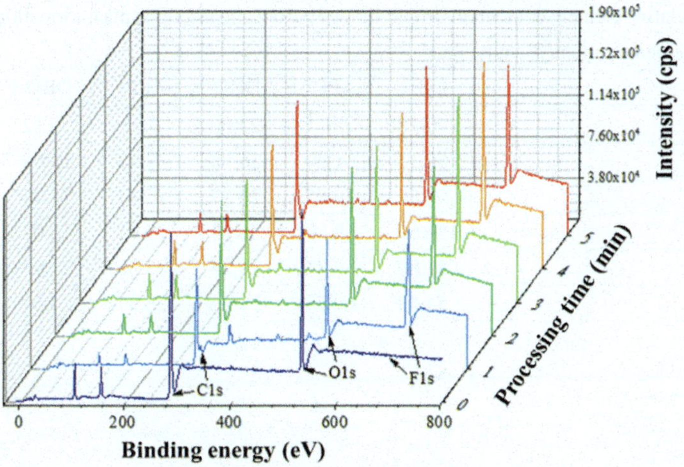

(a) DBD fluorination

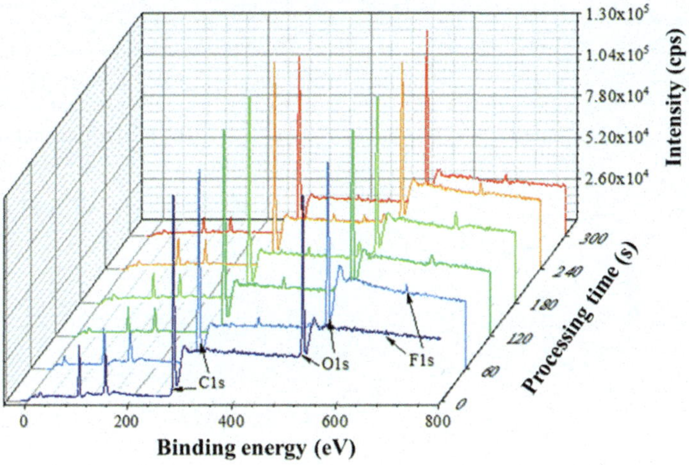

(b) APPJ fluorination

Fig. 28 XPS spectra of samples treated by plasma fluorination

to the total peak area is calculated and the summarized in Table 5. The analysis and summary results show that the carbon-containing functional groups on the surface of the untreated epoxy resin are mainly C–C bonds, together with some C–O bonds and O–C=O bonds. A large number of CF_x functional groups are added after plasma fluorination because the binding energy of the C–F bond (485 kJ/mol) and H–F bond (565 kJ/mol) is much higher than that of the F–F bond (153 kJ/mol) and C–H bonds (414 kJ/mol). When the high-energy plasma bombards the surface of the epoxy resin

Table 4 Relative element content of sample surface before and after plasma fluorination treatments

Relative element content/% (min)	C		O		F	
	DBD-F	APPJ-F	DBD-F	APPJ-F	DBD-F	APPJ-F
0	70.28		29.60		0.12	
1	60.33	61.29	26.58	37.43	13.09	1.28
2	57.89	70.85	26.17	26.43	15.94	2.72
3	57.69	68.89	26.29	25.97	16.02	5.14
4	51.82	69.84	26.24	25.36	21.93	4.80
5	54.43	69.78	21.26	26.37	24.31	3.86

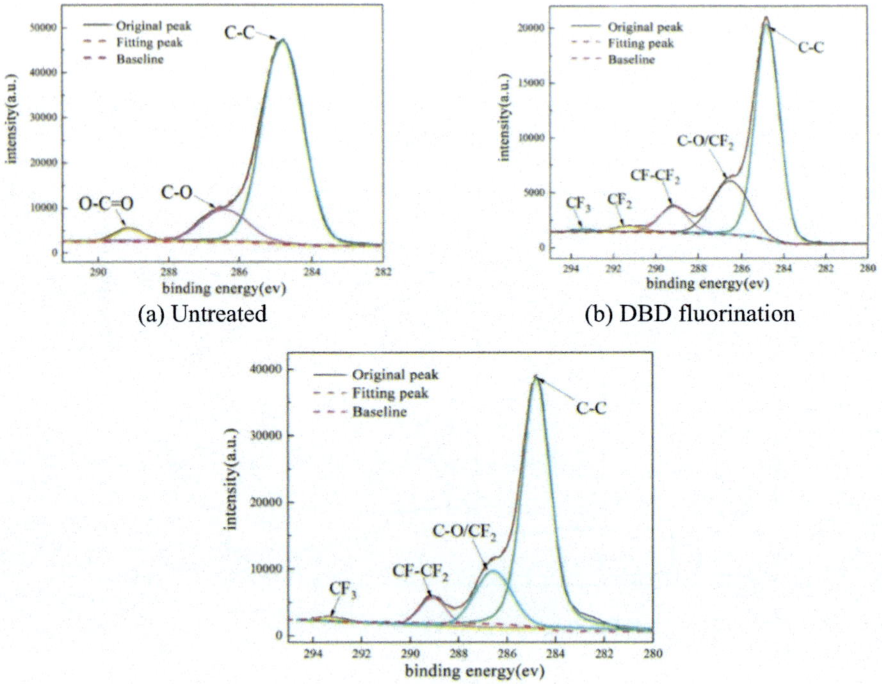

(a) Untreated

(b) DBD fluorination

(c) APPJ fluorination

Fig. 29 C1s XPS spectra of samples before and after plasma fluorination

to break the carbon-containing chemical bond, this position will become a reactive site, thus allowing the amalgamation with the free fluorine element in the plasma to form a more stable C–F bond—a process known as plasma grafting.

Further analysis of the composition of each functional group after plasma fluorination revealed that both DBD and APPJ contained C–CF_2 (286.6 eV), CF–CF_2 (289.3 eV) and CF_3 (293.0 eV) functional groups after fluorination. However, the relative amount of each functional group is different. The amount of CF_2-containing

Table 5 Ratios of chemical bond content before and after plasma fluorination

Approach	Relative element content/%					
	C–C	C–O/C–CF_2	O–C=O/C–F	CF–CF_2	CF_2	CF_3
Unprocessed	81.50	15.01	3.49	–	–	–
DBD fluorination	61.25	24.73	–	11.36	1.98	0.68
APPJ fluorination	73.09	20.15	–	5.95	–	0.82

functional groups (C–CF_2 and CF–CF_2) after DBD fluorination was significantly higher than that of APPJ fluorination and additional CF_2 (291.0 eV) functional groups were added to the surface of epoxy resin after DBD fluorination. When taking into account the considerable the discharge power of DBD fluorination as well as the relative presence of F element, it is suggested that the DBD structure can provides a more comprehensive plasma fluorination.

In addition, the F 1s peak in the XPS spectrum after fluorination of DBD is subjected to peak splitting and the obtained spectrum is shown in Fig. 30. It can be found that the F1s peak is divided into two peaks, corresponding to the organometallic fluorine peak (685 eV) and the organic fluorine peak (688 eV). The organic fluorine peak refers to the CF_x functional group formed by the free fluorine element and the carbon element on the surface of the epoxy resin. The metal fluorine peak is attributable to the $AlOF_x$ functional group, as C–H bonds are broken, leading to the generation of free protons reacted with the free hydrogen elements to form HF. Moreover, there are a certain amount of micron alumina particles on the surface of the epoxy resin, and more alumina is exposed after plasma fluorination. The aluminum oxide will react with hydrogen fluoride to give rise to $AlOF_x$ functional groups. Besides, hydrogen fluoride is corrosive, further improving the etching effect of plasma fluorination [57].

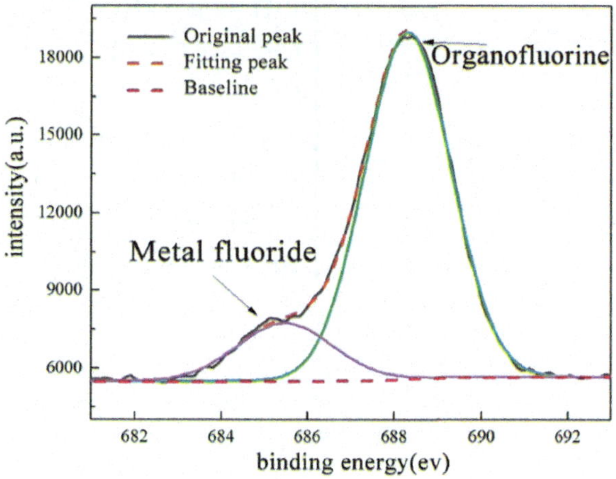

Fig. 30 F1s XPS spectrum of the epoxy resin after DBD fluorination

By studying the alterations in physical and chemical properties after plasma fluorination, the mechanism of the influence of plasma fluorination on the surface physical and chemical properties of epoxy resin can be obtained: The Ar/CF_4 mixed gas, up being excited by a high-frequency high-voltage power supply and transformed into a plasma state with tremendous energy, which internally ionizes many free fluorine elements. When the high-energy plasma bombards the surface of the epoxy resin, the high-energy plasma will break carbon-containing chemical bonds such as C–C bonds and C–H bonds and form reactive active sites. The reactive sites will combine with the free fluorine elements in the plasma to form strong C–F bonds. From the point of view of physical properties, the bond breaking and recombination of carbon chains lead to the destruction of the original surface morphology, and some alumina particles are exposed. Therefore, plasma fluorination can etch the surface of epoxy resin, the etching intensity strengthening with the extension of treatment time. From the perspective of chemical properties, plasma fluorination changes the original distribution of chemical elements on the surface of epoxy resin. A large amount of fluorine being grafted onto the surface of epoxy resin in the form of CF_x functional groups significantly increased the relative content of fluorine. In addition, hydrogen fluoride produced by C–H bond cleavage is corrosive. Hydrogen fluoride can react with alumina to generate $AlOF_x$ functional groups, further enhancing the plasma fluorination etching. From the above analysis we can conclude that the plasma fluorination method can regulate epoxy resin's surface element distribution, especially the fluorine elements' relative content. The reaction mechanism of plasma fluorination on the surface of epoxy resin is shown in formula (3). In addition, due to the higher discharge power of the DBD electrode, the plasma fluorination reaction can be more sufficient and the etching effect is more pronounced. Therefore, the surface of the epoxy resin is rougher after DBD fluorination, and more fluorine is grafted. Compared with APPJ, DBD modification can improve the effect of plasma fluorination.

$$\begin{cases} \text{R-H} \rightarrow \text{R}\cdot + \text{H}\cdot \\ \text{CF}_4 \rightarrow \text{CF}_3\cdot + \text{F}\cdot \\ \text{CF}_3\cdot \rightarrow \cdot\text{CF}_2\cdot + \text{F}\cdot \\ \cdot\text{CF}_2\cdot \rightarrow \cdot\text{CF}\cdot + \text{F}\cdot \\ \text{R}\cdot + \text{F}\cdot \rightarrow \text{RF} \\ \text{H}\cdot + \text{F}\cdot \rightarrow \text{HF} \\ \text{HF} + \text{Al}_2\text{O}_3 \rightarrow \text{AlOF}_x + \text{H}_2\text{O} \end{cases} \quad (3)$$

4.3 Modulation of Chemical Composition on the Surface of Epoxy Resin by Plasma SiO_x Deposition

Similar to the research steps of plasma fluorination, the XPS spectra of epoxy resin before and after plasma SiO_x deposition is shown in Fig. 31. It was found that the C1s peak, O1s peak and Si 2p peak changed after plasma SiO_x deposition. The peak value of C1s decreased with the extended plasma SiO_x deposition time, while the O1s peak and Si 2p peak increased with the increase of modification time. The relative content of main elements before and after plasma SiO_x deposition was also computed based on XPS spectra, as shown in Table 6. It was found that the untreated epoxy resin sample is still mainly composed of C element, but there is about 12% Si element. This is because a certain amount of silicon-containing coupling agent will be added to the micro-alumina filler during production, which can reduce the degree of agglomeration between the fillers and enhance the dispersion of the filler in the matrix, though a certain amount of Si element will be introduced. Unlike plasma fluorination, the relative content of C element decreased significantly after plasma SiO_x deposition. Instead, the content of Si element has increased, and the content of O element increased sharply. Especially after APPJ-SiO_x deposition, the content of O element accounts for 56% at most, which indicates that plasma SiO_x deposition not merely augmented the prevalence of Si element on the epoxy resin surface, but boosted the O element portion as well. The improvement effect changes slightly with the deposition time of plasma SiO_x. Furthermore, despite the DBD electrode encouraging the reaction, the content of O and Si elements in APPJ-SiO_x deposition is still higher than those in DBD-SiO_x deposition, which is contrary to the conclusion of plasma fluorination. This is because DBD-SiO_x deposition is carried out in the reactor, with the reaction atmosphere consisting of argon, while the reaction environment of APPJ-SiO_x deposition is air. Consequently, the plasma plume is exposed to the air and interacts with the epoxy resin sample, resulting in the surrounding oxygen participating in the reaction. Some studies have confirmed that by incorporating a appropriate amount of oxygen during plasma SiO_x deposition can promote the formation of thin films [47]. Therefore, the APPJ-SiO_x deployment can complete the reaction more efficiently.

Currently, it has been verified that plasma SiO_x deposition does indeed create a film on the surface of epoxy resin. In order to study the chemical composition of the film, the Si 2p in the XPS spectrum after plasma SiO_x deposition is processed by peak separation, and the new spectra is shown in Fig. 32. It can be found that the Si 2p peak before and after procedure can be divided into three peaks with the same peak value, but the peak areas of the three peaks corresponding to different methods are significantly different.

In order to ascertain the composition of the film on the surface of epoxy resin after plasma SiO_x deposition, each peak value was compared with the binding energy of typical functional groups in the XPS binding energy comparison table, and calculate the proportion of each peak area to the total peak area in each spectrum. The summarized results are shown in Table 7. It is found that the silicon-containing functional

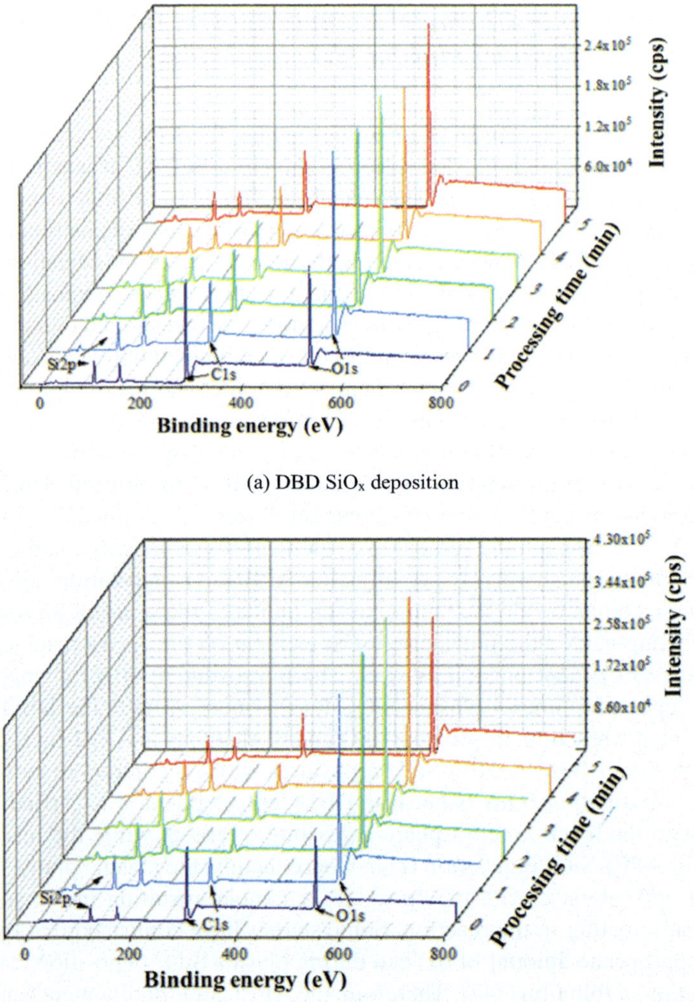

(a) DBD SiO_x deposition

(b) APPJ SiO_x deposition

Fig. 31 XPS spectra of samples treated by plasma SiO_x deposition

groups before and after plasma SiO_x deposition are Si ($–O_2$) (102.2 eV), Si($–O_3$) (102.8 eV) and Si($–O_4$) (103.4 eV). The relative content of Si($–O_2$) in untreated epoxy resin is up to 88%, while the relative content of Si($–O_3$) and Si($–O_4$) after plasma SiO_x deposition is significantly increased, especially in APPJ-SiO_x deposition, the relative content of Si($–O_4$) is up to 92%, while Si ($–O_2$) is only 5%. This is consistent with the significant increase of O element content after plasma SiO_x deposition, and the increase of the proportion of Si($–O_4$) attests to a higher degree of reaction completeness in plasma SiO_x deposition.

Table 6 Relative element content of sample surface before and after plasma SiO_x deposition

Relative element content/% (min)	C		O		Si	
	DBD-Si	APPJ-Si	DBD-Si	APPJ-Si	DBD-Si	APPJ-Si
0	64.37		23.41		12.22	
1	41.63	24.81	41.37	52.97	17	22.22
2	39.47	19.75	41.7	56.43	18.83	23.81
3	40.01	22.25	39.24	53.34	20.75	24.41
4	43.78	20.85	38.49	54.66	17.73	24.48
5	43.09	35.61	39.44	43.18	17.48	21.2

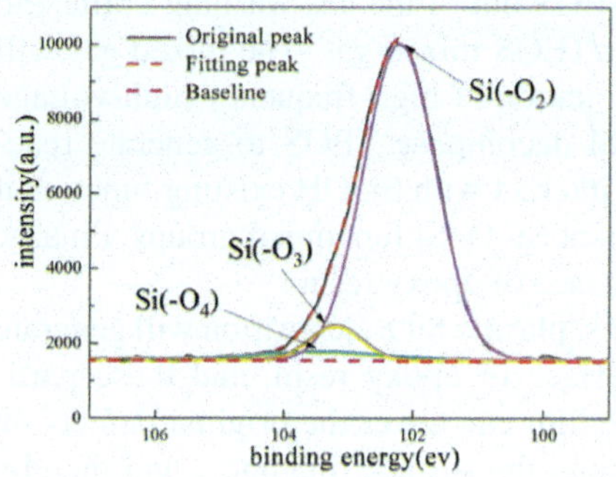

(a) Untreated

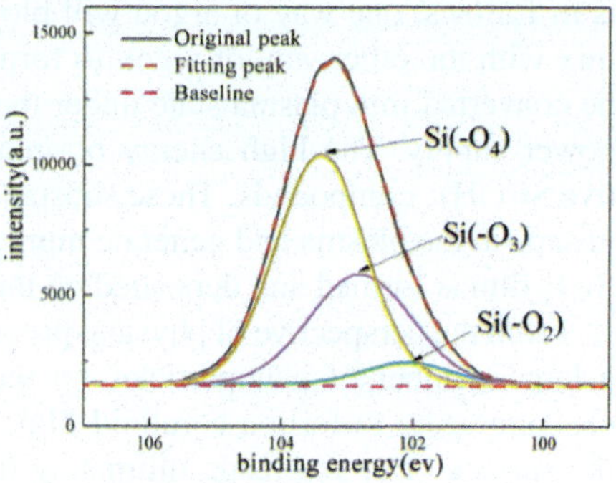

(b) DBD SiO_x deposition

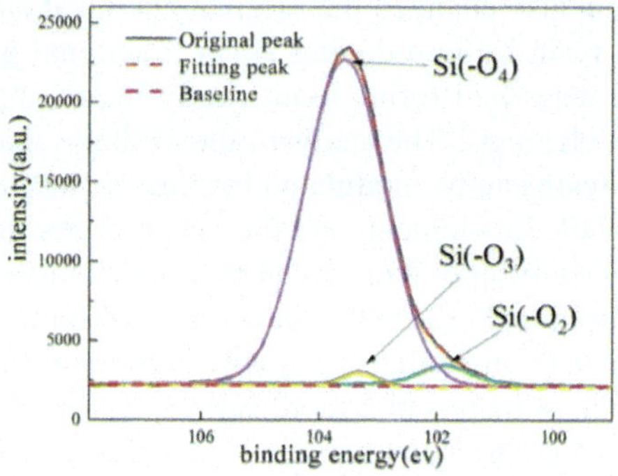

(c) APPJ SiO_x deposition

Fig. 32 Si 2p XPS spectra of samples before and after plasma SiO_x deposition

Table 7 Ratios of chemical bond content before and after plasma SiO_x deposition

Treatment method	Relative element content/%		
	$Si(-O_2)$	$Si(-O_3)$	$Si(-O_4)$
Untreated	88.79	7.19	4.01
DBD-SiO_x deposition	5.43	33.41	61.15
APPJ-SiO_x deposition	5.39	1.67	92.93

By comprehensive analysing the changes of physical and chemical properties after plasma SiO_x deposition, it can be concluded that the influence mechanism of plasma SiO_x deposition on the physicochemical properties of epoxy resin surface is as follows: one way of argon will blow TEOS out of the gas-washing bottle and mix with the other way of argon to form Ar/TEOS mixed gas. The mixed gas will be converted into plasma state under the excitation of high-frequency high-voltage power supply. The high energy plasma will decompose TEOS to generate reactive $Si(OH)_x$ compounds. These substances interact with Si–OH existing functional groups in the plasma and generate numerous of Si–O–Si functional groups. Finally, SiO_x film is formed and deposited on the surface of epoxy resin.

From the perspective of physical properties, plasma SiO_x deposition will generate a large number of nanoparticles on the surface of epoxy resin, and these particles aggregate to form a coralloid film. The film can cover the original defects on the surface of epoxy resin, ultimately reducing the surface roughness and thereby resulting in a more even surface. And the film coverage will continue to increase with the extension of treatment time. From the perspective of chemical properties, plasma SiO_x deposition also changed the original chemical element distribution on the surface of epoxy resin by introducing silica functional groups on the surface of epoxy resin in the deposited form, significantly improving the relative content of silicon and oxygen elements. The analysis above shows that the surface element distribution of epoxy resin can be modulated by plasma SiO_x deposition, especially the relative content of silicon element, and the reaction mechanism of plasma SiO_x deposition on the surface of epoxy resin can be obtained as shown in Formula (4) and Formula (5). In addition, because the reaction environment of APPJ-SiO_x deposition is air, the presence of ambient oxygen will accelerates the reaction process, so the APPJ electrode makes the reaction more sufficient; Because the APPJ electrode is more flexible, there are corresponding devices to ensure the uniformity of treatment, so the film produced by APPJ-SiO_x deposition is more even in comparison to DBD-SiO_x deposition.

$$Si(OC_2H_5)_n(OH)_{4-n} + e \rightarrow Si(OC_2H_5)_{n-1}(OH)_{4-n+1} + C_2H_4 + e \quad (4)$$

$$Si{-}OH + Si{-}OH \rightarrow Si{-}O{-}Si + H_2O \quad (5)$$

5 Modulation of Electrical Properties of Epoxy Resin Surfaces by Plasma Modification

It has been established that the surface physicochemical properties of insulating material are correlated to its electrical properties. Previous research has demonstrated that the physicochemical properties of the epoxy resin can be mortified by changing the experimental parameters. Therefore, it is necessary to further investigate into the effects of plasma modification methods on the electrical properties of epoxy resins, as well as the mechanism of electrical property changes.

This section presents the effect of different means on the flashover performance of epoxy resin surface, and analyze the influence of different methods on electrical parameters such as surface conductivity and surface charge aggregation characteristics. Additionally, the enhancement mechanism of various modification methods on the surface flashover performance will be explored, thereby establishing the correlation between plasma modification, physical and chemical properties of epoxy resin and electrical properties of epoxy resin. This enables the plasma modification to modulate the surface conductivity and surface charge dissipation rate of epoxy resin, as well as to enhance the surface flashover e performance.

5.1 Modulation of Surface Conductivity of Epoxy Resin by Plasma Modification

For the sub-atmospheric glow discharge treatment, the trend of surface conductivity with treatment time was illustrated in Fig. 33. It can be seen that the surface conductivity increase from 8×10^{-15} to 4×10^{-14} S/m when the treatment time is 1 min. The surface conductivity increased by a factor of 5. The surface conductivity continue to increase with the extension of the treatment time, but the increase in surface conductivity maintained a gradual yet slow rate. When the treatment time reaches 10 min, the surface conductivity of the sample is rose by a 10x. However, the bulk conductivity increase slightly after the modification, but the magnitude is much smaller than the surface conductivity increase. In summary, the glow discharge showed great efficacy in boosting the material surface conductivity by 10x.

Figure 34 shows the surface conductivity results of all samples before and after plasma fluorination and SiO_x deposition. The surface conductivity of the untreated epoxy resin samples is 3.11×10^{-18} S/m, and the surface conductivity is enhanced after various treatments. In general, the surface conductivity of the samples is monotonically increasing with the treatment time. In addition, the surface conductivity of the sample after plasma SiO_x deposition is higher than that of plasma fluorination. DBD-SiO_x deposition has the best effect in enhancing the electrical conductivity of epoxy resin surface up to 1.87×10^{-14} S/m. And the sample with the highest surface conductivity in plasma fluorination is DBD-F5, which is only 6.89×10^{-16} S/m. This result also shows that the surface conductivity of the epoxy resin can be

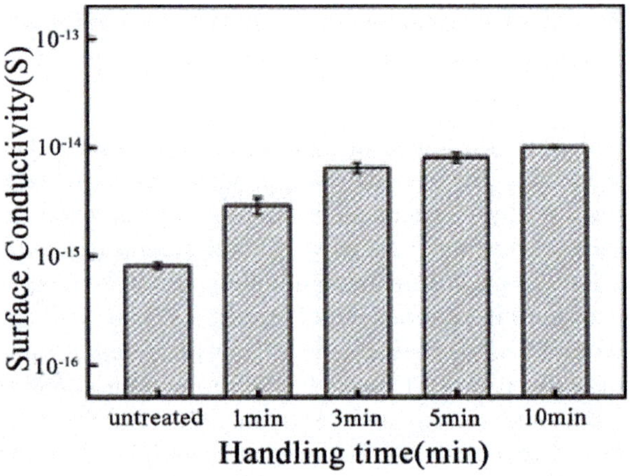

Fig. 33 Surface conductivity at different times of sub-atmospheric glow discharge

regulated between 10^{-18} and 10^{-16} S/m orders of magnitude using the plasma fluorination method described herein. However, plasma SiO_x deposition can modulate the surface conductivity of the epoxy resin between 10^{-17} and 10^{-14} S/m orders of magnitude.

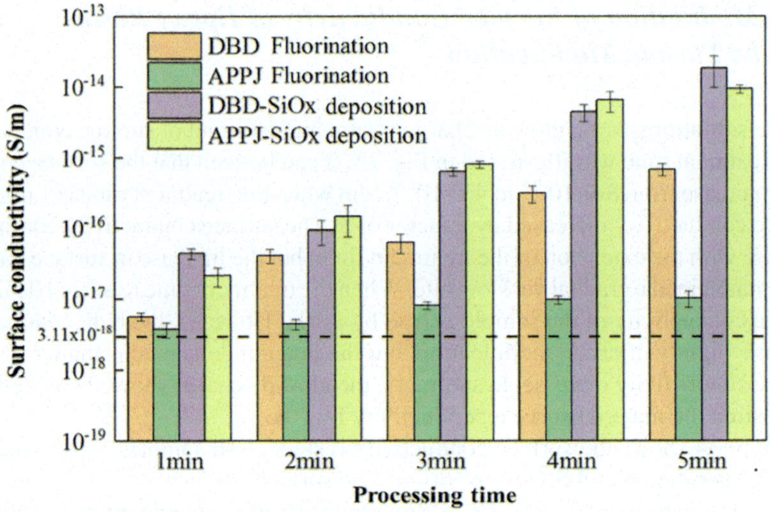

Fig. 34 Surface conductivity of epoxy resin before and after plasma modification

5.2 Modulation of Charge Dissipation Rate on Epoxy Resin Surface by Plasma Modification

Studies have shown that the surface flashover performance of an insulating material is influenced by the dissipation of the surface charge. And this rate can be approximated by the decay of surface potential [58]. With respect to the sub-atmospheric glow discharge, the Fig. 34 shows the sample surface potential decay curves at different modification times. It can be seen that These curves exhibit a double exponential decay form over the passage of time. In the initial moment of dissipation, an accelerated decline is observed on the surface potential, and the decay curve becomes slow and smooth as the dissipation time increases. Both the initial potential amplitude and the dissipated 600 s potential amplitude of the surface of the modified samples are lower than those of the untreated samples. It is also seen that the rate of decrease of surface potential rises with the increase of modification time from 0.5 to 3 min process. However, when the modification time continued to increase to 10 min, a saturation trend in the drop of surface potential was registered without any major variation in response to the prolonged modification duration.

For plasma fluorination and SiO_x deposition, the Fig. 35 shows the normalized surface potential decay from 0 to 30 min. The black dotted line in the figure denotes surface potential decay of the untreated epoxy resin. It can be found that the surface potential decay curves of most of the samples are below the black curve, indicating that the surface potential decay rate of these samples is faster than that of the untreated samples. In other words, the surface charge dissipation rate of these samples is faster than that of the untreated epoxy. From the perspective of processing time, the surface potential decay rates of all plasma-treated samples are positively correlated with the processing time. It is suggested that extending the processing time of plasma modification can accelerate the dissipation of charge on the surface of epoxy resin. From the perspective of the processing method, it is obvious for the decay of the epoxy resin surface potential after plasma SiO_x deposition. In particular, the surface potential of the APPJ-Si5 sample after 30 min is only 10% of the initial surface potential. And the degree of potential change of epoxy resin surface after plasma fluorination is not as good as plasma SiO_x deposition, especially APPJ fluorination. The surface potential of the APPJ-F5 sample with the highest decay is 83% of the initial surface potential at 30 min. Even the surface potential of the APPJ-F1 and APPJ-F2 samples decayed more slowly than that of the untreated samples. It indicates that both plasma fluorination and SiO_x deposition can accelerate the rate of charge dissipation on the epoxy resin surface. However, plasma SiO_x deposition is significantly more effective in dissipating the surface charge than plasma fluorination. From the perspective of plasma discharge electrodes, DBD fluorination is more competent in the decay of epoxy resin surface potential than APPJ fluorination. But APPJ-SiO_x deposition promotes the decay of epoxy resin surface potential better than DBD-SiO_x deposition. It indicates that the use of DBD electrode structure can enhance the dissipation rate of plasma fluorination on the surface charge of epoxy

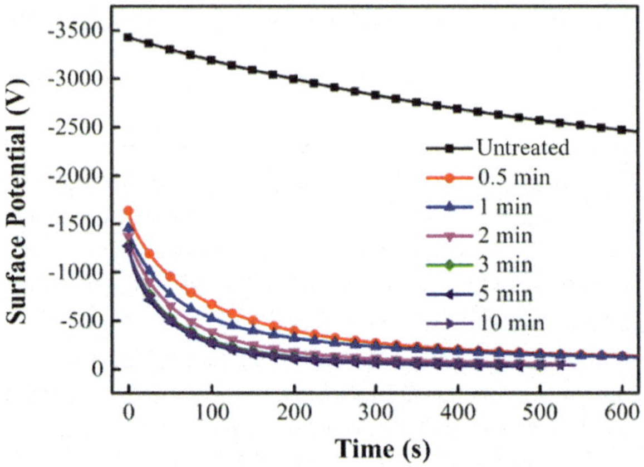

Fig. 35 Decay curve of surface potential under different modification time

resin, while the use of APPJ electrode structure can boosts the dissipation rate of plasma SiO_x deposition on the surface charge of epoxy resin.

According to the solid state energy band theory [59, 60], the ordered stacking of atoms in crystals creates an overlapping array of atomic orbitals, forming quasi-continuous molecular orbital energy levels. This results in multiple energy bands and an energy gap between them. In general, the highest energy band in which electrons can survive and exist stably in a crystal determines the properties of the crystal, and this energy band is called the valence band. Since the electron distribution follows the principle of lowest energy, the energy band, which is lower in energy than the valence band, has been filled with electrons and cannot conduct electricity. The energy band, which is higher in energy than the valence band, is only partially occupied by electrons, allowing for the free movement of electrons here and they can conduct electricity. This energy band is called the conduction band. One criterion in solid state physics to distinguish a crystal as a metal, semiconductor or insulator is the size of the energy gap between the valence band and the conduction band. With no gap present, a material is a metal, a gap greater than 4 eV makes it an insulator, and between 0 and 4 eV causing a semiconductor. The region between the valence band and the conduction band is called forbidden band, since it's where electrons cannot exist stably.

But in fact, the atomic arrangement of most insulators is disordered, especially the epoxy resin is a three-dimensional network structure formed by countless carbon chains cross-linked and intertwined, which is a typical amorphous, plus the intrinsic defects of the material or the role of additives, will form a localized state whose energy level is in the forbidden band, where the electrons will transition and be bound. This localized state is called a carrier trap. On the one hand, the carrier trap will restrict the movement of carriers and promote the accumulation of a large number of charges. On the other hand, the carrier trap will reduce the energy of electronic transition, but will promote the migration of carriers. Whether carrier traps bind or promote carrier migration depends on the energy level of the traps. Therefore, studying the depth and distribution of trap energy levels on the surface of epoxy resin before and after plasma modification can reveal mechanism of the effect of plasma modification on the charge dissipation rate on the surface of epoxy resin.

According to the Isothermal Surface Potential Decay (ISPD) theory [61], the surface potential decay of an insulating material satisfies a certain function with its trap energy level depth and trap density. Therefore, it is necessary to describe the surface charge decay curve by a functional relationship. Studies show that the double exponential function fit the surface potential decay curve of epoxy resin. The fitting formula is shown in Eq. (6):

$$\varphi_s(t) = Ae^{-\frac{t}{m}} + Be^{-\frac{t}{n}} \tag{6}$$

where $\varphi_s(t)$ is the surface potential, t is a moment of the surface potential decay curve, and A, B, m, and n are the fitting coefficients. Therefore, the trap energy level E_T and trap density Q_s of this sample are calculated as shown in Eqs. (7) and (8):

$$E_T = k_B T \ln(vt) \tag{7}$$

$$Q_s = t\frac{\varepsilon_0 \varepsilon_r}{eL} \cdot \frac{d\varphi_s(t)}{dt} \tag{8}$$

where k_B is the Boltzmann constant of 1.380649×10^{-23} J/K; T is the absolute temperature. The surface charge dissipation experiments of the group were conducted at a temperature of 26 °C. So the absolute temperature is taken as 299 K; v is the electron attempt escape frequency, taken as 4.17×10^{13} s^{-1}. e is the meta-charge, which is 1.6×10^{-19} C; L is the thickness of the sample, the epoxy resin samples prepared by the group are 3 mm thick, so the thickness of the sample is taken as 3 mm. ε_0 is the vacuum dielectric constant, which is 8.85×10^{-12} F/m; ε_r is the relative dielectric constant of the sample, and the study shows that the modification of the epoxy resin surface has less effect on its dielectric constant. Then the relative dielectric is taken as 4.5, which is the relative dielectric constant of the untreated epoxy resin.

For the sub-atmospheric glow discharge, Fig. 37 shows the surface charge trap energy level distribution curves of the specimens at varying modification times. It can be seen that the surface charge trap distribution curve of the untreated sample shows 2 characteristic peaks, corresponding to 2 different trap energy level depths and different peaks. After plasma treatment, the trap energy depths of the 2 characteristic peaks of the epoxy resin were gradually decreased. The second characteristic peak gradually disappears as the modification time increases. By further increasing the modification time, the depth of the trap energy level corresponding to the characteristic peak no longer decreases. The surface potential drop curves in Fig. 34 all show a double exponential drop pattern, implying that there are two trap mechanisms inside the epoxy resin. Corresponding to the pattern shown in the curves of Fig. 36, the peaks corresponding to the lower trap energy level depths in the figure represent shallow traps, while the peaks corresponding to higher trap energy level depths are associated with deep traps. It corresponds to 2 processes of rapid decrease of the specimen surface potential at the initial moment and slow decrease at the later stage [62]. With the extension of the modification time, the two trap energy levels become progressively shallower in depth, and the deep traps gradually decrease or even disappear. However, further increasing the modification time did not significantly change the trap energy depth. This was in line with the pattern reflected in Fig. 34, it can be seen that the trap energy level depth and peak of the surface charge trap energy level distribution curve can reflect the rate of surface potential drop.

For plasma fluorination and SiO_x deposition, Fig. 37 shows the trap energy level distribution curves for all samples. The trap energy level distribution curve of the epoxy resin was found to consist of a single peak or a double peak, and the depth of the trap energy level corresponding to the peak of the curve is the depth of the trap energy level mainly distributed in the sample. The trap energy level curve of the untreated epoxy is a single peak with a peak corresponding to a trap energy level depth of 1.08 eV. It indicates that the surface of untreated epoxy resin is mainly distributed with traps of energy level depth of 1.08 eV. This trap is deeper and is

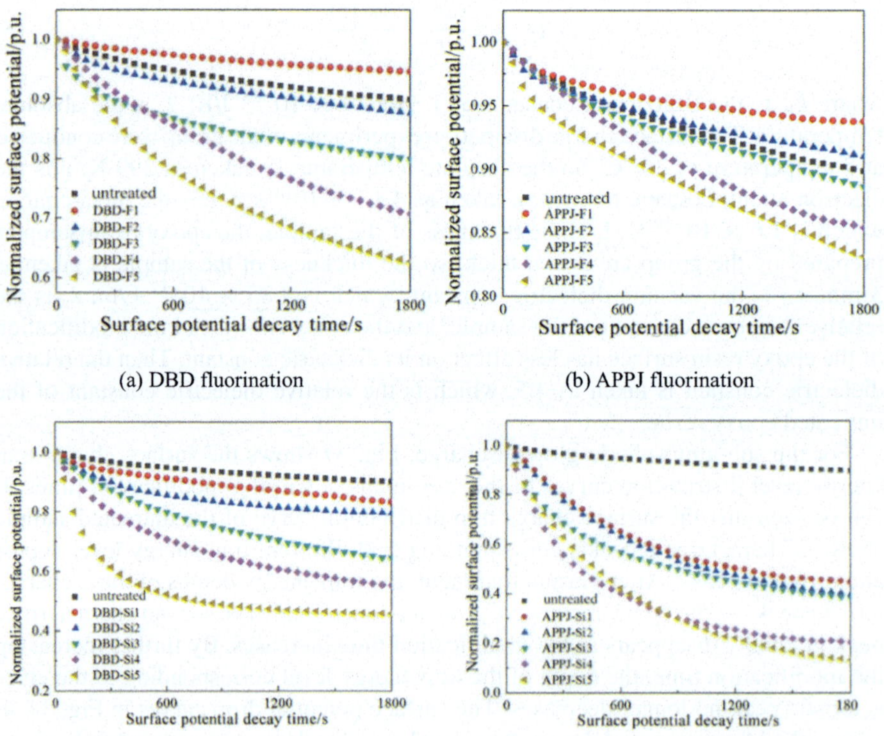

(a) DBD fluorination

(b) APPJ fluorination

(c) DBD SiOx deposition

(d) APPJ SiOx deposition

Fig. 36 Surface potential decay curve of epoxy resin before and after plasma modification

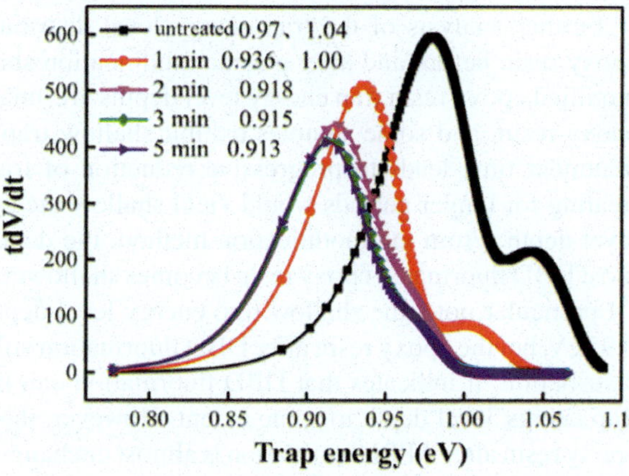

Fig. 37 Energy level distribution of surface traps under different modification times

generally referred to as a deep trap. By analogy, the trap energy level distribution curve of APPJ-Si1 sample shows not only a deep trap peak of 1.06 eV, but also a peak with a trap energy level depth of only 0.98 eV. This trap depth is generally referred to as a shallow trap, indicating that there is already a partial distribution of shallow traps on the surface of the APPJ-Si sample (Fig. 38).

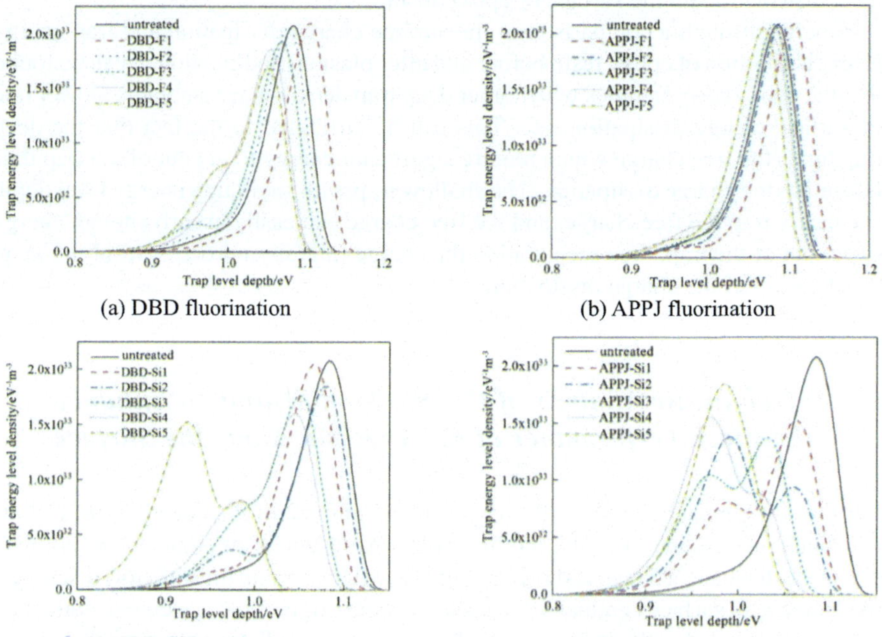

(a) DBD fluorination

(b) APPJ fluorination

(c) DBD SiO_x deposition

(d) APPJ SiO_x deposition

Fig. 38 Charge trap variation curve of epoxy resin before and after plasma modification

Further analysis of the trap energy level distribution curves on the surface of epoxy resin before and after plasma modification showed that most of the plasma-modified epoxy resin trap energy level depths are smaller than those of the untreated epoxy resin, and some samples exhibit shallow trap peaks. Increasing the plasma treatment time leads to progressive reduction of traps depths. This confirms that treating for longer periods would yield shallow the epoxy resin surface trap energy level depth. From the modification method, the depth of deep trap energy level of most DBD fluorinated epoxy resin becomes shallower, and then shallow traps appear. Of particular note, the shallow trap energy level depth of DBD-F3 sample reached 0.95 eV, but the epoxy resin after DBD fluorination still has deep trap as the main trap distribution. It indicates that DBD fluorination can lighten the epoxy resin surface trap energy level depth to some extent. However, the energy depth of deep traps in epoxy resin after APPJ fluorination is almost unchanged, and even if there are shallow traps, their proportion is much smaller than that of deep traps. It indicates that the effect of APPJ fluorination on the energy level distribution of epoxy traps is not significant. Plasma SiO_x deposition not only makes the depth of the deep trap energy level in the trap energy level distribution curve of the epoxy resin surface smaller, but also increases the area of the shallow trap peak. For example, the shallow trap peak height of DBD-Si5 and APPJ-Si5 samples has exceeded the deep trap. It shows that plasma SiO_x deposition can not only shallow the deep traps of epoxy resin, but also introduce more shallow traps, resulting in significant changes in the trap energy level distribution on the surface of epoxy resin.

Finally, through a comparison of the surface charge dissipation and trap energy level distribution of epoxy resin before and after plasma modification, it can be found that the epoxy resin samples with higher deep trap density have a notably slower rate of surface charge dissipation rate. This can be attributed to the fact that the deep trap binds the free charge which require significant energy to get out of the trap thus inhibiting the charge dissipation. The shallow trap with small trap energy level depth can easily trap the free charge, and the free charge can easily absorb enough energy to get out of the trap, thus accelerating the charge migration process, resulting in an acceleration of the charge dissipation.

5.3 Mechanistic Analysis of Plasma Modification to Enhance the Flash Performance of Epoxy Resin Along the Surface

For the sub-atmospheric glow discharge, the magnitude of the negative DC voltage resistance before and after the modification was tested at an atmospheric ambient electrode distance of 5 cm, as shown in Fig. 39. After a certain time of glow discharge, the increase in surface conductivity accelerates the original surface charge collection and slows down the effect of surface charge on electric field distortion. This leads to an increase in the DC flashover voltage. When the modification time is increased to 5 min, the increase in conductivity of the material surface reaches saturation, thus

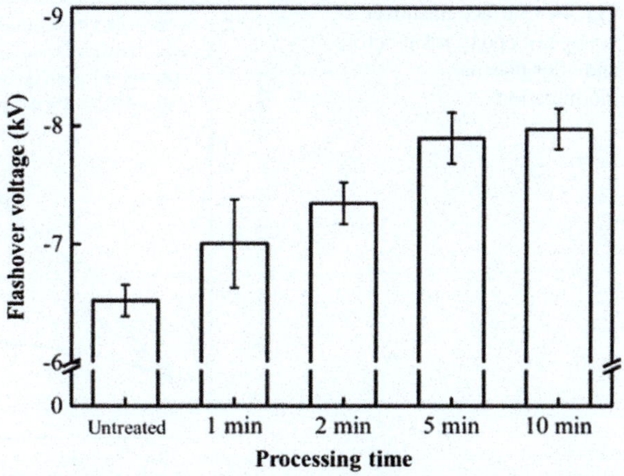

Fig. 39 Flashover voltage under different modification time

causing a cessation in surface charge dissipation rate growth and inhibiting further enhancement of flashover voltage. Therefore, with the extension of modification time, the flashover voltage sharply initially and then more slowly and the surface flashover voltage increased from −6.6 to −7.8 kV without treatment, with a maximum increase of about 17%.

For plasma fluorination and SiO_x deposition, the surface flashover voltages of all epoxies before and after modification are shown in Fig. 40. Since we use a negative polarity DC power supply to test the surface flash performance of all the samples, we take all the along-surface flash voltage values as absolute values for the sake of comparison. It can be discovered the surface flashover voltage of the untreated epoxy samples is 9.45 kV, and the along-surface flashover voltage of all the plasma-modified epoxies has increased. The DBD fluorinated epoxy resin has the best increase in flashover voltage up to 11.45 kV, which is 21.2% higher than the untreated sample. We observed the effect of various plasma modification methods on the surface flashover voltage with respect to the length of treatment time. We found that all plasma modification means caused a decrease in the flashover voltage at long modification time. Among them, the flashover voltage of APPJ-SiO_x deposition decreased severely after 3 min of modification, with the highest decrease from 11.28 kV at 3 min to 10.03 kV, which is a decrease of 11%. The plasma treatment time of 3 min and less was considered as short time treatment, and the enhancement effect of each plasma modification method on the epoxy resin along the surface flash in this interval was ranked as follows: DBD fluorination > APPJ-SiO_x deposition > DBD-SiO_x deposition > APPJ fluorination. Taking the plasma treatment time of 3 min and above as the long time treatment, the enhancement effect of each plasma modification method on epoxy resin along the surface flash in this interval was ranked as follows: DBD fluorination > DBD-SiO_x deposition > APPJ fluorination > APPJ-SiO_x deposition.

Based on a range of the experimental studies and analyses, we uncovered the correlation between the plasma modification method, physical and chemical properties of

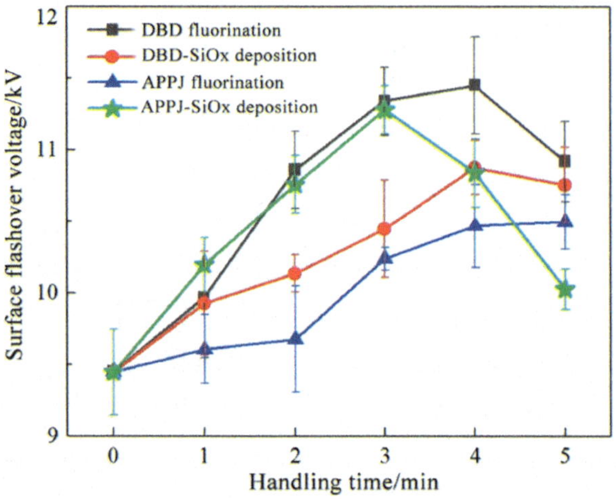

Fig. 40 Surface flashover voltage of epoxy resin before and after plasma modification

epoxy resin, and electrical properties of epoxy resin, and analyzed the mechanism of plasma modification to enhance the surface flash performance of epoxy resin.

To explain the process of flashing along the surface of the epoxy resin: when the negative polarity DC high-voltage power supply is continuously boosted, a large amount of surface charge is generated at the junction of the needle-pin electrode, the epoxy resin surface and the air. These charges migrate to the opposite electrode aided by the action of the external electric field. When the surface charge continuously migrates and forms a conductive channel through, the two electrodes conduct and flashover occurs along the surface. If surface charges accumulate in a large amount at the triple bonding point, these charges will seriously distort the external electric field, which will further promote the occurrence of flashover along the surface. Therefore, the flashover performance of epoxy resin mainly depends on the dissipation rate of surface charges. The surface conductivity of the untreated epoxy sample is only 10^{-18} orders of magnitude and the deep trap density is too high, which makes it difficult for surface charges to migrate across the surface. This leads to the gathering of substantial amount of charges at the triple bonding point, thereby distorting the external electric field and triggering surface flashover. The results of the surface flashover experiments further confirm that the untreated epoxy resin has the lowest surface flashover voltage.

Sub-atmospheric pressure glow discharge, on the one hand, the modification improves the surface conductivity of the material, and promotes the surface charge dissipation process along the surface. On the other hand, the surface trap level depth of the modified sample is lower than before, and the number of deep traps is significantly reduced, which accelerates the charge detrapping process. The improvement of surface conductivity and the shallowing of trap energy levels jointly promote the rapid dissipation of surface charges, thereby slowing down the effect of surface charges on electric field distortion and inhibiting the occurrence of discharge along the surface.

As for plasma fluorination and SiO_x deposition, we compared the surface flashover voltage of all modified epoxy resins, and APPJ fluorination has the smallest increase in the surface flashover voltage of epoxy resins. It was also found that APPJ fluorination improved the surface roughness of epoxy resin. However, only a small amount of fluorine was grafted on the surface of the epoxy resin, and the surface conductivity and surface charge dissipation rate were slightly improved, with surface trap energy level distribution changing little. It can be known from the literature [18] that an appropriate amount of surface roughness of insulating materials can increase the probability of surface charges being recaptured during surface migration, thereby increasing the rate of surface charge dissipation. Therefore, it can be concluded that the improvement mechanism of APPJ fluorination on the surface flashover performance of epoxy resin: Due to the lower discharge power and small treatment area of APPJ fluorination. Only a small amount of fluorine is grafted onto the surface of the epoxy resin, which has little effect on the chemical composition of the surface of the epoxy resin. Therefore, the surface roughness of the epoxy resin is mainly increased by the etching effect of the plasma, which increases the chances of the surface charge being recaptured when it migrates on the surface of the epoxy resin. At the same time, APPJ fluorination also slightly increased the surface conductivity of the epoxy resin. Thereby, the dissipation of the surface charge is accelerated, suppressing the distortion effect caused by the accumulation of the surface charge. So compared with untreated epoxy resin, APPJ fluorination can slightly improve the flashover performance along the surface.

This paper experimentally demonstrates that DBD fluorination of epoxy resin vastly supersedes APPJ fluorination in terms of improving the surface flashover voltage of epoxy resin. It is revealed that DBD fluorination produces a rough surface texture with a high concentration of fluorine atoms and groups, both of which enhance surface conductivity and charge dissipation rate. In addition, the deep traps on the surface of epoxy resin were shallowed and a small number of shallow traps appeared. These combined changes hold profound implications on the surface flashover performance of epoxy resin—that DBD fluorination improves the surface flashover performance of epoxy resin: due to the high discharge power and large treatment area of DBD fluorination, a large number of fluorine elements and fluorine-containing groups are grafted onto the surface of epoxy resin. And the etching effect of the high-energy plasma is used to greatly improve the surface roughness of the epoxy resin. On the one hand, the presence of fluorine-containing groups changes the original chemical composition of the epoxy resin surface. This leads to changes in the distribution of surface trap energy levels, that is, the depth of deep trap energy levels becomes shallower and a small number of shallow traps are introduced, thereby promoting the migration of surface charges on the surface of epoxy resin; On the other hand, a rougher surface will also increase the probability of surface charges being captured during surface migration, and the increase in surface conductivity will also accelerate surface charge migration. DBD fluorination guarantees greater charge dissipation compared to APPJ fluorination. It makes the surface charge more difficult to accumulate, so the effect of DBD fluorination on the surface flashover performance of epoxy resin is better than that of APPJ fluoridation.

The above analysis suggests that increasing the charge dissipation rate can improve the surface flashover performance. However, experiments show that the surface charge of APPJ-SiO_x deposited on the surface of epoxy resin dissipates the fastest, and its surface flashover voltage has an obvious downward trend when the treatment time is greater than 3 min. In addition, the lowest creeping flashover voltage is even close to that of untreated epoxy resin, and the surface charge dissipation rate of the DBD fluorinated 5 min sample is greater than that of other DBD fluorinated samples. However, the phenomenon of flashover voltage drop along the surface also appeared. These observations indicate that further explorations are required to understand the mechanism of flashover lifting along the surface. Further studying the changes in physicochemical properties and electrical properties of epoxy resin after APPJ-SiO_x deposition, we found that APPJ-SiO_x deposition enables TEOS to go through a condensation reaction, resulting in the formation and deposition of dense nanoscale SiO_x films over the epoxy resin surface. Through doing this, the existing material defects in the epoxy resin are covered up, thus flattening its surface. In addition, the deposited SiO_x films introduce an abundance of silicon-oxygen groups, which change the original chemical composition of the epoxy resin surface. The wax film is filled with a multitude of shallow traps, which also changes the trap energy level distribution on the epoxy resin surface and increases the surface conductivity of the epoxy resin by nearly four orders of magnitude. Even though the reduction in surface roughness decreases the rate of surface charge dissipation, the presence of a large number of shallow traps, on the contrary, greatly promotes the migration of surface charge. Therefore, the surface charge dissipation of the epoxy resin after APPJ-SiO_x deposition is the most rapid. However, due to excessive charge migrating to the opposite electrode, flashover may take place along the face. If the surface charge of epoxy resin is dissipated at a rapid speed, although the surface charge is not easy to accumulate in large quantities and distort the external electric field, it will provide seed charge for the development of flashover along the surface and promote the formation of penetrating conductive channels. This will make the surface flash performance of epoxy resin decline.

Reviewing the flashover voltage and surface charge dissipation along the surface of DBD fluorination and APPJ-SiO_x deposition, we found that the surface charge of epoxy resin dissipated too fast at 5 min of DBD fluorination and greater than 3 min of APPJ-SiO_x deposition. These findings indicated flashover voltage decrease along the surface. Therefore, it can be summarized that the mechanism of APPJ-SiO_x deposition to improve the flash performance of epoxy resin along the surface: APPJ-SiO_x deposition, through the polymerization effect of high-energy plasma, causes TEOS condensation reaction to generate SiO_x film and deposit on the surface of epoxy resin, covering the original defects of epoxy resin and reducing the surface roughness. At the same time the presence of SiO_x films introduces a large number of silicone oxygen groups on the epoxy resin surface, which significantly increases the surface conductivity while making the surface trap energy level distribution dominated by shallow traps. This greatly accelerates the dissipation of surface charge. Therefore, the short time of APPJ-SiO_x deposition also inhibits the accumulation of surface charges and improves the performance of flashover along the surface. However, prolonged

APPJ-SiO_x deposition causes too rapid dissipation of surface charge, which inhibits surface charge accumulation but significantly accelerates charge migration between electrodes and promotes the formation of penetrating conductive channels, which instead reduces the flashover performance along the surface.

The changes in the physicochemical properties and electrical properties of the epoxy resin surface after DBD-SiO_x deposition are basically the same as those of APPJ-SiO_x deposition, so the enhancement mechanism of DBD-SiO_x deposition on the flashing properties of the epoxy resin along the surface is similar to that of APPJ-SiO_x deposition. However, the effect of DBD-SiO_x deposition on the flashover voltage along the surface of epoxy resin is generally more limited due to the fixed the relative position of the DBD electrode, which makes it difficult to control the uniformity of the SiO_x film. This affects the effect of improving the electrical properties of the epoxy resin. The APPJ electrode structure is more flexible, and other devices can be configured to improve the uniformity of the deposited film. Moreover, the deposition of APPJ-SiO_x is carried out in the air, with oxygen further promoting the decomposition of TEOS and thus allowing for a better effect than DBD-SiO_x deposition.

Finally, a summary of the correlation between the plasma modification conditions, the physical and chemical properties of the epoxy resin surface, and the electrical properties of the epoxy resin surface can be made: the reaction precursor determines which plasma effect is used to modify the epoxy resin surface. Using CF_4 as a precursor facilitates both plasma etching and grafting effects to increase the surface roughness of the epoxy resin and graft fluorine-containing groups on the surface of the epoxy resin, thereby increasing the surface conductivity and reducing the depth of the original trap level. Ultimately, the surface charge dissipation rate is increased. Using TEOS as a precursor can make use of the plasma polymerization effect to cause TEOS to become polymerized and form a SiO_x film and deposit it on the surface of epoxy resin. This leads to reduction of the surface roughness of the epoxy resin and introduction of a large number of siloxane groups, thereby greatly improving the surface conductivity and significantly reducing the depth of the original trap level and forming an array of shallow traps—thus greatly increasing the surface charge dissipation rate.

The duration and the discharge electrode employed are significant factors that affect the efficacy the plasma modification. Prolonging the treatment time allows the plasma to modify the epoxy resin, and strengthen the effects of plasma etching, grafting and polymerization. This magnifies the advantages and disadvantages of plasma modification. The DBD electrode has a greater discharged power and a larger treated area, which enable more sufficient; but the structure of the DBD electrode is fixed, which leads to an uneven treatment effect, so it is not suitable for plasma SiO_x deposition. The APPJ electrode structure is flexible, and can be combined with appropriate apparatus to boost the evenness of modification, and the reaction can be carried out in the air. Therefore, when combined with the plasma SiO_x deposition method, the SiO_x film formed is uniform and dense; but the discharge power of the APPJ electrode is low, and the treatment area is small, which leads to poor effect of plasma fluorination.

6 Conclusion

This chapter systematically presents three plasma modification methods of epoxy resin, including sub-atmospheric glow discharge, plasma fluorination, and plasma SiO_x deposition. In addition, this chapter makes a comprehensive comparison of the effects of three methods on the surface properties of epoxy resins. Experiments attest to the effect of three plasma modification methods on the surface micro-morphology, chemical composition, surface conductivity and surface charge dissipation rate of epoxy resin. The relationship among the plasma modification method, the physical and chemical properties of the epoxy resin surface and the electrical properties of the epoxy resin surface was established. The mechanism of plasma modification to improve the surface flashover performance of epoxy resin is also elucidated. The main work and conclusions are as follows:

(1) Sub-atmospheric pressure glow discharge and plasma fluorination modification utilize the etching effect and grafting effect of high-energy plasma. The subatmospheric pressure glow discharge increased the oxygen content of the epoxy resin. In addition, the plasma fluorination enhanced the surface roughness and the relative content of fluorine-containing groups of the epoxy resin. Plasma SiO_x deposition modification utilizes the polymerization effect of high-energy plasma. Plasma SiO_x deposition modification reduces the surface roughness and increases the content of siloxane groups by depositing SiO_x film on the surface of epoxy resin.
(2) Increasing the surface roughness of epoxy resin will accelerate the dissipation rate of its surface charge. Introducing oxygen-containing and siloxy groups on the surface of epoxy resin will reduce the depth of trap energy levels and present many shallow traps, through which the charge can rapidly transit, leading to significantly improvement in the surface charge dissipation rate.
(3) Improving the surface charge dissipation rate is conducive to evacuating surface charges, thereby restraining the electric field distortion effect caused by a large amount of charge accumulation. Therefore, increasing the surface charge dissipation rate is beneficial to improving the surface flashover performance of epoxy resin. Nevertheless, the rapid dissipation of surface charge will promote the formation of penetrative conducting channels between electrodes, thus diminishing the surface flashover performance. To sum up, the surface charge dissipation rate needs should be appropriately regulated within a certain range.

References

1. Shao, T., Kong, F., Lin, H.F., et al.: Correlation between surface charge and DC surface flashover of plasma treated epoxy resin. IEEE Trans. Dielectr. Electr. Insul. **25**(4), 1267–1274 (2018)
2. Lu, X.P., Yan, P., Ren, C.S., et al.: Review on atmospheric pressure pulsed DC discharge. Sci. Sinica Phys. Mech. Astron. **41**(07), 801–815 (2011)

3. Wang, F., Fang, Z., Qiu, Y.C.: Study of charge accumulation on insulator surface in HVDC gas-insulated switchgear. J. Chin. Electr. Eng. Sci. **03**, 107–111 (2005)
4. Dai, D., Ning, W.J., Shao, T.: A review on the state of art and future trends of atmospheric pressure low temperature plasmas. Trans. China Electrotech. Soc. **32**(20), 1–9 (2017)
5. Xie, Q., Wang, Y., Liu, X., et al.: Characteristics of microsecond-pulse surface flashover on epoxy resin surfaces in SF6. IEEE Trans. Dielectr. Electr. Insul. **23**(4), 2328–2336 (2016)
6. Zhang, H.B., Chen, Q.: Recent progress of non-thermal plasma material surface treatment and functionalization. Phys. Sci. J. **70**(9), 22–38 (2021)
7. Zhang, C., Gu, J.W., Shao, T., et al.: Discharge mode in the repetitive nanosecond-pulsedischarge in atmospheric pressure air. High Power Laser Part. Beams **26**(4), 203–209 (2014)
8. Shen, Y., Wang, R.X., Zhang, C., et al.: Characterization of atmospheric pressure helium plasma jet driven by microsecond pulse. High Power Laser Part. Beams **28**(5), 112–118 (2016)
9. Ma, Y.F., Zhang, C., Li, C.Y., et al.: Experimental study of accelerating surface charge dissipation on polymer treated by repetitively pulsed discharge plasmas. J. Chin. Electr. Eng. Sci. **36**(6), 1731–1738 (2016)
10. Ishikawa, S., Yukimura, K., Matsunaga, K., et al.: Surface modification of poly (tetrafluoroethylene) film using dielectric barrier discharge of intermittent pulse voltage. Jpn. J. Appl. Phys. **39**(9R), 5223 (2000)
11. Sarani, A., Nikiforov, A.Y., Leys, C.: Atmospheric pressure plasma jet in Ar and Ar/H_2O mixtures: optical emission spectroscopy and temperature measurements. Phys. Plasmas **17**(6), 063504 (2010)
12. Shao, T., Wang, R.X., Zhang, C., et al.: Atmospheric-pressure pulsed discharges and plasmas: mechanism, characteristics and applications. High Volt. **3**(1), 14–20 (2018)
13. Mei, D.H., Fang, Z., Shao, T.: Recent progress on characteristics and applications of atmospheric pressure low temperature plasmas. J. Chin. Electr. Eng. Sci. **40**(4), 1339–1358+1425 (2020)
14. Liu, W.Z., Chai, M.L., Hu, W.L., et al.: Generation of atmospheric pressure diffuse dielectric barrier discharge based on multiple potentials in air. Plasma Sci. Technol. **21**(7), 074004 (2019)
15. Luan, P., Knoll, A., Wang, H., et al.: Model polymer etching and surface modification by a time modulated RF plasma jet: role of atomic oxygen and water vapor. J. Phys. D: Appl. Phys. **50**(3), 03LT02 (2016)
16. Dai, Y.C., Zhang, M., Li, Q., et al.: Separated type atmospheric pressure plasma microjets array for maskless microscale etching. Micromachines **8**(6), 173 (2017)
17. Seok, D.C., Yoo, S.R., Lee, K.I., et al.: Relation between etching profile and voltage–current shape of sintered SiC etching by atmospheric pressure plasma. Plasma Sci. Technol. **21**(4), 045504 (2019)
18. Shao, T., Zhou, Y.X., Zhang, C., et al.: Surface modification of polymethyl-methacrylate using atmospheric pressure argon plasma jets to improve surface flashover performance in vacuum. IEEE Trans. Dielectr. Electr. Insul. **22**(3), 1747–1754 (2015)
19. Shao, T., Liu, F., Hai, B., et al.: Surface modification of epoxy using an atmospheric pressure dielectric barrier discharge to accelerate surface charge dissipation. IEEE Trans. Dielectr. Electr. Insul. **24**(3), 1557–1565 (2017)
20. Shao, T., Zhang, C., Wang, R.X., et al.: Atmospheric-pressure pulsed gas discharge and pulsed plasma application. High Volt. Eng. **42**(3), 685–705 (2016)
21. Chen, S.L., Wang, S., Wang, Y.B., et al.: Surface modification of epoxy resin using He/CF_4 atmospheric pressure plasma jet for flashover withstanding characteristics improvement in vacuum. Appl. Surf. Sci. **414**, 107–113 (2017)
22. Zhang, C., Zhou, Y., Shao, T., et al.: Hydrophobic treatment on polymethylmethacrylate surface by nanosecond-pulse DBDs in CF_4 at atmospheric pressure. Appl. Surf. Sci. **311**, 468–477 (2014)
23. Zhan, Z.Y., Ruan, H.O., Lv, F.C., et al.: Plasma fluorinated epoxy resin and its insulation properties in C_4F_7N/CO_2 mixed gas. Trans. China Electrotech. Soc. **35**(8), 1787–1798 (2020)

24. Wan, F.C.: The Surface fluorination of Crosslinked Polystyrene and Its Vacuum Surface Flashover Performance Research. Wuhan University of Technology (2017)
25. Wan, F.C., Xu, F.F., Wei, W., et al.: Surface modification of cross-linked polystyrene by CF_4 plasma and its vacuum surface flashover performance. High Volt. Eng. **44**(12), 3857–3864 (2018)
26. Kang, Y.C., Yan, J.Y., Peng, C.K., et al.: Plasma dielectric barrier discharge fluorination modified epoxy resin and its ageing behavior. High Power Laser Part. Beams **33**(6), 161–171 (2021)
27. Peyroux, J., Dubois, M., Tomasella, E., et al.: Enhancement of surface properties on commercial polymer packaging films using various surface treatment processes (fluorination and plasma). Appl. Surf. Sci. **315**, 426–431 (2014)
28. Zhang, C., Ma, Y.Y., Kong, F., et al.: Atmospheric pressure plasmas and direct fluorination treatment of Al_2O_3-filled epoxy resin: a comparison of surface charge dissipation. Surf. Coat. Technol. **362**, 1–11 (2019)
29. Kong, F., Chang, C., Ma, Y.Y., et al.: Surface modifications of polystyrene and their stability: a comparison of DBD plasma deposition and direct fluorination. Appl. Surf. Sci. **459**, 300–308 (2018)
30. Matsubara, K., Danno, M., Inoue, M., et al.: Surface fluorination of polystyrene particles via CF_4 plasma irradiation using a barrel-plasma-treatment system. Surf. Coat. Technol. **236**, 269–273 (2013)
31. Xie, Q., Duan, Q.J., Shao, S., et al.: Effect of barium titanate nanofibers and plasma fluorination on surface insulation properties of epoxy resin composites. J. Chin. Electr. Eng. Sci. **40**(12), 4051–4063 (2020)
32. Ruan, H.O.: The Effect of Nano-Al_2O_3 and Nano-TiO_2 and Their Fluorination on the Surface Insulation Properties of Epoxy Resin. North China Electric Power University, Baoding (2019)
33. Ruan, H.O., Xie, Q., Lü, F.C., et al.: Filler fluorination of nanoTiO_2/ER composites and their surface insulation properties: a comparison of dielectric barrier discharge and chemical solution fluorination. J. Phys. D Appl. Phys. **53**(14), 145204 (2020)
34. Lv, F.C., Zhan, Z.Y., Zhang, L.G., et al.: Effect of plasma fluorinated modified micro-AlN filler epoxy resin on the insulation properties. Trans. China Electrotech. Soc. **34**(16), 3522–3531 (2019)
35. Lv, F.C., Liu, C.B., Zhan, Z.Y., et al.: Effect of plasma fluorinated nano-SiC/epoxy composite coating on the flashover characteristics. High Volt. Eng. **46**(7), 2444–2452 (2020)
36. Armenise, V., Milella, A., Fracassi, F., et al.: Deposition of thin films containing carboxylic acid groups on polyurethane foams by atmospheric pressure non-equilibrium plasma jet. Surf. Coat. Technol. **379**, 125017 (2019)
37. Kakiuchi, H., Higashida, K., Shibata, T., et al.: High-rate HMDSO-based coatings in open air using atmospheric-pressure plasma jet. J. Non-Cryst. Solids **358**(17), 2462–2465 (2012)
38. Yim, J.H., Rodriguez-Santiago, V., Williams, A.A., et al.: Atmospheric pressure plasma enhanced chemical vapor deposition of hydrophobic coatings using fluorine-based liquid precursors. Surf. Coat. Technol. **234**, 21–32 (2013)
39. Xie, Q., Lin, H.F., Zhang, S., et al.: Deposition of SiCxHyOz thin film on epoxy resin by nanosecond pulsed APPJ for improving the surface insulating performance. Plasma Sci. Technol. **20**(2), 025504 (2017)
40. Lin, H.F., Wang, R.X., Xie, Q., et al.: Rapid surface modification by plasma jet to promote surface charge decaying. Trans. China Electrotech. Soc. **32**(16), 256–264 (2017)
41. Lin, H.F.: Study of Effect of Film Deposition by Low Temperature Plasma on Surface Electrical Properties of Epoxy Resin. North China Electric Power University, Baoding (2018)
42. Lian, H.L., Yan, J.Y., Xie, J., et al.: Enhancement of surface insulation properties of alumina/epoxy composites through ZrO_2 film sprayed by plasma atomization. Polym. Compos. **41**(10), 4020–4030 (2020)
43. Ma, Y.Y., Zhang, C., Kong, F., et al.: Effect of plasma jet array assisted film deposition on epoxy resin surface electrical characteristics. High Volt. Eng. **44**(9), 3089–3096 (2018)

44. Zhou, Y.X., Shao, T., Zhang, C., et al.: Atmospheric pressure plasma jet array in helium driven by microsecond pulses. High Power Laser Part. Beams **26**(4), 68–72 (2014)
45. Wang, R.X., Li, W.Y., Zhang, C., et al.: Thin insulating film deposition on copper by atmospheric-pressure plasmas. Plasma Process. Polym. **14**(7), 1600248 (2017)
46. Zhang, C., Lin, H.F., Zhang, S., et al.: Plasma surface treatment to improve surface charge accumulation and dissipation of epoxy resin exposed to DC and nanosecond-pulse voltages. J. Phys. D Appl. Phys. **50**(40), 405203 (2017)
47. Wang, T.T., Zhang, C., Zhang, F.Z., et al.: Effect of oxygen concentration on improvement of surface pressure resistance of epoxy resin by atmospheric pressure plasma deposition. High Volt. Eng. **46**(10), 3708–3714 (2020)
48. Zhang, P.H., Zhang, S., Kong, F., et al.: Atmospheric-pressure plasma jet deposition of bumpy coating improves polypropylene surface flashover performance in vacuum. Surf. Coat. Technol. **387**, 125511 (2020)
49. Cui, C.C.: Study on the Inhibition of Micro Discharge by Surface Treatment of Atmospheric Pressure Plasma. Zhengzhou University (2018)
50. Cui, C.C., Zhang, C., Ren, H.Y., et al.: Surface modification of Cu by atmospheric pressure plasma jet for micro discharge inhibition. J. Chin. Electr. Eng. Sci. **38**(5), 1553–1561 (2018)
51. Hai, B., Zhang, C., Wang, R.X., et al.: Plasma depositing SiO_2-like film to suppress surface charge accumulation on epoxy resin. High Volt. Eng. **43**(2), 375–384 (2017)
52. Hu, D., Ren, C.Y., Zhang, C., et al.: Effect of deposited film on the surface insulation characteristics of FEP material by atmospheric pressure plasma jet. J. Chin. Electr. Eng. Sci. **39**(15), 4633–4641 (2019)
53. Kostov, K.G., Nishime, T.M.C., Castro, A.H.R., et al.: Surface modification of polymeric materials by cold atmospheric plasma jet. Appl. Surf. Sci. **314**, 367–375 (2014)
54. Zhang, C.Q.: The Surface Plasma Treatment and Its Study on Electrical Properties of Epoxy Resin. North China Electric Power University, Baoding (2020)
55. Yan, J.Y., Liang, G.S., Lian, H.L., et al.: Improving the surface flashover performance of epoxy resin by plasma treatment: a comparison of fluorination and silicon deposition under different modes. Plasma Sci. Technol. **23**(11), 115501 (2021)
56. Xue, J.Y., Wang, H., Chen, J.H., et al.: Effects of surface roughness on surface charge accumulation characteristics and surface flashover performance of alumina-filled epoxy resin spacers. J. Appl. Phys. **124**(8), 083302 (2018)
57. Riello, D., Zetterström, C., Parr, C., et al.: AlF_3 reaction mechanism and its influence on α-Al_2O_3 mineralization. Ceram. Int. **42**(8), 9804–9814 (2016)
58. Gao, Y., Wang, X.F., Li, N., et al.: Characterization method for carrier trap and the effect on insulation breakdown within polymer insulating materials: a review. High Volt. Eng. **45**(7), 2219–2230 (2019)
59. Lei, Q.Q., Liu, G.Y.: How to understand the two basic physical processes of polarization and conductance in engineering dielectrics and scientific principles and methods of their measurement. J. Chin. Electr. Eng. Sci. **38**(23), 6769–6789+7113 (2018)
60. Lei, Q.Q., Liu, G.Y.: Basic physical knowledge of engineering dielectrics and learning thinking method. J. Chin. Electr. Eng. Sci. **38**(16), 4898–4917+4995 (2018)
61. Simmons, J.G., Tam, M.C.: Theory of isothermal currents and the direct determination of trap parameters in semiconductors and insulators containing arbitrary trap distributions. Phys. Rev. B **7**(8), 3706 (1973)
62. Que, L.K., An, Z.L., Ma, Y., et al.: Improved DC flashover performance of epoxy insulators in SF6 gas by direct fluorination. IEEE Trans. Dielectr. Electr. Insul. **24**(2), 1153–1161 (2017)

Regulation Method and Mechanism of Functionalized Modified Nano Filler on Electrical Properties of Epoxy Resin

Haoou Ruan, Qing Xie, Qijun Duan, and Jun Xie

1 Introduction

Epoxy resin is a kind of insulating material with excellent electrical, thermal and mechanical properties. It is widely used in GIS, GIL, dry-type transformers, IGBT and other electrical equipment in high-voltage transmission [1, 2]. However, epoxy insulation components not only face higher requirements such as power resistance and heat resistance, but also have to endure extreme conditions and withstand long-term use. This high demand is intensified with the increasing construction of UHV transmission grids, extended transmission range, and complex operational environment. The insulation strength of general epoxy resin base material is far from sufficient to withstand the demand of high voltage or even ultra-high voltage power transmission. Therefore, the modification of epoxy curd materials has become a hotspot. Filler doping is an important modification method to improve the comprehensive properties of epoxy resin [3, 4]. The epoxy casting composite insulators in high voltage transmission engineering promote the overall withstand level, thermal and mechanical properties of the materials by doping with a large number of inorganic fillers. Epoxy composite insulating materials can synthesize insulating components with excellent electrical properties by making use of epoxy resin matrix as continuous phases and inorganic filler particles such as alumina and silicon oxide as dispersed phases (Figs. 1 and 2).

Nano-doping modification is a technology that uniformly fills nanoscale inorganic filler particles into the polymer or other matrix materials to synthesize composite materials with excellent properties. Originated in the early 1980s, nanocomposites are the composites material with the size of dispersed phase material at least within 100 nm in a one-dimensional direction. The common synthesis methods of polymeric

H. Ruan · Q. Xie (✉) · Q. Duan · J. Xie
Department of Electrical Engineering, North China Electric Power University, Baoding, China
e-mail: xq_ncepu@126.com

Q. Xie (ed.), *Electrical Materials*, Engineering Materials,
https://doi.org/10.1007/978-981-99-9050-4_4

Fig. 1 Gas-insulated switch

Fig. 2 Epoxy resin composite insulation

nanocomposites include the sol–gel method, blending method, in-situ polymerization method, and intercalation and polymerization method, etc. [5, 6]. Since the introduction of "nanocomposite dielectric" by Lewis in 1994, a lot of research in electrical composite materials has suggested that the unique high specific surface area and size effect of nano-inorganic fillers can effectively modify the insulating properties of composite materials. Furthermore, the characteristics of nano inorganic fillers render the electrical composite materials excellent properties such as dielectric response, bulk breakdown strength, corona resistance and electric trace resistance. All of this makes nanocomposites a necessary functional material in the field of high-voltage insulation [7]. Moreover, many scholars have provided molecular explanations and theoretical models with regard to nanocomposites' macroscopic effect and charge behavior. Lewis proposed an interface model to divide the association region between nano particles and epoxy resin into shielding and dispersion regions [7].

In 2005, Tanaka put forward a multi-core model which separated the filler interface from bonded, bound and loose layers [8]. The multi-core model further divides the interface between filler and matrix and also comprehensively describes the differences in charge behavior between different layers in the interface region (Figs. 3 and 4).

Although the introduction of nano filler has dramatically promoted the overall performance of epoxy resin systems, there remain two major practical issues. Firstly, the surface activity of nano filler is very high, which leads to the agglomeration in the process of resin curing. This phenomenon significantly affects the dispersion

Fig. 3 Interface model

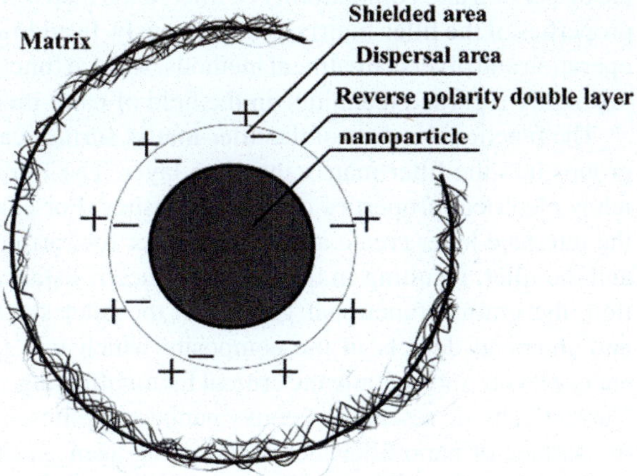

Fig. 4 Multicore model

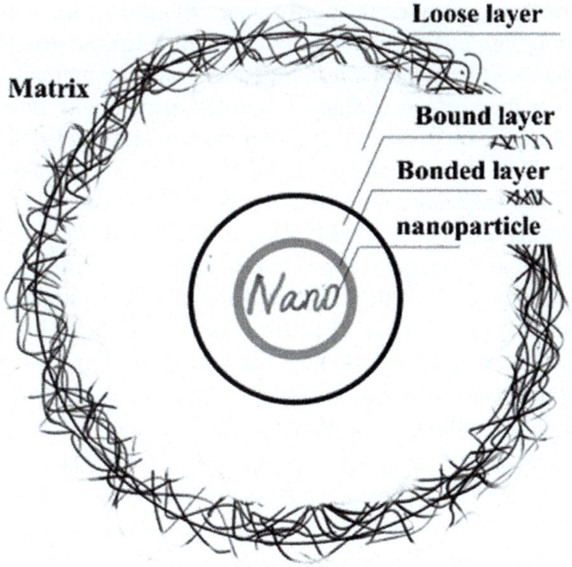

properties of epoxy resin system, consequently diminishing the nano-modification effect. Secondly, the surface of inorganic nanofillers often lack groups that can form durable bonds with the resin, resulting in poor compatibility between fillers and epoxy resin. To sum up, these two problems seriously restrict the modification of nano filler on the performance of epoxy resin composites, and have become the principal impediment to nano filler modification research. Therefore, modifying nano fillers has become a prevalent approach in the advancement of nano dielectrics. The modification methods can be categorized as physical morphology modification, surface functional modification, element or defect doping modification, etc. In these modification methods, the surface functional modification of nano fillers can optimize the dispersion and compatibility of filler in the matrix, as well as regulate the interface properties of the filler-matrix simultaneously. In addition, due to the relatively simple operation and diverse treatment methods, surface functional modification has become the highly regarded technique in the field of nano-modification.

The functionalization of the filler-matrix surface can introduce various functional groups into the filler-matrix interface layer. These functional groups have changed many electrical properties of the composites. For example, the nonpolar groups in the interface layer create an energy barrier for carrier transport between the matrix and the filler, resulting in the high interface resistance of charge transport. In addition, the grafted functional groups on the filler surface will cause many physical and chemical defects in the composite which will form charge traps of different energy levels and become the critical factor affecting their charge transfer ability [7]. Currently, in the research of epoxy nanocomposites, the methods for functionalizing the surface of nano filler include coupling agent grafting modification, fluorination modification, amination modification, hydroxylation modification, dopamine modification, etc. The functional grafting on the filler surface has a significant effect on further improving the performance of epoxy composites. Wang et al. used two silane coupling agents, octamethylcyclotetrasiloxane and hexamethyl-disilazane, to graft nano-SiO_2 and dope it to prepare composite materials. Subsequent analysis showed that the surface charge accumulation of epoxy nanocomposites was mitigated in comparison to pure epoxy after surface grafting [9]. Luca Valentini et al. proposed to graft fluorinated groups on the surface of single-walled carbon nanotubes and further functionalized amino with the help of fluorinated groups [10] which improved the dispersion of the modified filler in the epoxy resin matrix and enhanced the electrical and mechanical properties of the composite. Li Licheng and his team applied dopamine to modify h-BN and alumina microspheres and co-doped the two fillers to modify epoxy resin. The team found that the grafting modification of filler can adjust the interface interaction between the particle group and EP matrix. This modification can not only alleviate the ablation of insulating materials caused by surface discharge, but also enhance the apparent activation energy of the EP pyrolysis process [11].

This chapter focuses on the surface functionalization modification technology of nano filler in the research of epoxy composites which introduces in detail the methods of coupling agent grafting, fluorination grafting, and amination grafting in the surface functionalization of nanofiller and their influence rule on the properties of epoxy composites. In addition, the chapter analysis the mechanism of functionalization of nano fillers and its effect on the performance of epoxy composites by combining the interface model, molecular dynamics simulation, and Density Functional Theory (DFT) calculation.

2 Functional Modification of Filler and Preparation and Characterization of Composite

2.1 Functional Modification of Nanofiller

Surface functional modification of nanofiller is referred to the treatment of filler that grafting organic functional groups or certain element onto the filler surface, with the purpose of conferring the filler with certain property. Surface functional modification is commonly achieved through hydrothermal method, self-assembly method, plasma treatment and so on. In these methods, groups with functional elements like F and N or biomolecules such as dopamine are grafted to the nanofiller surface, in order to make up for the filler defects and enhance comprehensive performance of composite. The functional modification of nanofiller surface can ameliorate the dispersibility and compatibility of filler in matrix while also regulating the interface between filler and matrix, and pointedly improve the electrical properties of epoxy composite. This section introduces an overview of the experimental platforms and preparation methods involved in coupling agent modification, fluorination modification and amino modification of nanofiller for epoxy nanocomposite.

2.1.1 Platform and Method of Coupling Agent Grafting Modification

Coupling agents are frequently used in conventional synthesis of epoxy composite to graft and modify doped fillers. Regular coupling agents feature an inorganic group at one end and organophilic group on the other. Coupling agent is also called the surface modifier due to the improvement of the dispersibility and compatibility of filler in a resin matrix. The most commonly used coupling agent in epoxy composite is silane coupling agent. Generally, the processability, insulating property and thermal property of composite can be effectively improved by coupling agent with 0.5–2% of filler mass fraction. The technology that using coupling agent to modify nanofiller is relatively mature and convenient, which can be achieved by basic material preparation platform. A brief introduction is as follows.

(1) Platform of coupling agent grafting modification for nanofiller

The coupling agent treatment platform mainly includes magnetic stirrer, ultrasonic disperser and vacuum drying oven.

Magnetic stirrer is mainly applied to stir liquid with low viscosity or solid–liquid mixture. Utilizing the fundamental principle of magnetic field that the same polarity will repel each other, the magnetic stirrer inside container is pushed to rotate by continuously transforming two polarities at the base of the vessel and then rotate the samples in container by stirrer to achieve the purpose of uniformly mixing samples. Typically, coupling agent treatment regards ultrapure water or ethyl alcohol as dispersant, and filler will be added to container to stir for completely reaction after the coupling agent and solvent are uniformly mixed (Fig. 5).

Ultrasonic cleaner, one of the physical dispersion methods, utilizes ultrasonic dispersion technology to reduce nanoparticle agglomeration. As shown in Fig. 6, with stainless still liner and mechanical control heating inside, ultrasonic cleaner can autonomously control environment temperature during dispersion process and improve the dispersion efficiency of nanoparticle. The frequency of ultrasonic vibration head inside is 40 kHz, and the device can adjust the ultrasonic time according to specific experimental requirement. The ultrasonic dispersion of the mixture of filler and coupling agent contributes to complete contact and reaction between filler and coupling agent molecules.

Vacuum drying oven can provide stable reaction temperature and steady high-temperature environment to dry nanofiller. The temperature fluctuation is constantly lower than 1 °C. When processing the coupling agent grafting reaction of filler over

Fig. 5 Thermal collecting homoiothermic magnetic stirrer

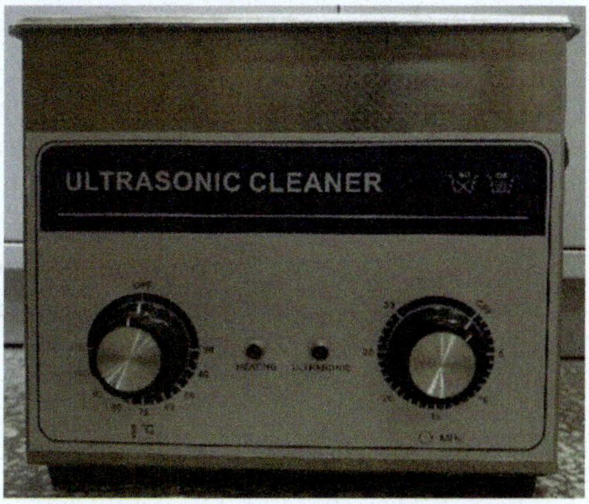

Fig. 6 Ultrasonic cleaner

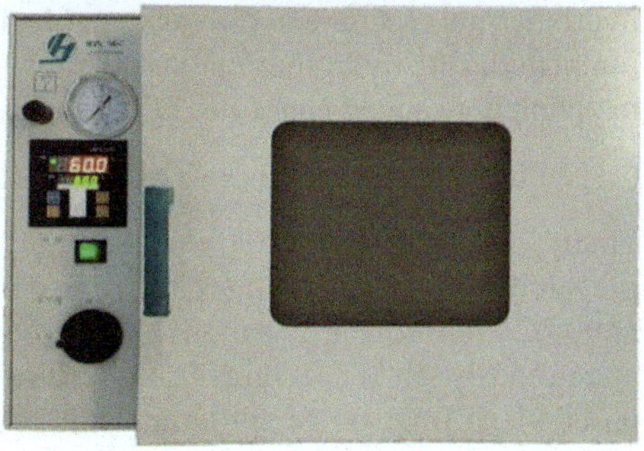

Fig. 7 Vacuum drying oven

a long time, the temperature can be controlled invariable using a vacuum drying oven. Furthermore, under the vacuum, it can also rapidly evaporate solvent such as ultrapure water and ethyl alcohol after reaction (Fig. 7).

(2) Method of coupling agent grafting modification for nanofiller

Firstly, required quality of coupling agent was weighed and added to a beaker, and then proper absolute alcohol was also added to the beaker followed by placing the beaker on a magnetic mixer and stirring for 30 min at 30 °C in order to thoroughly mix coupling agent and absolute alcohol. Secondly, nanoparticle to be processed was added to the uniformly-stirred mixture of coupling agent and absolute alcohol. The mass ratio of nanoparticle to coupling agent is 100:5. After that, stirring the mixture for another 30 min at 30 °C in magnetic mixer, and then placing the mixture into a ultrasonic cleaner to disperse for 10 min to completely mix filler and coupling agent solution. Finally, the samples mixing uniformly were placed in a vacuum drying oven to sufficiently react for 20 h, and the solvent was completely evaporated. The dried nanoparticle, presenting as lumpy and realizing coupling agent surface grafted, can be preserved to reserve after grinding in mortar.

2.1.2 Platform and Method of Fluorination Modification

Filler fluorination, one of the filler functionalization, can effectively improve the dispersibility of filler in matrix by grafting strongly electronegative fluorine-containing groups on the surface of the nanofiller. Simultaneously, the unique atomic structure of fluorine entrusts nanofiller fortissimo charge binding ability, which can improve the insulating property of composite significantly [12]. At present, the grafting methods of filler fluorination mainly include hydrothermal fluorination, fluorine-containing coupling agent grafting, direct process with fluorine gas, plasma fluorination, etc. This paper focuses on the hydrothermal fluorination.

(1) Platform of fluorinated grafting modification for nanofiller

As shown in Fig. 8, platform of hydrothermal fluorination mainly consists of high-temperature hydrothermal reactor, suction flask and vacuum pump. High-temperature hydrothermal reactor can provide a stable, high-temperature and high-pressure environment during hydrothermal fluorination. Suction flask and vacuum pump are used to obtain fluorinated nanofiller by suction filtration following hydrothermal reaction.

(2) Method of fluorinated grafting modification for nanofiller

Firstly, To begin with, 0.04 g/ml Sodium hydrogen fluoride ($NaHF_2$) solution was prepared by stirring $NaHF_2$ particle of 98% purity in deionized water for 30 min at 60 °C. Subsequently, nanofiller was weighed and added to the solution above to continuously stir for 10 min. The mixed solution was then added to the polytetrafluoroethylene liner of a high-temperature hydrothermal reactor. After that, the liner was hermetically sealed with a stainless steel high pressure reactor and placed into a vacuum drying oven to heat to 90 °C for full reaction for 24 h. After the reactor was cooled down to room temperature, it was removed from the drying oven and suction filtrated under vacuum. The resulting sample was cleaned by deionized water until the solution presents as neutral and was placed into the vacuum drying oven to dry thoroughly at 60 °C. Finally, the nanofiller after drying can be used in experiment after grinding by a mortar.

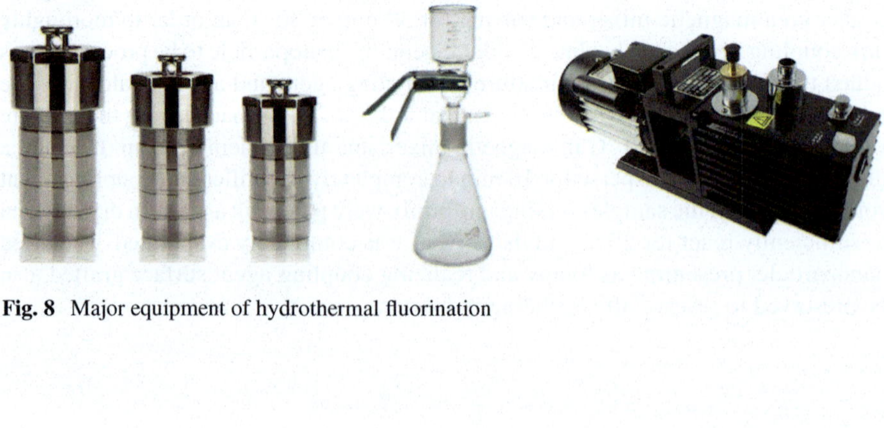

Fig. 8 Major equipment of hydrothermal fluorination

2.1.3 Platform and Method of Amination Modification

Amination modification of nanofiller is a new method to adjust the filler-matrix interface of epoxy composite insulation material. By grafting amino groups on the surface, filler can form cross-linked bonds with epoxy resin, which has a splendid effect on improving the compatibility of filler and matrix. Meanwhile, amino groups can accelerate the migration rate of composite space charge, which has a positive effect on inhibiting charge accumulation and field distortion effects in local insulating material.

(1) Platform of amination grafting modification for nanofiller

Taking the amino modification of BN nanosheet as an example, this paper introduces the platform of amino modification. The required experimental equipment includes planetary ball mill, suction filtration device, vacuum filtration pump, agate mortar, vacuum drying oven, etc. Planetary ball mill can co-ball mill the hexagonal boron nitride together with urea, and graft amino groups on the surface of boron nitride nanosheet during the ball milling stripping. Suction filtration device is used to clean and suction filter the modified filler in order to remove excess reagent. Drying oven and mortar can dry and grind the modified filler (Figs. 9 and 10).

(2) Method of amino grafting modification for nanofiller

The method to amination modification of Boron Nitride Nanosheets (BNNS) introduced in this book is a one-step fabrication method of urea co-ball milling [13].

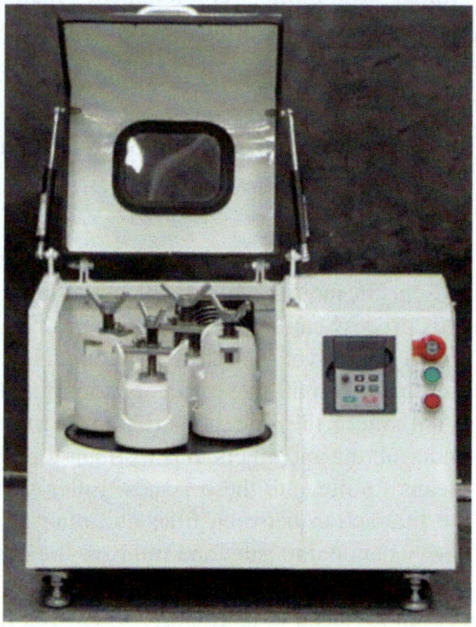

Fig. 9 Planetary ball mill

Fig. 10 Stainless steel ball mill tank

Initially, 1 g hexagonal boron nitride powder and 4 g urea were weighted and 25 ml isopropyl alcohol solution was measured before being homogenously combined and stirred in an agate jar of a planetary ball mill. Whereafter, the agate jar was placed into the planetary ball mill and the rotate speed of mill was set to 300 r/min. After continuously operating for 48 h, take out the mixture and place it in a beaker to for 2 h static. Then the upper suspension was extracted into a suction flask with an injector, and was suction filtered through a filtration membrane with 0.22 μm aperture. The samples on the filtration membrane were placed into a vacuum drying oven to dry for 12 h at 80 °C. After cooling at room temperature, the blocky boron nitride nanosheet grafted with amino were obtained. Finally, the powder of boron nitride nanosheet grafted with amino groups were obtained by grinding the nanosheet in a mortar.

2.2 *Platform and Method of Epoxy Nanocomposite Preparation*

Epoxy composite have been extensively utilized in high voltage insulation field at present. In industry, micro filler is often applied to dope and modify epoxy resin. This process can be achieved by conventional material mixing and curing reaction, which is relatively simple and mature. Nevertheless, due to the large specific surface area and the small size effect of nanofiller, the doping amount of nanofiller is limited. What's more, the process also has problems with easy aggregation, bubble abundance, undermixing and others. It is worth noting that filler functionalization is an important means to alleviate these issues. Filler surface grafting modification can facilitate the interaction between filler and matrix, reduce the actual synthesis difficulty of epoxy nanocomposite, and improve the comprehensive properties of material. This chapter introduces the preparation platform and technology of epoxy nanocomposite commonly used in laboratory, which can be applied to the synthesis of most epoxy nanocomposite.

2.2.1 Preparation Platform of Epoxy Nanocomposite

The preparation of epoxy nanocomposite can be divided into the steps of blending, degassing, casting, curing and demolding. The required equipment mainly includes balance, stirring device, dispersion device, vacuum drying oven, vacuum pump and mold, etc.

Machine mixer is mainly used in mixing. A high power, high-speed mechanical mixer is necessary to intensively mix all raw materials due to the viscosity of epoxy resin at normal temperature is high and the difficulty in dispersing nanofiller into the epoxy matrix by surface energy effect. Stirring paddle and rotating rod are made of polytetrafluoroethylene, which is a kind of natural release agents that can effectively alleviate the adhesion between epoxy resin and equipment. The stirring device is equipped with temperature-regulated electric heating equipment that can heat the epoxy resin mixture and reduce its viscosity, thereby promoting that all kinds of raw materials are mixed more homogeneously. At the same time, with glass reactor as its complement, it allows one to observe the mixing state of blends at any moment.

Shear machine is used to decompose and diffuse the solid or liquid which is incompatible with liquid medium, and form an even distribution within the medium. The principle is generating a strong centrifugal force field through the high-speed rotation of the rotor, and forming a negative pressure zone inside the knife head. Nanoparticle to be processed will enter the dispersion knife head along the axial direction, then are removed from the slots that arranged on the stator and rotor by the internal force, radially disperse in all directions. The forces that effect nanoparticle are strong shear force and axial force generated by high acceleration process. In addition, shear gradient generated between stator and rotor will form strong turbulence, which makes agglomerate particle to disperse and mix under high-speed impact.

Vacuum drying oven and vacuum pump are necessary equipment to prepare epoxy composite, which are applied for vacuum degassing of mixed material and material casted into the mold. During the epoxy resin preparation, on the one hand, lots of bubbles will mix in the blend when stirring, on the other hand, a great deal of air will be introduced in the process of casting which will form a large amount of defect sites inside the epoxy composite material, and seriously degrade the mechanical and electrical properties of cured product. The material vacuum degassing through vacuum drying oven and vacuum pump can effectively eliminate bubbles mixed in material and improve the comprehensive properties of material.

Aiming at the practical application and testing requirements, epoxy composite needs to be casted to solidified samples with different shapes. Therefore, different kinds of molds need to be pointedly customized. Metal mold possess impressive heat resistance, solidity and durability. Moreover, it has excellent manufacturability and convenient to process which makes it the most prevalent casting mold in industry. However, metal mold is easy to bond with epoxy resin, so releasing agents are usually coated on the iron mold to release epoxy material after curing in industry (Fig. 11).

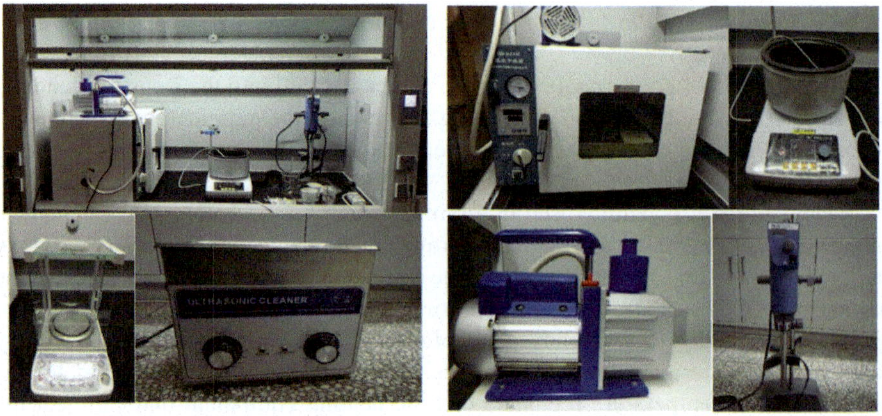

Fig. 11 Preparation platform and equipment of epoxy resin composites

2.2.2 Preparation Method of Modified Epoxy Composite

The preparation method of epoxy composite is shown in Fig. 12. Firstly, all raw materials were weighed in equal proportion according to established mass fraction and poured into a glass reactor. The heating temperature was 60 °C to reduce the viscosity of epoxy resin composite. About 5 min later, the height of agitating valve was adjusted to immerse the valve into the blend solution, and the rotate speed was 300 r/min, stirring at 60 °C. After about 10 min, the mixing could be considered uniform and the stirring could be stopped if the blend presents as faint yellow homogeneous liquid. Later, the blend after stir was placed into a vacuum drying oven at 60 °C and a vacuum pump was used to extract air until the oven is close to vacuum state. It could be observed through an inspection window that bubbles in blend continually rose and broke. About 30 min later, if there were no bubbles in the blend, the oven was returned to normal pressure and the blend was picked out to cast. The mold needed to be preheated before casting. Coated with mold release agents on the surface of metal mold and preheated the mold to 80 °C could prevent the huge temperature difference between mold and blend destroying properties of cured products during casting process. After preheating, the blend was slowly poured into the mold and degassed again. Subsequently, the mold was placed into the vacuum drying oven to heat to 80 °C and repumped to vacuum state until there were no bubbles in all the blend in mold. After degassing, the mold was placed into the drying oven, and the temperature in oven was adjusted to 120 °C for continuously curing for 2 h. During the process, the blend in mold would occur crosslinking reaction, eventually solidifying. The mold was removed from the vacuum drying oven after solidifying and left to cool to room temperature. Finally, the cured epoxy nanocomposite samples were obtained by releasing the mold after it was cooled to room temperature. The samples was placed into the vacuum drying oven at 120 °C to continuously solidify for 10 h, which can fully cross-link the epoxy resin and acid anhydride. This process

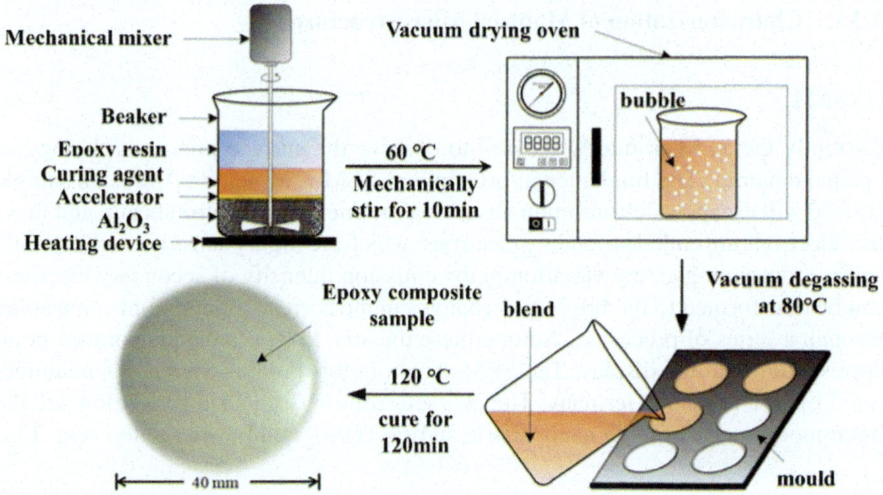

Fig. 12 Preparation process of epoxy resin composites samples

can guarantee that the largest crosslinking of epoxy composite condensate, and the uniform properties of different samples become more excellent.

2.3 *Characterization Method of Material*

Epoxy nanocomposite can optimize the properties of matrix materials with the help of nanofiller performance. The doping of inorganic filler on the macro-level effects insulation, mechanical property, heat resistance and other properties of material by adjusting the internal microstructure and chemical composition of composite. According to the interface model proposed by T. J. Lewis and the multicore model introduced by Tanaka, the independent properties and the action mechanism between interphases of composites at multiple scales should be focused depending on the characterization and joint analysis for components, morphology, structure and properties of material when analyzing the effect rule of nanofiller. Various characterization methods are available in current technology for experiment and research on epoxy composite materials. With the help of Scanning Electron Microscope (SEM), Atomic Force Microscope (AFM), Fourier Transform infrared spectroscopy (FTIR) and X-ray photoelectron spectroscopy (XPS), the comprehensive analysis of the micro-morphology and molecular organization of nanofiller and composite was conducted. In addition, various properties of material can be measured and characterized in accordance with different professional standards.

2.3.1 Characterization of Material Microstructure

(1) SEM

Currently the SEM primarily utilized to observe the surface micromorphology of opaque material. The fundamental principle of SEM is as below: Atoms on sample surface will escape electrons upon absorbing the energy of electron beam, and these free electrons are called secondary electrons, which are highly sensitive to the sample surface morphology. Any variation in the emission intensity of secondary electrons can be transformed to the brightness changes in the corresponding points on display through a series of processes. Among these the area to be scanned is point-to-point corresponding to the display. The SEM shown in this book is Sigma300, manufactured by The Zeiss in Germany. The characterization results showed below are the micromorphology of BFO nanoparticle and FBTO/EP composite section (Fig. 13).

(2) AFM

AFM can be used to study surface morphology, nanostructure and chain conformation of sample to obtain information on nanoparticle size, aperture, material surface roughness, material surface defects, etc. At the same time, it can also track the surface structure and morphology (vary with time, temperature and other conditions). The pictures below are stereograms and plans of surface morphology at flat of electrode edge characterized by AFM (Fig. 14).

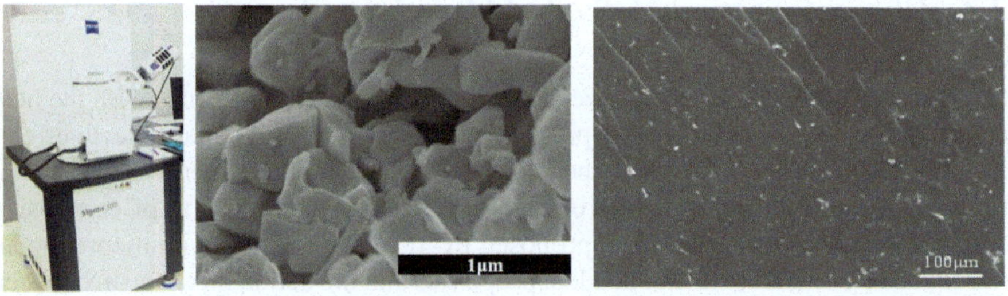

Fig. 13 SEM and characterization results

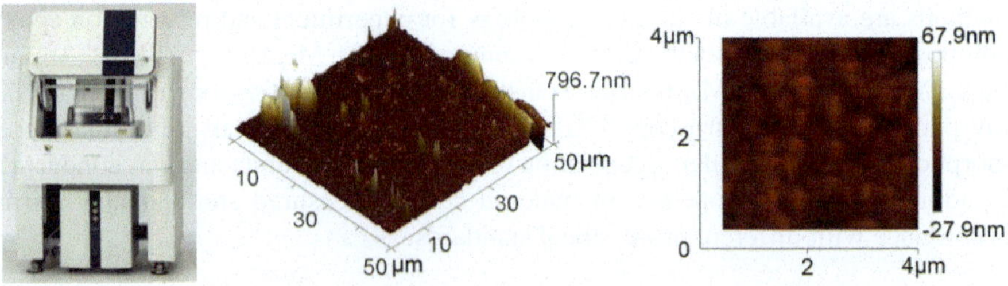

Fig. 14 AFM and characterization results

2.3.2 Characterization of Material Chemical Constituents

(1) XPS

At present, the content of chemical elements and functional groups on the material surface is usually measured by XPS. The principle is: When X-ray with certain energy is irradiated to sample surface, if the energy of X-ray is greater than the work function of sample surface, electrons in atoms at sample surface will break away from the nucleus and become photoelectrons. Now that the energy of X-ray is known, we can measure the kinetic energy of photoelectron, thereby obtaining the electron binding energy of the sample. Because the orbital electron binding energy of various atoms and molecules is constant, the elements composition in samples can be quantitatively known by measuring binding energy. The XPS shown below is Escalab250Xi by The Thermo Fisher Scientific, and the characterization results are XPS spectra of ZnO nanofiller before and after modification (Fig. 15).

(2) XRD

In order to characterize the crystal structure of sample and confirm information on physical phase, crystal structure and space groups of prepared materials, now X-ray diffraction (XRD) is frequently used. The fundamental is below: Crystals consist of lattice cells formed by regular arrangement of atoms. The interatomic distance arranged regularly has the same order of magnitude with the wavelength of incident X-ray. Consequently, when a beam of monochromatic X-ray is incident on crystal, the X-ray scattered by various atoms will interface with each other, resulting in strong diffraction in some special directions. The spatial distribution directions and intensity of diffracted ray is particularly related to the structure of crystal (Fig. 16).

(3) FTIR

There are many kinds of vibrations in molecule, and some of which can change the molecular dipole moment. When the frequency of these types of vibrations is same

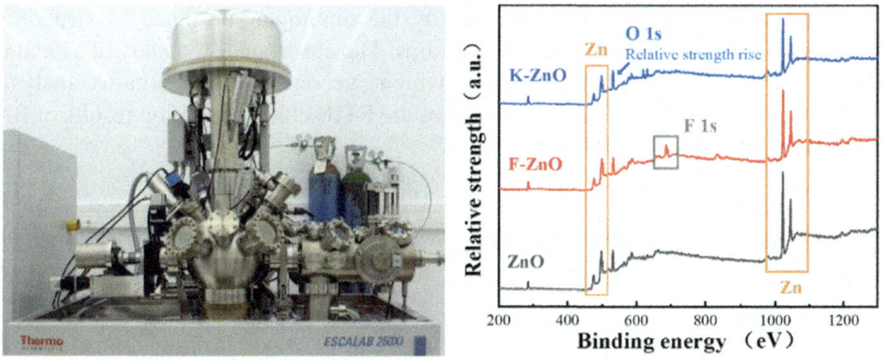

Fig. 15 XPS and characterization result spectra

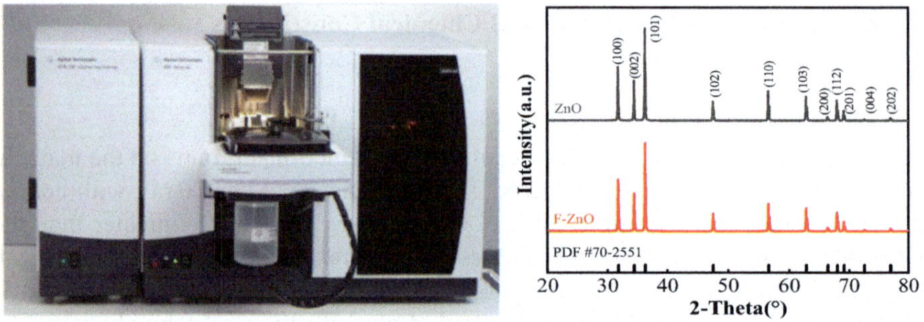

Fig. 16 XRD and characterization result spectra

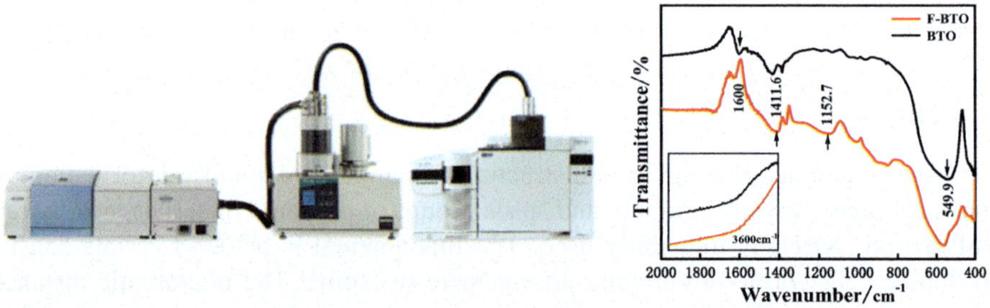

Fig. 17 FTIR and characterization result spectra

to that of infrared light, molecules can absorb the energy of infrared light to form infrared absorbing spectrum (IR). Different compounds have different characteristic peaks of infrared absorption spectrum due to their different molecular structures. FTIR analysis is a significant modern analytical method, which has been widely used for structure identification, quantitative analysis and study of chemical kinetic, etc. Its analysis can provide a large amount of information about functional groups. The position and intensity of infrared absorption peaks reflect characteristics of molecule structure, which can be applied to identify the unknown substance by structural composition or determine the chemical groups. The absorbance intensity of a band is related to the content of chemical groups, which can be used for quantitative analysis and confirming its purity. The below images are FTIR characterization results of BN before and after fluorination (Fig. 17).

2.3.3 Electrical Performance Test

(1) Dielectric

Relative permittivity is a physical parameter that characterizes the dielectric property or polarization property of dielectric materials. It is determined by the capacity ratio of the same size capacitors that are respectively made of predictive materials and vacuo as dielectric. And the value also reflects the characterization of material power storage. The picture showed below is impedance analyzer Agilent 4294A with a circle with radius of 5 mm effective measurement area and its frequency domain ranges from 40 Hz to 110 MHz (Fig. 18).

(2) Conductivity

The surface conductivity and volume conductivity can be measured by high precision ohmmeter. The device showed in picture below is Hioki SM-7100 ohmmeter. During measurement, samples are placed into a shielding box and three metal electrodes are built in electrode according to the three electrodes method. The ohmmeter can apply voltage to electrodes and detect trace leakage currents before and after application, thus it can calculate the surface conductivity and volume conductivity of samples to be measured (Fig. 19).

(3) Surface flashover voltage

The surface flashover test platform of epoxy resin includes high voltage DC power supply, metal cavity, experimental electrode, protection resistor, oscilloscope and

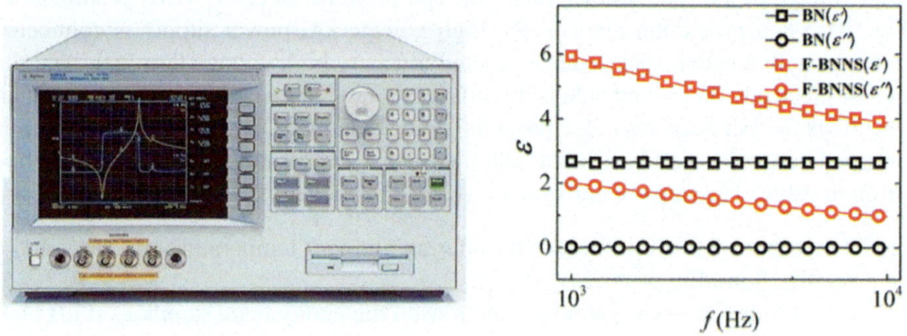

Fig. 18 Impedance analyzer and characterization results

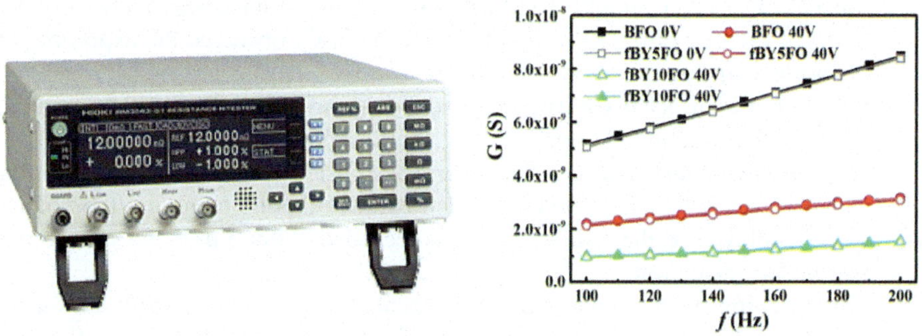

Fig. 19 Ohmmeter and characterization results

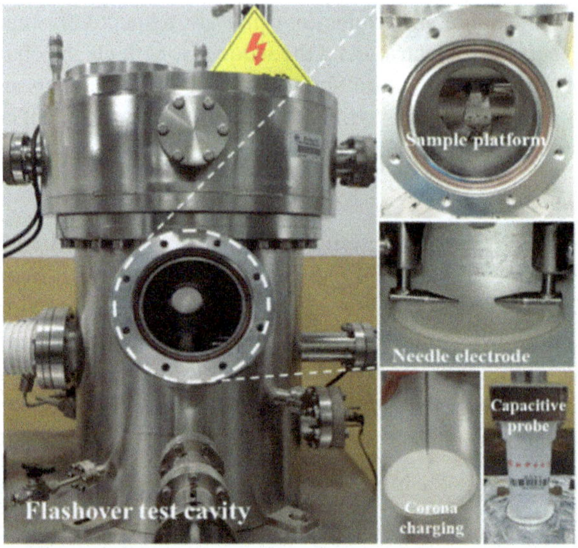

Fig. 20 Experimental platform of epoxy resin surface flashover

high voltage probe. The surface flashover test platform of epoxy resin is showed as Fig. 20. Via a protection resistor, the high voltage DC power supply is connected in series with a needle electrode connected to cavity high voltage terminal. And the needle electrode on ground side is directly grounded via the cavity ground terminal. The distance between two electrodes is 7 mm. One end of high voltage probe is propped on the protection resistor, and the other accesses channel 1 of oscilloscope through cable. The steps of epoxy resin surface flashover experiment are as follows:

(1) The experiment was conducted under an ambient temperature of 26 °C and a relative humidity of 30%.
(2) Opened the observing window and placed the epoxy resin samples on lift platform. After closing observing window, lifted those samples until their surface clings to the bottom of needle-needle electrode.
(3) The oscilloscope was turned on and set as rising edge trigger. The high voltage DC power supply was turned on and its voltage was increased at a uniform velocity of 0.5 kV/s, and then continuously observed the experimental phenomena in cavity.
(4) If the sample surface in cavity generated arcs along the needle-needle electrodes with discharge sound of "pop" and the oscilloscope detected rising edge waveform simultaneously, surface flashover was considered to have occurred in the sample. The source was turned off immediately this moment and the voltage value displayed on the oscilloscope was recorded. After 5 min, the sample was restored to insulation.
(5) After waiting, step (3) and (4) were repeated. 10 times was recorded on the voltage of the sample surface flashover, and the average was taken as the final surface flashover voltage of the sample.

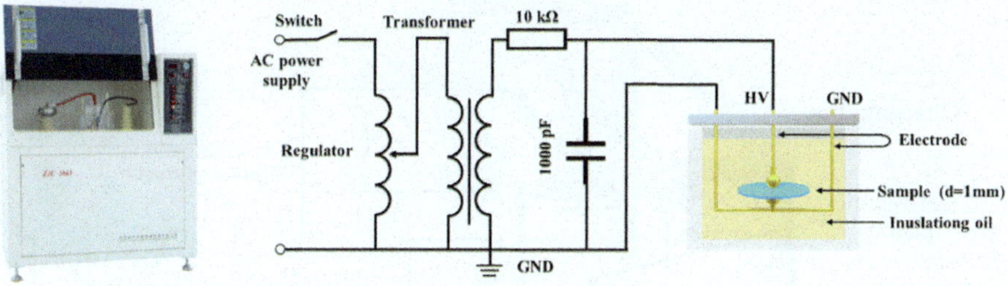

Fig. 21 Measurement of power frequency breakdown voltage

(6) After 10 flashovers, lifted the platform and separated the sample and electrode. Following this step, the observing window was opened to replace a new sample. And then step (2) was repeated.

(4) Breakdown strength measurement

Breakdown strength was measured according to the national standard GB/T 1408.1-2016 (sphere electrode). In order to prevent corona discharge from affecting the value of breakdown voltage, the whole electrode system was placed in transformer oil filtered clean and was polished by abrasive paper. Voltage booster was rotated to increase voltage with the speed of 2 kV/s until the sample was broke down, and recorded the voltage at breakdown. Breakdown strength is an important index that characterizes the materials insulating property which is defined as the ratio of voltage U to the thickness d of sample when the breakdown occurs. And the average breakdown field strength was represented by E. The computational formula is shown below in (1) (Fig. 21).

$$E = U/d \tag{1}$$

(5) Surface charge measurement

The experiment platform of interfacial charge aggregation and dissipation includes high voltage DC power supply, two-dimensional motion control device, charging pin, electrostatic probe, electrometer, data acquisition card and lift platform. The experiment platform of epoxy resin interfacial charge aggregation and dispersion is shown in Fig. 22. High voltage DC power supply is series connected to the charging pin. The charging pin and electrostatic probe are fixed to the X-axis of two-dimensional motion control device, respectively. Besides, stepping motors on X-axis and Y-axis are connected to the motion control device through data transmission line, and the motion control device is connected to computer by the same way. The outlet terminal of electrostatic probe accesses the electrometer, and the data transmission terminal of electrometer accesses computer by data acquisition card.

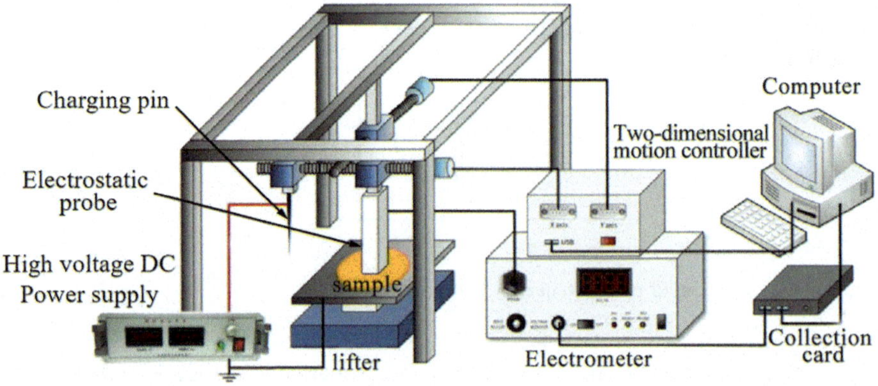

Fig. 22 Experiment platform of epoxy resin interface charge

The procedure of epoxy resin interfacial charge dissipation experiment are as follows:

(1) The experimental ambient temperature was at 26 °C, and the relative humidity was set as 30%. The motion control device and electrometer were turned on. Subsequently, the motion control software and data acquisition software were opened with the sampling frequency setting as 10 Hz.
(2) The sample to be processed was placed on the lift platform centre. After that, the charging pin was moved to 7 mm above the centre of sample by two-dimensional motion control device and lift platform.
(3) High voltage DC power supply was turned on and its voltage was uniformly increased to −7 kV with the speed of 0.5 kV/s. The voltage was kept constant for 60 s to charge the sample surface.
(4) The power was turned off and the charging pin was moved away after charging. Instead, the electrostatic probe was moved to 2 mm above the centre of sample by two-dimensional motion control device and lift platform. At the same time, electrostatic probe was switched to work by electrometer. Following preparation, the sample surface potential was measured rapidly with a sampling frequency of 10 Hz for 30 min.
(5) Upon the completion of the first interfacial charge dissipation experiment after sampling, the data acquisition card and electrostatic probe were turned off, and the sample was substituted by a new one to repeat step (2). To guarantee the accuracy of experiment, each sample was needed to be tested for 5 times. The average of 5 values at the same time was taken as the surface potential of sample at that time.
(6) In order to compare all samples directly, the data sampled completely were needed to be processed normalization, which means finding out surface potential with the largest absolute value and dividing all potential values by the maximum.

(6) Calculation of surface charge trap

According to the Isothermal Surface Potential Decay (ISPD) [14], surface potential decay of insulation material is a function of its trap level depth and trap density. As a consequence, a function equation can be used to describe the surface charge decay curves. Research shows that double exponential function can fit the epoxy resin surface potential decay curve. The fitting formula is as shown in Eq. (2):

$$\varphi_s(t) = Ae^{-\frac{t}{m}} + Be^{-\frac{t}{n}} \tag{2}$$

In Eq. (2), where $\varphi_s(t)$ is the surface potential, t is one moment of the surface potential decay curve. *A, B, m, n* are fitting coefficients. Therefore, the trap level E_T and trap density Q_s can be worked out by Eqs. (3) and (4) shown below:

$$E_T = k_B T \ln(vt) \tag{3}$$

$$Q_s = t\frac{\varepsilon_0\varepsilon_r}{eL} \cdot \frac{d\varphi_s(t)}{dt} \tag{4}$$

In the two equations, k_B is the Boltzmann constant that is 1.380649×10^{-23} J/K. T is the absolute temperature taken as 298 K for the reason that the interface charge dissipation experiment was conducted at 25 °C. Where v is the electron escape frequency taken as 4.17×10^{13} s^{-1}, e is the elementary charge quantity of 1.6×10^{-19} C. Besides, L is the thickness of samples of 3 mm since thickness of all the samples prepared in this paper is 3 mm. ε_0 is the vacuum permittivity of 8.85×10^{-12} F/m and ε_r is the relative permittivity of samples. According to the equations above, the samples trap level distribution curve can be worked out.

3 Regulation Electrical Properties of Epoxy Resin by Coupling Agent Modified Nano Filler

Coupling agent is an additive used to improve the organism and the filler interface performance during the preparation of composite materials. The coupling agents can be divided into four categories in terms of chemical composition: organic complexes, aluminate compound, titanates and silanes. Taking the silane coupling agent (KH550) treating nano-SiO_2 as an example, the reaction process is shown in Fig. 23.

The silane coupling agent can react with the hydroxyl groups on the surface of nano-SiO_2 after hydrolysis. The hydrophilic hydroxyl group is replaced by the lipophilic group, which makes the nano-SiO_2 particles change from hydrophilic to lipophilic, reducing the surface energy. This reaction enhances the compatibility of nano-particles in an organic environment, effectively decreasing the agglomeration of nano-particles and improving their dispersity in the matrix. In addition, one end of the coupling agent molecular chain is connected with the nano-SiO_2 surface and the

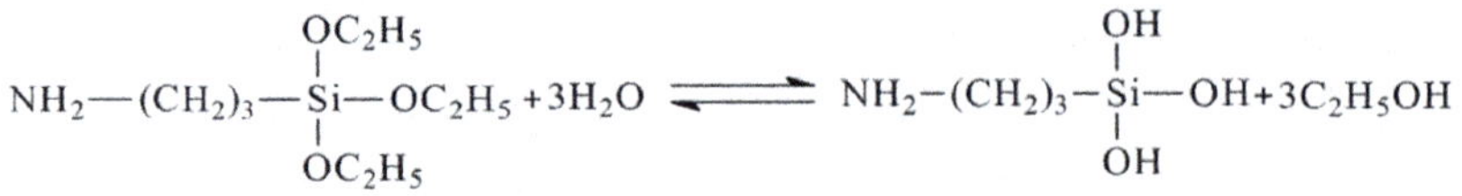

(a) Hydrolysis of coupling agent

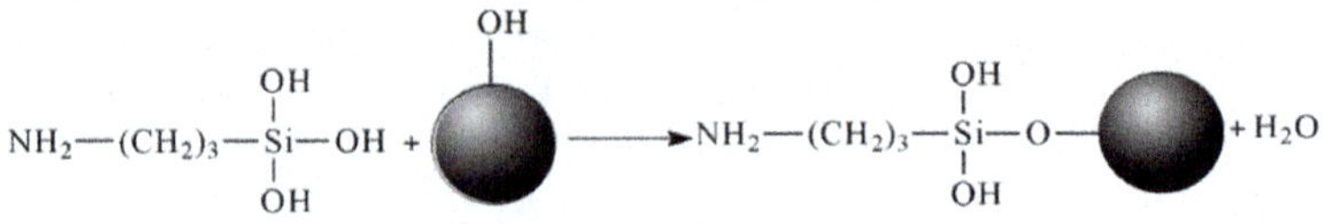

(b) Reaction of coupling agent with SiO_2

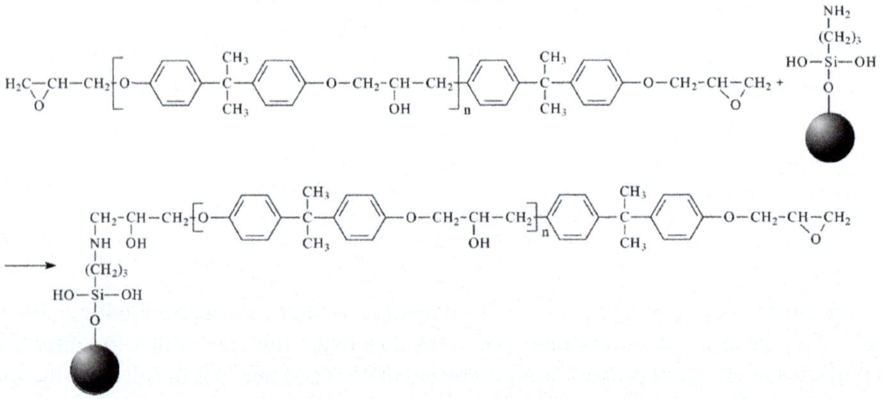

(c) Modified SiO_2 reacts with epoxy resin

Fig. 23 Action mechanism of silane coupling agent

other is associated with the matrix polymer chain, as shown in Fig. 23c. The nanoparticles are connected with the polymer chains through chemical bonds, improving the compatibility of the nanoparticles in the matrix. This chemical bond connection can also effectively avoid microphase separation inside the composite material under external factors such as electric, thermal, and force fields. In addition, it can also improve the stability of the composite material interface and increase the interaction between the two phases, so that the composite material can be used for a long time.

3.1 Effect of Different Kinds of Nano Filler Coupling Agent Grafting on Electrical Properties of Epoxy Resin

The coupling agent modified nano filler primarily regulates the interface area between the filler and the matrix, but there are different effects on different types of nano filler coupling agent. This paper provide an overview of the effect of silane coupling agents on the properties of composite materials by taking two insulating fillers of 100 nm SiO_2 and Al_2O_3 as examples.

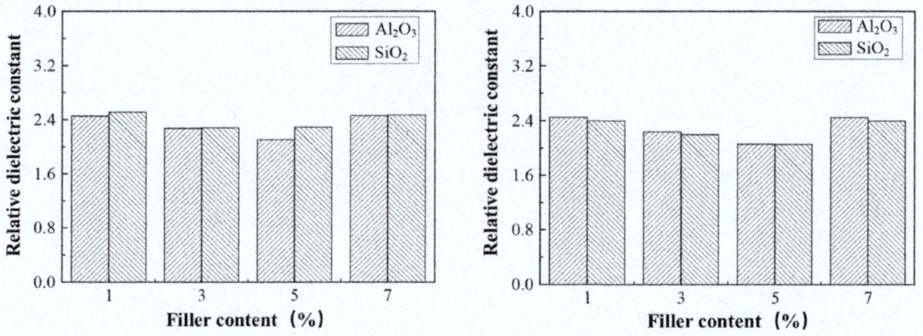

(a) Before coupling agent treatment (b) After coupling agent treatment

Fig. 24 Test results of composite materials dielectric constant

The agglomeration of nanoparticles will directly affect the interface effect between nanoparticles and the epoxy matrix. The more agglomeration phenomenon of particles, the wider the overlapping area of the interface. The overlapped interface layer loses its stable and compact structure, which impacts its dielectric constant, thus affecting the properties of the composite material. Simultaneously, particle agglomeration affects the interaction between the nano-particles and the epoxy resin matrix, reduces the resistance of the polar polymer chain segment or the side group turning at the interface. Consequently, particle agglomeration increases the dielectric constant of the composite.

After treatment with a coupling agent, the agglomeration of filler has been significantly improved and the overlapping range of the interface layer has been decreased. Moreover, the effect of coupling agent treatment on the interaction between nanoparticles and resin matrix is also reduced, resulting in a slight decrease in the dielectric constant. When the nano-particles are well dispersed, the composite can reflect the excellent performance of the filler itself (Fig. 24).

After being treated with coupling agent, the dispersion of nanoparticles in the epoxy resin composite system is significantly improved and the overlapping area of the interface layer is reduced. As a result, the charge high-conductivity channel is reduced, effectively improving the composite material insulation performance. At the same time, the trap energy level generated by the introduction of inorganic particles such as nano-SiO_2 and nano-Al_2O_3 will also limit the movement of space charges, thereby increasing the flashover voltage along the surface. It can be seen from the above that nano-doping will change the trap density as well as the energy level of the composite. However, under the same ratio, the number of nano-SiO_2 particles is larger than that of nano-Al_2O_3 at the same ratio, leading to the higher trap density of nano-SiO_2/epoxy resin composite compared to the nano-A12O3/epoxy resin composite. At the same time, it also makes the charge carrier more easily trapped and has a more vital binding ability to charge. Therefore, the flashover voltage of modified nano-SiO_2/epoxy resin composite is higher than that of nano-Al_2O_3/epoxy resin composite (Fig. 25).

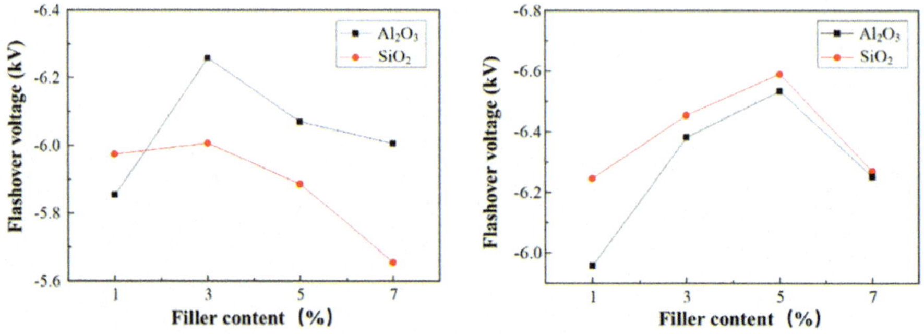

(a) Before coupling agent treatment

(b) After coupling agent treatment

Fig. 25 Test results of composite materials flashover voltage

3.2 Effect of Mass Fraction of Coupling Agent Modified Nano-filler on Electrical Properties of Epoxy Resin

The mass fraction of filler doping is one of the key factors affecting its modification effect. It is difficult to dope a large quantity of nano fillers at present when conducting a study of nanofillers doping modification as the fillers surface energy and small size effect play a role. It is an essential issue for the ongoing research on nano-modification of epoxy composites to Seek the optimal doping concentration and obtain a higher modification effect with a lower filling cost. This book introduces the change rules of epoxy composites electrical properties with nano filler mass fractions of 1%, 3%, 5%, and 7% respectively and studies the nano content impact on the properties of composites. The test results demonstrate that the dielectric constant of the composite initially decreases and then increases with the increase of the content. This is attributed to the low dielectric constant of the interface layer between the nano-particles and epoxy resin, and the interface area has a larger volume relative to the material. The interface region plays a leading role in the macroscopic properties, making the composite have a small dielectric constant. With the gradual increase of nano fillers, more interface layers will be introduced, further reducing the dielectric constant. Moreover, with the increase of nano fillers, the interaction between nano-particles and resin matrix will be intensified, further hindering the turning of polymer chain segments or side groups. Consequently, the dielectric constant of the composite will decrease as the content of nanofillers increases.

In addition, the dielectric constant of the composite will not decrease limitlessly with the increasing content of nano fillers. Once the content of the nano filler reaches to a point, excess particles per unit volume will lead to an intensified agglomeration of nanofiller as well as a larger overlapping area of the interface layer. At the same time, the increasing number of nanoparticles enhances the overlapping probability of the interface layer and further increases the overlapping range of the interface layer. The overlapping interface layer destroys the previous stable structure and affects the original small dielectric constant of the interface layer. Furthermore, the interaction

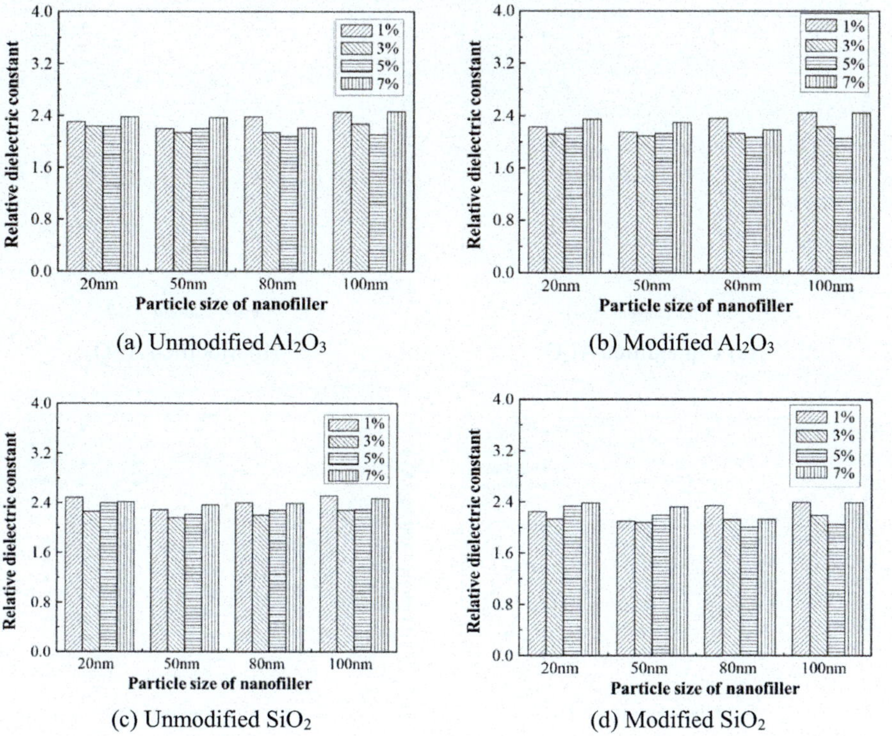

(a) Unmodified Al_2O_3 (b) Modified Al_2O_3

(c) Unmodified SiO_2 (d) Modified SiO_2

Fig. 26 Dielectric constant of composites with different nano content

between nano-particles and resin matrix has also changed, and the restriction on the polar polymer chain segment or side group turn in the interface area has weakened, therefore enhancing the dielectric constant of the composite. Consequently, the dielectric constant of the composite will increase with the increasing number of nano filler content when the nano filler content exceeds a specific limit (Fig. 26).

The results show that the trapped energy level will be generated after the ionization of the doped nano filler. Carriers flowing through the adjacent region will be attracted and bound within the interface region. At this time, the carriers are transported mainly by hopping conductivity and the primary process is trapped electrons—thermal excitation—conduction band electrons—reentry—trapped electrons. Due to the high trap energy level introduced by nano-particles, it is challenging for the trapped electrons to detrap through thermal excitation, thus severely restricting the charge transport between the dielectric body and the interface impeding the advancement of flashover. Therefore, creating conductive channels has been highly difficult, impeding the advancement of flashover. With the increase of nano content, more and more traps are introduced, which increases the trap density of composite materials and further decreases the mobility of carriers. At the same time, it lessens the concentration of free electrons participating in the conduction and further amplifies

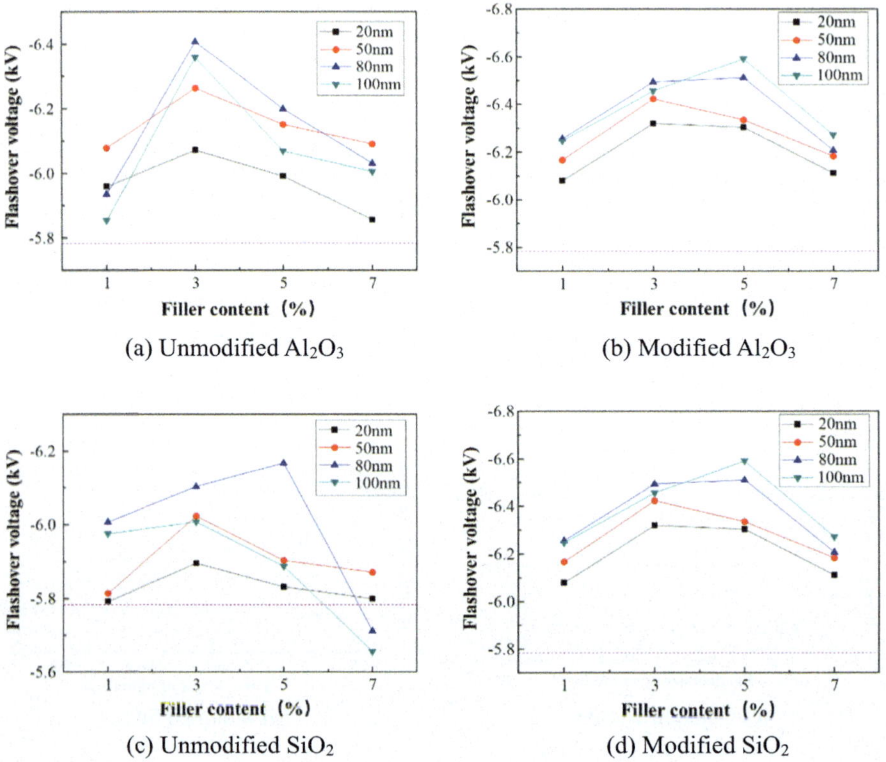

Fig. 27 Flashover voltage of composite materials with different nano content

the flashover voltage of the composite. As the concentration of nanoparticles per unit volume reaches a certain threshold, the agglomeration of nano-particles begins to occur. However, there are not many high conductivity channels generated as a result of the interface layers overlapping, the average mobility of carriers will not be improved significantly, thus having little effect the insulation performance of the composite. Concurrently, a slight increase in the conductivity leads to a rapid dissipation of the composite surface charge and mitigates the distortion of the space electric field, thus improving the flashover voltage of the composite further (Fig. 27).

3.3 *Effect of Particle Size of Nano-filler Modified by Coupling Agent on Electrical Properties of Epoxy Resin*

The particle size of nano filler will directly affect the interface effect between filler and matrix which further affects many macroscopic properties of the composite. In practical engineering applications, nanofillers with smaller sizes comes with greater preparation difficulty and higher production expense in practical engineering applications. Therefore, while utilizing the nano-size effect of fillers, it is also necessary to screen the particle size of fillers to obtain a cost-effective and high-performance epoxy nanocomposite preparation. In this paper, four types of nano fillers with particle sizes of 20, 50, 80 and 100 nm are used to analysis the effect of stuffing particle size on the performance of composite materials. In addition, the work also tests the dielectric constant and flashover voltage of composite materials.

The dielectric constant of the composite decreases first and then increases with the decrease of the filler particle size. With the filler particle size reduction, its specific surface area, reflecting the interfacial bonding degree between the nano-particles and the resin matrix and affecting its interfacial properties, gradually increases. The increase of specific surface area increases the interaction between nano-particles and epoxy resin and strengthens the restriction on the molecular chain movement of epoxy resin, and then makes it difficult for the main chain or side group to turn, thereby reducing the polarization inside the composite material and decreasing the dielectric constant with the decrease of the particle size of the filler.

When the particle size of nano filler is too small, although it provides a large specific surface area and reinforces the interfacial effect of epoxy resin, it also improves the adsorption of particles, leading to agglomerate more easily. The agglomeration of nano-particles impairs the interfacial bonding performance of nano-particles and epoxy resin and enhances the dielectric constant of the interface layer. In addition, agglomeration affects the interaction between nano-particles and resin matrix, leading to alleviation of the restrictions on the movement and direction of molecular chains in epoxy resin. At the meantime, the dielectric constant of the composite is improved in two ways. Therefore, when the particle size of the nano-particles is less than a certain value, the dielectric constant of the composite will increase with the diminution of the particle size (Fig. 28).

The flashover voltage of the composite increases first and then decreases with the decrease of the filler particle size. As the particle size of the filler continues to decline, its specific surface area and surface energy continues to increase which is easy to adsorb other particles and agglomerate. The aggregation of fillers makes the interface layer overlap between the adjacent nano-particles and the resin matrix, generating a high-conductivity charge channel, and thus promoting the carrier migration. However, the number of high-conductivity channels is small at this moment, and the mobility of carriers is not heightened, which does not affect the insulation performance of the composite. On the contrary, it can accelerate the surface charge dissipation of composite materials and reduce the space electric field distortion caused by charge accumulation. Thus, the surface flashover voltage of the composite is further optimized.

When the particle size of the filler decreases to a critical point, the agglomeration becomes serious. Owing to the interface layers overlapping, numerous high-conductivity channels of charge are established, leading to a remarkable augmentation in carrier mobility. The researches demonstrate that the small particle size of the filler curtails the formation of deep traps, and the smaller the particle size is, the stronger the inhibition ability will be. This phenomenon reduces the density

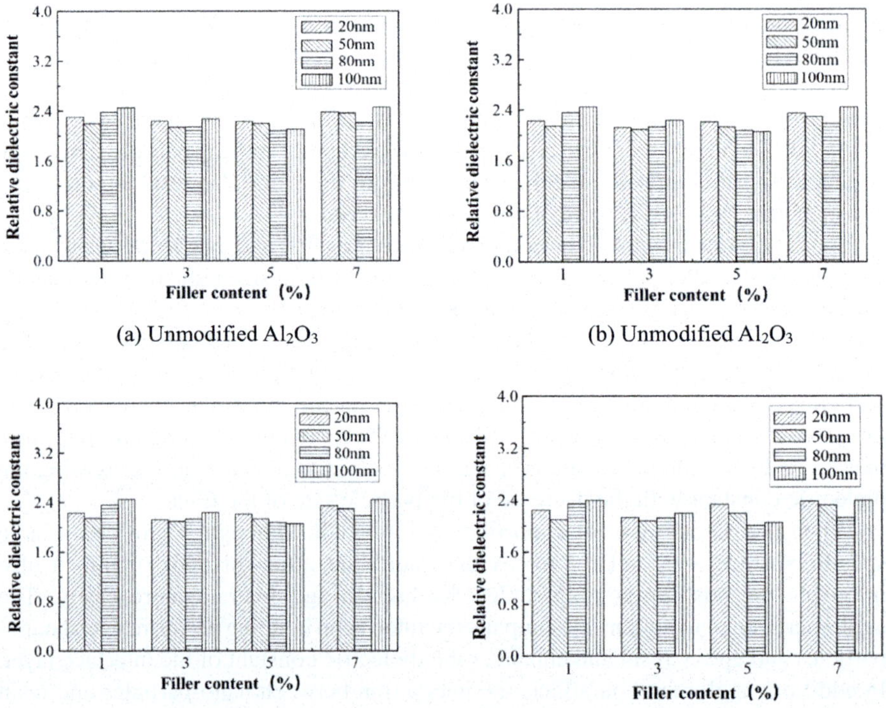

(a) Unmodified Al_2O_3 (b) Unmodified Al_2O_3

(c) Unmodified SiO_2 (d) Modified SiO_2

Fig. 28 Dielectric constants of composites with different nano-particle size

of deep traps and boosts the density of shallow traps. However, the lifetime of carriers in shallow traps is much shorter than that in deep traps. Therefore, carriers in shallow traps easily get stimulated and de-trapped, increasing carrier concentration. In summary, when the particle size of the filler is small to a certain degree, the flashover voltage of the composite decreases with the decrease in particle size of the filler (Fig. 29).

4 Regulation Electrical Properties of Epoxy Resin by Fluorinated Modified Nano-filler

Fluorination modification has proven to be a conducive approach to augmenting the insulation properties of epoxy composites. This method has been extensively acknowledged by industry experts as an area deserving of further exploration. Fluorine has extremely strong electronegativity, which can firmly bind valence electrons near its atomic nucleus and makes fluorine-containing groups have predictable inhibition ability to the development process of discharge [12]. The current research

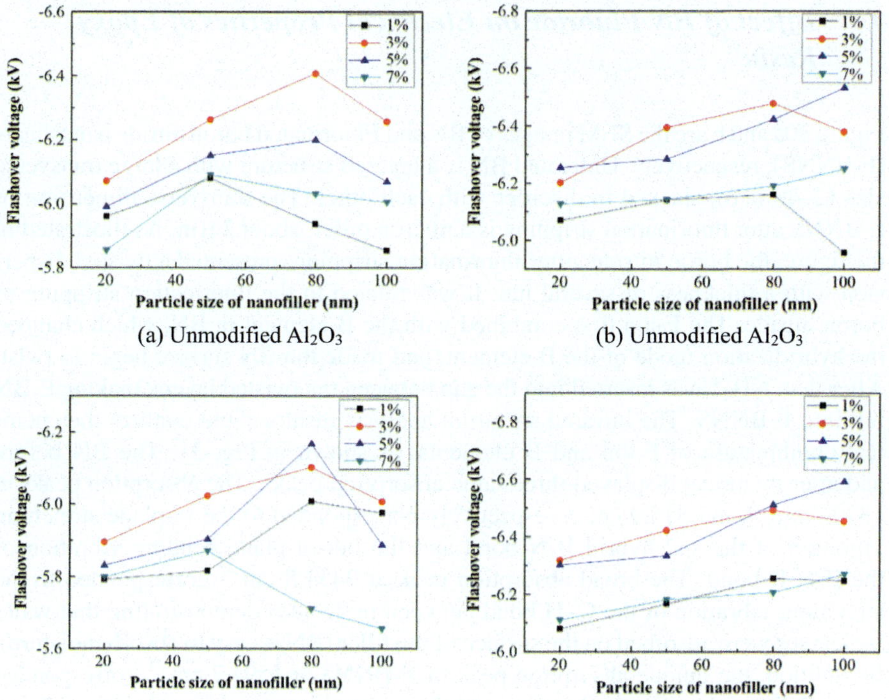

(a) Unmodified Al_2O_3 (b) Unmodified Al_2O_3

(c) Unmodified SiO_2 (d) Modified SiO_2

Fig. 29 Flashover voltage of composites with different nano-particles diameters

on fluorination modification primarily focus on surface fluorination. The fluorinated layer can inhibit the secondary electron emission and improve the trap distribution characteristics on the surface of epoxy composites, resulting in an improvement of the pressure resistance along the surface. Grafting fluorine-containing groups onto epoxy resin monomer can also effectively improve the insulation performance of cured epoxy. Fluorination of nano fillers is an innovation to introduce fluorine-containing groups into epoxy composites. The fluorine-containing group graft modification of nano filler functions as a coupling agent thereby enhancing the dispersion of filler in the polymer matrix and heightening the compatibility between the filler and matrix. Simultaneously, fluorine-containing groups are introduced at the interface between filler and matrix for grafting modification. This paper introduces the influence of fluorinated nano filler on the electrical properties of epoxy composites, taking BN nanofiller [15] and graphene nano filler [16] as examples.

4.1 *Effect of BN Fluoride on Electrical Properties of Epoxy Resin*

Figure 30a and b are the SEM images of BN and Fluorinated boron nitride nanosheets (F-BNNS), respectively. Untreated BN is a layered structure with a large transverse size (2–40 μm), stacked in disorder with each other. The transverse dimension of F-BNNS after fluorination stripping is a micron order, about 2 μm. As illustrated in the figure, the boron nitride, after fluorination stripping, presented a twisted dispersion with a thickness of several nm. It was related to the fluorination stripping of boron nitride. The F-ion first combined with the B element in BN, which changed the hybridization mode of the B element, and made the BN surface begin to twist. After that, NH_4^+ was inserted into the gap between the twisted layers, making F-BN fall into F-BNNS. The infrared spectrum analysis results could confirm the chemical combination of F-ion and B elements, as shown in Fig. 31. The BN before and after treatment displayed three same absorption peaks. The absorption peaks at 1372.8 cm^{-1} and 815.9 cm^{-1} respectively corresponded to the in-plane stretching vibration of the sp2 hybrid B-N bond and the out-of-plane bending vibration of the B-N-B bond. The broad absorption peak at 3431.8 cm^{-1} corresponded to the stretching vibration of the O–H bond between molecules demonstrating that water molecules were adsorbed on the surface of the filler (BN is easy to absorb moisture). In addition, the unique absorption peak of F-BNNS at 1594.7 cm^{-1} corresponded to the B-F covalent bond vibration, and the spectral intensity was changed. Therefore, it is believed BN after ammonium fluoride treatment has achieved exfoliation fluorination.

The law of surface charge dissipation of epoxy composites was explored by applying positive and negative corona to BN/EP and F-BNNS/EP composites. Normalized data were obtained by dividing all the measured potentials by the initial potential, as shown in Fig. 31. It was found that the trend of charge decay under positive and negative polarity is roughly similar. Over time, the total charge quantity revealed a decay law of initially fast and then slow, and the decay rate increases with filler concentration. Comparing the charge dissipation rate of the two materials, it

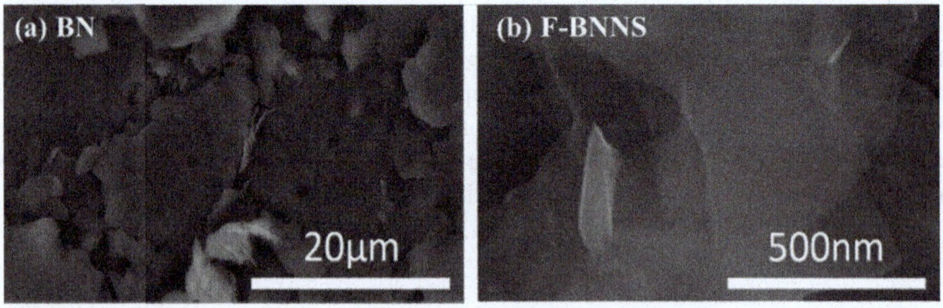

Fig. 30 Microstructure of BN nano fillers before and after fluorination

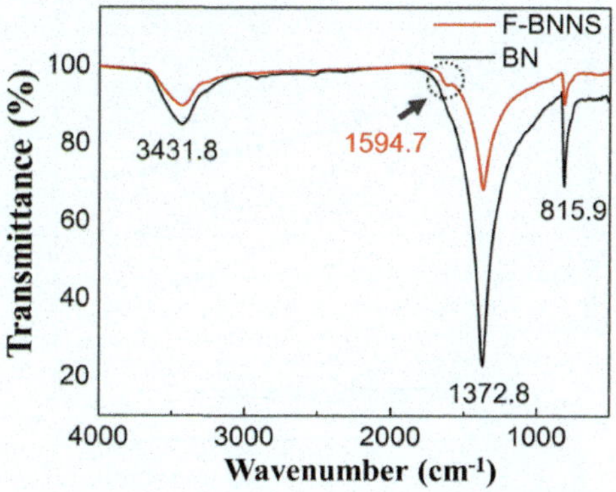

Fig. 31 FTIR spectra of BN nano fillers before and after fluorination

was observed that the charge dissipation rate of F-BNNS/EP nanocomposite was higher than that of BN/EP composite.

Material surface traps will trap corona charges. Upon removal of the charging pin, observed charge dissipation rate followed an arrangement of initially fast and then slow, which is closely related to the charge de-trapping. At the initial stage of corona charging, due to the large charge density on the material surface, many charges will de-trap under the excitation by its own electric field, and the charge will decay rapidly. Over time, the surface electric field of the sample decreases continuously due to the charge decay. Consequently, the rate of charge de-trapping also dwindles gradually (Fig. 32).

The flashover voltage of BN/EP and F-BNNS/EP composites increases with the filler mass fraction. There is a positive correlation between them, which is consistent with the charge dissipation rate. When the filler mass fraction is 8%, the flashover voltage of the BN/EP composite reaches the maximum value of 9.68 kV. Similarly, F-BNNS/EP nanocomposites have the highest flashover voltage of 10.5 kV when the mass fraction of filler is 8%. It is noteworthy that the flashover voltage of F-BNNS/EP nanocomposites has been substantially improved in comparison to the BN/EP composite.

The high flashover voltage of F-BNNS/EP nanocomposites can be attributed to the following reasons. First, by comparing the charge dissipation rate, it can be deduced that F-BNNS/EP nanocomposites with the same filler concentration possess a superior charge dissipation ability in comparison to BN/EP composites. This can be attributed to the fact that boron nitride fluoride can reduce its band gap width and make boron nitride change from insulator to the semiconductor. Therefore, compared with untreated BN/EP composites, F-BNNS filled with the same filler concentration accelerates the charge decay of the composites and alleviates the charges accumulation that induce surface flashover. In addition, the fluorine element on the surface of fluorinated boron nitride filler can react with epoxy groups with active chemical properties. Fluorinated nano-sheets with a high specific surface area have better adhesion with resin, which widens the charge dissipation channel inside the composite. Thus, the surface charge accumulation is further suppressed, and the flashover voltage is increased (Fig. 33).

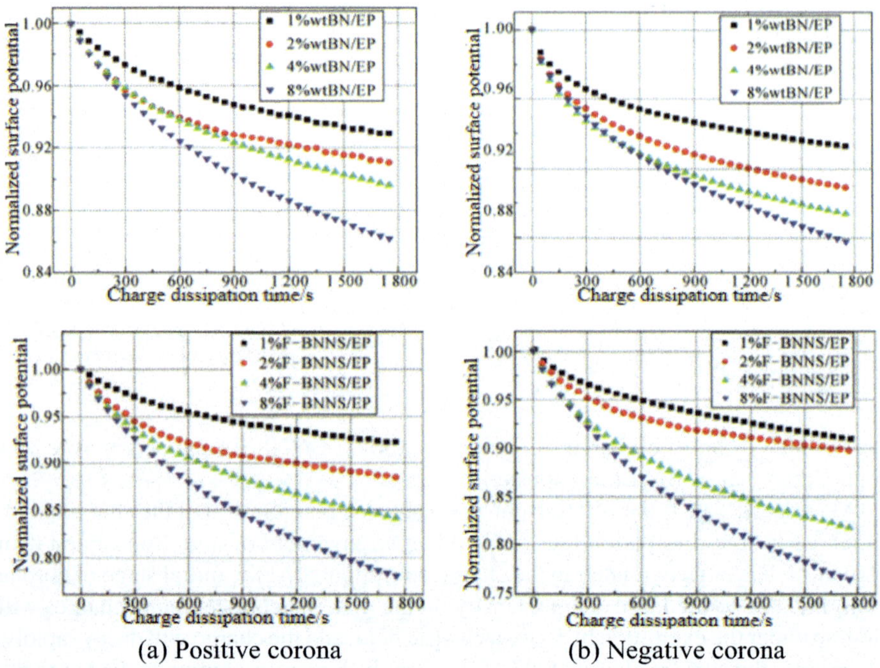

Fig. 32 Charge dissipation rate

Fig. 33 Surface flashover voltage

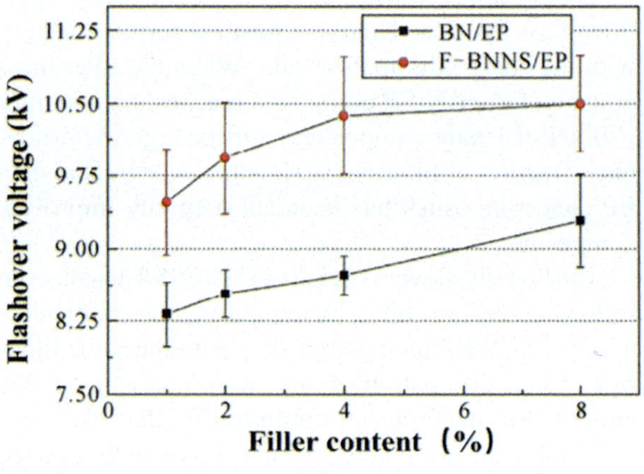

4.2 *Effect of Fluorographene Nano-sheets on Electrical Properties of Epoxy Resin*

Firstly, graphene nanosheets (GNs) were prepared by ball milling and fluorinated graphene nano-sheets (F-GNs) were prepared by grafting modification with fluorinated coupling agents. XPS characterization of graphene nano-sheets before and after modification was shown in Fig. 34. There were apparent F1s peaks at 689 eV in the GNs after fluorination treatment, belonging to the fluorine element grafted by surface treatment. Moreover, F-GNs had a Si 1 s peak at 835 eV, belonging to the silicon element in the fluorinated coupling agent. Further dividing the peaks of C 1 s shown a prominent peak corresponding to the C–F bond at 286.58 eV and a peak corresponding to $-CF_3$ at 291.98 eV. According to the FTIR examination results of GNs fillers before and after fluorination treatment, C–F peak and $-CHF_2$ peak appeared in F-GNs at about 803 cm^{-1} and 1265 cm^{-1}, respectively. The test demonstrates that the fluorinated group was successfully grafted on the GNs surface (Fig. 35).

Comparison with the surface micro-morphology of graphene nano-sheets before and after fluorination reveals that the thickness of GNs and F-GNs are 100 nm. The surface and edges of unmodified GNs appear to be relatively smooth. However, after the fluorination, slight protrusions can be observed on the surface and edge of F-GNs. It is believed that the fluorinated GNs are combined with the fluorinated coupling agent by bonding. Fluorine-containing long link branches construct a fluorine-containing coating on the surface of the filler, which increases the surface roughness of the material (Fig. 36).

The DC surface flashover voltage of EP composite with nanofiller was higher than that of pure EP through the DC surface flashover test of GNs before and after fluorination modification. With the increasing nanofiller doping, the flashover voltage of epoxy composites increased at first and then decreased. When the doping content

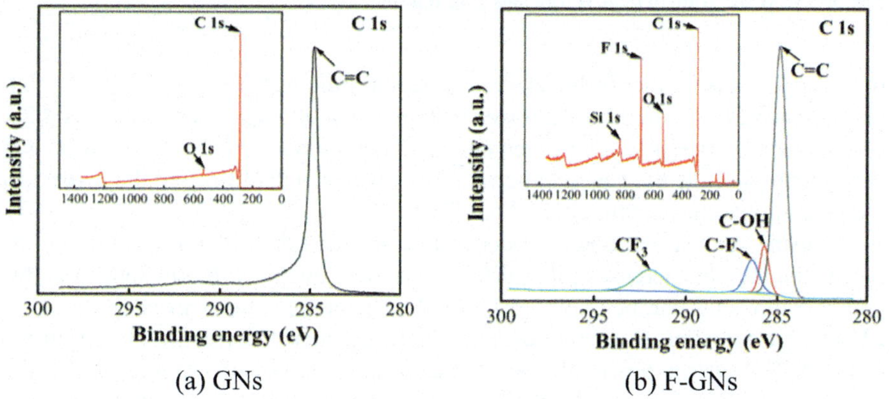

(a) GNs (b) F-GNs

Fig. 34 XPS spectra of **a** GNs and **b** F-GNs

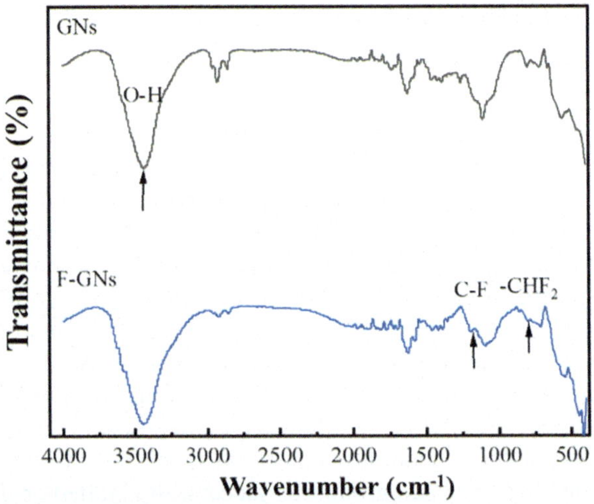

Fig. 35 FTIR spectrum of GNs before and after fluorination

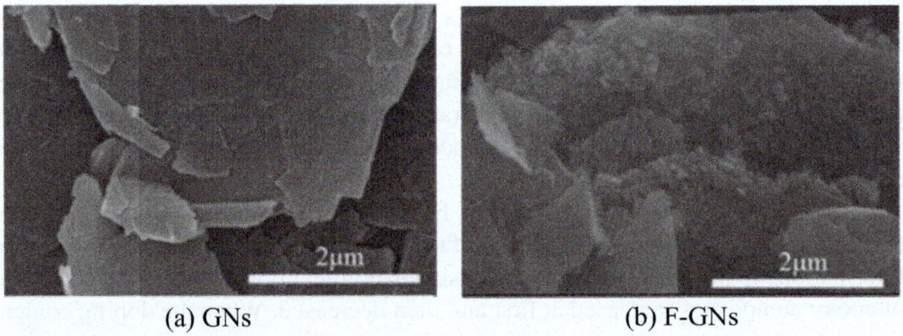

(a) GNs (b) F-GNs

Fig. 36 SEM spectrum of GNs before and after fluorination

of GNs/EP is 0.4%, the flashover voltage reaches the highest value of 9.77 kV. F-GNs/EP achieves its highest flashover voltage when doping concentration is 0.2%, which is up to 10.31 kV. It has been analyzed that there is a strong correlation with the formation of the fluorinated interface layer between the filler and the matrix and the percolation effect of the filler (Fig. 37).

By using the surface potential decay test and trap distribution calculation, the mechanism for the increase of the surface flashover voltage of fluorinated nano filled epoxy composites can be explored in depth. The charge curve revealed that GNs/EP surface potential decay rate dropped and then increased, while the decay rate of F-GNs/EP surface potential exhibited an upward trend with the increase of filler concentration. According to the calculation of charge traps on the surface of materials, the depth of surface traps of GNs/EP composites increased first and then decreased

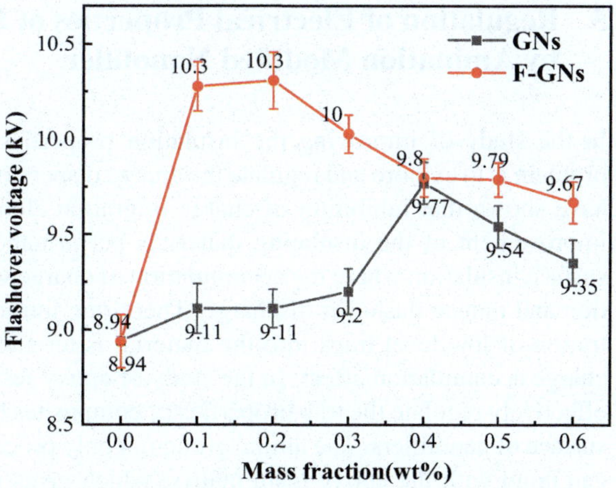

Fig. 37 Surface flashover voltage of EP composite

as the doping amount increase, however, the trap depth of the surface layer of the F-GNs/EP composite revealed a downward trend. When the filler concentration reached 0.6%, the trap depth of GNs/EP composites before and after fluorination decreased significantly. This phenomenon indicated that when the doping concentration of graphene nano-sheets is higher than 0.6%, the material has exceeded the percolation threshold. Due to the high conductivity of graphene nano-sheets, the electrical properties of epoxy composites have changed dramatically, and the surface charge dissipation rate has improved significantly (Fig. 38).

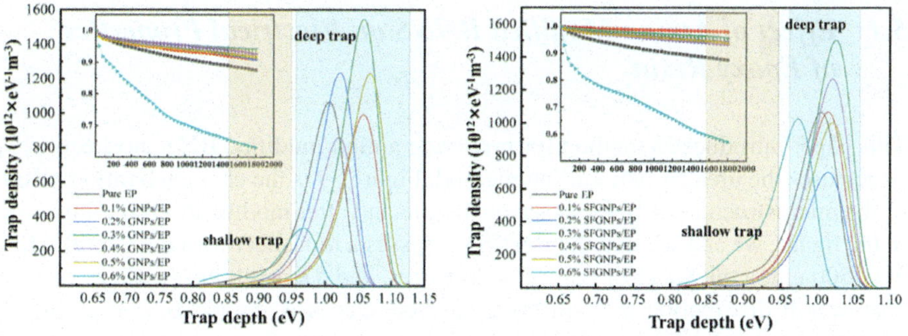

(a) GNs/EP (b) F-GNs/EP

Fig. 38 Surface charge dissipation characteristics and trap distribution curve

5 Regulation of Electrical Properties of Epoxy Resin by Amination Modified Nanofiller

In the study of improving the insulation properties of epoxy composites, a key problem is to explore and regulate the charge dispersion effect. Numerous researches have shown that inhibition of charge migration alone is ineffective in continuous improvement of the insulating materials performance, particularly in the case of surface insulation where the accumulation of charge will lead to electric field distortion and induce flashover discharge. Therefore, introducing an appropriate concentration of low-level traps into the material is an effective approach to control the charge accumulation effect. In the study of epoxy resin composites, it is possible to effectively regulate the trap properties of composites by grafting polar groups on the surface of nanofillers. The amino group not only processes a strong polarity, but also can bond with the epoxy resin matrix, which gives it distinct benefits. Amination modification of nanofillers can enhance the bonding with the matrix, make nanomaterials participate in the cross-linked network of EP, and alleviate the effect of nanofiller on EP cross-linked structure. At the same time, graft modification can adjust the bandgap characteristics and dielectric properties of nanofillers to a certain extent, and further affect the flashover performance of EP composites. In this chapter, we first grafted amino groups on BNNS and layered GNs [17], then we filled the amino-grafted two-dimensional nanofiller into EP to further enhance the electrical properties and investigate its flashover characteristics, charge dissipation, and trap characteristics.

5.1 Effect of Amino Modified BNNS on Electrical Properties of Epoxy Resin

This paper introduces a method for preparing amino-modified BN nanosheets [13], which uses the urea co-ball milling method. Figure 39 is the characterization results of the microstructure of BN nanosheets before and after modification. As illustrated in the figure, we can see that the BNNS possesses a thin sheet with the lateral size is 300–500 nm, a thickness of tens of nanometers, and a smooth, no-particles surface. The overall distribution is multi-layer stacked. Because aminodized boron nitride nanosheets (A-BNNS) is grafted with a large number of amino groups, there are many nano-scale particles attached to its surface and showed a layer of rough granular protrusions. Figure 39c, d are the cross-sectional morphologies of the composites after doping epoxy resin with the filler before and after modification. The unmodified BNNS in the figure displays a granular distribution in the epoxy matrix, and there is a marked agglomeration in partial areas, whereas the modified A-BNNS was well dispersed in the EP matrix, and the filler showed a dendritic distribution. Following analysis, we believe that this phenomenon may be caused by the bonding between the amino groups on the surface of BNNS and the EP matrix.

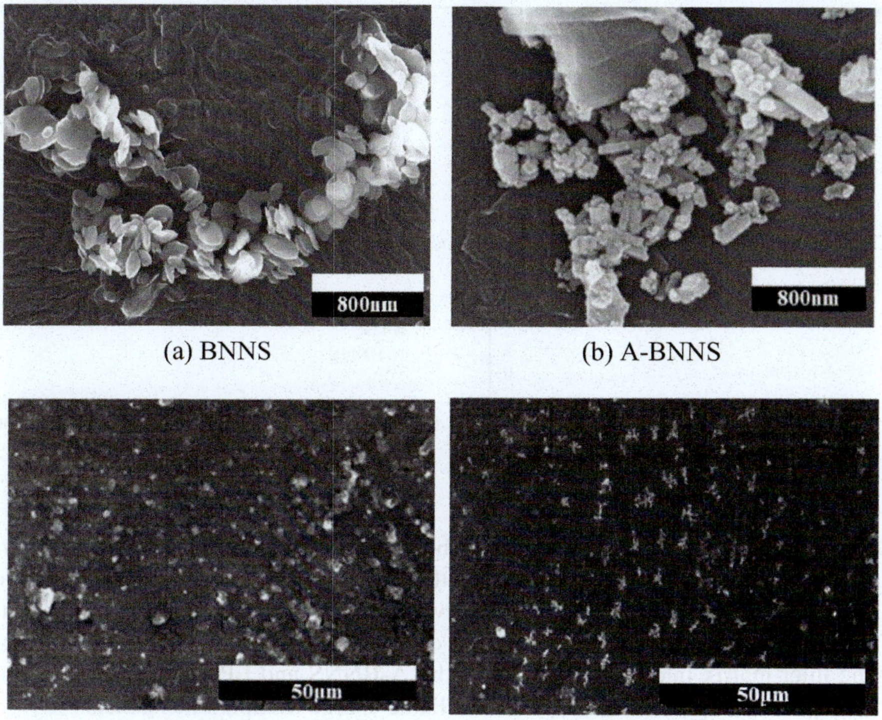

(a) BNNS (b) A-BNNS

(c) BNNS/EP (d) A-BNNS/EP

Fig. 39 SEM images of BNNS and its epoxy composites

In Fig. 40, FTIR analysis of BN nanosheets before and after amination modification can characterize the change of surface groups of fillers. Before and after modification, B-N bending vibration peaks and stretching vibration peaks all appeared at 806 and 1380 cm^{-1} for both fillers. A-BNNS has N–H absorption peaks at 3100–3500 cm^{-1}, and there are obvious double peaks of $-NH_2$ at 3208 and 3433 cm^{-1}, but BNNS does not appear the absorption peaks of these groups. This shows that after amination modification, NH_2 groups were successfully grafted onto the surface of BN nanosheets.

We doped BN nanosheets with different mass fractions of modified epoxy resins both before and after amination modification. Figure 41 presents the test result of its DC flashover voltage. The enhancement of surface flashover characteristics of EP composites is generally more prominent than that of A-BNNS. With the increase of filler content, the flashover voltage of BNNS and A-BNNS doped EP composites experienced a fluctuating tendency initially increasing, then decreasing and finally rising again. And the highest increase is achieved when the filler mass fraction is 0.5%. A-BNNS/EP increased by 11.3% compared with pure EP, and BNNS/EP increased by 10.77%. Upon conducting the analysis, we believe that the excellent insulating properties of BN can improve the overall surface flashover characteristics

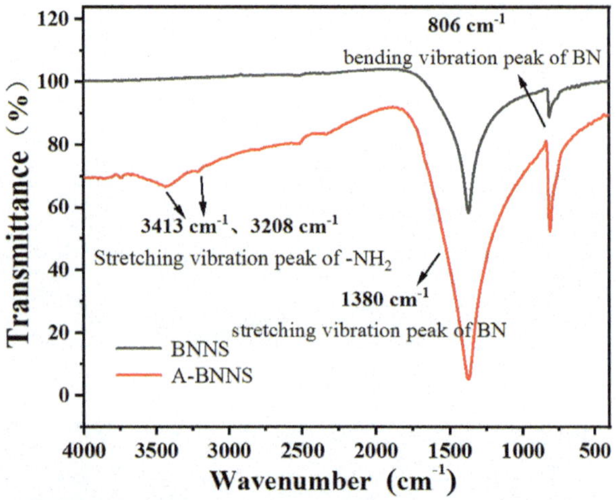

Fig. 40 FTIR of boron nitride nanosheets before and after amination modification

of EP composites, and at the same time, the interface region derived from nanofillers will also have a considerable impact on the surface pressure resistance of composites. There is less interfacial bonding between BNNS and the epoxy matrix, and its introduction will destroy a degree of the crosslinking network of the epoxy matrix. The filler-matrix interface layer formed by A-BNNS and EP through bonding has more polar groups, which can provide reliable charge dissipation channels.

The introduction of nano-fillers can regulate the charge dissipation and trap distribution characteristics of EP composites. The potential-time variation curve of EP composites is shown in Fig. 42. In the figure, we can see that the dissipation rate of the material surface charge presents a trend of climbing first and then sagging as the filler content increases. As a whole, the charge dissipation rate of the A-BNNS/EP composite demonstrates a higher value than the BNNS/EP composite. This is

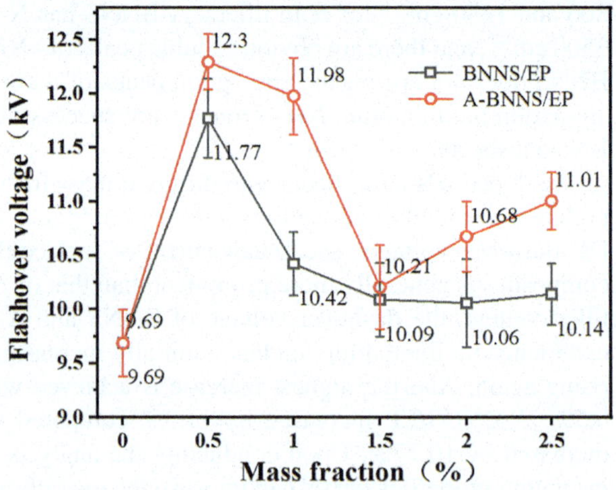

Fig. 41 Flashover voltage of EP composite materials

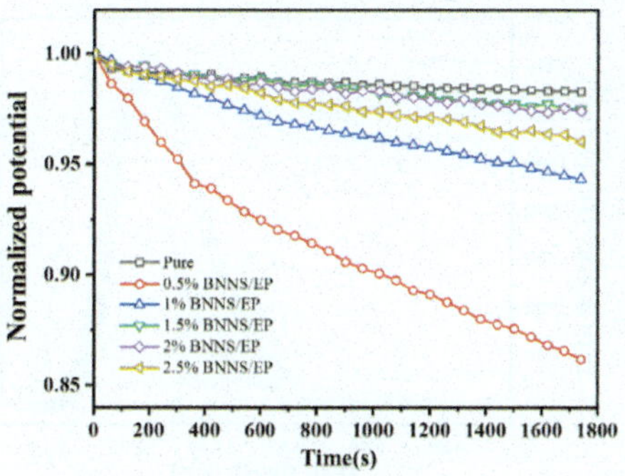

(a) BNNS/EP

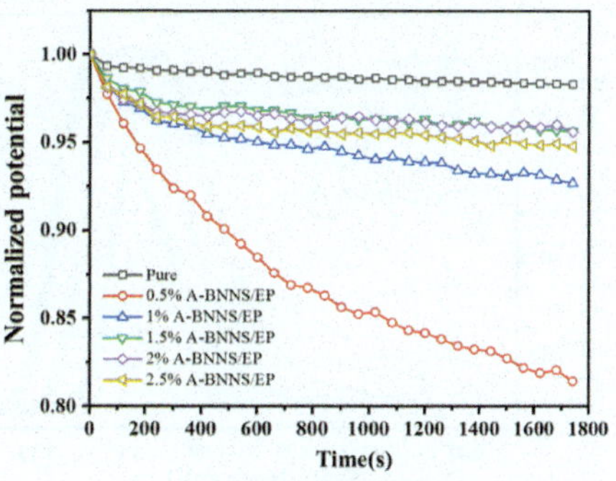

(b) A-BNNS/EP

Fig. 42 Surface potential decay characteristic curve of EP composite material

consistent with the change of the flashover voltage of the material, which shows that the accumulation effect of the surface charge of the EP composite material plays a major role in leading to the flashover. After A-BNNS was doped with modified EP, the interface layer between the filler and the matrix formed a charge transport network composed of dipoles, promoting the rate of space charge dissipation through the surface side and the solid side.

Figure 43 is the distribution characteristics of traps on the surface of EP composites. After adding nanofillers, the trap energy levels of EP composites can be shown that declining before rising. Moreover, the trap energy level of the A-BNNS/EP composite is lower than that of the BNNS/EP composite with the same filler mass fraction. As the EP matrix is a high polymer with a relatively high energy gap, approximately 8–10 eV, the introduction of BNNS into the EP matrix caused a formation of free-moving molecular segments in the interfacial region, which made defects in the formerly dense macromolecular network structure, thus introducing lower-energy traps. As the filler content increases, the interfacial regions begin to overlap and this effect is gradually weakened. At the same time, more BN with high energy gap is introduced into the matrix, which makes an upward trend in the trap energy level on the surface of the material. The surface of A-BNNS contains NH_2 groups, which upgrades the compatibility between the filler and the matrix and also introduces a more polar group which makes it easier for the EP composite to form polarized charges under the action of an electric field. Consequently, the trap energy level of the A-BNNS/EP composite is significantly lower than that of the BNNS/EP composite. The shallower trap energy level boosts the carrier mobility within the EP composite. Under the effect of an external electric field, the space charge injected into the material is hard to accumulate, which alleviates the electric field distortion in the local area. This makes it trouble for the initial electrons to be excited during the generation of the flashover discharge, and forming a large-scale electron avalanche is hard to do because of the inadequate of high-field-strength accelerated energy after the secondary electrons emission in the flashover development stage. Therefore, this

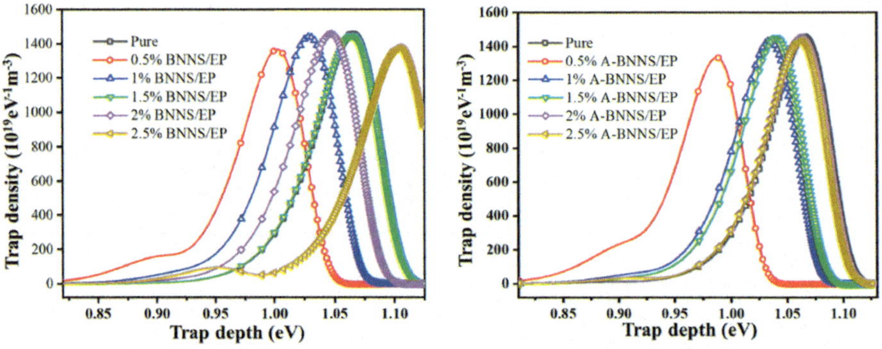

(a) BNNS/EP (b) A-BNNS/EP

Fig. 43 Energy level distribution of traps on the surface of EP composites

hinders the development of flashover discharge and is conducive to improving the surface flashover voltage of the material.

5.2 *Effect of Amino Modified GNs on Electrical Properties of Epoxy Resin*

The preparation process of GNs nanosheets and aminated GNs nanosheets is basically the same as that of BNNS, which is to obtain two-dimensional sheet materials by ball milling exfoliation. We used scanning electron microscopy to characterize the exfoliated nanosheets, and the result is shown in Fig. 44. Figure (a) depicts the microscopic morphology of GNs, which averages in size to around 500–800 nm and features a smooth nanosheet surface without any additional attached particles. The nanosheet thickness is estimated to be approximately 100 nm. Figure (b) shows the microscopic characterization of AGNs nanofillers. Its particle size is closely resembling that of GNs. Rough particles are attached to the surface of aminated GNs nanosheets, and amino groups are grafted on the surface of GNs through chemical bonds.

We filled EP with GNs before and after amination modification to prepare composite materials. Figure 45 displays the dispersion effect of fillers inside the matrix. Analysis of the characterization results reveals that the dispersion of GNs filled EP composites is relatively poor with many fillers experiencing serious agglomeration. The surface of AGNs is grafted with a large number of amino groups, which can participate in the curing crosslinking reaction of the epoxy resin/anhydride system, thus providing better compatibility with the matrix. Simultaneously, it is hard for a condensation reaction to proceed between two amino groups, resulting in minimal agglomeration of the filler particles. This makes it the optimal choice for dispersibility.

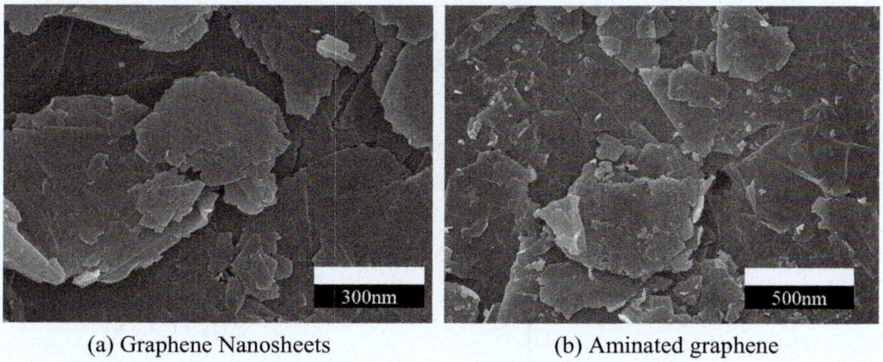

Fig. 44 Microscopic morphology of graphene and aminated graphene

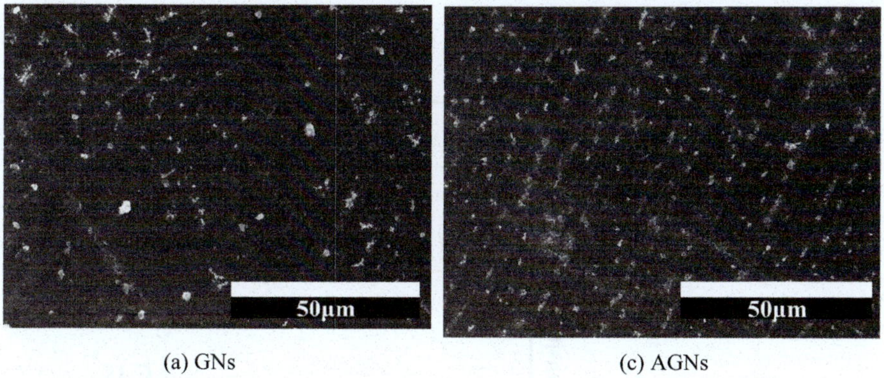

Fig. 45 Graphene nanofiller dispersion SEM

We performed FTIR and XPS analysis on graphene nanosheets before and after amination modification, using Nicolet Nexus 410 Fourier infrared spectrometer and ThermoFisher K-Alpha X-ray photoelectron spectrometer. Figure 46 is the analysis result. The FTIR spectrum of AGNs showed the characteristic peak of C–N bond at 1446 cm^{-1} and the characteristic peak of N–H bond representing amino group at 1592 cm^{-1}, which indicated a successful grafting of amino group. The scanning results of XPS displayed that the AGNs after amination modification exhibited N 1 s peak at 401.5 eV. By merging the results of the FTIR characterization, it can be concluded that the amination grafting modification effectively graft the amino groups onto the surface of GNs (Fig. 47).

Due to the high conductivity of graphene nanofillers, the mass fraction of fillers in EP composites needs to be controlled at a relatively low level in order to effectively promote the surface insulation performance. The flashover voltage of EP composite samples modified by graphene nanofillers was significantly higher than that of pure epoxy resin. The flashover voltage of the EP composite sample doped with GNs is

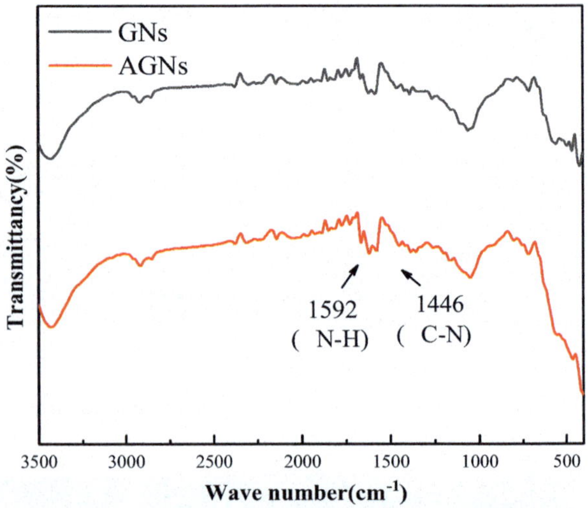

Fig. 46 FTIR of graphene and its modified filler

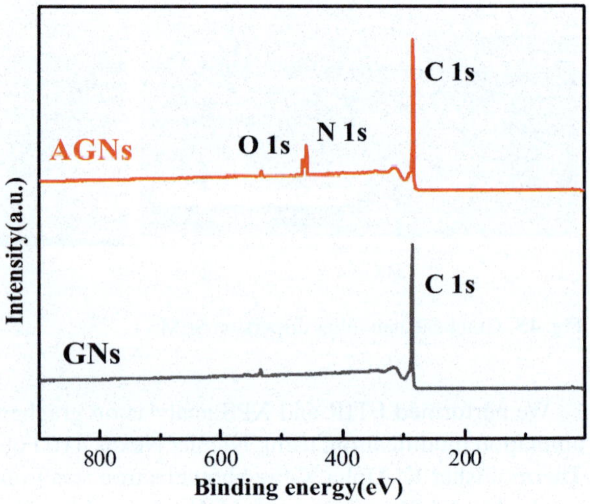

Fig. 47 Graphene and its modified filler XPS

demonstrated by the black curve. When the doping amount is 0.2 wt%, compared with the pure epoxy resin, the flashover voltage of the sample increased by 20.26%, reaching 11.22 kV, which is an obvious improvement; but as the mass fraction of GNs advanced to 0.4 wt%, the flashover voltage dropped to 9.6 kV, and then the flashover voltage gradually increased with the mass fraction increasing. The amination grafting of GNs can effectively strengthen the compatibility between the filler and the EP matrix, and reduce the physical and chemical defects during the preparation process. The blue curve in Fig. 48 is the flashover voltage change curve of the AGNs system. When 0.2 wt% of AGNs is dropped, the flashover voltage of the sample increases to 11.26 kV, an increase of 20.68% compared to that of the pure EP. Upon further

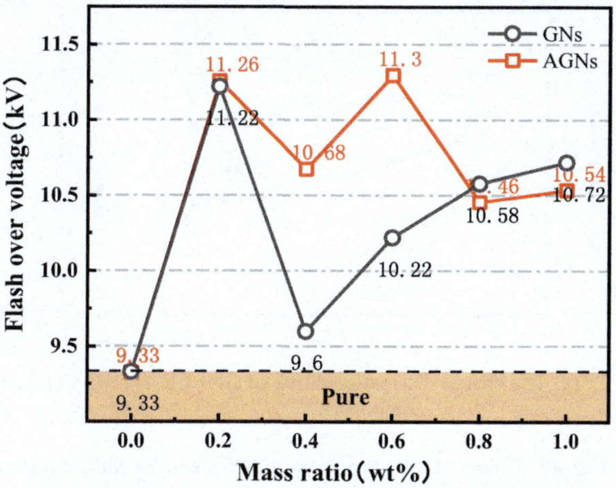

Fig. 48 Flashover voltage of graphene material modified EP sample

increasing the doping content of AGNs, the flashover voltage presents a "W"-shaped change pattern, exhibiting the optimal performance at 0.6 wt%, reaching 11.3 kV, 21.11% higher than that of pure EP. The contribution of AGNs to the improvement of flashover is primarily through accelerating charge dissipation, which is highly affected by the interface effect. The repetitive overlapping of multi-layer interface models results in the "N" type change law of the flashover voltage of the AGNs/EP system.

We can get the curve in Fig. 49 by testing the surface potential decay characteristics of the epoxy composite before and after modification. The charge dissipation rate generally presents a trend of first decreasing and then increasing as the doping concentration increases. After GNs are incorporated into the EP matrix, the compatibility with EP is poor, leading to difficulty in forming a stable chemical bond, which cause destroying the cross-linked structure of the EP matrix to a certain degree. The original conductive path is blocked and thus reduce the rate of charge dissipation. However, with the increasing doping amount of GNs nanofillers, the contact and overlap between the nanosheets are formed, and new conductive paths are generated. The GNs itself has a high carrier mobility, so the surface charge dissipation rate of GNs/EP composites continues to accelerate with the increase of doping amount. The AGNs can crosslink with the EP matrix due to the grafting of a considerable amount of amino groups on the surface. As the doping concentration rises, the superior dispersibility and matrix compatibility make AGNs nanosheets overlap with each other, adding a large number of new conductive paths, which makes a better charge dissipation rate of the system than that of GNs system at higher doping concentration.

Figure 50 shows the relationship between the trap concentration and energy level of epoxy composites before and after modification. Pure EP demonstrates a complete cross-linking structure and a lower concentration of shallow traps and thus there is a

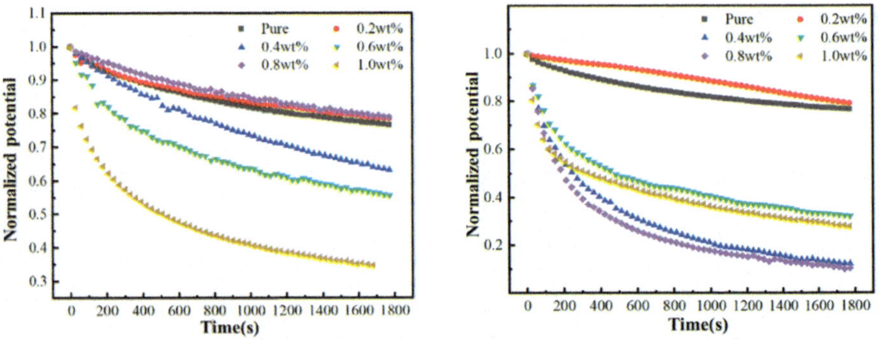

(a) The charge dissipation curve of GNs/EP composites

(b) The charge dissipation curve of graphene nanofiller doped EP

Fig. 49 Charge dissipation curve of graphene nanofiller doped EP

slow rate of charge dissipation. After doping GNs, with the increase of doping concentration, the concentration of shallow traps gradually increases, while the concentration of deep traps first climbing and then sagging. Following our analysis, we have determined that there exists a substantial number of charge traps within the interface region between the filler and the matrix, and the interface effect is closely related to the compatibility of materials and the width of the interface between particles. Once the doping concentration of nanofillers surpasses a specific value, overlap among the different layers of particles may be induced, which will further lead to changes in trap energy level and concentration. Moreover, upon reaching the percolation threshold of filler mass fraction, the high electrical conductivity of GNs would significantly accelerate the rate of charge transfer, which is also manifested as a shift in the trap energy level to a lower direction. As the doping concentration of the AGNs system increases, the trap distribution characteristics of the composite material develop in the direction of increasing the concentration of shallow traps and decreasing the concentration of deep traps. This is due to the fact that the amino groups grafted on the surface of the nanosheets can easily react with the EP matrix, the bonding between the filler and the matrix is stronger, and thus the promotion of GNs itself on trap shallowing can be better exerted.

6 The Regulation Mechanism of Filler Functional Modification on the Electrical Properties of Epoxy Resin

The modification of epoxy nanocomposites mainly is largely determined by the properties of filler particles and the interfacial phase formed by them. And the functionalization of nanofillers is primarily manifested in two aspects. On the one hand, it is to maximize the effect of nanoparticles within the composite material. On the other

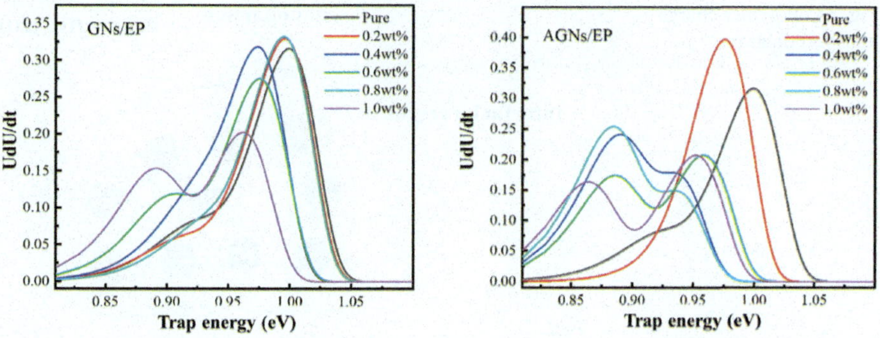

(a) Trap distribution of GNs/EP composites (b) Trap distribution of AGNs/EP composites

Fig. 50 Composite trap distribution characteristics

hand, regulate the interfacial interaction between the filler and the matrix in chemical methods. Tanaka proposed a multi-nuclear model in 2005 [8], which categorized the filler interface as a bond layer (the layer where the valence bond is formed), a bound layer (the layer where the high-energy trap is formed), a loose layer (the layer where the ion trap or low-energy trap is formed) and the electric double layer (the layer where the overlaps with the first three layers). The multi-nuclear model divides the filler-matrix interface more finely, and also adequately explains the difference in charge behavior between different layers in the interface region. With the deepening of the research and the continuous advancement of the current computer simulation technology, the researchers used simulation methods such as molecular dynamics, density functional theory to analyze the action principle of the filler-matrix interface region at the molecular scale. It provides a new theoretical reference for revealing the regulation mechanism of filler functional modification on the electrical properties of epoxy resin. This chapter illustrates the research results of the interaction law of the filler functional interface and its influence on the electrical properties of epoxy composites.

6.1 Interface Structure Between Filler and Matrix

The interfacial properties of nano-filler and matrix are closely related to the types of functional groups on the surface of the filler, the length of the bond segment, the bonding strength and other factors. The interface layer act as a conductive area, thus allowing the accumulation of electrons on the surface of the filler and between the bonding layer and the transition layer. An interface region is formed after doping with different fillers, which will alter the trap energy level and density inside the composite material. The carriers can only transition after gaining enough energy owing to the fact that there are potential barriers inside the bonding layer and the

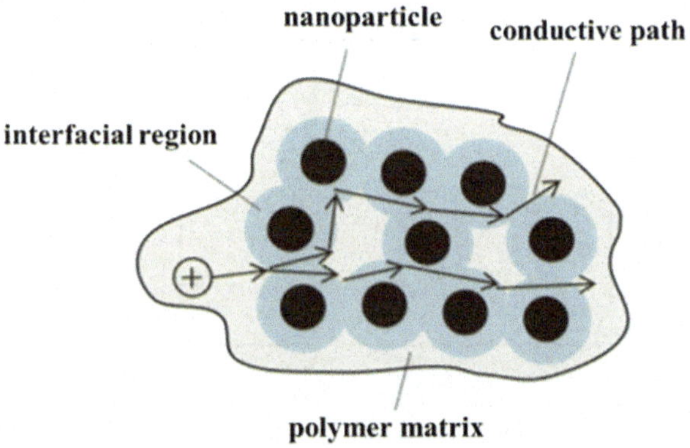

Fig. 51 Polymer charge transport diagram

transition layer. Due to the width of the transition layer being shorter than the mean free path of the carriers, when the amount of doped filler is small, the filler particles are isolated in the EP matrix, and the carrier migration is limited by the transition layer. As the amount of filler doping increases and the filler percolation threshold is reached, the interface regions will overlap to form percolated conductance channels, that is, conduction paths. As shown in the figure, it shows that the binding effect on the carriers is weakened, and the carriers in the composite material are easily transported through the conductive path (Fig. 51).

This section takes the fluorinated interface as an example to illustrate the impact of the interface structure on the material properties. The multi-nuclear model reveals that that the interface layer mainly consists of the bond layer and the transition layer. The former is where the surface groups of the nanofillers interact with the epoxy matrix and generally has a narrow width. This layer is composed of fluorinated reagent molecular chains coated on the surface of the filler or grafted fluorine-containing groups, and has a small width. The transition layer is mainly composed of molecules of the polymer matrix, which has a low density and a large width. The hydrothermal fluorination treatment destroyed the molecular structure of the surface of the nanofiller and grafted fluorine-containing groups on the surface of the filler. These groups manifested strong oxidizing properties, allowing for a chemical reaction with the epoxy matrix to realize the chemical bond connection in the interface region. Another fluorination technique is to utilize fluorine-containing coupling agent to fluorinate the nanofiller. One end of the molecular structure of the fluorinated coupling agent reacts with the filler surface to graft a longer fluorine-containing segment on the surface, which is difficult to form a bond with the epoxy matrix.

After the nanofiller is directly doped with the epoxy matrix, forming a stable bond with the matrix is hindered because the surface of the filler lacks active groups, resulting in the narrowness of the bonding layer. The interface layer forms a potential barrier due to Coulomb blockade effect, and the negative charge injected at the electrode is not easy to de-trap after being trapped, forming a deep trap, and thus the charge is not easy to dissipate. Following hydrothermal fluorination, the filler can be combined with the EP matrix through a small amount of covalent bonds and ionic

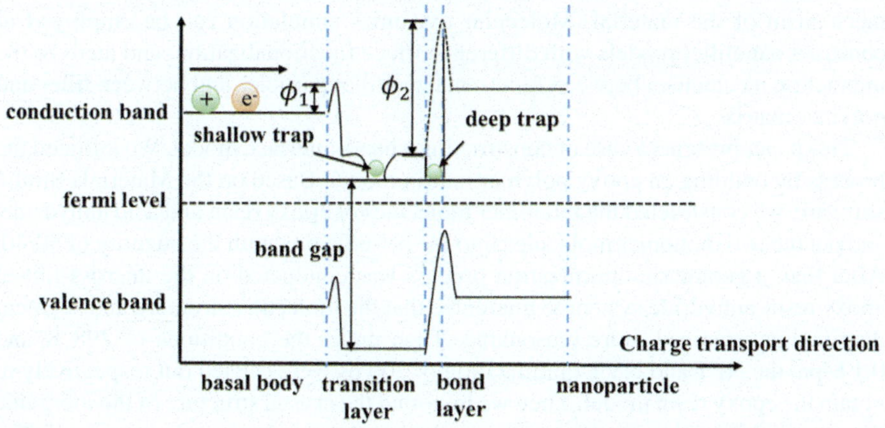

Fig. 52 The relationship between band gap characteristics and charge traps of materials

bonds, thus extending the breadth of the bonding layer to a slight degree. Electrons can fully participate in the charge transport through chemical bond migration, then the charge transport rate becomes faster, and the surface traps of the composite material converted to lower energy levels. As for the nanofiller modified by fluorine-containing coupling agent, its surface is grafted with fluorine-containing coupling agent molecules with longer chain segments, and the width of the bonding layer is larger. The filler and the matrix can only be connected through a longer fluorine-containing chain segment, which enhances the Helmholtz effect at the between the filler-matrix interface. The bound state charge centers of fluorine atoms seriously hinder the charges injected from the outside from transferring to the filler, causing a slowdown of the charge dissipation rate. Both the energy level and density of the traps showed an increasing trend (Fig. 52).

6.2 MD-Based Interfacial Interaction Mechanism Analysis

In the study of nano-modification of composite materials, the agglomeration effect of fillers is a crucial hindrance that affects the modification effect, especially in terms of insulation properties. Agglomeration will bring about uneven distribution of filler interface area and interface layer thickness, and the interaction between filler and matrix interface molecules will produce a special structure and properties, influencing multiple electrical properties of the material. Nanofillers, due to their relatively high surface energy, are often poorly compatible with organic matter, and tend to form agglomerates with mutual affinity within the polymer. The development of computer simulation technology has enabled researchers to explore the interaction between microscopic particles increasingly through methods such as molecular dynamics simulation. This can more effectively analyze the performance alteration mechanism of the material. Molecular dynamics simulation can be employed to construct nanofiller models with different surface functionalization, and analyze the interaction mechanism between functionalized filler particles and between filler and polymer matrix.

This paper presents a case of constructing a nano-interface model. We initiated the process by building an epoxy polymer matrix model. Based on the Materials Studio software, we constructed the monomer molecules of epoxy resin and acid anhydride, and put the two monomer molecules into the periodic box with the quantity of 80/40. After that, a geometric optimization process was conducted on the uncross-linked epoxy resin-anhydride system to guarantee that the molecular structure and segment distribution remained more reasonable. Then under the conditions of 298 K and 0.1 Mpa, the NPT and NVT optimization of 200 ps were carried out respectively to obtain the epoxy resin model. Then we imported the crystal structure of the inorganic filler, treated the unit cell with section method, and then construct the super cell. The following procedure was to utilize the build layer function to create the filler-epoxy resin interface model, and set a vacuum layer with a thickness of 30 Å above the model to eliminate the impact of periodicity. We then performed geometric optimization and 200 ps NVT optimization on the interface model with a 1 fs time step and a frame file being output every 1 ps. Subsequently, we calculated the interface binding energy of different models, selected the interface energy of the last 50 frames and took the average value as the final data. The interface interaction energy can be calculated by the following formula:

$$E_{interface} = E_{total} - (E_{Layer1} + E_{Layer2}) \tag{5}$$

As demonstrated in Fig. 53, the calculation result reveals that the interfacial interaction between fluorinated nanoparticles is significantly less than that of nanofillers with coupling agent modified and unmodified.

Fig. 53 Interface interaction energy

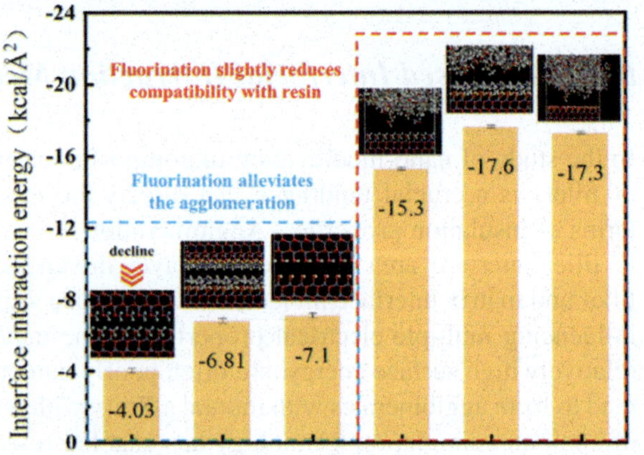

After analysis, we believe that fluorine-containing groups have been grafted onto the filler surface after fluorination treatment. The special structure of fluorine atoms makes it tightly bind the valence electrons near the nucleus, which endows it with extremely low surface energy. It enables the interaction energy between fluorine-containing groups much lower than that of general chemical groups. When two adjacent fluorinated fillers are in close proximity, the high surface energy and mismatching sites originally caused by the nano-size effect are covered by fluorine-containing groups. The nature between the filler particles transitions from affinity to mutual exclusiveness, leading to a significant enhancement in dispersion within the epoxy matrix. At the same time, we calculated the interaction energy between the three nanofillers and the uncross-linked epoxy resin to assess their compatibility with the epoxy matrix. It can be observed that the interaction energy between the fluorinated filler and the epoxy resin is slightly diminished, while the modification of the coupling agent can improve the affinity between filler and the epoxy matrix. This can be attributed to the introduction of fluorine-containing groups which reduce the affinity sites between the filler surface and epoxy, and the lower surface energy of fluorine-containing groups also weakens the interfacial interaction.

6.3 Influence Mechanism Analysis of Trap Characteristics Based on DFT

To further discover the influence mechanism of functional modification of fillers on the surface flashover characteristics of EP composites, three common treatment methods such as unmodified fillers, fluorinated fillers, and coupling agent grafted fillers were selected for study. And we took the molecular chain segment at the interface between the nanofiller and the epoxy matrix as the object, calculating the molecular orbital energy level of the interface chain segment by quantum chemical method. Then we used MS software to construct the models of the three interface molecular segments. Based on the Dmol3 module, we performed DFT calculations on it with the PBE as the exchange correlation functional and using the generalized gradient approximation. Taking into account the electronic polarization effect, the double numerical plus polarization (DNP) basis set was used for calculation and the optimal quality grid size was adopted as numerical integration [18]. In the calculation process, we firstly performed geometric optimization of the molecule, then engaged NVT optimization at 50 ps, and then analyzed the molecular orbital energy levels of the three chain segments. Figure 54 displays the calculation result. The orbital energy levels of molecular segments in the interface region include the lowest unoccupied orbital and the highest occupied orbital. The LUMO orbital acts as a trap site for capturing electrons inside the material, and the HOMO orbital acts as a trap site for electron holes. The midpoint of the two energy levels is the Fermi level. Electrons, after being excited, will have a transition from the HOMO orbital to the LUMO orbital to obtain conductivity, and the energy required for its transition is the energy

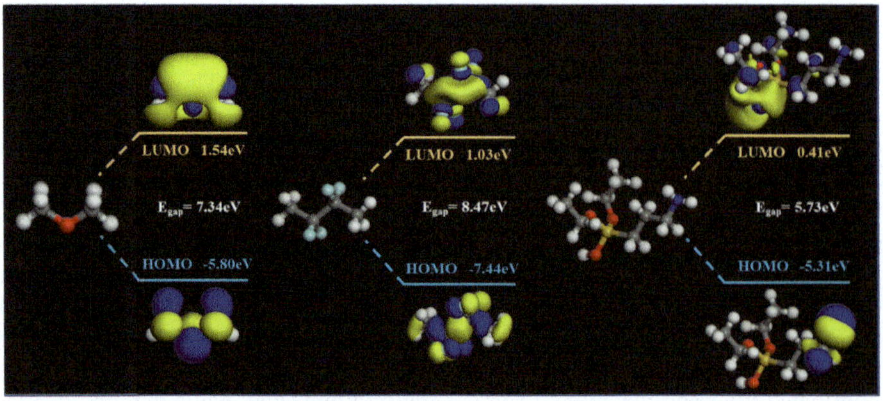

Fig. 54 DFT calculation results

gap. The energy gap between LUMO energy and HOMO energy can reflect the electronic transition ability. In the study of insulating materials, Egap is considered to be a form of trap existence [19]. The model frontier orbital level gap is:

$$E_{gap} = E_{LUMO} - E_{HOMO} \quad (6)$$

The calculation results indicate that the surface of the unmodified filler forms bonds with epoxy through a small amount of hydroxyl groups, and its energy gap is basically at 7.34 eV. Nevertheless, it is worth pointing out that the surface of the untreated filler contains only a tiny amount of hydroxyl groups and thus there are often more vacuum bands at its interface. This makes it tougher for carriers to migrate between the filler and the matrix. The filler treated with the coupling agent is linked to the epoxy via the coupling agent segment. Since there are more polar groups in the molecular chain segment, it can be easily to form polarized charges and transport carriers in the interface region. The energy gap of treated filler is much smaller than that of the untreated filler, leading in a significant increase in the density of shallow traps within the composite material. The fluorinated nanofiller is bonded to the epoxy matrix through the fluorine-containing chain segment, and the fluorine-containing group makes the energy gap of the chain segment significantly climbed. Thereby introducing a large number of deep traps at the interface. Meanwhile, the fluorine-containing chain segment also provides some bonding sites between the filler and the matrix, thus remaining a relatively stable density of the interface trap energy level, and the charge can still be transported through the chain segment. The energy required for carriers is still reduced compared to directly transitioning from the epoxy side to the filler side.

The functional modification of nanofillers has two main effects in ameliorating the surface pressure resistance of epoxy composites. On the one hand, it can improve the bonding state between the filler and the matrix, enhance the dispersibility and compatibility of the nanofiller in the resin matrix, thus better upgrading ability of the nanofiller itself in charge release and discharge suppression. On the other hand, it is to regulate many electrical properties of the filler-matrix interface region, especially the trap distribution characteristics. Conventional coupling agent treatment usually emphasize the first aspect, which can refine the modification effect of nano fillers by improving the dispersibility and compatibility of fillers. And treatments such as fluorination and amination modulation offer both of these capabilities simultaneously. Taking fluorination as an example, fluorination can not only enhance the dispersibility of fillers, but also introduce deep traps at the interface of fillers and inhibits the development of discharge and secondary electron emission at the interface. The dielectric parameters at the junction of filler-matrix-vacuum are significantly different, which will lead to obvious electric field distortion effect. This causes the triple junction to become a weak link in the flashover development, and easily excite a large number of secondary electrons under the initial electrons. Fluorinated nanofillers introduce substantial fluorine-containing groups in the interface region. These fluorine-containing groups can provide massive deep trap energy levels, which effectively inhibit the development of discharge in the interface region and improve the dielectric strength at the filler-matrix interface. On the other hand, the fluorine-containing groups often possess high reactivity, which facilitate their further reaction with the epoxy resin, thereby establishing a chemical bond between the filler and the resin. This eliminates the situation where the carriers cannot be transported due to the lack of effective bonding at the filler interface. Therefore, when the filler concentration reaches a certain threshold, the charge dissipation rate on the surface of the EP composite will also be improved obviously. This provides a good path for the dissipation of local charges on the surface of the composite material, alleviating the field distortion effect due to charge accumulation. The two mechanisms act on the filler side and the interface layer respectively and the surface pressure resistance of the composite is enhanced synergistically.

7 Conclusion

This chapter systematically and comprehensively introduces methods of nanofiller functional modification and their effect rule and mechanism on the electrical properties of epoxy composite. By assessing the current research regarding epoxy composite nanomodification research, methods of functional modification of nanofiller surface are proposed to improve the dispersity and compatibility of filler in matrix. Meanwhile, the interface properties of filler-matrix can be targeted adjusted by controlling the grafting of surface functional groups. Furthermore, the effect mechanism of nanofiller functional modification was analyzed and summarized by combining experiment results and simulation results of MD and DFT. The technology of nanofiller functional modification can enhance the modification effect of traditional inorganic nanoparticle on the insulation performance of epoxy composite, and optimize the preparation of epoxy nanocomposite, which has a broad industrial application prospect. The main conclusions of this chapter are as follows:

(1) This chapter introduced a systematic overview of the surface functional modification for nanofiller in epoxy composite insulation material research. And the platforms and methods of nanofiller surface coupling agent grafting, hydrothermal fluorination and one-step method of amination modification were elaborately discussed. Moreover, various types of functional inorganic nanofiller were prepared successfully.

(2) The experiments were conducted to study the influence law of nanofiller surface functional modification on microstructure and electrical properties of epoxy composite. The dispersity and compatibility of nanofiller in matrix were improved by coupling agent graft, fluorination and amination modification all, so as to increase the modification effect of nanofiller on insulating property of epoxy resin. In addition, fluorine-containing groups and amino groups could influence the trap distribution of composite surface and targeted adjust the aggregation and dispersion of material surface charge, which was essential for the improvement of epoxy composites surface insulation.

(3) Based on the nanointerface model and simulation results of MD and DFT, this study examined the effect mechanism of nanofiller functional modification on insulating property of epoxy composite. It could be considered that groups grafted on nanofiller surface can boost macro electric properties of epoxy composite by modifying filler- matrix interfacial properties. Filler surface fluorination modification could effectively inhibit the nanoparticle agglomeration effect. Furthermore, fluorine-containing groups could increase the energy gap of molecule chain at interface, thus high level deep traps were introduced to the internal of materials to improve the insulating property of materials.

References

1. Li, S., Li, J.Y.: Condition monitoring and diagnosis of power equipment: review and prospective. High Volt. **02**(02), 82–91 (2017)
2. Yang, S., Xiang, D., et al.: Condition monitoring for device reliability in power electronic converters: a review. IEEE Trans. Power Electron. (2010)
3. Fan, B.H., Zha, J.W., Wang, D., et al.: Size-dependent low-frequency dielectric properties in the $BaTiO_3$/poly (vinylidene fluoride) nanocomposite films. Appl. Phys. Lett. **100**(1), 012903 (2012)
4. Zha, J.W., Dang, Z.M., Li, W.K., et al.: Effect of micro-Si_3N_4-nano-Al_2O_3 co-filled particles on thermal conductivity, dielectric and mechanical properties of silicone rubber composites. IEEE Trans. Dielectr. Electr. Insul. **21**(4), 1989–1996 (2014)
5. Landry, C., Coltrain, B.K., Landry, M.R., et al.: Poly(vinyl acetate)/silica-filled materials: material properties of in situ vs fumed silica particles. Macromolecules **26**(14), 3702–3712 (1993)
6. Giannelis, E.P.: Polymer layered silicate nanocomposites. Adv. Mater. **8**(1), 1–9 (1996)
7. Lewis, T.J.: Interfaces: nanometric dielectrics. J. Phys. D Appl. Phys. **38**(2), 202 (2005)
8. Tanaka, T., Kozako, M., Fuse, N., et al.: Proposal of a multi-core model for polymer nanocomposite dielectrics. IEEE Trans. Dielectr. Electr. Insul. **12**(4), 669–681 (2005)

9. Heid, T., Fréchette, M., David, E.: Enhanced electrical and thermal performances of nanostructured epoxy/POSS composites. IEEE Trans. Dielectr. Electr. Insul. **23**(3), 1732–1742 (2016)
10. Valentini, L., Puglia, D., Carniato, F., et al.: Use of plasma fluorinated single-walled carbon nanotubes for the preparation of nanocomposites with epoxy matrix. Compos. Sci. Technol. **68**(3–4), 1008–1014 (2008)
11. Yin, Z., Sun, P., Sima, W., et al.: Synergistic enhancement of arc ablation resistance and mechanical properties of epoxy resin insulation. IEEE Trans. Dielectr. Electr. Insul. **27**(3), 748–756 (2020)
12. Jiyuan, Y.A.N., Liang, G., Hongliang, L., et al.: Effect of plasma step gradient modification on surface electrical properties of epoxy resin. Plasma Sci. Technol **23**(6), 064012 (2021)
13. Lei, W., Mochalin, V., Liu, D., et al.: Boron nitride colloidal solutions, ultralight aerogels and freestanding membranes through one-step exfoliation and functionalization. Nat. Commun. **6**, 8849 (2015)
14. Simmons, J., Tam, M.: Theory of isothermal currents and the direct determination of trap parameters in semiconductors and insulators containing arbitrary trap distributions. Phys. Rev. B **7**(8), 3706–3713 (1973)
15. Lv, F.C., Yin, K., Fu, K.X., et al.: Effect of fluorination and exfoliation filler on the surface flashover voltage of boron nitride/epoxy resin composites. High Volt. Eng. **43**(9), 8 (2017)
16. Xie, Q., Zhang, Y.J., Duan, Q.J., et al.: Effect of fluorinated modified graphene nanoplatelets on surface voltage resistance properties of epoxy resin composites. Proc. CSEE
17. Wang, Y.N., Lu, X., Yan, L.D., et al.: Effect of amino modified boron nitride nanosheets on surface insulation properties of epoxy resin. Insul. Mater. **54**(09), 48–54 (2021)
18. Zhang, S., Chen, W., Zhao, Y., et al.: Surface NH_2-functionalized by C doping of boron nitride nanotube to improve the thermal conductivity of epoxy composites. Compos. B Eng. **223**, 109106 (2021)
19. Li, J., Liang, H., Xiao, M., et al.: Mechanism of deep trap sites in epoxy/graphene nanocomposite using quantum chemical calculation. IEEE Trans. Dielectr. Electr. Insul. **26**(5), 1577–1580 (2019)

Research on the Influence of Nanoparticles and Surface Microstructure on the Hydrophobic and Electrical Properties of Silicone Rubber Materials

Peng Wang, Qizhi Chen, Jiaxuan Zhang, and Zinan Wang

1 Introduction

The concept of "fluorination pretreatment, nanoparticles, polyamide mesh micron-scale structure, and insulating substrate" is proposed in the design. The fluorinated nano-alumina/PA/SR composite material's surface was formed by connecting the fluorinated low surface energy nano-alumina to the surface of the polyamide mesh (PA) and fixing it to the surface of the silicone rubber material (SR). This paper tested and analyzed the surface properties of composite materials. A preparation method of superhydrophobic material with high flashover performance was determined by analyzing the properties of the samples before and after treatment from the aspects of monitoring the changes of functional groups by analyzing the chemical elements of the surface structure using surface morphology tests, x-photoelectron spectroscopy analysis, FTIR infrared spectroscopy analysis, and other means.

The influence of binary nano-structures on the surface characteristics of fluorinated nano-alumina/PA/SR composite materials is investigated in this work. Controlling the number of PA net regulations of the composite material surface micron structure, analyzing the different the composite material surface microstructure, surface roughness, hydrophobic properties, the change of the flash data characteristics, and further attempting to construct the surface flash model mechanism interpretation through surface potential attenuation and trap distribution.

The influence of nano-structure on the surface of fluorinated nano-alumina/PA/SR composites at the micron scale was investigated further. The nano-particle blending experiment was carried out by dissolving and resolidifying nano-fluoride Al_2O_3/

P. Wang (✉) · Q. Chen · J. Zhang · Z. Wang
School of Energy, Power and Mechanical Engineering, North China Electric Power University, Baoding 071000, China
e-mail: wang.peng.ncepu@foxmail.com

Q. Xie (ed.), *Electrical Materials*, Engineering Materials,
https://doi.org/10.1007/978-981-99-9050-4_5

MWCNT in the PA network structure. The effects of various mass fractions of nano-fluorinated Al_2O_3 and MWCNT on the hydrophobic characteristics of the composite surface were investigated, and a physical model was developed to attempt to explain the mechanism.

2 Preparation and Performance Characterization of Nano-alumina, Polyamide Mesh, Silicone Rubber Composites

This work attempted to build a macroscopic-micro dual-scale surface structure to achieve a composite material with both superhydrophobic and high flashover capabilities. According to research, the material surface morphology and composition have a significant impact on the flashover and hydrophobic properties of the material surface. As a result, this section was built by dissolving and resolidifying nano-alumina particles as filler and polyamide mesh as the skeleton of binary micro-nanostructures for composite surfaces. Based on this preparatory plan, this chapter will build on the discussion. Meanwhile, the surface characteristics of the composite were characterized using a scanning electron microscope surface morphology test, an X-ray photoelectron spectroscopy test, an FTIR Fourier infrared spectroscopy test, and the hydrophobic properties of the flashover properties, as well as the effect of modified nanoparticles on the surface physical and chemical properties of polyamide mesh/silicone rubber composite.

2.1 Preparation of Modified Nano-alumina, Polyamide Mesh, Silicone Rubber Composites

(1) Preparation of the experimental materials

The materials prepared for the experiment included: polyamide mesh (40 mesh, 100 mesh, 160 mesh, 300 mesh, 400 mesh, 400 mesh, 500 mesh, and 600 mesh, purchased from Jiangsu Honghao Silk Grid Co., Ltd.); silicone rubber (adhesive with back) purchased from Shanghai Taozhe Products Co., Ltd.; 1H, 1H, 2H, 2H-perfluorinatedtriethoxlosilane (C8F13H4Si(OCH2CH3)3, FAS) were purchased from Aladin reagent (Shanghai) Co., Ltd.; formic acid (85%), ethanol (99%), and acetone (99%) were purchased from Tianjin Comio Chemical Reagent Co., Ltd.; Al_2O_3 nanoparticles (about 30 nm in diameter, China Research Institute of Metal Metallurgy) and carbon nanotubes (TNM 7, Chengdu Organic Chemistry Co., LTD.) as nanofillers. All chemicals were analytical-grade reagents.

(2) Fluorination of the nanoparticles

Firstly, acetone and FAS mixed solutions (25 (wt)% FAS and 1 (wt)% Al_2O_3) were produced. The nano-alumina particles were added to the mixing solution and well stirred to ensure homogeneity. The modified superhydrophobic powder suspension was achieved after 16 h of 50 °C reaction. The upper solvent was removed after the suspension had been allowed to layer up and down. To obtain the superhydrophobic powder, the fluid was dried in an 80 °C drying oven for 24 h.

(3) Experimental preparation

A mixture of 20 g of formic acid and ethanol was prepared with a mass ratio of 3.5:6.5. After stirring for 10 min until uniform, modified nano-alumina powder and carbon nanotubes were added to the mixed solution. The mass percentage of the modified nano-alumina, carbon nanotube mass, and the mixed solution of formic acid and ethanol were defined as, respectively, γAl_2O_3 and γCNT. Before use, the suspension is stirred further for 60 min to uniformity.

Place the polyamide mesh in a sealed container with the suspension to ensure that it is completely immersed in the suspension. After sealing the container, set it in a drying oven at 60 °C for 1 h. Finally, the polyamide mesh was removed and allowed to air-dried. The air-dried polyamide mesh is directly adhered to the surface using the silicone rubber product's own glue (Fig. 1).

In the experiment, 40, 100, 160, 300, 400, 500, and 600 mesh and γAl_2O_3 = 5% were prepared, and various preparation degrees were prepared in the experimental procedure for further micron-level experimental examination. The surface of

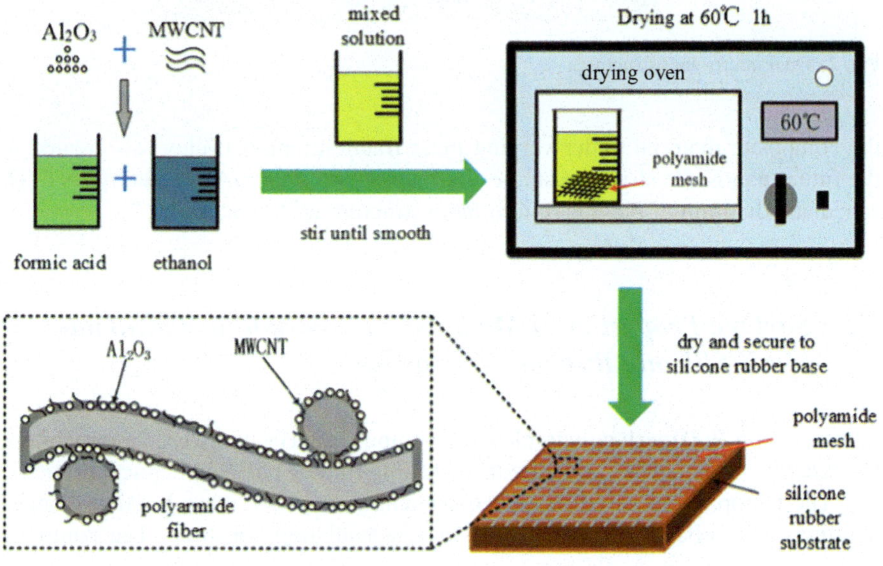

Fig. 1 Schematic diagram of the experimental sample preparation process

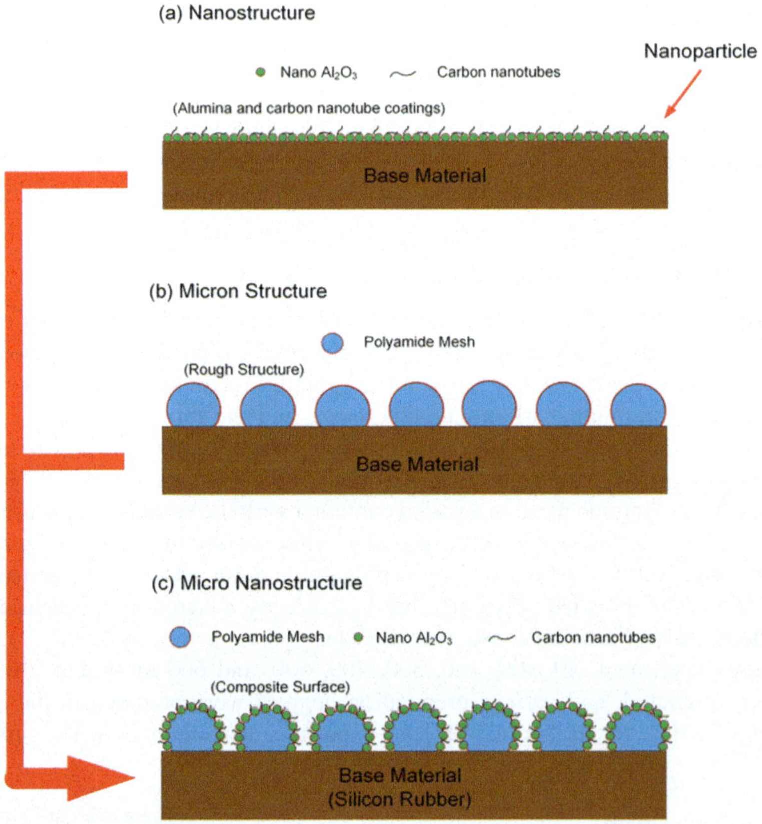

Fig. 2 Dual micro-nanostructures

the composite material under varying mass fraction nanoparticles was created for the future nanoscale structure studies by manipulating the γAl_2O_3 and γCNT. The schematic diagram of the dual micro-nano structure is shown in Fig. 2.

2.2 *Surface Properties of Modified Nano-alumina/Polyamide Mesh/Silicone Rubber Composites*

In this section, 300 mesh polyamide mesh composite surfaces are used as an example for experimental material fabrication and testing for key physicochemical properties [1]. The nanoparticles used in the experiment were a 5% mass fraction of nano-alumina, with silicone rubber serving as the insulating substrate. The composite

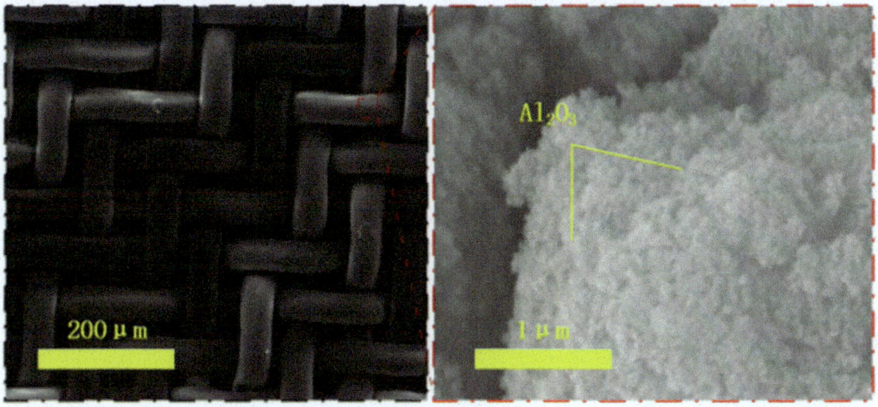

Fig. 3 Surface structure of 300 mesh polyamide mesh composite

surface's surface morphology, X-ray photoelectron, FTIR Fourier infrared spectroscopy, hydrophobic characteristics, and surface flashover qualities were also investigated.

(1) Surface morphology test

Figure 3 depicts the surface morphology of the polyamide mesh/silicone rubber composite surface. It depicts the electron microstructure of a sample of 300-mesh polyamide mesh composite. Low-power electron microscopy can reveal the interwoven structure of the polyamide mesh wire. Surface-attached aggregate nano-alumina particles can be seen using high-power electron microscopy.

(2) X-ray optoelectronic test

As shown in Fig. 4, the surface of the composite material was examined for chemical components, and the treated samples were evaluated using the XPS equipment. When dissolved and resolidified, a prominent peak develops at the binding energy 74.49 eV, which corresponds to the distinctive peak of Al 2p, which originates from the nano-alumina particles on the surface of the composite material, as shown in Fig. 4a. We achieve the outcomes depicted in Fig. 4b. The C1s on the sample surface became multimodal after surface fluorination, revealing four peaks at 284.50, 285.49, 286.94, and 291.32 eV, corresponding to the typical peaks of –CH, C–O, C–CF, and –CFx, respectively. This indicates that after FAS fluoridation, element F is successfully grafted on the surface of nanoparticles on the surface of composite materials. It is worth noting that C–CF and –CFx are crucial for providing low surface energy.

(3) The FTIR infrared spectroscopy test

The Fourier spectra of the several experimental control groups are shown in Fig. 5. We attempt to investigate the changes and effects of the composite surface during the experimental treatment by comparing the Fourier spectra of the composite surface in different experimental treatment phases. The three curves are the composite surface of

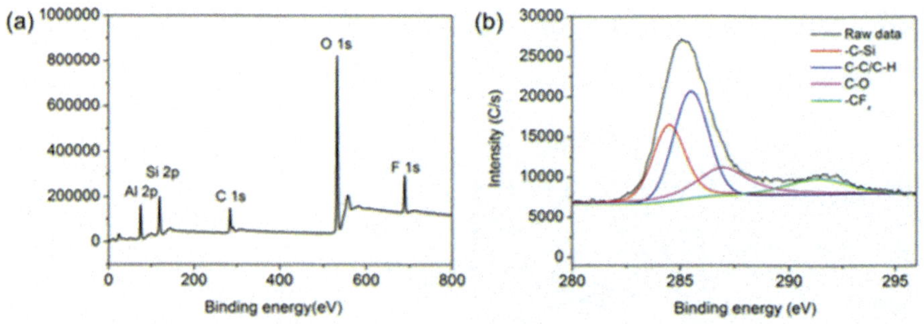

Fig. 4 An X-ray photoelectron test map of the experimental samples

untreated polyamide mesh and silicone rubber, the composite surface of polyamide mesh formed with silicone rubber that has been soaked in formic acid, and the composite surface of polyamide mesh and silicone rubber that has been dissolved and solidified after being soaked in formic acid. Experimental different samples prepared in the process of processing Fourier spectrum approximate, namely formic acid etching polyamide mesh and nano-alumina by dissolving solidification technology attached in the process of polyamide mesh, functional group basic no change, the two processes are given priority with physical change.

Fig. 5 The X-ray photoelectron test map of the experimental samples

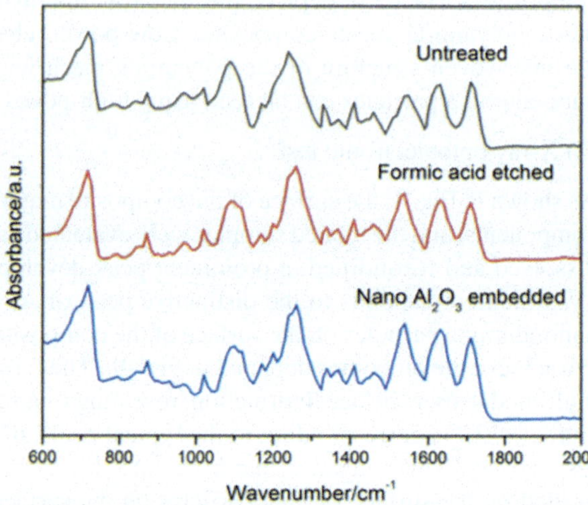

2.3 *Electrical Characteristics of Modified Nano-alumina/Polyamide Mesh/Silicone Rubber Composites*

Figure 6 depicts the surface hydrophobicity test results. The hydrophobic Angle of the composite surface increased from 105 to 152 after the experiment treatment, indicating that the superhydrophobic function of the composite surface can be effectively realized by constructing the micro-nano dual-scale surface topography and reducing the surface energy of the composite surface with fluorinated fillers.

The test results of flashover characteristics along the surface of silicone rubber material before and after treatment are given in Fig. 6. The flashover voltage along the surface of the composite material was increased by about 20% after the experimental treatment, demonstrating that the flashover performance could be improved to some extent by modifying the surface topography of the composite material and doping nanoparticles.

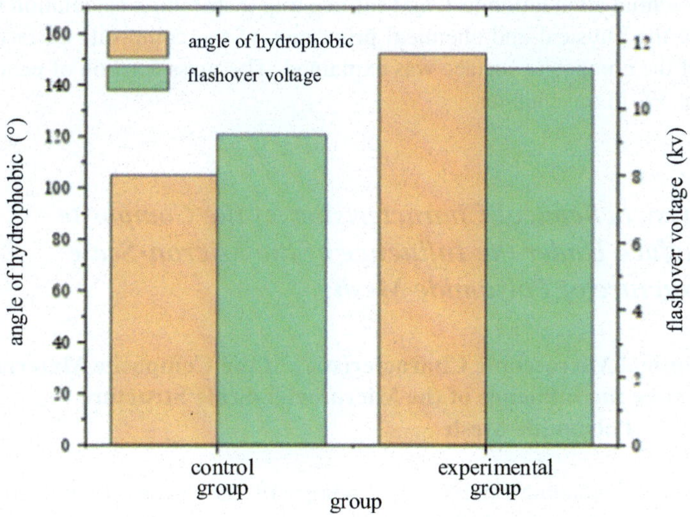

Fig. 6 The changes of hydrophobic characteristics and flashover characteristics before and after treatment of composite samples

3 Study on the Influence of the Micrometer-Scale Structure of the Polyamide Mesh on the Hydrophobic and Electrical Properties of the Composites

The experimental preparation idea of "small molecule nanostructure (nano-Al_2O_3 particles) + micron structure of polyamide mesh (300-mesh polyamide mesh) + insulating substrate (silicone rubber)" was preliminarily constructed in previous experimental studies, and the realization of the superhydrophobic properties of the composite surface and the improvement of the flashover properties along the surface were preliminarily proven. The surface characteristics of composite materials are explored using polyamide mesh of various mesh numbers based on the experimental investigation in the preceding chapter. The optical microscope observation test, water contact Angle test, and flashover performance test were performed on polyamide mesh/silicone rubber composite surfaces of 40 mesh, 100 mesh, 160 mesh, 300 mesh, 400 mesh, 500 mesh, and 600 mesh. Surface profile test, roughness test, surface potential attenuation test, and surface trap distribution calculation were used to analyze the physical and chemical properties of the composite surface, and the process of the composite surface was explained. The mass fraction of nano-alumina remains at 5% in this chapter.

3.1 Physicochemical Characteristics of the Composite Surface Under the Influence of the Micron-Scale Structure of Polyamide Mesh

3.1.1 Optical Microscopic Characteristics of the Composite Materials Under the Influence of the Micrometer-Scale Structure of the Polyamide Mesh

The optical microscopic observation findings of the surface of polyamide mesh composites with 40 mesh, 100 mesh, 160 mesh, 200 mesh, 300 mesh, 400 mesh, 500 mesh, and 600 mesh are presented in Fig. 7a–g, respectively. The diameter and spacing of 40 mesh polyamide mesh are the greatest among them. The mesh diameter and mesh spacing of the polyamide mesh reduced continually during the process from 40 to 600 mesh. Further measurement and analysis can yield the changing table of mesh diameter and mesh spacing of polyamide mesh with the number of mesh sizes displayed in Table 1.

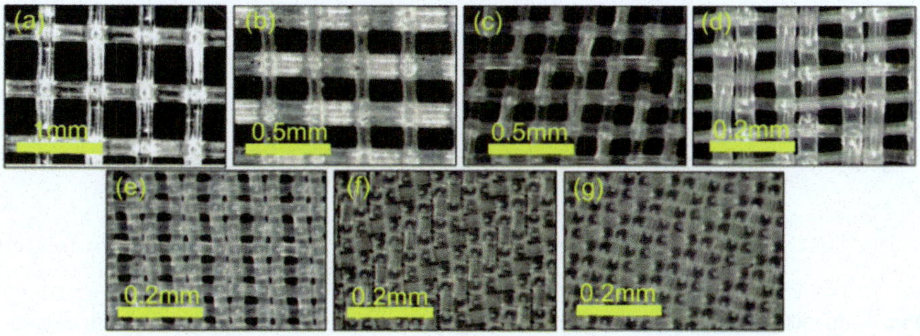

Fig. 7 40–600 polyamide mesh/silicone rubber composite test samples

Table 1 Table of diameter/spacing variation of 40–600 mesh polyamide mesh wire

Serial number	Number of grids	Average mesh diameter (μm)	Average wire spacing (μm)	Area fraction
1	40	195	410	0.566
2	100	88	171	0.654
3	160	70	113	0.679
4	300	48	52	0.815
5	400	35	30	0.890
6	500	–	–	–
7	600	–	–	–

3.1.2 Surface Profile Characteristics of the Composite Materials Under the Influence of the Micron-Scale Structure of the Polyamide Mesh

The surface profile of a 40–600 mesh polyamide mesh composite surface is evaluated to examine the surface morphology of composites qualitatively and quantitatively, and the two-dimensional and three-dimensional test results are presented in Figs. 8 and 9, respectively.

The surface roughness of the material surface is frequently expressed in experimental study by Sa (arithmetic mean height), and the basic calculation formula is stated in Eq. (1). Where A represents the computed surface roughness area, Z indicates the material's surface height, and x and y represent the material's position in the horizontal direction, respectively.

$$Sa = \frac{1}{A} \iint_A |Z(x, y)| dxdy \tag{1}$$

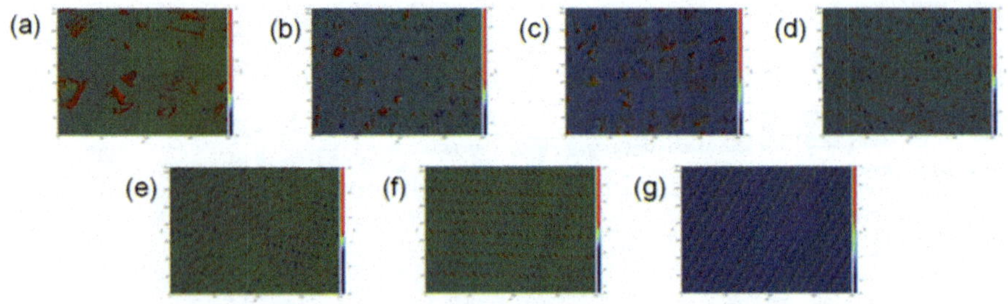

Fig. 8 40–600 2D surface profile of a mesh polyamide mesh/silicone rubber composite sample

The surface profiler can present the surface structure of polyamide mesh composite, as illustrated in Figs. 8 and 9, and the specific value of surface roughness of polyamide mesh composite is shown in Fig. 9h. The composite surface of 40-mesh polyamide mesh has the coarsest filament diameter and the greatest filament distance, while the material surface has the roughest and highest roughness of 9.8 μm. Simultaneously, as the number of polyamide meshes increases, the diameter of the mesh lowers, the distance of the mesh decreases, and the surface of the composite gradually smooths. The surface roughness of the composite is reduced to 6.8 μm when the number of polyamide meshes reaches 600. The results of the composite surface surface profile reveal that managing the composite surface morphology quantitatively may make the number of the composite surface simple and effective.

3.1.3 Water Contact Angle Characteristics of Composites Under the Influence of Micrometer Structure of Polyamide Mesh

Multiple linkages are used in the experiment to prepare the composite surface. In this paper, the contact angle of the polyamide mesh composite surface was measured for each production stage to clearly and quantitatively analyze the influence of the composite material's surface hydrophobicity [2]. "B-SR" is the blank silicone rubber test group; "B-PA" is the composite surface of the untreated polyamide mesh and silicone rubber substrate; "B-Al_2O_3" is the composite surface of the polyamide mesh and re-solidified; and "F-Al_2O_3" is the composite surface of fluorinated graft before "B-Al_2O_3" is dissolved and solidified. The following figure depicts the test results.

Among them, "hydrophobic crucial" is a 90° contact angle, a contact angle less than 90° is called hydrophilic material, a contact angle greater than 90° is called hydrophobic material, and a contact angle greater than 150° is called superhydrophobic material.

The silicone rubber substance itself has a strong hydrophobicity, according to the test, and its hydrophobic angle has been regularly recorded around 105° (Fig. 10). Although the polyamide mesh/silicone rubber composite material has a complex rough structure, the improvement in hydrophobic performance due to the better hydrophobic silicone rubber material is not visible. The polyamide mesh/silicone

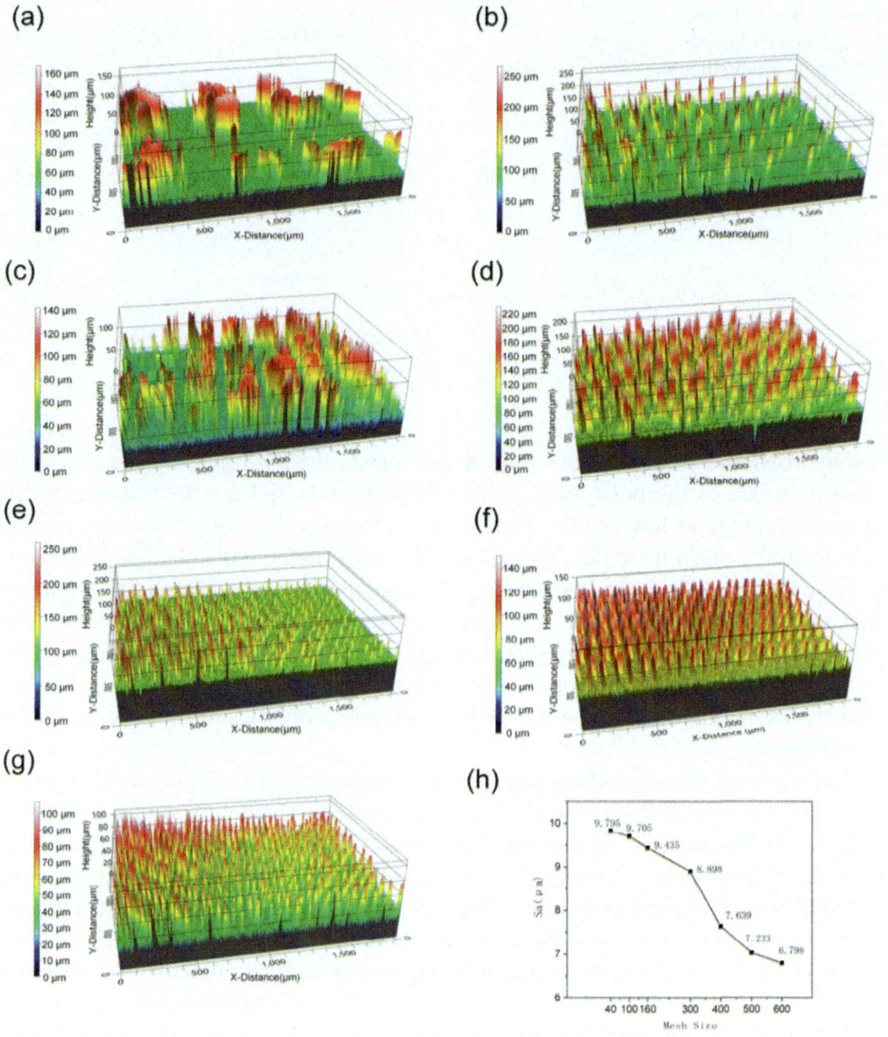

Fig. 9 **a–g** 40–600 mesh polyamide mesh/silicone rubber composite **h** surface roughness of 40–600 mesh polyamide/silicone rubber composite

rubber composite offers the best hydrophobic characteristics at 300 mesh. The hydrophobic performance of the 40 mesh polyamide mesh/silicone rubber composite is slightly worse, which is due to the rough texture of the composite material's surface. Because the 40-mesh polyamide mesh spacing is too widely apart and the mesh diameter is too high, the surface roughness of the composite material is the greatest at this time. At 400 mesh, the hydrophobic properties of the polyamide mesh composite material decrease sharply, and the hydrophobicity is lower than that of ordinary silicone rubber material; at 500 mesh, the composite material surface loses

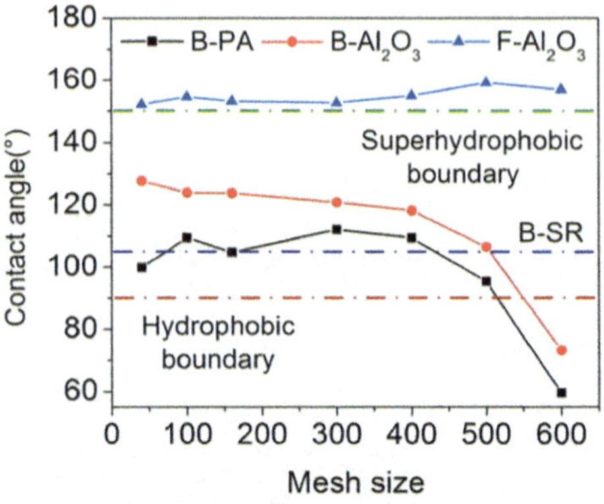

Fig. 10 40–600 water contact angle of mesh polyamide mesh/silicone rubber composite samples

the hydrophobic characteristics; and at 600 mesh, the hydrophobic characteristics on the surface of the polyamide mesh composite material are the lowest, and the contact Angle is as low as 60°. This is due to the fact that when the order number of polyamide mesh increases, the polyamide mesh filament distance decreases, the polyamide mesh dominates the surface of the composite material, and the polyamide mesh is overall relatively smooth. This experiment examined the surface hydrophobic properties of the polyamide mesh/silicone rubber composite to determine the effect of the micron-scale structure on the hydrophobic properties of the composite surface, which was not affected by the higher hydrophobic properties of the silicone rubber material itself.

Furthermore, the nano-alumina particles were connected to the polyamide mesh by using the recoagulation technique on the polyamide mesh/silicone rubber composite surface [3]. The experimental group with the addition of nano-alumina particles had a trend of decreasing contact angle with increasing polyamide mesh number, and the total contact angle was similar to the polyamide mesh/silicone rubber composite. However, the nanostructure of the nanoparticles improves the hydrophobic properties of the composite surface, resulting in an increasing contact angle of the composite surface following the introduction of the nanoparticles, and this promotion is more visible when the number of polyamide mesh is lower. The contact angle of the 40-mesh composite surface is the greatest at this point, reaching about 130°. However, at 400–600 mesh, the hydrophobic characteristics of the composite material will still fall dramatically. The contact angle of the composite surface remains at 106° for 500 mesh; at 600 mesh, the contact angle of the composite surface lowers to 70°. In this experiment, the binary micro-nanostructures significantly improved the hydrophobic characteristics of the composite surface, demonstrating that the binary micro-nanostructures can improve the hydrophobic characteristics of the composite and that the regulation can be adjusted by changing the number of polyamide mesh.

The surface of superhydrophobic materials should be further prepared, according to the experimental concept. The qualitative shift of the composite from hydrophobic to superhydrophobic is obtained by grafting the nano alumina particles and lowering the surface energy of the composite. The nanoparticles were fluorinated in this work using nano alumina particles/polyamide mesh/silicone rubber composites. The findings of the experimental tests demonstrate that the nanoparticles in the fluorinated composite have outstanding superhydrophobic characteristics. As shown in Fig. 10, after fluoride of nanoparticles, the number of polyamide mesh in 40–600 attained superhydrophobic function and displayed a shifting pattern with increasing polyamide mesh number and composite material surface contact angle. The contact Angle of the 40-mesh composite surface is the lowest, but still more than 150°, reaching 152°. The hydrophobic Angle of the 500–600 mesh composite surface is reaching 156°. The mesh spacing and mesh diameter become smaller as the number of polyamide meshes increases, giving the high-order polyamide mesh a higher specific surface area and more opportunities for fluorinated nanoparticles to adhere. The addition of more low surface energy nanoparticles improves the composite's hydrophobic characteristics [4, 5]. Based on the previous chapter's experimental concept of "small molecule nanostructure (nano-Al_2O_3 particles) + polyamide mesh micron structure (300 mesh polyamide mesh) + insulating substrate (silicone rubber)," the influence of binary micro-nanostructure composed of fluorine nanoparticles on the hydrophobic characteristics of the composite surface was investigated in this experiment.

3.2 *Electrical Characteristics of the Composite Surface Under the Influence of the Micron-Scale Structure of Polyamide Mesh*

3.2.1 Flashover Properties of Composites Under the Influence of Micrometer Structure of Polyamide Mesh

In this section, several techniques are involved in the surface preparation of polyamide mesh composites. The flashover characteristics of different sets of control groups were evaluated to conduct out analysis in order to dynamically monitor the influence of each process on the flashover characteristics of composite materials. "B-SR" was the untreated polyamide mesh and silicone rubber substrate composite surface; "B-PA" was the blank silicone rubber test group. After dissolving and resolidification of untreated nano-alumina, "B-Al_2O_3" is the composite surface of polyamide network and silicone rubber substrate; "F-Al_2O_3" was fluorinated grafted, and then B-Al_2O_3 was dissolved and coagulated on the composite surface of nano-alumina particles. The outcomes are as follows (Fig. 11).

The "B-SR" was the first to be tested in the experiment. The average flashover voltage after ten flashover tests was 9.2 kV. The flashover voltage on the surface of the "B-PA" polyamide mesh/silicone rubber composite is somewhat greater than that of the blank silicone rubber, fluctuating at around 9.5 kV. The flashover voltage increased with the number of polyamide mesh, slightly increasing and then gradually reducing, with the flashover voltage of 300 mesh polyamide mesh composite surface being the highest, reaching 9.8 kV.

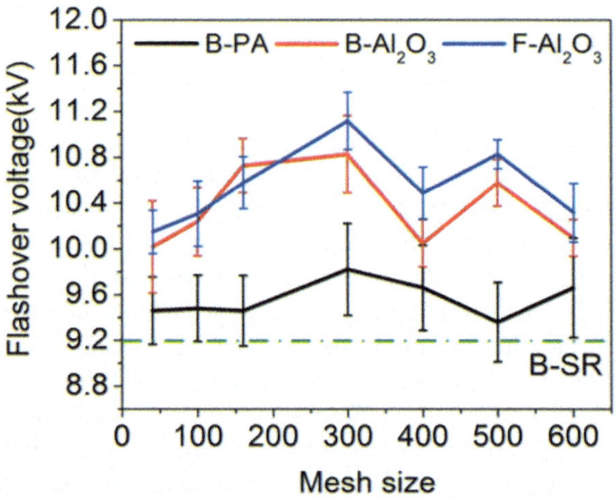

Fig. 11 Flashover properties of the 40–600 mesh polyamide mesh/silicone rubber composite samples

Non-conductive nanoparticles are commonly employed as fillers to increase the flashover voltage [6]. As a result, the surface flashover voltage of the composite material is dissolved and swiftly and dramatically raised after solidification during the composite material's surface treatment. The surface flashover voltage of the material corresponding to each mesh exceeded 10 kV in the test, the flashover voltage of the 300 mesh polyamide mesh remained the highest, reaching nearly 11 kV, and the flashover characteristic increased by more than 20% when compared to the untreated silicone rubber material. The overall variation pattern of the "B-Al_2O_3" experimental group's flashover voltage was comparable to that of the "B-PA," which displayed a shifting tendency of rising initially and then lowering. Simultaneously, the flashover voltage of different-order composite material surfaces changes more dramatically, which may be related to the alumina material's dissolution and resolidification efficiency and the change in discharge distance along the surface caused by the material's surface roughness.

Following fluorination of the composite material, the flashover voltage on the composite surface is increased once more. The overall flashover property of the test material continued to change with the number of polyamide mesh, first rising and then falling, but the increase in flashover voltage in the low-number test group was not obvious, whereas the increase in flashover voltage in the high-number polyamide mesh test group was more obvious. The flashover voltage of the composite surface of the 40 mesh polyamide mesh was 10.2 kV, whereas the mesh 300 experimental group's flashover value was 11.2 kV. This could be related to the fluoridation-induced trap modification and the improved attachment efficiency of nanoparticles by high order number polyamide mesh.

3.2.2 Surface Potential Attenuation Characteristics of the Composites Under the Influence of the Micrometer-Scale Structure of the Polyamide Mesh

This research continues to investigate the surface charge decay properties of the composite material before and after nanoparticle incubation using the surface charge decay test platform, which corresponds to the order polyamide mesh. "B-100," "B-300," and "B-600" correspond to the experimental groups of 100 mesh polyamide mesh, 300 mesh polyamide mesh, and 600 mesh polyamide mesh, respectively; "F-100," "F-300," and "F-600" correspond to the fluorinated group of nano-alumina particles of 100 mesh polyamide mesh, 300 mesh polyamide mesh, and 600 mesh polyamide mesh, respectively.

The charge dissipation rate in this study is defined as the ratio of the potential difference between the beginning and end, and the potential after the withdrawal (formula 2 and 3).

$$V = \frac{U_{(o)} - U_{(t)}}{U_{(o)}} \times 100\% \tag{2}$$

$$L = \frac{U_{(to)}}{U_{(o)}} \tag{3}$$

In formula (2), V represents the surface charge dissipation rate, $U_{(0)}$ represents the test zero surface charge, and $U_{(t)}$ represents the material's surface charge at the end of the test. The surface charge normalization results are examined here to emphasize the dissipation rate of the surface charges. At each time, the normalization is the ratio of the surface potential to the initial potential (such as formula 3).

When illustrated in Fig. 12, the surface charge decay of the nonfluorinated group ("B-100," "B-300," and "B-600") and the fluorinated group ("F-100," "F-300," "F-600") accelerates as the polyamide mesh number increases. When the surface charge stabilizes, the amount of residual charge decreases. However, the proportion of surface charge decreasing at the same moment is growing. At the same time, the surface potential decay of the fluorinated composite is faster than the surface potential decay of the nonfluorinated composite (e.g., "B-100" is faster compared to "F-100"). This could be due to the trapping effects of fluoride of nanoparticles on the composite surface.

3.3 Mechanism Analysis of Composite Surface Affected by the Micron Structure of Polyamide Mesh

This work primarily employs two ways of creating binary micro-nano surface structure and dissolving and resolidification of nanoparticles for the surface preparation of polyamide mesh composite. The surface structure of composite materials with binary

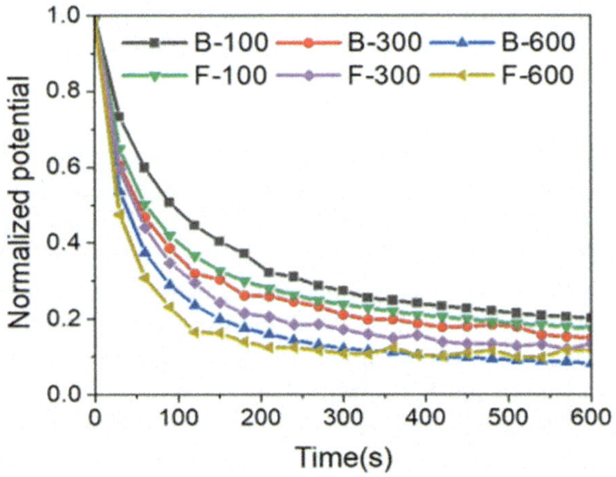

Fig. 12 Surface charge dissipation curve of 100–600 mesh polyamide mesh/silicone rubber composite samples

micro-nano structure has been tested in this paper. On the one hand, the optical fiber structure was used to measure the wire diameter and wire distance of polyamide mesh with varied mesh numbers. The results reveal that a lower polyamide mesh has a greater wire diameter and a longer wire distance. The surface profile test, on the other hand, assessed the surface roughness of the composite surface as it varied with the mesh number of polyamide mesh. The results reveal that a polyamide mesh with a low mesh number has a higher roughness, which corresponds to the optical microscopic observations. On the one hand, the length of the discharge channel along the composite surface flashover is lengthened. On the other hand, significant electric field distortion occurs. The surface trap calculation approach is utilized in this section to examine the effect of nanoparticle introduction on the surface of composite materials throughout the experimental preparation phase. In addition, try to develop a model to describe the mechanism of synergy.

3.3.1 Polyamide Mesh Structure's Influence on Surface Trap Distribution in Composites

The distribution of trap energy levels is determined using Eqs. (4) and (5) based on the surface potential decay curve and the isothermal surface potential decay theory (Isothermal Surface Potential Decay, ISPD) (5).

$$E_{\mathrm{t}} = k_B T \ln(V_{ATE} t) \tag{4}$$

$$Q_{\mathrm{s}}(t) = t \frac{\varepsilon_0 \varepsilon_t}{q_e L} \frac{d\varphi_s(t)}{dt} \tag{5}$$

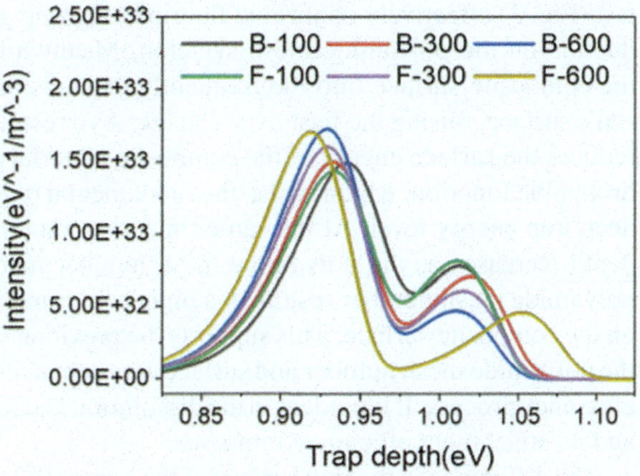

Fig. 13 Surface trap distribution curve of 100–600 mesh polyamide mesh/silicone rubber composite samples

Et is the trap energy level; Qs is the trap charge density; k_B is the Boltzmann constant; T is the temperature of the sample; νATE is the escape frequency of the trapped charge (taking 4.17 × 1013 s^{-1}); L is the thickness of the sample; $\varphi_s(t)$ is the surface potential.

Figure 13 depicts the composite surface's trap level study results. The deep trap energy levels and shallow trap energy levels on the surface of 100 mesh, 300 mesh, and 600 mesh polyamide mesh composites did not change appreciably in the non-fluorinated group. Deep trap energy levels are all about 1.02 eV; shallow trap energy levels decreased from 0.942 eV for 100 mesh to 0.928 eV for 600 mesh. The trap level of the surface of the polyamide mesh composite in different orders changed dramatically in the fluoride experiment group. The deep trap level of the polyamide mesh composite increases dramatically after fluoride, from 1.01 eV for 100 mesh to 1.053 eV for 600 mesh; the shallow trap level decreases marginally, from 0.934 eV for 100 mesh to 0.918 eV for 600 mesh.

It has been proposed that the change in trap depth on the composite surface is connected to the chemical bonding on the surface. The FTIR test in the material preparation process shows that there is no obvious peak change in the Fourier spectra in the process of soaking in formic acid mixed solution and in the process of dissolving and resolidification of nano-alumina particles, so physical changes are primarily used in these two processing processes. As a result, even if the number of polyamide meshes changes, the change in specific surface area caused by thinning the polyamide mesh and shorter wire distances only influences the dissolution efficiency of the polydactyl mesh and the efficiency of attaching the nano-alumina particles. It will not alter the surface chemical bond of the composite material. This results in a negligible change in trap depth on the composite surface.

On the one hand, the fluorinated group had a higher deep trap depth and a lower shallow trap density. On the other hand, there is a decrease in deep trap density as well as an increase in shallow trap density. The previous X-ray photoelectron spectroscopy test (Fig. 3) effectively confirmed fluorinated group grafting and nanoparticle introduction on the polyamide mesh skeleton. Meanwhile, past research indicates that the composite surface fluoride frequently expands the deep trap level on the material's surface, raising the flashover voltage. As a result, the fluorine-containing group reduces the surface energy of the composite material and accomplishes the superhydrophobic function, which is also the fundamental reason for the material's increased deep trap energy level. At the same time, the composite material surface deep trap depth increases as the polyamide mesh number increases, implying that a greater polyamide mesh number results in a higher fluorinated nanoparticle attachment rate on the composite surface. This supports the previously stated explanation: increasing the polyamide mesh number and surface area enhances each process in the processing efficiency process. It is evident in the dissolution and resolidification of nanoparticles, and its attachment efficiency improves.

The difference in the trap surface of the composite material of the polyamide mesh was primarily reflected in the trap density in the experimental test, as illustrated in Fig. 14. "B-D" is the non-fluoridated experimental group's deep trap density; "F-D" is the fluoridated experimental group's deep trap density; "B-S" is the non-fluoridated experimental group's shallow trap density; and "F-S" is the fluoridated experimental group's shallow trap density. The deep trap density of 100, 300, and 600 mesh polyamide mesh composites decreased successively in the non-fluoridated experimental group, with values of 8.31, 6.02, and 4.54 (all units are 10^{32} eV^{-1}/m^{-3}); the shallow trap density increased successively, with values of 1.44, 1.60, and 1.71 (all units are 10^{32} eV^{-1}/m^{-3}). The deep trap density of 100, 300, and 600 mesh composite reduced gradually, with values of 7.76, 6.93, and 4.71 (all units are 10^{32} eV^{-1}/m^{-3}); the shallow trap density grew successively, with values of 1.49, 1.51, and 1.73 (all units are 10^{32} eV^{-1}/m^{-3}). The trap density of the composite surface changed significantly during the experiment due to the change in polyamide mesh number.

The shallow trap density of the composite surface grew as the number of polyamide mesh increased, but the deep trap density fell dramatically. Previous research has found that shallow traps on the surface of composite materials are more affected by surface physical imperfections, while deep traps are frequently related to chemical defects. Two major adjustments are examined in this experiment: (1) On the one hand, when the number of polyamide meshes grows, the larger specific surface area contacts the polyamide more fully during the dissolution soaking process, generating more physical defects due to the increase in the density of shallow traps; (2) There were some researches found that the interfacial layer on the surface of the composite material is more prone to form deep traps. The nanofillers are partially incorporated in the polyamide fabric in this paper using a dissolving and resolidification technique. The interfacial layer of the nanoparticle polyamide mesh is subsequently created, introducing deep traps to the composite material's surface. When the polyamide mesh number is increased, the specific surface area increases, giving the nanoparticles more possibilities to attach. As a result, there is a possible tendency toward greater deep trap density in the composite surface. However, the solution concentration and material processing time continue to limit the embedding of the nanofillers. The

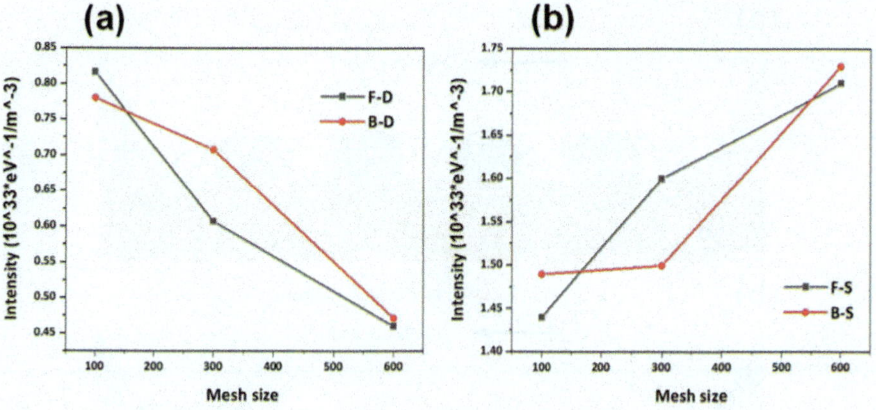

Fig. 14 Deep-depth trap density of 100–600 mesh polyamide mesh/silicone rubber composite samples

findings of the experimental tests demonstrate that the polyamide mesh is physically dissolved in the formic acid solution, and the process of generating a shallow trap takes precedence, resulting in an increase in shallow trap density and a decrease in deep trap density.

3.3.2 Flashover Model of Composite Material Under the Influence of the Micron-Scale Structure of Polyamide Mesh

To attempt to explain changes in the flashover properties of composite surfaces, a physical model is constructed (Fig. 15). The network structure of polyacrylamide is important at the macroscopic level. According to the roughness statistics (Fig. 15), decreasing the grid size causes an increase in roughness. The effect of roughness on material flashover is separated into two categories. On the one hand, high roughness causes lengthy discharge lengths along the surface, preventing the flashover from developing further and increasing the flashover voltage. Increased roughness, on the other hand, creates severe local electric field distortion, which causes a flashover along the surface and reduces the flashover voltage on the composite surface. As a result, the composite demonstrated the best surface flashover performance at the required roughness.

At the same time, the effect of nanoscale fillers (Al_2O_3 nanoparticles and carbon nanotubes) and surface modification is critical from a microscopic standpoint. Because they improve the thermal stability, mechanical characteristics, and corona resistance of composites, Al_2O_3 nanoparticles are frequently used in industrial applications [7, 8]. Further research revealed that a large number of unpaired atoms and flaws on the surface of the nano-packing can trap carriers and limit charge transport in the nanoscale polymer. The inclusion of CNTs raises surface conductivity. A

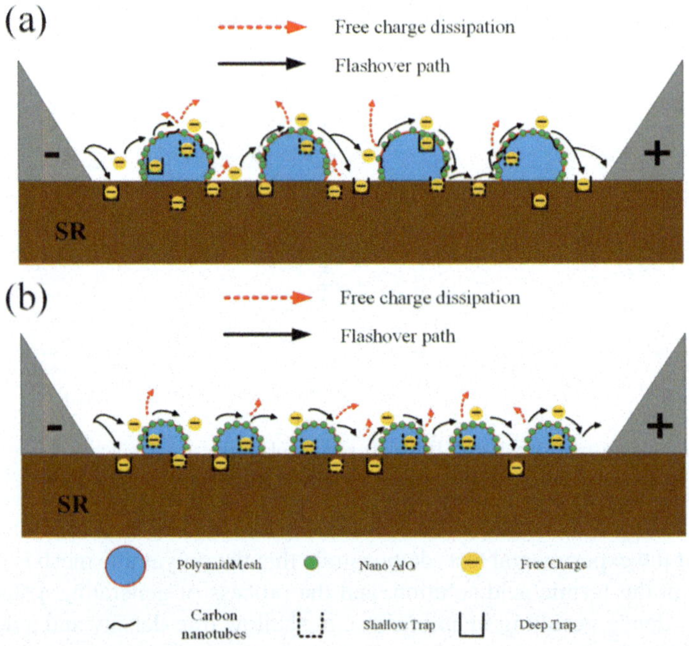

Fig. 15 Physical model of polyamide mesh/silicone rubber composites with different mesh sizes

proper increase in surface conductivity aids in the acceleration of charge dissipation, which in turn increases the flashover voltage. Previous research has demonstrated that correct fluoride adjustment can reduce surface charge accumulation and boost flashover tolerance voltage.

4 Study on the Influence of Nanostructures on the Surface Properties of Composite Materials Under the Mixed Modification of Nano Alumina and Carbon Nanotubes

Based on the preceding section of this chapter, the composite surface will be attempted to increase further electrical performance based on the ideal 300 mesh polyamide meshsilicone rubber composite surface. As a result, carbon nanotubes (CNTs) were injected as a modified filler during the dissolving and resolidification process to combine carbon nanotubes with nano-alumina particles. The water contact Angle test, flashover performance test, surface potential attenuation test, and surface trap distribution calculation of the composite surface were carried out by modifying the blending ratio of the two nano-fillers, and the method was attempted to explain.

In the experiment, 300 mesh polyamide mesh, fluorinated modified nano-alumina particles, and a silica rubber substrate were used. Nanoparticle doping studies were carried out with alumina particles of 0%, 2%, 4%, 4%, 6%, 8%, and 10% mass fraction and carbon nanotubes of 0%, 0.05%, 0.1%, 0.15%, and 0.2% mass fraction, respectively. In the experiment, 300 mesh polyamide mesh, fluorinated modified nano-alumina particles, and a silica rubber substrate were used. Nanoparticle doping studies were carried out with alumina particles of 0%, 2%, 4%, 4%, 6%, 8%, and 10% mass fraction and carbon nanotubes of 0%, 0.05%, 0.1%, 0.15%, and 0.2% mass fraction, respectively.

4.1 Effect of Nanostructures on the Hydrophobic Properties of Composite Surfaces

Figure 16 depicts the results of the hydrophobic properties test on nanoparticles. The doping of fluorinated nano alumina particles significantly affects the surface hydrophobic properties of the composite material, although the influence on the varied doping of carbon nanotubes is not obvious. The experimental group of nonfluorinated nano-alumina particles demonstrated the worst hydrophobic qualities of the composite surface, with a contact angle of roughly 120°, corresponding to the experimental group of mass fraction of each carbon nanotube. The surface hydrophobic characteristics of the composite material increased fast after the addition of partly fluorinated nano-alumina particles to the surface, surpassing 150°. When the fluorinated nano-alumina particles' mass fraction exceeds 4%, the surface of the composite material becomes stable and exhibits a superhydrophobic function.

The contact Angle of the composite material's surface varies depending on the type of nanoparticle. On the one hand, the doping amount of carbon nanotubes is thought to be modest, resulting in no substantial changes in the hydrophobic qualities

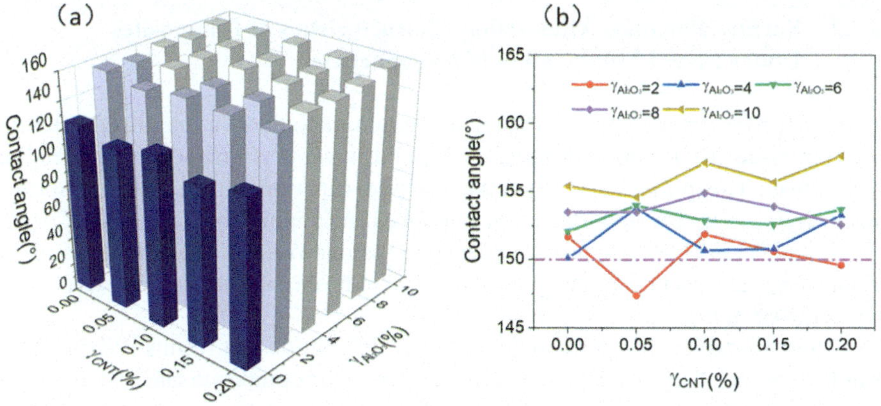

Fig. 16 Water contact angle characteristics of composite surface for nano-Al_2O_3 and MWCNT blend doping

of the material surface. Conversely, increasing the mass fraction of fluorinated nano-alumina enhances the ultimate adhesion amount of nano-alumina particles during the dissolving and resolidification process. The bigger the amount of nano alumina attachment, the lower the surface energy of the composite. As a result, for superior hydrophobic qualities, the composite surface with a greater mass fraction of fluorinated nano alumina is used, and can stably display superhydrophobic capabilities once the mass fraction exceeds 4%.

4.2 Effect of Nanostructures on the Electrical Characteristics of Composite Surfaces

4.2.1 Surface Flashover Properties of Composites Modified by Nanoparticle Blending

(1) Flashover test results of nanoparticle blend composite surface

The surface flashover test results of fluorinated nano-alumina particles on 300 mesh polyamide mesh composite surface and carbon nanotubes are illustrated in Fig. 17. The overall flashover property of the composite material increased with the amount of doping of both nanoparticles, with a trend of increasing initially and then decreasing. The surface flashover properties of a composite material made with 6% nanoscale alumina particles and 0.15% carbon nanotubes were the best among them. At this point, the flashover voltage on the composite material's surface reaches 12.3kV, the flashover performance improves by 10%, and the cumulative flashover performance improves by more than 30%.

4.2.2 Surface Potential Attenuation Characteristics of Composites Under Mixed Modification of Nanoparticles

Using 6% nano-alumina and 0.15% nano-carbon nanotubes in the mixed solution, the composite surface demonstrated the best flashover performance in the blending experiment. This paper investigated the charge dissipation of the composite surface with varying mass fractions of carbon nanotubes and alumina nanoparticles at $Al_2O_3 = 6\%$ and 0.15% CNT to further understand the microscopic characteristics of the composite surface and the changing mechanism of flashover qualities. The test results are depicted in Fig. 18.

Figure 18 depicts the normalized potential dissipation test results. When γCNT was 0.15%, the charge dissipation on the surface of the composite material displayed a rapid changing trend. When γAl_2O_3 reaches 6%, charge dissipation on the surface of the composite material first slows and then accelerates as the mass fraction of carbon nanotubes increases.

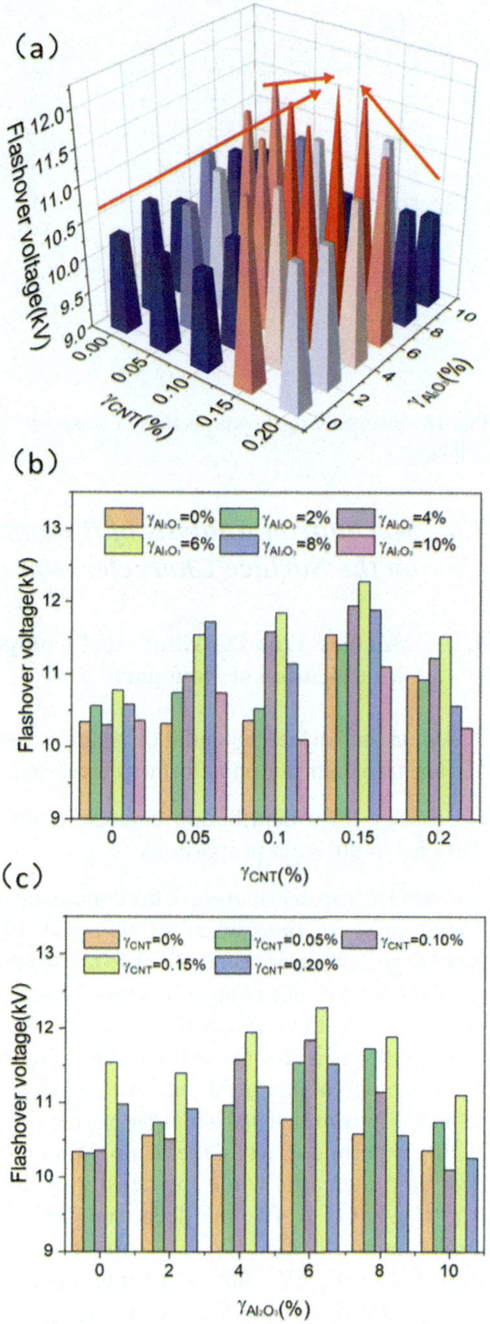

Fig. 17 Flashover properties of composite materials with nano-Al_2O_3 and MWCNT

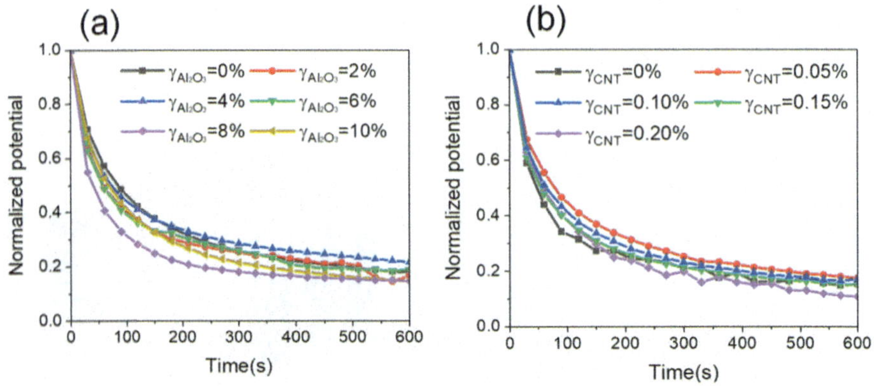

Fig. 18 Charge dissipation properties of composite materials at nano-Al_2O_3 and MWCNT blend doping

4.3 *Mechanism Analysis of the Influence of Nanostructure on the Surface Characteristics of Composite Materials*

4.3.1 Surface Trap Distribution of Composites Under Mixed Modification of Nanoparticles

Based on the surface potential dissipation test of the nanoparticle blend sample, the surface trap distribution is further calculated.

(1) Energy level distribution characteristics of the composite surface of alumina particles in different proportions.

The surface trap distribution of the composite material was under varied nano alumina doping mass fractions when γ CNT was 0.15% in Fig. 19a. Deep trap levels on the composite surface are centered at 1.02 eV, while shallow trap levels are concentrated around 0.93 eV. The change in nano-alumina was not significant.

Changes in the trap density of the composite surface are also obtained, as seen in Fig. 19b. It is discovered that the scenario is comparable to that which occurs when the polyamide mesh number changes. Surface trap modifications are mostly reflected in trap density when the surface of the composite is physically modified. In this experiment, when the nano-alumina mass fraction grows, the density of the deep trap increases while the density of the shallow trap drops. From $\gamma Al_2O_3 = 0\%$ to $\gamma Al_2O_3 = 4\%$, the deep trap density grew from 0.81 (10^{33} eV^{-1}/m^{-3}) to a maximum of 1.07 (10^{33} eV^{-1}/m^{-3}), whereas the shallow trap density declined from 1.48 (10^{33} eV^{-1}/m^{-3}) to a minimum of 1.22 (10^{33} eV^{-1}/m^{-3}). Subsequently, when the γAl_2O_3 was further increased to 10%, the deep trap density decreased to a minimum of 0.7 (10^{33} eV^{-1}/m^{-3}), the shallow trap density was increased to a maximum of 1.57 (10^{33} eV^{-1}/m^{-3}).

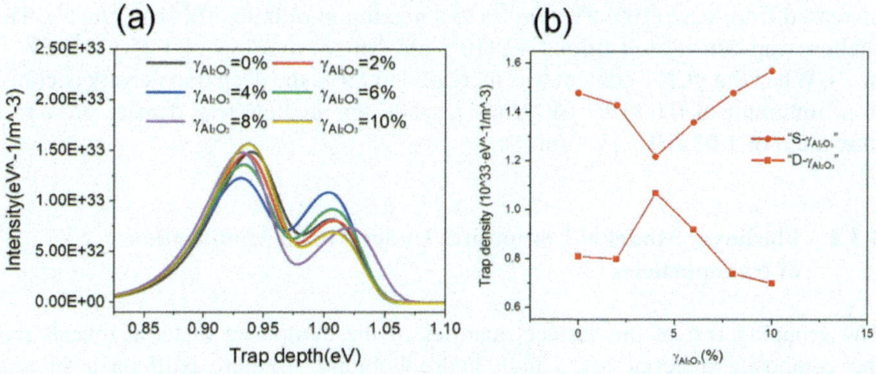

Fig. 19 Surface trap distribution of nano-Al_2O_3 composites with different mass fractions

(2) Energy level distribution characteristics of the composite surface of carbon nanotube particles with different proportions.

Figure 20a shows the distribution of composite surface traps at different nano-alumina doping mass fractions when γAl_2O_3 was 6%. The deep trap level on the composite surface is found to be concentrated at around 1.02 eV, whereas the shallow trap level is concentrated at around 0.93 eV. The corresponding change in CNT mass fraction did not occur.

Furthermore, as illustrated in Fig. 20b, the trap density fluctuates on the composite surface. It is discovered that the condition is identical to that which occurs when the mass fraction of nano alumina changes. Surface trap modifications are mostly reflected in trap density when the surface of the composite is physically modified. In this experiment, when the mass fraction grew, the density of the deep trap increased little at first, then rapidly declined. The change in shallow trap density on the material surface is the inverse. From $\gamma CNT = 0\%$ to $\gamma CNT = 0.05\%$, the deep trap density

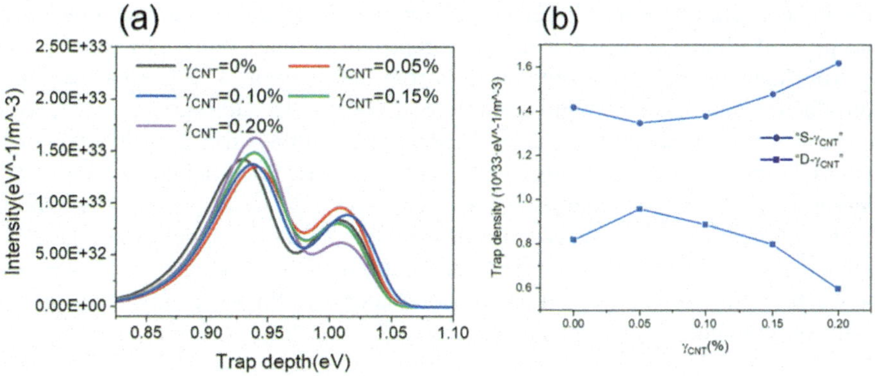

Fig. 20 Surface trap distribution for different mass fractions of MWCNT composites

increased from 0.82 (10^{33} eV^{-1}/m^{-3}) to a maximum of 0.96 (10^{33} eV^{-1}/m^{-3}). The shallow trap density fell from 1.42 (10^{33} eV^{-1}/m^{-3}) to a low of 1.35 (10^{33} eV^{-1}/m^{-3}). When the γCNT content was increased to 10%, the deep trap density declined to a minimum of 0.6 (10^{33} eV^{-1}/m^{-3}), while the shallow trap density grew to a maximum of 1.62 (10^{33} eV^{-1}/m^{-3}).

4.3.2 Flashover Model of Composites Under Mixed Modification of Nanoparticles

The grouping test of the surface qualities of the composite material reveals that the composite material has a high hydrophobicity structure built on a silicone rubber hydrophobic material substrate, polyamide mesh, and nanoparticles. Although increasing the mass fraction of fluorinated nano-alumina particles can improve the composite surface's superhydrophobic characteristics, the overall impact is small. The impact of nanoparticles on the surface properties of composite materials is most visible in the flashover properties along the surface.

Figure 21 depicts a tiny representation of the composite surface. Cheng K and others discovered the phenomena of nanoparticle introduction in their research. The researchers expected that introducing nanoparticles onto the surface of the composite material would result in the formation of an interface zone between distinct components [9]. Deep traps on the composite surface are mostly caused by thc interface region between these distinct media. Along with being connected to the composite material surface, the nano-alumina particles attached to the polyamide mesh/silicone rubber composite material surface initially created the interface area with the composite material, resulting in a significant number of deep traps. Further doping of nano-alumina results in the formation of a loose layer on the surface of the composite material made of nano-alumina particles, which is the primary cause of the decrease in the deep trap density of the composite surface in the second stage [10].

Concerning the influence of carbon nanotubes on the composite surface, the researchers believe that the high electrical conductivity of carbon nanotubes increases the pace of charge dissipation of the composite surface, which is related to the improvement of the composite surface's trap distribution [11]. Carbon nanotube introduction can be divided into three steps. To begin, the addition of a very modest amount of carbon nanotubes causes the deep trap on the surface of the composite material to rise. Carbon nanotubes with increasing mass fraction improve trap distribution on the surface of composite materials, accelerate surface charge dissipation, hinder the creation of surface flashover discharge route, and eventually raise flashover voltage [12]. As the number of carbon nanotubes increases, they lap to form a local or via bridging network, resulting in a dramatic drop in the composite's surface flashover voltage.

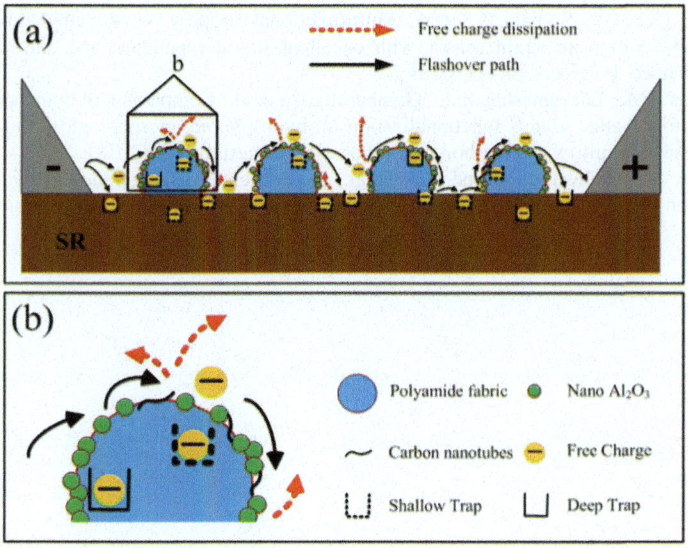

Fig. 21 Model of the flashover mechanism along the surface of the composite material

References

1. Touko, L.S.K.K., Vasuki, D., et al.: Protein adsorption and its effects on electroanalytical performance of nanocellulose/carbon nanotube composite electrodes. Biomacromolecules (2023)
2. Guannan, J., Lei, Z., Jiao, L., et al.: Robust metallic-based superhydrophobic composite with rigid micro-skeleton structure for anti-icing/frosting. J. Mater. Process. Tech. **316** (2023)
3. Assaedi, H.S., Olawale, D.M.: Impact of nano-alumina on the mechanical characterization of PVA fibre-reinforced geopolymer composites. J. Taibah Univ. Sci. **16**(1) (2022)
4. Jianwen, Z., Xinyu, P., Jinquan, H., et al.: Construction of hierarchical micro/nanostructured ZnO/Cu-ZnMOFs@SA superhydrophobic composite coatings with excellent multifunctionality of anticorrosion, blood-repelling, and antimicrobial properties. ACS Appl. Mater. Interfaces (2022)
5. Niu, L., Cai, L., Xinyu, X., et al.: Highly reliable transparent superhydrophobic composite by organosilane/denture base resin-modified alkylated silica nanoparticles against contaminants. Surf. Interfaces **35** (2022)
6. Qin, X., Luyao, W., Wenfang, F., et al.: Ratiometric electrochemical sensor based on polythionine/multiwalled carbon nanotube composite for butylated hydroxyanisole determination. J. Alloys Compd. **925** (2022)
7. Xiaoguang, Z., Dean, P., Gangqiang, Z.: Preparation and characterization of form-stable phase-change materials with enhanced thermal conductivity based on nano-Al_2O_3 modified binary fatty acids and expanded perlite. Energy Build. **271** (2022)
8. Ke, L., Chao, Y., Siyuan, Z., et al.: Laser direct writing of a multifunctional superhydrophobic composite strain sensor with excellent corrosion resistance and anti-icing/deicing performance. Mater. Des. **218** (2022)
9. Cheng, K., Yifan, F., Zheng, Z., et al.: Superhydrophobic composite coating with excellent mechanical durability. Coatings **12**(2) (2022)

10. Ke, L., Chao, Y., Siyuan, Z., et al.: Multifunctional slippery polydimethylsiloxane/carbon nanotube composite strain sensor with excellent liquid repellence and anti-icing/deicing performance. Polymers **14**(3) (2022)
11. Sansotera, M., Talaeemashhadi, S., Gambarotti, C., et al.: Comparison of branched and linear perfluoropolyether chains functionalization on hydrophobic, morphological and conductive properties of multi-walled carbon nanotubes. Nanomaterials **8**(3) (2018)
12. Mathur, A., Wadhwa, S., Byrne, A.J., et al.: Experimental demonstration of hydrophobicity variation in carbon nanotubes by surface modifications. e-J. Surf. Sci. Nanotechnol. **7** (2009)

Molecular Dynamics Simulation of the Effect of Water Intrusion on the Epoxy Resin/glass Fiber Interface of Composite Insulator Core

Jun Xie, Ke Chen, Qing Xie, and Fangcheng Lü

1 Introduction

1.1 Background of the Topic Selection and Research Significance

Composite insulators have the advantages of being lightweight, resistant to pollution, and exhibiting excellent flashover performance, making them widely used in power systems. According to the investigation, the number of standard composite insulators used in domestic grid lines in our country has exceeded 8 million. However, composite insulators are typical organic insulation devices that operate for long periods in high-temperature and high-pressure environments, often under harsh conditions. Various factors such as wind, sunlight, rain, ultraviolet radiation, and strong electric fields can cause aging in composite insulators, leading to a decrease in pollution flashover performance and even serious accidents like insulation breakdown and core rod fracture. Statistical data shows that dozens of 500 kV composite insulator stringing accidents have already occurred in the eastern coastal areas of our country.

Therefore, studying the aging mechanism of composite insulators under different environmental conditions is of great significance for improving their material properties and ensuring the safety and stability of the power system.

Currently, most of the research on the aging characteristics of composite insulators is based on artificial accelerated aging methods. Silicone rubber is an essential factor contributing to the excellent pollution flashover performance of composite insulators. The aging mechanism of silicone rubber materials is primarily studied through methods such as wet-heat aging, corona aging, and ultraviolet radiation

J. Xie (✉) · K. Chen · Q. Xie · F. Lü
Department of Electrical Engineering, North China Electric Power University, Baoding, China
e-mail: junxie@ncepu.edu.cn

Q. Xie (ed.), *Electrical Materials*, Engineering Materials,
https://doi.org/10.1007/978-981-99-9050-4_6

aging. The core rod is another important factor contributing to the excellent mechanical strength of composite insulators. However, factors like surface damage leading to moisture ingress can cause aging and overheating in the core rod, ultimately leading to accidents such as fractures [1].

In addition, besides studying silicone rubber and core rods, an increasing number of scholars have been focusing on the interface issues of composite insulating materials. The interface is a vulnerable link in composite insulation devices. In order to study the interface, Yang [2] investigated the impact of moisture ingress on the crimping interface of composite insulators, while Wang [3, 4] studied the boiling aging characteristics of the silicone rubber and cycloaliphatic epoxy resin interface in composite insulators and the effect of insulator mold joints on water diffusion tests.

The interface of composite insulators includes both macroscopic and microscopic interfaces. The macroscopic interfaces of composite insulators mainly include silicone rubber sheath/skirt, silicone rubber sheath/core rod, and end fittings/core rod. The representation of the microscopic interface is the interface between the core rod epoxy resin and glass fiber. The performance of the epoxy resin/glass fiber interface will significantly influence the overall performance of the composite insulator core rod. Therefore, conducting relevant research on the degradation characteristics of the epoxy resin/glass fiber interface is of great importance in understanding the aging mechanism of the core rod at the micro level.

Molecular simulation, as an important method for studying the microscopic properties of materials, has been widely applied in various fields. Compared to traditional experiments, molecular simulation methods have the advantages of high accuracy, low cost, and high efficiency. They are consistent with experimental results and can be flexibly and efficiently applied in polymer materials, synthetic chemistry, biology, and other fields. Molecular simulation methods integrated into the field of insulating materials can provide better insights into the material's changes and mechanisms in different environmental conditions.

Therefore, in order to clarify the degradation mechanism of the epoxy resin/glass fiber interface under moisture ingress, this study employs molecular simulation methods to analyze the impact characteristics of moisture ingress on the epoxy resin/glass fiber interface under different electric fields and environments. The research findings can contribute to a better understanding of the deterioration mechanism of composite insulators and provide theoretical and engineering guidance for enhancing the performance of composite insulators.

1.2 Research Status at Home and Abroad

1.2.1 Research Progress on Epoxy Resin/Fiber Interface Properties

Rain et al. [5] found that when moisture infiltrates the epoxy resin and penetrates into the quartz, it leads to the separation of interconnected epoxy resin and quartz, creating an interface. Other related studies have shown that moisture entering the interior of composite materials can cause damage and severe degradation of the epoxy resin/fiber interface [6], and may also result in matrix plasticization [7]. In the wet-heat aging process, the synergistic effect of moisture and temperature on the epoxy resin/fiber interface is primarily due to the high temperature causing the interface to fracture, and the post-curing reaction of the alkoxy material further deepens the fracture plane [8], resulting in a significant accumulation of moisture in the cross-sectional area. It is generally believed that the phenomenon of interface fracture in composite materials is mainly caused by the swelling or expansion stress resulting from moisture infiltration into the interface. Xu et al. [9–11], in their study of epoxy resin/fiber composite materials in a humid high-temperature and high-pressure environment and under the influence of an electric field, found that under conditions where other electrical parameters remain the same, the dielectric loss and dielectric constant of the composite material are directly proportional to the experimental treatment time and temperature, inversely proportional to the electric field intensity. Furthermore, with an increase in the experimental duration, the material's resistivity gradually decreases, indicating a deterioration in its insulating properties.

According to relevant studies, epoxy resin in composite materials may undergo hydrolysis reactions in a humid and hot environment, resulting in the rupture of the epoxy resin chain structure [12, 13]. FTIR analysis of the epoxy resin/fiber interface material after wet-heat aging reveals changes in the characteristic peaks of carboxyl and hydroxyl groups [14]. When composite materials absorb water, hydrolysis reactions occur between the epoxy resin and water, generating C–O bonds and N–CO–N bonds, accompanied by the cleavage of epoxy groups [15]. Furthermore, FTIR analysis of the water solution after immersing the composite material shows a characteristic peak at 1012 cm^{-1}, indicating the precipitation of some substances from the epoxy resin/fiber material after aging in a humid and hot environment, and this characteristic peak corresponds to Si–O bonds. However, the original composite material does not contain the mentioned chemical bonds, providing strong evidence of hydrolysis reactions occurring in composite materials under wet-heat conditions [16]. Additionally, the FTIR scan of the material after wet-heat aging shows a characteristic peak at 1735 cm^{-1}, indicating that the wet-heat environment can cause oxidation of the composite material [17]. Soles et al. [18] found that after wet-heat aging, samples develop polar hydroxyl groups that form hydrogen bonds. Since epoxy resin molecules do not contain amino groups, the formation of hydrogen bonds in the samples is likely due to hydroxyl groups interacting with ether groups or self-interactions.

In conclusion, researchers usually conduct simulated aging experiments on epoxy resin/fiber interfaces under conditions of wet-heat and applied external electric fields. However, the actual aging process of the interface is influenced by factors such as the ambient temperature during long-term operation, operating voltage, electric field, and electrochemical heating. The simulation experiments conducted by researchers only consider the effects of wet-heat and discharge, and the macroscopic experimental results cannot fully reflect the aging deformation process of the material's internal structure. Therefore, it is essential to explore characteristic values that indicate significant changes in the microscopic structure of epoxy resin/glass fiber materials in environments with moisture ingress. By using molecular dynamics methods, the macroscopic properties of the material can be organically linked to the microscopic structure and characteristics of the internal structure. This will provide a more comprehensive understanding of how the material behaves in the presence of moisture infiltration and its impact on the microstructure.

1.2.2 Molecular Simulation for Epoxy/Fiber Interface Properties

Lau et al. [19] conducted a molecular simulation analysis at the microscopic level to study the adhesion characteristics at the interface between fiber-reinforced resin composite materials and wood. They also analyzed the interfacial free energy. They found that the mechanism of hydrogen bond formation differs at different water molecule concentrations. At higher water molecule concentrations, water molecules tend to form larger clusters connected by hydrogen bonds, whereas at lower concentrations, smaller clusters connected by hydrogen bonds are formed with the composite organic materials. The entry of water molecules into the interface between epoxy resin and single-walled carbon nanotubes significantly affects the adhesion of the two molecules, leading to a decrease in the material's mechanical strength [20]. The internal water content of epoxy resin composite materials is positively correlated with their density but negatively correlated with the Young's modulus. As the water content increases, the Young's modulus initially increases and then decreases, with the lowest value being as low as 94.5% of the original performance. Other studies have found that the hydrogen bonding forces between water molecules and composite materials severely affect the movement of water molecules within the composite structure [21]. Additionally, through the analysis of root mean square displacement, volume expansion rate, and radial distribution function, it was observed that in the early stage of wet-heat aging, the moisture uptake is small, and the hydrogen bonding between water molecules and the composite material is the dominant form. As water molecules continue to increase, they bind together, enhancing their diffusion ability [22]. Due to the topological constraints of the spatial distribution of amino and hydroxyl groups, they are not affected by temperature limitations. However, the diffusion coefficient of water molecules shows a positive correlation with temperature [23]. Analyzing the simulation results, it was observed that water molecules enter the interior of the composite material and condense in the interface region between epoxy resin and carbon fiber. They form hydrogen bonds with the chemical bonds present in the interface, thereby reducing the surface free energy between epoxy resin and carbon fiber [24]. When the water content in the interface material is 0 wt%, the free volume accounts for 5.1% of the total volume. As the internal water content increases from 4 to 7 wt%, the proportion of free water decreases from 4.4 to 4.0%. This indicates that water molecules entering the interface interior occupy the free volume in the interface analysis [25, 26]. Xiao [27] studied water molecule infiltration into the epoxy resin/carbon fiber interface and found that water molecules enter the gaps between the epoxy resin and carbon fiber surfaces. Through intermolecular forces, they further expand the gaps between epoxy resin and carbon fiber molecules at the interface, resulting in reduced adhesion at the interface.

The basic properties of the composite material were analyzed using Material Studio software [28]. The simulation results obtained the glass transition temperature of the material, which showed a high degree of agreement with experimental results. However, the glass transition temperature calculated from the diffusion coefficient was found to be higher compared to experimental results [29–31]. Incorporating carbon nanotubes with high aspect ratios into the epoxy resin composite material can enhance the material's stiffness, leading to an early occurrence of tensile yield phenomena. Consequently, adding carbon nanotubes with high aspect ratios to the epoxy resin composite material significantly reduces its thermal stability [32, 33]. Research has indicated a negative correlation between the mechanical properties of epoxy resin and its moisture absorption and environmental temperature. However, the mechanical properties can be improved by increasing the crosslinking density of the epoxy resin [34, 35]. Therefore, in a humid and hot environment, epoxy resin composite materials tend to undergo accelerated aging, resulting in a decrease in their basic mechanical properties. The infiltration of water into the interior of composite materials leads to a reduction in their physical and chemical properties, subsequently affecting their fundamental insulating properties and mechanical strength [36–38].

In conclusion, scholars worldwide have mainly focused on studying the physicochemical properties of epoxy resin composite materials at the macroscopic level after water infiltration at the interface. There has been limited research on the microscopic structural changes of water molecules at the interface. Therefore, we need to shift our perspective and conduct comprehensive research from the perspectives of temperature and electric fields to study the changes in the epoxy resin/fiber interface after water infiltration in different environments. Subsequently, we can delve into the investigation and analysis of the movement patterns of microscopic molecules within the interface and their correlation with the macroscopic fundamental properties, as well as the variations and mechanisms of the interfacial bonding performance.

1.3 What This Article Examines

This paper aims to investigate the impact and failure mechanism of water infiltration on the epoxy resin/glass fiber interface using molecular dynamics methods. Considering the actual working conditions of composite insulators, this study focuses on the effects of temperature, electric fields, and their combined influence on the characteristics of water at the epoxy resin/glass fiber interface. The main tasks of this study are as follows:

(1) Based on the production process of composite insulator core rod epoxy resin/glass fiber interface, the composition ratios and reactions occurring during the preparation process are studied. The molecular-level model of the epoxy resin/

glass fiber interface is constructed and optimized, and water molecules are introduced as the working environment. After appropriate treatment, a stable conformational molecular model is obtained.

(2) Molecular simulations are performed on the epoxy resin/glass fiber interface in a water environment, and the results are compared with simulations conducted in a dry environment. Interface binding energy, ester group quantity, radial distribution function, and hydrogen bond quantity are selected as indicators to study the effects of water infiltration on the epoxy resin/glass fiber interface under the influence of temperature and electric fields.

(3) By simultaneously varying temperature and electric field parameters, the study investigates the impact of temperature/electric field synergistic effects on the epoxy resin/glass fiber interface in the presence of water infiltration. The variations in interface binding energy, ester group quantity, radial distribution function, and hydrogen bond quantity are analyzed to elucidate the degradation mechanism of the epoxy resin/glass fiber interface after water infiltration.

2 Establishment and Optimization of Epoxy/Glass Fiber Interfacial Molecular Models

Composite insulators commonly use glass fiber and epoxy resin as the core rod material, preparing fiber-reinforced composite materials with a certain proportion and processing method to combine the advantages of both materials. The adhesive performance of the epoxy resin/glass fiber interface in the core rod of composite insulators greatly affects the overall performance of the core rod. Therefore, establishing a simulation model for the epoxy resin/glass fiber interface and studying its changes in interface performance under different environments is of great significance.

In this chapter, we first analyze the composition and structure of the epoxy resin/glass fiber interface based on the production process of composite insulator core rods. Using the Material Studio simulation software, we construct a simulation model for the epoxy resin/glass fiber interface. By employing molecular dynamics theory, we perform energy optimization and relaxation operations on the interface model to establish a stable epoxy resin/glass fiber interface structure model. Subsequently, we place the model in environments with water molecules and dry conditions, laying the foundation for the study of interface performance under different environments.

2.1 Fundamentals of Molecular Dynamics Simulation

The molecular dynamics theory was initially proposed by Alder and Wainwright in the late 1950s. The earliest model of molecular dynamics was a simple hard sphere fluid model, which gradually developed into a mature theory through a series of optimization methods such as constant pressure and constant temperature simulations [39–42]. Using molecular dynamics methods, one can determine the molecular kinetic properties of materials during static equilibrium and physical deformation, and thus understand the macroscopic changes in material properties arising from the microscopic movements of molecules.

To simplify the calculations, the theory ignores the free movement of electrons and assumes that the forces acting on the atoms within the molecular material depend solely on the interatomic distances of neighboring nuclei. Therefore, the main calculation process of molecular dynamics involves calculating the acceleration of each atom based on Newton's second law under the same system constraints (as shown in Eq. 1). The potential energy of the molecular system is then calculated using fundamental mechanical parameters, and the first-order partial derivatives of the atomic coordinates are obtained for force analysis (as shown in Eq. 2). Multiple iterations are performed to compute the coordinates and velocities of all atoms in the system, resulting in the comprehensive trajectory of molecular motion.

$$\frac{\partial^2 r_i}{\partial t^2} = \frac{F_i}{m_i} = a_i \tag{1}$$

$$F_i = \frac{\partial U}{\partial r_i} \tag{2}$$

In the equations, t represents time; mi denotes the mass of atom i; ri is the coordinate of atom i; Fi represents the force acting on atom i; ai is the acceleration of atom i; and Ui stands for the potential energy of atom i.

During the calculation of molecular motion trajectories, the initial values of parameters directly affect the deviation between simulation results and actual results. To address this, ensemble theory can be used to adjust the overall computation process of atoms by controlling the system temperature and pressure. An ensemble refers to the sum of independent microscopic systems in different motion states under the same macroscopic conditions. The state of the atoms includes the coordinates, velocities, and momenta of all atoms. Common ensembles include the microcanonical ensemble (NVE), canonical ensemble (NVT), grand canonical ensemble (μVT), Gibbs ensemble (NPT), and isobaric-isenthalpic ensemble (NPH). The NVE ensemble has the constraint of a microsystem collection with the same number of molecules N, volume V, and temperature T, while the NPT ensemble has the constraint of a microsystem collection with the same number of molecules N, pressure P, and temperature T.

Furthermore, due to computational limitations, the scope of molecular dynamics research is limited to a collection of molecules in a finite space. Therefore, the boundary effects have a significant impact on the computational results. To reduce the boundary restrictions and computational complexity, the method of analyzing multi-atomic systems under force is often employed. Additionally, periodic boundary conditions are commonly used in molecular dynamics simulations to simplify computations and force analysis.

Taking the example of molecular dynamics simulations using the NVE ensemble with the same number of molecules, pressure, and temperature, the main computational steps are as follows:

(1) Determine the initial values of running parameters (number of molecules, velocities, structural density, integration time step, and running time).
(2) Initialize the system by determining the initial coordinates and velocities of atoms.
(3) Analyze the forces on atoms based on the system's initial set of basic mechanical parameters. Calculate the accelerations and potential energies of the atoms using Newton's second law and atomic nuclear distances.
(4) Calculate the velocities of atoms and iteratively integrate to calculate the velocities at the present time step based on the previous time step's velocities and forces on atoms.
(5) Calculate the system's kinetic energy and system temperature based on the forces on atoms, velocities, and potential energies. Optimize the running velocities of atoms accordingly.
(6) Calculate the coordinates of atoms and iteratively integrate to calculate the coordinates at the present time step based on the previous time step's coordinates and forces on atoms.
(7) Repeat steps 2–6 in a loop to compute the complete trajectories of atoms and the basic atomic structure of the system at its static equilibrium state. Analyze the statistical thermodynamic parameters of the system based on molecular dynamics theory.

2.2 *Establishment and Optimization of Molecular Models*

2.2.1 Establishment and Optimization of the Molecular Model for Epoxy Resin

Epoxy resin refers to a high molecular weight oligomer with two or more epoxy groups in its molecular structure and a backbone of aromatic, cyclic, or aliphatic compounds. It can form thermosetting products through the mutual reaction of epoxy groups. The epoxy resin used in composite insulator core rods is mainly composed of diglycidyl ether of bisphenol A (DGEBA) as the base resin and methyl tetrahydrophthalic anhydride (MTHPA) as the curing agent. The molecular structures of DGEBA and MTHPA are shown in Figs. 1 and 2, respectively.

The epoxy resin cured product obtained using anhydride exhibits excellent heat resistance and chemical stability. Its main cross-linking reaction mechanism is as follows:

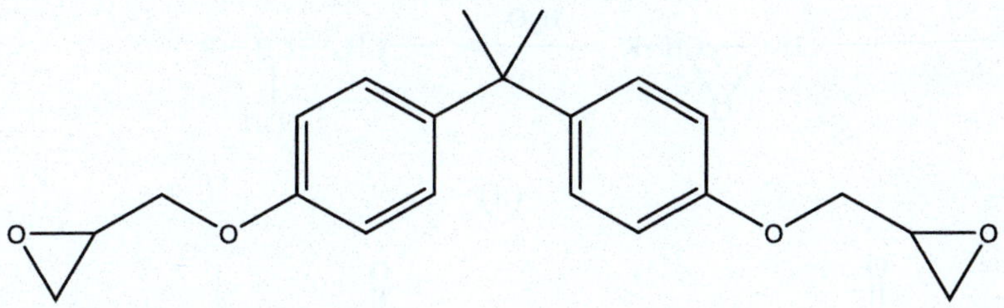

Fig. 1 Molecular structure of DGEBA

Fig. 2 MTHPA molecular structure

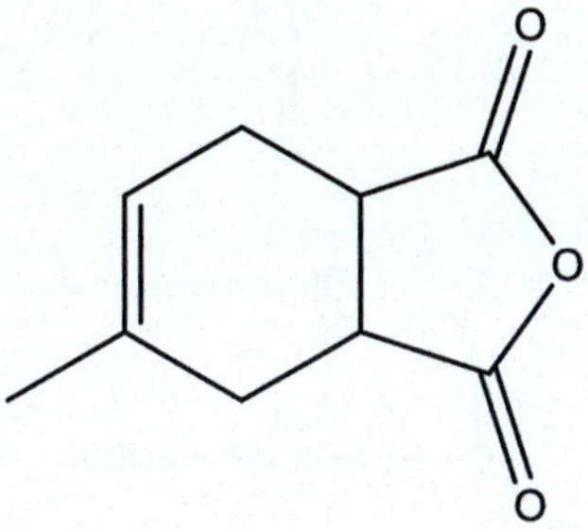

(a) Water molecules cause the epoxy resin's epoxy groups to open, generating hydroxyl groups.
(b) Anhydride groups react with hydroxyl groups in the epoxy resin, forming ester and carboxyl groups.
(c) Carboxyl groups undergo addition reaction with epoxy groups, producing new ester groups.
(d) In an acidic environment, epoxy groups undergo etherification reaction with hydroxyl groups.

The specific reaction mechanism is illustrated in Fig. 3.

Based on the above principles, the molecular configurations of DGEBA and MTPHA were established using Materials Studio simulation software, as shown in Fig. 4 and Fig. 5, respectively.

In the actual production process of curing the DGEBA resin with MTHPA, the mass ratio of the curing agent to epoxy resin is typically 80:100. Given the molecular masses of MTHPA (166) and DGEBA (330), the required quantity ratio of DGEBA to MTHPA is calculated as 83:132. Considering the computational limitations of the computer, the final epoxy resin model in this study includes 35 DGEBA molecules and 56 MTHPA molecules.

Based on the cross-linking mechanism of epoxy resin, a cross-linking program was written in Perl language to facilitate the automated cross-linking of DGEBA and MTHPA using the Amorphous Cell module of Materials Studio software. The resulting cross-linked model is shown in Fig. 6.

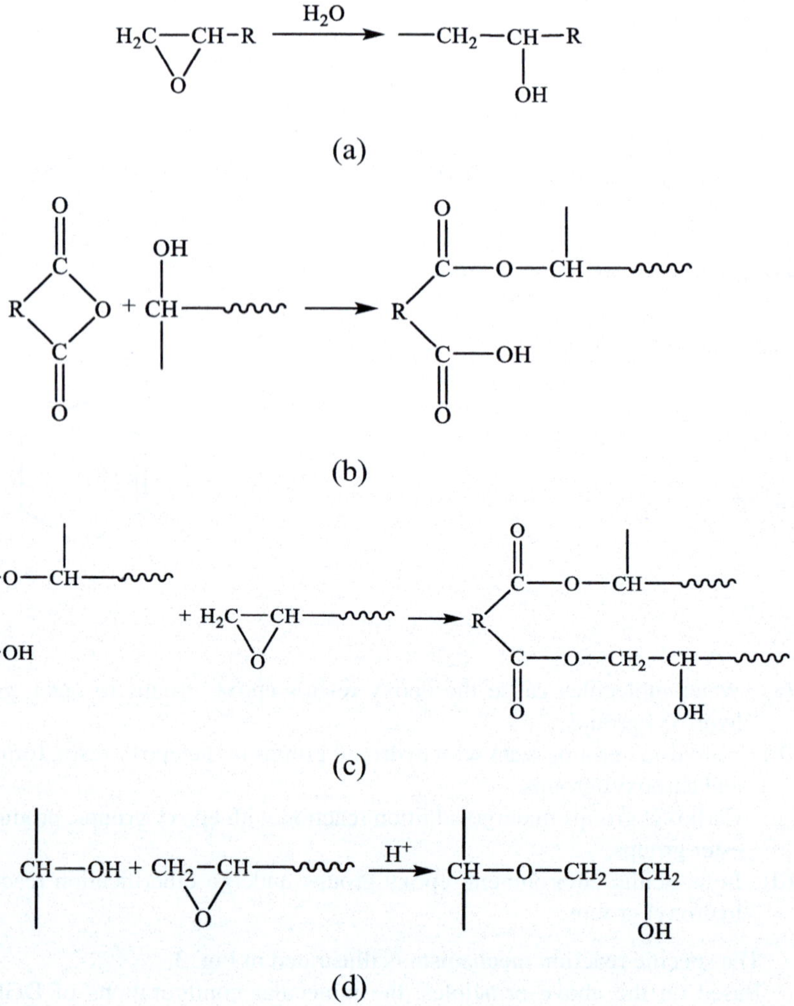

Fig. 3 Crosslinking reaction mechanism of DGEBA and MTHPA

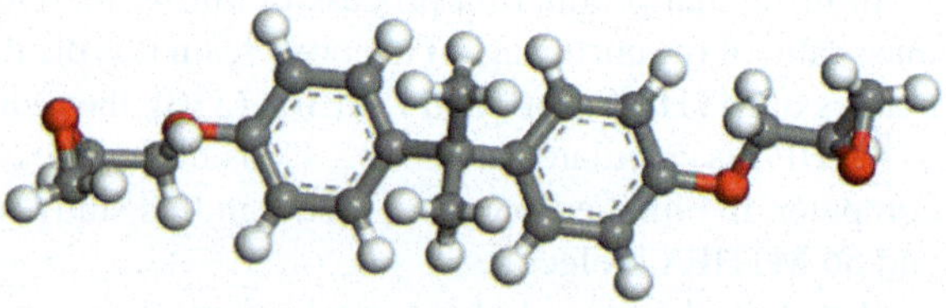

Fig. 4 DGEBA molecular model

Fig. 5 MTHPA molecular model

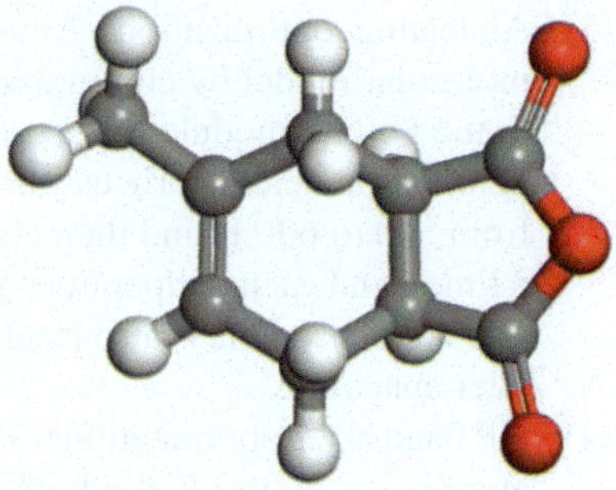

Fig. 6 Molecular model after crosslinking of DGEBA and MTHPA

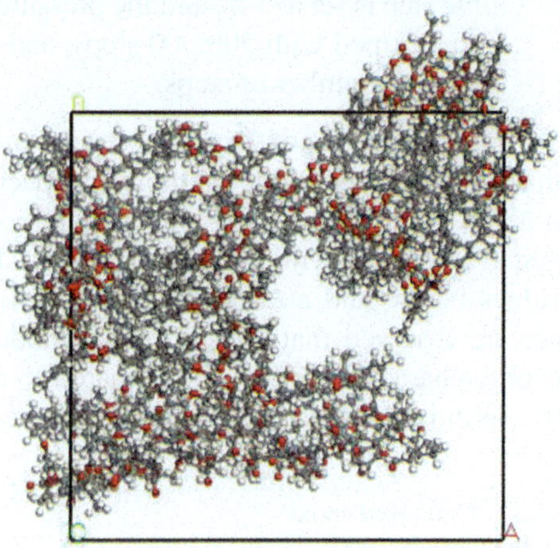

To ensure that the cross-linked model aligns with the real situation, it needs to be optimized to obtain the lowest and most stable molecular structure. The optimization process of the model includes the following steps:

(1) Energy minimization optimization: The cross-linked epoxy resin is structurally optimized using the "Geometry Optimization" function in the Forcite module of Materials Studio. The COMPASS force field is employed, and the Smart algorithm is iterated for 50,000 steps to achieve energy minimization optimization of the model.

(2) High-temperature relaxation optimization: After performing energy minimization optimization on the molecular model, the COMPASS force field and NPT ensemble are used to simulate the system at an elevated temperature of 600 K. The simulation is iterated for 500,000 steps with a time step of 1 fs. The pressure is controlled using the Berendsen method, and the temperature is controlled using the Nose method. The molecular configuration is obtained based on Newton's second law and iterative calculations after 500.0 ps of simulation at high temperature.

(3) Annealing optimization: Annealing is performed to gradually stabilize the molecular model by continuously heating and cooling it. The Anneal function in the Forcite module is used for this optimization, with the COMPASS force field. The annealing process includes cycles of gradually raising the temperature from 300 to 600 K and then cooling it back to 300 K. The simulation is cycled 5 times, and each temperature gradient is simulated with 10,000 dynamic steps. The NPT ensemble is also used. The model's structure becomes relatively stable after annealing.
(4) NPT and NVT optimizations: The COMPASS force field is used, and the temperature is set to 300 K for both optimizations to ensure that the model is at its lowest and most stable energy state under normal temperature conditions. The time step is set to 1 fs, and the pressure is set to 0.1 MPa. The NPT optimization is performed with 500,000 steps, and the NVT optimization is performed with the same number of steps.

The optimized epoxy resin molecular model is shown in Fig. 7. It can be observed that the optimized model is more compact compared to the model just after cross-linking, with a significant reduction in voids, indicating a lower energy state and better alignment with actual conditions. The comparison of energy characteristic values before and after molecular model optimization is presented in Table 1. It can be observed that the optimized model exhibits considerable improvement in energy characteristic values compared to the non-optimized model, indicating that the optimized molecular model has reached a more stable state.

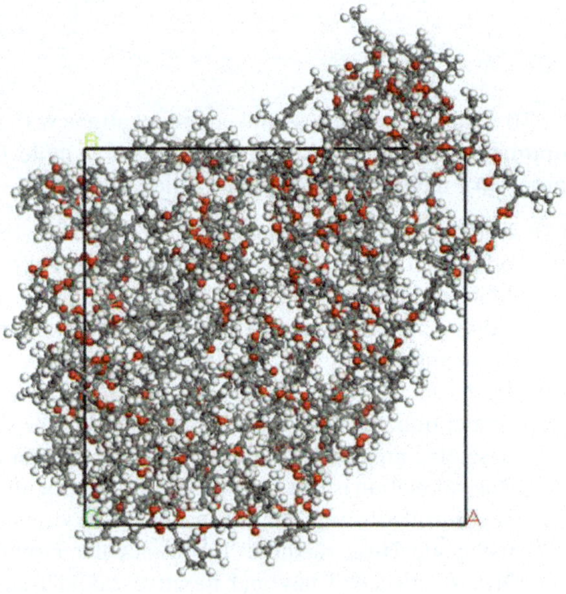

Fig. 7 Optimized epoxy resin molecular model

Table 1 Comparison of energy characteristic values before and after epoxy resin molecular model optimization

Energy (kcal/mol)	Before optimization	After optimization
Total energy	12,585.512	7447.213
Bond angle bending energy	3096.446	2036.045
Dihedral angle torsion energy	1032.854	896.637
Bond length stretching energy	2296.233	1432.258
Non-bonded interaction energy	856.414	422.174

2.2.2 Establishment of Glass Fiber Molecular Model

Glass fibers exhibit excellent corrosion resistance, heat resistance, and insulation properties. They are mainly made from natural minerals through processes such as high-temperature melting, drawing, spinning, and weaving. In molecular simulations, to simplify computations, the silica (SiO_2) structure can be used to model the surface of glass fibers.

The SiO_2 structure is retrieved from the Materials Studio software library and then trimmed to a size of 42.789742.789710.5853 Å^3 with a density of 2.254 g/cm^3. To enhance the simulation's authenticity and reliability, the SiO_2 surface undergoes hydrogenation treatment to simulate the oxidation process of SiO_2. The unsaturated O atoms on the SiO_2 surface are paired with H atoms, and the unsaturated Si atoms are bonded with –OH, resulting in the specific molecular model shown in Fig. 8.

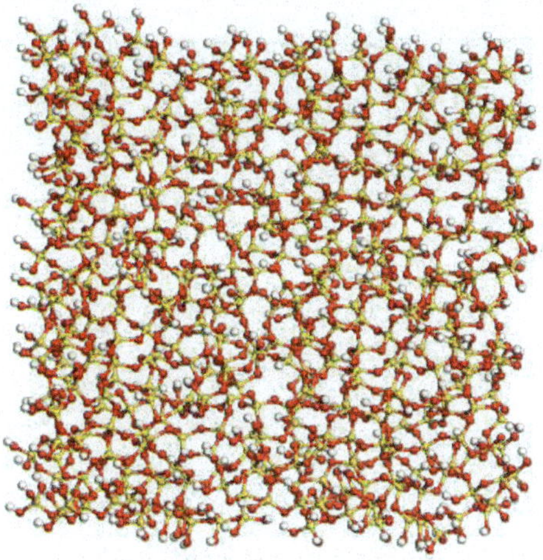

Fig. 8 SiO_2 amorphous molecular model

2.2.3 Establishment of Epoxy/Glass Fiber Molecular Model

The optimized epoxy resin and SiO_2 models are combined into a layered structure using the Amorphous Cell module in Materials Studio software. To eliminate the unwanted interface binding energy due to periodicity, a vacuum layer of 30 Å is added above the epoxy resin molecules. The final box size is set to 42.789742.789790.2477 $Å^3$, with SiO_2 at the bottom, epoxy resin in the middle, and a vacuum layer at the top. The initial model's interface binding performance is not satisfactory and does not represent the actual situation accurately. Hence, further optimization of the model is required. Following the epoxy resin optimization method and parameters, the interface model is optimized. The optimized Epoxy Resin/SiO_2 molecular model before and after optimization is shown in Fig. 9. It can be observed that the optimized epoxy resin molecules are closer to the SiO_2 interface, indicating improved interface binding performance, laying the foundation for studying the effect of water molecules on interface binding performance.

Composite insulator core rods often work in humid environments. Therefore, it is crucial to investigate the basic mechanical properties and adhesion interface free energy changes of epoxy resin/glass fiber composites under humid conditions. To simulate the water molecule's action mechanism at the epoxy resin/glass fiber interface, water molecules are randomly added inside the aforementioned simulation model. Considering the hydrophobic nature of glass fiber and the adsorption behavior of epoxy resin towards water molecules, according to the research results in reference [42], as shown in Fig. 10, the moisture absorption process of epoxy resin can be divided into several stages, and the saturation moisture absorption rate can generally

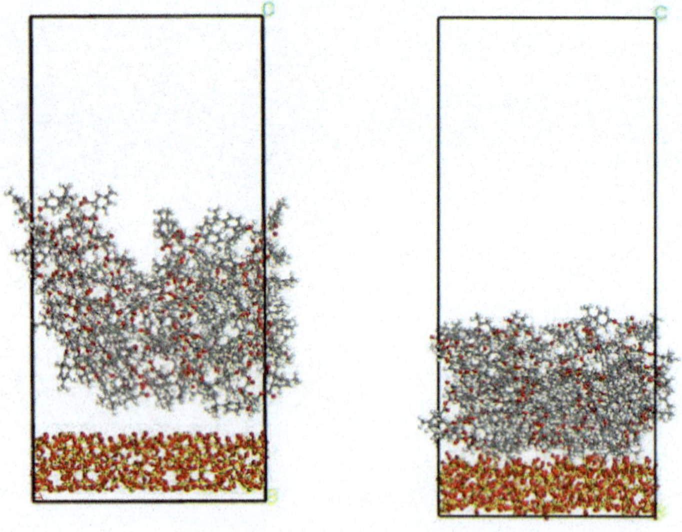

Fig. 9 Epoxy/SiO_2 interface model before and after optimization

reach 2.4% [42]. Based on these results, the model is randomly supplemented with 56 water molecules, and the water molecule model follows the TIP4P model proposed by Abascal, Jose LF, and Carlos Vega. Specific parameters of the water molecule model are shown in Table 2.

Following the above-mentioned method, the molecular model of the epoxy resin/glass fiber interface was established in the presence of water molecules. Subsequently, the model was optimized using the previously described approach. The optimized model is shown in Fig. 11.

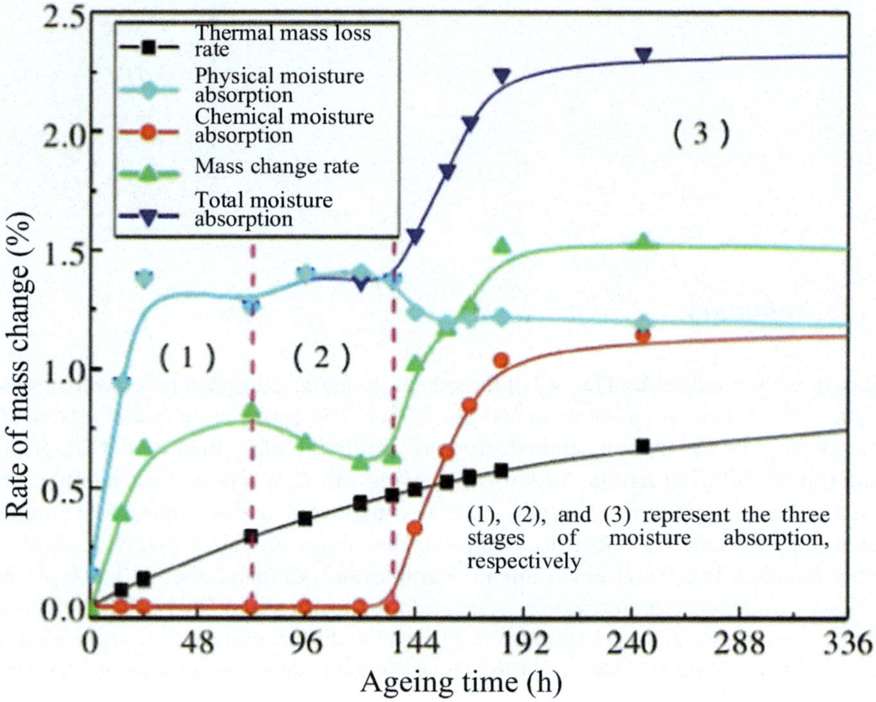

Fig. 10 Change law of moisture absorption of epoxy resin [42]

Table 2 TIP4P water model parameters

Oxygen atom mass	Hydrogen atom mass	Oxygen atom charge	Hydrogen atom charge	Hydroxyl bond length (A)	HOH bond angle (deg)	OM distance (A)
15.9994	1.008	−1.0484	−1.0484	0.9572	104.52	0.1250

Fig. 11 Represents the molecular model of the epoxy resin/SiO_2 interface in the presence of water molecules

2.3 Summary

Based on the fundamental theory of molecular dynamics, this chapter establishes the molecular models of epoxy resin and glass fiber. The epoxy resin model represents the DGEBA/MTPHA cross-linked structure, while the glass fiber model adopts an amorphous SiO_2 structure. Additionally, using the Amorphous Cell module, the interface model of epoxy resin/glass fiber is constructed, and an appropriate amount of water molecules is added to create the molecular model of epoxy resin/glass fiber interface in a water environment. Furthermore, to match the models with the real system, energy minimization, relaxation, annealing, and NVT/NPT molecular dynamics optimizations are conducted. The establishment of these various molecular models in this chapter serves as a foundation for the research in subsequent chapters.

3 Considers the Influence of Water Intrusion on the Epoxy Resin/Glass Fiber Interface Under Temperature Effect

The high altitude and cold environment in some regions of China have significantly different requirements for transmission equipment and lines compared to other regions, which is also a major challenge in the construction of ultra-high voltage projects in China. In the case of moisture intrusion, the composite insulator core rod may experience abnormal heating, and in extreme cases, partial discharge may occur, leading to a further increase in the temperature around the core rod. Different environmental temperatures may cause varying degrees of damage to the interface after water intrusion. Therefore, it is important to study the microstructure and interfacial bonding performance changes of the epoxy resin/glass fiber interface under different temperature conditions. Therefore, this chapter mainly studies the impact of water molecules invading the epoxy resin/glass fiber interface in different temperature environments, summarizes and summarizes the simulation results, and finally obtains the factors and mechanisms of water molecules affecting the interface, providing reference for the causes of composite insulator core rod failures at the molecular level.

3.1 Temperature Selection Basis and Parameter Settings

In terms of temperature selection, this study mainly relies on the following two criteria.

(1) According to the Meteorological Bureau, extreme weather below −40 °C has occurred in Inner Mongolia and Heilongjiang Province in northeast China. The lowest winter temperature in the northernmost Heihe region of China can reach −45 °C.
(2) Epoxy resin undergoes hydrolysis reactions in water molecular environments, with hydrolysis reactions and epoxy resin degradation occurring at temperatures ranging from 400 to 500 °C. The air gap between the sheath and the core rod will be generated due to aging, and the air gap is easy to conduct due to the invasion of moisture, which will lead to partial discharge. In serious cases, it will lead to the carbonization of epoxy resin. Generally, this reaction occurs at 400–580 °C [43].

Considering the above environmental temperature and the additional temperature caused by the partial discharge phenomenon in the core rod area and other factors, the possible temperature range of the composite insulator core rod is analyzed, and a certain margin is reserved. The simulation temperature range is 213–853 K, and the interval is 60 K. Simulate and calculate the changes in the internal microstructure and basic microscopic properties of the epoxy resin/glass fiber interface when water invades at different environmental temperatures. The simulation takes the standard atmospheric pressure of 0.1 MPa. According to the constraint theory of NVT ensemble, the number, volume, and temperature of molecules in the system are kept constant. The integration step size is 1 fs, and the number of iterations is set to 20,000, totaling 20.0 ps. Considering that the interface model contains C, H, O, and Si atoms, and that the Dreiding universal force field is suitable for calculating the structure and other basic properties of polymer polymers, this simulation uses the GULP/Dreiding module to simulate and statistically calculate the interface model. This also establishes a foundation for the subsequent study of the GULP module on the impact of water molecule invasion on the epoxy resin/glass fiber interface in temperature and electric field synergistic environments.

3.2 *Analysis of the Influence of Water Intrusion on the Epoxy Resin/Glass Fiber Interface Under Temperature Effect*

3.2.1 Interface Structure Changes

This article first uses molecular simulation technology to analyze the changes in the interface structure of epoxy resin/glass fiber under the action of water molecules at different temperatures.

From the results of Fig. 12, it can be seen that in a water molecule environment, at low temperatures (213–293 K), the epoxy resin/glass fiber interface is tightly bonded, and the number of water molecules entering the interface through epoxy resin is very small; As the temperature increases, the tightness of the epoxy resin structure decreases, and holes gradually form inside. A small amount of water gradually enters the epoxy resin/glass fiber interface through the epoxy resin; As the temperature continues to rise, the bonding degree of epoxy resin becomes more and more loose, and the number of holes in the top epoxy resin increases. When the temperature rises to 453 K, the epoxy resin molecular chain in the water molecule environment undergoes fracture, and the bottom epoxy resin shows slight bulging, and the number of water molecule accumulation at the epoxy resin/glass fiber interface increases significantly; As the temperature continues to rise, when the temperature exceeds 613 K, the epoxy resin begins to decompose and disperse around; When the temperature reaches 693 K, water molecules gather on the surface of the decomposed epoxy resin group, and the interface has been completely destroyed.

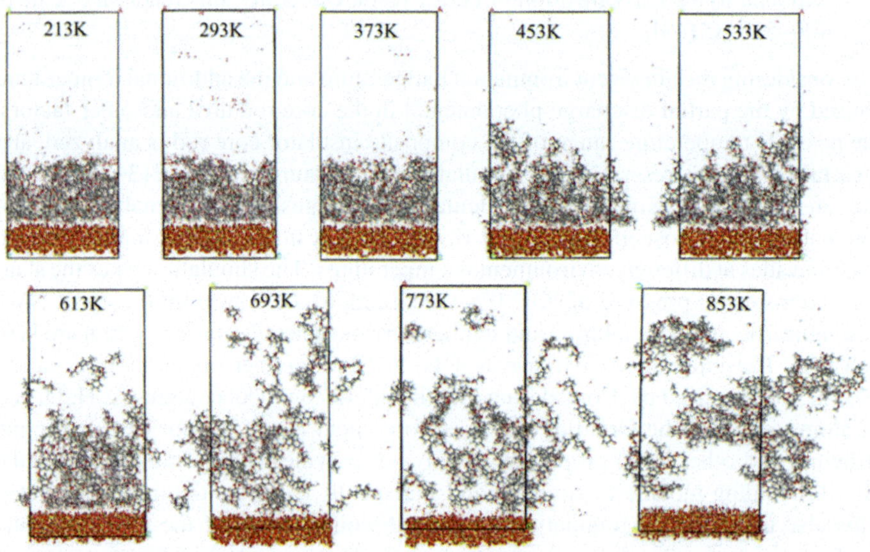

Fig. 12 Changes in interface structure of epoxy resin/glass fiber under water molecular environment

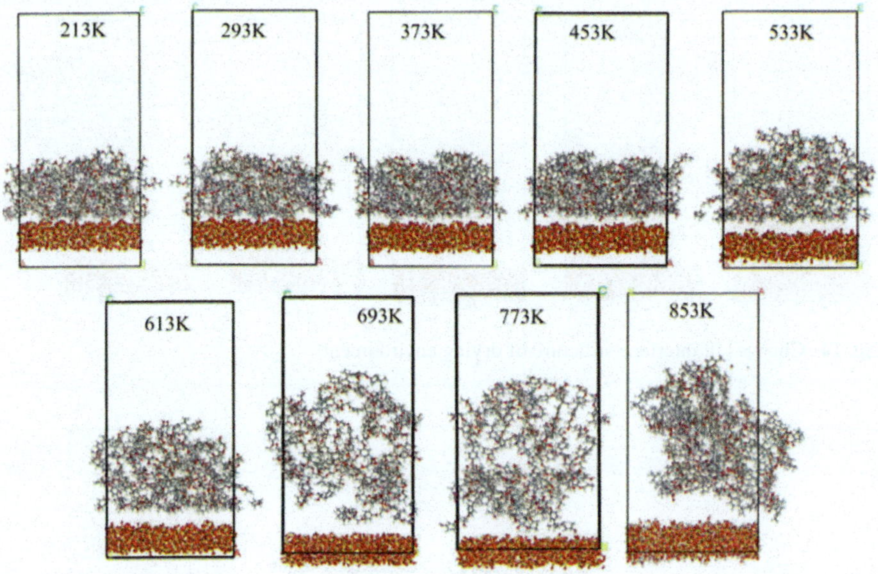

Fig. 13 Changes in the interface structure of epoxy resin/glass fiber in a dry environment

In order to further illustrate the destructive effect of temperature on the epoxy resin/glass fiber interface, this article also presents the results of the structural changes of the epoxy resin/glass fiber interface in a dry environment under temperature changes, as shown in Fig. 13.

From the results in Fig. 13, it can be seen that in a dry environment, as the temperature increases, the tightness of the epoxy resin structure gradually decreases, and the number of internal holes in the epoxy resin gradually increases. But when the temperature reached 853 K, there was no significant fracture of the epoxy resin molecular chain, and the epoxy resin/glass fiber interface was not completely destroyed.

Figure 14 shows the change of molecular structure of epoxy resin/glass fiber interface model with time at 853 K, in water molecular environment and dry environment. From the results of this figure, it can be further observed that spontaneous separation occurs at the epoxy resin/glass fiber interface at high temperatures, as shown in Figs. 14 and 15; The presence of water molecules will break the Chemical bond of epoxy resin and decompose the epoxy resin, thus further promoting the migration of water to the epoxy resin/glass fiber interface, and ultimately accelerating the destruction process of the interface.

In summary, after water intrusion, water molecules will enter the epoxy resin/glass fiber interface, promoting interfacial debonding; When the temperature increases, the epoxy resin/glass fiber also exhibits debonding behavior; When the temperature increases, the tightness of the epoxy resin molecular chain decreases, and the internal

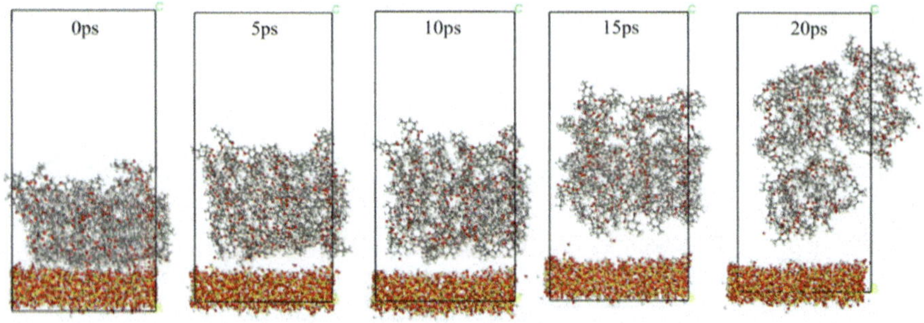

Fig. 14 Changes in interface structure of drying environment

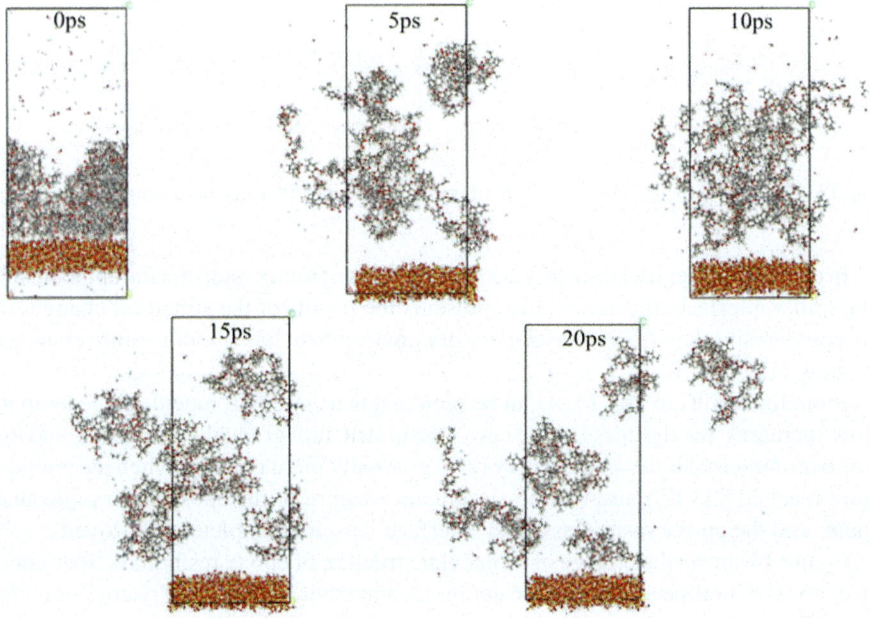

Fig. 15 Changes in interface structure under water molecular environment

holes gradually increase, accelerating the migration process of water towards the interface. In addition, under the combined action of high temperature and moisture, the molecular chains of epoxy resin break, further damaging the bonding degree of the interface.

3.2.2 Interface Adhesion Ability

Interface binding energy refers to the energy released to make the system more stable when atoms or other particles combine with each other. It is a parameter of the interaction energy of different components in a mixed system. Based on the changes in interface structure in last section, it can be seen that the interface between epoxy resin and silicon dioxide has undergone separation and damage in different environments. Studying interface binding energy can quantitatively explore the specific changes in interface binding performance. The calculation method for interfacial binding energy in water molecular environments is as follows:

$$E_{Epoxy\&SiO_2} = \begin{bmatrix} E_{Total} - \left(E_{Epoxy} + E_{SiO_2+H_2O}\right) + E_{Total} \\ -\left(E_{SiO_2} + E_{Epoxy+H_2O}\right) - E_{Total} + \left(E_{H_2O} + E_{Epoxy+SiO_2}\right) \end{bmatrix} \tag{3}$$

$E_{Epoxy\&SiO_2}$ Interface Binding energy of epoxy resin and silica;
E_{Epoxy} Epoxy resin energy;
E_{SiO_2} Silicon dioxide energy;
E_{H_2O} Water molecular energy;
$E_{SiO_2+H_2O}$ The total energy of water molecules and silica;
$E_{Epoxy+H_2O}$ Total energy of water molecules and epoxy resin;
$E_{Epoxy+SiO_2}$ Total energy of silica and epoxy resin;
E_{Total} Total energy of the system.

Before calculating the interface Binding energy, the above parameters were calculated by deleting the epoxy resin, silica and water molecules in the model. The interfacial Binding energy of epoxy resin/glass fiber interface changing with temperature under the water molecular environment was calculated by introducing the formula.

See Eq. 4 for the calculation formula of Binding energy of interface in dry environment. Similarly, delete the epoxy resin and silica from the model separately, and calculate the relevant parameters in the formula.

$$E_{Epoxy\&SiO_2} = \left(E_{Epoxy} + E_{SiO_2}\right) - E_{Total} \tag{4}$$

$E_{Epoxy\&SiO_2}$ Interface Binding energy of epoxy resin and silica;
E_{Epoxy} Epoxy resin energy;
E_{SiO_2} Silicon dioxide energy;
E_{Total} Total energy of the system.

Based on the above calculation method, the changes in interfacial binding energy of epoxy resin/glass fiber in water molecular environment and dry environment were calculated at different temperatures. The results are shown in Figs. 16 and 17, respectively.

From the results in Fig. 16, it can be seen that in the water molecular environment, the interfacial binding energy of epoxy resin/glass fiber shows a trend of first increasing and then decreasing. The interfacial binding energy is relatively small in

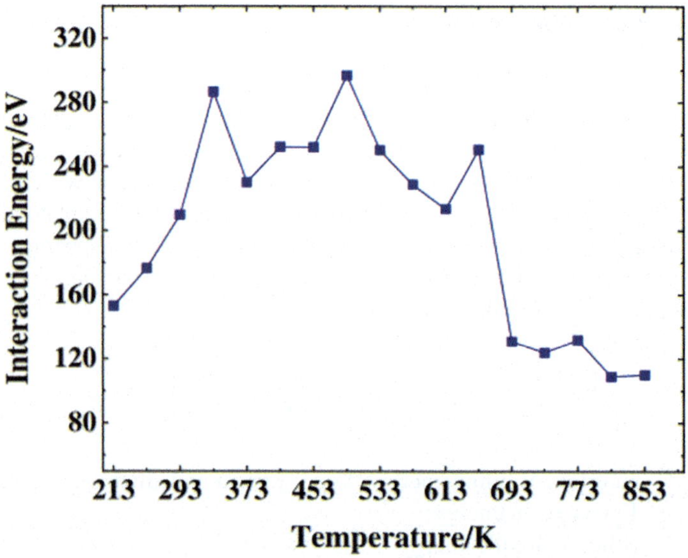

Fig. 16 Change of interface binding energy in water molecular environment

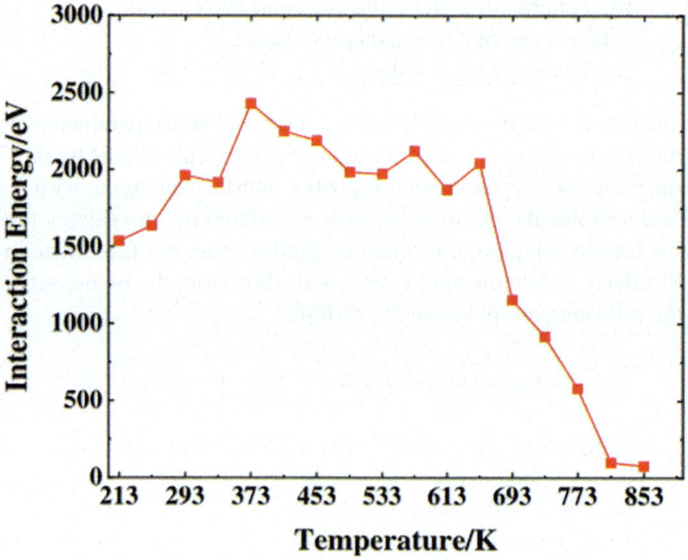

Fig. 17 Interface binding energy change in dry environment

low temperature environments, and increases rapidly with increasing temperature. When the temperature is between 373 and 653 K, the interfacial binding energy shows a fluctuating pattern and remains basically stable. When the temperature exceeds 653 K, the interfacial binding energy rapidly decreases. Afterwards, due to the severe damage of the epoxy resin/glass fiber interface, as the temperature further increased, the binding energy of the interface remained at a lower level without significant change.

It can be seen from the results in Fig. 17 that the change trend of the Binding energy of the epoxy resin/glass fiber interface in the dry environment is similar to that in the water molecular environment, but the value of the Binding energy of the interface is significantly higher than that in the water molecular environment at the same temperature, so the water molecule has an obvious role in destroying the interface. In addition, in dry environment, when the temperature is higher than 373 K, the fluctuation of interfacial Binding energy decreases, and when the temperature is higher than 653 K, the interfacial Binding energy decreases significantly.

In summary, as the temperature increases, the interface binding energy will show a trend of first increasing and then decreasing. Based on the changes in the interface structure mentioned earlier, it can be seen that when the temperature is around 373 K, the epoxy resin will expand due to the influence of temperature and water molecules entering the holes, causing the epoxy resin in the interface area to approach SiO_2, making the interface bonding more tight, and thus increasing the interface binding energy. This indicates that the increase of temperature does not necessarily lead to the decrease of interface bonding ability. Only when the temperature is higher than a certain limit value, will the decrease of interface Binding energy be caused in the anhydrous environment, which will reduce the degree of interface bonding. When water is present, it will significantly reduce the interfacial binding energy and affect the degree of interfacial bonding.

3.2.3 Microstructural Changes

Composite insulators often operate in humid environments, and the epoxy resin in the core rod spontaneously absorbs water in the humid environment. During the operation of the composite insulator core rod, it may encounter high-temperature environments, where the epoxy resin undergoes hydrolysis reactions in the water molecule environment. From Fig. 12, it can be observed that epoxy resin molecules undergo cracking when the temperature reaches a certain value. The specific hydrolysis reaction equation is shown in Fig. 18.

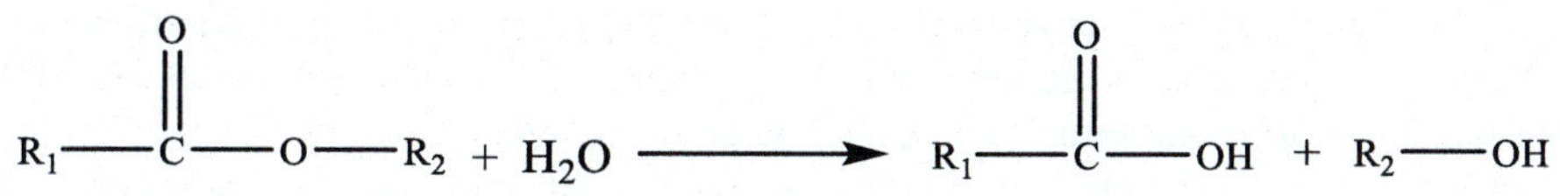

Fig. 18 Epoxy resin hydrolysis reaction equation

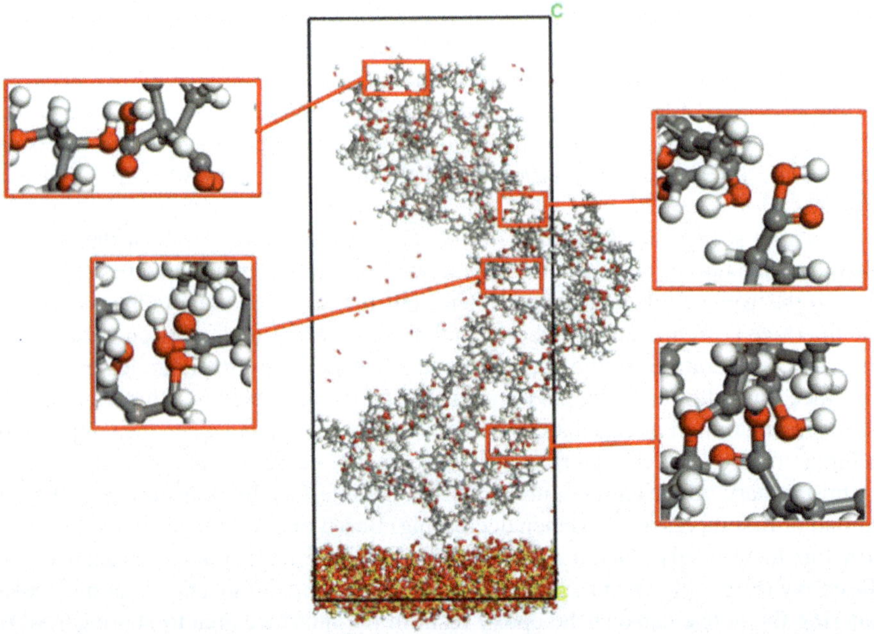

Fig. 19 693 K interface model ester group fracture diagram

From the previous analysis results, it can be seen that in the water molecule environment, as the temperature increases, the epoxy resin undergoes chemical bond breakage and gradually generates holes. Figure 19 shows an enlarged view of the local structure inside the epoxy resin at a temperature of 693 K. It can be seen from the figure that the ester group inside the epoxy resin breaks and generates carboxyl and hydroxyl groups. The above hydrolysis reaction did indeed occur.

From Fig. 19, it can be seen that the ester groups formed by cross-linking in the epoxy resin components will react with water molecules to generate carboxyl and hydroxyl groups, leading to the breaking of cross-linking bonds and the overall cracking of the epoxy resin. From Fig. 19, it can be observed that the degree of molecular cleavage of epoxy resin varies under different temperature environments. Therefore, the degree of hydrolysis reaction can be determined by counting the number of ester groups inside the epoxy resin.

According to statistics, the previously established molecular model contains a total of 56 ester groups. Due to the absence of water molecules in the model in a dry environment and the absence of conditions for hydrolysis reactions to occur, this statistical analysis only covers the ester group breakage in a humid environment. The statistical results of the number of ester groups in the interface model of water molecules at different temperatures are shown in Fig. 20.

From Fig. 20, it can be seen that there is no significant change in the number of ester groups at temperatures below 373 K; When the temperature is greater than 373

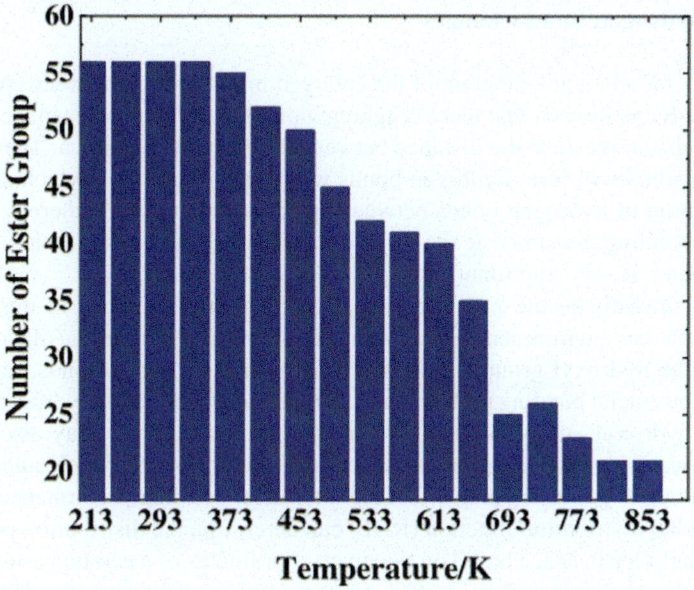

Fig. 20 Changes in the number of ester groups

K, the number of ester groups begins to decrease; When the temperature exceeds 613 K, the number of ester groups rapidly decreases; When the temperature reaches 693 K, the number of ester groups remains basically unchanged.

Based on the changes in the interface model at different temperatures mentioned above, it can be concluded that when the temperature is below 373 K, the number of ester groups does not change, indicating that this temperature environment does not comply with the temperature at which the epoxy resin hydrolysis reaction occurs, and the epoxy resin does not undergo significant hydrolysis. When the temperature reaches 373 K, reaching the hydrolysis temperature of the epoxy resin, the number of ester groups gradually decreases. As the temperature further increases, the molecular structure of the epoxy resin gradually relaxes, and water molecules are more likely to enter the epoxy resin to participate in the reaction. Therefore, as the temperature increases, the number of ester groups rapidly decreases. As the temperature further increases, the broken epoxy resin molecules drift freely in the system, causing water molecules to be too far away from the epoxy resin and making it difficult to react. Therefore, the ester base amount is relatively stable at this time.

In summary, based on the previous analysis results, it can be seen that the epoxy resin/glass fiber interface begins to undergo hydrolysis reaction at 373 K, leading to a decrease in interfacial binding energy. As the temperature continues to rise, a large number of water molecules enter into the epoxy resin molecules, leading to more intense hydrolysis reaction, which eventually causes the epoxy resin molecules to gradually crack and disperse, leading to a sharp drop in the interface binding energy, further damaging the interface.

3.2.4 Hydrogen Bond Changes

Figure 21 is a schematic diagram of the enlarged interface area between epoxy resin and silica. It can be seen that there is aggregation of water molecules in the interface area, which can increase the distance between epoxy resin and silica. The hydroxyl groups in silica will form hydrogen bonds with water molecules, which may affect the formation of hydrogen bonds between epoxy resin and silica, thereby affecting interface bonding. So studying the formation of hydrogen bonds in water molecular environments is very important.

Firstly, investigate the possible formation of hydrogen bonds in the interface region. In a dry environment, the oxygen atoms in the epoxy resin/silica interface area and the hydroxyl groups in the silica may form hydrogen bonds, resulting in stronger interfacial bonding performance. In humid environments, hydrogen bonding between hydroxyl atoms in silica and oxygen atoms in water may also occur at the interface. In humid environments, there are numerous water molecules, and the hydrogen bonds generated may affect the interfacial bonding performance.

The radial distribution function (RDF) can determine the distribution probability of other particles in space based on the given coordinates of a certain particle. Based on RDF, the probability density of another molecule appearing at a distance of r from a certain molecule can be achieved. By using the RDF method, the distance between atoms that may form hydrogen bonds at the epoxy resin/glass fiber interface and the overall distance between epoxy resin molecules and silica molecules can be analyzed.

Based on the possible forms of hydrogen bonds mentioned above, the main research objects of this section are as follows:

(1) Hydrogen bonds formed by O_{epoxy}—$H_{silicon}$ in water molecular environments;
(2) Hydrogen bonds formed by O_{water}—$H_{silicon}$ in water molecular environments;
(3) Hydrogen bonding formed by O_{epoxy}—$H_{silicon}$ in a dry environment.

The distance at which the peak appears in the RDF spectrum represents the fixed distance from one of the studied atoms, find the distance corresponding to the highest probability of finding another atom. Figure 22 shows the RDF analysis of hydrogen atoms in silicon dioxide and oxygen atoms in water molecules at a temperature of 253 K. It can be seen from the figure that the probability of oxygen atoms appearing is the

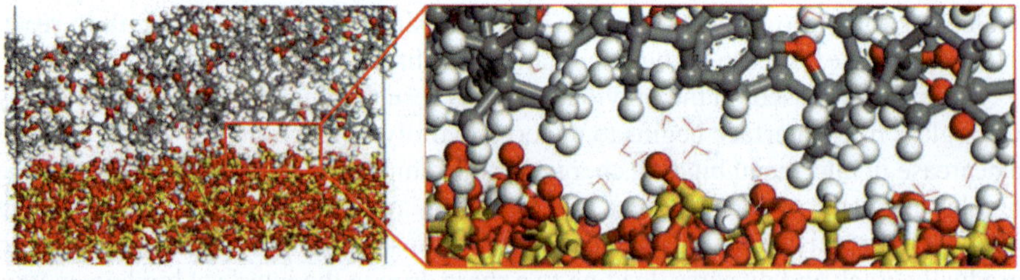

Fig. 21 Schematic diagram of water molecule aggregation in the interface area

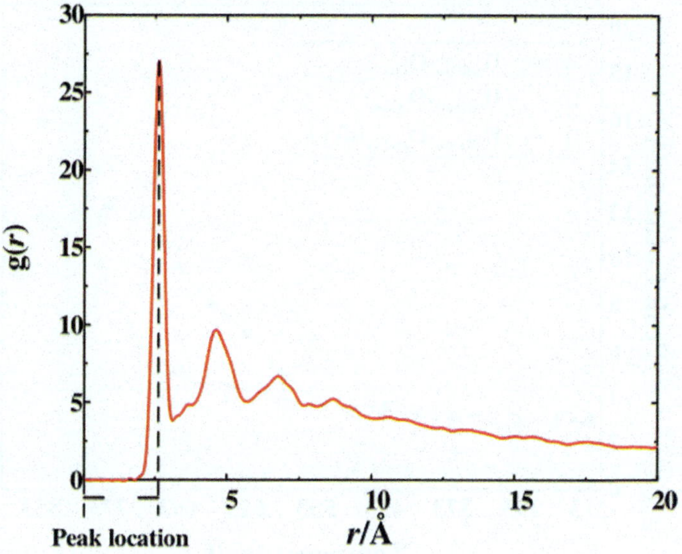

Fig. 22 RDF diagram of $H_{silicon}$—O_{water} at 253 K

highest at a distance of 2.5 Å from the hydrogen atom. Therefore, the most important parameter in the RDF study in this section is the distance from the peak, which is the abscissa of the peak. By studying the distance between peaks, it is possible to study both the distance between interfaces and the sequence of hydrogen bond formation between interfaces.

The research statistical results are shown in Fig. 23. It can be seen that the RDF peak distance of $H_{silicon}$—O_{epoxy} under different environments shows an overall upward trend with the increase of temperature, and the increasing speed of the peak distance increases when the temperature is greater than 653 K. Among them, at temperatures ranging from 373 to 653 K, the distance from the peak of $H_{silicon}$—O_{epoxy} to the peak of $H_{silicon}$—O_{water} is generally larger than that of $H_{silicon}$—O_{water}, indicating that the formation of hydrogen bonds between water molecules and silica occurs earlier than that between epoxy resin and silica, which inevitably leads to a decrease in the mechanical properties of the interface between epoxy resin and glass fiber. As the temperature continues to rise, water molecules disperse in high-temperature environments, leading to a rapid increase in the distance from the peak. In a dry environment, the distance between $H_{silicon}$—O_{epoxy} first remains stable and then rapidly increases, indicating that even without the presence of water, the hydrogen bonding between epoxy resin and glass fiber will weaken as the temperature increases.

From the changes in interface structure mentioned earlier, it can be seen that water will gradually migrate through epoxy resin to the epoxy resin/glass fiber interface. The accumulation of water molecules at the interface will lead to the formation of hydrogen bonds between water molecules and silicon dioxide molecules, hindering

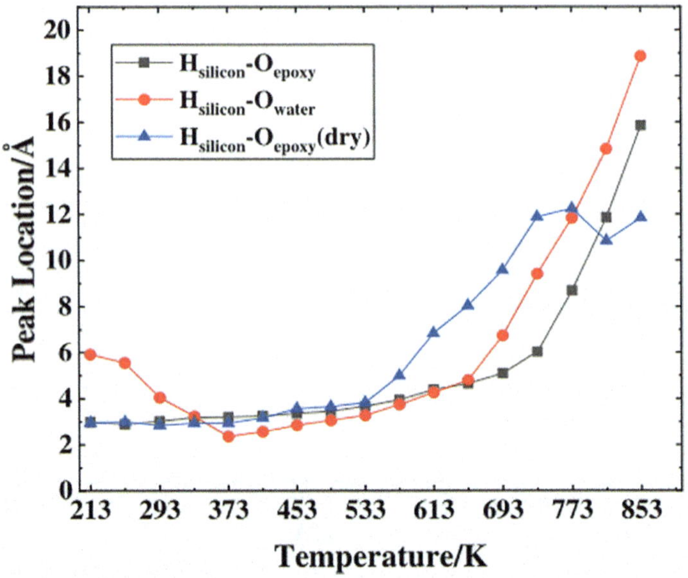

Fig. 23 RDF peak occurrence distance

the formation of hydrogen bonds between epoxy resin molecules and silicon dioxide molecules, thereby affecting the interface bonding performance.

The phenomenon of water accumulation at the interface indicates that water molecules and silica adhere to each other, which leads to more hydrogen bonds forming in the interface area than in dry environments. Therefore, counting the number of hydrogen bonds has an important impact on explaining the changes in interface bonding performance. Using Perl language to write a script for counting the number of hydrogen bonds, this script combines the trajectory file of the interface under water molecules with temperature changes to obtain the average value of hydrogen bonds generated by the interface over time, which can accurately determine the number of hydrogen bonds in the system. The statistical results are shown in Fig. 24.

From Fig. 24, it can be seen that in the water molecule environment, the number of hydrogen bonds increases at 373 K, and begins to decrease when the temperature exceeds 493 K. The number of hydrogen bonds decreases significantly when the temperature exceeds 653 K, and remains stable at 693 K.

Based on the changes in interface structure mentioned earlier, it can be seen that when the temperature is below 373 K, water molecules do not move violently and cannot enter the interface area due to the lower temperature. As the temperature increases (above 373 K), epoxy resin molecules begin to hydrolyze, causing damage to the overall structure of epoxy resin molecules, causing water molecules to enter the internal holes of epoxy resin, leading to epoxy resin expansion, making the epoxy resin in the interface area slightly move downwards, making the interface fit more

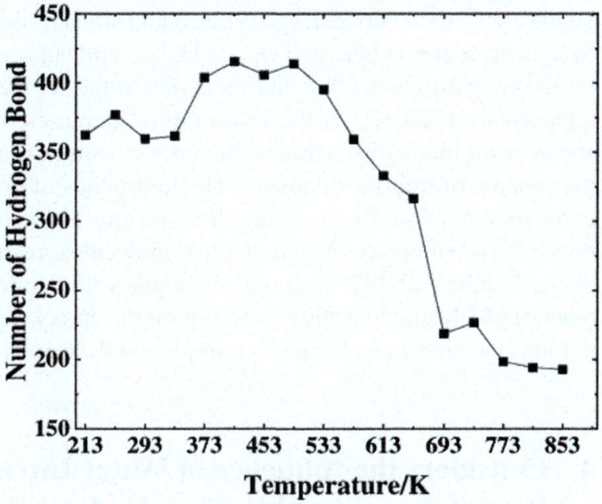

Fig. 24 Changes in the number of hydrogen bonds

tightly, leading to an increase in hydrogen bonding. Afterwards, as the temperature continues to rise, water molecules adhere to the surface of SiO_2 under the action of temperature, forming hydrogen bonds with them. At this time, the hydrogen bonds formed by O_{epoxy}-$H_{silicon}$ break, and the hydrogen bonds formed by O_{water}-$H_{silicon}$ remain stable at this temperature. As the temperature continues to rise (greater than 653 K), water molecules begin to break free from the binding of hydrogen bonds and leave the surface of silica under the influence of temperature, resulting in a sharp decrease in the number of hydrogen bonds. When the temperature is above 693 K, the distance between each component in the system is relatively far, and hydrogen bonds cannot be formed, resulting in a stable number of hydrogen bonds.

In summary, epoxy resin undergoes expansion and cracking under the influence of temperature and hydrolysis reaction, causing water molecules to pass through the epoxy resin to reach the interface area and form hydrogen bonds with silica molecules, hindering the formation of hydrogen bonds between epoxy resin and silica molecules, resulting in a decrease in interfacial binding energy. This is manifested in a decrease in interfacial binding energy at temperatures ranging from 493 to 613 K.

3.3 *Summary*

In this chapter, the influence of water intrusion on the epoxy resin/glass fiber interface under the consideration of temperature effect was studied, and the changes of interface structure, interface Binding energy, ester group number and hydrogen bond were emphatically analyzed. From the results of this chapter, it can be seen that as the temperature gradually increases, the molecular structure of epoxy resin gradually relaxes, water molecules are easy to enter the internal and interface areas of epoxy resin molecules. After water molecules invade the interface, hydrogen bonds are formed between epoxy resin and silica molecules, hindering the formation of hydrogen bonds between O_{epoxy}-$H_{silicon}$ and subsequently hindering the bonding of epoxy resin/glass fiber interface. When the temperature reaches the hydrolysis temperature (373 K), as the temperature increases, the hydrolysis reaction of the epoxy resin intensifies, causing the epoxy resin to gradually crack and accelerating the process of interface damage. The hydrolysis of epoxy resin promotes the easier entry of water into the interface area, further hindering the formation of hydrogen bonds between epoxy resin and silica molecules, resulting in a significant decrease in the bonding ability of epoxy resin/glass fiber interface. When the temperature reaches 653 K, the hydrolysis reaction of the epoxy resin is basically completed, and the interface has been basically completely destroyed.

4 Considers the Influence of Water Intrusion on the Epoxy Resin/Glass Fiber Interface Under Electric Field Effect

Under working conditions, composite insulators bear the effect of electric fields. In addition, due to water intrusion and other reasons, the interface between the core rod and the sheath and the inside of the core rod are prone to field strength distortion. Due to the polarity of water molecules, their migration behavior within the core rod is inevitably influenced by the strength of the electric field. In view of this, this chapter focuses on studying the changes in the epoxy resin/glass fiber interface after water intrusion under the electric field effect.

4.1 Selection Basis and Parameter Settings for Electric Field Strength

According to the research results of Ref. [44], due to stray capacitance and other reasons, the axial potential and electric field distribution of composite insulators exhibit significant non-uniformity, with a prominent manifestation being the high field strength on the wire side, which can reach a value of 3.9 kV/cm. The field strength in the middle of the insulator is the lowest, with a value of approximately 0.8 kV/cm. The field strength in the middle is relatively low. In addition, due to the infiltration of moisture at the interface of the core rod/sheath, the 15–20 cm electric field at the defect site will undergo distortion. When the field strength distortion occurs on the high-voltage side, the maximum field strength can reach 33.14 kV/cm.

According to the above principles, when conducting molecular simulation, the electric field intensity range is set to 1 kV/cm–33 V/cm, with an interval of 2 kV/cm. According to the research conclusion in last chapter, when the temperature is higher than 373 K, the epoxy resin will undergo a hydrolysis process, causing the interface performance to gradually decay. Therefore, during the simulation in this chapter, the temperature is still set at 373 K.

In summary, the parameter settings in this chapter are: temperature is 373 K; the air pressure is standard atmospheric pressure (0.1 MPa); the integration step size is 0.1 fs; 20,000 iterations; the total simulation time is 20.0 ps; the simulation module is GULP/Dreiding.

In addition, due to the extremely short molecular simulation time, usually calculated in femtoseconds (fs), which is much smaller than the alternating period of the AC electric field, the electric field can be equivalent to a constant electric field during simulation.

4.2 Analysis of the Influence of Water Intrusion on the Epoxy Resin/Glass Fiber Interface Under the Consideration of Electric Field Effect

4.2.1 Interface Structure Changes

This article first uses molecular simulation technology to analyze the changes in the interface structure of epoxy resin/glass fiber in a water molecular environment under the action of an electric field. The results are shown in Fig. 25.

From the results of Fig. 25, it is not difficult to find that when the electric field intensity is low (1 kV/cm), the number of water molecules entering the epoxy resin

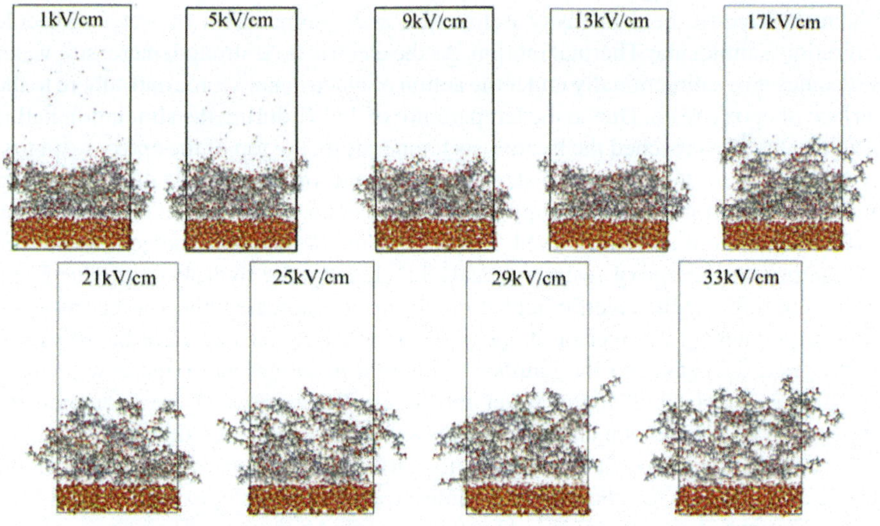

Fig. 25 Changes in the interface structure of epoxy resin/glass fiber in water molecular environments under different electric fields

is relatively small, and there is no significant change in the structure at the interface. When the electric field is 5–9 kV/cm, the number of water molecules attached to the epoxy numerical surface gradually increases, but there are still no obvious holes inside the epoxy resin, and the structure at the epoxy numerical/glass fiber interface remains unchanged. When the electric field strength exceeds 9 kV/cm, the top of the epoxy resin undergoes splitting and creates holes, through which water molecules can enter the interior of the epoxy resin. As the electric field strength further increases, the splitting phenomenon of epoxy resin intensifies, and the number of internal holes in epoxy resin increases. The number of water molecules entering the epoxy value and entering the interface through epoxy resin gradually increases. When the field strength reaches 29 kV/cm, the vast majority of water molecules have reached the interface area and attached to the SiO_2 surface.

To further analyze this phenomenon, this article presents the evolution results of the interface structure between epoxy resin and glass fiber in a water molecular environment at a field strength of 29 kV/cm, as shown in Fig. 26. During the 0–5 ps process, water molecules shift from random distribution to directional movement towards epoxy resin molecules, with a large number of water molecules adhering to the surface of the epoxy resin and some entering the interior of the epoxy resin. At 10 ps, the water molecules in the space have completely entered the interior of the epoxy resin, and the epoxy resin above begins to split, creating holes, and some water molecules reach between the epoxy resin and silica. At 15 ps, the number of water molecules at the interface increases, and the epoxy resin at the interface also decomposes. At 20 ps, a large number of water molecules have entered the interface area and the interface has been basically destroyed.

After analyzing the above rules, it can be concluded that after water intrusion, water molecules will gradually migrate towards the epoxy resin/glass fiber interface. When the electric field intensity is low, the main inducement of water molecule migration is molecular Thermal motion. As the electric field strength increases, water molecules move directionally under the action of electric field force and adhere to the surface of epoxy resin. Due to the temperature of 373 K during the simulation in this article, which has reached the hydrolysis temperature, the top of the epoxy resin first hydrolyzes under the action of water molecules and forms holes. At this time, water molecules are prone to pass through the holes and enter the interior of the epoxy resin under the action of an electric field. However, some of the water molecules entering the interior of the epoxy resin are easily blocked by the hydrolysis products and stop migrating. As the electric field strength further increases, the number of water molecules entering the interior of epoxy resin increases, the degree of hydrolysis of epoxy resin deepens, and the number of internal holes gradually increases. Under the strong electric field force, water molecules are more likely to overcome the obstruction of epoxy resin and its hydrolysis products and reach the interface.

To compare and illustrate the destructive effect of moisture on the interface, this article also presents the changes in the interface structure of epoxy resin/glass fiber under different electric fields in a dry environment. As shown in Fig. 27.

From the results in Fig. 27, it can be seen that due to the fact that both epoxy resin and silica are non-polar molecules and have no functional groups available for

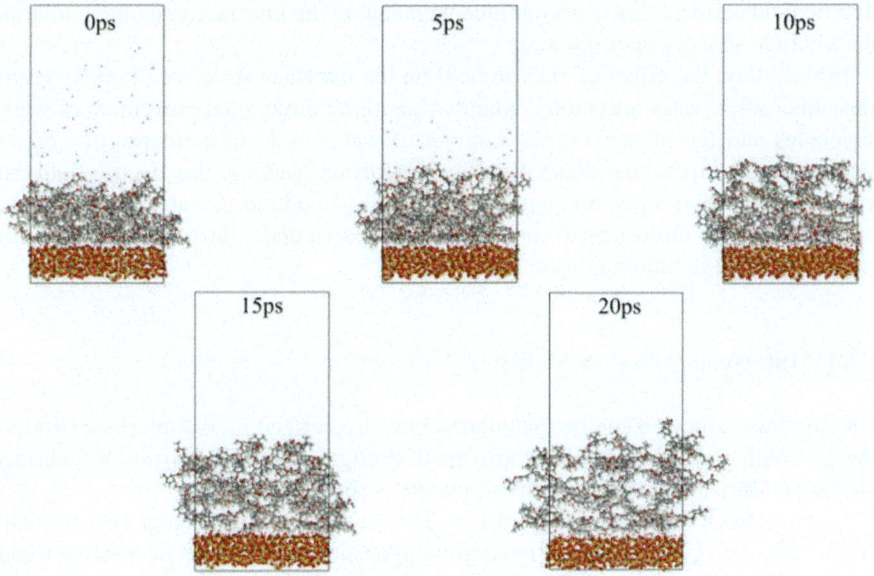

Fig. 26 Changes in interface model under a 29 kV/cm electric field environment

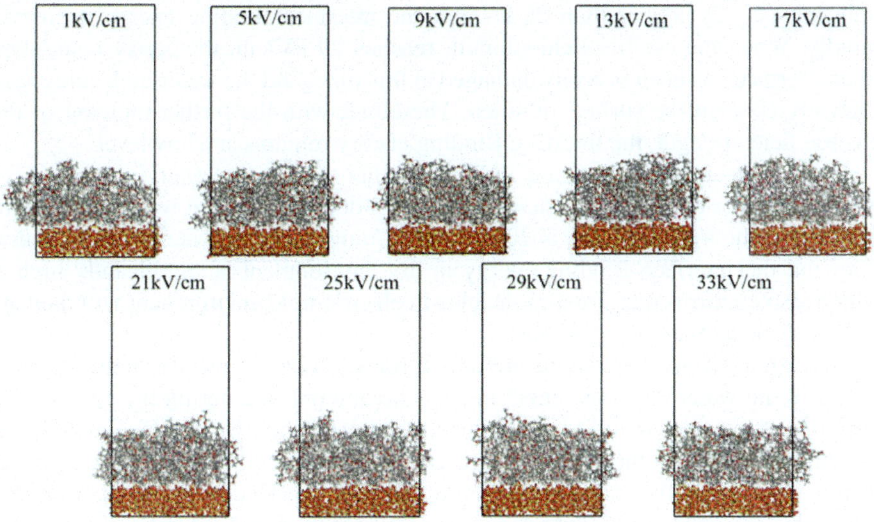

Fig. 27 Changes in the interface structure of epoxy resin/glass fiber in dry environments under different electric fields

electric field action, there is no significant change in the interface structure under the electric field in a dry environment.

In summary, the effect of electric field on the interface structure of epoxy resin/glass fiber after water intrusion is mainly due to the directional movement of water molecules and the increase in the number of water molecules entering the epoxy resin. At the temperature allowed by the hydrolysis reaction, the electric field can trigger the hydrolysis reaction and generate holes. In addition, water molecules can cause more water molecules to enter the interface area under the action of an electric field, affecting the binding of the interface.

4.2.2 Interface Adhesion Ability

The interface adhesion can be quantitatively characterized by the interface Binding energy. With the increase of electric field strength, the change rule of interface Binding energy under different environments is shown in Fig. 28.

It can be seen from the results of Fig. 28a. In the water molecular environment, the Binding energy of the epoxy resin/glass fiber interface shows a decreasing trend. When the electric field intensity is low (1–9 kV/cm), the interface Binding energy value is high and basically stable. When the electric field intensity is 9–25 kV/cm, the interface Binding energy shows a gradual decline trend. When the electric field intensity is greater than 25 kV/cm, the interface Binding energy decreases rapidly. When the electric field strength reaches 29 kV/cm, the epoxy resin/glass fiber interface has been severely damaged at this time, and the water molecules have basically covered the surface of silica. Therefore, with the further increase of the electric field strength, the interface Binding energy remains at a low level.

It can be seen from the results in Fig. 28b that the Binding energy of the epoxy resin/glass fiber interface is basically stable in a dry environment with small fluctuations, and the average value is 2017.98 eV. Compared with water molecule environment, the interface Binding energy in dry environment is significantly higher, which also confirms that water molecules in electric field environment will damage the epoxy resin/glass fiber interface.

According to the analysis of the above rules, when the electric field intensity is low in the water molecule environment, the amount of entering the epoxy resin interior and interface is still small, at this time, the interface Binding energy remains at a high value. With the increase of electric field strength, water molecules begin to move directionally and enter into the epoxy resin. Under the action of water molecules, the epoxy resin undergoes hydrolysis reaction and generates holes. Water molecules entering the interface also increase, and the interface Binding energy gradually decreases. With the further increase of the electric field strength, the water entering the epoxy resin gradually becomes more and more, the degree of hydrolysis of the epoxy resin deepens and the holes increase. Under the strong electric field, water molecules are more likely to break through the barriers of the epoxy resin and its hydrolysates and enter the interface, which greatly reduces the interface Binding energy and severely damages the epoxy resin/glass fiber interface.

Fig. 28 Electric field intensity

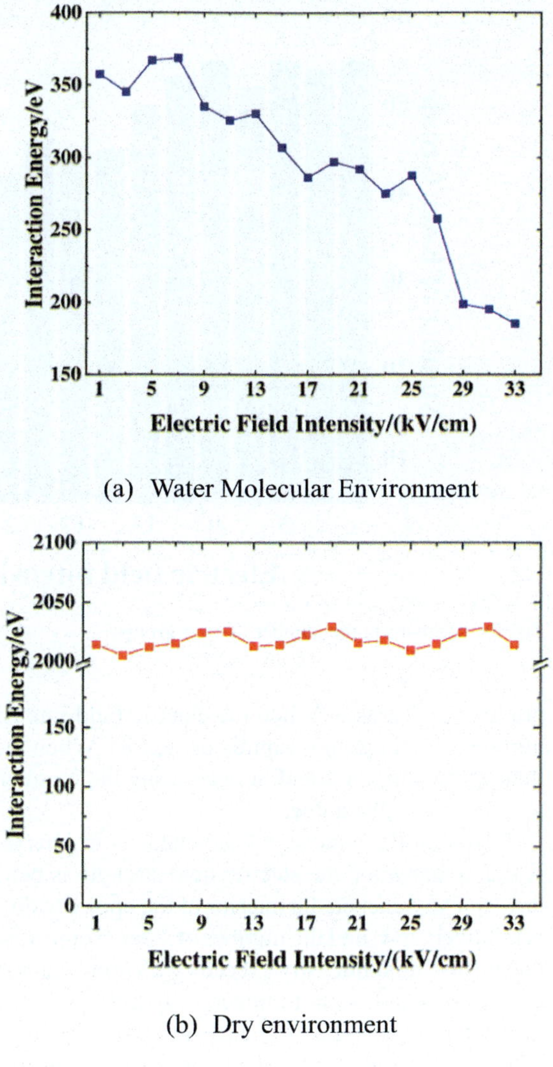

4.2.3 Microstructural Changes

As the electric field strength increases, the internal moisture of the epoxy resin gradually increases and hydrolysis occurs. Therefore, this section follows the method in last chapter and characterizes the degree of hydrolysis reaction based on the number of ester groups. The results are shown in Fig. 29.

From Fig. 29, it can be seen that as the electric field intensity increases, the overall number of ester groups shows a decreasing trend. When the electric field intensity is less than 9 kV/cm, there is no significant change in the number of ester groups. When the electric field intensity is greater than 9 kV/cm, the number of ester groups

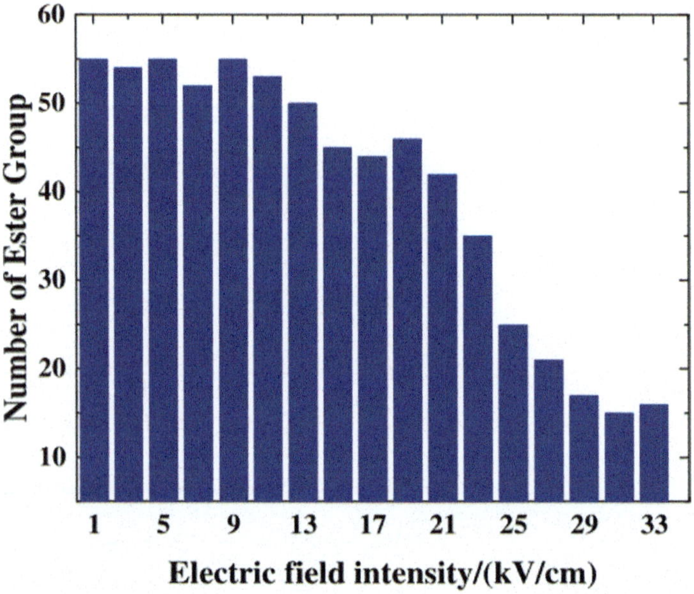

Fig. 29 Changes in the number of ester groups

gradually decreases. When the electric field intensity is greater than 19 kV/cm, the number of ester groups rapidly decreases. When the electric field intensity is greater than 25 kV/cm, the trend of decreasing the number of ester groups slows down and remains basically stable.

Based on the analysis of the changes in interface results in the previous text, it is found that when the electric field strength is below 9 kV/cm, the water molecules have not yet reached the interior of the epoxy resin, the degree of hydrolysis reaction is relatively low, and the number of ester groups remains basically unchanged. When the electric field intensity exceeds 9 kV/cm, water molecules reach the interior of the epoxy resin and begin to undergo hydrolysis, resulting in a decrease in the number of ester groups. When the electric field is greater than 19 kV/cm, a large amount of water molecules enter the interior of the epoxy resin, causing a large number of ester groups to fracture. As the electric field intensity continues to increase, when it reaches 25 kV/cm, most of the water molecules have reached the surface of silica with the strong electric field effect, so the ester base amount is relatively stable.

4.2.4 Hydrogen Bond Changes

According to the analysis results in Sect. 4.2.1, when the electric field strength is high, a large number of water molecules pass through the epoxy resin and reach the interface area. According to the analysis of hydrogen bond formation, oxygen atoms in water molecules are prone to form hydrogen bonds with hydrogen atoms on the

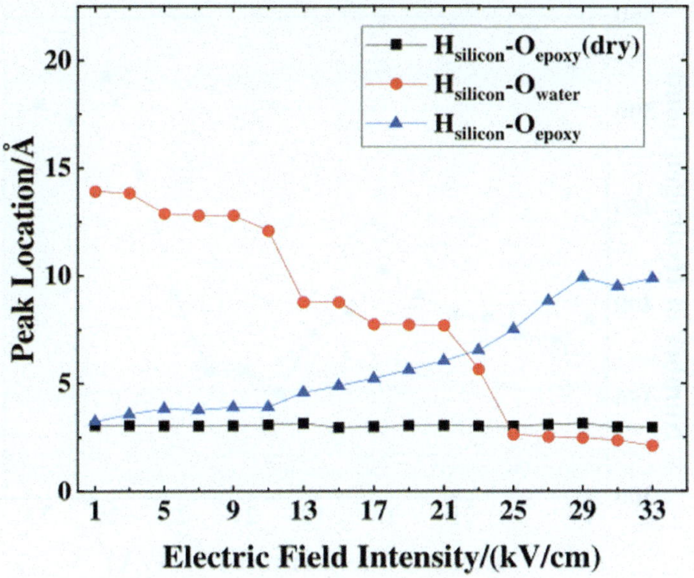

Fig. 30 RDF peak distance

surface of silica, hindering the formation of hydrogen bonds between epoxy resin and silica, and damaging the interface bonding. Therefore, consistent with the method in Sect. 3, RDF analysis can be performed on the relevant atoms in the interface region under electric field environment. This section also conducts RDF analysis on the relevant atoms, and the results are shown in Fig. 30.

From the results in Fig. 30, it can be seen that the distance from the peak RDF of $H_{silicon}$-O_{epoxy} (dry) in a dry environment remains basically unchanged with the change of electric field intensity. This indicates that the distance between epoxy resin and silica in a dry environment does not change significantly. This is consistent with the conclusion that electric field intensity has no significant effect on the epoxy resin/glass fiber interface in a dry environment when analyzing the changes in interface structure in the previous section. In the water molecule environment, as the electric field intensity increases, the distance from the peak RDF of $H_{silicon}$-O_{epoxy} gradually increases, while the distance from the peak RDF of Hsilicon-O_{water} gradually decreases. In addition, in the water molecule environment, when the electric field intensity exceeds 25 kV/cm, the RDF peak distance of $H_{silicon}$-O_{water} is already smaller than that of $H_{silicon}$-O_{epoxy}. This indicates that in the water molecule environment, when the electric field intensity exceeds 23 kV/cm, hydrogen bonds will preferentially form between the water molecule and silicon dioxide.

In summary, as the electric field strength increases, water molecules gradually accumulate at the interface and preferentially form hydrogen bonds with SiO_2, thereby hindering the formation of hydrogen bonds between SiO_2 and epoxy resin, resulting in a decrease in the degree of bonding between epoxy resin and glass fiber.

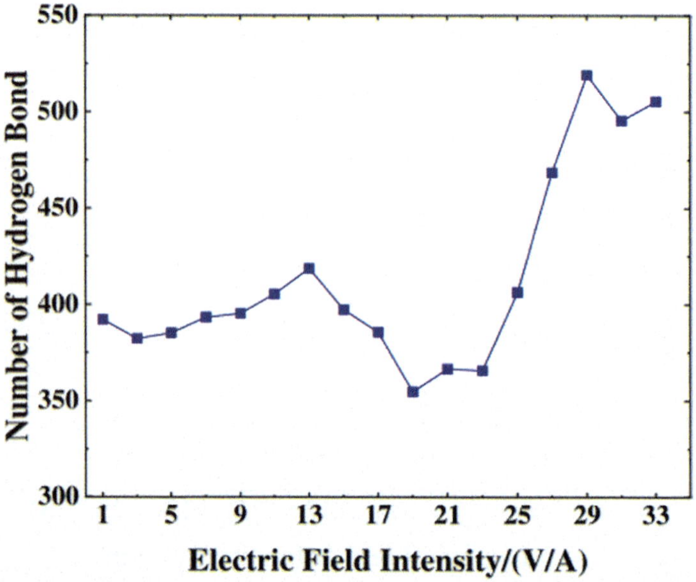

Fig. 31 Changes in the number of hydrogen bonds at the interface under different environments

Therefore, analyzing the changes in the number of hydrogen bonds can help clarify the destructive effect of water intrusion on the epoxy resin/glass fiber interface under the action of an electric field. Figure 31 show the changes in the number of hydrogen bonds when the electric field changes.

From Fig. 31, it can be seen that in the water molecular environment, the overall number of hydrogen bonds shows a trend of first decreasing and then increasing. When the electric field intensity is between 1–13 kV/cm, the number of hydrogen bonds in the interface remains basically stable. When the electric field intensity is greater than 13 kV/cm, the number of hydrogen bonds decreases. When the electric field intensity is greater than 19 kV/cm, there is a significant increase in the number of hydrogen bonds. When the electric field intensity is greater than 29 kV/cm, the number of hydrogen bonds reaches its peak and remains stable.

Based on the analysis of the above rules, it can be concluded that when the electric field strength is low, the main form of hydrogen bonding at the interface is O_{epoxy}-Hsilicon. However, there is less O_{epoxy} near the interface area, resulting in less hydrogen bonding with SiO_2. Additionally, there are fewer water molecules at the epoxy resin and interface, which do not participate in the formation of hydrogen bonds. As the electric field strength increases, water molecules enter the interior of the epoxy resin. At this point, a small amount of water molecules reach the bottom epoxy resin and undergo hydrolysis reactions with it, breaking some hydrogen bonds. Most water molecules are blocked by the hydrolysis products of the epoxy resin and cannot continue to move. When the electric field strength continues to increase, water molecules break through the barrier of epoxy resin hydrolysis products and reach the surface of silica, forming hydrogen bonds with them. The higher the electric field strength, the more water molecules reach the surface of silica, and the more hydrogen bonds can be formed. The final manifestation is that there are more hydrogen bonds formed when the electric field intensity is high than when the electric field intensity is low.

In summary, when the environmental temperature reaches the hydrolysis temperature of the epoxy resin, water molecules migrate directionally to the interior of the epoxy resin under the action of an electric field, causing the epoxy resin to undergo hydrolysis. When water molecules reach the interface area and form hydrogen bonds with silica, the formation of hydrogen bonds between epoxy resin and SiO_2 is hindered, which leads to the reduction of interface Binding energy.

4.3 Summary

In this chapter, the influence of water intrusion on the epoxy resin/glass fiber interface under the effect of electric field was studied, and the change laws of interface structure, interface Binding energy, ester group number, hydrogen bond number, etc. were emphatically analyzed. From the results of this chapter, it can be seen that: (1) In a dry environment, the electric field has no significant effect on the epoxy resin/glass fiber interface; (2) In the environment of water molecules, an electric field will cause water molecules to migrate directionally through epoxy resin to the interface, and when the environmental temperature reaches the hydrolysis temperature of epoxy resin, it will cause the epoxy resin to undergo hydrolysis and generate holes; (3) The hydrolysis of epoxy resin and the generation of holes promote the migration process of water molecules, and under strong electric fields, water molecules are more likely to break through the binding of epoxy resin and its hydrolysis products and migrate to the interface; (4) As the electric field strength increases, the moisture at the interface gradually increases, promoting the hydrolysis of epoxy resin and hindering the formation of hydrogen bonds between epoxy resin and silica molecules. Under the combined effect of the above effects, the epoxy resin/glass fiber interface gradually damages.

5 Effect of Moisture Intrusion on Epoxy Resin/Glass Fiber Interface Considering Synergistic Effects of Temperature and Electric Field

The temperature and electric field at the insulator rod are variable due to environmental factors and local defects. Therefore, in this chapter, based on the previous paper, molecular simulation techniques will be used to investigate the characteristics of the moisture intrusion on the epoxy resin/glass fiber interface characteristics when considering the synergistic effect of temperature and electric field. Combining the principles of temperature and electric field settings in Sects. 3 and 4, this chapter determines that the temperature range of the simulation experiment is 213–853 K, with an interval of 80 K; the electric field strength range is 1–33 kV/cm, with an interval of 4 kV/cm.

5.1 Considering the Synergistic Effect of Temperature and Electric Field, the Influence of Water Intrusion on the Interface of Epoxy Resin/Glass Fiber Was Analyzed

5.1.1 Interface Structure Changes

Firstly, the epoxy resin/glass fiber interface in a water molecule environment is analyzed when the temperature and electric field change synergistically. Figure 32 shows the results of the interfacial structure changes in the water molecule environment at temperatures of 213 K, 453 K, 613 K, and 853 K with electric field strengths of 1 kV/cm, 13 kV/cm, 21 kV/cm, and 33 kV/cm, respectively.

The results in Fig. 32 can be summarized as follows:

At low temperatures (typically 213 K), with the increase of electric field strength, water molecules can migrate to the surface of epoxy resin, but the structure of epoxy resin and epoxy resin/glass fiber interface do not change significantly.

At medium and high temperatures (typically 431 K), the hydrolysis of the epoxy resin gradually appears, resulting in the expansion of the epoxy resin structure, and along with the increase in the electric field strength, water is more likely to migrate to the interior of the epoxy resin and the interface, resulting in the destruction of the structure of the epoxy resin and the interface between the epoxy resin and the glass fibers.

At high temperatures (typically 613 K), the internal and interface structures of epoxy resin are basically similar under different electric fields, and the effect of the electric field on the epoxy resin/glass fiber interface is relatively small, even if the strength of the electric field is low, the moisture exists more around the epoxy resin and at the epoxy resin/glass fiber interface.

For comparison, this paper gives the temperature and electric field in Fig. 32, dry environment, epoxy resin/glass fiber interface structure change rule, the results are shown in Fig. 33:

In the dry environment, the electric field strength had no significant effect on the epoxy resin/glass fiber interface at different temperatures.

In dry environments, as the temperature rises, the holes in epoxy resin increase gradually.

In summary, when moisture intrusion occurs, the epoxy resin/glass fiber interface structure will be co-influenced by the electric field and the temperature. At lower

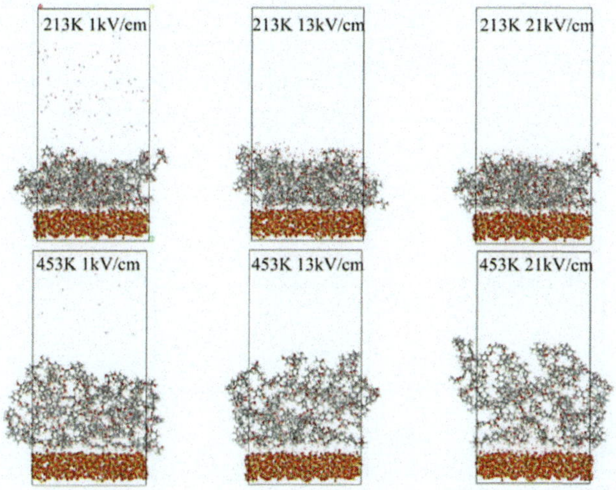

(a) 213K-453K

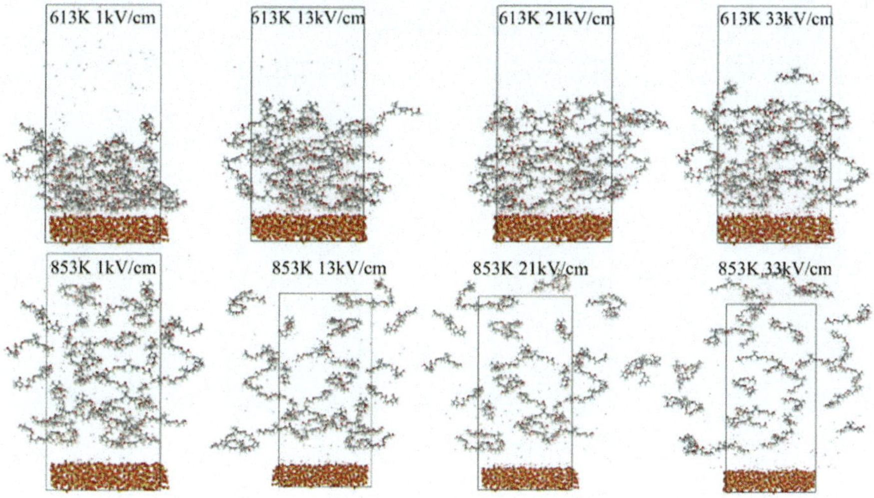

(b) 213K-853K

Fig. 32 Interfacial structure change underwater molecule environment

temperatures (below the hydrolysis temperature), the electric field has no significant effect on the interfacial structure. When the temperature is higher than the hydrolysis temperature, the electric field will make the water molecules move directionally and promote the migration process of water molecules inside the epoxy resin, therefore, the electric field promotes the destruction process of water on the interface at this temperature. When the temperature is further increased, the molecular thermal movement is intensified, the migration of water will be mainly affected by the molecular thermal movement, and the hydrolysis of the epoxy resin is also stronger, at the

(a) 213K-453K

(b) 613K-853K

Fig. 33 Changes in the interfacial structure under a dry environment

same time, the thermal effect under the action of the epoxy resin itself will produce cavities, the structure of the epoxy resin expands, resulting in a lower temperature when the interfacial region of the epoxy resin is closer to the interface. And when the temperature is further increased, at this time the epoxy resin can epoxy resin/glass fiber interface will be extremely easy to be damaged, at this time the electric field on the epoxy resin/glass fiber interface structure influence effect will become weaker.

5.1.2 Interface Adhesion Ability

In this paper, we analyze the adhesion capacity of epoxy resin/glass fiber interface under a water molecule environment at different temperatures and electric fields. Consistent with the analytical approach in Sects. 3 and 4, the interface adhesion capacity is still characterized by calculating the interface binding energy. Also, for comparison, the epoxy resin/glass fiber interface bonding energies are given together in this paper for dry environments.

As can be seen in Fig. 34, the interface binding energy in the water molecule environment decreases at different temperatures and with the increase of electric field strength. Among them, the interface bonding energies at low and high temperatures (213, 693, 773, 853 K) are lower than those at other temperatures. At 453–613 K, the decrease of interface binding energy with the increase of electric field strength is larger, while the decrease of interface binding energy with the increase of electric field strength is smaller at 213 K and 693–853 K, and the higher the temperature, the slower the decrease of interface binding energy with the increase of electric field.

Combined with the analysis of the interfacial structure change in Sect. 5.1.1, it can be seen that the effect of electric field on the interface bonding energy is mainly due to the directional movement of water molecules by the electric field, which will cause water molecules to enter the internal cavities of the epoxy resin and lead to its expansion, and also trigger the hydrolysis of the epoxy resin and the formation of hydrogen bonds with silica. Therefore, the interface binding energy tends to decrease as the electric field strength increases. The reason for the smaller magnitude of change in interface binding energy at low temperatures may be that the

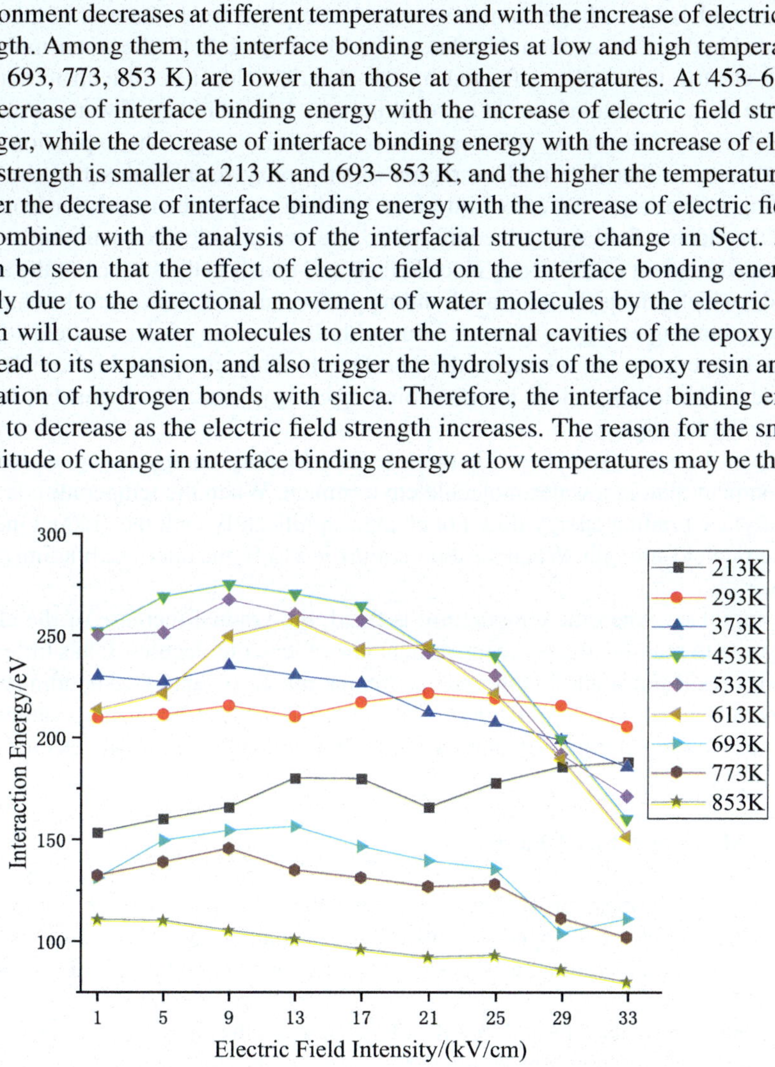

Fig. 34 The change of interface binding energy in the water molecular environment

epoxy molecules are more compact in this environment due to the lower temperatures, and thus it is difficult for water molecules to pass through the epoxy to enter the interfacial region even at higher electric fields. The movement of water molecules is mainly affected by temperature under high temperature environment, the electric field for water molecules is small, resulting in high-temperature environments interface binding energy with the increase in the electric field does not change much. In the temperature range 453–613 K, water molecules are similarly affected by temperature and electric field, and are able to diffuse with the increase of temperature and move along the direction of electric field. In this temperature environment and the electric field strength is low, a large number of water molecules into the interior of the epoxy resin and hydrolysis reaction, but due to the low electric field, the water molecules by the electric field is less affected by the electric field, it is difficult to continue to move to the surface of the SiO_2, resulting in a large number of water molecules in the interior of the epoxy resin cavities build up and trigger the expansion of the epoxy resin, so that the interface region of the epoxy resin is closer to the SiO_2, resulting in the improvement of interface bonding properties in this environment, when the electric field strength continues to rise, water molecules continue to move in the environment of the strong electric field and attach to the surface of the silicon dioxide, which affects the interface bonding performance. It shows that the change of interface binding energy is more obvious with the increase of electric field strength in this temperature environment, the interface binding energy is higher when the electric field is lower, and the interface binding energy decreases obviously when the electric field is higher.

As can be seen from Fig. 35, the interface binding energy is higher in a dry environment than in a water molecule environment. When the temperature is fixed, the interface binding energy does not change significantly with the gradual increase of electric field strength. When the temperature is 213 K, the interface binding energy is lower.

In summary, when the temperature is fixed, the gradual increase of the electric field strength mainly affects the movement law of water molecules. It has little effect on epoxy resin and silica. At low and high temperatures, the interface bonding energy is generally lower but less affected by the electric field; at other temperatures, the interface bonding energy is generally higher but more affected by the electric field.

5.1.3 Microstructure Change

To further analyze the hydrolysis reaction law of the epoxy resin when the temperature and electric field were changed synergistically, the number of ester groups inside the system at each temperature and electric field was counted, and the results are shown in Fig. 36.

Comparison by the 213 and 293 K in Fig. 36 when the change rule of ester group can be seen; in the low temperature, the number of ester group in the system is basically stable, and the value is higher, and with the electric field strength is not related. This analysis shows that at low temperatures, there is no obvious hydrolysis

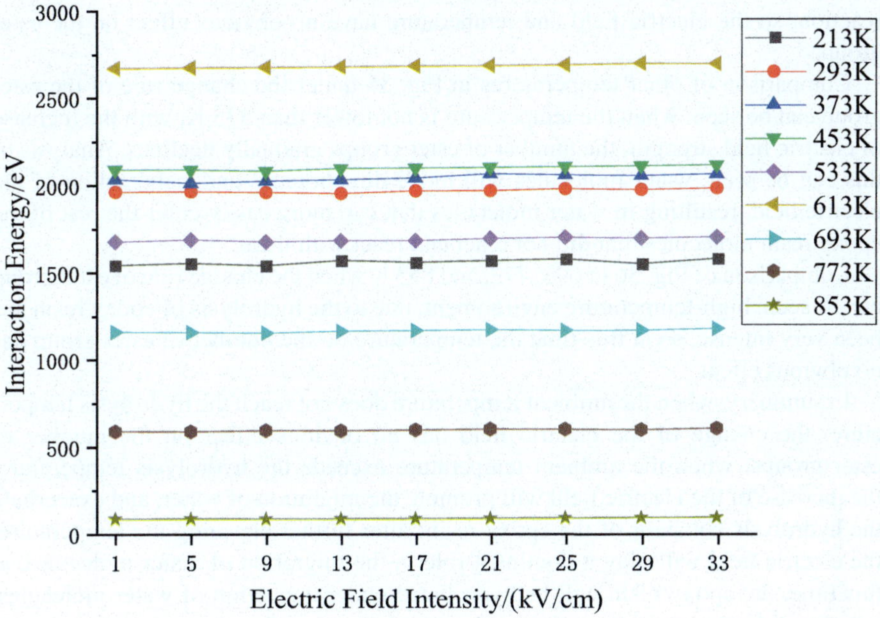

Fig. 35 The change of interface binding energy in a dry environment

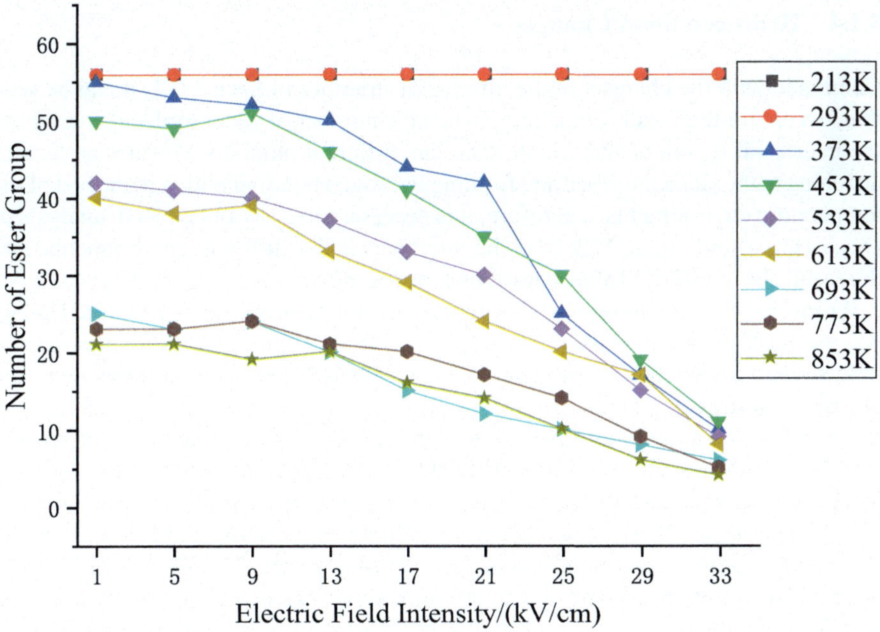

Fig. 36 Changes in the number of ester groups

reaction, so the electric field and temperature have no obvious effect on the ester group.

Comparison of other temperatures in Fig. 36 under the change rule of the ester group can be seen, when the temperature is not lower than 373 K, with the increase of electric field strength, the number of ester groups gradually declines. Analysis of this can be seen, water molecules will move directionally under the action of the electric field, resulting in water molecules that can more easily enter the rest of the epoxy resin molecules that did not react and react with them.

Comparison of Fig. 36 in 693, 773, and 853 K when the change rule of ester group can be seen, high-temperature environment, due to the hydrolysis of epoxy resin has been very intense, so at this time the temperature on the number of ester group has no obvious effect.

In summary, when the ambient temperature does not reach the hydrolysis temperature, the change of the electric field has no obvious effect on the number of ester groups; when the ambient temperature exceeds the hydrolysis temperature, the increase of the electric field will promote the migration of water, and exacerbate the hydrolysis behavior of the epoxy resin; after further elevating the temperature, the electric field will play a dominant role in the migration of water molecules, at this time, the epoxy resin will be hydrolyzed under the action of water molecules, which will lead to the destruction of the interface.

5.1.4 Hydrogen Bond Changes

Combined with the changes in the interfacial structure in Sect. 5.2.1, it can be seen that, except for the lower temperature when the number of water molecules reaching the interfacial region is very small, there are different numbers of water molecules attaching to the silica interface and forming hydrogen bonding with it in the rest of the temperature environments. Therefore, it is necessary to carry out an RDF analysis of O_{water}-$H_{silicon}$ and O_{epoxy}-$H_{silicon}$ at the interfaces under different environments, and the results of the RDF analysis are shown in Fig. 37.

From Fig. 37a, it can be seen that in the low-temperature environment (213–293 K), the peak distance of O_{epoxy}-$H_{silicon}$ is stable with the increase of the electric field strength; when the temperature is greater than 293 K, with the increase of the electric field strength, the increase of the peak distance of O_{epoxy}-$H_{silicon}$ is faster and faster; when the temperature is greater than 773 K, the peak distance is increasing rapidly. As can be seen from Fig. 37b, the O_{epoxy}-$H_{silicon}$ peak distance at a gradually increasing distance and fluctuates with increasing temperature. As can be seen in Fig. 37c, the O_{water}-$H_{silicon}$ peak distance at temperatures below 293 K at a gradually decreasing distance to near 8 Å and then remains stable. The O_{water}-$H_{silicon}$ peak distance decreases to around 2.5 Å at temperatures greater than 373 K and then stabilizes, but the O_{water}-$H_{silicon}$ peak distance varies at lower field strengths: at 373–453 K, the peak distance decreases from the higher value to around 2.5 Å; at 533–693 K, the peak distance reaches around 2.5 Å after a small decrease; at 773–853 K, the peak distance does not start to decrease until the electric field strength is at 13 kV/cm,

Fig. 37 RDF analysis of interfaces in different environments

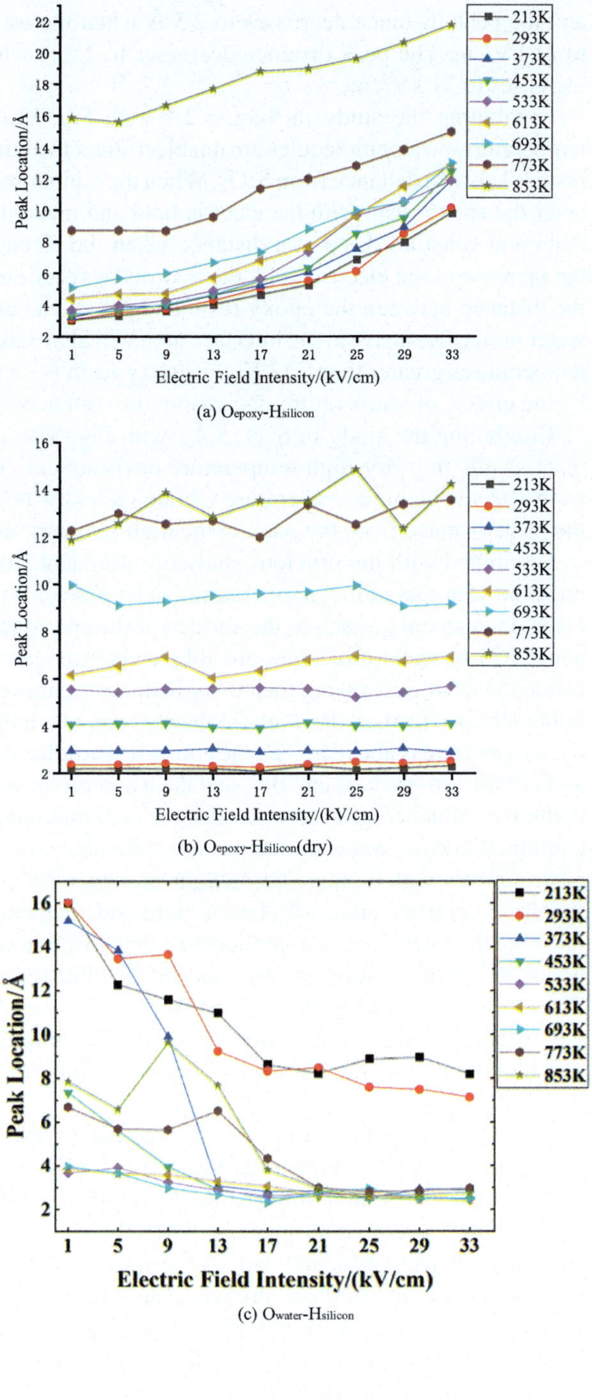

(a) O_{epoxy}-$H_{silicon}$

(b) O_{epoxy}-$H_{silicon}$(dry)

(c) O_{water}-$H_{silicon}$

and the peak distance decreases to 2.5 Å when the electric field strength increases to 21 kV/cm. The peak distance decreases to 2.5 Å when the electric field strength increases to 21 kV/cm.

Combining the study in Sect. 5.2.1 with Fig. 37a, it can be seen that at low temperatures, water molecules are unable to enter the interfacial region, and the epoxy resin is kept at a distance from SiO_2. When the temperature rises, water molecules can enter the epoxy resin with the electric field and make it crack and expansion, epoxy resin and silica are keep at a distance when the electric field strength is low, with the increase of the electric field, the hydrolysis reaction is more intense, at this time, the distance between the epoxy resin and the silica has increased, and some of the water molecules arrive at the interface area will also make the interface separation. At temperatures greater than 773 K, the epoxy resin is dispersed into the spatial region by the effects of temperature, increasing the distance to the peak significantly.

Combining the study in Sect. 5.2.1 with Fig. 37b, it can be seen that the epoxy resin swells in a dry high-temperature environment. The groups inside the epoxy resin affected by high temperature vibrate violently, resulting in large fluctuations in the peak distance from the peak in the high-temperature environment.

Combined with the previous study of interfacial structure analysis and Fig. 37c, it can be seen that at low temperature, water molecules can not enter into the epoxy resin, and can only attach to the surface of the epoxy resin. As the temperature rises above 373 K, water molecules are able to enter the hydrolyzed epoxy resin and stay inside the cavities, causing the epoxy resin to expand as a whole, and as the electric field increases further, the water molecules can reach the surface of the silica. When the temperature is above 533 K, the epoxy resin molecules are subjected to hydrolysis and swelling to produce cavities, and the water molecules reach the interfacial region under the influence of a stronger electric field and temperature. As the temperature continues to rise, water molecules are only able to reach the interfacial region in environments with higher electric field strengths.

The synergistic effect of electric field and temperature on the water molecules invading the interface is a complicated mechanism, in which the change of hydrogen bonds affects the bonding performance of the interface under different environments. Therefore, in this paper, the method of counting the number of hydrogen bonds in the interface is also used to investigate the bonding degree of the interface under different environments. The variation in the number of hydrogen bonds is shown in Fig. 38.

As can be seen from Fig. 38, the number of hydrogen bonds did not change significantly when the temperature was 213–293 K; when the temperature was 373–613 K, the trend of the number of hydrogen bonds with the increase of the electric field strength was a small increase when the electric field strength was 9–13 kV/cm, and then decreased, and then increased, and then reached the lowest point at 21 kV/cm. As the electric field strength continues to increase, there is a large overall decrease in the number of hydrogen bonds; as the temperature increases, the curve of the number of hydrogen bonds with the electric field strength flattens out.

Combined with the structural change of the interface in Sect. 5.2.1 and the RDF analysis, it can be seen that in the lower temperature environment, there is no water

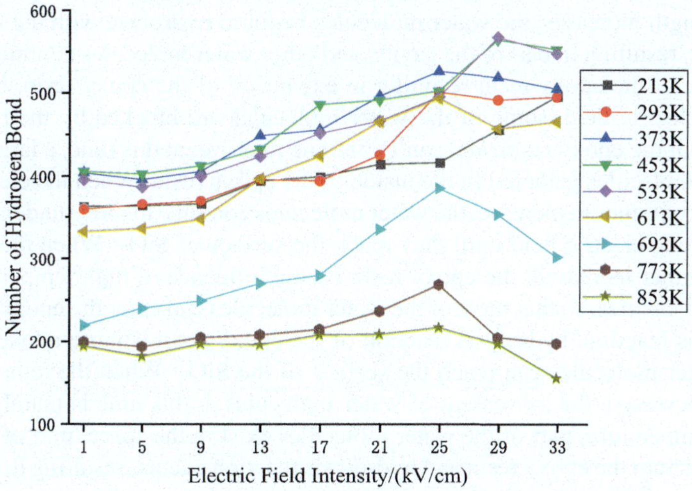

Fig. 38 The change of the number of hydrogen bonds at the interface under the environment of water molecules

molecule entering into the interface region, so the number of hydrogen bonds has not changed significantly. The temperature continued to increase above 373 K. At the lower electric field strength, water molecules hydrolyzed with the top epoxy resin, resulting in the expansion of the epoxy resin after the creation of cavities at the top of the epoxy resin, and the epoxy resin in the interface region bulged out to the SiO_2, and hydrogen bonds was formed, at which time the number of hydrogen bonds increased slightly. With the increase of the electric field, a small number of water molecules into the bottom of the epoxy resin area and hydrolysis reaction with it, destroying part of the hydrogen bond, most of the water molecules by the epoxy resin hydrolysis products blocking, can not reach the SiO_2 surface, at this time, the number of hydrogen bonds show a gradual decrease in the trend. When the electric field strength is large, the water molecules break through the barrier in the strong electric field environment and continue to move to the surface of SiO_2 and form hydrogen bonds with it, at which time the trend of the number of hydrogen bonds shows a gradual increase. When the temperature is high, epoxy resin due to hydrolysis reaction and high temperature, gradually cracked and dispersed to the space region, at this time, the number of hydrogen bonds drops dramatically, the interface is damaged.

In summary, the water molecules in the electric field and temperature under the synergistic effect of the law of motion and distribution of the main: in the low temperature and low electric field environment, water molecules are mainly distributed in the space region; when the electric field is raised, the water molecules move directionally to the surface of the epoxy resin and cannot enter the inside of the epoxy resin. When the temperature is raised to 373 K and above, the water molecules still exist in the space region under the low electric field environment. When the electric field strength increases, the water molecules begin to hydrolyze with the top epoxy resin first, resulting in part of the cavity, and other water molecules continue to enter the inside of the epoxy resin resulting in expansion of the epoxy resin, but due to the low electric field, some of the water molecules are blocked by the hydrolysis products of the epoxy resin, and can't continue to move, at this time, a large number of water molecules gathered in the inside of the epoxy resin. When the electric field strength continues to increase, the water molecules continue to move under the action of the strong electric field until they reach the surface of SiO_2. When the temperature is further increased, the epoxy resin by the influence of high temperature first generated cavities, so that most of the water molecules can enter the epoxy resin and hydrolysis reaction, hydrolysis reaction of the epoxy resin is more relaxed, so that more water molecules can reach the surface of the SiO_2. When the temperature is further increased, the movement of water molecules at this time is mainly affected by the temperature, part of the water molecules exist in the space, part of the water molecules into the epoxy resin and hydrolysis reaction occurs, resulting in the epoxy resin is cracked and dispersed in the space region, and a part of the water molecules attached to the surface of SiO_2.

5.2 *Destruction Mechanism of Moisture on Epoxy Resin/ Glass Fiber Interface of Composite Insulator Rod*

The process of moisture intrusion at the epoxy/fiberglass interface of composite insulator mandrels considering the synergistic effects of temperature and electric field is summarized.

As the temperature rises, the epoxy resin expands and creates cavities, while the water molecules move irregularly in response to the heat. When the temperature reaches 373 K, the epoxy resin and water molecules will produce a hydrolysis reaction, the hydrolysis reaction further caused by the epoxy resin structure, resulting in an increase in the number of cavities within the epoxy resin, and leading to a decrease in the interfacial bonding properties. In addition, due to the increase of cavities inside the epoxy resin, the migration process of water molecules to the epoxy resin/glass fiber interface can be facilitated, and the water molecules at the interface will first form hydrogen bonds with SiO_2 and hinder the formation of hydrogen bonds between epoxy resin-SiO_2. Under the combined effect of hydrolysis and hydrogen bonding disruption, the bonding ability of the epoxy resin/glass fiber interface decreases, and the interface is gradually destroyed.

The migration process of water inside the epoxy resin is affected by both temperature and electric field. When the temperature is low, the water molecules can be directed to move inside the epoxy resin by increasing the strength of the electric field, but because the temperature is too low to reach the hydrolysis conditions of the epoxy resin, and the structure of the epoxy resin is relatively dense, it is difficult for the water molecules to destroy the epoxy resin and the structure of the interface even under the action of the electric field, so there is no significant effect of the electric field on the interface of the epoxy resin/fiberglass when the temperature is low.

With the increase of temperature, when the temperature reaches the hydrolysis temperature of the epoxy resin (the simulation result in the paper is 373 K), under the action of low electric field, the water molecules are still dominated by the irregular thermal movement, and the number of water molecules entering the interior of the epoxy resin is relatively small. However, as the electric field strength increases, the electric field force on the water molecules gradually increases, which leads to more water molecules attaching to the surface of the epoxy resin or entering into the interior of the epoxy resin, and hydrolyzes the epoxy resin. Under the combined effect of hydrolysis reaction and spontaneous expansion of epoxy resin at high temperature, the number of cavities inside the epoxy resin increases. The increase of cavities and promote the migration of water inside the epoxy resin, but the epoxy resin and its hydrolysis products on the migration of water molecules have a certain impediment to the water molecules, when the water molecules are not enough to break through the above impediment to the electric field, the number of water molecules into the interface at the interface did not significantly increase, so the interface bonding energy decreases slowly at this time. As the field strength is further increased, the majority of water molecules can break through the epoxy resin and its hydrolysis products and reach the interface. Hydrolysis of the epoxy resin at the interface, and the presence of water molecules destroy the formation of hydrogen bonds between the epoxy resin and SiO_2, which further reduces the degree of interface bonding. Therefore, at this time, the increase of electric field strength has an accelerating effect on the destruction of the epoxy resin/glass fiber interface.

With the further increase of temperature, the irregular thermal movement of water molecules is the dominant factor for its migration, and the degree of hydrolysis damage to the epoxy resin has been more serious, the interface can be basically destroyed, so at this time to increase the electric field on the epoxy resin/glass fiber interface bonding degree has no significant effect.

In summary, when considering the synergistic effect of electric field and temperature, the damage mechanism of epoxy resin/glass fiber interface by moisture intrusion is analyzed from the molecular nanoscopic level as follows:

Temperature causes the epoxy resin to expand, and as the temperature rises, there are more cavities within the epoxy resin, which affects the bonding properties of the epoxy resin/glass fiber interface;

The destructive effect of moisture on the epoxy resin/glass fiber mainly exists: moisture makes the epoxy resin undergo hydrolysis, and moisture at the interface affects the formation of hydrogen bonds between the epoxy resin/glass fiber;

The electric field itself has no significant effect on the epoxy resin/glass fiber interface, but the electric field will cause the moisture to migrate directionally. When the temperature reaches the hydrolysis temperature of the epoxy resin, increasing the electric field can accelerate the migration process of the moisture and aggravate the damage process of the moisture on the body of the epoxy resin and the interface of the epoxy resin/glass fiber;

When the temperature is high enough, the hydrolysis reaction of the epoxy resin itself has been more obvious, and at this time increasing the electric field on the epoxy resin/glass fiber interface damage process has no obvious effect.

5.3 Summary

This section investigates the electro-thermal synergetic environment at the epoxy resin/fiberglass interface based on the studies in Sects. 3 and 4. At the beginning of this chapter, the selection of parameters is first described. Subsequently, the study of interfacial structure change, interface binding energy, number of ester groups, radial distribution function, number of hydrogen bonds, and the damage mechanism of moisture intrusion into the epoxy/fiberglass interface are summarized. The results show that the interface under an electrothermal synergistic environment undergoes a hydrolysis reaction with increasing temperature epoxy resin and the interface separates under high temperature environment. Under the influence of an electric field, water molecules begin to enter into the epoxy resin in large quantities in the internal and interfacial areas, which promotes the expansion and hydrolysis of the epoxy resin. From the molecular structure microscopic changes, the mechanism of the synergistic effect of the electric field and temperature on the interface and water molecules can basically be considered as the superposition of the separate effects of the electric field and temperature.

References

1. Xie, J., Chen, K., Yan, M., et al.: Effect of temperature and water penetration on the interfacial bond between epoxy resin and glass fiber: a molecular dynamics study. J. Mol. Liq. **350**, 118424 (2022)
2. Yang, C., Xu, C., Yan, N., et al.: The impact of water intrusion into the crimping interface on composite insulators. Power Grid Technol. **43**(5), 1841–1848 (2019)
3. Wang, Y., Liu, Y., Fan, H., et al.: Water boiling aging characteristics of the interface between silicone rubber and alicyclic epoxy resin composite insulators. High Volt. Technol. 1–8
4. Fan, H., Liu, Y., Wang, L., et al.: The effect of epoxy resin insulator mold joint on water diffusion test. Power Grid Technol. 1–8
5. Rain, P., Brun, E., Guillermin, C., et al.: Experimental model of a quartz/epoxy interface submitted to a hygrothermal ageing: a dielectric characterization. IEEE Trans. Dielectr. Electr. Insul. **19**(1), 343–351 (2012)
6. Li, Y., Li, R., Huang, L., et al.: Effect of hygrothermal aging on the damage characteristics of carbon woven fabric/epoxy laminates subjected to simulated lightning strike. Mater. Des. **99**, 477–489 (2016)
7. Deng, S.: Study on the Assembly Behavior and Interfacial Properties of Grafted Polymer Chains at the Fiber/Resin Interface. East China University of Science and Technology, Shanghai (2013)
8. Patel, S.R., Case, S.W.: Durability of hygrothermally aged graphite/epoxy woven composite under combined hygrothermal conditions. Int. J. Fatigue **24**(12), 1295–1301 (2002)

9. Xu, R., Cui, C., Wang, J., et al.: The effect of high-temperature and high-pressure humid heat treatment on the electrical properties of epoxy resin composite materials. Fiber Compos. Mater. **25**(1), 32–34 (2008)
10. Huang, Y., Yi, L.: Effect of thermal aging on the dielectric properties of epoxy resin in dry-type transformers. Insul. Mater. **60**(009), 53–56 (2016)
11. Yin, Q., Gao, W., Ye, K., et al.: The influence of different effects on the development of electrical tree branches in insulating materials. High Volt. Technol. (4), 766–771 (2009)
12. Bao, L.R., Yee, A.F.: Moisture diffusion and hygrothermal aging in bismaleimide matrix carbon fiber composites-Part I: uni-weave composites. Compos. Sci. Technol. **62**(16), 2099–2110 (2002)
13. Bao, L.R., Yee, A.F.: Effect of temperature on moisture absorption in a bismaleimide resin and its carbon fiber composites. Polymer **43**(14), 3987–3997 (2002)
14. Haghighi-Yazdi, M., Lee-Sullivan, P.: FTIR analysis of a polycarbonate blend after hygrothermal aging. J. Appl. Polym. Sci. **132**(3) (2015)
15. Xiao, G.Z., Delamar, M., Shanahan, M.E.R.: Irreversible interactions between water and DGEBA/DDA epoxy resin during hygrothermal aging. J. Appl. Polym. Sci. **65**(3), 449–458 (2015)
16. Sun, B., Li, Y.: Study on the damp heat aging behavior and durability prediction of composite materials. Fiberglass/Compos. Mater. (4), 28–34 (2013)
17. Wang, X., Liang, G., Zhang, W., et al.: The effect of damp heat aging on the properties of high performance composite materials. Solid Rocket Technol. **29**(4), 301–304 (2006)
18. Soles, C.L., Yee, A.F.: A discussion of the molecular mechanisms of moisture transport in epoxy resins. J. Polym. Sci. Part B: Polym. Phys. **38**(5), 792–802 (2015)
19. Zhou, A., Tam, L.H., Yu, Z., Lau, D.: Effect of moisture on the mechanical properties of CFRP-wood composite: an experimental and atomistic investigation. Compos. Part B: Eng. **71**, 63–73 (2015)
20. Tam, L.H., Wu, C.: Molecular mechanics of the moisture effect on epoxy/carbon nanotube nanocomposites. Nanomaterials **7**(10), 324–343 (2017)
21. Tam, L.H., Lau, D.: Moisture effect on the mechanical and interfacial properties of epoxy-bonded material system: an atomistic and experimental investigation. Polymer **57**, 132–142 (2015)
22. Pandiyan, S., Krajniak, J., Samaey, G., et al.: A molecular dynamics study of water transport inside an epoxy polymer matrix. Comput. Mater. Sci. **106**, 29–37 (2015)
23. Lee, S.G., Ji, I.C., Koh, W., et al.: Effect of temperature on water molecules in a model epoxy molding compound: molecular dynamics simulation approach. IEEE Trans. Compon. Packag. Manuf. Technol. **1**(10), 1533–1542 (2011)
24. Xiao, Y., Xian, G.: Effects of moisture ingress on the bond between carbon fiber and epoxy resin investigated with molecular dynamics simulation. Polym. Compos. **1**, 1–10 (2017)
25. Lee, S.G., Jang, S.S., Kim, J., et al.: Distribution and diffusion of water in model epoxy molding compound: molecular dynamics simulation approach. IEEE Trans. Adv. Packag. **33**(2), 333–339 (2010)
26. Li, H., Fan, X., Yue, S., et al.: Molecular dynamics simulation of water diffusion in epoxy resin. Comput. Appl. Chem. **31**(06), 696–700 (2014)
27. Xiao, Y.: Molecular Dynamics Simulation of the Interfacial Bonding Effect of Water on Carbon Fiber/Epoxy Resin. Harbin Institute of Technology, Harbin (2017)
28. Zhang, W., Li, Z., Wu, T., et al.: Molecular simulation research and progress on the crosslinking structure and macroscopic properties of epoxy resin and its composite materials. J. Compos. Mater. **36**(02), 269–276 (2019)
29. Wang, Z., Lv, Q., Chen, S., et al.: Glass transition investigations on highly crosslinked epoxy resins by molecular dynamics simulations. Mol. Simul. **41**(18), 1515–1527 (2015)
30. Zhang, M., Wang, R., Lin, Y., et al.: Molecular dynamics simulation of glass transition temperature and mechanical properties of crosslinked fluorosilicone rubber. Polym. Mater. Sci. Eng. **31**(04), 68–72 (2015)

31. Yang, Q., Sui, G., Xu, D., et al.: Molecular simulation research on the structure and properties of epoxy resins. In: Proceedings of the 17th National Composite Materials Academic Conference (Composite Materials and Raw Materials Sub Forum), Chinese Aerospace Society: Beijing AVIC Era Culture Communication Co., Ltd., p. 4 (2012)
32. Sul, J.H., Prusty, B.G., Kelly, D.W.: Molecular dynamics study on effects of aspect ratio of carbon nanotubes in thermosetting epoxy based nanocomposites including modeling of crosslinking process. Adv. Manuf. Polym. Compos. Sci. **9**(2), 94–104 (2015)
33. Chen, S., Sun, S., Gwaltney, S.R., et al.: Molecular dynamics simulation of the interaction between carbon nanofibers and epoxy resin monomers. J. Polym. Sci (10), 1158–1164 (2015)
34. Han, Q.: A molecular dynamics investigation on the compression of cross-linked epoxy resins. Mol. Simul. **41**(18), 1509–1514 (2015)
35. Chen, X.: Experimental and Molecular Dynamics Simulation of Thermodynamic Properties of Epoxy/SiO_2. Wuhan University, Wuhan (2017)
36. Lin, Y.C., Chen, X.: Moisture sorption-desorption-resorption characteristics and its effect on the mechanical behavior of the epoxy system. Polymer **46**(25), 11994–12003 (2005)
37. Li, D.: Performance Evolution and Molecular Dynamics Simulation of Epoxy Resin Under Water Alkali Immersion Conditions. Harbin Institute of Technology, Harbin (2015)
38. Lin, J.: Simulation of Mechanical Properties of Polymer Based Nanocomposites Based on Materials Studio. Jinan University, Guangzhou (2013)
39. Alder, B.J., Wainwright, T.E.: Studies in molecular dynamics. I. General method. J. Chem. Phys. **31**, 459–466 (1959)
40. Parrinello, M., Rahman, A.: Crystal structure and pair potentials: a molecular-dynamics study. Phys. Rev. Lett. **45**, 1196 (1980)
41. Nose, S.: A unified formulation of the constant temperature molecular dynamics methods. J. Chem. Phys. **81**, 511–519 (1984)
42. Liu, Y.: Study on the Aging Characteristics of Epoxy Resin in Humid and Hot Environments. Chongqing University, Chongqing (2018)
43. Liang, X., Gao, Y.: Research on composite insulator crumb fracture (1): main characteristics, definition, and criteria of crumb fracture. Chin. J. Electr. Eng. **36**(17), 4778–4786 (2016)
44. Xie, C., Liu, S., Liu, Q., et al.: The influence of internal defects in AC 500 kV composite insulators on axial electric field distribution. High Volt. Technol. **38**(04), 922–928 (2012)

Preparation and Thermal–Mechanical Property Evaluation of Cellulose Insulation Paper with Differing Nano-SiC Contents

Yiyi Zhang, Chuqi Xu, Shuangxi Nie, and Junwei Zha

1 Introduction

Oil-immersed transformers play an important role in power systems and their insulating properties are determined by the internal insulation consisting of insulating oil and insulating paper [1–3]. Because of its intrinsic mechanical and electrical qualities, cellulose pressboard is utilized as transformer insulation paper. During long-term operation, cellulose insulation paper will cause local thermal ageing due to uneven heat dissipation, which will lead to irreversible deterioration of mechanical strength and insulation properties, and greatly reduce the service life of insulation paper. In the actual operation of the transformer, the insulation oil can be updated by changing the oil, but the insulation paper cannot be replaced. As a result, the service life of the cellulose insulation paper directly determines the service life of transformers [4].

Although cellulose insulation paper has excellent properties in comparison to other insulation materials, it is still deficient in some specific aspects. In recent years, many experts have studied different aspects of improving the properties of cellulose insulation paper. Among them, the combination of polymers and nanoparticles is the preferred method for improving material properties. Tang et al. [5] prepared modified insulation paper doped with nano-Al_2O_3, the results showed that the addition of nano-Al_2O_3 had a positive effect on AC breakdown voltage and improved the thermal ageing performance of cellulose insulating paper. Liao et al. [6] have found that the addition of nano-TiO_2 changed the contact interface between the insulation paper and oil, created a large number of shallow traps, the AC breakdown strength of

Y. Zhang (✉) · C. Xu · S. Nie
School of Electrical Engineering, Guangxi University, Nanning, China
e-mail: zdsizyy@126.com

J. Zha
School of Chemistry and Biological Engineering, University of Science & Technology Beijing, Beijing, China

Q. Xie (ed.), *Electrical Materials*, Engineering Materials,
https://doi.org/10.1007/978-981-99-9050-4_7

oil-impregnated insulation paper increased by 20.8% when the nano-TiO_2 content reached 3%. Chen et al. [7] prepared transformer insulation paper using cellulose nanocrystals modified by citric acid crosslinking and melamine amidation, found that the tensile strength of the modified paper increased by 17.30% and the breakdown strength in oil increased by 40.28%. Sequeira et al. [8] prepared Nano-SiO_2 composite cellulose paperboard, the result showed that when the content was 40–60% SiO_2 in the paperboard, the hydrophilicity was reduced, but the thermal stability is greatly improved.

In comparison to other nanoparticles, nano-SiC particles have a high thermal conductivity and thermal stability, as well as a low dielectric constant and dielectric loss. So far, it has been an excellent filler. Park et al. [9] found that liquid silicone resin composites doped with 20% SiO_2 and 3% SiC had a more evenly distributed space charge than materials with other contents. Pu et al. [10] enhanced the properties of 7075Al alloy with nano-SiC particles, demonstrating that nano-SiC particles can strengthen the tensile properties of Al when containing a less amount.

Experimental approaches for polymer modification are prone to being influenced by experimental circumstances and equipment, and they can be time-consuming and costly. Molecular simulation technology has become widely used in the calculation and investigation of molecular micro-mechanisms as computer technology has advanced [11]. Yang et al. [12] established the pure cellulose model and the modified cellulose with three kinds of polyhedral oligomeric silsesquioxane derivatives by molecular dynamics. The addition of polyhedral oligomeric silsesquioxane derivatives can effectively reduce the polarizability of cellulose, and can improve the electric field distribution in the oil-paper insulation system. Du et al. [13] investigated the effects of carbon nanowires with three different structures on the thermodynamic properties of cellulose insulation paper by molecular dynamics simulation technology. The simulation results showed that the addition of carbon nanowires can increase the number of intermolecular hydrogen bonds and improve the thermodynamic performance.

Although the properties and applications of composite modified materials have received a lot of attention in recent years, there are few studies on the effects of nano-SiC particles on the properties of insulating paper.

In summary, in this paper, nano-SiC particles were doped into cellulose, and molecular dynamics was used to establish SiC/cellulose composite models with different contents to determine the feasibility of the experiment. Simultaneously, nano-SiC composite papersheets were prepared, the thermal conductivity, thermal stability, and mechanical properties of cellulose with different contents were studied; and the modification mechanism of nano-SiC on cellulose was investigated by using microscopic parameters such as free volume and mean square displacement. The aim is to investigate a method of enhancing the performance of cellulose insulation paper by combining theoretical calculations with practical experimental results.

2 Methods and Characterizations

2.1 Methods

In this paper, in order to study the effect of different contents of nano-SiC on the thermal–mechanical properties of cellulose insulation paper. Firstly, a pure cellulose model and nano-SiC modified models were established by molecular dynamics, and the changes of the thermal–mechanical properties of the models at different contents were simulated; then, according to the simulation results, the nano-SiC modified insulation papersheets at different contents were prepared experimentally and the properties were tested.

2.2 Characterizations

2.2.1 Mechanical Properties

The generalized Hooke's law [14] can be used to define the stress–strain relationship of solid materials, as shown in Eqs. (1) and (2).

$$\sigma_i = \boldsymbol{C}_{ij}\varepsilon_j \tag{1}$$

$$\begin{bmatrix} \sigma_1 \\ \sigma_2 \\ \sigma_3 \\ \sigma_4 \\ \sigma_5 \\ \sigma_6 \end{bmatrix} = \begin{bmatrix} C_{11} & C_{12} & C_{13} & C_{14} & C_{15} & C_{16} \\ C_{21} & C_{22} & C_{23} & C_{24} & C_{25} & C_{26} \\ C_{31} & C_{32} & C_{33} & C_{34} & C_{35} & C_{36} \\ C_{41} & C_{42} & C_{43} & C_{44} & C_{45} & C_{46} \\ C_{51} & C_{52} & C_{53} & C_{54} & C_{55} & C_{56} \\ C_{61} & C_{62} & C_{63} & C_{64} & C_{65} & C_{66} \end{bmatrix} \cdot \begin{bmatrix} \varepsilon_1 \\ \varepsilon_2 \\ \varepsilon_3 \\ \varepsilon_4 \\ \varepsilon_5 \\ \varepsilon_6 \end{bmatrix} \tag{2}$$

In Eqs. (1) and (2), σ_i is the stress, $\boldsymbol{C}_{ij}$ is the calculated elastic stiffness coefficient matrix, and ε_j is the strain tensor. λ and μ are lame constants. As shown in Eqs. (3) and (4).

$$\lambda = \frac{1}{3}(C_{11} + C_{22} + C_{33}) - \frac{2}{3}(C_{44} + C_{55} + C_{66}) \tag{3}$$

$$\mu = \frac{1}{3}(C_{44} + C_{55} + C_{66}) \tag{4}$$

Mechanical properties are important indicators of macroscopic strength that can be reflected by various moduli. Elastic modulus E is the ratio of stress to strain, the greater the value, the stronger the stiffness and deformation resistance of the material;

shear modulus G is the ratio of shear stress to strain, the higher the value, the better the ability to resist shear stress and the less likely it is to be damaged under external stress; bulk modulus K, a physical quantity describing the connection between strain and average stress, can reflect the macroscopic mechanical properties of materials.

The elastic modulus, shear modulus, and bulk modulus are calculated by molecular dynamics, and the calculation formulas are as follows:

$$E = \frac{\mu(3\lambda + 2\mu)}{\lambda + \mu} \tag{5}$$

$$G = \mu \tag{6}$$

$$K = \lambda + \frac{2}{3}\mu \tag{7}$$

In the experiment, a tensile strength tester (Lorentzen and Wettre, Stockholm, Sweden) was used to measure the tensile index of papersheets according to the GB/T453-1989 standard.

2.2.2 Thermal Conductivity

Molecular dynamics has been used to calculate the thermal conductivity of polymeric materials in many ways [15, 16]. The reverse non-equilibrium molecular dynamics (RNEMD) method has been widely used because of its faster convergence time of temperature gradient [17].

The calculation of the thermal conductivity is based on Fourier's Law of Heat Conduction:

$$\lambda = -\frac{Q}{\mathrm{dT}/\mathrm{dx}} \tag{8}$$

where Q is the energy flux in the x-direction, dT/dx is the temperature gradient, and the negative sign indicates that the energy flux is in the direction opposite to the gradient.

In the RNEMD method, the hot and cold regions are formed in the simulated system by simulating the exchange of atomic center-of-mass velocities in different regions of the system in the ideal state. Then, by averaging the kinetic energy exchange between the coldest particle in the hot layer and the hottest particle in the cold layer several times, a stable temperature gradient and energy flux in the system is achieved, and the thermal conductivity is finally obtained by calculation. The energy flux is imposed by exchanging the energy ΔE between two fixed layers in the system at each time interval Δt, the equation is as follows:

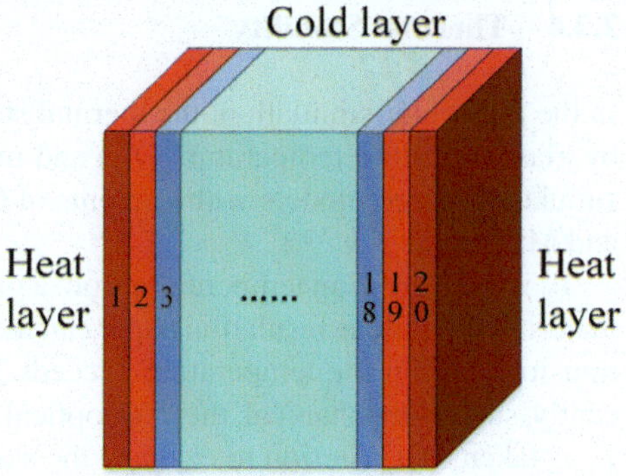

Fig. 1 Model diagram of RNEMD method

$$Q = \frac{1}{2A}\frac{\Delta E}{\Delta t} \tag{9}$$

where A is the area perpendicular to the flux direction and the factor *2* is due to the periodic boundary condition.

As shown in Fig. 1, the RNEMD method divides the model into 20 regions in one direction. The hot layer is in the left and right boundary layers, and the cold layer is in the middle. Heat is transferred by kinetic energy exchange between the hot and cold layers. After multiple exchanges, a stable temperature gradient is finally formed in the system. To calculate the thermal conductivity in one direction, the model is first simulated for 100 ps in the NVT ensemble and then 100 ps in the NPT ensemble.

The RNEMD method can only calculate thermal conductivity by balancing the temperature gradient in a single direction. To calculate the overall thermal conductivity of the composite material, it is necessary to calculate the thermal conductivity of the model in the x, y, and z directions, and take the average value as the thermal conductivity of the model [18].

In the experiment, the thermal conductivity of the papersheets was measured using LFA467HT HyperFlash (Netzsch Ltd.).

2.2.3 Microscopic Topography

In the experiment, the microstructure and morphology were observed by the field emission scanning electron microscope (GeminiSEM 300), with a beam accelerating voltage of 10 kV. Energy-dispersive X-ray spectroscopy (EDS) is an accessory used on scanning electron microscope for micro-area analysis of components energy spectrum. The specimens were broken in liquid nitrogen and the cross-sections were sputtered with gold before observation.

2.2.4 Thermal Stability

In the molecular simulation, the thermal stability of a material can be characterized by glass transition temperature (T_g) and mean square displacement (MSD). In this simulation, three models with contents of 0, 3, and 6% are used to calculate the T_g and MSD.

T_g is the transition temperature from a glass state to a high elastic state of polymer materials. Cellulose insulation paper can meet engineering requirements before glass transition, when the temperature exceeds T_g, many properties will change significantly, such as mechanical, thermal, optical, and electrical properties [19]. Therefore, T_g is taken as a criterion to evaluate the stability of materials.

The molecular dynamics simulation of stage cooling treatment was carried out at temperatures ranging from 650 to 200 K, with each step lowering the temperature by 50 K. The starting conformation for the next stage was taken from the final equilibrium conformation simulated in the previous stage. First simulated for 100 ps at each temperature in an NPT ensemble, then for 200 ps under 0.0001 GPa in an NPT ensemble is performed for data analysis and calculation of various properties.

The temperature is plotted against the specific volume derived from the NPT molecular dynamics simulation. Taking the inflection point of the specific volume as the boundary point, linear fitting was performed before and after the boundary point respectively. The boundary point is the glass transition point.

The entire movement of the center of the molecular chain can be described by MSD. The stronger the molecular chain motion, the higher the MSD value [20]. Its calculation is expressed as:

$$MSD = \left\langle |r_i(t) - r_i(0)|^2 \right\rangle \tag{10}$$

Among them: $r_i(t)$ is the position of i molecule or atom at t time; $r_i(0)$ represents the position of i molecule or atom at 0 time; $< >$ represents the average of all atoms in the group.

In the experiment, a simultaneous thermal analyzer (NETZSCH STA 449F3) was used for thermogravimetric experiments. The samples were cut into 1 mm * 1 mm particles with a mass of about 5 mg. The heating rate of the sample was 10 °C/min, the temperature was raised from 20 to 800 °C, the purge gas and the protective gas were nitrogen.

2.2.5 Internal Structure

In the simulation, according to the free volume theory, the free volume (V_f) is the gap between molecules, and the occupied volume is the space occupied by the molecular chains (V_0). The free volume provides space for the movement of molecular chains. The smaller the free volume fraction (FFV), the smaller the inner space of the material, the smaller the tendency of molecular chains to move at the molecular level, and the relatively stable structure of the material [21].

The calculation of FFV is expressed as follows:

$$FFV = \frac{V_f}{V_0 + V_f} \tag{11}$$

In the experiment, the pore size distributions and porosities of the papersheets were measured using Mercury Intrusion Porosimeters (MicroActive AutoPore V9600, Micromeritics Instrument Corporation). The sample of no less than 2 g was degassed under vacuum degree less than 1 Pa to remove the adsorbed substance before the experiment.

2.2.6 Thermal Ageing Resistance

In the experiment, papersheets with the nano-SiC content of 9% and unmodified papersheets are used for thermal ageing experiments. Karamay NO. 25 naphthenic mineral oil was chosen as the insulation oil. Dry the oil for 48 h under vacuum at 90 °C. The papersheets are immersed for 48 h in degassed and dewatered insulating oil.

Regular sampling is done to measure the degree of polymerisation (DP) of cellulose in papersheets during an accelerated heat ageing experiment at 130 °C. The viscosity method is used to measure the DP of paper in the laboratory. The fibers are dissolved in copper ethylene diamine, and the outflow time of the obtained solution in the capillary determines the DP of the paper.

3 Results and Discussions

3.1 Model Establishment and Parameter Setting

Cellulose models with a DP of 10, an initial density of 0.6 g/cm^3, and spherical nano-SiC particles with a radius of 5 Å were established in Materials Studio [22]. In the Amorphous cell module, nano-SiC/cellulose composite models with different contents of 0%, 2%, 3%, 4%, 5% and 6% were established, which were recorded as 0%-SiC, 2%-SiC, 3%-SiC, 4%-SiC, 5%-SiC and 6%-SiC respectively.

Geometry optimization of the models is carried out by using Forcite module for energy minimization calculation, followed by annealing treatment with six cycles between 300 and 900 K. After such treatment, the internal stress of the models is decreased, the simulation system is in a stable state, and the distribution of voids in the models is closer to the real material. Then, the models were simulated by dynamics, which included simulations in an NVT ensemble for 100 ps, an NPT ensemble for 100 ps under 1 GPa, and an NPT ensemble for 100 ps under 0.0001 GPa successively. Compass force field [23], Nose temperature control method [24], Ewald electrostatic

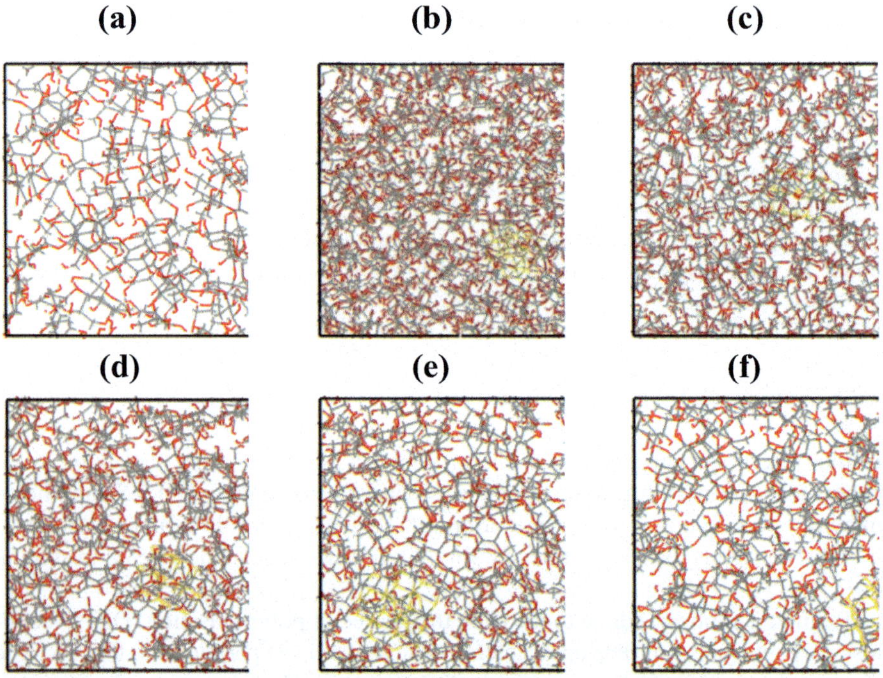

Fig. 2 Models of OAB surface **a** 0%-SiC/cellulose model. **b** 2%-SiC/cellulose model. **c** 3%-SiC/cellulose model. **d** 4%-SiC/cellulose model. **e** 5%-SiC/cellulose model. **f** 6%-SiC/cellulose model

method [25], Atom Based method for Vander Waals, and Berendsen method for pressure control are employed in the process of energy optimization and molecular dynamics simulation. After the above optimizations, the density and energy of the models can converge, and the models can be considered as close to the real cellulose. The models are shown in Fig. 2.

3.2 Sample Preparation

In this paper, the laboratory papersheet forming method was used. T4 pressboards (manufactured by Taizhou Weidmann High Voltage Insulation Co., Ltd., Taizhou, China) were employed to prepare Laboratory papersheets. Nano-SiC was produced by Beijing Xingrongyuan Technology Co., Ltd. with an average particle size of 50 nm and a purity of 99.99%. First, soak T4 pressboard and tear it into small pieces, the small pieces were stirred by a Messmer pulp disintegrator (Mavis Engineering Ltd., London, UK) at 3000 rpm for 10 min.

In order to investigate the effect of different nano-SiC contents on the variation of insulating paper properties, five groups of papersheets with 0%, 3%, 6%, 9%, and 12% were prepared based on the molecular simulation results, 0%-SiC, 3%-SiC, 6%-SiC, 9%-SiC, 12%-SiC were recorded as P0, P1, P2, P3 and P4, respectively.

Different contents of nano-SiC particles were added to deionized water and stirred magnetically. Then, different contents of insulating pulp and nano-SiC particles were mixed and diluted with deionized water to reach a final concentration of 1 wt%. Then the mixture was filtered, moulded, pressed, and dried to obtain papersheets in a normalized Rapid-Köthen papersheet former (PTI, Vorchdorf, Austria) according to ISO 5269/2 (2004). Finally, the papersheets were hot-pressed at 15 MPa and 90 °C to achieve densification and internal bonding. The papermaking process is shown in Fig. 3.

Finally, circular papersheets with an average thickness of 0.27 mm, a basis weight of 240 g/m^2, and a diameter of 16 cm were produced. After the papermaking was completed, they were dried at 90 °C for 48 h, and then sealed in a sample bag for storage.

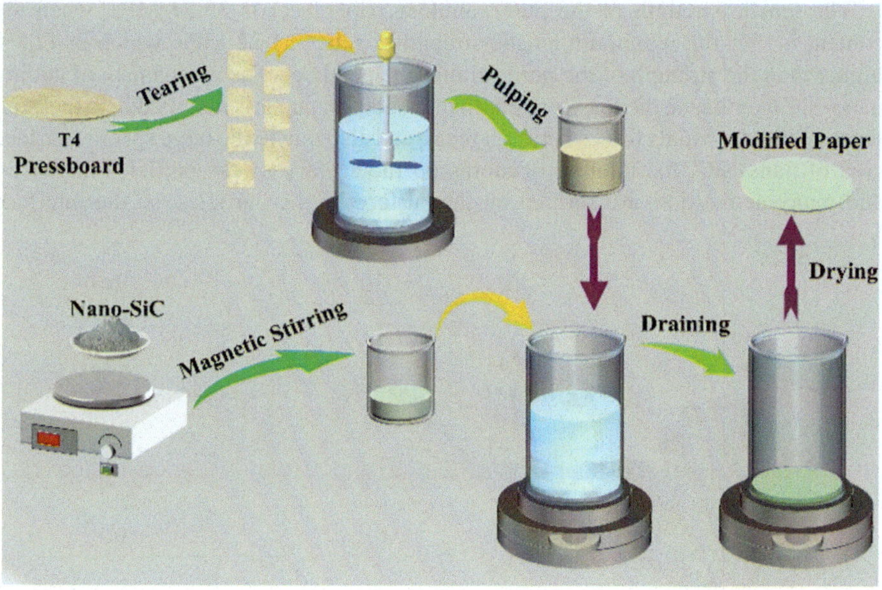

Fig. 3 Schematic diagram of the preparation process

3.3 Performance Analysis

3.3.1 Mechanical Properties Analysis

In the simulation, the relationships between bulk modulus, shear modulus, elastic modulus, and different contents of nano-SiC in the models are obtained, as shown in Fig. 4a.

The moduli of the modified models are higher than those of the unmodified model, the addition of nano-SiC particles can improve the bulk modulus, shear modulus, and elastic modulus, indicating an increase in stiffness and resistance to deformation, with the greatest increase in properties occurring at a content of 3%. When the content exceeds 3%, the mechanical properties begin to decline.

In the experiment, the tensile strength of pure cellulose papersheet and nano-SiC composite papersheets was studied, the curves of the stress–strain and the maximum tensile strength are shown in Fig. 4b, c, respectively.

The tensile strength of the pure cellulose papersheet is 35.87 MPa, when the content is 9%, the maximum tensile strength reaches 51.22 MPa, which is 42.8% higher than the strength of the pure cellulose papersheet. The mechanism of adding nano-SiC to enhance the tensile strength of insulation paper is mainly due to the existence of van der Waals forces between them, and also due to the large specific surface area of nano-SiC, the number of atoms on the surface of the particles increases, and many active dots are formed on the surface of the particles, so the interface

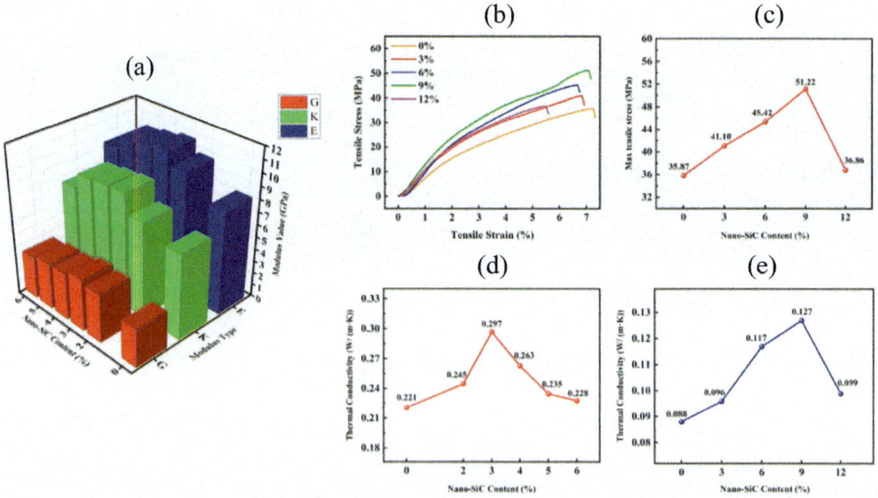

Fig. 4 **a** Bulk modulus (G), shear modulus (K) and elastic modulus (E) of SiC/cellulose models with different nano-SiC contents. **b** Stress–strain curves of papersheets with different nano-SiC contents. **c** Max tensile stress of papersheets with different nano-SiC contents. **d** Thermal conductivity of SiC/cellulose models with different nano-SiC contents in the simulation. **e** Thermal conductivity of papersheets with different nano-SiC contents in the experiment

between nano-SiC and cellulose is well bonded, thus improving the tensile strength of insulation paper.

With the further increase of nanoparticles, when the nano-SiC content exceeds 9%, it can be considered that the nano-SiC particles appear to agglomerate, at which time the particle size of nano-SiC becomes larger, the specific surface area becomes smaller, the unpaired atoms on its surface become fewer, the surface activity becomes lower, and the bond with cellulose becomes relatively weaker, which leads to a decrease in the mechanical properties of the insulation paper, which is the reason for the peak in tensile strength, which is also consistent with the trend obtained from molecular simulation.

Wrapping and winding copper conductors to act as electrical insulation is an important application for cellulose insulation paper in transformer insulation. If the mechanical properties of the cellulose insulation paper deteriorate, it will be liable to fracture when the conductor coil is subjected to mechanical stress, which affects the insulation performance of the transformer. As a result, improving the mechanical properties of cellulose insulation paper can enhance the insulation performance from a different perspective, extending the service life of transformers.

3.3.2 Thermal Conductivity Analysis

The normal operating temperature of power transformers is around 343 K, but due to various factors such as operating environment and load, the temperature of certain parts inside the transformer can sometimes soar to around 393 K, so 363 K was chosen in this paper to study the thermal conductivity of cellulose insulation paper.

Figure 4d shows the simulation results of the thermal conductivity at 363 K. It can be seen from Fig. 4d, with the addition of nano-SiC, the thermal conductivity first increases and then decreases. When the content is 3%, the thermal conductivity of cellulose pressboard increases the most, which is 34.4% higher than that of unmodified model.

In the experiment, the thermal conductivity of composite papersheets with different nano-SiC contents at 363 K was measured, the results are shown in Fig. 4e. It can be seen from the figure that the thermal conductivity of pure cellulose papersheet is 0.088. When the content of nano-SiC is 9%, the thermal conductivity reaches 0.127, which is 44.3% higher than unmodified papersheet. With the increase of nano-SiC content, the thermal conductivity first increases and then decreases, which is also the same as the simulation trend.

According to heat conduction pathway theory, with the addition of thermally conductive fillers, because of the high thermal conductivity of SiC, and some thermally conductive fillers begin to contact each other and form local thermal conductive chains or networks, accelerating the heat diffusion within the paper, resulting in an increase in the thermal conductivity. When the addition of fillers exceeds 9%, as the number of fillers continues to increase, the distribution of thermally conductive fillers in the paper becomes wider, the probability that the fillers are distributed on the surface gradually increases. The addition of thermally conductive fillers increases the surface roughness of papersheets, which increases the thermal resistance of the insulating paper interface, generates more phonon scattering, and hinders the propagation of phonons inside the insulating paper, so it results in a decrease in the thermal conductivity of papersheets.

3.3.3 Microscopic Topography Analysis

Figure 5a–e shows the SEM image of the cellulose/nano-SiC. Figure 5f, g shows the types of trace elements in cellulose, and the presence of silicon is found, indicating that nano-SiC was added successfully.

It can be roughly seen that the paper fibers are intertwined in a disorderly manner to maintain a certain mechanical strength. At the same time, it can be seen that the small particles (nano-SiC) are attached to the surface of the fibers, filling the voids in the cellulose network, and can form a network structure with the fibers, which enhances the mechanical properties of the composite paper.

When the content is small, the nanoparticles are dispersed in the fiber surface without obvious agglomeration. When the content is 12%, it can be seen that the

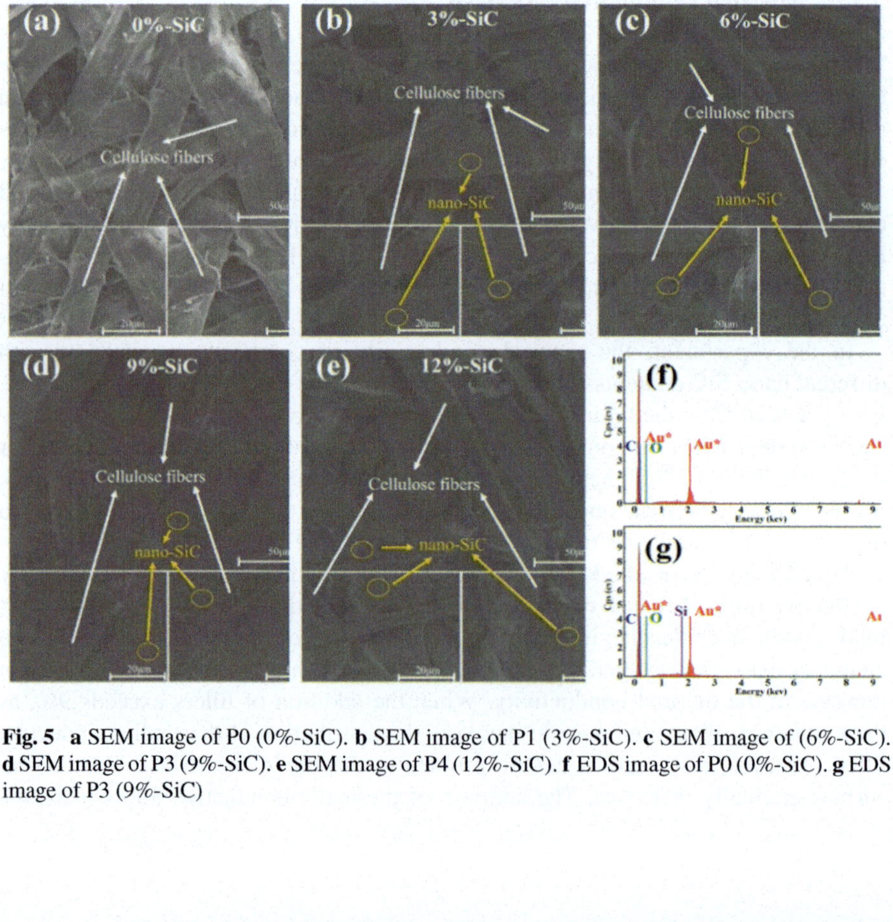

Fig. 5 **a** SEM image of P0 (0%-SiC). **b** SEM image of P1 (3%-SiC). **c** SEM image of (6%-SiC). **d** SEM image of P3 (9%-SiC). **e** SEM image of P4 (12%-SiC). **f** EDS image of P0 (0%-SiC). **g** EDS image of P3 (9%-SiC)

distribution range of the globules attached to the fiber surface becomes wider, and the particle size becomes larger. It shows that when the content is 12%, slight agglomeration occurs. The degradation of thermal–mechanical properties of insulation paper may be caused by the agglomeration of nanoparticles.

3.3.4 Thermal Stability Analysis

The simulation curves of T_g are shown in Fig. 6a–c. T_g of the unmodified model is 439 K, and that of the pure cellulose model obtained by Wang et al. is 448 K [26]. The result of this paper is not much different from 448 K, which verifies the accuracy of the models.

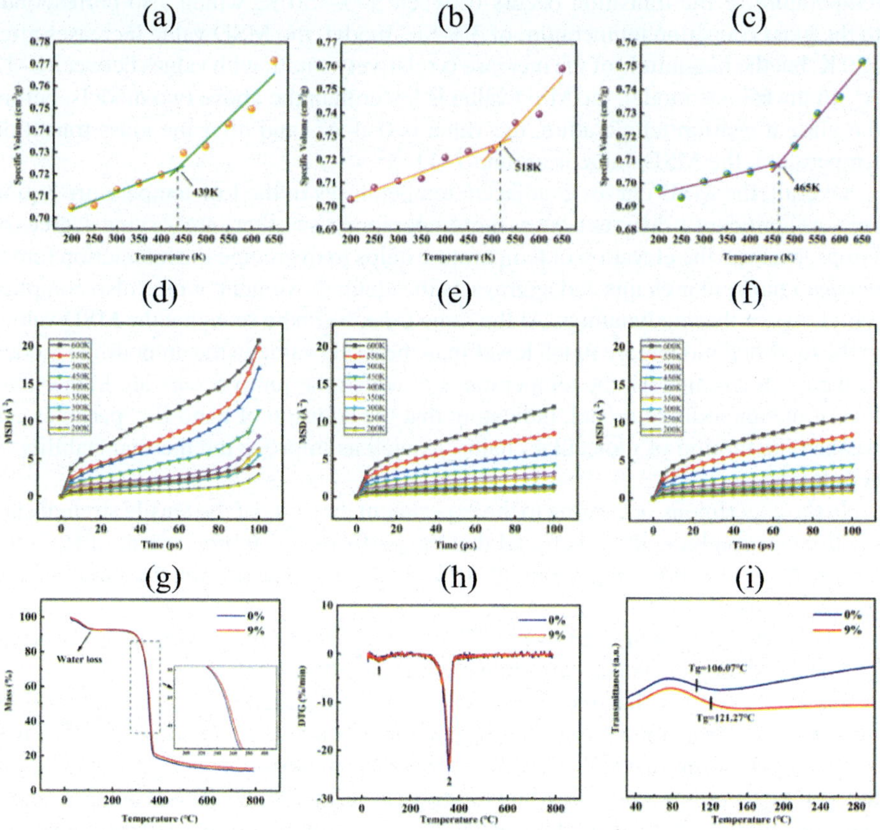

Fig. 6 **a** Glass transition temperature of 0% SiC unmodified model. **b** Glass transition temperature of 3% SiC modified model. **c** Glass transition temperature of 6% SiC modified model. **d** MSD curve of 0% unmodified model. **e** MSD curve of 3% modified model. **f** MSD curve of 6% modified model. **g** TGA curves of P0 (0%-SiC) and P3 (9%-SiC). **h** DTG curves of P0 (0%-SiC) and P3 (9%-SiC). **i** T_g curves of P0 (0%-SiC) and P3 (9%-SiC)

T_g of the modified model with 3%-SiC is 518 K and that of the modified model with 6%-SiC is 465 K. The T_g of the modified model with 3%-SiC is 79 K higher than that of the unmodified model, and T_g of the modified model with 6%-SiC is 26 K higher than that of the unmodified model. It shows that adding nano-SiC particles to cellulose insulation paper can increase T_g. The mechanical characteristics and thermal stability of composites are inextricably linked to T_g. Therefore, nano-SiC can improve the thermal–mechanical properties of cellulose insulation paper.

The MSD curves of cellulose chains are shown in Fig. 6d–f. As shown in Fig. 6d–f, the MSD value in 0%-SiC model is relatively small before 400 K, around 0–8 Å^2, and the temperature at which the MSD value jumps is between 400–450 K, this temperature range corresponds to the glass transition temperature of 0%-SiC model, after 450 K, the MSD value increases rapidly, reaching a maximum of 21 Å^2. While in the 3%-SiC model, the MSD value is lower before 500 K, around 0–6 Å^2, the temperature of the transition occurs between 500–550 K, which also corresponds to the glass transition temperature of 3%-SiC model, the MSD value increases after 550 K, but the magnitude of the increase is relatively small, with values between 8–12 Å^2. In the 6%-sic model, the MSD value is lower than the above two models, before the glass transition temperature, the value is 0–4 Å^2, and after the glass transition temperature, the MSD value is around 7–11 Å^2.

Overall, the MSD curves at adjacent temperatures in the low-temperature region are not significantly different, but as the temperature rises, there are varying degrees of jumps, because the elevation of temperature helps to overcome the interaction forces between molecular chains and aggravates the chain movement, which also complies with Laws of thermodynamics. At the same time, it can be seen that the MSD values in the modified models are much lower than the MSD value in the unmodified model. In the modified models, the temperatures at which the jump occurs are higher than that of the unmodified model, indicating that the addition of nano-SiC particles can weaken the motion of molecular chains, as well as improve the thermal stability of cellulose.

In the experiment, according to the experimental results of the tensile strength and the thermal conductivity, it is found that the performance improvement of P3 is the largest Therefore, P0 (0%-SiC) and P3 (9%-SiC) were measured by the simultaneous thermal analyzer.

The thermogravimetric Analysis (TGA), derivative thermogravimetry (DTG), and T_g curves of the papersheets are shown in the Fig. 6g–i.

Figure 6g shows that the samples have two significant weight losses in the measurement temperature range, corresponding to the two peaks in Fig. 6h. The first weight loss is around 100 °C, which corresponds to the loss of water. After 100 °C, all the water evaporates, corresponding to the area "1" in Fig. 6h. The primary reaction stage of cellulose thermal cracking is the second weight loss stage, and the main reaction temperature range for cellulose thermal cracking is 300–400 °C. In this process, the glycosidic bonds of cellulose are broken by ring-opening, small-molecule volatile compounds are generated, resulting in a significant decrease in quality at this stage, and the reaction rate reaches the maximum, corresponding to the "2" area in Fig. 6h.

Comparing the TGA curves of 0 and 9%, it can be seen that the final residual amount of P0 is 11.73%, while the final residual amount of P3 is 13.03%.

Figure 6h shows that the maximum reaction rate of P0 is higher than that of P3, indicating that the degradation rate of the papersheets becomes slower after adding nano-SiC, and the thermal stability of cellulose is improved.

In this paper, the glass transition temperature of the sample was measured by differential scanning calorimetry, and the results are shown in Fig. 6i. It can be seen that the T_g of P0 is 106.07 °C (379.22 K), while the T_g of P3 is 121.27 °C (394.42 K), which is 15.2 °C higher than that of the unmodified papersheet. It shows that adding nano-SiC to cellulose can increase the glass transition temperature, which is consistent with the conclusion drawn from the simulation.

3.3.5 Internal Structure Analysis

At 298 K, the free volume distributions of different models under the OAB surface are shown in Fig. 7. The blue area is V_f and the gray area is V_0. The *FFV* of each model under different contents is shown in Table 1.

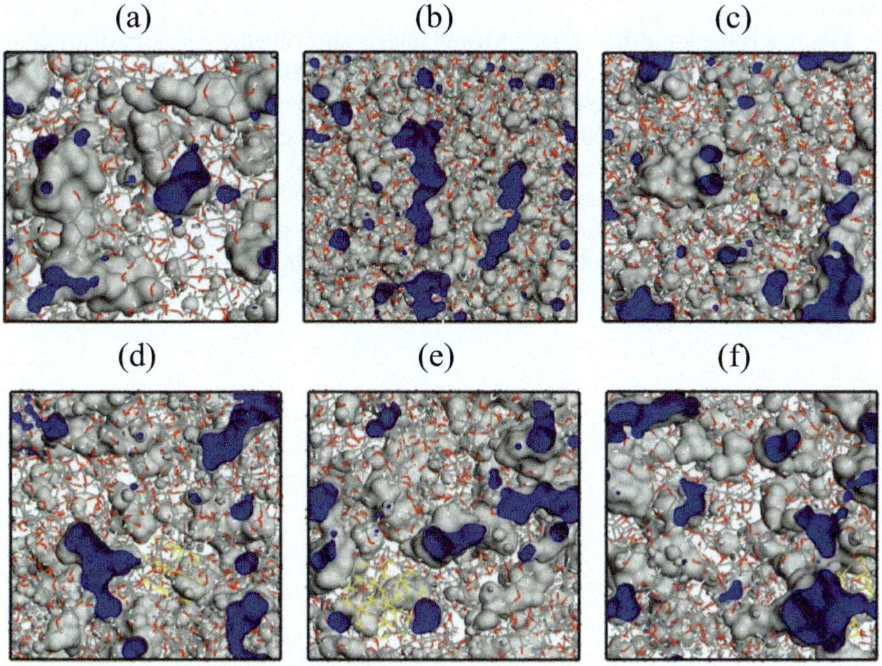

Fig. 7 The free volume distribution of OAB surface in models **a** free volume distribution of 0% unmodified model. **b** Free volume distribution of 2% modified model. **c** Free volume distribution of 3% modified model. **d** Free volume distribution of 4% modified model. **e** Free volume distribution of 5% modified model. **f** Free volume distribution of 6% modified model

Table 1 The proportion of the free volume of SiC/cellulose models under different nano-SiC content

Content	0%	2%	3%	4%	5%	6%
V_0	18,528	49,145	31,790	24,195	19,101	16,291
V_f	4578	9804	5258	3867	3569	2785
FFV	0.198	0.166	0.142	0.138	0.158	0.146

It can be seen from the data in Table 1, with the increase of nano-SiC particles gradually, the *FFV* of cellulose first decreases, then fluctuates up and down. When the content is between 3–4%, the *FFV* is relatively low.

Due to the addition of nano-SiC particles, the whole molecular structures of cellulose are combined more closely, and the gaps between molecule chains can be filled, resulting in a reduction in free volume and effectively weakening the movement of molecular chains. As a result, adding nano-SiC in cellulose can slow down the movement of molecular chains.

From the perspective of thermal conductivity and tensile strength, when the content of nano-SiC is 9%, the performance improvement is the largest. Therefore, in this paper, the pore distribution and porosity of P0 (0%-SiC) and P3 (9%-SiC) were measured.

Figure 8 is the log-differential intrusion image, showing the pore size distribution and quantity of papersheets. In Table 2, the porosity of P0 is 40.96%, the average pore size is 2599.38 nm, and the pore size distribution is obvious (361 and 2 μm). After adding nano-SiC particles, the porosity and average pore size of cellulose decreased to 40.23% and 2203.41 nm, respectively, and the pore size distribution decreased at 361 μm and 2 μm. It can also be seen from the Table 2 that the total pore area has decreased from 6.359 to 4.506 m^2/g. All show that the nano-SiC particles were attached to the surface of cellulose, filled the voids of the fiber network and reduced the number of pores.

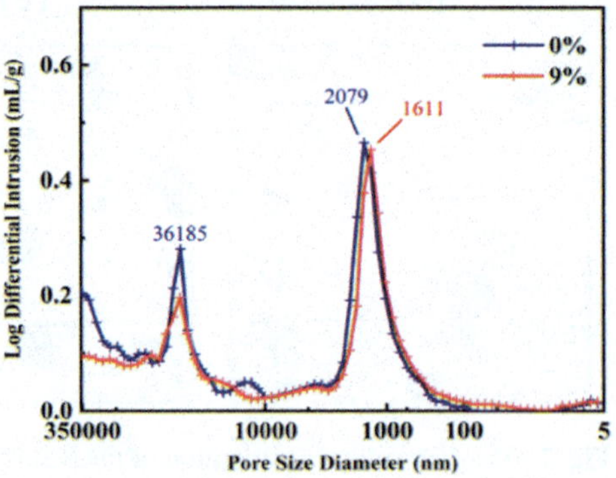

Fig. 8 Log differential intrusion image of P0 (0%-SiC) and P3 (9%-SiC)

Table 2 Porosity, average pore diameter and total pore area of papersheets with 0%-SiC and 9%-SiC

Sample	Porosity (%)	Average pore diameter (nm)	Total pore area (m^2/g)
P0	40.96	2599.38	6.359
P3	40.23	2203.41	4.506

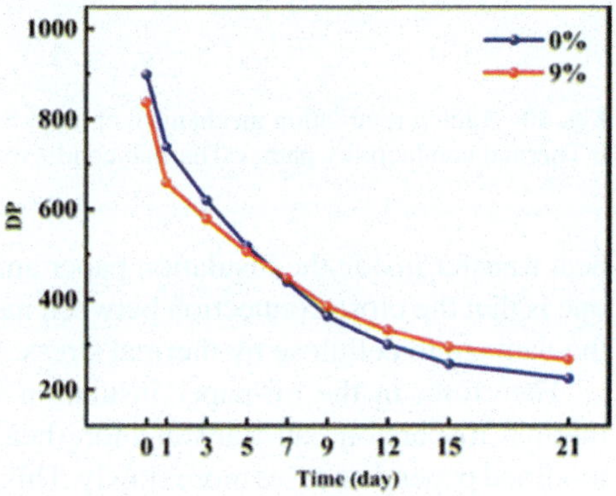

Fig. 9 The change curve of the DP of P0 (0%-SiC) and P3 (9%-SiC) with ageing time

3.3.6 Thermal Ageing Analysis

The variations of the DP of modified and unmodified paper with ageing time are shown in Fig. 9.

It can be seen from Fig. 9 that the DP of the modified papersheet is slightly smaller than that of the unmodified papersheet when they are not aged. This may be due to the fact that the same mass is taken from each sample when measuring the DP. The addition of nano-SiC makes the relative content of cellulose lower than that of the unmodified paper, resulting in a smaller viscosity and DP.

Before ageing for seven days, the DP of the unmodified papersheet was higher than that of the modified papersheet. As the ageing continued, the DP of the modified papersheets began to be gradually higher than that of the unmodified papersheets, indicating that the nano-SiC modified papersheets had excellent thermal stability.

The mechanism by which the addition of nano-SiC can enhance the anti-ageing performance is analyzed. In Fig. 10, the yellow lines represent cellulose molecules and the green spheres represent nano-SiC. Figure 10a shows fibers in insulation paper that are haphazardly intertwined. Figure 10b shows the schematic of the addition of a small number of nanoparticles, it can be seen that the small amount of added nano-SiC adheres to the surface of the cellulose and forms a thermal conductivity pathway. From Fig. 10c, there forms a thermally conductive network inside the cellulose with the further addition of nano-SiC particles.

A comprehensive analysis shows that there are two reasons why nano-SiC can improve thermal ageing resistance: One is that the addition of nano-SiC can fill the intermolecular gaps and the thermal conductivity network can greatly accelerate the

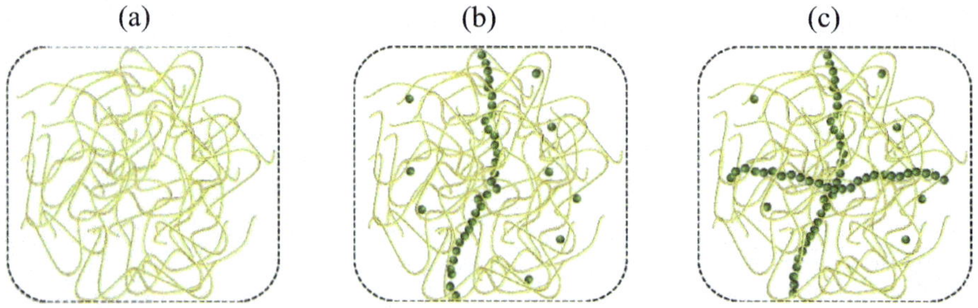

Fig. 10 Ageing retardation mechanism of nano-SiC modified insulation paper **a** Cellulose model. **b** Thermal conductivity path. **c** Thermal conductivity network

heat transfer inside the insulation paper and improve the local overheating. Another one is that the close connection between nano-SiC and cellulose molecules weakens the damage to cellulose by thermal stress.

Therefore, in the oil-paper insulation system, the insulation paper with faster thermal dispersion can transfer more heat to the insulation oil, so the DP of the modified paper decreases more slowly. This also indicates that the nano-SiC-modified insulation paper has better anti-ageing properties and can meet the requirements of long-term operation of transformers, which has potential application significance.

4 Conclusions

In summary, we have shown that the thermal–mechanical properties of cellulose insulation paper could be improved by adding nano-SiC particles. In comparison with that of unmodified papersheet, the tensile strength and thermal conductivity can be increased by 42.8% and 44.3%, respectively. The performance of cellulose insulating paper was simulated by molecular dynamics. Based on the simulation results, nano-silicon carbide modified insulating paper with different contents was prepared experimentally, and its performance was tested to verify the accuracy of the simulation. Because of the free settling of nanoparticles and cellulose during the experiment, and part of nanoparticles flow out with water, the optimal content of nano-SiC in the experiment is larger than that in the simulation. In the case of ignoring the experimental error, when the nano-SiC content is 9%, the thermal–mechanical performance can be improved the most. Our research provides a reference for the practical application of nanomaterials in the improvement of cellulose insulation paper.

References

1. Zhang, Y., et al.: Preparation and thermal-mechanical property evaluation of cellulose insulation paper with differing nano-SiC contents. High Volt. **8**(3), 599–610 (2023)
2. Liao, R., et al.: Influence of the copper ion on aging rate of oil-paper insulation in a power transformer. IET Electr. Power Appl. **3**(5), 407–412 (2009)
3. Oommen, T.V., Prevost, T.A.: Cellulose insulation in oil-filled power transformers: part II maintaining insulation integrity and life. IEEE Electr. Insul. Mag. **22**(2), 5–14 (2006)
4. Huang, M., et al.: Influence of voltage reversal on space charge behavior in oil-paper insulation. IEEE Trans. Dielectr. Electr. Insul. **21**, 331–339 (2014)
5. Tang, C., et al.: Molecular simulation and experimental analysis of Al_2O_3-nanoparticle-modified insulation paper cellulose. IEEE Trans. Dielectr. Electr. Insul. **24**(2), 1018–1026 (2017)
6. Liao, R., et al.: The insulation properties of oil-impregnated insulation paper reinforced with nano-TiO_2. J. Nanomater. **4**, 373–959 (2013)
7. Chen, Q., et al.: Effect of melamine modified cellulose nanocrystals on the performance of oil-immersed transformer insulation paper. Cellulose **27**, 7621–7636 (2020)
8. Sequeira, S., et al.: Synthesis and characterisation of cellulose/silica hybrids obtained by heteropoly acid catalysed sol–gel process. Mater. Sci. Eng. C **27**(1), 172–179 (2006)
9. Park, J.-J., Lee, J.-Y.: Effects of SiC nanoparticles on space charge behaviors of LSR/ nano-SiO_2/nano-SiC composites for insulating material of HVDC cable joints. Mater. Chem. Phys. **270** (2021)
10. Pu, B., et al.: Effect of SiC nanoparticles on the precipitation behavior and mechanical properties of 7075Al alloy. J. Mater. Sci. **55**, 6145–6160 (2020)
11. Tang, C., et al.: Thermal stability of polyphenylsilsesquioxane-modified meta-aramid insulation paper. High Volt. **5**(3), 264–269 (2020)
12. Yang, L., et al.: Cellulose insulation paper with high thermal stability and low polarizability: influence of different substituents on POSS modified cellulose insulating paper. Cellulose **28**, 6023–6033 (2021)
13. Du, D., et al.: Molecular simulation on the mechanical and thermal properties of carbon nanowire modified cellulose insulating paper. Compos. Struct. **261**, 113283 (2021)
14. Tanaka, F., Okamura, K.: Characterization of cellulose molecules in bio-system studied by modeling methods. Cellulose **12**, 243–252 (2005)
15. Jund, P., Jullien, R.: Molecular-dynamics calculation of the thermal conductivity of vitreous silica. Phys. Rev. B **59**(21), 13707 (1999)
16. Yang, X., et al.: Atomic-scale insights into interface thermal resistance between epoxy and boron nitride in nanocomposites. Int. J. Heat Mass Transf. **159**, 120105 (2020)
17. Liu, X., Rao, Z.: Molecular dynamics simulations on the heat and mass transfer of hyper-crosslinked shell structure of phase change nanocapsules as thermal energy storage materials. Int. J. Heat Mass Transf. **132**, 362–374 (2019)
18. Xie, Y., et al.: Research progress of thermal conductive epoxy resin matrix composites for electronic packaging. Eng. Plast. Appl. **46**(12), 143–147 (2018). J-GLOBAL ID:201902241552662274
19. Chang, K., et al.: Free volume and alcohol transport properties of PDMS membranes: insights of nano structure and interfacial affinity from molecular modeling. J. Membr. Sci. **417–418**, 119–130 (2012)
20. Yi, Z., et al.: Derivation of mean-square displacements for protein dynamics from elastic incoherent neutron scattering. J. Phys. Chem. B **116**(16), 5028–5036 (2012). https://doi.org/10.1021/jp2102868
21. Yin, F., et al.: Effect of moisture on mechanical properties and thermal stability of meta-aramid fiber used in insulating paper. Polymers **9**(10), 537 (2017)
22. Tang, C., et al.: Molecular dynamics simulations of the effect of shape and size of SiO_2 nanoparticle dopants on insulation paper cellulose. AIP Adv. **6**, 125106 (2018). https://doi.org/10.1063/1.4971280

23. Theodorou, D.N., Suter, U.W.: Detailed molecular structure of a vinyl polymer glass. Macromolecules **18**(7), 1467–1478 (1985)
24. Nose, S.: Constant temperature molecular dynamics method. Prog. Theor. Phys. Suppl. (103), 1–46 (1991)
25. Berendsen, H.J.C., et al.: Molecular dynamics with coupling to an external bath. J. Chem. Phys. **81**, 3684–3690 (1984).https://doi.org/10.1063/1.448118
26. Wang, Y.Y., et al.: Molecular dynamics simulation of glass transition in amorphous region of insulating paper. Gaodianya Jishu/High Volt. Eng. **38**(05), 1199–1206 (2012)

The manufacturer's authorised representative in the EU is Springer Nature Customer Service Centre GmbH, Europaplatz 3, 69115 Heidelberg, Germany. If you have any concerns regarding our products, please contact ProductSafety@springernature.com

Printed and bound by CPI Group (UK) Ltd, Croydon, CR0 4YY
01/10/2026
02225868-0006

Contents

Designed cover image: © Getty Images

First edition published 2027
by CRC Press
2385 NW Executive Center Drive, Suite 320, Boca Raton FL 33431

and by CRC Press
4 Park Square, Milton Park, Abingdon, Oxon, OX14 4RN

CRC Press is an imprint of Taylor & Francis Group, LLC

ISBN: 978-1-041-03998-3 (hbk)
ISBN: 978-1-041-04002-6 (pbk)
ISBN: 978-1-003-62633-6 (ebk)

DOI: 10.1201/9781003626336

Typeset in Times
by codeMantra

Weight of Evidence for Environmental Assessment

Glenn W. Suter II and Susan M. Cormier

CRC Press
Taylor & Francis Group
Boca Raton London New York

CRC Press is an imprint of the
Taylor & Francis Group, an **informa** business

Weight of Evidence for Environmental Assessment

Laws and regulations in the United States and some other nations require weight of evidence, as do any assessments that use multiple pieces of evidence. However, environmental assessors are frequently unaware of the need for defensible practices. This book provides the first clear framework for weight of evidence for environmental protection based on USEPA guidelines, as well as comparisons with alternative methods. Real-world case studies illustrate the implementation and application of weight of evidence for making diverse inferences and decisions that influence our environment and our lives.

Key Features

- Is written by experts who developed guidance on environmental assessment and weight of evidence for the United States government.
- Explains how weight of evidence is applied to different types of environmental and health decisions.
- Includes alternative frameworks and practices from other organizations that may be appropriate in particular cases.
- Uses an engaging style and draws on the authors' decades of practical experiences.
- Expands on the USEPA guidelines that have been a basis for weight-of-evidence guidance in Europe, Australia, and elsewhere.

This book is for senior and graduate students in environmental sciences and management and for assessment practitioners who want to understand the use of weight of evidence to achieve the purposes of environmental assessments.

PART III Using Weight of Evidence

Preface: What Kind of Book Is This?

Impetus: A few years ago, I (Suter II) tried to find a book on weight of evidence that was relevant to environmental assessors. However, all of the weight-of-evidence books that I found related to the law. The goddess of justice still holds her scales, and legal practitioners still take the weighing of evidence seriously. Since I was retired with relatively little to do, Susan Cormier suggested that we collaborate on a book on weight of evidence for environmental scientists. She and I have worked together for decades on methods and applications for weight of evidence, and we have enjoyed and benefited from those collaborations. I contacted my editor and proposed a book with the following scope.

Definition: By assessment, we mean the process of generating and presenting scientific information to inform an environmental regulatory or management decision. By environmental, we mean having to do with entities and processes occurring in the natural world, including modified environments such as reservoirs and croplands. Environmental assessments address pollutants, contaminants, introduced species, land modification, resource management, and other agents and activities.

Audience: This book is primarily written for environmental assessment practitioners. It is also appropriate for advanced students who want to be well prepared for a career in environmental science.

Goal: Like many other scientific texts, our goal is to define the state of knowledge and practice with respect to our field. Much is new in environmental toxicology, chemistry, and environmental science, and knowledge will continue to advance. However, the fundamental fact that most environmental assessment problems will require identifying, rating, and integrating multiple pieces of evidence will not change. The fundamentals of inference will not change. For example, statistical associations will never prove causation, no matter what novel statistical methods are deployed. Similarly, evidence will continue to be evaluated with respect to defined considerations, as it has been for 60 years. As a result, we believe that this account of the state of weight of evidence will continue to be useful.

As we make clear in Part I, there are many approaches and methods for assessment by weight of evidence and its alternatives. In Part II, we present an approach that we believe is best for environmental assessments, which is an update of the U.S. Environmental Protection Agency's guidelines. It is intended to provide clear and defensible results by explicitly addressing issues that tend to be glossed over in other approaches. However, as a result, it is relatively complex. If you do not have the time or resources to implement it, you can reduce the process to the essential steps for your case. In Part III, we discuss the use of weight of evidence in different assessment tasks.

This book is also a review of the literature and practices of weight of evidence in environmental assessments. The review can serve to guide you to alternative approaches.

We believe that our approach, as presented in this book and in the U.S. Environmental Protection Agency guidelines, is the most generally useful method. In addition, at least in the United States, using a standard method increases the credibility of an assessment. However, with the literature review you can explore alternative methods. In any case, you can be assured that weight of evidence is an accepted and common practice that is even mandated in some contexts.

Acknowledgments

We wish to acknowledge the former Office of Research and Development (ORD) of the U.S. Environmental Protection Agency and our many excellent ORD scientist colleagues. We and many others considered it to be the most important environmental research institution in the world. It is now being dismantled, and our friends have been forced out of the EPA or transferred to non-research positions. We both spent decades in that institution doing research on methods to improve environmental assessment and applying those methods to important problems in environmental protection. That work would not have been possible anywhere else.

Authors

Glenn W. Suter II is a retired environmental assessor. During his 20 years at the U.S. Environmental Protection Agency, he served as the Science Advisor to the Director of the National Center for Environmental Assessment in Cincinnati and as Chairman of the Risk Assessment Forum's Ecological Oversight Committee. Before that, he worked for 23 years in the Environmental Sciences Division of Oak Ridge National Laboratory. He has authored more than 300 publications, including five authored books and four edited books. Among his recognitions, he received the SETAC Founder's Award, the USEPA's Level One Scientific Achievement Award (with Susan Cormier and Sue Norton), and the Gold Medal for Exceptional Service. He is a fellow of the AAAS and SETAC. He has worked on the development of toxicity test protocols; environmental impact assessment for nuclear, geothermal, and coal energy technologies; the National Acid Precipitation Assessment Program; recovery of an endangered species; and ecological risk assessments of contaminated sites and of the mining of copper and coal. He participated in the development of guidance for ecological risk assessment, causal assessment, weight of evidence, field-based water quality criteria, assessment endpoints, ecosystem services, integrated environmental assessment, species sensitivity distributions, read-across, and other topics.

Susan M. Cormier, now retired, has been a university professor and a senior scientist and science advisor in the Office of Research and Development at the U.S. Environmental Protection Agency. Although trained as a cell biologist, she was drawn to championing environmental causes, even in high school, working to address nutrient enrichment in a neighborhood lake. As a university professor, she investigated and exposed epidemic levels of hepatocellular carcinoma in Atlantic tomcod in the Hudson River. She was recruited to join the EPA, and, in a moment of patriotism, she resigned her faculty position and committed herself for the next 33 years to developing effective ways to assess and protect aquatic resources. She developed USEPA methods, guidance, and applications for condition, causal, and criterion research and assessments, all of which depend on weight of evidence. Much of her work was co-authored with her long-time colleague and coauthor, Glenn Suter II, with whom she is excited to share this book.

Prologue: Our History of Weight of Evidence

We (Suter II and Cormier) have learned in our long careers in environmental assessment that weight of evidence is an essential but under-appreciated method. When I (Suter II) was working on contaminated site assessments in the 1980s–1990s, I discovered that a weight-of-evidence approach was necessary to deal with the diversity of studies being performed on-site and in the laboratory. I developed a general method based on weight-of-evidence tables and published multiple versions of it (Suter and Loar 1992; Suter 1996; Suter et al. 2000).

When I moved to the U.S. Environmental Protection Agency (USEPA), I joined Susan Cormier and Sue Norton on a project for the Office of Water. While the Office of Water had convinced many states to use biological as well as chemical data to identify impaired waters, they had no method to determine the cause of biological impairments. No cause, no remediation. When we reviewed the literature on causation, it became clear that, once again, a weight-of-evidence method was needed. That work led to a guidance document, a web-based support site, and a book, *Ecological Causal Assessment* (USEPA 2000, 2007; Norton, Cormier, and Suter 2014). Our approach has since been applied to many causal assessments of ecological impairments.

Next, Susan used field data to derive water quality criteria and other benchmark values. Her method used weight of evidence to determine whether a contaminant of concern is the cause of observed effects, whether other agents confound the observed causal relationship, and whether the results could be extrapolated within ecoregions (Cormier and Suter 2012, 2013; Suter and Cormier 2013; Cormier et al. 2018). Other USEPA projects in which we used our weight-of-evidence approach include the Bristol Bay watershed assessment (USEPA 2014) and the development of a new method for read-across in health risk assessments (Lizarraga et al. 2022; Suter and Lizarraga 2022). An assessment that used causal characteristics to distinguish natural from anthropogenic nutrient levels resulted in new standards for nutrients on the Grand Portage Reservation (Watkins 2023). Causal characteristics were also used to assess the decline of brook trout from exposure to 6PPD-quinone, a tire additive already infamous for killing coho salmon (Cormier, Wang, and Wharton 2025).

The USEPA work on weight of evidence led to the development of a general method and culminated in the development, by the Risk Assessment Forum, of Agency-wide guidelines (USEPA 2016; Suter and Cormier 2011). The guidelines document and its predecessors influenced European and Australian guidance on weight of evidence for ecological and human health assessments (EFSA 2017; Australia and New Zealand 2017).

This self-regarding prologue is intended to convince you that our long history of methods development and application in weight of evidence implies sufficient experience and expertise to justify this book. We believe that the concepts in this book can usefully be adopted and adapted to your assessments. We are quite aware that many others have contributed to the large and diverse body of weight-of-evidence methods. Those methods are considered and compared in this text.

As we are in the midst of writing this book, the Trump administration is in the midst of trying to gut the USEPA, the United States Fish and Wildlife Service, the National Oceanic and Atmospheric Administration, the United States Geological Survey, the Agency for Toxic Substances and Disease Registry, and other relevant environmental and public health agencies. In particular, websites are being taken down or diminished. Some of them are being archived by individuals inside and outside the agencies. Hence, you may have difficulty finding some webpages or reports that are cited here. We hope that, with the help of a good search engine, you will be able to find them. Also, the European REACH procedures for regulating chemicals in the environment are currently being revised to simplify the process. The proposed simplification has received many comments including the recommendation that all peer-reviewed evidence be "weighted consistently and transparently" (Wintermantel et al. 2026).

REFERENCES

Australia and New Zealand. 2017. *Weight of Evidence, Guidelines for Fresh and Marine Water Quality.* Canberra: Governments of Australia and New Zealand.

Cormier, S. M., and G. W. Suter. 2012. "A method for assessing causation of field exposure-response relationships." *Environmental Toxicology and Chemistry* 32: 272–276.

Cormier, S. M., and G. W. Suter. 2013. "A method for assessing causation of field exposure-response relationships." *Environmental Toxicology and Chemistry* 32 (2): 272–276.

Cormier, S. M., Y.-C. Wang, and C. Wharton. 2025. "Screening causal assessment of brook trout occurrence and road runoff." *Environmental Toxicology and Chemistry* 44 (11): 3375–3386.

Cormier, S. M., L. Zheng, G. W. Suter, and C. M. Flaherty. 2018. "Assessing background levels of specific conductivity using weight of evidence." *Science of the Total Environment* 628–629: 1637–1649.

EFSA. 2017. "Guidance on the use of the weight of evidence approach in scientific assessments." *EFSA Journal* 15 (8): 1–69.

Lizarraga, L. E., G. W. Suter, J. C. Lambert, G. Patlewicz, J. Q. Zhao, J. L. Dean, and P. Kaiser. 2022. "Advancing the science of a read-across framework for evaluation of data-poor chemicals incorporating systematic and new approach methods." *Regulatory Toxicology and Pharmacology* 137: 105293.

Norton, S. B., S. M. Cormier, and G. W. Suter. 2014. *Ecological Causal Assessment.* Boca Raton, FL: CRC Press.

Suter, G. W. 1996. *Risk Characterization for Ecological Risk Assessment of Contaminated Sites.* Oak Ridge, TN: Oak Ridge National Laboratory.

Suter, G. W., and S. M. Cormier. 2011. "Why and how to combine evidence in environmental assessments: Weighing evidence and building cases." *Science of the Total Environment* 409: 1406–1417.

Suter, G. W., and S. M. Cormier. 2013. "A method for assessing the potential for confounding applied to ionic strength in central Appalachian streams." *Environmental Toxicology and Chemistry* 32 (2): 288–295.

Suter, G. W., R. A. Efroymson, B. E. Sample, and D. S. Jones. 2000. *Ecological Risk Assessment for Contaminated Sites.* Boca Raton, FL: Lewis Publishers.

Suter, G. W., and L. E. Lizarraga. 2022. "Clearly weighing the evidence in read-across can improve assessments of data-poor chemicals." *Regulatory Toxicology and Pharmacology* 129: 105111.

Suter, G. W., and J. M. Loar. 1992. “Weighing the ecological risks of hazardous waste sites: The Oak Ridge case.” *Environmental Science and Technology* 26 (3): 432–438.

USEPA. 2000. *Stressor Identification Guidance Document.* Washington, DC: U.S. Environmental Protection Agency.

USEPA. 2007. *Causal Analysis/Diagnosis Decision Information System (CADDIS).* Washington, DC: U.S. Environmental Protection Agency.

USEPA. 2014. *An Assessment of Potential Mining Impacts on Salmon Ecosystems of Bristol Bay, Alaska.* Washington, DC: U.S. Environmental Protection Agency.

USEPA. 2016. *Weight of Evidence in Ecological Assessment.* Washington, DC: Risk Assessment Forum, U.S. Environmental Protection Agency.

Watkins, M. 2023. *Technical Support Document for Numeric Site-Specific Criteria for Nutrients, Aluminum, and Specific Conductance within the Reservation of the Grand Portage Band of Lake Superior Chippewa.* Grand Portage, MN: Departments of Biology and Environment. Grand Portage Band of Lake Superior Chippewa.

Wintermantel, Dimitry, Denicia Kassie, Lindsey Hendricks-Franco, Jessica Knapp, Annika Jahnke, Tom D Breeze, Mark JF Brown, Carsten A Brühl, Justine Dewaele, and Manon Fievet. 2026. “EU Omnibus proposal increases pesticide risks.” *Science* 392 (6805): 1348–1351.

Part I

Background of Weight of Evidence

1 History, Scope, and Definition of Weight of Evidence

All crimes shall cease, and ancient fraud shall fail,
Returning Justice lift aloft her scale.

Alexander Pope, from Messiah

Why did the kit fox population decline on the Elk Hills Naval Petroleum Reserve in the 1980s? I (Suter) was brought in to determine whether contaminants from the oil field were to blame. The answer was no (Suter et al. 1992). The Department of Energy said that drought was the cause, but I thought that was wrong too. It was only when I applied a structured weight-of-evidence analysis to all potential causes that I could confidently say that increased predation by coyotes caused the decline (Suter and O'Farrell 2009). We are all aware of weight of evidence, but that experience impressed on me the importance of taking weight of evidence seriously and doing it right. This section explains what weight of evidence is and why it is important.

1.1 WHAT IS WEIGHT OF EVIDENCE?

The weighing of evidence is derived from the ancient Greek goddess of justice, Themis, who used her scales to weigh evidence. Any type of evidence can go on the scales of justice: victim testimony, expert testimony, physical evidence, forensic test results, etc. Her judgments, or those of her judicial acolytes, are determined by the weight of evidence on each side of the case, which is expressed as the relative heights of the pans. Some pieces of evidence are more influential (weightier), and one side may have more pieces of evidence. Of course, that is a metaphor, but it is a powerful one. For thousands of years, the practice of drawing judicial conclusions from diverse evidence has been represented by Themis and her scales.

Weight of evidence is not just a legal concept. Ever since David Hume, it has been an important component of logic. He argued in *Of Miracles* that the only way to judge an empirical claim or compare claims is by weighing the evidence (Hume 1748). He discussed factors that increase weight, including reliability, manner, and number of witnesses.

The weighing of evidence has even been described as a basic element of human thought (Griffin and Tversky 1992). Fundamentally, weight of evidence is how the brain works. Multiple sensory inputs result in repeated nerve stimulation, each of which weights a hypothesis concerning the perceived reality (Gold and Shadlen 2002, 2007). Similarly, stimulation of multiple neural pathways becomes conscious

DOI: 10.1201/9781003626336-2

perception based on weight of evidence (Dennett 1991). Research on chimpanzees shows that they make choices by weighing conflicting evidence, including giving more weight to some types of evidence based on quantity and quality, and by updating their beliefs through integrating new evidence (O'Grady 2025; Schleihauf et al. 2025; Hare 2025).

More importantly, the weighing of evidence is recognized as essential to reliable inference in all fields. Steinzor (2009) concluded that "The only truly scientific decisions are ones reached on the basis of the weight of the evidence." More broadly, Boyd (2025) wrote that "Truth is defined as a state of knowledge verifiable by multiple independent methods applied by disinterested observers."

Puzzles provide another metaphor for weight of evidence. In crossword puzzles, the answer to one clue (i.e., a piece of evidence) constrains answers to others. So, although multiple words may be consistent with a clue for a row in the puzzle, the solutions to clues for intersecting columns determine which is correct. As suggested by this metaphor, to accept a hypothesis, pieces of evidence should not contradict each other.

In practice, weight of evidence is an inferential process that considers all available and potentially relevant evidence, assigns weights to the evidence, and integrates the body of evidence to generate a well-supported conclusion. The inference can take two forms.

1. In inductive inference, we use evidence to determine the best hypothesis. The most common output of induction is the relative degree of support provided by a body of evidence for a hypothesis. That is, weight of evidence can determine which hypothesis is best supported by the evidence. This induction begins with evidence such as injured plants, soil chemistry, dryness, plant symptoms, etc., and relates the evidence to hypotheses.
2. In abductive inference, we use hypotheses to explain evidence. Abductions begin with evidence that must be explained, (dying plants, dry soil, wilting, etc.), form hypotheses that might explain the evidence (soil contamination, air pollution, drought, herbicide drift, etc.), and determine the best explanatory hypothesis. From a pragmatic point of view, we infer that the best hypothesis is the one that makes the best predictions. For example, multiple hurricane models (hypotheses) may project different landfalls, but if a hurricane faithfully follows model C (evidence), then it makes sense to evacuate residents at model C's predicted landfall first. The focus on explaining the evidence of interest means that you will evaluate properties of the hypotheses. The properties of hypotheses may be duplicative of the explanatory virtues (e.g., simplicity, falsifiability) or theoretical virtues (e.g., coherence, consilience) (Sections 3.5 and 8.2.2), or you may base them on weighting attributes of the hypotheses that explain the evidence (e.g., drought explains dry soil and wilting).

Inference by weight of evidence, in either form, is an essential method in environmental assessment, as it is in science generally. In the interest of open science in regulatory environmental assessments, we need explicit weight-of-evidence frameworks and transparent implementation (Brock et al. 2021). Also, thanks to skepticism driven by social media, artificial intelligence, and anti-science politics, people no longer take an expert opinion as fact. Social media comments increasingly ask for sources and further evidence to back up claims, which is a good thing.

Weight of evidence is particularly important to the assessment of causation. Causation is essential to science and our everyday functioning in the world, but it is not countable or measurable (Norton, Cormier, and Suter 2014). It has many definitions and methods for its characterization (Susser 1991; Suter 2015), so it is best determined by plural approaches (Cartwright 2007).

Many practitioners have defined weight of evidence without the weighting of evidence such as Burton, Chapman, and Smith (2002). All that they required is that multiple pieces of evidence be combined. A nice definition of that sort is: Weight of evidence "describes the process to collect, analyze, and evaluate a combination of different qualitative, semi-quantitative, or quantitative lines of evidence to make an overall assessment of water/sediment quality and its associated management" (Warne et al. 2018). You can substitute whatever assessment problem you face for water/sediment quality. However, we believe that most nominal weight-of-evidence applications at least implicitly give some evidence more weight than others. Inferential methods that combine pieces of evidence without weighting them may be called combinations of evidence.

It is better to make weighting explicit, or state explicitly that all evidence is equally influential. For example, if the entire body of evidence for an assessment of toxic effects on fish consists of three standard acute lethal concentrations (LC50s) for zebrafish, there is no point in weighting those values. They are just replicates.

1.2 WHAT IS WEIGHT?

When Good (1991) redefined weight of evidence, he made a joke: "*I shall not use the expression to mean the weight of the paper on which some evidence is printed.*" That joke provides a warning that you should be careful not to mistake voluminous evidence for weighty evidence. There are two common definitions of weight that are useful for assessments. They are not inconsistent. Rather, they are different ways of thinking about the weight of your evidence.

- Weight is a unit of confidence. That is, it is the degree to which evidence instills confidence in a hypothesis or the degree of confidence that a hypothesis explains the evidence.
- Weight may also be a unit of importance. That is, weight is the importance of evidence to reaching the conclusion of an inference.

Either definition works, but it is helpful to choose a definition so that, as you weigh the evidence, the weights are conceptually consistent. That is, for a piece of information to constitute evidence, either:

- it gives you more or less confidence in the hypothesis, or
- it moves you closer to or further from a conclusion concerning the hypothesis.

We usually find confidence to be a more natural way to think of weight, but you may find importance to be more useful.

BOX 1.1 CONFIDENCE AND UNCERTAINTY

Confidence and uncertainty may seem to simply be inverse concepts, but they are conceptually distinct (SPC 2000; Mastrandrea et al. 2010; Oppenheimer et al. 2019; Suter 2023).

Confidence is an expression of degree of belief. In weight-of-evidence practices, the weight of supporting evidence is an expression of confidence in a hypothesis, given the evidence. In quantitative analysis, you may derive the interval in which you have some percent confidence that the true value occurs. In risk assessments, confidence is often expressed as the margin of exposure. That is, if the difference between the level of exposure and a toxicological benchmark value is large in either direction, you are confident in the risk.

Uncertainty represents the lack of knowledge about some information that is desired and could be obtained. Uncertainties are often listed in assessments and may be categorized in terms of their perceived importance. For example, could they change the conclusion? If the uncertainty is due to the absence of measurements needed to estimate a model parameter, its importance may be estimated by sensitivity analysis.

In an assessment, you should transparently reveal your uncertainty along with your confidence derived from the weight of evidence.

1.3 WHY DO WE NEED WEIGHT OF EVIDENCE?

Environmental assessors commonly face the questions: how do I use multiple pieces and types of evidence to assess a hypothesis or derive a value? Ecological assessors who are trying to determine aquatic benchmark values may have biological survey data from contaminated sites, tests of effluents, acute and chronic tests of multiple species exposed to individual chemicals, and chemical concentrations in streams. Human health assessors trying to determine whether a chemical is a human carcinogen may have rodent carcinogenicity tests, in vitro mutagenicity tests, and observational human data. What should you do with such mixtures of information? You guessed it, apply a weight-of-evidence method.

Some assessors integrate evidence by choosing one "best" piece of evidence, and too often it is the evidence that they generated. As has long been recognized, all types of evidence have limitations, and each piece of evidence has weaknesses. No one type of evidence reveals all the complexities of real-world cases. These differences are not just sampling error or other random variance that can be corrected by statistically combining results. They are qualitatively different, proverbial apples and oranges. Therefore, a logical weight-of-evidence process must be used to understand how the pieces of evidence relate to each other and to the hypothesis. Simply put, weight of evidence is more likely to give the right answer.

Multiple arguments may be made for making inferences by weighing evidence. The following list of 15 arguments is adapted from Suter and Cormier (2011).

1. If all evidence is not considered, false conclusions are more likely. In particular, the wrong hypothesis may be chosen because of biases:
   a. assessors are likely to choose their own evidence, their favorite type of evidence, or the most easily obtained evidence, and
   b. assessors are likely to avoid evidence that is unfamiliar, difficult to understand, controversial, or contrary to preconceptions or self-interest.
2. In effect, weighing multiple pieces of equivalent evidence can amount to evidence replication. Such replication can reveal the variance among pieces of evidence of the same type, among studied systems, and among investigators. One piece of evidence may compensate for the limitations of another.
3. If you cannot obtain evidence from replicate systems, you can study the same system using multiple methods (Cavalli-Sforza 2000). Weighing the evidence from the multiple methods as a body of evidence may reveal the best-supported hypothesis.
4. Different types of evidence have different implications, so they tell you something different about the hypothesis and the situation being assessed. For example, one piece of evidence may tell you whether a putative cause preceded the effect, while another tells you whether the cause was of sufficient magnitude.
5. The mistakes made in generating one piece of evidence may not be repeated in others, and random errors tend to balance out.
6. When whole-animal tests are not allowed, weighing a large set of sub-organismal tests may provide equivalent results (Svingen et al. 2024).
7. It is seldom the case that one piece or type of evidence is superior to all others in all respects.
8. Even if one piece of evidence is superior to all others, supporting evidence from other sources can increase confidence in the results from the superior study.
9. To not assemble and weigh all relevant evidence is to waste a limited resource—scientific evidence.
10. Including and weighting multiple risk estimates allows decision makers to make better-informed decisions (Gray 1994).
11. Weighing evidence makes assessments consistent with legal practice (Krimsky 2005; Walker 1996), and environmental assessments often enter legal proceedings.
12. Including all relevant evidence reassures stakeholders and decision makers that evidence is not being ignored or covered up (NRC 1994).
13. Weighing evidence is consistent with natural cognitive processes (Section 1.1).
14. Weighing evidence is recommended by agencies, organizations, and authoritative panels (Section 1.9).
15. When communicating contentious science, such as the safety of vaccines, weight of evidence and consensus are more effective than alternative strategies (Whitehead et al. 2023).

Most of these reasons are specific examples of epistemological pluralism, the philosophy that more types or sources of information or reasoning support better judgments.

Pluralism originated with the pragmatic philosophers Charles Sanders Peirce and William James (Suter and Cormier 2013). They equated multiple pieces of evidence with multiple strands of wire braided into a cable.

Although most of the confidence gained by weighing evidence comes from combining multiple pieces and categories of evidence, it is possible to weight a single piece of evidence when only one piece is available for a hypothesis. This is done by weighting that piece with respect to the same properties as multiple pieces. It provides a consistent evaluation of evidence and may highlight the need for more information if the weight is low.

1.4 DO WE REALLY NEED TO WEIGH EVIDENCE?

We must acknowledge that there are arguments against considering all relevant evidence (Linkov et al. 2015; de Zwart et al. 2009; Newman, Zhao, and Carriger 2007). There are also responses to those arguments.

1. *It is unnecessary because, in most cases, one piece of evidence is obviously right, or one type is clearly best. Therefore, no weight should be given to the others.* However, some sort of explicit or implicit evidence-weighting process must be used to identify that one best piece of evidence and distinguish it from the rest.
2. *It is qualitative and subjective.* Some authors who advocate particular quantitative methods for inference criticize weight of evidence as less rigorous than their equivalent method. Although quantitative methods using one type of evidence may be more rigorous, they are not more likely to be correct because they often discard important information that does not fit their quantitative methods.
3. *It is too difficult or time-consuming.* There is no need to use a weight-of-evidence method if it will not improve the assessment or if a rapid response is required.
4. *It is too complicated and confusing.* Clear presentation of methods and results can eliminate, or at least reduce, confusion in complicated assessments.
5. *It dilutes good evidence with poor evidence.* This is the reason that evidence is explicitly weighted. Good evidence has high weight, and poor evidence has low weight.
6. *In some cases, standard sets of information and interpretations are specified by guidelines.* In such cases, follow the rules, but you may also provide a weight of evidence in the discussion for additional context and support.

Some assessors and stakeholders have rejected inference by weight of evidence. If you are confronted by someone with that opinion, ask them if they are willing to consider more than one type of evidence. If they say yes, ask them if they believe that some evidence should have more influence than other evidence. If they say yes to that as well, they have endorsed weight of evidence, and perhaps the rejection is a matter of semantics or a particular process. In any case, it is an opportunity for discussion.

1.5 HOW MIGHT YOU COME TO HAVE MULTIPLE PIECES OF EVIDENCE?

Most of our justifications of weight of evidence presume that you have multiple pieces of evidence. That assumption may hold because you are dealing with a problem that has been extensively investigated, such as the aquatic toxicity of copper. On the other hand, you may have designed case-specific studies that provide multiple pieces of evidence that meet the needs of your assessment. In that case, you may use a weight-of-evidence method like the sediment quality triad (Section 2.7), which specifies a standard dataset. You may do both. That is, begin by searching for existing information and then fill in the information gaps with results of new studies. In some cases, one type of evidence comes from the literature (e.g., toxicological benchmark values), and another is inevitably case-specific (e.g., in situ toxicity).

As a result of these various circumstances, you may have:

- Multiple types of evidence
- Multiple studies of the same type
- Multiple endpoint effects in a type
- Multiple analytical methods for a single dataset

1.6 WHY DOES WEIGHT OF EVIDENCE REQUIRE EXPERT JUDGMENT?

Although evidence is objective (i.e., based on observations or measurements), judgments of its properties, such as reliability, are not. Further, most applications of weight of evidence in environmental assessment involve types of evidence that cannot be objectively integrated. Also, different pieces of evidence have different relevance, reliability, and strength, and those properties should be weighed and combined into an overall weight of the body of evidence. Combining the types of evidence and their properties is inherently judgment-based. Even when you use standard weights (e.g., using rules such as when relating measurements of stream biota and chemical concentrations, a correlation coefficient greater than 0.75 is evidence for a strong association), the choice of the standard weights was subjective, based on experience. Rather than denying or ignoring subjectivity, recognizing when, how, and why it influences arguments within a formal weight-of-evidence method engenders trust in your work. That is, it makes your judgments explicit, clear, consistent, and justified (Part II).

The subjective judgments in weight of evidence are, at least, explicit, which puts you in cognitive system 2 (Kahneman 2011). Cognitive system 1 is fast and effortless and likely to make false inferences. For example, if you see a dead fish floating in water with a rainbow sheen, under system 1, you immediately will think that spilled oil killed the fish. You might immediately weight that observation as strong evidence of the co-occurrence of cause and effect. Cognitive system 2 involves concentration, effort, and analysis. System 2 thinking would remind you that association is not causation, which might lead you to take samples, analyze the water and fish, and examine symptoms in the fish. The weight of that evidence may show that a pesticide in an oil carrier was the cause. A major theme of this book is to make sure your judgments are framed in the context of a structured inferential system that keeps you firmly in system 2.

Practices that make your judgments less questionable include working in a team, transparently presenting the bases for judgments, and enlisting qualified peer reviewers. Awareness of your values, expertise, and experiences that could influence your judgments, and consideration of the appropriateness of those values to the legal mandate, are also helpful (Wandall 2004).

1.7 WHO WEIGHS THE EVIDENCE?

To weigh evidence well, an assessor should have experience in generating and applying evidence for the problem being assessed. Ideally, that person would also have experience in assessments that use the weight-of-evidence approach being applied to the case. That is a lot to ask of any individual. It is part of the reason that we recommend weighing evidence in teams. The teams should be drawn from different areas of expertise to cover the range of scientific evidence and assessment experience. Some organizations insist on outside experts to weigh evidence, but, in my experience, most outside experts have scientific expertise but little or no expertise in weight of evidence in environmental assessment.

Teams may function in different ways. They may meet and discuss the evidence until they reach a consensus, weight the evidence independently and then meet to resolve differences, or use a formal interactive consultative process like the Delphi method (Hsu and Sandford 2007; Hasson, Keeney, and McKenna 2025). Even if there is specific guidance for assigning weights, a weight-of-evidence team may be needed to interpret the guidance with respect to the case and evidence at hand.

1.8 WHAT IS WEIGHT OF EVIDENCE USED FOR?

Weight of evidence is a sort of Swiss Army knife for assessments. In assessments, you can use it for summary, inference, and estimation. Health Canada (2022) stated that "The nature and scope of risk assessments may vary, but all have a WoE component."

In evidence summaries, weight of evidence brings the evidence together to elucidate what is known concerning a particular problem or issue of fact. Consider, for example, a literature review concerning the herbicide glyphosate. Weight of evidence may organize the evidence by type, location, medium, or other attribute, and then summarize the information that is relevant to estimating risks. Summaries may simply be reference works, or they may be a prelude to inference or estimation. Some have argued that we assessors should not proceed beyond summarization to the derivation of a conclusion (Gray 1994). They would leave conclusions to the decision maker. However, even in a summary, some weighing is involved in suggesting which evidence is most credible or sufficiently credible and whether the body of evidence is consistent and coherent. And, in our experience, decision makers want a conclusion or recommendation with reasons for them.

Inference is the derivation of a conclusion from evidence by reasoning. It is the most common use of weight of evidence in environmental assessment.

The most obvious examples are questions of causation. Hazard assessments for chemicals answer the question: What effects are potentially caused by the chemical (Chapter 16)? For example, does atrazine cause deformities, or what effects can atrazine cause?

Predictive assessments, such as risk, criteria, and management assessments, answer questions about potential future events. Will a proposed use of atrazine cause deformities? How much can safely be applied? How can it be applied to minimize harm? Causal assessments answer questions about observed effects such as, were the observed deformities caused by atrazine? These are all questions about qualities (Chapter 8). By weighing the evidence, you can infer which answer is most credible for any of those questions and whether the weight of evidence is sufficient to support the conclusion. Estimation derives quantities from evidence (Chapter 12). These may be quantities that are used in an assessment, such as the biodegradation rate of glyphosate or a benchmark value that drives decisions, such as a water quality criterion for glyphosate. Weight of evidence can determine the best estimate of a quantity from among the estimates provided by different types of evidence.

1.9 WHAT PROVIDES THE IMPETUS FOR WEIGHT OF EVIDENCE IN ENVIRONMENTAL ASSESSMENTS?

You may use weight of evidence simply because it is the most appropriate summary, inferential, or estimation method for your case. However, additional impetus is provided by laws, regulations, and guidelines. Some environmental legislation in the U.S. explicitly requires weight of evidence to ensure that all available evidence is used. The most recent U.S. law concerning pollution is the Lautenberg Act, which is an update of the Toxic Substances Control Act. It calls for weight of evidence six times. In particular, the law contains the following brief subsection: "WEIGHT OF SCIENTIFIC EVIDENCE—The Administrator shall make decisions under Sections 4, 5, and 6 based on the weight of the scientific evidence." In a review of European Union environmental laws, four of nine laws and six of nine guidance documents mentioned weight of evidence (Ågerstrand and Beronius 2016). The Canadian Environmental Protection Act of 1999 and its 2023 amendments require a weight-of-evidence approach in assessments and reviews of decisions. Systematic review, meta-analysis, decision analysis, Bayesian networks, or other tools that have been presented as alternatives to weight of evidence are not mandated by U.S., Canadian, or EU legislation.

Weight of evidence is often prescribed in the regulations that determine how laws are implemented. For example, the regulations that implement the U.S. Clean Water Act require that all existing and readily available information should be evaluated when determining whether a water body is impaired (40 CFR §130.7(b)(5)).

Guidance and guidelines provided by regulatory agencies or other organizations also encourage the weighing of evidence. Besides recommending methods, they also indicate when weight of evidence should be used. Some examples are (USEPA 1998, 2016, 2025; EFSA Scientific Committee 2017; Menzie et al. 1996, 2024; Australia and New Zealand 2017; ECHA 2010; NRC 2009; Bates, Massey, and Wood 2018; SCHEER 2024; Hall et al. 2017).

In industry, weight of evidence can be part of an assessment intended to inform responsible decision making, prepare for regulation, or respond to regulation. In particular, when regulatory assessments are based on worst-case exposure scenarios and the most sensitive toxicological effect, industries sometimes respond with a weight-of-evidence analysis that may give more realistic but less precautionary results.

Further, environmental assessors weigh evidence to be presented in legal proceedings. That is, you provide the weighted evidence that loads Themis's scales. When you serve as an expert witness or prepare materials for an expert witness, you should be aware of legal expectations and requirements. In the U.S., under the Daubert rule, the judge is empowered to evaluate expertise and the admissibility of evidence (Supreme Court of the United States 1993). Evidence must be admissible, authentic, complete, reliable, and believable. Otherwise, it may be excluded from the body of evidence to be weighed.

In sum, in a National Research Council workshop on Environmental Medicine Decision Making, Murray (2009) provided this conclusion: *It is widely agreed that these [public policy] decisions should be based on an unbiased evaluation of the weight of scientific evidence.*

1.10 BEST AVAILABLE SCIENCE AND WEIGHT OF EVIDENCE

While many laws and regulations call for weight of evidence, an alternative legal standard for judging evidence is "best available science." This may have multiple interpretations. It may mean that you must use a single best study. It may mean the result that is supported by the best set of studies that give a common result. It may require meeting certain criteria, such as having been peer reviewed, derived using standard methods, or representing the most recent science. Some discussions of the issue have equated determining best science with weight of evidence, and we believe that is the best solution. You can weigh evidence to identify the best (weightiest) piece of scientific evidence, or the hypothesis or quantitative result that is supported by the greatest weight of evidence. Best science may be judged on the basis of quality, which implies the reliability property (Section 7.1.2). An explicit, structured weight-of-evidence approach could resolve many of the disagreements concerning best available science. Instead, why not conclude a comprehensive weight of evidence and then explain how that is the best science?

1.11 WEIGHT OF EVIDENCE FOR RESEARCH

Although the subject of this book is environmental assessment, weight of evidence can make important contributions to scientific research. Science has become controversial, particularly biomedical and public health science, and scientific results and recommendations may be unclear due to inconsistent and irreproducible results. It has been argued that, rather than replication, the best strategy for preventing or correcting scientific controversies is to weigh multiple types of evidence for a scientific hypothesis (Munafò and Smith 2018). Simply replicating an experiment may only replicate flawed designs and execution.

REFERENCES

Ågerstrand, M., and A. Beronius. 2016. "Weight of evidence evaluation and systematic review in EU chemical risk assessment: Foundation is laid but guidance is needed." *Environment International* 92: 590–596.

Australia and New Zealand. 2017. "Weight of evidence, Guidelines for fresh and marine water quality." Accessed August 16, 2021.

Bates, M. E., O. C. Massey, and M. D. Wood. 2018. *Weight-of Evidence Concepts: Introduction and Application to Sediment Management.* Washington, DC: U.S. Army Corps of Engineers.

Boyd, I. (2025). *Science and Politics*. Boston, NY: Polity Press.

Brock, T. C. M., K.C. Elliot, A. Gladbach, C. Moermond, J. Romeis, T.-B. Seiler, K. Solomon, and G. P. Dohmen. 2021. "Open Science in regulatory environmental risk assessment." *Integrated Environmental Assessment and Management* 17 (6): 1229–1242.

Burton, G. A., P. M. Chapman, and E. P. Smith. 2002. "Weight-of-evidence approaches for assessing ecosystem impairment." *Human and Ecological Risk Assessment* 8 (7): 1657–1673.

Cartwright, N. 2007. *Hunting Causes and Using Them: Approaches in Philosophy and Economics*. Cambridge, UK: Cambridge University Press.

Cavalli-Sforza, L. 2000. *Genes, Peoples, and Languages*. New York: North Point Press.

Dennett, D. C. 1991. *Consciousness Explained.* Boston, MA: Little, Brown & Co.

ECHA. 2010. *Practical Guide 2: How to Report Weight of Evidence.* Helsinki: European Chemicals Agency.

EFSA Scientific Committee. 2017. "Guidance on the use of the weight of evidence approach in scientific assessments." *EFSA Journal* 15 (8): 1–69.

Gold, J. I.. and M. N. Shadlen. 2002. "Banburismus and the brain: Decoding the relationship between sensory stimuli, decisions, and reward." *Neuron* 36: 299–308.

Gold, J. I., and M. N. Shadlen. 2007. "The neural basis of decision making." *Annual Review of Neuroscience* 30: 535–574.

Good, I. J. 1991. "Weight of evidence and the Bayesian likelihood ratio." In *The Use of Statistics in Forensic Science*, edited by C. C. D Aitken and D. A. Stoney, 85–106. New York: Ellis Horwood.

Gray, G. M. 1994. "Complete risk characterization." *Risk in Perspective* 2 (4): 1–2.

Griffin, D., and A. Tversky. 1992. "The weighing of evidence and the determinants of confidence." *Cognitive Psychology* 24: 411–435.

Hall, A. T., S. E. Belanger, P. D. Guiney, M. Galay-Burgos, G. Maack, W. S., and O. Martin. 2017. "New approach to weight-of-evidence assessment of ecotoxicological effects in regulatory decision-making." *Integrated Environmental Assessment and Management* 13 (4): 573–579.

Hare, B. 2025. "Chimpanzees are natural scientists." *Science* 390 (6772): 444–445.

Hasson, F., S. Keeney, and H. McKenna. 2025. "Revisiting the Delphi technique – Research thinking and practice: A discussion paper." *International Journal of Nursing Studies* 168: 105119.

Health Canada. 2022. "Application of weight of evidence and precaution in risk assessment." *Fact Sheet Series: Topics in Risk Assessment of Substances under the Canadian Environmental Protection Act*, 1999.

Hsu, C., and B. A. Sandford. 2007. "The Delphi technique: Making sense of consensus." *Practical Assessment, Research, and Evaluation* 12 (1).

Hume, D. 1748. *An Inquiry Concerning Human Understanding*. Amherst, NY: Prometheus Books.

Kahneman, D. 2011. *Thinking Fast and Slow.* New York: Farrar, Straus, and Giroux.

Krimsky, S. 2005. "The weight of scientific evidence in policy and law." *American Journal of Public Health* 95: S129–S135.

Linkov, I., O. Massey, J. Keisler, I. Rusyn, and T. Hartung. 2015. "From "weight of evidence" to quantitative data integration using multicriteria decision analysis and Bayesian methods." *ALTEX* 32 (1): 3–8.

Mastrandrea, M. D., C. B. Field, T. F. Stocker, O. Edenhofer, K. L. Ebi, D. J. Frame, H. Held, E. Kriegler, K. J. Mach, P. R. Matschoss, G.-K. Plattner, G. W. Yohe, and F. W. Zwier. 2010. *Guidance Note for Lead Authors of the IPCC Fifth Assessment Report on Consistent Treatment of Uncertainties.* Geneva, Switzerland: Intergovernmental Panel on Climate Change.

Menzie, C. A., P. D. Guiney, S. E. Belanger, K.-C. Lee, G. Arts, B. O. Opeolu, and H. C. Silva de Assis. 2024. "Incorporating a weight-of-evidence approach into a tiered assessment for chemicals management, with emphasis on program development and applications in developing countries and emerging economies." *Integrated Environmental Assessment and Management* 20 (6): 2312–2325.

Menzie, C., M. H. Henning, J. Cura, K. Finkelstein, J. Gentile, J. Maughan, D. Mitchell, S. Petron, B. Potocki, S. Svirsky, and P. Tyler. 1996. "Special report of the Massachusetts weight-of-evidence workgroup A weight-of-evidence approach for evaluating ecological risks." *Human and Ecological Risk Assessment* 2 (2): 277–304.

Munafò, M. R., and G. D. Smith. 2018. "Repeating experiments is not enough." *Nature* 553 (7689): 399–401.

Murray, T. H. 2009. "General observations concerning conflict of interest." In *Environmental Health Sciences Decision Making: Risk Management, Evidence, and Ethics*, edited by Y. Coble, C. Coussens, and K. Quinn, 35–38. Washington, DC: National Academies Press.

Newman, M. C., Y. Zhao, and J. F. Carriger. 2007. "Coastal and estuarine ecological risk assessment: The need for a more formal approach to stressor identification." *Hydrobiolgia* 577: 31–40.

Norton, S. B., S. M. Cormier, and G. W. Suter II. 2014. *Ecological Causal Assessment.* Boca Raton, FL: CRC Press.

NRC. 1994. *Science and Judgment in Risk Assessment.* Washington, DC: National Research Council, National Academy Press.

NRC. 2009. *Science and Decisions: Advancing Risk Assessment.* Washington, DC: National Research Council, National Academy Press.

O'Grady, C. 2025. "New evidence? No problem. Chimps can weigh conflicting clues, just like humans." *Science.* doi: 10.1126/science.zt0mo02

Oppenheimer, M., N. Oreskes, D. Jamieson, K. Brysse, J. O'Reilly, M. Shindell, and M. Wazeck. 2019. *Discerning Experts: The Practices of Scientific Assessment for Environmental Policy.* Chicago, IL: University of Chicago Press.

SCHEER. 2024. *Memorandum on Weight of Evidence and Uncertainties, Revision 2024.* Luxembourg: European Commission Scientific Committee on Health, Environmental and Emerging Risks.

Schleihauf, H., E. M. Sanford, B. D. Thompson, S. Zhang, J. Rukundo, J. Call, E. Herrmann, and J. M. Engelman. 2025. "Chimpanzees rationally revise their beliefs." *Science* 390: 521–526.

SPC. 2000. *Risk Characterization Handbook.* Washington, DC: Science Policy Council, U.S. Environmental Protection Agency.

Steinzor, R. 2009. "The use and misuse of science in decision making." In *Environmental Health Sciences Decision Making: Risk Management, Evidence, and Ethics*, edited by Y. Cable, C. Coussens, and K. Quinn, 28–29. Washington, DC: National Academies Press.

Supreme Court of the United States. 1993. *Daubert v. Merrell Dow Pharmaceuticals,* 509 U.S. 579, 113 S.Ct. 2786, 125 *L.Ed.* 2d 469, *(U.S. Jun 28, 1993) (NO.* 92–102*).* Washington, DC: Supreme Court of the United States.

Susser, M. 1991. "What is a cause and how do we know one? A grammar for pragmatic epidemiology." *American Journal of Epidemiology* 133 (7): 635–648.

Suter, G. W. 2015. "What is a cause?" In *Ecological Causal Assessment*, edited by S. B. Norton, S. M. Cormier, and G. W. Suter, 17–31. Boca Raton, FL: CRC Press.

Suter, G. W. 2023. *Fundamentals of Environmental Assessment*. Boca Raton, FL: CRC Press.

Suter, G. W., and S. M. Cormier. 2011. "Why and how to combine evidence in environmental assessments: Weighing evidence and building cases." *Science of the Total Environment* 409: 1406–1417.

Suter, G. W., and S. M. Cormier. 2013. "Pragmatism: A practical philosophy for environmental scientists." *Integrated Environmental Assessment and Management* 9: 181–184.

Suter, G. W., and T. P. O'Farrell. 2009. *Analysis of the Causes of a Decline in the San Joaquin Kit Fox Population on the Elk Hills, Naval Petroleum Reserve #1, California*. Cincinnati, OH: U.S. Environmental Protection Agency.

Suter, G. W., A. E. Rosen, J. J. Beauchamp, and T. T. Kato. 1992. *Results of Analysis of Fur Samples from the San Joaquin Kit Fox and Associated Water and Soil Samples from the Naval Petroleum Reserve No. 1, Tupman, California*. Oak Ridge, TN: Oak Ridge National Laboratory.

Svingen, T., A.-M. Andersson, J. Angelova, M. Axelstad, J. Bakker, L. Baumann, A. Beronius, N. Bouftas, F. Chalmel, and S. Christiansen. 2024. "Enhanced identification of endocrine disruptors through integration of science-based regulatory practices and innovative methodologies: The MERLON Project." *Open Research Europe* 4 (68): 68.

USEPA. 1998. *Guidelines for Ecological Risk Assessment*. Washington, DC: Risk Assessment Forum, Washington, DC: U.S. Environmental Protection Agency.

USEPA. 2016. *Weight of Evidence in Ecological Assessment*. Washington, DC: Risk Assessment Forum, Washington, DC: U.S. Environmental Protection Agency.

USEPA. 2025. *Guidelines for Carcinogen Risk Assessment*. Washington, DC: Risk Assessment Forum, Washington, DC: U.S. Environmental Protection Agency. https://www.epa.gov/risk/guidelines-carcinogen-risk-assessment.

Walker, V. R. 1996. "Risk characterization and the weight of evidence: Adapting gatekeeping concepts from the courts." *Risk Analysis* 16: 793–799.

Wandall, B. 2004. "Values in science and risk assessment." *Toxicology Letters* 152 (3): 265–272.

Warne, M., G. E. Batley, R. A. van Dam, J. C. Chapman, D. R. Fox, C. W. Hickey, and J. L. Stauber. 2018. *Revised Method for Deriving Australian and New Zealand Water Quality Guideline Values for Toxicants*. Brisbane: Department of Science, Information Technology and Innovation.

Whitehead, H. S, C. E. French, D. M. Caldwell, L. Letley, and S. Mounier-Jack. 2023. "A systematic review of communication interventions for countering vaccine misinformation." *Vaccine* 41 (5): 1018–1034.

de Zwart, D., L. Posthuma, M. Gevrey, P. von der Ohe, and E. de Dekere. 2009. "Diagnosis of ecosystem impairment in a multiple-stress context—How to formulate effective river basin management plans." *Integrated Environmental Assessment and Management* 5: 38–49.

2 Weight of Evidence Methods

The concept of weight of evidence is based on two principles: decisions should be based on all potentially relevant evidence, and different evidence may have different degrees of influence. The weight of evidence concept has been implemented in various ways, which has inspired many reviews of weight of evidence methods for human health and ecological assessments (Weed 2005; Linkov et al. 2009; Dorne et al. 2016; Rhomberg et al. 2013; Martin et al. 2018; USEPA 2016; Burton, Chapman, and Smith 2002; Chapman, McDonald, and Lawrence 2002; Roth and Ciffroy 2016; EFSA Scientific Committee 2017; Moore et al. 2023). These reviews are dissimilar in their organization and description of the practice. Our review of methods in this chapter is determined by our taxonomy of methods and our view of what makes for good weight of evidence practices.

Evaluation of the methods depends on the answers to two questions. First, does it qualify as weight of evidence? As stated in Section 1.1, weight of evidence methods should:

- be an inferential process and not just a review,
- consider all available and potentially relevant evidence,
- assign weights to evidence to indicate how influential it should be, and
- integrate the evidence to generate a conclusion.

Second, does it result in high-quality assessments? This question may be answered by considering whether the methods meet Douglas's (2012) criteria for good weight of evidence approaches, condensed and paraphrased as follows.

- **Completeness (Evidential Inclusiveness)**: the method should incorporate all types of evidence.
- **Rigor (Epistemic Constraint)**: weighing evidence should constrain our inferences and not allow an ad hoc response to the evidence.
- **Transparency**: other experts should be able to understand and critique the weight of evidence.
- **Communicability**: in addition to transparency, the method should be communicable to decision-makers, stakeholders, and the interested public.
- **Scope**: the weight-of-evidence approach should be applicable to a wide range of uses so that we need not develop and justify a new method for each application.
- **Practicality**: weight-of-evidence processes should be neither interminable nor excessively burdensome.

 DOI: 10.1201/9781003626336-3

2.1 NARRATIVE WEIGHT OF EVIDENCE

The original method for weight of evidence is to gather evidence and then describe and evaluate it in a prose narrative (i.e., an essay). A pure narrative weight of evidence has no structured inferential method or explicit weighting of the evidence. It makes sense of the evidence or explains the evidence and then writes the story that the evidence tells, or it tells the story of the hypothesis and fits the evidence into that narrative.

The conclusion of a narrative weight of evidence is based on a subjective evaluation of the narrative. The expertise of the reviewer is assumed to be sufficient to assure a correct conclusion without the use of an explicit method or criteria.

Because the methods of traditional narrative weight of evidence are implicit, the results are often distrusted. However, in some cases, a narrative is appropriate or at least sufficient to guide a decision. If you use narrative weight of evidence, consider following these recommendations from (USEPA 2016).

- When using evidence from the literature, design and carry out a literature search that will find the information needed in a thorough and unbiased manner (Chapter 6). Define both the search terms and the criteria for screening the results to obtain relevant evidence.
- Even in a narrative approach, provide a table summarizing the selected evidence and indicate what was excluded and why.
- Avoid narrative reviews that describe one piece of evidence after another and then present a conclusion. Instead, logically structure the narrative. At minimum, present the evidence for a hypothesis and the evidence against it, and explain why one side has greater weight.
- Even without a formal weighting and weighing of evidence, discuss the relative importance of the evidence.
- Make the narrative organization clear. Use lists and subsection headings to help the reader understand the logical structure.
- Present the results clearly. Which hypothesis is best supported by the evidence, or which hypothesis best explains the evidence, and how much confidence can be placed in the conclusion?

A few organizations have their own guidance for narrative synthesis of evidence, such as (CEE 2018).

Subjective evaluations may benefit from clear organization of the evidence using these recommendations. A good example is the weight-of-evidence-based narrative review of evidence to determine whether imposex in marine snails is caused by tributyltin's endocrine-disrupting properties (Box 2.1).

BOX 2.1 NARRATIVE WEIGHING OF EVIDENCE FOR TRIBUTYLTIN-MEDIATED ENDOCRINE DISRUPTION IN MOLLUSKS

Imposex in marine snails, including penises and reduced fecundity in females, was first reported in 1970 in Plymouth Sound. Matthiessen and Gibbs (1998) published a narrative weight-of-evidence review of the hypothesis that the imposex was caused by endocrine-disrupting effects of tributyltin, which was applied to ships as an antifoulant. They summarized the evidence in terms of three implications of the available evidence.

Scale and Extent—Imposex was identified in more than 100 species of marine snails worldwide and is associated with tributyltin in harbors (e.g., frequent and regular association).

Specificity—Imposex is specific to tributyltin and even similar organotins do not induce it.

Mechanisms—Tributyltin's mechanism of action is unclear, but the weight of evidence suggests that TBT acts as a competitive inhibitor of cytochrome P450-mediated aromatase. Meager evidence suggests a secondary mechanism. Tributyltin may also inhibit the formation of sulfur conjugates of testosterone and its active metabolites, thus interfering with its excretion.

A narrative account of this body of evidence was sufficient to confirm the hypothesis that organotin from antifoulant treatments of ships causes imposex in marine snails by endocrine disruption.

Narrative weight of evidence has the advantage of being highly flexible and does not require learning a formal method. Because of its lack of an explicit method, narrative weight of evidence is being replaced in regulatory programs. However, an advantage of narrative methods is that they have little to argue over, while explicit methods can result in methodological controversies. For example, the newest method for the USEPA's IRIS program, which replaced a narrative method, required ten years of iterative development, review, and revision before finalization (USEPA 2022).

The biggest disadvantage of narrative weight of evidence is the narrative fallacy. We are all told that if you want to convince your audience, you should present a good story (Alda 2017). You may choose to follow that advice and present the hypothesis that allows you to tell the most convincing story. You may even fall prey to the narrative fallacy yourself and believe the hypothesis with the best story. However, true hypotheses may have a more complicated, less compelling, or less easily understood story.

If you do not have all of the relevant evidence, you may, inappropriately, choose a narrative that does not require the missing evidence. This is called the what-you-see-is-all-there-is fallacy (Kahnman 2011). An explicit, structured inferential method applied to good evidence should lead you to the best-supported hypothesis, thereby avoiding the narrative fallacy.

2.2 CONSIDERATIONS-GUIDED WEIGHT OF EVIDENCE

The first and probably most important advance in weight of evidence is to (1) list aspects of causal evidence to be considered, and (2) perform a narrative analysis of each consideration. That advance was made by the U.S. Surgeon General Terry and A.B. Hill (Box 2.2).

BOX 2.2 LUTHER TERRY, BRADFORD HILL, AND WEIGHT OF EVIDENCE CONSIDERATIONS (ADAPTED FROM SUTER ET AL. 2020)

In the early 1900s, it became apparent that cigarette smoking had changed lung cancer from being a rare disease to being a common cause of death. Several studies had shown an association between smoking and the disease, but since they were observational studies, they were not convincing. Supporters of tobacco, including the famous statistician Ronald Fisher, strongly questioned the causality of the association. Two advocates of weight of evidence, U.S. Attorney General Luther Terry and Austin Bradford Hill, successfully responded to that criticism (USDHEW 1964; Hill 1965).

The Surgeon General's report has properties that Hill's famous paper does not. First, he recognized that even with the considerations, the weighing of evidence was subjective. Therefore, he assembled an Advisory Committee on Smoking and Health consisting of ten unbiased experts. They had no connection to the tobacco industry, no research on the topic, no expressed opinion on the issue, and no stakeholders objected to them. Second, he recognized that selection of evidence could bias the results, so they performed a proto-systematic review that found more than 7,000 relevant studies. The studies were reviewed for signs of bias. Third, they consistently calculated the strength of association in each study and performed a proto-meta-analysis, which determined that the relative risk was between 9 and 10. Finally, Hill focused on cigarette smoking and lung cancer, but the Surgeon General assessed all diseases that had an apparent association with tobacco smoking.

Hill's list of considerations (also called viewpoints) was presented to the Royal Society of Medicine and published in its journal as a solution to the problem of causation. Like the Surgeon General, Hill clearly stated that statistical analyses could not demonstrate causation. Hill's list is the standard for considerations-based weight of evidence. That is in part because of his international reputation as a biomedical statistician and as the father of randomized clinical trials. In addition, he presented his considerations in a clear and concise six-page paper rather than a several-hundred-page report. Finally, Hill relied on his individual expertise expressed as short narratives for the nine considerations. That simple approach has been more appealing to many than the Surgeon General's carefully vetted committee, formal literature reviews, and debiasing measures.

The Surgeon General's five considerations are:

- **Consistency of Association**: An association is consistent if it has been repeatedly observed in different studies, populations, conditions, and times.
- **Strength of the Association**: This is usually expressed as a correlation or the magnitude of the difference between exposed and unexposed subjects. For example, the death rate of smokers relative to nonsmokers.

- **Specificity of the Association**: A general effect has many potential causes, but a specific effect has few or one cause, so an association with a specific effect is more likely to be causal. For example, death rate versus death rate due to lung cancer.
- **Temporal Relationship of the Association**: Causes must occur before effects. This is the one necessary condition. Tobacco smoking precedes lung cancer in individuals, and smoking rates increase before lung cancer rates in populations.
- **Coherence of the Association**: Evidence is coherent if the body of evidence consistently indicates causation or lack of causation and is consistent with biological, chemical, and physical science.

Hill adopted those considerations and added four more:

- **Biological Gradient**: The magnitude of effects should be related to the magnitude of exposure in a manner consistent with the hypothesized causal relationship. This was true of the rate and duration of smoking and lung cancer rates.
- **Plausibility**: If the causal nature of the association is plausible given biological knowledge, it is supported. The Surgeon General made this part of coherence. Hill emphasized that the lack of plausibility may be misleading since some true causes are implausible in the beginning. However, if the association makes no mechanistic sense, that can be strong negative evidence.
- **Experiment**: Hill was referring to natural experiments. For example, if one religious group forbids smoking but other religious groups in the same population are not forbidden, those conditions create treatment and control groups. However, many assessors have interpreted experiment to include controlled experiments.
- **Analogy**: If other similar associations are known to be causal, that evidence may be applied by analogy to the case at hand. For example, the now well-known causal relationship of tobacco smoking to lung cancer may be applied by analogy to the smoking of other herbs.

Terry and Hill were successful in convincing most of the public and most public health professionals that cigarette smoking causes lung cancer. The method of organizing evidence in terms of considerations has been widely adopted by epidemiologists, but not by all. In part, this is because the considerations were often treated as criteria and even as a checklist, although they are not appropriate for such uses. For that reason, the famous epidemiologist Kenneth Rothman strongly criticized Hill's considerations (Rothman, Greenland, and Lash 2008). Although, as Hill emphasized, the considerations are not a set of criteria for causation, they are often called Hill's criteria and are treated as such. Some of the individual considerations have been said to be misleading or useless (Ioannidis 2016; Rothman, Greenland, and Lash 2008). If the considerations are to be used, more guidance is needed concerning how they should be interpreted and whether they include everything that should be considered.

When you begin to interpret information in terms of the considerations, their incoherence becomes apparent. Neither Hill nor the Surgeon General considered the roles of the considerations in inference by weight of evidence (Suter and Cormier 2011; USEPA 2016). Hill did not acknowledge that he was mixing types of evidence (e.g., experiment), characteristics of causal relationships (e.g., temporality), properties of evidence (e.g., strength), and properties of bodies of evidence (e.g., consistency) (Cormier, Suter, and Norton 2010). Once you recognize that the considerations are examples of a few of the various attributes of evidence (Box 2.3), it becomes clear that the list of considerations is incomplete and inadequate to derive fully justified conclusions.

BOX 2.3 CONSIDERATIONS VERSUS CRITERIA

Considerations, such as those developed by A.B. Hill, are simply things to think about when weighing evidence. They are neither necessary nor sufficient.

Criteria, such as those of Forbes and Calow (Section 2.3) or Koch and Chapman (Section 2.5), are said to be necessary properties of a body of evidence. For example, Chapman's sediment quality triad requires that the sediment contain at least one chemical in toxic amounts, which is a criterion for toxicological causality. If a body of evidence includes a complete set of the criteria (necessary properties), it is considered to be sufficient.

Both sets of considerations begin with the assumption that the association of the effect with the putative cause is established. Their goal was to determine whether the association is causal or is contributing, indirect, confounded, or otherwise not the cause. In the case of cigarettes and lung cancer, an association was established by correlation (Doll and Hill 1954). Having clearly defined that association (e.g., it includes both sexes), the degree of strength, consistency, temporality, coherence, etc. can be weighed.

A second weakness is the difficulty of inference when applying separate narrative discussions of each consideration to complex causes or to multiple causal hypotheses (Rothman and Greenland 1998). In complex assessments, conceptual network models and tabulation of evidence for the networks are necessary.

A third weakness is that the considerations were specifically for assessing causation in epidemiological studies. The considerations are not applicable to many assessments because they lack epidemiological studies and include evidence that applies to attributes beyond the considerations.

A fourth weakness is that there is no method for weighing the body of evidence to infer causation. Conclusions come from an unstructured expert judgment after reviewing the considerations.

Despite these weaknesses, versions of Hill's considerations are still commonly used to weigh evidence, including in regulatory assessments. Examples include the USEPA's guidelines for carcinogen assessment (USEPA 2005), IRIS assessments of chemicals (https://www.epa.gov/iris), and Integrated Science Assessments for air quality standards (USEPA 2015). Peculiarly, the National Academies committee reviewing the IRIS program recommended against using Hill's considerations, but recommended the GRADE system, which uses Hill's "criteria" to grade evidence (Committee to Review the IRIS Process 2014; Guyatt et al. 2011; NRC 2014). Some authors have revised Hill's considerations for their specific use but have not acknowledged their fundamental weaknesses. For example, Fox (1991) developed a set of seven considerations for ecoepidemiology derived from Hill and Susser. Meek et al. (2014) modified Hill's epidemiological considerations to incorporate molecular evidence when identifying mode of action. Similarly, the USEPA's IRIS method for human health assessments uses a highly adapted list including confidence in study findings (risk of bias and sensitivity), consistency across studies or experiments, dose/exposure-response gradient, strength (effect magnitude) of the association, directness of outcome or endpoint measures, and coherence (USEPA 2022). Some have applied numerical scores to the considerations (Swaen and van Amelsvoort 2009). Others used Hill's considerations but have called their use "evidence-based toxicology" and denied that it is weight of evidence (Guzelian et al. 2005; James et al. 2015). Still others have retained Hill's considerations but encouraged new interpretations based on new types of evidence (Fedak et al. 2015) or have made them more transparent (Shimonovich et al. 2024). Johnson, Sumpter, and Depledge (2021) made a case for using Hill's considerations in ecological causal assessments but acknowledged that other weight-of-evidence approaches, including ours, are also suitable. None of these adaptations made the considerations more coherent. An extensive review of the uses and modifications of Hill's considerations is provided by Koterov and Ushenkova (2022) and Koterov (2019).

Arguably, the best use of Hill's considerations is presented by Ward (2009). He proposed that they would supplement statistical associations by serving as an informal test of the potential causality of the association. The inference uses the considerations to make abductive inferences to the best explanation of the evidence. He proposed that Hill had that use in mind but did not know about abduction. However, Ward also insisted that satisfying Hill's "criteria" does not justify claims of true causation. Similarly, Phillips and Goodman (2006) described causal considerations as "guideposts on the road to common sense." Even these modest uses would be better served by weighting the properties of evidence rather than considering the considerations.

Because Hill's considerations were developed for assessing causation, they have been developed and used primarily for that purpose. However, other sets of considerations can be used and have been used for other purposes. For example, considerations are provided for determining the utility of effluent toxicity tests (Norberg-King et al. 2018).

Ad hoc modification of Hill's considerations has been used to obtain a desired assessment outcome (Shrader-Frechette and ChoGlueck 2017). Basing inference on a loose set of things to consider almost invites such behavior. For that reason, it is important to make any modifications to methods up front, before the evidence is collected and analyzed.

You may think that we are being unduly hard on Hill and Terry. However, the continued routine use of their considerations, despite their flaws and limited applicability, has proven to be a hard habit for assessors to break. The considerations were an advance over unstructured narratives, but it is past time to move on to methods that structure the weighing of evidence in terms of a logical inferential process.

2.3 BEST EVIDENCE SYNTHESIS

Some assessors have been concerned that systematic review and meta-analysis may dilute the influence of evidence from the best studies by mixing it with evidence from poor studies. They have developed or advocated approaches that identify "best evidence" and use it in the inference (Slavin 1995; Cook, Mulrow and Haynes 1997; Alton-Lee 2004; Treadwell et al. 2012).

The distinguishing feature of best evidence synthesis is, as the name implies, prioritization to identify the best piece or perhaps the best type of evidence, rather than identifying the hypothesis supported by the weightiest body of evidence. Its method has been identified as "intellectual weighing of evidence" (Letzel 1995), but, in practice, multiple methods of weighing the pieces of evidence have been employed. They include narratives, considerations-guided narratives, and tabulation and scoring, but any method might do if it focused on the task of identifying the best evidence. Treadwell et al. (2012), who focused on medical trials, proposed a set of properties to be considered: risk of bias, applicability, replication, and conclusiveness.

Some guidance documents recommend using the best evidence. For example, risks from chemical mixtures may be assessed using mixture testing, component chemical testing, and data analysis. USEPA human health guidance for assessment of mixtures recommends choosing the best evidence derived from those alternative types of evidence (Risk Assessment Forum 2000).

2.4 REALIST SYNTHESIS

Realist synthesis was developed in response to the limitations of meta-analysis in systematic reviews. It focuses on understanding the mechanisms and circumstances by which interventions such as medical treatments succeed or fail (Pawson et al. 2004; Rycroft-Malone et al. 2012; Wong et al. 2013). Realist synthesis has focused on interventions in systems such as medical systems rather than proximate interventions such as a drug. Rather than asking what works, it asks what works for whom, in what circumstances, in what respects, and how (Pawson et al. 2004). The analogy to human and natural communities is obvious. The purpose is sense-making for the decision-maker rather than providing the answer. Conclusions are reached in consultation with decision-makers and stakeholders. The method of synthesis is a logical narrative built around the evidence and a theory of the system. Structured guidance for performing realist synthesis exists, but it is focused on how to perform the systematic review, given the purpose (Wong et al. 2013).

2.5 SEQUENTIAL CRITERIA WEIGHT OF EVIDENCE

In most weight-of-evidence methods, the inference is based on a body of evidence assembled from weighted pieces or types of evidence. However, some methods make inferences sequentially from a series of pieces or categories of evidence. They are effectively dichotomous keys for causes. For example, Forbes and Calow (2002) developed a sequential method for determining the cause of existing environmental impairments in general. Burkhardt-Holm and Scheurer (2007) adapted the same

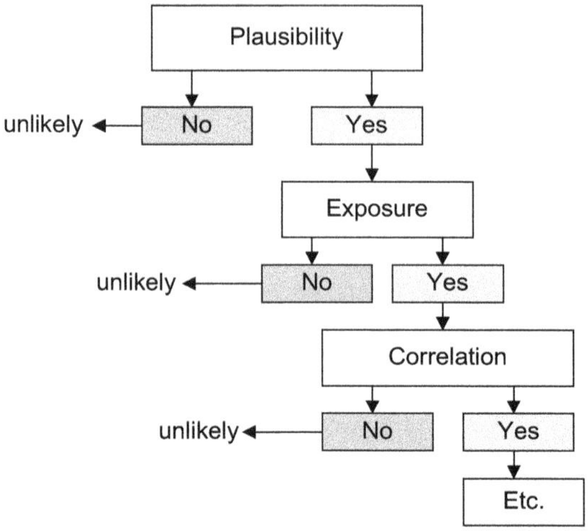

FIGURE 2.1 A diagram of sequential inference based on the first three steps of Burkhardt-Holm and Scheurer (2007).

approach to determine the cause of a decline in brown trout. Henry et al. (2020) further adapted it to determine the cause of disease in bobwhite quail.

The sequential approach is illustrated in Figure 2.1. The steps in the sequences are referred to by these authors as questions: does the hypothesized cause make sense logically and scientifically, is there evidence of exposure to the causal agent, etc.? The answers are provided by the implications of the evidence (Section 6.5). For example, measured body burdens of a potentially causal chemical imply exposure. Johnson, Sumpter, and Depledge (2021) independently developed a sequential approach based on Hill's considerations for chemical effects on wildlife in general. Another sequential, tiered approach was advocated for wildlife risk assessment. It includes: (1) concentration of chemicals in relation to levels reported to be harmful; (2) bioassays or toxicity studies to define dose-response relationships; and (3) field studies of population or community responses (Fairbrother 2003). Bridges and Solomon (2016) used a sequential method based on a hierarchy of biological organization: receptor, biochemical or physiological pathways, organ or tissue, organism, and population. In their system, if a link is broken in that sequence by the absence of supporting evidence, the chain is broken, and no risk to a population can be inferred.

The obvious advantage of this approach is efficiency. Assessment of a hypothesis may be completed in one step with one piece of negative evidence. For that reason, in decision analysis, such diagrams are called quick-exit decision trees (Cartwright and Hardie 2012). Another advantage is that the logic is clear, which can make the inference transparent to decision-makers and stakeholders.

A limitation of the sequential approach is that it is designed to consider one potential cause. In many cases, there are multiple possible causes, but the sequential approach has no comparative inference step. Also, there is no rating of the evidence to determine whether the evidence is sufficiently relevant or reliable. All that matters is whether the evidence supports the hypothesis or not. However, a sequential approach could be elaborated to include options for more structured weight of evidence.

The steps in the sequence should be necessary criteria. That is, to reject the hypothesis, the steps must necessarily be attributes of the evidence. They cannot be simply things to consider, as in Hill's considerations. Also, you should have evidence to evaluate each criterion as it appears in the sequence. For example, if the hypothesis is plausible, but you have no evidence concerning exposure, you are stuck, even though you may have decisive evidence for other criteria. For that reason, it is important to hierarchically sequence the criteria so that you begin with criteria that are most likely to have evidence and to be decisive. You may skip the no-data results and go to the next step, but that works only if a later step will reject the hypothesis (Burkhardt-Holm and Scheurer 2007). This elimination approach is well developed in quick-exit decision trees (Cartwright and Hardie 2012).

You will note that the decisive implications of evidence are all negative. This is because disproof of a hypothesis is more logically powerful and convincing than proof (Popper 1968). However, there is an exception in weight of evidence. If the effect is diagnostic, the occurrence of that effect is definitive evidence (Burkhardt-Holm and Scheurer 2007; USEPA 2000). That is, if the effect is so specific to an agent that its occurrence proves causation by that agent, the sequence ends with a yes. Such diagnostic effects are uncommon in environmental assessments.

In the end, if all the dichotomous decisions are "yes," the hypothesis is accepted if the series of decisions addresses a complete set of criteria. Otherwise, the inference is indeterminate. If any decisions cannot be made because there are no relevant data, the decision process is also indeterminate. In either case, some sort of weighing of the body of evidence must occur. The approach used by sequential systems is to consider what can logically be inferred from the totality of results (i.e., from the body of evidence—Section 7.2).

Alternatively, if not every step in the sequence is decisively dichotomous, the sequence may begin with dichotomous decisions that eliminate some causes and then switch to weighting and weighing the remaining evidence. An example is provided by the determination of the cause of biological impairment of Clear Fork and its tributaries in West Virginia (Figure 2.2).

Another sequential approach elaborates the process beyond causation to the entire analytical phase of an assessment process for contaminated sites (Hull and Swanson 2006). The sequences include determining the magnitude of direct toxic effects, indirect effects, habitat-mediated effects, and causation of identified effects, all leading to a decision concerning the need for remediation.

2.6 TABULATING AND SCORING

The lack of inferential structure and transparency when using Hill's considerations led the epidemiologist, Mervyn Susser, to create a new weight-of-evidence method (Susser 1986). He defined five properties of evidence of causation: time order, strength of association, specificity, consistency on replication, and coherence. For each of these, he defined criteria, possible effects on the hypotheses, and ratings for each combination of criterion and effect. Using his table for time order (Table 2.1), you can see that if the time order is incompatible with causation (i.e., the effect precedes the hypothesized cause), then the evidence strongly detracts from the hypothesis (++).

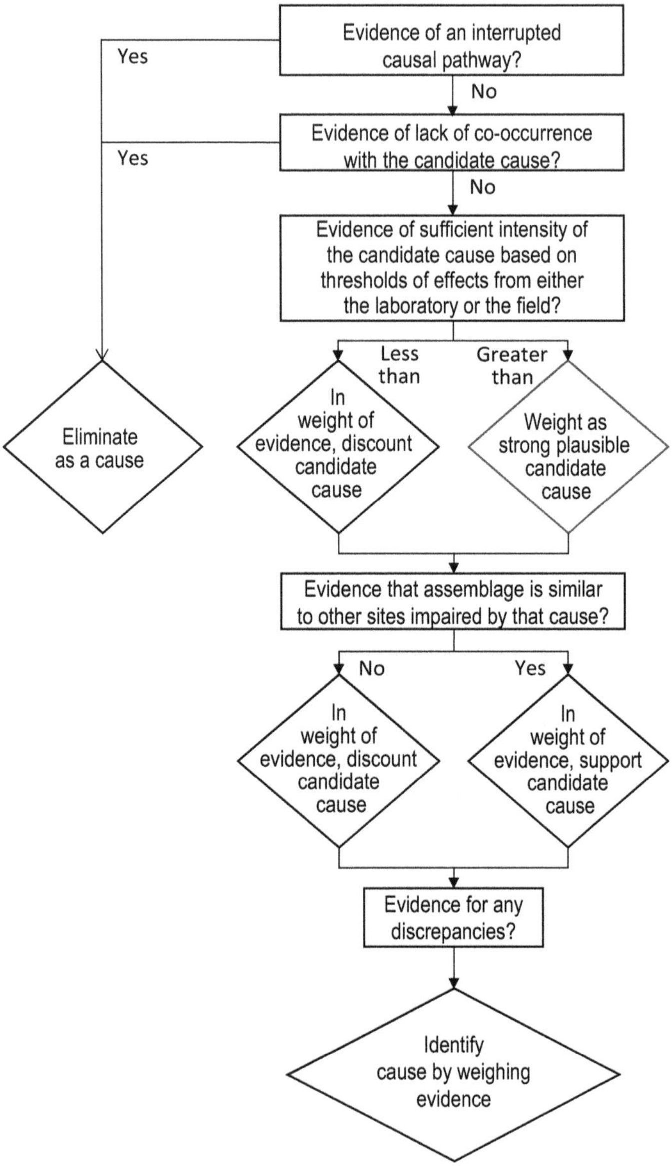

FIGURE 2.2 The sequential process used in the causal assessment for Clear Fork, WV (Zheng, Gerritsen, and Cormier 2014; USEPA 2010).

If the evidence is compatible with the hypothesis (i.e., the cause precedes the effect), the association could be causal, but not necessarily, so the effect on the hypothesis is indeterminate and weak (+). If the evidence is uncertain, the effect on the hypothesis is indeterminate (+). Equivalent logic is applied to the other properties of causal evidence.

TABLE 2.1
A Table for Rating Possible Evidence of Time Order with respect to its Effect on the Causal Hypothesis

	Effect on Hypothesis		
Criterion	**Supports**	**Detracts**	**Indeterminate**
Time Order			
Compatible			+
Incompatible		++	
Uncertain			+

Adapted from Susser (1986).
Each property of evidence has an equivalent table.

The logic and transparency of Susser's tables and qualitative rating system inspired our weight-of-evidence tables and ratings for contaminated sites (Suter and Loar 1992; Suter 1996; Suter et al. 2000) and those of Fox (1991) for ecoepidemiology. They include tables summarizing the results for each type of evidence and tables scoring and integrating the weight of evidence. Such tables may be sufficient alone to make judgments. They are certainly an advance over Hill's considerations. They may be a tool in a structured weight of evidence that includes additional inferential tools; see the methodology in Part II and (USEPA 2016). Susser did not propose a method for integrating weights across properties except to say that negative evidence is more influential and that you should be judicious.

2.7 EVIDENTIAL CRITERIA, CHECK LISTS, AND LOGIC TABLES

Real criteria (versus considerations) define necessary evidence. If you do not have evidence for each criterion, the hypothesis is not sufficiently supported. Often, to obtain a full set of evidence that meets those criteria, you must generate case-specific data. Inference based on criteria may use checklists, logic tables, or a structured weight of evidence.

Consider the criteria for convicting a criminal. As a prosecutor, you must demonstrate that the defendant had the means, a motive, and an opportunity to commit the crime. If the defendant was outside the country when the local bank was robbed, there was no opportunity, and he would be acquitted.

The original set of evidence criteria for the biological sciences is Koch's postulates for determining whether a pathogen causes a disease. The postulates were adapted in a weakened form (no experiments) for epidemiological studies of human diseases (Yerushalmy and Palmer 1959; Hackney and Linn 1979). They were also adapted to ecological effects of pollutants (Woodman and Cowling 1987; Suter 1993; Adams 1963). Koch's postulates are not always applicable, even for their original purpose of identifying causal pathogens. However, they serve as a model for defining sets of evidence that will establish causation.

My version of Koch's postulates (Suter 1993) is applicable to determining the cause of environmental toxic effects in humans and nonhuman organisms.

- The injury, dysfunction, or other putative effects of the agent should be regularly associated with exposure to the agent and any contributory causal factors.
- Indicators of exposure to the agent should be found in the affected organisms.
- The effects should be seen when normal organisms are exposed to the agent under controlled conditions, and any contributory factors should contribute in the same way during the controlled exposures.
- The same indicators of exposure and effects should be identified in the controlled exposures as in the field.

The criteria have been treated as a checklist. That is, the causal hypothesis is accepted if all four criteria are supported. That logic is reasonable if the evidence for each criterion unambiguously supports or refutes the causal hypothesis. However, the relevance, strength, and reliability of the evidence may be weak or ambiguous.

For tobacco and lung cancer, three types of evidence were required to convincingly demonstrate causation: observational population studies, animal experimentation, and cancer-causing chemicals in cigarette smoke (Proctor 2012). Note that this is a different approach from Hill's considerations because Hill was concerned with assessing the epidemiological evidence. The animal studies available to Hill were weak, and the cancer-causing chemicals had not been characterized. The addition of animal studies and a chemical agent makes Proctor's criteria much stronger than Hill's considerations.

Some standard sets of evidence types result from experience with inference in practice. For example, a study of medical doctors found that they believed that a treatment was effective if there was a clinical or epidemiological association between the hypothesized cause and effect and if there was a demonstrated mechanism (Russo and Williamson 2007).

In many cases, the results are a mixture of accepted and rejected criteria. This can result from ambiguity in the evidence. Peter Chapman developed a set of criteria that are analogous to Koch's. His sediment quality triad, and logic table 2.2 served to interpret possible results (Chapman et al. 1997). To determine whether contamination caused impairment of a sediment community by his method, you need three types of evidence from the site.

1. Chemical analyses of the sediment, to be compared to ecotoxicological benchmark values, determine whether the sediment contains any chemicals in toxic amounts.
2. Toxicity test results of the contaminated sediment are compared to results for similar but uncontaminated sediment to determine whether the sediment, as a whole mixture, is toxic.
3. Biological surveys of contaminated sediment and of reference sediment are compared to determine whether the biotic community is affected.

TABLE 2.2
Logic Table for the Sediment Quality Triad (Chapman 1990)

Situation	Chemicals Present	Toxicity	Community Alteration	Possible Conclusions
1	+	+	+	Strong evidence for pollution-induced degradation
2	−	−	−	Strong evidence that there is no pollution-induced degradation
3	+	−	−	Contaminants are not bioavailable or are present at nontoxic levels
4	−	+	−	Unmeasured chemicals or conditions exist with the potential to cause degradation
5	−	−	+	Alteration is not due to toxic chemicals
6	+	+	−	Toxic chemicals are stressing the system but are not sufficient to significantly modify the community
7	−	+	+	Unmeasured toxic chemicals are causing degradation
8	+	−	+	Chemicals are not bioavailable, or alteration is not due to toxic chemicals

Each type of sediment quality evidence is scored as positive or negative, and each possible combination of scores has a prescribed interpretation (Table 2.2). Ideally, the types of evidence in the triad should be consistent—all plus or all minus. The table provides interpretations for the inconsistencies. For example, if the contaminated sediment is toxic and has a depauperate biota, but no chemical is detected in toxic amounts, an unmeasured chemical or physical property is causing the effects (situation 6). Unmeasured chemicals are the designated source of ambiguity in this line of the table. However, the set of possible conclusions is incomplete. For example, the possibility of unmeasured chemicals is included, but not unmeasured toxicity effects or unmeasured community alterations. This triad has also been applied to other media, including water and soil (Kim et al. 2024; Jensen and Mesman 2006; Chapman and COA Sediment Task Group 2008), and it has been expanded to a quadrad by including biomarkers or tissue analyses (COA Sediment Task Group 2008).

While Chapman's approach deals with some of the ambiguity of the evidence, it does not address relevance, strength, or reliability. That is, the results may be false positives or false negatives because of limitations in the data. For example, situation 6 may also occur because the reference site habitat does not match the exposed site, or because the community measurement methods or endpoints were insensitive. Hence, because of limitations of the evidence, there are many ways in which the logic table can fail. That concern also applies to the use of other sets of criteria as checklists.

Case-specific logics may be developed that use the attributes of the case to identify the possible conclusions. We suggest that, in most cases, the evidence for each criterion should be clearly identified and weighted, and the body of evidence should be weighed to derive conclusions.

2.8 INDEPENDENT APPLICABILITY OF EVIDENCE

The opposite of a necessary and sufficient evidence set is the independent applicability of evidence. That is, if adequate evidence from a relevant and reliable study shows an impairment, that is sufficient to declare a system to be impaired, independent of other evidence. In the U.S., a stream is impaired if any of the following is true:

- a water quality criterion is exceeded,
- a toxicity test of the ambient water or diluted effluent is positive, or
- biological surveys demonstrate that an exposed biotic community is diminished.

If any of these three types of evidence is generated by approved methods and is positive, that is sufficient, given the policy that mandates protection. This method has been criticized by industry consultants (Lee and Jones-Lee 1995; Chapman 2000), but its logic has stood-up to legal challenges.

Negative evidence (a finding of no effect) is not decisive because each of these types of evidence has weaknesses that could result in false-negative results.

- Failure to exceed a water quality criterion may be because (1) the causal chemical does not have a criterion value, (2) the causal chemical may not be measured, or (3) the effects may be due to the combined effects of multiple chemicals or other agents.
- Tests of ambient waters or effluents may not show toxicity because they do not include sensitive species, life stages, or responses.
- Biological surveys may not detect effects because they do not include sensitive species or responses, or the reference sites may be inappropriate.

Of course, evidence may give a false-positive result, but that is judged to be less likely. Poor methods and sloppy implementation are more likely to fail to detect an effect than to detect a false effect. This is because bad practices tend to introduce extraneous variance, fail to measure endpoint responses in all relevant taxa and life stages, have insufficient duration, or miss the impaired locations.

Independent applicability is also justified by the fact that different types of evidence address different impairments. If any one type of evidence identifies an effect, then the community is impaired in that way. If another type of evidence shows no effect, then the community is not impaired in that way. For example, if the fish toxicity test is negative but the biological survey shows that the invertebrate community is depauperate, then it is appropriate to protect the invertebrates. It is not appropriate to integrate those two pieces of evidence.

Australia and New Zealand specify independent applicability as one of three acceptable weight-of-evidence methods in their water quality guidelines (ANZG 2018). While one type of evidence is sufficient to conclude that water or sediment is impaired, other types of evidence are required to specify the nature and extent of the impairment.

This method of inference is known in the European Water Framework Directive as the "One Out–All Out" (OOAO) principle (European Parliament 2000). The directive distinguishes good chemical status, good ecological status, and good hydrological status. All three types of good status must be achieved, so no approved measure of any of the three types of status can fail. The OOAO method is described as an implementation of the precautionary principle. Although the OOAO method has been criticized for being inherently overly pessimistic, others argue that poor evidence makes it overly optimistic, and weight of evidence may be needed to determine whether any evidence is sufficiently reliable to prompt remedial actions (Cunningham 2012).

Independent applicability weighs evidence to determine that it has acceptable relevance and reliability. Then, all types of evidence are considered and weighted dichotomously. Results that indicate impairment are given absolute weight, but no-effect results are given no weight. The inferential rule is clear. The results are negative (i.e., no hazard detected) only if no adequate body of evidence identifies an effect and some adequate evidence finds no effect. Finally, conventional weight-of-evidence methods can be used to determine why the types of evidence for impairment do not agree.

2.9 OTHER RULE-BASED METHODS

Independent applicability is the simplest rule for inference from multiple pieces and types of evidence, but other rules have been used. One such rule was developed by the European Center for Ecotoxicology and Toxicology of Chemicals for weighing evidence from animal and human data in human health risk assessments of chemicals (ECETOC 2009). The type of evidence with the highest quality (greatest weight for reliability) is used in the assessment, but if the quality of animal and human data is similar, the hypothesized effect is assumed to occur if either animal or human evidence shows the effect. This logic could be used with more than two types of evidence.

Other rule-based approaches are more complex, but all have inferences of the form, if the available evidence has characteristics $X_1 \ldots X_n$, then conclude Y. An example is the cancer risk assessment guidelines for categorizing potential carcinogens (USEPA 2005). The advantage of rule-based weight of evidence is consistency, transparency, and minimization of ad hoc judgments. The disadvantage is a lack of flexibility, which may result in inappropriate conclusions. As bodies of evidence and inferences become more complex, particularly due to evidence concerning mechanisms, weight-of-evidence guidance documents have become less rule-based and more general (USEPA 2016).

2.10 THE EXPLANATORY/ABDUCTIVE APPROACH

The explanatory approach attempts to address the problem that different experts interpret a piece of evidence differently (Douglas 2012). You begin by gathering evidence and developing different explanatory accounts for the evidence (i.e., hypotheses). For example, an epidemiological association may be interpreted as representing a causal relationship, a confounded relationship, or biased sampling. Second, you decide which explanatory interpretations are adequate and which should be discarded. Douglas's criteria for adequate explanations are internal consistency, empirical competency, and predictive potential. Third, you weigh the evidence to judge which of the remaining explanations is best.

The explanatory approach is an example of abductive inference by weight of evidence (Peirce 1955). Rather than weighing the evidence with respect to its support for a hypothesis, you weigh explanatory hypotheses with respect to their ability to explain the evidence (Section 1.1). Your choice depends on how the assessment is framed. You must decide whether the evidence serves the hypothesis, or the hypotheses serve the evidence (i.e., the explanatory approach). Both approaches should give the same result if applied well.

A variant of the explanatory approach is hypothesis-based weight of evidence (Lutter et al. 2015; Rhomberg 2013, 2015). It was developed for assessing the qualitative risk (i.e., hazard) of chemicals. It focuses on developing well-formulated hypotheses (e.g., the chemical causes cancer by inducing cell proliferation) and evaluating alternative hypotheses with respect to the evidence. A practical advantage of Rhomberg's hypothesis-based weight of evidence, relative to Douglas's approach, is that it has been applied to several chemicals, which can serve as models (Bailey, Prueitt, and Rhomberg 2012; Mihaich et al. 2017; de Peyster and Mihaich 2014; Mihaich and Borgert 2018). However, it is a weak inferential method in that it relies heavily on narrative without explicitly weighting the evidence or the hypotheses relative to the evidence.

2.11 FRANKLIN'S BALANCING METHOD

Benjamin Franklin weighed evidence by balancing considerations for and against a particular decision. It is presented in a letter to Joseph Priestly, which I quote here in its entirety (Franklin 1772). If you add transparency to Franklin's method by documenting which considerations balance against which, it could be a simple and transparent weight-of-evidence method, or as Franklin called it, a weight of reasons.

London, September 19, 1772

Dear Sir,

In the Affair of so much Importance to you, wherein you ask my Advice, I cannot for want of sufficient Premises, advise you *what* to determine, but if you please I will tell you *how*. When these difficult Cases occur, they are difficult chiefly because while we have them under Consideration all the Reasons *pro* and *con* are not present to the Mind at the same time; but sometimes one Set present themselves, and at other times another, the first being out of Sight. Hence the various Purposes or Inclinations that alternately prevail, and the Uncertainty that perplexes us. To get over this, my

Way is, to divide half a Sheet of Paper by a Line into two Columns, writing over the one *Pro*, and over the other *Con*. Then during three or four Days Consideration I put down under the different Heads short Hints of the different Motives that at different Times occur to me for or against the Measure. When I have thus got them all together in one View, I endeavour to estimate their respective Weights; and where I find two, one on each side, that seem equal, I strike them both out: If I find a Reason *pro* equal to some two Reasons *con*, I strike out the three. If I judge some two Reasons *con* equal to some three Reasons *pro*, I strike out the five; and thus proceeding I find at length where the Ballance lies; and if after a Day or two of farther Consideration nothing new that is of Importance occurs on either side, I come to a Determination accordingly. And tho' the Weight of Reasons cannot be taken with the Precision of Algebraic Quantities, yet when each is thus considered separately and comparatively, and the whole lies before me, I think I can judge better, and am less likely to make a rash Step; and in fact, I have found great Advantage from this kind of Equation, in what may be called *Moral* or *Prudential Algebra*. Wishing sincerely that you may determine for the best, I am ever, my dear Friend, Yours most affectionately.

B Franklin

2.12 ALGORITHMIC WEIGHT OF EVIDENCE

Some organizations have developed rule-based algorithmic processes for weight of evidence in specific assessment contexts. Such methods are characterized by being prescriptive. For example, the USEPA IRIS program recently published a process for identifying the hazards to humans posed by environmental chemical contaminants and the toxicity values for cancer and noncancer effects in humans (USEPA 2022). It includes numerous process flow charts and tables directing the interpretation, rating, and integration of information and evidence for hazards, and dose-response modeling for toxicity values. Other examples are the GRADE and Cochrane methods (Schünemann et al. 2013; Higgins et al. 2023). Such algorithmic systems are designed to increase consistency and defensibility by minimizing process variance. The difficulty is that the authors of the algorithm must be very wise and anticipate novel kinds and combinations of evidence and novel properties of new chemicals.

2.13 STRUCTURED CONTAMINATED SITE WEIGHT OF EVIDENCE

Contaminated sites may be assessed by comparing contaminant concentrations in site media to benchmark concentrations. This is usually the case with human health assessments. However, ecological assessments may gather data on the condition of biota on the site, tissue or whole-organism concentrations, biomarkers of exposure and effects, and toxicity tests of the media. These studies provide evidence that can be weighed to answer the question: Is the biotic community impaired, and if so, what is causing the impairment? The sediment quality triad (Section 2.5) was developed for this purpose, but its standard evidence and interpretations have been too constraining for most assessments. Weight-of-evidence frameworks and approaches have been developed for contaminated sites that allow various site studies and various approaches to inference (Burton et al. 2002; Grapentine et al. 2002; USEPA

2000; Suter 1996; Suter et al. 2000). They have in common the idea that you can use site-specific information, supported by site-independent information, to characterize contamination, toxicity, exposure, and effects. Integrating that body of evidence can clarify impairment and its cause.

2.14 STRUCTURED GENERIC WEIGHT OF EVIDENCE

Structured generic methods are, we believe, the most useful weight-of-evidence methods for environmental science and assessments. They provide clear guidance for performance of the process that makes applications transparent, consistent, and credible. However, they are sufficiently flexible to be broadly applied.

Over more than two decades, we have developed and applied a structured generic weight-of-evidence approach that provides the heart of this book (Part II). It has three properties:

- it includes the entire process from obtaining evidence to explaining the results,
- it is sufficiently structured to guide the weight-of-evidence process, but
- it is not so prescriptive that it is limited to one type of assessment problem.

The steps in the method identify evidence, assign the evidence to types that have implications for the hypotheses, assign weights to properties of the evidence, and integrate the weighted body of evidence. If assessors follow the process, the logic should be clear, and the conclusions should be defensible. For reviewers, decision-makers, and stakeholders, formal structured methods provide transparency that allows understanding of the basis for the conclusions and the possibility of replication and sensitivity analysis. The structured process is applicable to assessments ranging from hazard identification for new chemicals to determining the cause of ecosystem impairments or cancer clusters.

Structured weight-of-evidence methods have been developed by several organizations. Ours appears first in USEPA guidance for identifying causes of environmental impairments and then in generic guidelines for weight of evidence in ecological assessments (USEPA 2000, 2016). Although the 2016 guidelines were written for ecological assessments, the general structure is also applicable to human health assessments. Others have recommended structured weight-of-evidence systems that are specifically for chemical assessments (EFSA Scientific Committee 2017; Ågerstrand et al. 2011; OECD 2019; USEPA 2022; Hall et al. 2017).

A structured weight-of-evidence approach for the protection and management of water resources has been developed for Australia and New Zealand (ANZG 2018). "It is the central platform for water/sediment quality assessments in the Water Quality Guidelines." The weight-of-evidence-based guidelines include management plans, planning and data analysis of monitoring studies, derivation of guideline values, and assessments of baselines, multiple pressures and stressors, waste discharges, and unexpected events. They emphasize the site specificity of the assessment process.

Formal structured methods have many advantages. They provide consistency, transparency, procedural guidance, credibility, and a reduced likelihood of self-deception.

2.15 ALTERNATIVE TERMINOLOGY FOR WEIGHT OF EVIDENCE

Although weight of evidence has been a prominent concept for thousands of years and currently appears in environmental laws and regulations, some have rejected the term while accepting the concept.

A National Research Council committee opined that weight of evidence is vague and recommended that it be replaced by integration of evidence (NRC 2014). This is bizarre in that, unlike weight of evidence, which implies weighing the evidence for a hypothesis, integration does not imply a method. Hence, integration is vaguer. In addition, even though they stated that weight of evidence is vague, they were certain that the weight-of-evidence analogy "neglects to account for the total weight on either side (that is, the scope of evidence available) and captures only where the balance stands" (NRC 2014). That is not true in general. Unfortunately, some assessors have interpreted this NRC comment as requiring the rejection of the practice of weighing evidence rather than suggesting a change in terminology. The committee was reviewing the IRIS program, and IRIS management accepted the recommendation. However, the IRIS management has recently recognized that they are legally required to weigh the evidence and have acknowledged that evidence integration is another term for weight of evidence (USEPA 2022). By either name, the IRIS method is an algorithmic structured weight-of-evidence approach that emphasizes systematic literature review, as in the right leg of Figure 4.3.

Evidence-based risk assessment is another assessment term that describes systematic review followed by structured weight of evidence or meta-analysis for assessment of risks from chemicals in the environment (Krewski et al. 2022). The term is borrowed from assessments of risks from medical practices, particularly by insurance companies (Milano 2001).

Some assessment programs have been so focused on gathering the evidence that they call the entire process of assembling evidence and making inferences systematic review. This dates to the Cochrane assessments of medical treatments, which seem to assume that, after the long, hard process of systematically finding and reviewing the literature and extracting the data and metadata, inference is a subordinate task. However, combining and drawing inferences by weighing multiple pieces and types of evidence is more complex than systematic review.

The combining of multiple pieces or types of evidence is sometimes referred to as triangulation, particularly for qualitative judgments in medical or public health practices (Carter et al. 2014; NAS 2022). Triangulation is not a particular method and is often an ad hoc and informal inference presented as a narrative. Triangulation is generally narrower than weight of evidence in that it focuses on bias. "Triangulation encompasses thoughtful and explicit consideration of the different types of biases that pull results in different directions in key studies" (NAS 2022).

2.16 CHOOSING OR DEVELOPING A WEIGHT-OF-EVIDENCE METHOD

As shown in this section, there are many approaches and methods for weight of evidence. Your situation will determine whether you choose a standard method, develop a method for a particular assessment, or develop a new standard method for a particular type of problem or organization. In any case, your weight-of-evidence method should be suited to its purpose.

If you are in an organization that performs environmental assessments, you are likely to have a general approach and may have one or more standard weight-of-evidence methods. Use the standard methods that are appropriate. For example, a standard method for assessing the effects of individual chemicals would not be appropriate for assessing the risks from a contaminated site.

Why use standard methods?

- An explicit, standardized method increases the transparency of your assessment by relating it to the method guidance document.
- The overall assessment is more reliable and compelling because the evidence, inferences, and results have defined consistency, quality, and credibility. Thus, it is more likely to be accepted by the decision-maker.
- A standard method may be required by the relevant agency.

If you can choose or develop a nonstandard method for your assessment, remember that simple cases may be assessed by simple methods. For example, weighing three standard rodent carcinogenicity studies for the hypothesis that a chemical is a human carcinogen is relatively simple. All studies are equally relevant to humans, and weighting the reliability of standard tests is straightforward. If your case is complex, having multiple hypotheses and many types and pieces of evidence, you should clearly lay out all the relevant information, judgments, and justifications. For example, a study to determine the cause of decline in a population of San Joaquin kit foxes included eight causal hypotheses and 19 types of evidence (Suter and O'Farrell 2009). All types of evidence from five sites were presented and rated for each hypothesis, the body of evidence for each hypothesis was evaluated, and an integrated rating was derived and explained. While a prior narrative weight of evidence blamed climate, the structured weight of evidence convincingly showed that the cause was predation by coyotes. Thinking your way through a formally structured weight of evidence can keep you from jumping to conclusions, particularly if you have an initial bias.

In the interest of achieving both transparency and efficiency while reaching a justified conclusion, follow these steps in choosing, adapting, or developing your weight-of-evidence method (Cormier 2008; Donnelly et al. 2018).

- Clearly define the type of hypothesis to be evaluated or the value to be estimated.
- For the assembly of evidence, identify what information concerning the agents, conditions, and hypothesized relationships would be informative for developing evidence.

- Consider alternative weight-of-evidence methods.
- Use a method that is as formally structured as the case requires.
- Adopt prescriptions for the sake of guidance, transparency and reproducibility, if the prescriptions are relevant.
- Procedural consistency is generally desirable. Inconsistent procedures may lead to inconsistent quality of results and to charges that the procedures are arbitrary and capricious. On the other hand, requirements to adhere to a procedure that does not fit the case can lead to errors or an inability to inform a decision.
- For ease of implementation and understanding by decision makers, make your method clear and well organized.
- For feasibility, ensure that the method can be implemented on time, within budget, and with available staff. For example, a prescribed method may use techniques that require types of data and information that are not routinely available, uncommon expertise or training, or the use of tools that are not generally available or are difficult to implement.
- If information must be assembled and conclusions drawn quickly, consider a consensus workshop of available experts, taking advantage of online meeting capabilities or independent web-based surveys of the experts.

The Cochrane assessments of medical treatments have been held up as a model for how systematic and rigorous a method should be. However, clinical trials of a treatment in humans provide a relatively consistent and conceptually simple problem for evidence synthesis because there is only one type of evidence (clinical trials), one species, and the exposure conditions and effect endpoint are well defined. Assessments of environmental causes and risks are usually complex and heterogeneous, and they require a more flexible approach.

When deciding how to implement a chosen method, to reduce self-deception, reduce the scope for error, allow review and replication, and avoid leaping to a conclusion, follow the steps in the method, and make the judgments specific and well documented.

It is important to be unbiased in any assessment, but it is particularly important for WoE because of the importance of expert judgment (Suter and Cormier 2015).

REFERENCES

Adams, D. F. 1963. "Recognition of the effects of fluorides on vegetation." *Journal of the Air Pollution Control Association* 13: 360–362.

Ågerstrand, M., A. Küster, J. Bachmann, M. Breitholtz, I. Ebert, B. Rechenberg, and C. Ruden. 2011. "Reporting and evaluation criteria as means towards a transparent use of ecotoxicity data for environmental risk assessment of pharmaceuticals." *Environmental Pollution* 159 (10): 2487–2492.

Alda, A. 2017. *If I Understood You, Would I Have This Look on My Face?* New York: Penguin Books/Random House.

Alton-Lee, A. 2004. *Guidelines for Generating a Best Evidence Synthesis Iteration 2004.* Ministry of Education (Wellington, New Zealand).

ANZG. 2018. *Weight of evidence.* Australian and New Zealand Governments, Canberra

Bailey, L. A., R. L. Prueitt, and L. R. Rhomberg. 2012. "Hypothesis-based weight-of-evidence evaluation of methanol as a human carcinogen." *Regulatory Toxicology and Pharmacology* 62: 278–291.

Bridges, J., and K. R. Solomon. 2016. "Quantitative weight-of-evidence analysis of the persistence, bioaccumulation, toxicity, and potential for long-range transport of the cyclic volatile methyl siloxanes." *Journal of Toxicology and Environmental Health, Part B* 19 (8): 345–379.

Burkhardt-Holm, P., and K. Scheurer. 2007. "Application of the weight-of-evidence approach to assess the decline of brown trout (*Salmo trutta*) in Swiss rivers." *Aquatic Science* 69: 51–70.

Burton, G. A., G. E. Batley, P. M. Chapman, V. E. Forbes, E. P. Smith, T. Reynoldson, D. E. Schlekat, P. J. den Besten, A. J. Bailer, A. S. Green, and R. L. Dwyer. 2002. "A weight-of-evidence framework for assessing sediment (or other) contamination: Inproving certainty in the decision-making process." *Human and Ecological Risk Assessment* 8: 1675–1696.

Burton, G. A., P. M. Chapman, and E. P. Smith. 2002. "Weight-of-evidence approaches for assessing ecosystem impairment." *Human and Ecological Risk Assess* 8: 1657–1673.

Carter, N., D. Bryant-Lukosius, A. DiCenso, J. Blythe, and A. J. Neville. 2014. "The use of triangulation in qualitative research." *Methods & Meanings Number* 541 (5): 545–547.

Cartwright, N., and J. Hardie. 2012. *Evidence-Based Policy*. New York: Oxford University Press.

CEE. 2018. *Guidelines for Systematic Reviews in Environmental Management. Version 5.0.* Collaboration for Environmental Evidence, Houston, Texas.

Chapman, P. M. 1990. "The sediment quality triad approach to determining pollution-induced degradation." *Science of the Total Environment.* 97/98: 815–825.

Chapman, P. M. 2000. "Whole effluent toxicity testing—Usefulness, level of protection, and risk assessment." *Environmental Toxicology and Chemistry* 19 (1): 3–13.

Chapman, P. M., B. Anderson, S. Carr, V. Engle, R. Green, J. Hameedi, M. Harmon, P. Haverland, J. Hyland, C. Ingersoll, E. Long, J. Rodgers, M. Salazar, P. K. Sibley, P. J. Smith, R. C. Swartz, B. Thompson, and H. Windom. 1997. "General guidelines for using the sediment quality triad." *Marine Pollution Bulletin* 34 (6): 368–372.

Chapman, P. M., and COA Sediment Task Group. 2008. *Canada-Ontario Decision-Making Framework for Assessment of Great Lakes Contaminated Sediments.* Environment Canada and Ontario Ministry of the Environment, Ottawa.

Chapman, P. M., B. G. McDonald, and G. S. Lawrence. 2002. "Weight-of-evidence issues and frameworks for sediment quality (and other) assessments." *Human and Ecological Risk Assessment* 8: 1489–1515.

COA Sediment Task Group. 2008. *Canada-Ontario Decision-Making Framework for Assessment of Great Lakes Contaminated Sediments*. Environment Canada and Ontario Ministry of the Environment, Ottawa.

Committee to Review the Iris Process. 2014. *Review of EPA's Integrated Risk Information System (IRIS) Process*. Washington, DC: National Research Council. National Academies Press.

Cook, D. J., C. D. Mulrow, and R. B. Haynes. 1997. "Systematic reviews: Synthesis of best evidence for clinical decisions." *Annals of Internal Medicine* 126 (5): 376–380.

Cormier, S. M. 2008. "A synopsis of immediate and deliberate environmental assessments." In *Real-Time and Deliberative Decision Making: Application to Emerging Stressors*, edited by Igor Linkov, Elizabeth Ferguson and Victor S. Magar, 21–29. Springer, New York, NY.

Cormier, S. M., G. W. Suter, and S. B. Norton. 2010. "Causal characteristics for ecoepidemiology." *Human and Ecological Risk Assessment* 16 (1): 53–73.

Cunningham, R. 2012. *Glass Half Full or Half Empty? Why 2009 Water Framework Directive Classification Results Are Over-Optimistic about the State of Rivers Despite the One-out, All-Out Rule.* Sandy, UK: Royal Society for Protection of Birds.

Doll, R., and A. B. Hill. 1954. "The mortality of doctors in relation to their smoking habits." *British Medical Journal* 1 (4877): 1451.

Donnelly, C. A., I. Boyd, P. Campbell, C. Craig, P. Vallance, M. Walport, C. J. M. Whitty, E. Woods, and C. Wormald. 2018. "Four principles for synthesizing evidence." *Nature* 558 (7710): 361–364.

Dorne, J. L., B. Bottex, C. Merten, A. Germini, N. Georgiadis, E. Aiassa, L. Martino, L. Rhomberg, H. J. Clewell, M. Greiner, G. W. Suter, M. Whelan, A. D. M. Hart, D. Knight, P. Agarwal, M. Younes, J. Alexander, and A. R. Hardy. 2016. "Weighing evidence and assessing uncertainties." *EFSA Journal* 14: e00511.

Douglas, H. 2012. "Weighing complex evidence in a democratic society." *Kennedy Institute of Ethics Journal* 22 (2): 139–162.

ECETOC. 2009. *Framework for the Integration of Human and Animal Data in Chemical Risk Assessment.* Brussels: European Centre for Ecotoxicology and Toxicology of Chemicals.

EFSA Scientific Committee. 2017. "Guidance on the use of the weight of evidence approach in scientific assessments." *EFSA Journal* 15 (8): 1–69.

European Parliament. 2000. *Directive 2000/60/EC of the European Parliament and of the Council Establishing a Framework for Community Action in the Field of Water Policy.* Directorate-General for Environment, Brussels, Belgium.

Fairbrother, A. 2003. "Lines of evidence in wildlife risk assessment." *Human and Ecological Risk Assessment* 9: 1475–1471.

Fedak, K. M., A. Bernal, Z. A. Capshaw, and S. Gross. 2015. "Applying the Bradford Hill criteria in the 21st century: How data integration has changed causal inference in molecular epidemiology." *Emerging Themes in Epidemiology* 12: 1–9.

Forbes, V. E., and P. Calow. 2002. "Applying weight-of-evidence to retrospective ecological risk assessment when quantitative data are limited." *Human and Ecological Risk Assessment* 8: 1625–1639.

Fox, G. A. 1991. "Practical causal inference for ecoepidemiologists." *Journal of Toxicology and Environmental Health* 33: 359–373.

Franklin, B. 1772. "From Benjamin Franklin to Joseph Priestley, 19 September 1772." Founders Online. National Archives.

Grapentine, L., J. Anderson, D. Boyd, G. A. Burton, C. DeBarros, G. Johnson, C. Marvin, D. Milani, S. Painter, T. Pascoe, T. Reynoldson, L. Richman, K. Solomon, and P. M. Chapman. 2002. "A decision making framework for sediment assessment developed for the Great Lakes." *Human and Ecological Risk Assessment* 8 (7): 1641–1655.

Guyatt, G., A. D. Oxman, E. A. Akl, R. Kunz, G. Vist, J. Brozek, S. Norris, Y. Falck-Ytter, P. Glasziou, and H. DeBeer. 2011. "GRADE guidelines: 1. Introduction—GRADE evidence profiles and summary of findings tables." *Journal of Clinical Epidemiology* 64 (4): 383–394.

Guzelian, P. S., M. S. Victoroff, N. C. Halmes, R. C. James, and C. P. Guzelian. 2005. "Evidence-based toxicology: A comprehensive framework for causation." *Human and Experimental Toxicology* 24: 161–201.

Hackney, J. D., and W. S. Linn. 1979. "Koch's postulates updated: A potentially useful application to laboratory research and policy analysis in environmental toxicology." *American Revies of Respiratory Disease* 1119: 849–852.

Hall, A. T., S. E. Belanger, P. D. Guiney, M. Galay-Burgos, G. Maack, W. Stubblefield, and O. Martin. 2017. "New approach to weight-of-evidence assessment of ecotoxicological effects in regulatory decision-making." *Integrated Environmental Assessment and Management* 13 (4): 573–579.

Henry, C., M. Z. Brym, K. Skinner, K. R. Blanchard, B. J. Henry, A. L. Hay, J. L. Herzog, A. Kalyanasundaram, and R. J. Kendall. 2020. "“Weight of evidence” as a tool for evaluating disease in wildlife: An example assessing parasitic infection in Northern bobwhite (*Colinus virginianus*)." *International Journal for Parasitology: Parasites and Wildlife* 13: 27–37.

Higgins, J. P. T., J. Thomas, C. M. Chandler, T. Li, M. J. Page, and V. A. Welch. 2023. "Cochrane handbook for systematic reviews of interventions version 6.4." Cochrane Collaborative. Accessed 1/6/2023. https://www.training.cochrane.org/handbook.

Hill, A. B. 1965. "The environment and disease: Association or causation?" *Proceedings of the Royal Society of Medicine* 58: 295–300.

Hull, R. N., and S. Swanson. 2006. "Sequential analysis of lines of evidence—An advanced weight-of-evidence approach for ecological risk assessment." *Integrated Environmental Assessment Management* 2: 302–311.

Ioannidis, J. P. A. 2016. "Exposure-wide epidemiology: Revisiting Bradford Hill." *Statistics in Medicine* 35 (11): 1749–1762.

James, R. C., J. K. Britt, N. C. Halmes, and P. S. Guzelian. 2015. "Evidence-based causation in toxicology: A 10-year retrospective." *Human & Experimental Toxicology* 34 (12): 1245–1252.

Jensen, J., and M. Mesman. 2006. *Ecological Risk Assessment of Contaminated Land*. Bilthoven: National Institute of Public Health and the Environment (RIVM).

Johnson, A. C., J. P. Sumpter, and M. H. Depledge. 2021. "The weight-of-evidence approach and the need for greater international acceptance of its use in tackling questions of chemical harm to the environment." *Environmental Toxicology and Chemistry* 40 (11): 2968–2977.

Kahnman, D. 2011. *Thinking Fast and Slow*. New York: Penguin Books.

Kim, D., J. I. Kwak, T.-Y. Lee, L. Kim, H. Kim, S. H. Nam, W. Hwang, J. Wee, Y. H. Lee, and S. Kim. 2024. "TRIAD method to assess ecological risks of contaminated soils in abandoned mine sites." *Journal of Hazardous Materials* 461: 132535.

Koterov, A. N. 2019. "Causal criteria in medical and biological disciplines: History, essence, and radiation aspect. Report 1. Problem statement, conception of causes and causation, false associations." *Biology Bulletin* 46 (11): 1458–1488.

Koterov, A. N., and L. N. Ushenkova. 2022. "Causal criteria in medical and biological disciplines: History, essence, and radiation aspects. Report 4, part 1: The post-Hill criteria and ecological criteria." *Biology Bulletin* 49 (12): 2423–2466.

Krewski, D., P. Saunders-Hastings, R. A. Baan, T. S. Barton-Maclaren, P. Browne, W. A. Chiu, M. Gwinn, T. Hartung, A. D. Kraft, and J. Lam. 2022. "Development of an evidence-based risk assessment framework." *Altex* 39 (4): 667.

Lee, G F., and A. Jones-Lee. 1995. "Independent applicability of chemical and biological criteria/standards and effluent toxicity testing." *National Environment Journal* 5 (1): 60–67.

Letzel, H. 1995. "Best-evidence synthesis: An intelligent alternative to meta-analysis": Discussion. A case of "either-or" or "as well." *Journal of Clinical Epidemiology* 48 (1): 19–21.

Linkov, I., E. B. Long, S. M. Cormier, F. K. Satterstrom, and T. Bridges. 2009. "Weight-of-evidence evaluation in environmental assessment: Review of qualitative and quantitative approaches." *Science of the Total Environment* 497: 5199–5205.

Lutter, R., L. Abbott, R. Becker, C. Borgert, A. Bradley, G. Charnley, S. Dudley, A. Felsot, N. Golden, G. Gray, D. Juberg, M. Mitchell, N. Rachman, L. Rhomberg, K. Solomon, S. Sundlof, and K. Willett. 2015. "Improving weight of evidence approaches to chemical evaluations." *Risk Analysis* 35 (2): 186–192.

Martin, P., C. Bladier, B. Meek, O. Bruyere, E. Feinblatt, M. Touvier, L. Watier, and D. Makowski. 2018. "Weight of evidence for hazard identification: A critical review of the literature." *Environmental Health Perspective* 126 (7): 076001-1–076001-15.

Matthiessen, P., and P. E. Gibbs. 1998. "Critical appraisal of the evidence for tributyltin-mediated endocrine disruption in mollusks." *Environmental Toxicology and Chemistry* 17 (1): 37–43.

Meek, M. E., C. M. Palermo, A. N. Bachman, C. M. North, and R. J. Lewis. 2014. "Mode of action human relevance (species concordance) framework: Evolution of the Bradford Hill considerations and comparative analysis of weight of evidence." *Journal of Applied Toxicology* 34 (6): 595–606.

Mihaich, E., M. Capdevielle, D. Urbach-Ross, and B. Slezak. 2017. "Hypothesis-driven weight-of-evidence analysis of endocrine disruption potential: A case study with triclosan." *Critical Reviews in Toxicology* 47 (4): 263–285.

Mihaich, E. M., and C. J. Borgert. 2018. "Hypothesis-driven weight-of-evidence analysis for the endocrine disruption potential of benzene." *Regulatory Toxicology and Pharmacology* 100: 7–15.

Milano, A. F. 2001. "Evidence-based risk assessment." *Journal of Insurance Medicine-New York* 33 (3): 239–250.

Moore, D. W., B. Ruffle, A. McQueen, S. Thakali, and D. Edwards. 2023. "Frameworks for screening and risk management of chemicals and advanced materials: A critical review." *Integrated Environmental Assessment and Management* 19 (5): 1192–1206.

NAS. 2022. *Triangulation in Environmental Epidemiology for EPA Human Health Assessments: Proceedings of a Workshop.* Washington, DC: National Academies of Sciences, Engineering, and Medicine, National Academies Press.

Norberg-King, T. J., M. R. Embry, S. E. Belanger, T. Braunbeck, J. D. Butler, P. B. Dorn, B. Farr, P. D. Guiney, S. A. Hughes, and M. Jeffries. 2018. "An international perspective on the tools and concepts for effluent toxicity assessments in the context of animal alternatives: Reduction in vertebrate use." *Environmental Toxicology and Chemistry* 37 (11): 2745–2757.

NRC. 2014. *Review of EPA's Integrated Risk Information System (IRIS) Process.* Washington, DC: National Research Council of the National Academies of Science, Engineering, and Medicine, National Academies Press.

OECD. 2019. *Guiding Principles and Key Elements for Establishing a Weight of Evidence for Chemical Assessment.* Paris, France: Organization for Economic Cooperation and Development, Environmental Directorate.

Pawson, R., T. Greenhalgh, G. Harvey, and K. Walshe. 2004. "Realist synthesis – An introduction." *ESRC Research Methods Program* 2: 55.

Peirce, C. S. 1955. "Abduction and induction." In *Philosophical Writings of Peirce*, edited by J. Buchler, 150–156. New York: Dover Publications, Inc.

de Peyster, A., and E. Mihaich. 2014. "Hypothesis-driven weight of evidence analysis to determine potential endocrine activity of MTBE." *Regulatory Toxicology and Pharmacology* 69 (3): 348–370.

Phillips, C. V., and K. J. Goodman. 2006. "Causal criteria and counterfactuals; nothing more (or less) than scientific common sense." *Emerging Themes in Epidemiology* 3 (1): 5.

Popper, K. R. 1968. *The Logic of Scientific Discovery.* New York: Harper and Row.

Proctor, R. N. 2012. "The history of the discovery of the cigarette–lung cancer link: Evidentiary traditions, corporate denial, global toll." *Tobacco Control* 21 (2): 87–91.

Rhomberg, L. 2015. "Hypothesis-based weight of evidence: An approach to assessing causation and its application to regulatory toxicology." *Risk Analysis* 35 (6): 1114–1124.

Rhomberg, L. R. 2013. *Hypothesis-Based Weight of Evidence: An Approach to Assessing Causation and Its Application to Regulatory Toxicology.* Cambridge, MA: Harvard Center for Risk Analysis.

Rhomberg, L. R., J. E. Goodman, L. A. Bailey, R. L. Prueitt, N. B. Beck, C. Bevan, M. Honeycutt, N. E. Kaminski, G. Paoli, L. H. Pottenger, R. W. Scherer, K. C. Wise, and R. A. Becker. 2013. "A survey of frameworks for best practices in weight-of-evidence analyses." *Critical Reviews in Toxicology* 43 (9): 753–84.

Risk Assessment Forum. 2000. *Supplementary Guidance for Conducting Health Risk Assessment of Chemical Mixtures*. Washington, DC: U.S. Environmental Protection Agency.

Roth, N., and P. Ciffroy. 2016. "A critical review of frameworks used for evaluating reliability and relevance of (eco)toxicity data: Perspectives for an integrated eco-human decision-making framework." *Environment International* 95: 16–29.

Rothman, K. J., and S. Greenland. 1998. *Modern Epidemiology*, Second edition. Philadelphia, PA: Lippincott, Williams & Wilkins.

Rothman, K. J., S. Greenland, and T. L. Lash. 2008a. *Modern Epidemiology*, Third edition. Philadelphia, PA: Lippincott, Williams & Wilkins.

Russo, R., and J. Williamson. 2007. "Interpreting causality in the health sciences." *International Studies in the Philosophy of Science* 21: 157–170.

Rycroft-Malone, J., B. McCormack, A. M. Hutchinson, K. DeCorby, T. K. Bucknall, B. Kent, A. Schultz, E. Snelgrove-Clarke, C. B. Stetler, and M. Titler. 2012. "Realist synthesis: Illustrating the method for implementation research." *Implementation Science* 7: 1–10.

Schünemann, H., J. Brożek, G. Guyatt, and A. Oxman. 2013. "The GRADE handbook." Cochrane Collaboration. Accessed 12/29/2023. https://gdt.gradepro.org/app/handbook/handbook.html#h.svwngs6pm0f2.

Shimonovich, M., H. Thomson, A. Pearce, and S. V. Katikireddi. 2024. "Applying Bradford Hill to assessing causality in systematic reviews: A transparent approach using process tracing." *Research Synthesis Methods* 15: 826–838.

Shrader-Frechette, K., and C. ChoGlueck. 2017. "Pesticides, neurodevelopmental disagreement, and Bradford Hill's guidelines." *Accountability in Research* 24 (1): 30–42.

Slavin, R. E. 1995. "Best evidence synthesis: An intelligent alternative to meta-analysis." *Journal of Clinical Epidemiology* 48 (1): 9–18.

Susser, M. 1986. "Rules of inference in epidemiology." *Regulatory Toxicology and Pharmacology* 6 (2): 116–128.

Suter, G. W. 1996. *Risk Characterization for Ecological Risk Assessment of Contaminated Sites*. Oak Ridge, TN: Oak Ridge National Laboratory.

Suter, G. W., II. 1993. *Ecological Risk Assessment*. Boca Raton, FL: Lewis Publishers.

Suter, G. W., and S. M. Cormier. 2011. "Why and how to combine evidence in environmental assessments: Weighing evidence and building cases." *Science of the Total Environment* 409: 1406–1417.

Suter, G. W., and S. M. Cormier. 2015. "Bias in the development of health and ecological assessments and potential solutions." *Human and Ecological Risk Assessment* 22 (1): 99–115.

Suter, G. W., R. A. Efroymson, B. E. Sample, and D. S. Jones. 2000. *Ecological Risk Assessment for Contaminated Sites*. Boca Raton, FL: Lewis Publishers.

Suter, G. W., and J. M. Loar. 1992. "Weighing the ecological risks of hazardous waste sites: The Oak Ridge case." *Environmental Science and Technology* 26 (3): 432–438.

Suter, G. W., and T. P. O'Farrell. 2009. *Analysis of the Causes of a Decline in the San Joaquin Kit Fox Population on the Elk Hills, Naval Petroleum Reserve #1, California*. Cincinnati, OH: U.S. Environmental Protection Agency.

Suter, G. W., J. Nichols, E. Lavoie, and S. M. Cormier. 2020. "The origins of classic WoE and SR, supplemental material for: Systematic review and weight of evidence are integral to ecological and human health assessments: They need an integrated framework." *IEAM* 16 (5): 718–728.

Swaen, G., and L. van Amelsvoort. 2009. "A weight of evidence approach to causal inference." *Journal of Clinical Epidemiology* 62 (3): 270–277.

Treadwell, J. R., S. Singh, R. Talati, M. L. McPheeters, and J. T. Reston. 2012. "A framework for best evidence approaches can improve the transparency of systematic reviews." *Journal of Clinical Epidemiology* 65 (11): 1159–1162.

USDHEW. 1964. *Smoking and Health: Report of the Advisory Committee to the Surgeon General.* Washington, DC: US Department of Health Education and Welfare.

USEPA. 2000. *Stressor Identification Guidance Document.* U.S. Environmental Protection Agency, National Center for Environmental Assessment, Cincinnati, OH.

USEPA. 2005. *Guidelines for Carcinogen Risk Assessment.* Washington, DC: Risk Assessment Forum, U.S. Environmental Protection Agency.

USEPA. 2010. *Inferring Causes of Biological Impairment in Clear Fork Watershed, West Virginia.* Washington, DC: U.S. Environmental Protection Agency.

USEPA. 2015. *Preamble to the Integrated Science Assessments (ISA).* Washington, DC: U.S. Environmental Protection Agency. www.epa.gov/isa.

USEPA. 2016. *Weight of Evidence in Ecological Assessment.* Washington, DC: Risk Assessment Forum, U.S. Environmental Protection Agency.

USEPA. 2022. *ORD Staff Handbook for Developing IRIS Assessments.* Washington, DC: Office of Research and Development, U.S. Environmental Protection Agency.

Ward, A. C. 2009. "The role of causal criteria in causal inferences: Bradford Hill's "aspects of association." *Epidemiologic Perspectives and Innovations* 6: 2.

Weed, D. L. 2005. "Weight of evidence: A review of concept and methods." *Risk Analysis* 25 (6): 1545–1557.

Wong, G., G. Westhorp, R. Pawson, and T. Greenhalgh. 2013. "Realist synthesis." In *RAMESES Training Materials*. London: The RAMESES Project.

Woodman, J. N., and E. B. Cowling. 1987. "Airborne chemicals and forest health." *Environmental Science and Technology* 21: 120–126.

Yerushalmy, J., and C. E. Palmer. 1959. "On the methodology of investigations of etiologic factors in chronic diseases." *Journal of Chronic Diseases* 10 (1): 27–40.

Zheng, L., J. Gerritsen, and S. M. Cormier. 2014. "Clear fork watershed case study: The value of state monitoring programs." In *Ecological Causal Assessment*, edited by S. B. Norton, S. M. Cormier, and G. W. Suter II, 353–383. Boca Raton, FL: CRC Press.

3 Alternatives to Weight of Evidence

Most of the alternatives to weight of evidence are nominally quantitative. Quantitative alternative methods are appealing because they give an impression of objectivity and precision. However, in general they require either that only a fraction of the evidence is included in the analysis or that some or all of the numerical values are pseudoquantities or statistical degrees of belief. Otherwise, you would not be able to quantitatively combine heterogeneous evidence, like combining fish species richness with frequencies of fish deformities when determining stream impairment. These may seem to be the proverbial apples and oranges, but they are both evidence that a fish community is affected by contaminants.

3.1 VOTE COUNTING

The simplest approach to making inferences from multiple studies is vote counting. That is, determine the number of studies that support each alternative hypothesis, and the hypothesis with the most "votes" wins. It is the quickest and easiest approach to inference from multiple pieces of evidence. However, vote counting fails to consider the relevance or reliability of the evidence. For an example, see Rohr and McCoy (2010).

3.2 STATISTICAL METRICS

I.J. Good was Alan Turing's main statistical assistant on the team that broke the German Enigma code. After the war, he made important contributions to Bayesian statistics (Good 1983). One was a Bayesian metric for the weight of evidence (Good 1950). His weight-of-evidence metric is the log base 10 of Bayes' factor. Bayes' factor is the quotient of the Bayesian likelihood that a hypothesis is true divided by the likelihood that it is false (or any other alternative hypothesis). The likelihoods are the probabilities of the evidence given the hypotheses. Bayes' factor is also the standard metric for Bayesian hypothesis testing. The log conversion makes the weight of evidence additive, which is analogous to adding weights to Themis' scales. Good's examples include the weight of evidence for guilt provided by a lie detector result and for a disease provided by a medical test result. That is, it is based on frequencies of associations, such as the frequency of a disease in patients with positive outcomes on a test. It has been advocated for environmental weight of evidence (Smith, Lipkovich, and Ye 2002), but it has not caught on. Good (1991) modestly provided a warning that, "Because probabilities are often difficult to estimate, the formula does not solve all problems of evaluating evidence…To emphasize the difficulties of probability estimation, the formula could be called 'semi-quantitative'."

DOI: 10.1201/9781003626336-4

Good's Bayesian weight of evidence provides a measure of the property, strength of evidence (Section 7.1.3). It does not address the relevance or reliability of evidence. Other statistical metrics that have been used to express the relative strength of evidence for hypotheses include correlation, relative risk, and odds ratios. Those metrics, unlike Good's, are not additive and are not claimed to be measures of the weight of a body of evidence for a hypothesis.

3.3 META-ANALYSIS

Meta-analysis refers to statistical methods for combining the quantitative results of multiple studies of the same type (Glass 2015; Cumming and Calin-Jageman 2016). It is commonly used to test hypotheses of the efficacy of medical treatments and to estimate levels of treatment effects and their confidence intervals by analyzing results of multiple clinical trials. Because meta-analysis combines results of multiple studies, it is sometimes referred to as a weight-of-evidence method. However, it is limited to combining studies of the same type, and it does not weigh properties of the evidence, so it does not account for differences in the reliability of studies. The only generally accepted weighting in meta-analysis is inverse variance weighting. Meta-analysis does not address causality, so it is limited largely to studies in which the causality of interest is baked into an experimental study design, such as randomized clinical trials (Higgins et al. 2023). However, it can be used as evidential input to a weight-of-evidence analysis to estimate the strength and consistency of evidence (Section 4.2) (Weed 2000).

3.4 DECISION ANALYSIS

Conventional decision analysis, based on multi-attribute utility theory, can mathematically combine heterogeneous values and sets of evidence by converting them to utiles, a 0–1-scaled unit of utility. Utiles may be combined in a utility function that expresses the desirability of alternative actions based on the utilities of their attributes. For example, a remedial option for a contaminated site may be to remove contaminated soil and thereby destroy a forest, and the other option may be to leave the forest to naturally biodegrade the contaminants. Utiles could be the common metric for the loss of an area of forest and reproductive toxic effects. Based on the utilities and the probabilities of the estimated effects under each alternative, utiles may be calculated for each alternative. Utiles would combine the gains and losses in the various attributes of forests and streams due to the remedial alternatives and the costs of implementing them. The decision analysis determines the utility of alternative decisions by multiplying the utility of each outcome by its probability of occurrence. Conventionally, the probabilities of outcomes and the utiles of individual attributes are subjectively judged by the decision-maker.

Many environmental applications of decision analysis have been proposed. For example, the selection of a sediment management decision may have four criteria: (1) cost in dollars, (2) public acceptance in acres impacted, (3) human health in number of exposure pathways and maximum cancer risk, and (4) ecological health in number of pathways and maximum hazard quotient (Kiker et al. 2005; Linkov et al. 2006).

Other examples include choosing the best fish test for endocrine-active chemicals (Crane et al. 2010), choosing a pesticide fate model (Quilbé et al. 2006), and managing multiple air pollutants (Stahl and Cimorelli 2013).

Decision analysis has been described as a weight-of-evidence method (Crane 2016; Hristozov et al. 2014), but it is usually considered a distinct inferential approach. Some advocates argue that decision analysis is preferable to weight of evidence because it is quantitative (Linkov et al. 2015). However, in general, the preferences in decision analysis are as subjective as weights of evidence.

There are many alternative decision analysis methods, including Bayesian methods. Decision criteria may or may not be weighted for their importance. The common weakness of decision analysis is that, in our experience, most decision-makers do not want to make their decision criteria explicit and quantitative. Also, they will listen to stakeholders' preferences but do not want them to determine the decision. Decision analysis, however, may have a heuristic function. A decision analysis workshop may create a common understanding among stakeholders of the nature of a decision.

3.5 EVIDENCE SYNTHESIS

"Evidence synthesis is the process of identifying, compiling, and combining relevant knowledge from multiple sources so it is readily available for decision-makers" (Pullin et al. 2016). It compiles evidence and distributes it to assessors and decision-makers to be analyzed and interpreted in their decision context (Cook et al. 2017). It may serve as the back end of a systematic review or the front end of an assessment by weight of evidence (Section 3.6).

3.6 EXPLANATORY VIRTUES

Inference to the best explanation of the evidence is most commonly conducted by weighing the evidence for the potentially explanatory hypotheses (Section 1.1). However, an alternative is to evaluate the hypotheses directly using explanatory virtues (Mckay 2023; Zukerman 2024). There are many sets of explanatory virtues. These seven are commonly used.

1. **Explanatoriness**: Explanations must explain all the observed facts.
2. **Depth**: Explanations should not raise more questions than they answer.
3. **Power**: Explanations should apply in a range of similar contexts, not just the current situation in which the explanation is being offered.
4. **Falsifiability**: Explanations should be falsifiable—it must be possible for there to be evidence that would show that the explanation is incorrect.
5. **Modesty**: Explanations should not claim any more than is needed to explain the observed facts. Any details in the explanation must relate to explaining one of the observed facts.
6. **Simplicity**: Explanations that posit fewer entities or processes are preferable to explanations that posit more entities or processes. All other things being equal, the simplest explanation is the best.

7. **Conservativeness**: Explanations that force us to give up fewer well-established beliefs are better than explanations that force us to give up more well-established beliefs.

The virtues of hypotheses are evaluated in narratives like Hill's considerations. They are much more commonly used in philosophy, psychology, and cognitive science than in the natural sciences, where evidence is the focus of inference. Besides being a stand-alone inferential method, the virtues may be used as a tool of critical thinking when the weights of bodies of evidence are inconclusive (Section 8.2.2).

3.7 INDICES

Indices are ratios of numbers or of sets of numbers that are said to be indicative of some endpoint characteristic. Examples in medicine include indices of periodontal disease and the Health Index for England (Dhingra and Vandana 2011; ONS 2022). An early use of environmental indices was to combine concentrations of multiple chemicals into an indicator of poor air or water quality (Ott 1978). Indices in ecological assessment are also ways of summarizing data to estimate undefined properties such as ecosystem health, biological condition, and biological integrity. These multimetric indices may be based on combined levels of the metrics from field surveys that are thought to represent levels of acceptability. Others are based on the use of multivariate statistics such as cluster analysis to identify indices that distinguish acceptable from unacceptable sites. An early index for ecological assessment was the Index of Biotic Integrity (IBI) for fish in the Midwestern U.S. (Karr 1981). It combined 12 metrics into categories: fish species richness and composition, trophic composition, abundance, and condition. The numeric index of these heterogeneous fish metrics was created by assigning scores on a one-to-five scale.

More recently, multimetric indices have been used to evaluate contaminated media and sites. These include soil (Dagnino et al. 2008) and sediment (Benedetti et al. 2012; Semenzin et al. 2008; Bombardier and Bermingham 1999).

These indices have conceptual and operational flaws, including ambiguity, eclipsing, arbitrary combining functions, arbitrary variance, unreality, post hoc justification, being nondiagnostic, nonsense results, loss of information, and disconnect from testing and modeling (Ott 1978; Suter 1993; Burton, Chapman, and Smith 2002). However, they have been widely used in the US and elsewhere to assess condition and for regulatory applications. For example, the USEPA has attempted to enable comparative condition assessments by using the Biological Condition Gradient (USEPA 2016).

3.8 NETWORK MODELS

Network models use the connections between components of the system being assessed to analyze the causal relationships among variables. The original version is path analysis, which was developed by Sewall Wright and was first used in environmental assessment by Westman (1981) to determine the likely path of causation in the decline of native vegetation cover. A more powerful version of path analysis is structural equation modeling (Shipley 2000). These models are parametric and typically require abundant evidence. Bayesian belief networks have become a more popular form of network modeling. They analyze networks using conditional probabilities (Spirtes, Glymour, and Scheines 2000; Pearl 2000). All of these network models are more likely to be causal than unstructured statistics like multiple regression because they are directional, like causation. However, they are all associative and therefore not inherently causal. Although network models have been advocated as an alternative to weight of evidence, they are still just models of statistical associations. This is illustrated by the integration of multiple types of evidence of fish toxicity in noncausal Bayesian networks (Lillicrap et al. 2020; Belanger et al. 2023). In those Bayesian models, toxicity to *Daphnia* is statistically associated with toxicity to fish, but it is not a cause of toxicity to fish. Although such networks have been claimed to be weight of evidence, they are just structured multivariate statistical models.

3.9 MATHEMATICAL MODELS

Mathematical simulation models have been claimed to be a type of weight of evidence because they combine multiple pieces of evidence in the form of equation structures and parameter estimates. While that is certainly true, mathematical models of environmental contamination, population and ecosystem response models, and human toxicokinetic and toxicodynamic models are very different from anything recognizable as weight of evidence. They do not combine evidence to support a hypothesis, and they do not assign weights to evidence. The creators of AQUATOX recognized that, "Like any model, it [AQUATOX] is best used as one of several tools in a weight-of-evidence approach" (USEPA 2021). Most commonly, a chemical transport and fate model can contribute, along with measured contaminant levels, to the estimation of exposure and observed effluent plumes in determining the source of exposure.

Although simulation modeling is not a weight-of-evidence method, weight of evidence may be used to determine the best of multiple options for a model structure or a parameter value. Potential values may be weighted for their relevance and reliability, and the best value may be chosen, or multiple high-weight values may be averaged (Chapter 12). Similarly, weight of evidence may be used to determine the best-supported functional form or model structure.

3.10 WEIGHTING STATISTICAL PARAMETERS

Parameters of statistical or mathematical models may be weighted for statistical or conceptual reasons, but weighting parameters is not the same as weighting evidence. For example, weighting of species in a species sensitivity distribution has been suggested as a way of compensating for underrepresentation and overrepresentation of taxa in the dataset (Warne et al. 2018). That may be appropriate, but it is not weight of evidence.

3.11 CRUCIAL EXPERIMENTS

Histories of science have promoted the idea of a crucial experiment that will determine the acceptance of a hypothesis, particularly by rejecting all credible alternatives (Popper 1968). In theory, this is the antithesis of weight of evidence. However, even here there is a role for weighing evidence. First, it is important to determine whether the nominally crucial experiment is in fact sufficiently relevant and reliable. Second, a crucial experiment may be preceded by a weighing of existing evidence to determine that the available evidence is insufficient and to identify the crucial missing information.

Because of complexities and ambiguities, "It is very rare to find determinative experiments in environmental health sciences" (Krimsky 2005). An example is provided by a test of developmental effects of Atrazine on clawed frogs (*Xenopus laevis*) performed for Syngenta at Michigan State University (Hecker et al. 2003). It was intended to be crucial based on its standard protocol, which was agreed to by Syngenta and the USEPA. It did not show the gross exposure-related developmental effects seen in earlier tests of frogs, but there were issues with the conduct of the test. Confidence in nominally crucial tests should be determined by weighing their relevance and reliability. This test did not convince those who had found developmental effects in other anuran species. Atrazine is banned in the European Union but is still permitted in the U.S.

REFERENCES

Belanger, S. E., A. D. Lillicrap, S. J. Moe, R. Wolf, K. Connors, and M. R. Embry. 2023. "Weight of evidence tools in the prediction of acute fish toxicity." *Integrated Environmental Assessment and Management* 19 (5): 1220–1234.

Benedetti, M., F. Ciaprini, F. Piva, F. Onorati, D. Fattorini, A. Notti, A. Ausili, and F. Regoli. 2012. "A multidisciplinary weight of evidence approach for classifying polluted sediments: Integrating sediment chemistry, bioavailability, biomarkers responses and bioassays." *Environment International* 38 (1): 17–28.

Bombardier, M., and N. Bermingham. 1999. "The SED-TOX index: Toxicity-directed management tool to assess and rank sediments based on their hazard – Concept and application." *Environmental Toxicology and Chemistry* 18 (4): 685–698.

Burton, G. A., P. M. Chapman, and E. P. Smith. 2002. "Weight-of-evidence approaches for assessing ecosystem impairment." *Human and Ecological Risk Assessment* 8: 1657–1673.

Cook, C. N., S. J. Nichols, J. A. Webb, R. A. Fuller, and R. M. Richards. 2017. "Simplifying the selection of evidence synthesis methods to inform environmental decisions: A guide for decision makers and scientists." *Biological Conservation* 213: 135–145.

Crane, M. 2016. "In Response: Challenges when weighing evidence about environmental risks—An industry perspective." *Environmental Toxicology and Chemistry* 34 (11): 2444–2445.

Crane, M., M. Gross, P. Matthiessen, G. T. Ankley, S. Axford, P. Bjerregaard, R. Brown, P. Chapman, M. Dorgeloh, and M. Galay-Burgos. 2010. "Multi-criteria decision analysis of test endpoints for detecting the effects of endocrine active substances in fish full life cycle tests." *Integrated Environmental Assessment and Management* 6 (3): 378–389.

Cumming, G., and R. Calin-Jageman. 2016. *Introduction to the New Statistics: Estimation, Open Science, and Beyond.* Milton Park: Routledge.

Dagnino, A., S. Sforzini, F. Dondero, S. Fenoglio, E. Bona, J. Jensen, and A. Viarengo. 2008. "A "weight of evidence" approach for the integration of environmental "triad" data to assess ecological risk and biological vulnerability." *Integrated Enviromental Assessment and Management* 4 (3): 314–326.

Dhingra, K., and K. L. Vandana. 2011. "Indices for measuring periodontitis: A literature review." *International Dental Journal* 61 (2): 76–84.

Glass, G. V. 2015. "Meta-analysis at middle age: A personal history." *Research Synthesis Methods* 6 (3): 221–231.

Good, I. J. 1950. *Probability and the Weighing of Evidence*. New York, NY: Hafner Press.

Good, I. J. 1983. *Good Thinking: The Foundations of Probability and Its Applications*. Minneapolis: University of Minnesota Press.

Good, I. J. 1991. "Weight of evidence and the Bayesian likelihood ratio." In *The Use of Statistics in Forensic Science*, edited by C. C. D Aitken and D. A. Stoney, 85–106. New York, NY: Ellis Horwood.

Hecker, M., K. K. Coady, D. L. Villeneuve, M. B. Murphy, P. D. Jones, and J. P. Giesy. 2003. *Response of Xenopus laevis to Atrazine Exposure: Assessment of the Mechanism of Action of Atrazine*. East Lansing, MI: Sponsor: Syngenta Crop Protection, Inc., Aquatic Toxicology Laboratory, Michigan State University, National Food Safety and Toxicology Center.

Higgins, J., J. Thomas, M. Cumpston, J. Chandler, T. Li, M. J. Page, and V. A. Welch. 2023. Cochrane Handbook for Systematic Reviews of Interventions Version 6.4. Cochrane Collaborative. Accessed 1/6/2023. www.training.cochrane.org/handbook.

Hristozov, D. R., A. Zabeo, C. Foran, P. Isigonis, A. Critto, A. Marcomini, and I. Linkov. 2014. "A weight of evidence approach for hazard screening of engineered nanomaterials." *Nanotoxicology* 8 (1): 72–87.

Karr, J. R. 1981. "Assessment of biotic integrity using fish communities." *Fisheries* 6 (6): 21–27.

Kiker, G. A., T. S. Bridges, A. Varghese, T. P. Seager, and I. Linkov. 2005. "Application of multicriteria decision analysis in environmental decision making." *Integrated Environmental Assessment and Management: An International Journal* 1 (2): 95–108.

Krimsky, S. 2005. "The weight of scientific evidence in policy and law." *American Journal of Public Health* 95: S129–S135.

Lillicrap, A., S. J. Moe, R. Wolf, K. A. Connors, J. M. Rawlings, W. G. Landis, A. Madsen, and S. E. Belanger. 2020. "Evaluation of a Bayesian network for strengthening the weight of evidence to predict acute fish toxicity from fish embryo toxicity data." *Integrated Environmental Assessment and Management* 16 (4): 452–460.

Linkov, I., O. Massey, J. Keisler, I. Rusyn, and T. Hartung. 2015. "From "weight of evidence" to quantitative data integration using multicriteria decision analysis and Bayesian methods." *ALTEX* 32 (1): 3–8.

Linkov, I., F. K. Satterstrom, G. Kiker, T. P. Seager, T. Bridges, K. H. Gardner, S. H. Rogers, D. A. Belluck, and A. Meyer. 2006. "Multicriteria decision analysis: A comprehensive decision approach for management of contaminated sediments." *Risk Analysis* 26: 61–78.

Mckay, N. D. 2023. "Explanatory virtues and reasons for belief." *Analysis* 83 (4): 701–707.

ONS. 2022. *Health Index Contents and Definitions*. Office for National Statistics.

Ott, W. R. 1978. *Environmental Indices – Theory and Practice*. Ann Arbor, MI: Ann Arbor Science Publishers.

Pearl, J. 2000. *Causality: Models, Reasoning, and Inference*. Cambridge: Cambridge University Press.

Popper, K. R. 1968. *The Logic of Scientific Discovery*. New York: Harper and Row. Reprint.

Pullin, A., G. Frampton, R. Jongman, C. Kohl, B. Livoreil, A. Lux, G. Pataki, G. Petrokofsky, A. Podhora, and H. Saarikoski. 2016. "Selecting appropriate methods of knowledge synthesis to inform biodiversity policy." *Biodiversity and Conservation* 25 (7): 1285–1300.

Quilbé, R., A. N. Rousseau, P. Lafrance, J. Leclerc, and M. Amrani. 2006. "Selecting a pesticide fate model at the watershed scale using a multi-criteria analysis." *Water Quality Research Journal* 41 (3): 283–295.

Rohr, J. R., and K. A. McCoy. 2010. "A qualitative meta-analysis reveals consistent effects of atrazine on freshwater fish and amphibians." *Environmental Health Perspectives* 118 (1): 20–32.

Semenzin, E., A. Critto, M. Rutgers, and A. Marcomini. 2008. "Integration of bioavailability, ecology and ecotoxicology by three lines of evidence into ecological risk indexes for contaminated soil assessment." *Science of the Total Environment* 389 (1): 71–86.

Shipley, B. 2000. *Cause and Correlation in Biology*. Cambridge: Cambridge University Press.

Smith, E. P., I. Lipkovich, and K. Ye. 2002. "Weight-of-evidence (WOE): Quantitative estimation of probability of impairment for individual and multiple lines of evidence." *Human and Ecological Risk Assessment* 8: 1585–1596.

Spirtes, P., C. Glymour, and R. Scheines. 2000. *Causation, Prediction, and Search*. Cambridge, MA: MIT Press.

Stahl, C., and A. Cimorelli. 2013. "A demonstration of the necessity and feasibility of using a clumsy decision analytic approach on wicked environmental problems." *Integrated Environmental Assessment and Management* 9 (1): 17–30.

Suter, G. W. 1993. "A critique of ecosystem health concepts and indices." *Environmental Toxicology and Chemistry* 12: 1533–1539.

USEPA. 2016. *A Practitioner's Guide to the Biological Condition Gradient: A Framework to Describe Incremental Change in Aquatic Ecosystems*. Washington, DC: U.S. Environmental Protection Agency.

USEPA. 2021. *AQUATOX: Linking Water Quality and Aquatic Life*. Washington, DC: U.S. Environmental Protection Agency.

Warne, M., G. E. Batley, R. A. van Dam, J. C. Chapman, D. R. Fox, C. W. Hickey, and J. L. Stauber. 2018. *Revised Method for Deriving Australian and New Zealand Water Quality Guideline Values for Toxicants*. Brisbane: Department of Science, Information Technology and Innovation.

Weed, D. L. 2000. "Interpreting epidemiological evidence: How meta-analysis and causal inference methods are related." *International Journal of Epidemiology* 29 (3): 387–390.

Westman, W. E. 1981. "Factors influencing the distribution of species of California coastal sage scrub." *Ecology* 62: 439–455.

Zukerman, I. 2024. "A Brief Summary of Explanatory Virtues." arXiv preprint arXiv:2411.16709.

4 The Context of Weight of Evidence

Weight of evidence is used in all environmental assessments. It may also be used with systematic reviews to make them more powerful tools.

4.1 WEIGHT OF EVIDENCE IN THE CONTEXT OF ENVIRONMENTAL ASSESSMENT

Environmental assessment is a process of generating and presenting scientific information to inform environmental regulatory or management decisions. Four broad types of environmental assessments have been identified (Figure 4.1) (Cormier and Suter 2008). Weight of evidence may serve different purposes in these types of assessments. For example, the roles of weight of evidence in protecting water quality were laid out in a USEPA guidance document (Table 4.1).

- Condition assessments determine the condition of the environment, potentially including concentrations of contaminants, abundance of species, diversity of biotic communities, and health of human populations. Assessments to determine whether environmental quality standards are being violated are the most common examples.
- Causal assessments determine causes of observed impairments. The impairments may have been identified by a condition assessment or may be manifest, as in fish kills or a large numbers of illnesses in people swimming in a lake.

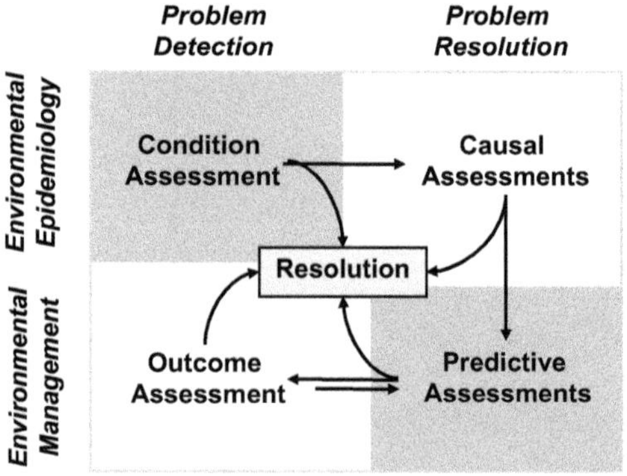

FIGURE 4.1 A framework for the relationships among types of environmental assessments (Cormier and Suter 2008).

 DOI: 10.1201/9781003626336-5

TABLE 4.1
Uses of Weight of Evidence in Assessments for Water Quality Protection under the U.S. Clean Water Act (Ridley et al. 2024)

Assessment Type	Clean Water Act Program	Example Question	Evidence That Could Be Combined to Answer the Example Question
Condition assessment	33 U.S.C. § 1315(b) - State reports on water quality	Is a specific water body biologically impacted?	Multiple biological endpoint measurements (e.g., algal biomass, macroinvertebrate index, fish abundance)
Causal assessment	33 U.S.C. § 1313(d) - TMDLs	What is the likely cause of impairment?	Observational field evidence showing time order and biological sufficiency; experimental evidence showing dose-response relationships; literature evidence showing causal relationships under similar circumstances
Predictive assessment	33 U.S.C. § 1314(a) - Criteria development	What level of the cause will reduce impairment? What level of the cause will ensure attainment of designated uses?	Reference condition; stressor-response for multiple endpoints; literature evidence showing levels under similar circumstances
Outcome assessment	33 U.S.C. § 1315(b) - State reports on water quality	Does reaching the predicted level of the cause result in recovery?	Multiple biological endpoint measurements

- Predictive assessments, such as risk assessments or impact assessments, predict the nature and magnitude of future exposures or effects (ecological, health, economic, etc.) from actions such as permitting the use of a chemical or mining an ore body.
- Outcome assessments determine the positive and negative results of management actions such as remediating a site, upgrading a water treatment plant, or permitting pesticide use.

Each of these types of assessment can fit the same general framework (Figure 4.2). All are initiated by some organization with an environmental mandate, which may be a legal or regulatory requirement, an industry's good environmental practices, or an advocacy group's desire to support its position. The initiators of the assessment define the scope of the assessment, including the question to be answered, the resources and time allowed, and constraints such as guidelines to be followed or actions to be evaluated. Assessments end by informing decision-makers who will take an action or make some other decision.

- The assessment is preceded by an initiator, which provides the question to be answered, guidelines that must be followed, deadlines, resources, and other constraints.
- The first assessment step is a planning process. It may have different names, but in the USEPA's risk assessment process, it is called problem formulation, and we will use that term here (Chapter 5). In all assessment frameworks, the planning process includes gathering information, which is the essential start of a weight-of-evidence process (Chapter 6). Other products may include assessment endpoints, conceptual models, alternative hypotheses, and an analysis plan.
- The second step is the analysis step. In conventional risk assessments, it includes an analysis of exposure and an analysis of the exposure-response relationship. Analysis includes summary statistics, mathematical and statistical modeling, and logical interpretation of the results.
- The third step is synthesis. In conventional risk assessments, it is called risk characterization, and it solves the exposure-response model for the exposure level to estimate effects. In a causal assessment, it is the final step of the weight of evidence. It summarizes and interprets the body of evidence to determine the result.
- Decisions and actions are informed by the assessment. Typically, assessors provide the assessment, including an executive summary, and they may brief decision makers and stakeholders. Less commonly, assessors may be involved in decision-making and implementation.

The weight-of-evidence process may be used to make inferences in any of the three assessment steps, or the entire assessment may be a weight-of-evidence process (left and right arrows in Figure 4.2). Assessments that are single, extended weight-of-evidence processes include assessments of individual chemicals or defined mixtures, such as IRIS and TOSCA assessments (Chapter 15), and many causal assessments

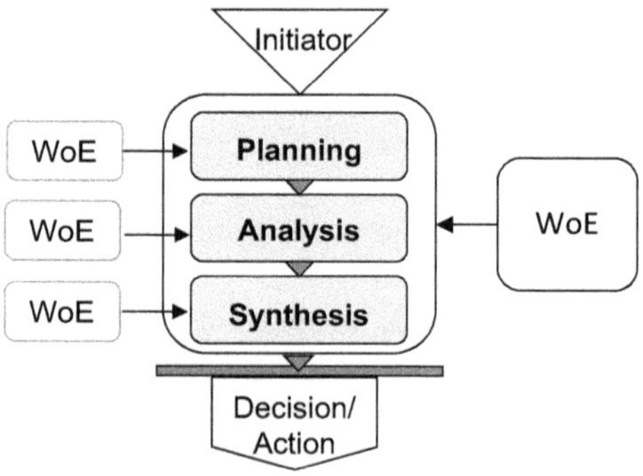

FIGURE 4.2 A generic framework for all types of environmental assessments, from Cormier and Suter (2008). Weight of evidence can contribute to inferences in any step of the process (WoE on the left) or can be the basis for a single inference in an entire assessment (WoE on the right).

(Chapter 13) (USEPA 2018, 2022; Norton, Cormier, and Suter 2015). In the planning phase of a risk assessment, weight of evidence may be used to determine the appropriate assessment endpoints or sources of exposure. In the analysis phase, weight of evidence may be used to select models, estimate parameters, and choose assumptions. In the synthesis phase, weight of evidence may be used to derive answers, such as the best estimate of risk or whether an ecosystem is impaired.

4.2 WEIGHT OF EVIDENCE IN THE CONTEXT OF SYSTEMATIC REVIEW

During World War II, the British physician Archie Cochrane became upset by the lack of reliable scientific support for medical treatments (Cochrane 1972; Suter et al. 2020a). Cochrane promoted randomized clinical trials as the only reliable guide to treatment, and he encouraged the compilation of findings from such trials. The compilation of trials of treatments has been systematized and is performed by the Cochrane Collaboration. They developed a systematic review procedure that finds relevant clinical trials, determines their suitability, and extracts the data (Higgins et al. 2023). It is distinguished from traditional literature reviews by procedures to thoroughly search for relevant literature, screen out inappropriate studies using explicit criteria, extract data and metadata into a database, and do it all transparently. Some investigators have advocated systematic review as a complete inferential method. However, as Rhomberg (2015) stated, "Systematic review of the data will not by itself make the appropriate conclusions evident, and the challenges of the evaluation and integration component need to be addressed."

It has been used by the Cochrane Collaboration and others to provide the data to combine results of multiple studies by meta-analysis and identify effective treatments (Section 3.3). In some contexts, meta-analysis is considered part of systematic review, but it is a statistical inferential method and not a review.

Because of the need to avoid bias in literature reviews, systematic review has been adopted in many research and assessment contexts. In particular, it has been used for literature reviews and information assembly for weight-of-evidence analyses in environmental assessments (Rooney et al. 2014; USEPA 2018, 2022). The Cochrane handbook (Higgins et al. 2023) has even been recommended as a model for toxicological weight-of-evidence guidance (Mandrioli and Silbergeld 2016). However, because it was designed for clinical trials and inference by meta-analysis, the Cochrane handbook is too conceptually simple for environmental toxicological assessments, or environmental assessments in general. Guidance for reviews of epidemiological studies has been added to the Cochrane handbook, but they are assessed separately from clinical trials. In that way, Cochrane reviewers avoid weighting and integrating multiple types of evidence.

A framework for combining systematic review, meta-analysis, and weight of evidence is presented in Figure 4.3. Beginning with the assessment question, the process proceeds to systematically reviewing the literature and incorporating other sources of information, including studies performed for the assessment and surveys of experts. If multiple quantitative studies of one type answer the question, you can proceed to a meta-analysis and, finally, the conclusion (left leg). It is important to remember that

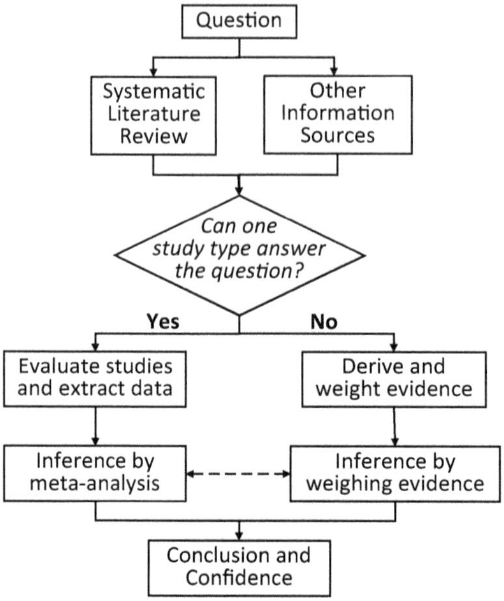

FIGURE 4.3 A diagram of merged systematic review, meta-analysis, and weight-of-evidence approaches to infer a quality from multiple studies or pieces of evidence. (Adapted from Suter et al. 2020b.)

statistical tests such as meta-analysis simply quantify the strength of association, so they do not alone establish causation. And yet, many assessors have not remembered this (Weed 2010). Causation may be inferred when the studies are highly relevant and experimental and, therefore, inherently causal. An example is pond mesocosm tests of the effects of atrazine on algal production (Moore et al. 2017). If the meta-analysis cannot provide results because no evidence is relevant and based on reliable experiments, weight of evidence provides the conclusion (right leg).

The dashed line is the complicated part. If the meta-analysis might answer the question, but other evidence could provide more weight or even a different answer, then the results of the meta-analysis might be added to the body of evidence to be weighed (dashed arrow to the right). In such cases, you can use meta-analysis to estimate the strength and consistency of the body of evidence (Weed 2000). If you are weighing the evidence, the body of evidence may contain multiple pieces of quantitative evidence of the same type, which may be meta-analyzed (dashed arrow to the left). For example, if the question concerns a chronic exposure, but there are four acute studies, it may be worthwhile to use meta-analysis to reduce that set of studies to one piece of acute toxicity evidence. That is because the acute studies are peripheral, and it is not worthwhile to weigh their relative merits. Finally, if the studies in a meta-analysis vary in their reliability, weight of evidence may be used to down-weight less reliable studies (dashed arrow to the left).

The National Research Council, in its review of the IRIS program, suggested that meta-analysis and weight of evidence are two possible modes of analysis for the evidence from a systematic review (NRC 2014). The integration of systematic review, meta-analysis, and weight of evidence in Figure 4.3 is also equivalent to the broad approach advocated as evidence-based toxicology (Martin et al. 2019; Stephens et al. 2016). However, none of those groups provided a framework for integrating systematic review, weight of evidence, and meta-analysis as in Figure 4.3.

REFERENCES

Cochrane, A. L. 1972. *Effectiveness and Efficiency: Random Reflections on Health Services.* UK: The Newfield Provincial Hospitals Trust.

Cormier, S. M., and G. W. Suter II. 2008. "A framework for fully integrating environmental assessment." *Environmental Management* 42 (4): 543–556.

Higgins, J. P. T., J. Thomas, M. Cumpston, J. Chandler, T. Li, M. J. Page, and V. A. Welch. 2023. Cochrane Handbook for Systematic Reviews of Interventions Version 6.4. London, UK: Cochrane Collaborative. Accessed 1/6/2023.

Mandrioli, D., and E. K. Silbergeld. 2016. "Evidence from toxicology: The most essential science for prevention." *Environmental Health Perspectives* 124 (1): 6–11.

Martin, O. V., J. Adams, A. Beasley, et al. 2019. "Improving environmental risk assessments of chemicals: Steps towards evidence-based ecotoxicology." *Environment International* 128: 210–217.

Moore, D. R. J., C. D. Greer, G. Manning, K. Wooding, K. J. Beckett, R. A. Brain, and G. Marshall. 2017. "A weight-of-evidence approach for deriving a level of concern for atrazine that is protective of aquatic plant communities." *Integrated Environmental Assessment and Manag*ement 13 (4): 686–701.

Norton, S. B., S. M. Cormier, and G. W. Suter II. 2015. *Ecological Causal Assessment.* Boca Raton, FL: CRC Press.

NRC. 2014. *Review of EPA's Integrated Risk Information System (IRIS) Process (2014).* Washington, DC: National Research Council of the National Academies of Science, Engineering, and Medicine, National Academies Press.

Rhomberg, L. 2015. "Hypothesis-based weight of evidence: An approach to assessing causation and its application to regulatory toxicology." *Risk Analysis* 35 (6): 1114–1124.

Ridley, C. E., S. D. Kaylor, S. S. Lee, J. N. Miller, S. Penry, and K. A. Schofield. 2024. *Application of Weight-of-Evidence Methods for Transparent and Defensible Numeric Nutrient Criteria.* Washington, DC: U.S. Environmental Protection Agency.

Rooney, A. A., A. L. Boyles, M. S. Wolfe, J. R. Biucher, and K. A. Thayer. 2014. "Systematic review and evidence integration for literature-based environmental health science assessments." *Environmental Health Perspectives* 122: 711–718.

Stephens, M. L., K. Betts, N. B. Beck, V. Cogliano, K. Dickersin, S. Fitzpatrick, J. Freeman, G. Gray, T. Hartung, and J. McPartland. 2016. "The emergence of systematic review in toxicology." *Toxicological Sciences* 152 (1): 10–16.

Suter, G. W., J. Nichols, E. Lavoie, and S. M. Cormier. 2020a. "The origins of classic WoE and SR, supplemental material for: Systematic review and weight of evidence are integral to ecological and human health assessments: They need an integrated framework." *Integrated Environmental Assessment and Management* 16 (5): 718–728.

Suter, G. W., J. Nichols, E. Lavoie, and S. M. Cormier. 2020b. "Systematic review and weight of evidence are integral to ecological and human health assessments: They need an integrated framework." *Integrated Environmental Assessment and Management* 16 (5): 718–728.

USEPA. 2018. *Application of Systematic Review in TSCA Risk Evaluations.* Washington, DC: U.S. Environmental Protection Agency.

USEPA. 2022. *ORD Staff Handbook for Developing IRIS Assessments.* Washington, DC: Office of Research and Development, U.S. Environmental Protection Agency.

Weed, D. L. 2000. "Interpreting epidemiological evidence: How meta-analysis and causal inference methods are related." *International Journal of Epidemiology* 29 (3): 387–390.

Weed, D. L. 2010. "Meta-analysis and causal inference: A case study of benzene and non-Hodgkin lymphoma." *Annals of Epidemiology* 20 (5): 347–355.

Part II

Process for Weight of Evidence

5 Frameworks for Weight of Evidence

Krimsky wrote in (2005) that, "Without an explication of how evidence is 'weighed' or 'weighted,' the claim WOE seems to be coming out of a 'black box' of scientific judgement." The framework presented in this section and elaborated in the following sections is intended to open up Krimsky's black box, show the mechanics for weighing evidence, and explain how to operate it. Many authors have called for standardization of weight-of-evidence methods to provide consistent and unbiased assessment practices and results (Martin et al. 2019). However, that is a difficult demand because, even within an agency like the USEPA, differences in legal mandates, tasks, and traditional practices make it difficult to impose common methods. Our goal in this section is to present a framework for a general approach that breaks down the weight-of-evidence process into basic components, provides consistent terminology, explains the processes, and relates it to other relevant weight-of-evidence approaches. The framework is based on a guidance document that we and other members of the USEPA Risk Assessment Forum created and that was extensively reviewed and approved by the Office of the Administrator as Agency-wide guidelines (USEPA 2016). Although it was developed by ecological risk assessors, the framework is also applicable to human health.

Remember that this weight-of-evidence framework may provide the inferential process for the entire assessment or for any discrete inference in any step of an assessment (Figure 4.2). For example, the scale of an assessment, the exposure or effect endpoints, or a statistical approach might be chosen by a weight of evidence. In other words, you may perform a weight of evidence within a larger weight of evidence encompassing the assessment. The need for weight of evidence is determined by a planning process that defines the goals of the assessment, the assessment process, and the endpoints of the assessment (Figure 5.1). Most importantly, planning defines the question to be answered by the assessment (Appendix A). The weight-of-evidence process is inserted where needed in the overall assessment process. The results of one or more weight-of-evidence processes may contribute to the synthesis phase of the assessment. For example, a human health hazard assessment may include weights of evidence for various effects such as cancer, reproductive effects, and hepatotoxic effects.

5.1 A BASIC FRAMEWORK FOR WEIGHT OF EVIDENCE

The framework for the weight-of-evidence process to assess evidence for qualities such as hazard, causation, and impairment or to choose the best quantity is shown in Figure 5.1. The assessment begins with the directions provided by the planning phase. Then the weight-of-evidence process within the assessment begins with

DOI: 10.1201/9781003626336-7

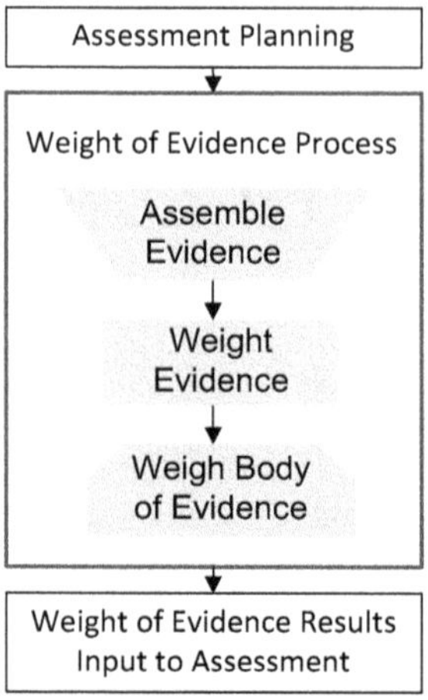

FIGURE 5.1 The basic weight-of-evidence framework from (USEPA 2016).

assembling the evidence (Chapter 6). This step may involve searching the literature, performing new studies, implementing a model, or polling expert judgments. Then the information from those sources must be screened to determine that it meets minimal relevance and reliability criteria, its implications for the hypotheses must be identified, and the data extracted to generate useful evidence. Next, the pieces of evidence are weighted (i.e., evaluated and rated) to determine their influence on the inference (Chapter 7). Finally, the body of evidence for each hypothesis is weighed to derive the conclusion (Chapter 8). The weighing involves integrating the assigned weights of the individual pieces or types of evidence. Then the body of evidence for each hypothesis is weighed with respect to collective properties, and finally, the weighed evidence for each hypothesis is interpreted to determine the conclusion. An extension of this framework to derive quantities is presented in Chapter 8.

Although weight-of-evidence methods vary, they all need these three basic steps because:

- All assessments require useful evidence,
- The pieces and types of evidence are likely to have greater or lesser importance (i.e., weightiness), which influences the final weighing of the evidence, and
- Finally, combining and interpreting evidence for hypotheses (i.e., weighing it) is necessary to inform the conclusion and help explain the rationale for that conclusion.

The second step, the explicit assigning of weights, is skipped by many weight-of-evidence methods, even though it is usually apparent that evidence is not all equally important. If you do assume that all evidence is equal, that assumption should be stated and justified. For example, a statement might be included: "Because all evidence was generated from well-designed and well-conducted case-specific studies, it is all highly relevant and reliable."

5.2 AN ONTOLOGICAL FRAMEWORK FOR WEIGHT OF EVIDENCE CONCEPTS

Weight of evidence requires a set of terminology that describes the component concepts of the process and their relationships. This is called an ontology because it determines what exists in the system. It is described and illustrated by a conceptual diagram (Figure 5.2). The terms in boxes are basic weight-of-evidence concepts, which are conceptually linked by processes. This diagram differs from the assessment framework (Figure 5.1), in which the boxes are procedural steps. The ontology is important because weight-of-evidence concepts and the terminology used to express them have not been standardized.

Questions, hypotheses, and qualities: Assessments begin with a question that you answer by performing the assessment. The same is true of a weight-of-evidence process within an extensive assessment. The question implies certain potential answers, which are the hypotheses of the assessment. In some cases, the hypotheses are simply reframed questions. For example, if the question is, "Is the chemical a teratogen?", the alternative hypotheses are that it is a teratogen and that it is not a teratogen. However, if the question is, "What is the cause of the decline in a trout population?", there are many plausible hypotheses, such as harvesting, disease, water temperature, reduced flow, or sewage causing the decline. In such cases, the gathering of information may suggest additional hypotheses beyond those that are implied by the question and background knowledge.

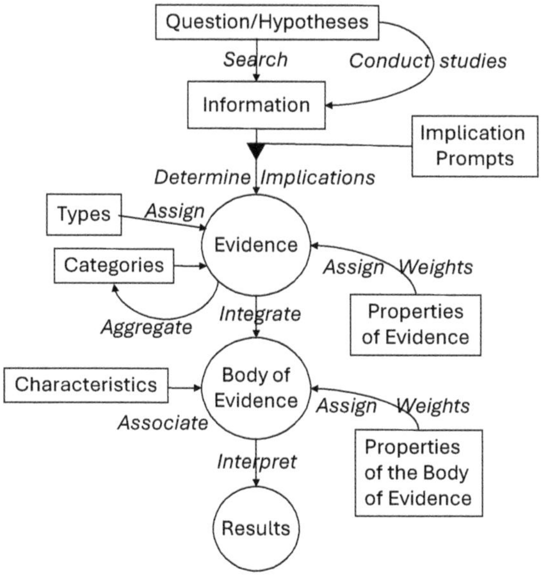

FIGURE 5.2 A conceptual diagram of the ontological framework for our weight-of-evidence method.

The hypotheses involve qualities such as teratogenicity and causation, which may have defined characteristics (Appendix B).

Information: Data, observations, model results, and other facts constitute information if they contribute to evidence. Information may be obtained by searching the literature and databases or by conducting studies.

Implications: Information is evidence if it has some implication for a hypothesis or question. For example, very similar chemical structures (the information) imply similar biodegradation pathways (an implication) with respect to the hypothesis that the biodegradation rate of one chemical can approximate the rate of the other. Implications are assigned to information.

Implication prompts: Implications of information may be based on prompts. Prompts include the characteristics of causation or of other qualities. For example, if organisms co-occur in space and time with a chemical spill, that information is evidence of the causal characteristic, co-occurrence. Conceptual models are also useful prompts. That is, the information may imply that one of the processes or states in the conceptual model exists in the real-world case being assessed.

Evidence: Information that has implications for the hypotheses is evidence or may contribute to evidence. The implications may be self-evident or based on implication prompts.

Types of evidence: Some pieces of evidence are of the same type in the sense that they are from studies with the same design and subject. There are conventional types, such as benthic invertebrate surveys, rat acute lethality tests, population demographic models, effluent toxicity tests of fish larvae, and quantitative structure-activity relationships.

Categories of evidence: Pieces of evidence may be aggregated into categories to reduce duplication during the weighting of evidence and to identify sets of evidence that, in some sense, are effectively replicates. Evidence may be categorized by type, implication, or characteristic. For example, all evidence for the co-occurrence of a potential cause and the effect of interest may be aggregated, even though the pieces of evidence are not of the same methodological type. Finally, a series of pieces of evidence that together constitute a line of reasoning may be considered a line of evidence.

Properties of evidence: Properties are features of evidence that determine how much weight it should be assigned. They may be properties of pieces, types, or other categories of evidence. The most commonly used properties of evidence are relevance and reliability.

Body of evidence: All the evidence that is applicable to a question or hypothesis is integrated into a body of evidence.

Properties of bodies of evidence: Bodies of evidence have collective properties, such as coherence and diversity.

Characteristics of the hypothesized quality: Qualitative weight of evidence makes inferences concerning a hypothesized quality that could be supported by evidence. Examples of qualities include causation, impairment, remediation, carcinogenicity, and protection. These qualities are defined by certain characteristics that may be supported or diminished by evidence. For example, sufficiency is a characteristic of causation.

Results: An interpretation of the weighted bodies of evidence for each hypothesis determines which is the best-supported hypothesis. The weight of a body of evidence may also be compared to a standard to determine whether a hypothesis is sufficiently supported.

REFERENCES

Krimsky, S. 2005. "The weight of scientific evidence in policy and law." *American Journal of Public Health* 95: S129–S135.

Martin, O. V., J. Adams, A. Beasley, et al. 2019. "Improving environmental risk assessments of chemicals: Steps towards evidence-based ecotoxicology." *Environment International* 128: 210–217.

USEPA. 2016. *Weight of Evidence in Ecological Assessment.* Washington, DC: Risk Assessment Forum, U.S. Environmental Protection Agency.

6 Assembling Evidence

Because many judgments must be based on limited information, it is critical that all reliable information be considered.

The Presidential/Congressional Commission on Risk Assessment and Risk Management (1997)

Weight of evidence depends on assembling evidence by finding it in the literature or by generating it in new studies (Figure 6.1). We define evidence as information that can contribute to evaluating a hypothesis (Box 6.1). In environmental assessments, evidence is often:

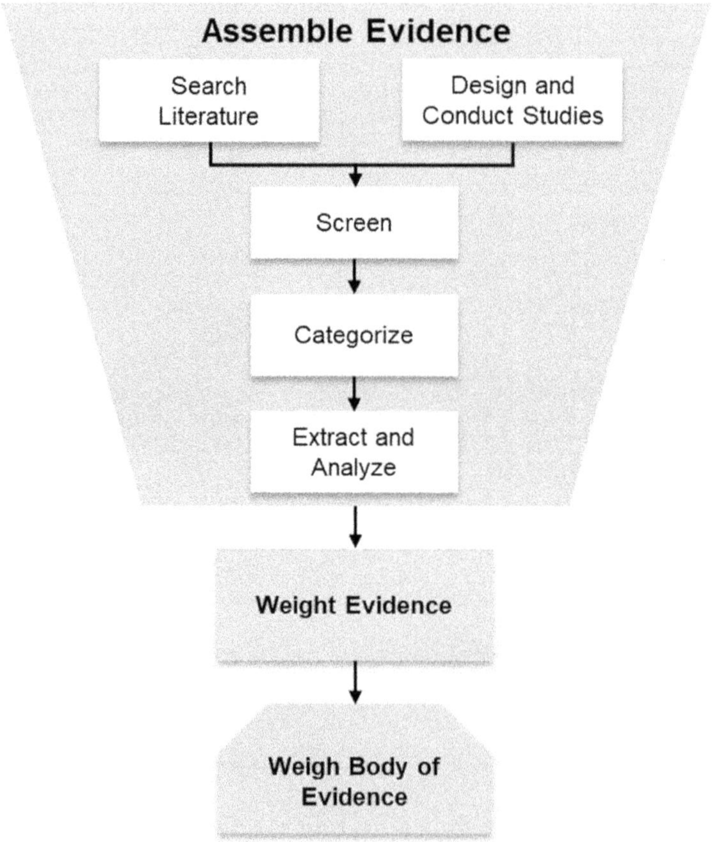

FIGURE 6.1 An elaboration of the process for assembling evidence, the first step of weight of evidence. (Adapted from USEPA 2016.)

DOI: 10.1201/9781003626336-8

- multiple (e.g., three mouse acute lethality tests),
- heterogeneous (e.g., three acute mouse tests, a rat reproductive test, a cellular assay, and four human case reports),
- produced by multiple disciplines (e.g., molecular genetics, toxicology, and epidemiology), or
- methodologically diverse (e.g., concentrations measured by multiple analytical methods (Toselli and Sgolastra 2020)).

Evidence may be empirical or mechanistic, observational or experimental, site-specific or generic, quantitative or qualitative, measured or modeled, and objective or subjective.

As a result of this diversity, searching for relevant data or qualitative information can be a challenge. To be evidence, information must have an implication and be relevant, and determining whether a piece of information is sufficiently relevant may require considerable expertise. Carnap's (1947) principle of total evidence guards against inappropriate exclusion by defaulting to inclusion. In utilitarian terms, we should use all available evidence, provided that the cost of doing so is not excessive (Good 1966). Weight-of-evidence methods make this inclusiveness reasonable by assigning weights even to evidence with little relevance or reliability (Chapter 7). However, in most weight-of-evidence systems, evidence is screened to some degree before weighting it (Section 6.6) because further analysis of irrelevant or poor-quality evidence is time-consuming and may distract attention from relevant evidence.

BOX 6.1 WHAT IS EVIDENCE?

The definition of evidence used here is a pragmatic one. Evidence is information that is useful for an inference. However, there are other definitions.

Philosophers of science have addressed the nature of evidence at length but without the pragmatic point of view. They want to know how to define evidence in such a way that they can prove it is evidence. A good summary of evidential concepts from a philosopher of science is found in (Achinstein 2001, 2005). Achinstein lists four types of theories of evidence and their founders: inductivist (Isaac Newton and John Stuart Mill), explanationist (William Whewell), falsificationist (Richard Feynman and Karl Popper), and Bayesianist/probabilist (various Bayesians). That is, evidence may inductively support a hypothesis, evidence may be the information to be explained by a hypothesis, it may show that a hypothesis is untrue, or the relationship between evidence and hypothesis may be a subjective and probabilistic.

Induction is most familiar, but assessors and other scientists practice abduction (inference to the best explanation) in their work, often without awareness of the pragmatic philosophy of science. Notably, artificial intelligence systems use abductive algorithms to, for example, determine the best diagnostic explanation (hypothesis) for a set of symptoms (evidence).

6.1 EVIDENCE NEEDS AND PROMPTS

Assessments of qualities begin with an assessment question and then frame hypotheses concerning the answer to the question. Evidence is applied to hypotheses. Therefore, it is important to clearly specify the question and hypotheses so that you know what evidence is needed. Then you must translate them into a set of desired evidence.

6.1.1 PICO/PECO/PECOTS

If the assessment is routine and the process is standardized, the question to be assessed may have a standard format. For example, in the GRADE framework for assessing medical treatments, the questions are framed by the PICO acronym. It specifies four components: Population, Intervention, Comparison, and Outcome (Schünemann et al. 2013). The USEPA and others who assess human chemical exposures have changed PICO to PECO (Thayer, Angrish, et al. 2022; Morgan et al. 2018).

- **Population**: a human population or life stage (e.g., infants) or an experimental animal species and life stage.
- **Exposure**: chemical, form, route, and dosage.
- **Comparator**: a comparison or reference population exposed to lower levels (or no exposure or exposure below detection limits), exposure for shorter periods of time, or cases versus controls.
- **Outcome**: health outcomes (e.g., cancer and noncancer).

The PECO components can serve as prompts for information to be obtained when assembling evidence. They also serve as guides for formulating hypotheses.

An ecological PECO question might be: In a **P**opulation of honeybees, **E**xposed by contact with deltamethrin in rapeseed fields, and using a **C**omparator of unsprayed fields, is the **O**utcome increased death? Ideally, a literature review for an assessment to answer this question would find studies that meet all four PECO criteria. However, one might also extrapolate from laboratory toxicity tests of individual bees to field-exposed bees.

The PECO system is not applicable to all assessments or the needs of all organizations, so it is modified. The U.S. Office of Health Assessment and Translation (OHAT) added two components, Time and Setting, to create PECOTS (OHAT 2019). Because PECO is intended for the assessment of individual chemicals in humans, it is often inappropriate for ecological assessments. The exposed entities may be communities, ecosystems, or even regions, not populations. Also, PECO and similar questions may not fit your assessment. For example, causal assessments (Chapter 15) require results of models or monitoring studies to estimate real-world exposures, but chemical hazard identifications (Chapter 17) do not require such estimates. When deriving a simple assessment question, a template like the PECO acronym might help ensure that the evidence is adequately informative. However, complex assessments, like assessing risks from a proposed mine in the Alaskan wilderness, cannot be encompassed by simple PECO questions.

6.1.2 Conceptual Models

A conceptual model is a diagram of the components of a system and the causal processes that connect them (Figure 6.1). They are created in risk assessment problem formulations or other assessment planning processes. Their content depends on the type of assessment. For contaminated sites, the basic model begins with one or more sources (e.g., wastewater pits, contaminated soil from spills, or groundwater plumes) and extends through transport pathways to endpoint entities (e.g., fish populations, plant communities, or stream ecosystems). However, they can include complexities such as secondary effects due to the loss of habitat structure or food species. Some are simple, such as an assessment of the hazard posed by a chemical that may hypothetically contaminate water and may be taken up by the gills of fish. At the other extreme, assessments of projects may have very complex conceptual models. For example, an assessment of the proposed Pebble Mine included digging a pit; piling waste rock; processing the ore; disposing of tailings; using, contaminating, treating, and releasing water; building roads, pipelines, power plants, and other infrastructure; accidents; and all the direct and indirect ecological and cultural effects of those actions. The associated conceptual models go on for pages (USEPAR10 2014).

One approach for associating evidence with the needs of an assessment is to use the conceptual model as a scaffold on which to hang the evidence (Norton and Schofield 2018). From that, you can see which cause-effect relationships (arrow-and-box pairs) have evidence and which do not.

For example, the conceptual model of accidents as a cause of San Joaquin kit fox mortality (Figure 6.2) showed that evidence concerning the causes of deaths was needed. Those causes were determined by collecting dead radio-collared foxes and

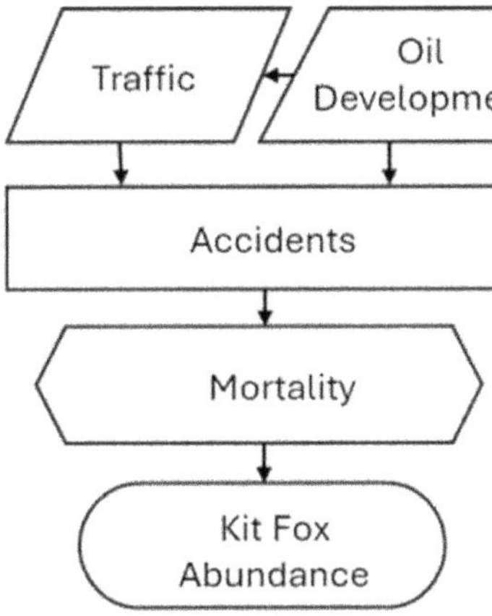

FIGURE 6.2 Section of the conceptual model related to lethal accidents as a candidate cause of kit fox population decline on the Elk Hills, California. (Redrawn from Suter and O'Farrell 2009.)

performing necropsies. Evidence was also needed to determine that mortality from accidents was sufficient to cause the decline in abundance. Anthropogenic accidental mortality was determined by demographic modeling to be insufficient to cause the decline in abundance (Suter and O'Farrell 2009).

Note that, to accept a hypothesis, it is not necessary to have positive evidence for every causal link in your conceptual model. Evidence for missing links may be assumed, and it is sufficient to have, at the end of the assessment, a body of evidence that supports a conclusion.

6.1.3 Standard Evidence Sets

Some methods for weight of evidence specify standard, mandatory sets of evidence (Section 2.7). For example, the sediment quality triad requires chemical concentrations in sediment, toxicity test results for sediment, and surveys of sediment biotic communities. Koch's postulates require a set of four results from field and laboratory studies. Also, some regulatory and management programs specify a minimum set of evidence. See, for example, the requirements for pesticide registration in the United States (USEPA 2024b). The Australia and New Zealand water quality guidelines call for weighing evidence for each step in the pressure–stressor–ecosystem receptor causal pathway (ANZG 2018). Standard evidence requirements usually imply the generation of assessment-specific data.

Standard evidence may imply a standard assessment methodology. For example, the European Union's OSIRIS weight-of-evidence approach is designed to analyze nonanimal test data from integrated testing strategies for specific endpoints (Buist et al. 2013; Vermeire et al. 2013).

6.1.4 Missing Evidence

People tend to focus on what is known given the evidence. We tend to neglect the missing evidence that would be influential in the weighing of evidence but was not found. That is why we have devoted so much space to defining the evidence set. Otherwise, you may be unaware of what is missing. It is important to track the missing evidence so that, when making an inference, you will consider how the missing evidence might change the results if it were available.

Missing evidence most often results from a lack of information in the literature, but it may also result from an incomplete specification of the system structure. That is, parts of your conceptual model may be missing, so you did not search for the evidence. For example, when assessing risks to birds from pesticide applications, you might assess direct toxic effects and leave out the indirect effects due to toxic effects on insects in the birds' diets (Mineau and Kern 2023).

Having identified missing evidence, you may want to decide whether it is worth obtaining in terms of time, effort, cost, and influence on the decision. If you want to be formal about it, you can use value-of-information analysis (Keisler et al. 2014; Wilson 2015).

6.2 SOURCES OF EVIDENCE

Having determined what evidence is required or desired, you can proceed to obtain it. The most obvious and common source is to review the literature, but you can also search databases and repositories, collect new empirical evidence, generate relevant results from models, or elicit expert judgments.

6.2.1 Literature Review

All of us have performed literature reviews for our research or assessments. However, prior to the development of systematic reviews in the 1990s, the process was informal, incomplete, inconsistent, and subject to bias (Section 4.2). Systematic reviews were developed as the principal tool of evidence-based medicine and have led to evidence-based toxicology and evidence-based ecotoxicology (Hartung 2010; Guzelian et al. 2005; Martin, Adams, Beasley, Belanger et al. 2019; Harris et al. 2014; OHAT 2019). In addition to addressing risks from chemicals and other agents, systematic review has been adapted to environmental conservation practices (CEE 2018). The primary purposes of systematic review are to find all or nearly all the information concerning a topic, screen out information that is not worthy of consideration, and assemble the rest.

Although systematic review for the assessment of chemicals in the environment has focused on the hazardous properties of chemicals, the USEPA's Office of Pollution Prevention and Toxics has broadened its application to include reviews of attributes of exposure and environmental context (USEPA 2025). They also more clearly link systematic review to risk assessment. From our perspective, those differences are advantageous. Systematic reviews for environmental assessments may address agents other than chemicals, such as hydropeaking, which is the rapid rise and fall in discharges of water from hydroelectric plants (Melcher et al. 2017).

The advantages of systematic reviews are that their methods and results are transparent, they make more complete use of the available literature than unsystematic reviews, they minimize bias, and they document the review results, which allows review and provides information for future assessments. For example, they document the criteria for eliminating toxicity test data (Table 6.1). Preferred Reporting Items for Systematic Review and Meta-Analysis Protocols (PRISMA-P) provides a detailed general protocol for systematic reviews to ensure the acceptability of your review methods (Moher et al. 2015). Guidance is also available on systematic reviews specifically for environmental evidence (Livoreil et al. 2017). Some regulatory agencies have provided guidance on systematic reviews (USEPA 2018, 2022; EFSA 2010). Despite these advantages, medical systematic reviews have often been found to be misleading or conflicted (Ioannidis 2016). Therefore, caution is advised when adopting systematic reviews performed by and for others. Biases may be difficult to detect without replicating the published review.

The criteria used to accept or eliminate studies can be a source of controversy because they can determine the outcome of assessments. See, for example, the conflict over the toxicity of atrazine to aquatic vertebrates that turned on the criteria for

TABLE 6.1
Circumstances and Types of Toxicity Data for Which Guideline Values Should Not Be Calculated (Warne et al. 2018)[a]

Test Characteristic	Conditions Excluded
Experimental design	Where the test concentrations differ by a large amount (for example ≥10-fold differences such as 1, 10, 100, 1,000, and 10,000 μg/L)
Duration of exposure	If not stated or does not conform with Table 1 [in (Warne et al. 2018)].
Toxicological endpoint	If not stated and/or endpoints other than lethality, immobilization, growth, population growth, or the equivalent unless the endpoint has been proven to be ecologically relevant
Aqueous solubility	If toxicity values are greater than twice the aqueous solubility

[a] Except where the data are of particular significance and a strong justification for their inclusion can be provided and is deemed to be of net benefit to the derivation of the guideline value.

eliminating studies (Section 6.6). Ideally, standard criteria for acceptance and elimination (e.g., Table 6.1) should be used (USEPA 2011a). Alternatively, case-specific criteria can be defined prior to the review. You may perform an exploratory phase by, say, reviewing 10% of the papers and then setting the exclusion criteria based on that experience.

The disadvantages of systematic reviews are that they become mired in prescribed processes for finding literature, screening it, extracting and documenting data and metadata, and documenting the searching and screening processes. Systematic reviews can take years to complete and consume significant financial and human resources. They can result in missed deadlines, blown budgets, or incomplete assessments.

A solution for excessively burdensome systematic reviews is rapid reviews (Moons, Goossens, and Thompson 2021; Garritty et al. 2021; Collins et al. 2015). These reviews are intended to come to the right conclusion or, to a limited conclusion, within the available time and resources. Rapid evidence assessment is an alternative name for this process. By either name, this abbreviated process may be accomplished by reducing the scope of the question to be answered, setting a time limit, limiting yourself to only a few reliable sources or types of studies, using strict eligibility criteria for including studies, or limiting the amount of documentation of the review. The USEPA has developed a decision-support tool for rapid assessments that includes exposure and toxicity information, as well as graphic and analytical tools (Lambert et al. 2025) (https://comptox.epa.gov/rapidtox/auth/login). It is unclear what shortcuts rapid assessments can take without missing essential evidence or introducing bias (Garritty et al. 2021). In any case, acknowledge any shortcuts that you take.

Systematic literature reviews and rapid reviews may be used for purposes other than supplying evidence to answer an assessment question. They may also summarize the state of knowledge with respect to an issue and identify knowledge gaps, types of evidence, weaknesses, trends, and clusters (subsets of evidence that address specific issues) (Figure 6.3). These reviews are sometimes called systematic mapping or evidence mapping. Conceptual maps have been proposed for environmental

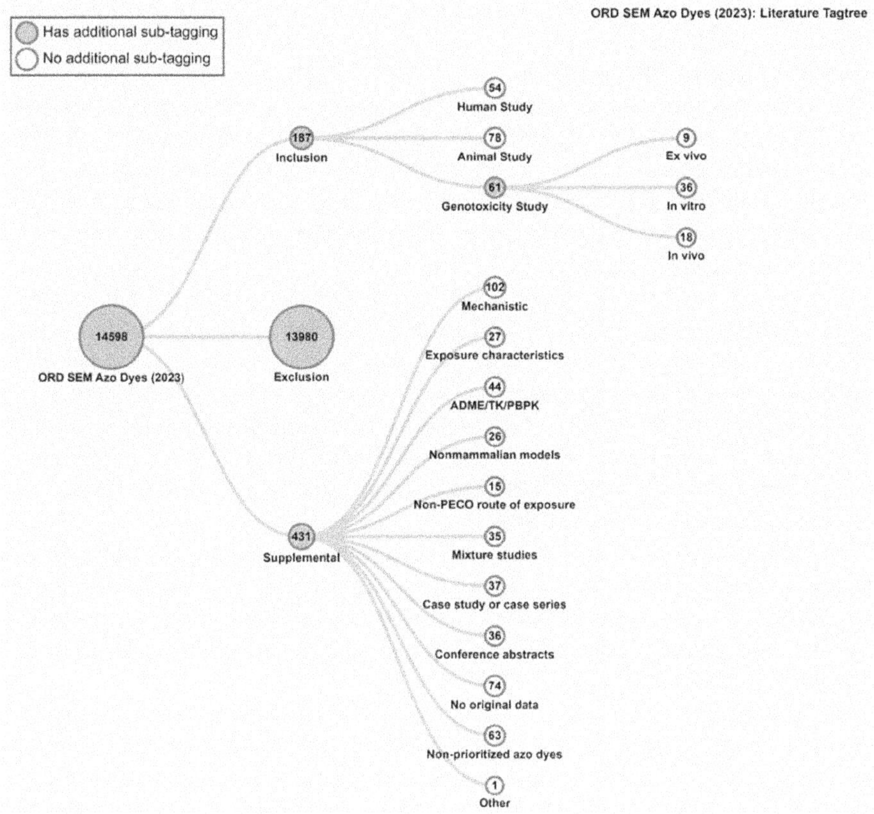

FIGURE 6.3 A systematic map for assessing human health risks from azo dyes, created using the Health Assessment Workspace Collaborative (HAWC) open software for the USEPA (Keshava et al. 2023).

sciences (James, Randall, and Haddaway 2016; Thayer, Shaffer, et al. 2022). They are used in organizing evidence, planning a research or modelling program, or determining how much support the literature provides for a hypothesis or intervention. They may also be published as review papers for reference by assessors and other scientists. For example, a large systematic review of interactions of hazardous agents in freshwater has organized and summarized 2396 multiple-stressor experiments (Orr et al. 2024). Its database is open access, so it could serve as a starting point for weighing evidence concerning a particular interaction. Similarly, an evidence bank is being developed to compile reviews of how well particular policies have worked to control climate change (Pearson 2024).

6.2.2 Databases and Repositories

Databases and repositories have been created specifically to support human health and ecological assessments. They are particularly useful for rapid reviews and assessments. They can also be used to validate a systematic review. That is, the systematic review process for searching the literature should find the relevant information in an applicable database. However, it is wise to check the date of the last revision. Examples include the following.

ECOTOX is a database of ecotoxicological information on single chemicals developed by the USEPA (https://cfpub.epa.gov/ecotox/). It includes aquatic and terrestrial data. An interactive feature allows exploration and visualization of results during a search.

The Pesticide Properties DataBase includes human toxicity, ecotoxicity, and physicochemical properties for pesticides, biopesticides, and veterinary treatments https://sitem.herts.ac.uk/aeru/ppdb/.

The Organisation for Economic Co-operation and Development's (OECD) eChemPortal (https://www.echemportal.org/echemportal/) provides information on physicochemical properties, ecotoxicity, toxicity, environmental fate and behavior, classification and labeling, exposure, and use (OECD 2024). It collects chemical hazard and risk information prepared for governmental chemical programs.

The Water Quality Portal (https://www.waterqualitydata.us/) is a compilation of water quality datasets for the United States from the USGS, USEPA, USDA, and state, tribal, and local agencies (USGS 2019).

The eWater Toolkit, created at the University of Melbourne and University of Canberra, includes Eco Evidence (https://toolkit.ewater.org.au/tools/eco-evidence). It is a database of associations between causal agents and related effects, along with associated metadata, for water pollution. It also includes analysis software to weight the evidence for the causal link.

The Toxicity Forecaster (ToxCast) (https://www.epa.gov/comptox-tools/toxicity-forecasting-toxcast) program makes *in vitro* medium- and high-throughput screening assay data publicly available for prioritization and hazard characterization of thousands of chemicals. It stores, manages, curve-fits, and visualizes data and populates a relational database.

Papyrus is a publicly available dataset of ligand–protein bioactivity data (Bequignon et al. 2023). It is an example of the new datasets being created for specific applications in new computational and mechanistic methods.

Repositories are sites in which information can be deposited for sharing. For example, the AOP Wiki is a repository for adverse outcome pathways (https://aopwiki.org/).

The CADDIS Literature Database (CADLink) illustrates a major problem with databases (https://www.epa.gov/caddis/cadlink). It is difficult to maintain support for updating them. CADLink is a component of the CADDIS system for ecological causal assessment. It provides data concerning evidence to support causal linkages in the environment, but it has not been updated in years. Although the web page is still accessible, the database has been taken down.

Other USEPA data can be found at: Where to Find EPA Data I US EPA – https://www.epa.gov/data/where-find-epa-data.

6.2.3 New Studies

Sometimes, the available information on a topic is not sufficient to support an assessment. In extreme cases, such as the causal assessment that determined that 6PPDq was the cause of Coho salmon mortality in Seattle, there was no relevant literature or compiled data to review. Instead, ad hoc studies generated the evidence needed to identify the cause (Feist et al. 2017; Tian et al. 2021, 2022). More often, new studies are needed to obtain a particular piece of information. For example, targeted laboratory and field studies are conducted to characterize exposure in a contaminated site assessment, effluent composition for a permit, or pathology for a wildlife disease cluster.

Standard methods are available from regulatory agencies and the OECD for toxicity testing, chemical analyses, epidemiology, biotic community characterization, and other information needs. Data generated from those protocols will be acceptable as evidence in assessments. Methods developed for research or ad hoc studies may or may not be acceptable (OECD 2025).

When conducting new studies, consider best practices for reporting results. Guidance for evidence-based toxicology and ecotoxicology provides directives for reporting studies (Martin, Adams, Beasley, Belanger, et al. 2019; Hlisnikova 2024; NRC 2011).

6.2.4 Analogies

Analogies are sources of information if you are lacking information for your case, but know of one or more similar (i.e., analogous) cases. For example, the chemical or toxicological properties of a poorly known chemical may be estimated from the properties of similar chemicals, a process called read-across (Chapter 20).

The ultimate analogy is between your current case and prior cases. For example, if you are assessing an open-pit copper mine, previous such mines are valuable sources of information. The information can come from risk assessments for licensing prior mines, from monitoring reports, condition assessments during operation and remediation, and reviews of effects from copper mines or their components (e.g., tailings dams and waste rock piles) (USEPAR10 2014). The assessments for the prior cases are applicable by analogy, and the weight given to the resulting evidence depends on the similarity of the cases.

6.2.5 Models

In assessments of chemical risks, the most common models for estimating unmeasured attributes are statistically derived empirical models. Of these, quantitative structure–activity relationships (QSARs) are the most commonly used (Chinen and Malloy 2022), and they are readily available in the QSAR Toolbox (https://qsartoolbox.org). Similarly, the Web-based Interspecies Correlation Estimation (Web-ICE) application estimates the toxicity of a chemical to a species, genus, or family from the known toxicity of the chemical to surrogate species (Lilavois, Nelson, and Raimondo 2024). These empirical models have been extended to include chronic toxicity and a wider range of input data (Douziech et al. 2024). Estimates of a parameter from multiple QSARs and extrapolations from toxicity tests may be combined by weighting evidence and averaging (Thomas, Bicherel, and Bauer 2019).

Mathematical simulation models use mechanistic understanding to estimate system attributes. Their most common use in environmental assessment is estimating exposure given attributes of sources and receiving environments (Di Guardo et al. 2018). Mathematical simulations are also useful for estimating population or community responses to exposures from organism-level effects. For example, fish populations and honeybee colonies responses have been estimated from organism-level responses of fish and bees (Preuss et al. 2022; Barnthouse, Suter, and Rosen 1990). Such models can be more relevant than tests or observations, but their reliability is often doubted. To counter this mistrust of models, the properties of the models may be weighted to determine the appropriate degree of confidence in the model results as evidence. This weight-of-evidence process may be used during model development or to weigh an existing model as evidence. An approach for weighting population models as evidence is provided by Spromberg et al. (2025).

6.2.6 Expert Elicitation

When evidence is insufficient or unobtainable, expert elicitation may be used to substitute for empirical evidence. It may be a consensus-building exercise in which experts are brought together and express their consensus opinions after discussion. Alternatively, experts may express their opinions independently, and then you synthesize them. It is important to identify the credentials of the selected experts and their relevance to the assessment so that you know how much weight their opinions should bear. In some cases, local stakeholders may be the most relevant experts.

It is important to distinguish whether you are eliciting expert opinions or evidence from the experience of the experts. For example, to estimate the size of buffer zones needed to avoid disturbance of nest sites of certain raptor species, I (Suter) sent questionnaires to investigators who had studied the reproduction of those species (Suter and Jones 1981). I asked them how close they could get before the birds flushed from their nests. Hence, I was eliciting unpublished observations rather than their opinions on buffer zones. Since then, formal methods for expert elicitation have proliferated. Structured methods of elicitation have been recommended that include weighting the evidence from expert responses (Krewski et al. 2022; Cooke 2015; Slottje, Sluijs, and Knol 2008). Courts of law also rely on expert elicitation to identify, weight, and weigh scientific evidence.

6.2.7 Indigenous Knowledge

Elicitation of Indigenous knowledge is a variant of expert elicitation, but it is treated differently. Indigenous peoples are experts based on having been users and stewards of natural resources for millennia. They have a particular interest in management decisions. Their values, attitudes, and knowledge differ from those of other people, and because of their use of natural resources for food, drinking water, and cultural purposes, protective benchmark values for a general population may not be protective of them and their uses. In addition, because of cultural differences, standard information elicitation techniques may be ineffective and even offensive. In the United States, guidance is available from different agencies and organizations (USEPA 2011b), and an Indigenous liaison can be very helpful. When I (Suter) visited Alaskan Indigenous villages in the course of an assessment, I was accompanied by an Indigenous USEPA employee to keep me from being inadvertently offensive. Indigenous environmental experts ranged from a woman who had been fishing and gathering for over 80 years and spoke no English to a young man with a degree in fish and wildlife management but still a distinctly Indigenous attitude. Some Indigenous knowledge is appropriate to weight and integrate with the knowledge of scientific experts, and some is not, particularly knowledge based on traditional values (Chapman 2007). Some tribes have data from their own environmental monitoring programs that comply with the same quality standards as those of state agencies. Their data can be treated like any other scientific information.

6.2.8 Machine Learning and Artificial Intelligence

Machine learning is a subset of artificial intelligence that learns from data and improves over time without being explicitly reprogrammed. It is being explored as an alternative to statistically derived empirical models such as QSARs (Gasser et al. 2024). Artificial intelligence, which is a broader field, may have uses that are not determined at this time. These potential tools require very large datasets, which limit their applicability. Also, they do not have the background knowledge that a human ecological assessor has. However, in the form of large language models, they may be acquiring it. Sometimes their responses are spot on, and at other times they are so erroneous that their results are called hallucinations. Artificial intelligence is highly dependent on the query format and the quality of information extracted by the algorithm. The old modeling adage still applies: garbage in, garbage out.

6.3 TYPES OF EVIDENCE

It is important to organize your evidence in terms of a typology that may be conceptualized as ontologies and expressed as evidence maps. That will help you get a handle on how many of the various types of evidence you have and what types are missing. Types of evidence may be defined broadly (acute toxicity tests) or narrowly (the OECD standard 96-hour LC50 for fathead minnows).

Types of evidence have three aspects. The following aspects are defined for an example hypothesis that toxicity is causing the loss of fish species.

- **The Type of Study**: There are conventional study types or methods such as single-chemical acute lethality tests, surveys of fish pathologies or specific standard methods (Box 6.2).
- **The Assessment Endpoint Addressed**: Fish species richness in a stream.
- **The Implication of the Study**: Acute lethality of a chemical to fish could cause the loss of a species in a contaminated stream reach.

BOX 6.2 METHODS AS TYPES

In most assessments, types of evidence are derived from multiple applications of a method. The type is weighted based on its role in an assessment. The weight may be different for evidence used to determine the cause of an impairment in the field versus evidence used in the hazard assessment for a chemical. In some cases, the evaluation is of a method per se rather than the utility of evidence from the method in a particular case. The method may be assessed to determine whether it is appropriate for a particular regulatory purpose. Should it replace a current standard method? Is it a useful addition to an institution's set of standard methods?

Properties of proposed new methods or seldom used methods may be weighted to determine confidence in them as generators of evidence rather than in the particular evidence generated by the method. Van Der Zalm et al. (2022) proposed a framework that evaluates methods for five properties: fitness for purpose, human biological relevance, technical characterization, data integrity and transparency, and independent review. Rattner et al. (2024) evaluated methods for toxicological assessment with respect to three properties: relevance, reliability, and readiness.

If a method has been weighted and weighed, assessors who are using results from its use as a type of evidence may refer back to a prior weighting of the method for guidance.

Consideration of these aspects of evidence ensures that the pieces of evidence within a type are equivalent. If they are, you can aggregate them and weigh them in aggregate to determine the weight of evidence provided by the type to the body of evidence.

In human health risk assessments for chemical hazards, typically, the primary division of types is between direct evidence and indirect (supplemental) evidence (Thayer, Angrish, et al. 2022). Direct evidence may include epidemiological studies or animal tests that are sufficiently relevant to the exposures and effects of concern. Indirect evidence may include pharmacokinetic and toxicokinetic information, animal studies that are not sufficiently relevant for direct evidence, information on susceptible human populations, exposure studies with no effects, mixture models, and case reports.

Specific types of assessments require particular types of evidence. For example, an assessment of site remediation by natural attenuation used three types of evidence (Tullis 2008). Primary evidence included historical groundwater or soil analyses that showed decreasing, increasing, or stable concentrations. Secondary evidence included site data that determined the type and rate of attenuation processes. Tertiary evidence included data from field or mesocosm studies demonstrating microbial activity.

The types of evidence that are available can determine which assessment approaches are applicable. For example, if you have a biotic community assessment endpoint but have only organism-level toxicity data, the assessment tool could be a species sensitivity distribution. However, if you have monitoring data for exposed biotic communities, multiple metrics and weight-of-evidence approaches are available (Rode et al. 2025).

In ecological assessments, the initial division is usually laboratory versus field. Observational versus experimental is also an essential division. Finally, evidence types may directly address the assessment endpoint and conditions that comprise the hypothesis or may relate to mechanisms of exposure or effects. Evidence from each of these broad types has advantages and disadvantages for an assessment that can be complementary.

- Laboratories lend themselves to well-designed experiments with nearly identical replicates of treatments and precise quantification of effects. Replication and randomization assure that differences in effects are caused by the treatments. However, the relevance and reliability may be low.
- Field observational studies directly engage with the real world, so they are realistic. When they are performed for a contaminated or disturbed site assessment, they are completely relevant. However, the observed relationships are not necessarily causal due to confounding. That is, the lack of randomization and replication of the cause and effect (i.e., an effluent and an invertebrate community) leaves these studies open to false associations or multiple associations. Unless the confounding is addressed, it can make the evidence highly unreliable. Multi-site studies, such as a statewide biological survey, are replicated but not randomized. Therefore, they reveal relationships that are more likely to be causal than those from single-site studies, but they can still be confounded.
- Field experimental studies include mesocosms, such as constructed streams or ponds, organisms confined to a location with a particular exposure, such as fish caged above and below an outfall or plants placed at locations with different levels of soil contamination, or sites that are treated with the agent of interest, such as terrestrial plots treated with different pesticide doses. These studies fall between laboratory experiments and field observations in terms of relevance.
- Empirical models represent relationships or processes by statistically analyzing data. Obvious examples in assessments are exposure-response relationships or relationships between distance from a source and contaminant concentrations.
- Simulation models use mechanistic understanding of the structure and function of a system to create a mathematical representation. In environmental assessments, simulation models are largely limited to estimating the transport and fate of hazardous agents. However, in ecological assessments, they may also be used to estimate the population- or community-level consequences of organism-level effects.

New approach methods (i.e., toxicological methods that do not use whole-animal tests) include in vitro, in chemico, and in silico techniques. In vitro methods are used to generate genomics, metabolomics, proteomics, or transcriptomics data. Weight-of-evidence approaches have been developed for the use of those data in weight of evidence (Cecchetto et al. 2023; Bridges et al. 2017; Sauer et al. 2017). In silico (i.e., computational) techniques range from empirical QSARs to models that simulate chemical effects based on the structural interactions of chemicals with biological receptor proteins (Thomas, Bicherel, and Bauer 2019). The USEPA's Comptox program's in silico system is SeqAPASS (USEPA 2024c). In general, they will have less weight in an assessment than the whole-animal tests that they substitute for. However, they may be better than a low-relevance or low-reliability conventional test.

Sub-organismal analyses of field samples, known as biomarkers, are used in observational studies of contaminated or impaired sites. They can be useful in causal or risk assessments to demonstrate that organisms have been exposed to a particular agent or that effects are related to a particular mode of action or to a particular causal agent. They are increasingly important contributors to weight-of-evidence-based assessments (Hook, Gallagher, and Batley 2014).

6.4 DATA ANALYSES

You perform data analyses when primary data are generated for your assessment or when secondary data have not been appropriately analyzed for your assessment. Data in itself is not evidence, and the results of irrelevant data analyses are not evidence for your hypothesis. Inappropriate analyses are weaker evidence than the data could potentially provide. The choice of data analysis method depends on the hypothesis, the nature of the data, and the potential implication of the evidence to be generated. However, some generalizations are possible.

- The most generally useful result of an analysis is a statistical model of the relationship between two or more variables. Such models identify potentially causal relationships which are the foundation of any environmental assessment. Examples include exposure-response relationships and relationships between contaminants of concern and waste discharges.
- Estimates of values are obviously useful, but they lack the power of a potentially causal relationship. Examples include concentrations in media or tissues, numbers of species at a site, and contaminated area. Confidence intervals for estimates should be provided.
- The least useful analyses are statistical hypothesis tests. They tell you nothing about biological significance or the magnitude and nature of effects (Suter 1996; 2023). They are also biased in favor of the null hypothesis, which, in environmental assessments, is a bias against finding adverse effects. Relative risks and odds ratios are more defensible for conveying importance.

In addition to changing the analysis method, reanalysis can involve editing the original dataset. For example, you might analyze only data with a circumneutral pH, only aquatic insects, or only oral exposures.

Reanalysis of secondary data is often valuable, in part because so many results in the literature are expressions of statistical significance and in part to catch errors. Reanalysis is more likely now that many journals and reports provide the primary data. The desire for transparency has caused many governmental agencies to provide internet access to monitoring data (e.g., state Clean Water Act data) and research data (e.g., the USEPA Enterprise Data Catalog) (https://www.epa.gov/data/where-find-epa-data).

6.5 IMPLICATIONS OF EVIDENCE

Considering the implications of information determines whether you should use the information, how much weight it should be assigned, and how it contributes to a sufficient body of evidence (Figure 5.2). Implications have two components: (1) an attribute of the information that relates it to the hypothesis and (2) a determination that it strengthens or weakens the hypothesis.

BOX 6.3 IMPLICATIONS AND A THEORY OF EVIDENCE

The need for evidence to have implications for a hypothesis is implicit in Achinstein's (2001) theory of evidence. He defines evidence by the following condition: "for a fact *e* to be evidence for a hypothesis *h, e* must provide a good reason to believe *h*. Good reasons to believe must include implications for the truth or falsity of the hypothesis."

6.5.1 Deriving Implications

Implications may be identified by examining the information and asking yourself, "What, if anything, does this tell me that might support or diminish confidence in the hypothesis?" In most cases, it will be evident whether a particular result of a study has implications. For example, in assessments of the hypothesis that a chemical causes cancer in humans, any mammalian carcinogenicity test of the chemical has implications. However, two heuristic tools may aid in identifying implications: characteristics of the type of hypothesis being assessed and causal networks.

6.5.1.1 Characteristics

It is advantageous to use consistent and logical sets of implications, such as characteristics of the type of hypothesis being assessed. For example, when weighing evidence for a hypothesized cause of an environmental impairment, the implication could be that the information provides evidence for one of the characteristics of causation (Section 14.2, Appendix B). These characteristics of evidence are particularly useful because they explicitly identify the relevance of the evidence to the causal relationship. Associations of the implications of information with characteristics are also useful when weighting evidence and weighing bodies of evidence (Chapters 7 and 8).

6.5.1.2 Causal Networks

If conceptual models or other causal networks were created in a planning process and used to determine what information should be found or generated (Section 6.1.2), they may then be used to define the implications of that information. For example, if the conceptual model indicates that exposure of bass to mercury occurs due to dietary bioaccumulation, then information concerning concentrations in prey organisms can imply that the hypothesized route of exposure has or has not occurred. Therefore, it is evidence concerning risks to bass and to bass-eating wildlife and humans.

The IRIS handbook provides an explanation of the use of adverse outcome pathways, a type of toxicological causal pathway model, in assessments of substances (USEPA 2022).

Adverse outcome pathways have become functional and versatile tools for use in the risk assessment process. They organize the sequential connections of empirically measured key events between a single molecular initiating event and an adverse outcome. They are focused on the molecular initiating event rather than exposure to a specific chemical, although they establish, when possible, a quantitative understanding of the key event relationships that describe the progression from one key event to the next (Ankley et al. 2010). They provide a clear graphical representation and integrative construct for organizing the more complex relationships and associations described in a mechanism of action. Thus, the outcomes of other exposures with similar molecular initiating events or key events can be predicted from measurable upstream events. Approaches are available for using weight of evidence to develop or evaluate adverse outcome pathways in ecotoxicological assessments (Fay et al. 2017; Collier et al. 2016; Becker et al. 2015; OECD 2013, 2018).

6.5.2 Particular Uses of Implications

Implications are particularly important in inference by analogy because not all analogies are relevant. For example, consider the extrapolation by analogy between a well-characterized chemical and a poorly characterized chemical in the read-across method (Figure 20.1) (Lizarraga et al. 2022; Suter and Lizarraga 2022). When weighing evidence in read-across, it is common to collect all information that is available for both chemicals concerning structural, physicochemical, distribution, metabolism, excretion, toxicity, or mechanism attributes. However, it is important to consider the chemical property of interest and distinguish information that has clear implications for that property from information that does not. If the property is carcinogenicity, information concerning mutagenicity has clear implications because mutations are a common mechanism of carcinogenicity. If the information is vapor pressure, that is a physicochemical property, but it has no clear implication for carcinogenicity. Giving weight to vapor pressure would reduce the quality of the analogy.

Implications are also an important issue when using biomarkers as evidence. Biomarkers are measurable substances in organisms that indicate an abnormal condition and may be indicative of exposure to an agent. They are not effects of concern and are not measures of exposure, such as internal concentrations. Therefore, if they are to contribute to an inference concerning environmental risks, their implications must be identified. For example, a biomarker may be an enzyme that is induced by the agent of interest, or it may be a metabolite of that agent.

Considerations also appear in the legal weighing of evidence. For example, the considerations for weighing evidence to determine whether a person is disabled include five considerations that support greater weight and eight that support lesser weight (20 CFR 220.14 – Weighing of Evidence). For example, evidence that implies "inconsistency between reports of pain and functional impact" has negative weight with respect to the contention that a person is disabled by pain from an injury.

One approach to implications is termed hypothesis-based weight of evidence (Rhomberg 2015). It establishes a detailed hypothesis, such as "chemical x causes cancer in humans by a particular mode of action" (Merrington, Nowell, and Peck 2024).

6.6 SCREENING EVIDENCE

Information is screened to determine whether it should be retained as evidence. Screening is usually based on its reliability (quality), but it may also consider relevance (Hulzebos et al. 2010). In addition, under REACH, evidence is screened for the validity of methods and regulatory need (ECHA 2022). In general, relevance and reliability are accounted for by weighting the evidence (Section 7.1.2). However, some information is not worth retaining to be weighted (Ågerstrand, Breitholtz, and Rudén 2011). Studies with no controls or reference sites, for example, are not worth considering. Information from secondary sources is also screened out in most systems. Screening criteria are often specified in regulatory systems. For example, many agencies specify detailed standards for the acceptance of species sensitivity distributions used to derive water quality benchmarks (Belanger et al. 2017).

Screening information from the literature is an important component of systematic reviews (Section 6.2.1), but it is also applicable to other information sources, including your own studies. Many systems are available for assessing and weighting relevance and reliability (Section 7.1). For example, the CRED system for evaluating ecotoxicity data and the CREED system for evaluating exposure data include relevance as well as reliability (Moermond et al. 2016; SETAC 2024; Bertram et al. 2025). Some screening systems are devoted to the completeness of recording the study procedures (e.g., is the acclimation and feeding protocol reported?) rather than to the weight of the evidence (Hanson et al. 2017).

Some systems substitute a detailed screening process for weighting the evidence. For example, the CREED system evaluates exposure data with respect to six prescreening criteria and then evaluates relevance with respect to 11 criteria and reliability with respect to 19 criteria (Merrington, Nowell, and Peck 2024; SETAC 2024; Hladik et al. 2024). Each study or dataset is evaluated with respect to those criteria and rated as fully met, partly met, not met, inappropriate, not reported, or not applicable. Those ratings are integrated to obtain an overall rating for relevance and reliability:

- Reliable or relevant without restrictions,
- Reliable or relevant with restrictions,
- Not reliable or relevant, or
- Not assignable.

This screening process is similar to weight-of-evidence processes, but its purpose is to determine whether a study or dataset may be used or should be used with qualifiers. It does not determine whether the data should be influential in the inference.

This example illustrates the ambiguity of the distinction between screening and weighting. Consider whether you are applying a screening criterion or assigning a weight.

Most of these screening criteria address technical considerations. However, for regulatory assessments, it is also important to screen for suitability, also termed "regulatory suitability," "regulatory need," or "fitness for purpose" (Franklin et al. 2019; ECHA 2022). It is a property that determines whether a piece of evidence is suitable for a particular case or class of cases, such as the derivation of an air quality standard (Section 14.5). Regulations and guidelines provide more specific definitions of goals that indicate what constitutes regulatory suitability. For example, a physiological response suitable for causal assessments (USEPA 2000) may not be suitable for deriving water quality criteria for aquatic life (Stephan et al. 1985). The USEPA water quality criteria use only population-relevant responses: survival, growth, and reproduction. Other endpoints are screened out. Further, some limitations are placed on exposure chemistry, such as tests with extreme salinity, pH, or temperature. Similarly, some human health assessments screen out nonmammalian data.

The criteria used to accept or eliminate studies can be a source of controversy because they can determine the outcome of assessments. For example, a strict systematic review of information concerning the effects of atrazine on fish and amphibians by Rohr and McCoy (2010) found consistent effects at relevant concentrations in the accepted studies. They stated that the finding of no relevant effects by Solomon et al. (2008) was due to the acceptance of studies without defined inclusion criteria. In turn, Rohr and McCoy were accused of eliminating papers based on statistical criteria that were too strict (Van Der Kraak et al. 2014). Ideally, standard criteria for acceptance and elimination (e.g., Table 6.1) should be used, or case-specific acceptance criteria should be defined a priori.

For the sake of transparency, it is valuable to provide an appendix to your assessment that lists studies and endpoints that have been screened out, along with the reason for their elimination. This can minimize the sort of contention that occurred with respect to competing atrazine assessments.

6.7 SUFFICIENCY OF EVIDENCE

Having assembled the evidence, you should consider whether it is sufficient to proceed with a weight-of-evidence-based assessment. Information gaps may be filled by techniques such as QSARs and read-across, or new conventional studies may be conducted. If there are no evidence standards or no opportunity to conduct new studies, you must proceed with whatever you have. Otherwise, the application of criteria for the sufficiency of the evidence should be demonstrated.

Many regulatory organizations have explicit criteria for sufficiency. For example, the U.S. Food, Drug, and Cosmetics Act requires that applicants submit sufficient evidence of the effectiveness and safety of drugs. Originally, the Food and Drug Administration (FDA) interpreted that as two clinical trials conducted by experts qualified by scientific training and experience. However, the FDA now accepts applications with only one human trial if a sufficient weight of evidence is provided by related studies (Cormier 2011).

Exceptions are often possible. For example, the USEPA's ambient water quality criteria guidelines have explicit and demanding requirements for sufficient evidence (Stephan et al. 1985), but they allow criteria to be set without the standard dataset when necessary to protect endangered, commercially important, or otherwise important species. Further, the Clean Water Act regulations allow for "other scientifically defensible methods" and not just the standard numerical guidelines (Chapter 15). However, USEPA management has been reluctant to use nonstandard methods. For example, 6PPD is highly toxic to coho salmon, causing mass mortalities. That is clearly a sensitive and important effect that is observed in the field and supported by laboratory data for that species. Nevertheless, the Agency derived only acute screening values for 6PPD and 6PPDq, because the standard data requirements were not met (USEPA 2024a).

REFERENCES

Achinstein, P. 2001. *The Book of Evidence*. Oxford: Oxford University Press.

Achinstein, P., ed. 2005. *Scientific Evidence, Philosophical Theories and Applications*. Baltimore, MD: The Johns Hopkins University Press.

Ågerstrand, M., M. Breitholtz, and C. Rudén. 2011. "Comparison of four different methods for reliability evaluation of ecotoxicity data: A case study of non-standard test data used in environmental risk assessments of pharmaceutical substances." *Environmental Sciences Europe* 23: 1–15.

Ankley, G. T., R. S. Bennett, R. J. Erickson, D. J. Hoff, M. W. Hornung, R. D. Johnson, D. R. Mount, J. W. Nichols, C. L. Russom, and P. K. Schmieder. 2010. "Adverse outcome pathways: A conceptual framework to support ecotoxicology research and risk assessment." *Environmental Toxicology and Chemistry* 29 (3): 730–741.

ANZG. 2018. *Weight of Evidence*. Australian and New Zealand Governments. Canberra.

Barnthouse, L. W., G. W. Suter, and A. E. Rosen. 1990. "Risks of toxic contaminants to exploited fish populations: Influence of life history, data uncertainty, and exploitation intensity." *Environmental Toxicology and Chemistry* 9: 297–311.

Becker, R. A., G. T. Ankley, S. W. Edwards, S. W. Kennedy, I. Linkov, B. Meek, M. Sachana, H. Segner, B. Van Der Burg, D. L. Villeneuve, H. Watanabe, and T. S. Barton-Maclaren. 2015. "Increasing scientific confidence in adverse outcome pathways: Application of tailored Bradford-Hill considerations for evaluating weight of evidence." *Regulatory Toxicology and Pharmacology* 72 (3): 514–537.

Belanger, S., M. Barron, P. Craig, S. Dyer, M. Galay-Burgos, M. Hamer, S. Marshall, L. Posthuma, S. Raimondo, and P. Whitehouse. 2017. "Future needs and recommendations in the development of species sensitivity distributions: Estimating toxicity thresholds for aquatic ecological communities and assessing impacts of chemical exposures." *Integrated Environmental Assessment and Management* 13 (4): 664–674.

Bequignon, O. J. M., B. J. Bongers, W. Jespers, A. P. IJzerman, B. van der Water, and G. J. P. van Westen. 2023. "Papyrus: A large-scale curated dataset aimed at bioactivity predictions." *Journal of Cheminformatics* 15 (1): 3.

Bertram, M. G., M. Ågerstrand, E. S. J. Thoré, et al. 2025. "EthoCRED: A framework to guide reporting and evaluation of the relevance and reliability of behavioural ecotoxicity studies." *Biological Reviews* 100 (2): 556–585.

Bridges, J., U. G. Sauer, R. Buesen, L. Deferme, K. E. Tollefsen, T. Tralau, B. van Ravenzwaay, A. Poole, and M. Pemberton. 2017. "Framework for the quantitative weight-of-evidence analysis of 'omics data for regulatory purposes." *Regulatory Toxicology and Pharmacology* 91: S46–S60.

Buist, H., T. Aldenberg, M. Batke, S. Escher, R. K. Entink, R. Kühne, H. Marquart, E. Pauné, E. Rorije, and G. Schüürmann. 2013. "The OSIRIS weight of evidence approach: ITS mutagenicity and ITS carcinogenicity." *Regulatory Toxicology and Pharmacology* 67 (2): 170–181.

Carnap, R. 1947. "On the application of inductive logic." *Philosophy and Phenomenological Research* 8 (1): 133–148.

Cecchetto, M., L. Peruzza, E. Giubilato, I. Bernardini, G. D. Rovere, A. Marcomini, F. Regoli, L. Bargelloni, T. Patarnello, and E. Semenzin. 2023. "An innovative index to incorporate transcriptomic data into weight of evidence approaches for environmental risk assessment." *Environmental Research* 227: 115745.

CEE. 2018. Guidelines for Systematic Reviews in Environmental Management. Version 5.0. Collaboration for Environmental Evidence. www.environmentalevidence.org/information-for-authors.

Chapman, P. M. 2007. "Traditional ecological knowledge (TEK) and scientific weight of evidence determination." *Marine Pollution Bulletin* 54: 1839–1840.

Chinen, K., and T. Malloy. 2022. "Multi-strategy assessment of different uses of QSAR under REACH analysis of alternatives to advance information transparency." *International Journal of Environmental Research and Public Health* 19 (7): 4338.

Collier, Z. A., K. A. Gust, B. Gonzalez-Morales, P. Gong, M. S. Wilbanks, I. Linkov, and E. J. Perkins. 2016. "A weight of evidence assessment approach for adverse outcome pathways." *Regulatory Toxicology and Pharmacology* 75: 46–57.

Collins, A. M., D. Coughlin, J. Miller, and S. Kirk. 2015. *The Production of Quick Scoping Reviews and Rapid Evidence Assessments, A How to Guide*. London: Joint Water Evidence Group, UK Department for Environment, Food, and Rural Affairs.

Cooke, R. M. 2015. "The aggregation of expert judgment: Do good things come to those who weight?" *Risk Analysis* 35 (1): 12–15.

Cormier, J. W. 2011. "Advancing FDA's regulatory science through weight of evidence evaluations." *Journal of Contemporary Health Law and Policy* 28: 1–22.

Di Guardo, A., T. Gouin, M. MacLeod, and M. Scheringer. 2018. "Environmental fate and exposure models: Advances and challenges in 21 st century chemical risk assessment." *Environmental Science: Processes and Impacts* 20 (1): 58–71.

Douziech, M., S. A. Oginah, L. Golsteijn, M. Z. Hauschild, O. Jolliet, M. Owsianiak, L. Posthuma, and P. Fantke. 2024. "Characterizing freshwater ecotoxicity of more than 9000 chemicals by combining different levels of available measured test data with in silico predictions." *Environmental Toxicology and Chemistry* 43 (8): 1914–1927.

ECHA. 2022. *Guidance on Information Requirements and Chemical Safety Assessment, Part B*. Helsinki: European Chemicals Agency. https://echa.europa.eu/guidance-documents/guidance-on-information-requirements-and-chemical-safety-assessment.

EFSA. 2010. "Application of systematic review methodology to food and feed safety assessments to support decision making." *EFSA Journal* 8 (6): 1–90.

Fay, K. A., D. L. Villeneuve, C. A. LaLone, Y. Song, K. E. Tollefsen, and G. T. Ankley. 2017. "Practical approaches to adverse outcome pathway development and weight-of-evidence evaluation as illustrated by ecotoxicological case studies." *Environmental Toxicology and Chemistry* 36 (6): 1429–1449.

Feist, B. E., E. R. Buhle, D. H. Baldwin, J. A. Spromberg, S. E. Damm, J. W. Davis, and N. L. Scholz. 2017. "Roads to ruin: conservation threats to a sentinel species across an urban gradient." *Ecological Applications* 27 (8): 2382–2396. https://doi.org/10.1002/eap.1615.

Franklin, P., R. Stoffels, J. Clapcott, D. Booker, A. Wagenhoff, and C. W. Hickey. 2019. *Deriving Potential Fine Sediment Attribute Thresholds for the National Objectives Framework*. Hamilton, NZ: National Institute of Water & Atmospheric Research. https://environment.govt.nz/assets/Publications/Files/deriving-potential-fine-sediment-attribute-thresholds-for-the-national-objectives-framework.pdf.

Garritty, C., G. Gartlehner, B. Nussbaumer-Streit, V. J. King, C. Hamel, C. Kamel, L. Affengruber, and A. Stevens. 2021. "Cochrane Rapid Reviews Methods Group offers evidence-informed guidance to conduct rapid reviews." *Journal of Clinical Epidemiology* 130: 13–22.

Gasser, L., C. Schür, F. Perez-Cruz, K. Schirmer, and M. Baity-Jesi. 2024. "Machine learning-based prediction of fish acute mortality: Implementation, interpretation, and regulatory relevance." *Environmental Science: Advances* 3 (8): 1124–1138.

Good, I. J. 1966. "On the principle of total evidence." *British Journal of Philosophy of Science* 17 (4): 319–321.

Guzelian, P. S., M. S. Victoroff, N. C. Halmes, R. C. James, and C. P. Guzelian. 2005. "Evidence-based toxicology: A comprehensive framework for causation." *Human & Experimental Toxicology* 24: 161–201.

Hanson, M. L., B. A. Wolff, J. W. Green, M. Kivi, G. H. Panter, M. St. J. Warne, M. Ågerstrand, and J. P. Sumpter. 2017. "How we can make ecotoxicology more valuable to environmental protection." *Science of the Total Environment* 578: 228–235.

Harris, C. A., A. P. Scott, A. C. Johnson, G. H. Panter, D. Sheahan, M. Roberts, and J. P Sumpter. 2014. "Principles of sound ecotoxicology." *Environmental Science and Technology* 48 (6): 3100–3111.

Hartung, T. 2010. "Evidence-based toxicology: The toolbox of validation for the 21st century?" *ALTEX* 27: 253–263.

Hladik, M. L., A. Markus, D. Helsel, L. H. Nowell, S. Polesello, H. Rüdel, D. Szabo, and I. Wilson. 2024. "Evaluating the reliability of environmental concentration data to characterize exposure in environmental risk assessments." *Integrated Environmental Assessment and Management* 20: 981–1003.

Hlisnikova, H. 2024. "SciRAP - Science in Risk Assessment and Policy." Karolinska Institute. Accessed12/02/2024.https://ki.se/en/imm/scirap-science-in-risk-assessment-and-policy.

Hook, S. E., E. P. Gallagher, and G. E. Batley. 2014. "The role of biomarkers in the assessment of aquatic ecosystem health." *Integrated Environmental Assessment and Management* 10 (3): 327–341.

Hulzebos, E., S. Gunnarsdottir, J.-P. Rila, Z. Dang, and E. Rorije. 2010. "An integrated assessment scheme for assessing the adequacy of (eco)toxicological data under REACH." *Toxicology Letters* 198: 255–262.

Ioannidis, J. P. A. 2016. "The mass production of redundant, misleading, and conflicted systematic reviews and meta-analyses." *The Milbank Quarterly* 94 (3): 485–514.

James, K. L., N. P. Randall, and N. R. Haddaway. 2016. "A methodology for systematic mapping in environmental sciences." *Environmental Evidence* 5: 1–13.

Keisler, J. M., Z. A. Collier, E. Chu, N. Sinatra, and I. Linkov. 2014. "Value of information analysis: The state of application." *Environment Systems and Decisions* 34 (1): 3–23.

Keshava, C., S. Nicolai, S. V. Vulimiri, F. A. Cruz, N. Ghoreishi, S. Knueppel, A. Lenzner, P. Tarnow, J. T. Vanselow, B. Schulz, A. Persad, N. Baker, K. A. Thayer, A. J. Williams, and R. Pirow. 2023. "Application of systematic evidence mapping to identify available data on the potential human health hazards of selected market-relevant azo dyes." *Environment International* 176: 107952.

Krewski, D., P. Saunders-Hastings, R. A. Baan, T. S. Barton-Maclaren, P. Browne, W. A. Chiu, M. Gwinn, T. Hartung, A. D. Kraft, and J. Lam. 2022. "Development of an evidence-based risk assessment framework." *Altex* 39 (4): 667.

Lambert, J. C., J. Brown, H. Gong, C. Kilburn, J. Krysa, B. Kuntzelman, J. Lee, A. Luke, J. Powell, and A. Rashid. 2025. "RapidTox: A decision-support workflow to inform rapid toxicity and human health assessment." *Computational Toxicology*: 100377.

Lilavois, C. R., S. A. Nelson, and S. Raimondo. 2024. *Web-Based Interspecies Correlation Estimation (Web-ICE) for Acute Toxicity: User Manual. Version 4.0.* Gulf Ecology Division, Office of Research and Development, Gulf Breeze, FL: U.S. Environmental Protection Agency.

Livoreil, B., J. Glanville, N. R. Haddaway, H. Bayliss, A. Bethel, F. Flamerie de Lachapelle, S. Robalino, S. Savilaakso, W. Zhou, G. Petrokofsky, and G. Frampton. 2017. "Systematic searching for environmental evidence using multiple tools and sources." *Environmental Evidence* 6 (1): 23.

Lizarraga, L. E., G. W. Suter, J. C. Lambert, G. Patlewicz, J. Q. Zhao, J. L. Dean, and P. Kaiser. 2022. "Advancing the science of a read-across framework for evaluation of data-poor chemicals incorporating systematic and new approach methods." *Regulatory Toxicology and Pharmacology* 137: 105293.

Martin, O., J. Adams, A. Beasley, S. E. Belanger, R. Breton, T. Brock, V. Buonsante, M. Burgos, J. Green, and P. Guiney. 2019. "Improving environmental risk assessments of chemicals: Steps towards evidence-based ecotoxicology." *Environment International* 128: 210–217.

Melcher, A. H., T. H. Bakken, T. Friedrich, F. Greimel, N. Humer, S. Schmutz, B. Zeiringer, and J. A. Webb. 2017. "Drawing together multiple lines of evidence from assessment studies of hydropeaking pressures in impacted rivers." *Freshwater Science* 36 (1): 220–230.

Merrington, G., L. H. Nowell, and C. Peck. 2024. "An introduction to Criteria for Reporting and Evaluating Exposure Datasets (CREED) for use in environmental assessments." *Integrated Environmental Assessment and Management* 20 (4): 975–980.

Mineau, P., and H. Kern. 2023. "Neonicotinoid insecticides: Failing to come to grips with a predictable environmental disaster." *American Bird Conservancy*. Washington, DC.

Moermond, C. T., R. Kase, M. Korkaric, and M. Ågerstrand. 2016. "CRED: Criteria for reporting and evaluating ecotoxicity data." *Environmental Toxicology and Chemistry* 35 (5): 1297–1309.

Moher, D., L. Shamseer, M. Clarke, D. Ghersi, A. Liberati, M. Petticrew, P. Shekelle, L. A. Stewart, and Prisma-P Group. 2015. "Preferred reporting items for systematic review and meta-analysis protocols (PRISMA-P) 2015 statement." *Systematic Reviews* 4: 1–9.

Moons, P., E. Goossens, and D. R. Thompson. 2021. "Rapid reviews: The pros and cons of an accelerated review process." *European Journal of Cardiovascular Nursing* 20 (5): 515–519.

Morgan, R. L., P. Whaley, K. A. Thayer, and H. J. Schünemann. 2018. "Identifying the PECO: A framework for formulating good questions to explore the association of environmental and other exposures with health outcomes." *Environment International* 121 (Pt 1): 1027.

Norton, S. B., and C. L. Schofield. 2018. "Conceptual model diagrams as evidence scaffolds for environmental assessment and management." *Freshwater Science* 36 (1): 231–239.

NRC. 2011. *Guidance for the Description of Animal Research in Scientific Publications*. Washington, DC: National Research Council, National Academies Press.

OECD. 2013. *Guidance Document on Developing and Assessing Adverse Outcome Pathways. Series on Testing and Assessment No. 184*. Paris: Organization for Economic Cooperation and Deeloment.

OECD. 2018. "Users' handbook supplement to the guidance document for developing and assessing adverse outcome pathways." *OECD Series on Adverse Outcome Pathways*. Paris.

OECD. 2024. *eChem Portal*. Paris: Organization for Economic Cooperation and Development. https://www.echemportal.org/echemportal/.

OECD. 2025. *OECD Guidance Document on the Generation, Reporting and Use of Research Data for Regulatory Assessments*. Paris: Organization for Economic Cooperation and Development, Environment Directorate, Chemicals and Biotechnology Committee.

OHAT. 2019. *Handbook for Conducting a Literature-Based Health Assessment Using OHAT Approach for Systematic Review and Evidence Integration*. Research Triangle Park, NC: Office of Health Assessment and Translation, National Toxicology Program.

Orr, J. A., S. J. Macaulay, A. Mordente, B. Burgess, D. Albini, J. G. Hunn, K. Restrepo-Sulez, R. Wilson, A. Schechner, and A. M. Robertson. 2024. "Studying interactions among anthropogenic stressors in freshwater ecosystems: A systematic review of 2396 multiple-stressor experiments." *Ecology Letters* 27 (6): e14463.

Pearson, H. 2024. "'Evidence bank' could help scientists to tackle climate change." *Nature* 630: 540–541.
Preuss, T. G., A. Agatz, B. Goussen, V. Roeben, J. Rumkee, L. Zakharova, and P. Thorbek. 2022. "The BEEHAVEecotox model—Integrating a mechanistic effect module into the honeybee colony model." *Environmental Toxicology and Chemistry* 41 (11): 2870–2882.
Rattner, B. A., T. G. Bean, V. R. Beasley, P. Berny, K. M. Eisenreich, J. E. Elliott, M. L. Eng, P. C. Fuchsman, M. D. King, and R. Mateo. 2024. "Wildlife ecological risk assessment in the 21st century: Promising technologies to assess toxicological effects." *Integrated Environmental Assessment and Management* 20 (3): 725–748.
Rhomberg, L. 2015. "Hypothesis-based weight of evidence: An approach to assessing causation and its application to regulatory toxicology." *Risk Analysis* 35 (6): 1114–1124.
Rode, O., M. Mather, D. Oliver, K. Nelson, V. Reed, T. Moore, and S. Pratap. 2025. "A framework tool that applies weight-of-evidence integration to the analysis of existing datasets to guide freshwater conservation." *Frontiers in Freshwater Science* 3: 1520312.
Rohr, J. R., and K. A. McCoy. 2010. "A qualitative meta-analysis reveals consistent effects of atrazine on freshwater fish and amphibians." *Environmental Health Perspectives* 118 (1): 20–32. https://doi.org/10.1289/ehp.0901164.
Sauer, U. G., L. Deferme, L. Gribaldo, J. Hackermüller, T. Tralau, B. van Ravenzwaay, C. Yauk, A. Poole, W. Tong, and T. W. Gant. 2017. "The challenge of the application of 'omics technologies in chemicals risk assessment: Background and outlook." *Regulatory Toxicology and Pharmacology* 91: S14–S26.
Schünemann, H., J. Brożek, G. Guyatt, and A. Oxman. 2013. *GRADE Handbook*. GRADE Working Group. McMaster University, Canada.
SETAC. 2024. *Criteria for Reporting and Evaluating Exposure Data Sets (CREED).* Washington, DC: Society of Environmental Toxicology and Chemistry.
Slottje, P., J. P. van der Sluijs, and A. B. Knol. 2008. *Expert Elicitation: Methodological Suggestions for Its Use in Environmental Health Impact Assessments*. National Institute for Public Health and the Environment (RIVM), Bilthoven, The Netherlands.
Solomon, K. R., J. A. Carr, L. H. Du Preez, J. P. Giesy, R. J. Kendall, E. E. Smith, and G. J. Van Der Kraak. 2008. "Effects of atrazine on fish, amphibians, and aquatic reptiles: A critical review." *Critical Reviews in Toxicology* 38 (9): 721–772.
Spromberg, J. A., S. A. Hecht, C. A. Laetz, T. Hawkes, and D. H. Baldwin. 2025. "Evaluation process for matching population models to regulatory decisions regarding threatened or endangered species by considering model risk." *Integrated Environmental Assessment and Management* 21 (2): 384–395.
Stephan, C. E., D. I. Mount, D. J. Hanson, J. H. Gentile, G. A. Chapman, and W. A. Brungs. 1985. *Guidelines for Deriving Numeric National Water Quality Criteria for the Protection of Aquatic Organisms and Their Uses*. Washington, DC: U.S. Environmental Protection Agency.
Suter, G. W. 1996. "Abuse of hypothesis testing statistics in ecological risk assessment." *Human and Ecological Risk Assessment* 2: 331–349.
Suter, G. W. 2023. *Fundamentals of Environmental Assessment*. Boca Raton, FL: CRC Press.
Suter, G. W. and J. L. Jones. 1981. "Criteria for golden eagle, ferruginous hawk, and prairie falcon nest site protection." *Raptor Research* 15 (1): 12–18.
Suter, G. W., and T. P. O'Farrell. 2009. *Analysis of the Causes of a Decline in the San Joaquin Kit Fox Population on the Elk Hills, Naval Petroleum Reserve #1, California*. Cincinnati, OH: U.S. Environmental Protection Agency.
Suter, G. W., and L. E. Lizarraga. 2022. "Clearly weighing the evidence in read-across can improve assessments of data-poor chemicals." *Regulatory Toxicology and Pharmacology* 129: 105111.
Thayer, K. A., M. Angrish, X. Arzuaga, L. M. Carlson, A. Davis, L. Dishaw, I. Druwe, C. Gibbons, B. Glenn, and R. Jones. 2022. "Systematic evidence map (SEM) template: Report format and methods used for the US EPA Integrated Risk Information System

(IRIS) program, Provisional Peer Reviewed Toxicity Value (PPRTV) program, and other "fit for purpose" literature-based human health analyses." *Environment International* 169: 107468.

Thayer, K. A., R. M. Shaffer, M. Angrish, X. Arzuaga, L. M. Carlson, A. Davis, L. Dishaw, I. Druwe, C. Gibbons, and B. Glenn. 2022. "Use of systematic evidence maps within the US Environmental Protection Agency (EPA) Integrated Risk Information System (IRIS) program: Advancements to date and looking ahead." *Environment International* 169: 107363.

The Presidential/Congressional Commission on Risk Assessment and Risk Management. 1997. *Risk Assessment and Risk Management in Regulatory Decision-Making*. Washington, DC: Government Printing Office.

Thomas, P. C., P. Bicherel, and F. J. Bauer. 2019. "How in silico and QSAR approaches can increase confidence in environmental hazard and risk assessment." *Integrated Environmental Assessment and Management* 15 (1): 40–50.

Tian, Z., M. Gonzalez, C. A. Rideout, H. N. Zhao, X. Hu, J. Wetzel, E. Mudrock, C. A. James, J. K. McIntyre, and E. P. Kolodziej. 2022. "6PPD-Quinone: Revised toxicity assessment and quantification with a commercial standard." *Environmental Science and Technology Letters* 9 (2): 140–146.

Tian, Z., H. Zhao, K. T. Peter, et al. 2021. "A ubiquitous tire rubber–derived chemical induces acute mortality in coho salmon." *Science* 371 (6525): 185–189.

Toselli, G., and F. Sgolastra. 2020. "Seek and you shall find: An assessment of the influence of the analytical methodologies on pesticide occurrences in honey bee-collected pollen with a systematic review." *Chemosphere* 258: 127358.

Tullis, D. S. 2008. *Issues Associated with the Natural Attenuation*. Washington, DC: Office of Solid Waste and Emergency Response, U.S. Environmental Protection Agency.

USEPA. 2000. *Stressor Identification Guidance Document*. Washington, DC: U.S. Environmental Protection Agency.

USEPA. 2011a. *Evaluation Guidelines for Ecological Toxicity Data in the Open Literature*. Washington, DC: U.S. Environmental Protection Agency, Office of Pesticide Programs.

USEPA. 2011b. *Integration of Traditional Ecological Knowledge (TEK) in Environmental Science, Policy and Decision-Making*. Washington, DC: U.S. Environmental Protection Agency.

USEPA. 2016. *Weight of Evidence in Ecological Assessment*. Washington, DC: Risk Assessment Forum, U.S. Environmental Protection Agency.

USEPA. 2018. *Application of Systematic Review in TSCA Risk Evaluations*. Washington, DC: U.S. Environmental Protection Agency.

USEPA. 2022. *ORD Staff Handbook for Developing IRIS Assessments*. Washington, DC: Office of Research and Development, U.S. Environmental Protection Agency.

USEPA. 2024a. *Acute Aquatic Life Screening Value for 6PPD and 6PPD-Quinone in Freshwater*. Washington, DC: U.S. Environmental Protection Agency, Office of Water.

USEPA. 2024b. *Data Requirements for Pesticide Registration*. Washington, DC: Office of Pesticide Programs, U.S. Environmental Protection Agency.

USEPA. 2024c. *Sequence Alignment to Predict Across Species Susceptibility (SeqAPASS) Resource Hub*. Washington, DC: U.S. Environmental Protection Agency.

USEPA. 2025. "Procedures for Chemical Risk Evaluation Under the Toxic Substances Control Act (TSCA)." *Federal Register* 90 (182): 45690–45717.

USEPAR10. 2014. *An Assessment of Potential Mining Impacts on Salmon Ecosystems of Bristol Bay, Alaska. EPA 910-R-001A*. Seattle, WA: U.S. Environmental Protection Agency, Region 10.

USGS. 2019. *Water Quality Portal*. Washington, DC: U.S. Geological Survey.

Van Der Kraak, G. J., A. J. Hosmer, M. L. Hanson, W. Kloas, and K. R. Solomon. 2014. "Effects of atrazine in fish, amphibians, and reptiles: An analysis based on quantitative weight of evidence." *Critical Reviews in Toxicology* 44 (sup5): 1–66.

Van Der Zalm, A. J., J. Barroso, P. Browne, W. Casey, J. Gordon, T. R. Henry, N. C. Kleinstreuer, A. B. Lowit, M. Perron, and A. J. Clippinger. 2022. "A framework for establishing scientific confidence in new approach methodologies." *Archives of Toxicology* 96 (11): 2865–2879.

Vermeire, T., T. Aldenberg, H. Buist, S. Escher, I. Mangelsdorf, E. Pauné, E. Rorije, and D. Kroese. 2013. "OSIRIS, a quest for proof of principle for integrated testing strategies of chemicals for four human health endpoints." *Regulatory Toxicology and Pharmacology* 67 (2): 136–145.

Warne, M., G. E. Batley, R. A. van Dam, J. C. Chapman, D. R. Fox, C. W. Hickey, and J. L. Stauber. 2018. *Revised Method for Deriving Australian and New Zealand Water Quality Guideline Values for Toxicants.* Brisbane: Department of Science, Information Technology and Innovation.

Wilson, E. C. F. 2015. "A practical guide to value of information analysis." *Pharmacoeconomics* 33 (2): 105–121.

7 Weight the Evidence

Weighting the evidence is the extra step beyond vote counting (Section 3.1) that addresses Griffin and Tversky's (1992) objection to combining evidence without weighting: "We propose that people's confidence is determined by the balance of arguments for and against competing hypotheses, with insufficient regard for the weight of evidence." By weighting, you assign weight to a piece, type, or other category of evidence to indicate the influence that it should have. How is that different from weighing evidence? Weighing the evidence (Chapter 8) is the process of combining the weighted evidence into a body of evidence and deriving a combined weight. Picture Themis's scales. In weighting, you imbue each piece or type of evidence with a small, medium, or large weight. In weighing the evidence, you load the weighted evidence into the pans for each hypothesis and then check the relative heights of the pans. Once again, the metaphor is not perfect, and some interpretation will be involved (Figure 7.1).

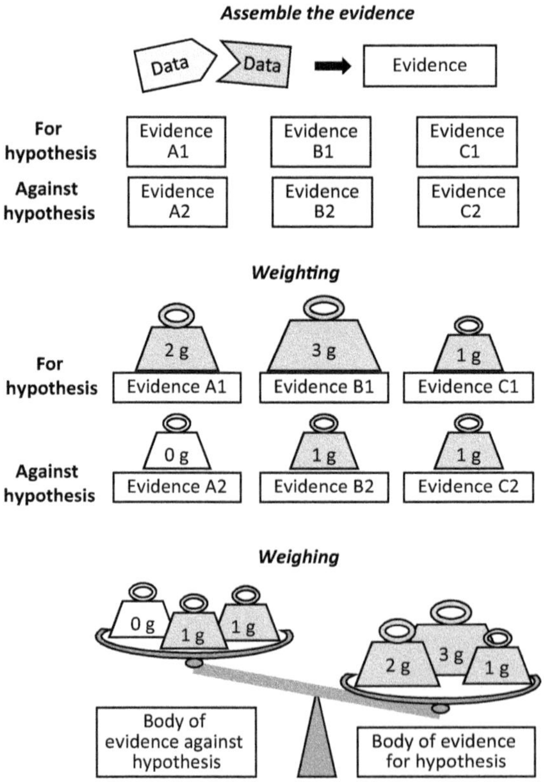

FIGURE 7.1 The scales metaphor for the weight-of-evidence process.

DOI: 10.1201/9781003626336-9

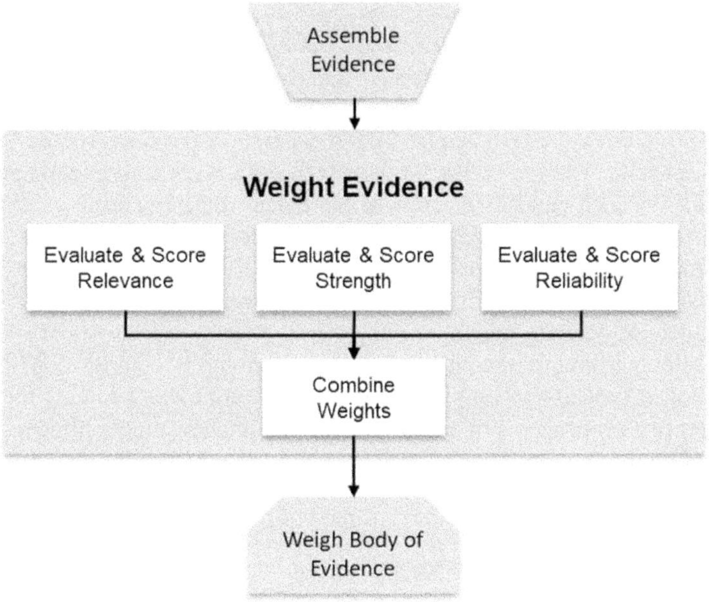

FIGURE 7.2 An elaboration of the process for weighting evidence, the second step in a weight-of-evidence process. (Adapted from USEPA 2016.)

Weighting has three steps that answer three questions (Figure 7.2). First, with respect to what is the evidence weighted? That is, what is the source of the weight? It is the properties of the evidence with respect to the hypothesis. You choose properties and sub-properties that are appropriate to the evidence and the inference. Two properties, relevance and reliability, are used in many contexts. In the U.S., judges determine that evidence from expert testimony is admissible based on its relevance and reliability. These properties are also used to assign weights to evidence in environmental assessments by various agencies and organizations (USEPA 2016; EFSA Scientific Committee 2017; OECD 2019; ECHA 2008, 2011) and by non-institutional authors (Moermond et al. 2017; Rudén et al. 2017; Ågerstrand et al. 2011; Brock et al. 2021; Hall et al. 2017; Ingre-Khans et al. 2019; Klimisch, Andreae, and Tillmann 1997). Other properties are considered in specific assessment contexts (Sections 7.1.3–7.1.5).

The second question is, what expression of the evidence do you weight?

In the simplest case, you weight each piece of evidence. This is appropriate if there are few pieces of evidence or if the pieces are dissimilar.

If there are multiple similar pieces of evidence, you may aggregate them by type and weight the types (Section 6.3). Types may be defined by specific methods, and the methods may be weighted. For example, seven methods for assessing toxicity to terrestrial vertebrates have been qualitatively weighted with respect to relevance, reliability, and readiness (Box 6.2). You may also aggregate pieces of evidence in terms of their implications for the hypothesis (Section 6.5). For example, aggregate all evidence of exposure of a population to an agent. Similarly, you might aggregate pieces of evidence in terms of characteristics of the qualities to be inferred, such as characteristics of causation, impairment, or remediation (Appendix B). Aggregating in terms of implications or characteristics and then rating those aggregates is not as simple as rating pieces or types of evidence, but it may give you more useful input to the weighing of evidence (Chapter 8). An example is provided by the Clear Fork causal assessment, which weights the evidence with respect to each characteristic of causation for each candidate cause of biological impairment; see Table 23.4 in Zheng, Gerritsen, and Cormier (2014). For example, the candidate cause, low pH, is rated + for the causal characteristic antecedence because sources were observed, and sources are necessarily prior to effects. However, other causal characteristics, co-occurrence, sufficiency, and alteration, are rated – or – – because pH was equal to statewide reference and threshold values and to community composition.

Third, how do you assign weights? This is usually done by evaluating the evidence with respect to each property of evidence to determine how much influence it should have. Based on the evaluation, what rating should the evidence be assigned to codify the evaluation?

7.1 PROPERTIES OF EVIDENCE

Evidence has properties that are judged to determine how much weight it should have when judging a hypothesis. In this section, we present two properties that are commonly used in weight-of-evidence contexts: relevance and reliability. Then we follow the USEPA guidelines by including the strength of evidence as an often-useful property. We have also added suitability as a potentially useful property that can be important when regulations, guidance, convention, or precedents constrain the types of evidence that are accepted. We believe that these are the best properties in general for environmental assessments, but others have been recommended. For example, in the Australia/New Zealand method for applying weight to evidence, weight incorporates judgments about the quality, quantity, relevance, and congruence of the data contained in the different types of evidence (Warne et al. 2025). Some recommended systems are limited to evaluating laboratory tests and are not applicable to field-derived evidence (Klimisch, Andreae, and Tillmann 1997; Schneider et al. 2009; Breton et al. 2009).

Some systems for evidence evaluation rate evidence with respect to the completeness of reported data and metadata for each included study (Section 6.2.3). We believe that such reporting is helpful in evaluating the properties of evidence, but reporting in itself is not a property of evidence that increases or decreases confidence in the hypotheses being evaluated.

7.1.1 RELEVANCE

Relevance is a general property of evidence that expresses the degree of correspondence between the evidence and the hypothesis or endpoint to which it is applied. In the law, evidence is relevant if: "(a) it has a tendency to make a fact more or less probable than it would be without the evidence, and (b) the fact is of consequence in determining the action" (Federal Rules of Evidence, Rule 401). For an environmental example, in a site-specific assessment, evidence concerning effects of interest on a population of concern living on the contaminated site is highly relevant. Laboratory tests of a distantly related species are far less relevant. For that reason, we use a gradient of weights for relevance. The PECO criteria, conceptual model nodes, and other evidence prompts that help determine what information should be gathered (Chapter 6) can also be used here to help weight the relevance of evidence. A single relevance weight may be applied, but it is usually helpful to define and weight more specific sub-properties of relevance: biological, physical/chemical, and environmental relevance (Box 7.1). One might even define and weight more specific sub-properties of those three sub-properties. In any case, you will combine them into an overall relevance weight (Section 7.2.6). Weighting sub-properties can provide more transparency concerning the basis for the relevance weight.

According to the Oxford English Dictionary, relevance is the "quality of being… closely connected or appropriate." That is, relevant evidence is related to the case being considered. Evidence is relevant to the system that is the subject of the hypothesis if it is, in some sense, related to that system from which the evidence is derived. For example, applying data from one ecosystem to another requires determining the relevance of the biota, geology, and climate of the ecosystem that is the source of the evidence to the ecosystem being assessed (Section 20.4). Similarly, test results from mice are less relevant to humans than clinical trials but more relevant than results from zebrafish.

Relevance is usually defined with respect to the hypotheses and the assessed system. Because relevance is hypothesis-specific, weights of evidence may be hypothesis-specific. For example, hypotheses of endocrine disruption are biochemically specific (e.g., the chemical is an estrogen agonist). Relevance weights may be specified for such biochemical endpoints. For example, increases in vitellogenin in male fish are a rank 1 endpoint for estrogen agonists in the system of Borgert et al. (2014). It is assigned that rank because it is sensitive for the hypothesized mechanism of action, is interpretable without ancillary data, and is rarely confounded. Similar standard weighting systems can be derived for individual adverse outcome pathways.

Biological Relevance: Correspondence between the biological properties of the evidence (the taxa, life stages, level of organization, mode of exposure, and responses measured or observed) and the assessment endpoint. For example, a *Daphnia* acute lethality test does not correspond well to the life cycles of insect populations, so its relevance to insect-dominated stream communities may be low. Relevance is a particularly important property for evidence from cellular or subcellular tests. That is, responses from such tests must be shown to be predictive of the responses of assessment endpoint organisms. When the assessment is for a generic endpoint such as biotic integrity, relevance may come from the inclusion of a diverse set of species from common taxa. When sensitivity is known, relevance may come from the inclusion of sensitive taxa, life stages, and responses.

Physical/Chemical Relevance: Correspondence between the chemical or physical agent tested in a laboratory or measured at the studied site and the chemical or physical agent constituting the agent of concern. For example, pyrene may be used to represent a polycyclic aromatic hydrocarbon mixture, but, depending on the mixture composition, relevance may be low. Other considerations include ionization, complexation, metabolites, and formulations. If actual or potential exposure levels are known in a case, evidence of effects from exposures in the same range is relevant. Physical/chemical relevance usually refers to properties of agents that make them hazardous. However, it may also refer to properties related to exposure, such as persistence or solubility.

Environmental Relevance: Correspondence between test conditions and environmental conditions at the assessed site or between the environmental conditions in a studied ecosystem and conditions in the region of concern determine relevance. For example, a pond mesocosm does not correspond well to a stream for many environmental parameters, so its relevance may be low.

BOX 7.1 RELEVANCE OF A PIECE OR TYPE OF EVIDENCE (ADAPTED FROM NORTON, CORMIER, AND SUTER (2014) AND USEPA (2016))

The property, relevance of evidence, may be broken down into three basic sub-properties for ecological assessments: biological, physical/chemical, and environmental relevance.

Broad or vague hypotheses may invite bias in weighting relevance. For example, if the hypothesis is, "the stream is unimpaired," the evidence for impairment may be that the fish community includes an introduced fish species (e.g., brown trout in America) or the evidence against impairment may be that the invertebrate community is typical of undisturbed reference streams. Which, if either, is more relevant to impairment? Do we give less weight to the introduced trout as relevant to impairment because it is a popular sport species and not a carp? Do we consider the fact that it outcompetes native brook trout? Clearly, relevance may be hard to judge when the endpoint is vague or value-laden.

Some weight-of-evidence systems have also included regulatory relevance to weight the evidence with respect to appropriateness for regulatory processes and restrictions (Rudén et al. 2017). However, we believe those considerations are distinct and should be used in screening of the evidence (Section 6.6) or in the weighting for suitability (Section 7.2.4). Still others have included statistical significance or the strength of an exposure-response relationship as relevance (Van Der Kraak et al. 2014), but we believe that those properties are not related to relevance and belong in strength of evidence (Section 7.2.3). For reviews of alternative lists of sub-properties of relevance for ecotoxicological studies, see Rudén et al. (2017), Ågerstrand, Breitholtz, and Rudén (2011), and Ingre-Khans et al. (2019).

7.1.2 Reliability

The wise lend a very academic faith to every report which favours the passion of the reporter.

David Hume (1748)

Reliability is a property of evidence that gives it credibility. It is determined by the degree of belief in the quality of the evidence, based on whether it is well designed, well performed, well reported, and otherwise inspires confidence. Some weight-of-evidence systems refer to reliability as "quality" (Bridges and Solomon 2016; Samuel et al. 2016) or "credibility" (Conrad and Becker 2011).

Reliability in U.S. law is based on the expertise of expert witnesses (Federal Rules of Evidence, Rules 702–705). The proponent demonstrates to the court that it is more likely than not that:

- the expert's scientific, technical, or other specialized knowledge will help the trier of fact understand the evidence or determine a fact in issue;
- the testimony is based on sufficient facts or data;
- the testimony is the product of reliable principles and methods; and
- the expert's opinion reflects a reliable application of the principles and methods to the facts of the case.

In scientific assessments, reliability should be based on properties of the evidence itself and not those of the expert who is the source of the evidence. An exception is questions of bias or fraud, which involve the character of the investigator.

Reliability differs from the other properties of evidence in that it is frequently used alone (Moermond et al. 2017; Bilotta, Milner, and Boyd 2014; Ågerstrand, Breitholtz, and Rudén 2011; Moermond et al. 2016; Samuel et al. 2016). In such cases, rather than being part of a process for inference by weight of evidence, it is used to determine whether the evidence is acceptable for use. In that role, it is used as part of assembling evidence (Section 6.6). However, the concepts and approaches in such reliability assessments may also be relevant to weighting your evidence.

The various weight-of-evidence systems have different ideas of what makes evidence reliable. Standard reliability sub-properties are available for toxic effects on humans (Klimisch, Andreae, and Tillmann 1997; Dekant and Bridges 2016) and ecological endpoints (Moermond et al. 2017; USEPA 2016; Ågerstrand, Breitholtz, and Rudén 2011; Ågerstrand et al. 2011; Klimisch, Andreae, and Tillmann 1997). The Klimisch system is often used in Europe and in Organization for Economic Cooperation and Development (OECD) guidance, and there is a tool for its implementation (Schneider et al. 2009; Segal et al. 2015). It is criticized for emphasizing the use of standard tests and the completeness of reporting rather than the quality of the study design and implementation. However, it is applicable to both human health and ecological assessments (ECHA 2011). Some systems for environmental assessments also emphasize reporting as part of reliability (Ågerstrand et al. 2011).

The Klimisch system and some other systems use a hierarchy of types of evidence to rate reliability. For example, the following are adapted from Mupepele et al. (2016). The highest rank (four) is indicative of the best evidence.

1. Studies without underlying data (individual expert opinion or mechanism-based reasoning).
2. Observational studies (inferential studies with statistical testing, descriptive studies without statistical testing, or multiple lines of weak evidence).
3. Studies with a reference or control (case-control, before-after control-impact, or multiple lines of moderate evidence).
4. Reviews (systematic reviews with meta-analysis or weight of evidence).

Some weight-of-evidence systems weight epidemiological or ecological field studies more highly than other types of evidence because the results of these studies are inherently realistic (Chapman and Anderson 2005). However, they may have low reliability due to lack of control of exposure, confounding by co-occurring agents, low replication, lack of appropriate reference sites or populations, or imprecise or inaccurate field measurement techniques. So, do not assume that any type of evidence is best without weighing its properties!

Properties of evidence that may suggest that evidence is reliable in environmental assessments are listed and defined in Box 7.2. Sub-properties of reliability that require more explanation than the definitions in the box are discussed in the following text.

BOX 7.2 RELIABILITY SUB-PROPERTIES OF EVIDENCE (ADAPTED FROM NORTON, CORMIER, AND SUTER (2014), USEPA (2016)).

Design and Execution: Evidence generated with a good study design that is well performed and statistically well analyzed is more reliable.

Reporting: Complete reporting of results gives greater confidence, permits replication, and allows checking of statistical analyses.

Abundance: Evidence derived from more numerous data is more reliable because it reflects greater replication or resolution.

Minimized Confounding: Evidence is more reliable when the sampling design or analysis controls extraneous correlates.

Specificity: Evidence specific to one cause or a few related causes is more reliable. For example, a characteristic symptom or set of symptoms is reliable evidence of causation.

Potential for Bias and Fraud: Evidence from a study that is not funded by an interested party, is not produced for advocacy, and is not produced by an investigator with conflicts of interest is more reliable.

Standardization: A standard method decreases the likelihood that the evidence is biased or analyses are inaccurate. Standard methods have been shown to be more replicable. Standard test methods, field sample collection and analysis methods, and good laboratory practices are available from the USEPA and other agencies and from international bodies, particularly the OECD. They may be inflexible and may minimize relevance by limiting the range of taxa, life stages, responses, and conditions. However, they provide a base level of reliability.

Corroboration: Findings that have been replicated by other studies or have made verified predictions are more reliable.

Transparency: Complete descriptions of methods, inferential logic, statistical analyses, and data provide the means to check the results and are presumed to increase reliability by reducing the likelihood of hidden faults.

Peer Review: Independent peer review of a study increases its reliability as a source of information. More reviewers and review workshops increase weight.

Consistency: The degree to which evidence does not vary in repeated instances within a study (e.g., across years, locations, sampling teams, or methods) is an indicator of the reliability of a piece of evidence. When weighting types of evidence, consistency across studies of the same type is an indicator of the reliability of the type. However, inconsistency may not be important if there is extraneous variance, such as collecting field samples in different seasons.

Consilience: Evidence shown to be consistent with scientific knowledge and accepted theory, particularly with respect to underlying mechanisms, is more reliable.

Study Design and Execution: This sub-property is given the most attention in many weight-of-evidence systems, particularly those developed for human health (USEPA 2022). Study design depends on the type of study and on case-specific constraints. It is advisable to develop a checklist of design features for each type of study that you commonly use. For example, a checklist of considerations for laboratory and field studies of sediment chemistry, toxicology, and community structure was developed by Batley et al. (2002). Statistical methods should be designed to avoid misleading results (Section 6.4). For example, evidence may be statistically insignificant but provide biologically or chemically important evidence. In some weight-of-evidence systems, nearly all of reliability weighting is related to study design and reporting (Moermond et al. 2017). In general, evidence from secondary sources should not be used, but if it is, the review method for those sources should be systematic and well designed (Section 6.2.1). Bad study designs (e.g., lack of randomization, calibration, or blinding) can bias the results of a study, but this is different from personal bias, discussed below.

Study design is not only important during weight of evidence: inattentiveness to design, execution, and reporting can lead to the exclusion of studies and the ineffectiveness of research efforts. For example, behavioral ecotoxicity studies are abundant (Bertram et al. 2022). However, Ågerstrand et al. (2020)

found just six cases in European Union chemical regulation where behavioral studies were considered usable, and only three were found to provide key or supporting evidence. They recommended that behavioral toxicologists should conduct more robust and well-reported studies.

Standardization: In general, evidence from standard test methods, survey methods, and analytical methods has been considered to be more reliable than evidence from nonstandard methods. In particular, regulatory bodies have relied on national and international Good Laboratory Practices (GLPs). The most important of these are the OECD standards (oecd.org/chemicalsafety/testing/good-laboratory-practiceglp.htm). However, GLPs may not be best in some situations. If standard tests do not include the endpoint effects, taxa, and life stages, you lose relevance. For example, the effects of major ions in streams were not recognized because the standard tests for aquatic invertebrates use daphnids, and the sensitive species are insects, particularly mayflies. Similarly, the U.S. Food and Drug Administration rejected 157 nonstandard studies of bisphenol A, which included the only studies that showed effects at low doses, and accepted 4 standard studies that showed effects only at high doses (Cornwall 2017). Some have argued that giving greater weight to GLP studies prioritizes studies by industry over studies by academics who prefer innovative methods. Some contend that GLPs emphasize bookkeeping at the expense of relevant study design (Borgert et al. 2016). Others contend that GLP tests are meant only for screening (Cornwall 2017). Still others argue that peer review provides better reliability than GLPs (vom Saal and Myers 2010). However, McCarty, Borgert, and Mihaich (2012) argued that neither GLPs nor peer review is a sufficient criterion for reliability, and weight of evidence should be applied. We agree.

Bias and Fraud: The most controversial aspect of reliability is the biases of the authors of studies, which may result in fraud or misleading conclusions. Many authors involved in health and environmental science insist that assessors should judge the science, not the scientists. However, fraud and bias based on employment, funding, or ideology are well documented (Mandrioli and Silbergeld 2016; Suter and Cormier 2014; Mebane et al. 2019; Higgins and Thomas 2023; Wagner, Fisher, and Pascual 2018). Because of apparent funding bias, evidence-based toxicology is considered by some to be tainted by the overt anti-regulation bias of its founders and some of its practitioners (Rudén and Hansson 2008). The potential for bias due to conflicts of interest should be revealed by the cited authors and accounted for in the assessment (Higgins and Thomas 2023). Patterns of bias can be detected if there is a large body of evidence from multiple sources. For example, in studies of new drugs, 5% of those funded by the manufacturer, compared to 38% of independently funded studies, were unfavorable (Friedberg et al. 1999). Simply revealing conflicts of interest has not proven to be sufficient to guarantee that the authors will not bias their science (Elliott 2008). In one case, an author of a paper that attacked a USEPA method approached me (Suter) at a meeting and said that he knew that we were right, but he needed to provide something for his client. Some fraudulent publications are simply due to the desire to professionally advance the author's career or to successfully complete a study by inventing data. Scientists in some nations have much higher rates of retractions than others (see *Retraction Watch*). However, it could be difficult to defend down-weighting evidence from those nations. When differences in results are associated with funding source or laboratory, the evidence should be particularly carefully reviewed to determine whether there is a methodological difference that would account for the difference and affect the relevance or reliability. Systems for minimizing the influence of bias are reviewed by Conrad and Becker (2011) and Bolland, Avenell, and Grey (2024). Scientific papers that have been challenged or found to be fraudulent can be found in the *Retraction Watch* database (https://retractionwatch.com) and the comments in PubPeer (https://pubpeer.com). Artificial intelligence that detects fake papers from paper mills, plagiarism, invented data, or manipulated images is under development (Liverpool 2023). However, advanced AI may also be designed to produce undetectable fraud. When detected, fraudulent sources of evidence must be weighted as zero. Sources that are biased but not detectably fraudulent are more difficult to address. Evidence may be biased by practices that are not malicious. For example, many authors do not publish negative results, and journals are reluctant to publish them because they are considered unimportant. However, publishing effects but not publishing no-effects results is a clear source of bias in a body of evidence.

Journals: Papers in predatory, hijacked, and scam journals are likely to be unreliable because they publish papers that legitimate journals reject, and they are poorly reviewed or not reviewed at all. Other journals have biases that justify special attention, such as the pro-industry bias of *Regulatory Toxicology and Pharmacology* (Michaels 2008; Chrisman and Weislogel 2003; Velicer, St Helen, and Glantz 2017). A prominent case that would have sorely benefited from careful review of papers in questionable journals is presented in Box 7.3. High-quality journals can publish fraudulent papers, but it is less likely.

Specificity: It is reasonable to think that if an association between a putative cause and effect is specific, it is more likely to be causal. Specificity may apply to effects, exposures, or susceptibility (Weiss 2002). If an effect has only one or a few causes, that effect is less likely to be due to a confounding cause. For example, cranial deformities in fish populations are specifically caused by selenium toxicity and by deficiencies in calcium and phosphorus. On the other hand, low abundance of fish has many causes. Specificity may apply to exposures. That is, an association with a specific pesticide is more likely to be causal than an association with agriculture. Finally, if a particular type of organism is known to be susceptible to a particular agent, that specific association of those organisms and the agent is evidence for causation. For example, unionid mussels are highly susceptible to ammonia, but fish and arthropods are less so. That susceptibility is evidence of ammonia as the cause of the loss of mussels where fish, insects, and crustaceans are not lost.

BOX 7.3 THE GLYPHOSATE SCANDAL

This morning (1/4/2026), the Sunday New York Times had a headline, "Study's Retraction Renews Weedkiller Concerns: EPA Facing Pressure to Reassess Roundup" (Tabuchi 2026). This is the latest episode of a long-running story, which is an embarrassment to the USEPA, *Regulatory Toxicology and Pharmacology*, Monsanto (now Bayer), and the scientific community (Kaurov and Oreskes 2025). In 2000, a paper that apparently cleared glyphosate of human carcinogenicity was ghostwritten by Monsanto staff but published by authors from two universities and a consulting firm (Williams, Kroes, and Munro 2000). Relying heavily on this paper, the USEPA found in 2016 that glyphosate is "not likely to be carcinogenic to humans," while the International Agency for Research on Cancer (IARC) found in 2015 that it is "probably carcinogenic to humans." The primary difference was found to be the USEPA's greater reliance on industry-funded studies and, secondarily, the failure to adequately consider formulations or exposures other than dietary (Benbrook 2019). The ghostwriting was revealed when Monsanto was sued for causing cancer, and it provided documents known as the "Monsanto Papers," which were released in 2017. A review of Monsanto-funded glyphosate papers found that two were ghostwritten and that the content of the others was controlled by Monsanto (Matheson 2023). *Regulatory Toxicology and Pharmacology* finally retracted the Williams et al. (2000) paper after eight years (Williams et al. 2025). In the 2017–2025 interval, hundreds of peer-reviewed papers approvingly cited the fraudulent paper, and the USEPA failed to change its regulatory finding (Kaurov and Oreskes 2025). However, a court has ordered the USEPA to revise its assessment this year (2026). The lessons for us are: pay attention when industry-funded studies consistently differ from others, and pay attention when a substance is being or has been assessed by the courts.

7.1.3 Strength

Strength is a property of evidence determined by the degree of differentiation from randomness or from control, background, or reference conditions. A strong signal is better differentiated from noise than a weak signal, so a strong signal should be given more weight. Strength can be determined by (1) the magnitude of difference among variables, such as species richness of exposed and unexposed communities, (2) the degree of association between potential causes and effects, such as an exposure-response correlation, and (3) the number of elements in the evidence for a hypothesis, such as the number of streams surveyed (Box 7.4). Strong evidence shows a large magnitude of differences, high levels of association, and numerous instances of the evidence in space or time.

Strength is a property of the study results, not the type of evidence or study method. The metrics for strength of evidence are familiar. Magnitude is typically represented by absolute and relative differences (e.g., body burdens were 20 times higher than at reference sites). Association is typically represented by correlation coefficients or slopes (e.g., the regression of species richness on dissolved oxygen had a slope of 7). Number is commonly represented by the number of elements or frequency of occurrences (e.g., a fish kill has occurred at snowmelt every year for the past 6 years). Although strength occurs less frequently as a property in weight-of-evidence systems than relevance and reliability, some have emphasized strength (Chapman 2007).

BOX 7.4 STRENGTH SUB-PROPERTIES FOR A PIECE OR TYPE OF EVIDENCE

The strength of a piece or type of evidence is the degree of differentiation from control, reference, or randomness (modified from Norton, Cormier, and Suter 2014 and USEPA 2016).

Magnitude: The relative or absolute difference between the amount of response at affected sites and at reference sites, between treatments and controls, between degrees of exposure, or other relevant differences in the evidence.

Association: The degree to which a variable is related to one or more other variables, most commonly expressed as a correlation coefficient. Associations of categorical variables can be expressed as relative risk or odds ratios.

Number: The number of elements in a set of evidence (e.g., symptoms, overt effects in a response, or steps in a causal pathway) that are reported to be observed or the number of occurrences.

Strength is typically quantitative, but it may be qualitative. For example, a stream receiving treated sewage may be described as "highly eutrophic" (strong descriptive evidence). However, that evaluation may not provide reliable evidence if the source of the description was an inexperienced investigator, who may not have used a standard descriptive scale or may have been describing a photograph.

Strength differs from relevance and reliability in that it can be inappropriate for derivations of quantitative endpoints. For example, if the question is, what proportion of species are lost when a pesticide is applied to soil, and a biological survey of contaminated soil estimates that species richness is reduced by 60%, that is an estimate of the endpoint, and not an strength of evidence property of toxicity. If you are weighing multiple estimates from four studies, such as 25%, 55%, and 60% reduction, the qualitatively strongest evidence of a causal relationship in terms of magnitude is 60%. However, it is not appropriate to conclude that it is the weightiest estimate based on strength because that weight biases the choice of quantitative result toward the highest value. When estimating quantities from multiple pieces of evidence, consider whether it is more appropriate to combine the estimates or to weigh the evidence based on appropriate properties (Chapter 12).

7.1.4 Suitability

Suitability is the property of being appropriate for the purpose of the assessment. Most criteria for suitability are binary and are used in the screening process (Section 6.6). However, some suitability criteria are not yes/no. A toxicity test result expressed as statistical significance has traditionally been considered suitable by the USEPA Office of Water for deriving chronic water quality criteria, but more recently, it has not been enough. It may be given a low suitability weight because it does not specify the endpoint effect, e.g., growth or reproduction. The Office of Water now prioritizes exposure-response-based test endpoints such as a tenth centile effective concentration (EC_{10}). We present no set of suitability sub-properties because suitability is very context-specific. In many cases, weighting for suitability is not an issue. However, you should consider whether suitability is appropriate in your case to ensure that your efforts are not wasted.

7.1.5 Properties for Evidence from Models

The properties described so far were developed with empirical evidence in mind, but they can be applied to mechanistic models and to their results as evidence.

Both the models and their input data may be weighted for their relevance and reliability.

- Model relevance depends on the degree to which the modeled system corresponds to the system being assessed. For example, a lake model has low relevance for a river.
- Model reliability depends on validity. Valid models have been shown to be free of logical flaws and internal inconsistencies (Oreskes, Shrader-Frechette, and Belitz 1994).
- Model reliability may also be increased if it is verified by comparing its results to attributes of the system being assessed (Oreskes, Shrader-Frechette, and Belitz 1994). The model verification should be based on observations that are independent of both the input data and the observations that were used in the model design. Various statistical models may be used to quantify the relationship between the estimated and observed values of the output variable.
- Data relevance depends on the degree to which the data correspond to the parameter. For example, the bioaccumulation factor for one polycyclic aromatic hydrocarbons may be used as an estimate for the bioaccumulation factor for a different PAH, but the relevance will be low.
- Data reliability depends on the quality of the method used to estimate the parameter.

Standard weights may be developed for particular parameters. Table 7.1 presents EFSA's standard ratings for uncertainties that may increase or decrease consumption

TABLE 7.1
Examples from a Larger Table of Qualitative Evaluation of the Influence of Uncertainties on an Assessment of Ochratoxin Exposure for High Consumers of Foods Contaminated with This Substance (EFSA 2007)

Sources of Uncertainty	Direction and Magnitude of Effect
Moderate under reporting of consumption is known to occur	– –
The possibility of large regional and temporal differences in OTA levels may lead to underestimation of exposure for consumers eating locally-produced food	– –
Use of broad food categories causes overestimation of exposure (e.g., assuming all juice is contaminated at the same level as grape juice; all hot drinks at the same level as coffee)	+++
Extrapolation from food consumption during 7-day survey overestimates long term consumption for high consumers	+++
Extrapolation of food consumption from France, Italy, and Sweden to the other 22 EU Member States	++/ – –
Qualitative evaluation of overall effect of identified uncertainties: estimates for high consumers are likely to overestimate adult exposure by a moderate amount, but might be underestimates for regional populations consuming locally-produced food and are probably underestimates for children	++ adults +/– local populations – – children

In these scores, the positive and negative refer to whether the evidence supports or weakens the hypothesis and the number of signs the rating of that uncertainty.

of a contaminant in their human dietary exposure model. Although these ratings are framed in terms of uncertainties, they could also be framed as weights for evidence reliability. That is, evidence would have low weight if it is highly uncertain. These uncertainty tables have been used by others (UK Department of the Environment, Food and Rural Affairs 2000), but we believe it is preferable to consistently use weight of evidence.

7.2 ASSIGNING WEIGHTS

The Judge will give greater weight to specific, signed, factual evidence than to general, unsupported allegations or opinions.

Code of Federal Regulations § 134.1011 Weight of evidence.

Weights are assigned for various reasons.

- Weight is assigned to evidence based on its properties. How relevant, how reliable, and how strong is it? This is the essential process of weight of evidence.

- A few assessors weight the properties as well as the evidence. That is, what is the relative importance of relevance, reliability, and strength or of their sub-properties?
- Some assessors weight the types of evidence. For example, some assessors give greater weight to field biological surveys than to toxicity tests.
- Finally, some assessors weight assessment endpoints. For example, when determining whether a stream is impaired, what is the relative importance of algal production, invertebrate species richness, and fish abundance?

Weight-of-evidence systems that assign weights to the properties of evidence base the weights on the premise that some properties are more important than others. For example, if you believe that taxonomic similarity is always more important than environmental conditions, you may weight biological relevance twice as highly as environmental relevance. This is not recommended by the USEPA (2016) guidelines or most other systems. In a review of methods for judging the reliability of toxicological studies, Moermond et al. (2017) found only three that weighted properties. They recommend weighting sub-properties of reliability, which they call criteria, depending on the specifics of the case: "For example, frequent measures of exposure concentrations in a test are more important for compounds that degrade rapidly than those that do not." That is a legitimate consideration, but it could be addressed when weighting the evidence from the tests rather than weighting properties. While there is nothing wrong with weighting the properties as well as the evidence, we believe that it adds unnecessary complexity. However, it may be helpful when the interpretation of the weight of evidence for hypotheses is ambiguous (Section 8.2.2).

Because weighting of properties, types of evidence, and endpoints is uncommon, we will focus on weighting evidence, which is the necessary heart of weight of evidence. However, if, for whatever reason, you believe that your assessment would benefit from weighting properties, types of evidence, or endpoints, go ahead.

7.2.1 Evaluating Evidence

Weighting of evidence begins with evaluation, to determine how influential/weighty it should be. It may involve either of two approaches. First, it may involve unguided expert judgment, an unstructured method constrained only by an assessor's understanding of the evidence and case. Such traditional ad hoc narrative weighting has become less acceptable, particularly in regulatory contexts.

Second, it is desirable to constrain expert judgment by prescribing a weighting method or at least providing some guidance. For example, if the applicant for a permit to market a pesticide and the regulatory agency obtain different assessment results, or if regulators from different nations obtain different results, a method for assigning weights that is prescribed for both parties would make it possible to determine where and why the differences occurred. As valuable as such prescriptive guidance is in obtaining consistent evidence evaluation, it may not always be applicable to weighting the particular evidence in your particular case. Evidence evaluation can benefit from judgment by multiple assessors.

An example of systems that assign weights in a very prescriptive manner is provided by Dekant and Bridges (2016). They present a table of criteria that assigns each of 14 aspects of the quality of animal test data to one of five scores from a row of the 70-cell scoring matrix. For example, the quality, "accessibility of raw data," receives a score of 0 for "data provided very limited" and extends to a score of 4 for "complete access to all raw data." They provide details for other components of their system, such as weighting different test endpoints and scoring the relevance of evidence. Applications of such prescriptive systems should be replicable, and they minimize opportunities for biased judgments. However, each piece of evidence requires its own scoring table, so the tables must be derived for each narrowly defined issue and type of evidence.

A more common prescriptive approach is to make a table of types of evidence and their potential outcomes with corresponding ratings (Stevenson and Chapman 2017; USEPA 2000; Hobbs, Warne, and Markich 2005). For example, temporality of a piece of evidence may be compatible (+), uncertain (0), or incompatible (– – –) (USEPA 2000). Incompatible temporality gets a high negative rating because effects cannot precede their causes. In fact, negative evidence is generally stronger because it can disprove a hypothesis (Popper 1968). This approach results in consistency and transparency, but assessors may find such prescriptions to be ambiguous, not applicable to their specific case. Prescriptive weighting systems may be particularly useful when consistency is especially important, as in regulatory assessments involving consistent cases (e.g., only aquatic effects of insecticides). Prescribed weights may be treated as mandates, suggestions or defaults.

One solution is for the assessment team to develop tables of standard weights for their particular assessment up front. That way, the outcomes are relevant to the case, but the ratings are still prescriptive and not biased by the assessor's weighting decisions.

However weights are assigned, they should be justified. This can be done with a brief explanation in the evidence-weighting table or a more expansive explanation in the text. This provides transparency and can increase the credibility of the weights.

7.2.2 Designating and Representing Weights of Evidence

Explicit weights have two forms: verbal and symbolic. Verbal weights express the assessors' evaluation of the evidence with terms such as "strongly supports" or "refutes." The low, moderate, and high scale is common. The symbols, such as +, 0, and –, correspond to the verbal weights. They facilitate the interpretation of evidence-weighting tables. A table with cells containing words does not show clear patterns of weights, but it is easy to see patterns of symbols. For example, you can quickly see which rows and columns have mostly + symbols. The standard expressions of weight in the USEPA guidelines are:

+++, – – –	Convincingly supports or weakens
+ +, – –	Strongly supports or weakens
+, –	Somewhat supports or weakens
0	No effect (neutral or ambiguous)
NE	No evidence

These standard weights express the degree to which the evidence supports a hypothesis. However, depending on the inference to be performed by weighing evidence, you may weight the degree to which the evidence is explained by a hypothesis (Box 7.5).

BOX 7.5 DEFAULT EXPRESSIONS OF WEIGHT FOR INDUCTIVE AND ABDUCTIVE INFERENCE

The standard weights of evidence in environmental assessments in this text and in the USEPA guidelines are based on the assumption that you are signifying the confidence in a hypothesis instilled by the evidence. That is, you are inferring by induction (going from specific information to a general hypothesis). For example, if you are assessing the hypothesis that atrazine causes deformities in frogs, atrazine in the blood of deformed frogs would be strongly supporting evidence (+ +).

However, in some cases, you are assessing the ability of hypotheses to explain the evidence or determining which hypothesis best explains the evidence. That is, you are inferring by abduction (going from general hypotheses to specific information/evidence). For example, if frogs develop deformities in ponds where atrazine is applied, that could be explained by the hypothesis that atrazine causes frog deformities (+ +). If no atrazine was found in the blood of deformed frogs, that would be unexplained by the hypothesis (– –).

+++, – – –	Convincingly explained by or unexplained by
+ +, – –	Strongly explained by or unexplained by
+, –	Somewhat explained by or unexplained by
0	No effect (neutral or ambiguous)
NE	No evidence

These symbolic expressions of weight are reasonably intuitive and commonly used (Staveley et al. 2014; Ridley et al. 2024; Rooney et al. 2014; OHAT 2019). Up and down arrows ↑↓ may be equivalent to + and – (Stevenson and Chapman 2017), or they may also indicate whether effects increase or decrease. Some assessors use colors (Cecchetto et al. 2025). Others use numbers for the symbols, but that creates a false impression of quantitative analysis when the weights are qualitative.

Some systems include only evidence that supports the hypothesis, including the first formal weighting of causal evidence in environmental epidemiology (Susser 1986). However, that requires ignoring negative or neutral evidence, which may be essential to the inference.

The system shown here has three types of weights (positive, negative, and none) and three levels of weight (convincing/high, strong/medium, and somewhat/low), plus the no-effect (zero) weight. This degree of differentiation of weights has been

TABLE 7.2
Standard Weighting of the Strength of Co-occurrence Measured as Linear Correlations (Absolute Value of *r*) for Evidence from Stream Biological Surveys (USEPA 2011; Cormier and Suter 2013)

Assessment	Strength	Rating
The sign of the correlation coefficient depends on the relationship. For toxic relationships such as the correlation between conductivity and number of *Ephemeroptera*, the sign should be negative. Weak or positive correlations weaken the case for that candidate cause.	$\lvert r \rvert > 0.75$	+ +
	$0.75 > \lvert r \rvert > 0.25$	+
	$0.1 < \lvert r \rvert < 0.25$	0
	$\lvert r \rvert < 0.1$	−
	r has the wrong sign	− −

useful in practice. Its seven levels (3 negative, 3 positive, and 1 neutral) correspond to the 7-option Likert scale, which is the number of options that people can usefully distinguish (Dawes 2008). Others have advocated simple binary weights (+/−) for accept/reject, supports/weakens, or equivalent binary options (Hope and Clarkson 2014).

In most cases, you evaluate the evidence, decide how it influences the hypothesis and to what extent, and assign the appropriate weight and its representation (i.e., its rating). However, you may use standard criteria for assigning weights. Strength metrics lend themselves to the standardization of weights. For example, correlations between potential environmental causes and extirpation of aquatic invertebrate genera in Central Appalachian streams were assigned weights based on typical correlation coefficients in such regional biological surveys (Table 7.2). No regional correlations were considered convincing, so none received three + or −. Absolute values of coefficients greater than 0.75 were considered strongly supportive evidence. Coefficients with the wrong sign, such as positive correlations for hypothesized toxic effects, were considered strongly weakening.

Standard weights may also be assigned to possible outcomes of particular types of evidence. Table 7.3 presents a portion of a table of standard weights for potential findings from two types of evidence. They were developed for weighting evidence concerning a hypothesized cause of the impairment of an aquatic community. The full table may be used as found in CADDIS, or this sample may serve as a model for standard weighting of evidence. Tables of standard weights may be found for other specific weight-of-evidence cases, such as microcosm and mesocosm studies of herbicides (Moore et al. 2017).

7.2.3 Arithmetically Combining Numerical Weights

Some authors arithmetically combine numerical weights and refer to the results as quantitative weight of evidence. Some even assign plus and minus ratings to evidence, recognizing the weights are qualitative, but then count them to obtain "quantitative" weights (Becker et al. 2017). Dekant and Bridges (2016) defined strength of evidence as a score that "is derived by multiplying the final relevance/effects score by the

TABLE 7.3
Table of Standard Weights for Two Example Types of Evidence Out of 15 Types in CADDIS (http://www.epa.gov/caddis/)

Type of Evidence	Finding	Interpretation	Rating
Spatial/temporal co-occurrence in site surveys	The effect occurs where or when the candidate cause occurs, **OR** the effect does not occur where or when the candidate cause does not occur.	This finding *somewhat supports* the case for the candidate cause, but is not strongly supportive because the association could be coincidental.	+
	Whether the candidate cause and the effect co-occur is uncertain.	This finding *neither supports nor weakens* the case for the candidate cause because the evidence is ambiguous.	0
	The effect does not occur where or when the candidate cause occurs, **OR** the effect occurs where or when the candidate cause does not occur.	This finding *convincingly weakens* the case for the candidate cause because causes must co-occur with their effects.	−−−
Laboratory tests of site media	Laboratory tests with site media show clear biological effects that are closely related to the observed impairment.	This finding *convincingly supports* the case for the candidate cause.	+++
	Laboratory tests with site media show ambiguous effects, **OR** show clear effects that are not closely related to the observed impairment.	This finding *somewhat supports* the case for the candidate cause.	+
	Laboratory tests with site media show uncertain effects.	This finding *neither supports nor weakens* the case for the candidate cause.	0
	Laboratory tests with site media show no toxic effects that can be related to the observed impairment.	This finding *somewhat weakens* the case for the candidate cause but is not strongly weakening because test species, responses, or conditions might be inappropriate relative to field conditions.	−

Each type of evidence is explained in its own CADDIS page.

quality/reliability score for particular study." Then, their weight of evidence is the sum of the scores for all studies. This can be misleading because the scores are not quantities, and there is no basis in mathematics or logic to add or multiply them. Two strongly supporting laboratory tests (++ and ++) are not equal to four weakly supporting field tests (+, +, +, and +), but if the weights for laboratory tests are represented as 2 and 2, addition is invited. For a test result, a − rating for study design and a + rating for replication of the test do not sum to 0 because they are not commensurable. They simply signify different results for the different qualitative properties.

More practically, if there are many low-reliability studies, they will dominate the combined score and suggest that there is no good evidence for the hypothesized effect, even if there are one or a few high-reliability studies. This "quantitative" method was developed for human health risks, but some ecological assessments have adapted it (Bridges and Solomon 2016). Based on mathematical principles and the goals of weighing evidence, and to avoid erroneous conclusions, we recommend that you avoid summing scores.

7.2.4 Presenting and Combining Weights

The primary tool for presenting and combining weights across properties is evidence-weighting tables. The tables can take many forms, but the components are pieces or types of evidence, ratings, and explanations. The tables may combine ratings for sub-properties to obtain a rating for a higher-level property (Table 7.4), combine ratings for pieces or types of evidence to obtain an overall rating for each property (Table 7.5), or combine ratings for each property to obtain an overall rating of pieces or types of evidence (Table 7.6). You might even find it helpful to combine ratings across rows and down columns in one table.

The basic weighting table presents the ratings of the sub-properties for a piece, type, or other category of evidence. For example, Table 7.4 shows a hypothetical rating table for the relevance of a biological survey for determining the cause of a fish kill.

TABLE 7.4
An Example Table for Weighting the Sub-properties of Relevance of Evidence from a Biological Survey to Determine the Cause of a Hypothetical Fish Kill

Property	Rating	Explanation
Biological relevance	+++	The surveyed community was altered, and the relative sensitivities of present and missing taxa provided evidence of the cause.
Chemical relevance	++	The proposed causal agent was present at the time of the kill, but was diminished at the time of the survey.
Environmental relevance	+++	The habitat properties were largely unchanged.
Overall relevance	+++	Although changes occurred between the times of the kill and the survey, the affected stream is the best source of evidence.

TABLE 7.5
Evidence Weighting Table for the Hypothesis That a Diesel Spill into a Salmon Spawning Stream Will Harm the Biotic Community

Type of Evidence (Exposure/ Exp-Resp)	Relevance	Strength	Reliability		Explanation
			Exp	Exp-Resp	
Dissolved hydrocarbons (Model/laboratory acute tests)	+	+	0	+	Modeled dissolved diesel concentrations are clearly lethal to invertebrates and approximately lethal to trout.
Dissolved hydrocarbons (Model/laboratory–based standard)	+	++	0	++	Modeled dissolved diesel concentrations greatly exceed the State standard
Dispersed hydrocarbons (Diesel-to-water ratio/ laboratory acute tests)	+	++	0	+	Diesel oil/water ratios in the spills and in tests suggest lethality to invertebrates and trout.
All routes in actual spills (Amount spilled/ observed effects)	+	++	+	+	Diesel spills in other streams cause acute biological effects.
Combined Weight by Property	+	++	0	+	The effects by four types of evidence are consistent and the observed effects are strong. The greatest uncertainty is the relation of laboratory to field exposures.
Overall Weight of the Evidence		++			The individual types of evidence only somewhat support the hypothesis, but the consistency of the body of evidence gives strong support.

Reliability is rated separately for exposure (Exp) and exposure/response (Exp/Resp) relationships. Adapted from the Bristol Bay Watershed Assessment (USEPA 2014) where the evidence and results are described.

The ratings are combined to derive a rating for overall relevance. That relevance rating would carry forward to Table 7.6, first column, fourth row. Tables may present more detailed sub-properties, such as rating different taxa within biological relevance.

Table 7.6 presents a basic form in which all of the high-level properties are rated for each type of evidence and combined. In this example, explanations of the ratings are presented in the text rather than in the table. An overall weight is presented for

TABLE 7.6
A Weighting Table with Ratings of Conventional Types of Evidence with Respect to Properties and a Combined Weight for the Type

Types of Evidence	Relevance	Strength	Reliability	Overall Weight
Laboratory toxicity tests of the agent	+	++	+++	+
Effluent toxicity tests				
Ambient media toxicity tests				
Field biological surveys				
Field biomarkers and organ or whole-body concentrations				
Field symptoms				

Source: Taken from USEPA (2016).
This table would be used to weight the evidence that a chemical is the cause of an aquatic ecological impairment. The first line is hypothetically completed for illustrative purposes.

each type of evidence by combining the ratings for the properties. The hypothetical weights in the first line show why the ratings should not be added or averaged. If the case involves a kill of trout in a stream, conventional *Daphnia* and fathead minnows have low (but still some) relevance. Therefore, the overall weight is positive but low despite strong results from reliable standard tests. If we had averaged the numbers of + ratings, the result would be ++, a moderate weight for laboratory tests. In general, negative evidence has more influence than strong positive evidence. Disproofs are more influential than proofs of hypotheses (Kuhn 1997; Popper 1968), and that generalization extends to the greater influence of negative evidence relative to positive evidence.

There is a simple alternative to weighting the evidence with respect to its properties and then combining the weights to get an overall weight. You may go directly to the overall weight by asking yourself: If I had only this one piece of evidence to support the hypothesis, how confident would I be in accepting it? If the evidence is negative, ask: If I had only this one piece of evidence, how confident would I be in rejecting the hypothesis? We do not recommend that method because it is too opaque and subjective. However, asking those questions after you have weighted the properties can act as a self-check. That is, it can be a replicate weighting method to determine whether the original weighting makes sense.

If you weight multiple sub-properties within a property, it is generally desirable to combine them before combining the properties. This is particularly important for reliability, which has 12 suggested sub-properties (Box 7.2). If you combine 4 relevance sub-properties with 12 reliability sub-properties, reliability could inappropriately overwhelm relevance.

However you form your evidence-weighting tables, you will carry them forward to the weighing of the body of evidence (Chapter 8).

TABLE 7.7
Standardized Scoring for Assigning Weights to Evidence of Association When Assessing Causation (USEPA 2011)

Rationale	Description	Rating Assignment
Logical implication	Registers that the evidence is relevant and either supports or discounts the causal relationship.	+ or –
Especially strong or logically compelling	The association was quantitatively strong or predicted from first, principles of chemistry or physics, or logically excludes or confirms the relationship.	Increase score
Corroborated	An independent dataset corroborated the evidence.	Increase score

7.2.5 An Alternative—Accumulating Ratings

The standard method of assigning ratings to evidence based on its properties is presented above. An alternative approach is to increase the rating by one qualitative unit for each consideration. In the example in Table 7.7, evidence for a causal association of an exposure and effect is assigned a + or – for supporting or discounting the association. It then may be given another + or – for the strength of the association or for consistency with principles of chemistry, physics, or logic. Finally, a third rating of + or – may be assigned based on whether independent data corroborate the association. A functionally equivalent method is used by the GRADE system to weight evidence quality (Schünemann et al. 2013). One or two up arrows are applied for each factor that potentially increases the quality of evidence, and one or two down arrows are applied for each factor that potentially reduces the quality of evidence. In both versions of this approach, the ratings are additive. The number of up or down arrows, or of pluses or minuses, indicates the weight of the evidence. In GRADE, four up arrows and two down arrows yield a rating of two up arrows. In Table 7.7, if the logical implication of the evidence makes it supportive (+), then the other ratings may increase or decrease the support. If the implication of the evidence is to weaken the hypothesis (–), the other ratings may increase or decrease the weakening.

7.3 CONCLUSIONS FOR WEIGHTING EVIDENCE

One of the important distinctions in the USEPA's weight-of-evidence guidelines is between weighting evidence and weighing the body of evidence. Many methods and applications do not make that distinction and, as a result, they are unclear about what the weights are being applied to. In this chapter, we have presented examples of how we and others have weighted pieces and types of evidence and have organized the evidence and weights in weight-of-evidence tables to serve as input to weighing the body of evidence in Chapter 8.

REFERENCES

Ågerstrand, M., K. Arnold, S. Balshine, T. Brodin, B. W. Brooks, G. Maack, E. S. McCallum, G. Pyle, M. Saaristo, and A. T. Ford. 2020. "Emerging investigator series: Use of behavioural endpoints in the regulation of chemicals." *Environmental Science: Processes and Impacts* 22 (1): 49–65.

Ågerstrand, M., M. Breitholtz, and C. Rudén. 2011. "Comparison of four different methods for reliability evaluation of ecotoxicity data: A case study of non-standard test data used in environmental risk assessments of pharmaceutical substances." *Environmental Sciences Europe* 23: 1–15.

Ågerstrand, M., A. Küster, J. Bachmann, M. Breitholtz, I. Ebert, B. Rechenberg, and C. Ruden. 2011. "Reporting and evaluation criteria as means towards a transparent use of ecotoxicity data for environmental risk assessment of pharmaceuticals." *Environmental Pollution* 159 (10): 2487–2492.

Batley, G. E., G. A. Burton, P. M. Chapman, and V. E. Forbes. 2002. "Uncertainties in sediment quality weight-of-evidence (WOE) assessments." *Human and Ecological Risk Assessment* 8: 1517–1547.

Becker, R. A., V. Dellarco, J. Seed, J. M. Kronenberg, B. Meek, J. Foreman, C. Palermo, C. Kirman, I. Linkov, and R. Schoeny. 2017. "Quantitative weight of evidence to assess confidence in potential modes of action." *Regulatory Toxicology and Pharmacology* 86: 205–220.

Benbrook, C. M. 2019. "How did the US EPA and IARC reach diametrically opposed conclusions on the genotoxicity of glyphosate-based herbicides?" *Environmental Sciences Europe* 31 (1): 1–16.

Bertram, M. L. G., J. M. Martin, E. S. McCallum, L. A. Alton, J. A. Brand, B. W. Brooks, D. Cerveny, J. Fick, A. T. Ford, and G. Hellström. 2022. "Frontiers in quantifying wildlife behavioural responses to chemical pollution." *Biological Reviews* 97 (4): 1346–1364.

Bilotta, G. S., A. M. Milner, and I. L. Boyd. 2014. "Quality assessment tools for evidence from environmental science." *Environmental Evidence* 3 (1): 14.

Bolland, M. J., A. Avenell, and A. Grey. 2024. "Publication integrity: What is it, why does it matter, how it is safeguarded and how could we do better?" *Journal of the Royal Society of New Zealand* 55 (2): 267–286.

Borgert, C. J., R. A. Becker, B. D. Carlton, M. Hanson, P. L. Kwiatkowski, M. S. Marty, L. S. McCarty, T. F. Quill, K. Solomon, and G. Van Der Kraak. 2016. "Does GLP enhance the quality of toxicological evidence for regulatory decisions?" *Toxicological Sciences* 151 (2): 206–213.

Borgert, C. J., L. D. Stuchal, E. M. Mihaich, R. A. Becker, K. S. Bentley, J. M. Brausch, K. Coady, D. R. Geter, E. Gordon, and P. D. Guiney. 2014. "Relevance weighting of tier 1 endocrine screening endpoints by rank order." *Birth Defects Research Part B: Developmental and Reproductive Toxicology* 101 (1): 90–113.

Breton, R. L., G. Gilron, R. Thompson, S. Rodney, and S. Teed. 2009. "A new quality assurance system for the evaluation of ecotoxicity studies submitted under the new substances notification regulations in Canada." *Integrated Environmental Assessment and Management* 5: 127–137.

Bridges, J., and K. R. Solomon. 2016. "Quantitative weight-of-evidence analysis of the persistence, bioaccumulation, toxicity, and potential for long-range transport of the cyclic volatile methyl siloxanes." *Journal of Toxicology and Environmental Health, Part B* 19 (8): 345–379.

Brock, T. C. M., K. C. Elliott, A. Gladbach, C. Moermond, J. Romeis, T.-B. Seiler, K. Solomon, and G. Peter Dohmen. 2021. "Open Science in regulatory environmental risk assessment." *Integrated Environmental Assessment and Management* 17 (6): 1229–1242.

Cecchetto, M., E. Giubilato, M. Picone, A. Volpi Ghirardini, C. Bettiol, F. Corami, I. Bernardini, M. Milan, T. Patarnello, and V. Matozzo. 2025. "An extended quantitative weight of evidence with uncertainty evaluation for the risk assessment of dredged sediment." *Environmental Sciences Europe* 37 (1): 169.

Chapman, P. M. 2007. "Determining when contamination is pollution—Weight of evidence determinations for sediments and effluents." *Environment International* 33 (4): 492–501.

Chapman, P. M., and J. Anderson. 2005. "A decision-making framework for sediment contamination." *Integrated Environmental Assessment and Management* 1: 163–173.

Chrisman, K., and P. Weislogel. 2003. "Correspondence about publication ethics and regulatory toxicology and pharmacology." *International Journal of Occupational and Environmental Health* 9 (4): 386–391.

Conrad, J. W., and R. A. Becker. 2011. "Enhancing credibility of chemical safety studies: Emerging consensus on key assessment criteria." *Environmental Health Perspectives* 119 (6): 757–764.

Cormier, S. M., and G. W. Suter. 2013. "A method for assessing causation of field exposure-response relationships." *Environmental Toxicology and Chemistry* 32 (2): 272–276.

Cornwall, W. 2017. "Rules of evidence." *Science* 355: 564–567.

Dawes, J. 2008. "Do data characteristics change according to the number of scale points used? An experiment using 5-point, 7-point and 10-point scales." *International Journal of Market Research* 50: 61–77.

Dekant, W., and J. Bridges. 2016. "A quantitative weight of evidence methodology for the assessment of reproductive and developmental toxicity and its application for classification and labeling of chemicals." *Regulatory Toxicology and Pharmacology* 82: 173–185.

ECHA. 2008. "Guidance on information requirements and chemical safety assessment. Chapter R.10: characterization of dose [concentration]-response for environment." European Chemicals Agency (Heksinki, Finland). https://echa.europa.eu/guidance-documents/guidance-on-information-requirements-and-chemical-safety-assessment.

ECHA. 2011. "Guidance on information requirements and chemical safety assessment, Chapter R.4: Evaluation of available information." European Chemicals Agency (Helsinki, Finland).

EFSA. 2007. "Opinion of the Scientific Committee related to uncertainties in dietary exposure assessment." *EFSA Journal* 5 (1): 438.

EFSA Scientific Committee. 2017. "Guidance on the use of the weight of evidence approach in scientific assessments." *EFSA Journal* 15 (8): 1–69.

Elliott, K. C. 2008. "Scientific judgment and the limits of conflict-of-interest policies." *Accountability in Research* 15 (1): 1–29.

Friedberg, M., B. Saffran, T. J. Stinson, W. Nelson, and C. L. Bennett. 1999. "Evaluation of conflict of interest in economic analyses of new drugs used in oncology." *JAMA* 282 (15): 1453–1457.

Griffin, D., and A. Tversky. 1992. "The weighing of evidence and the determinants of confidence." *Cognitive Psychology* 24: 411–435.

Hall, A. T., S. E. Belanger, P. D. Guiney, M. Galay-Burgos, G. Maack, W. Stubblefield, and O. Martin. 2017. "New approach to weight-of-evidence assessment of ecotoxicological effects in regulatory decision-making." *Integrated Environmental Assessment and Management* 13 (4): 573–579.

Higgins, J., and J. Thomas. 2023. *Cochrane Handbook for Systematic Reviews of Interventions Version 6.4*. London, England: Cochrane Collaboration.

Hobbs, D. A., M. St J. Warne, and S. J. Markich. 2005. "Evaluation of criteria used to assess the quality of aquatic toxicity data." *Integrated Environmental Assessment and Management* 1: 174–180.

Hope, B. K., and J. R. Clarkson. 2014. "A strategy for using weight-of-evidence methods in ecological risk assessments." *Human and Ecological Risk Assessment* 20 (2): 290–315.

Hume, D. 1748. *An Inquiry Concerning Human Understanding*. Amherst, NY: Prometheus Books.

Ingre-Khans, E., M. Ågerstrand, A. Beronius, and C. Rudén. 2019. "Reliability and relevance evaluations of REACH data." *Toxicology Research* 8 (1): 46–56.

Kaurov, A. A., and N. Oreskes. 2025. "The afterlife of a ghost-written paper: How corporate authorship shaped two decades of glyphosate safety discourse." *Environmental Science & Policy* 171: 104160.

Klimisch, H. J., M. Andreae, and U. Tillmann. 1997. "A systematic approach to evaluating the quality of experimental toxicological and ecotoxicological data." *Regulatory Toxicology and Pharmacology* 25: 1–5.

Kuhn, T. S. 1997. *The Structure of Scientific Revolutions*. Vol. 962. Chicago, IL: University of Chicago Press.

Liverpool, L. 2023. "AI intensifies fight against 'paper mills' that churn out fake research." *Nature* 618 (7964): 222–223.

Mandrioli, D., and E. K. Silbergeld. 2016. "Evidence from toxicology: the most essential science for prevention." *Environmental Health Perspectives* 124 (1): 6–11.

Matheson, A. 2023. "The "Monsanto papers" and the nature of ghostwriting and related practices in contemporary peer review scientific literature." *Accountability in Research* 31 (8): 1152–1181.

McCarty, L. S., C. J. Borgert, and E. M. Mihaich. 2012. "Information quality in regulatory decision making: Peer review versus good laboratory practice." *Environmental Health Perspectives* 120 (7): 927–934.

Mebane, C. A., J. P. Sumpter, A. Fairbrother, T. P. Augspurger, T. J. Canfield, W. L. Goodfellow, P. D. Guiney, A. LeHuray, L. Maltby, and D. B. Mayfield. 2019. "Scientific integrity issues in environmental toxicology and chemistry: Improving research reproducibility, credibility, and transparency." *Integrated Environmental Assessment and Management* 15 (3): 320–344.

Michaels, D. 2008. *Doubt Is Their Product*. New York: Oxford University Press.

Moermond, C., A. Beasley, R. Breton, M. Junghans, R. Laskowski, K. Solomon, and H. Zahner. 2017. "Assessing the reliability of ecotoxicological studies: An overview of current needs and approaches." *Integrated Environmental Assessment and Management* 13 (4): 640–651.

Moermond, C. T. A., R. Kase, M. Korkaric, and M. Ågerstrand. 2016. "CRED: Criteria for reporting and evaluating ecotoxicity data." *Environmental Toxicology and Chemistry* 35 (5): 1297–1309.

Moore, D. R. J., C. D. Greer, G. Manning, K. Wooding, K. J. Beckett, R. A. Brain, and G. Marshall. 2017. "A weight-of-evidence approach for deriving a level of concern for atrazine that is protective of aquatic plant communities." *Integrated Environmental Assessment and Management* 13 (4): 686–701.

Mupepele, A.-C., J. C. Walsh, W. J. Sutherland, and C. F. Dormann. 2016. "An evidence assessment tool for ecosystem services and conservation studies." *Ecological Applications* 26 (5): 1295–1301.

Norton, S. B., S. M. Cormier, and G. W. Suter. 2014. *Ecological Causal Assessment*. Boca Raton, FL: CRC Press.

OECD. 2019. *Guiding Principles and Key Elements for Establishing a Weight of Evidence for Chemical Assessment*. Paris: Organization for Economic Cooperation and Development, Environmental Directorate. https://www.oecd.org.

OHAT. 2019. *Handbook for Conducting a Literature-Based Health Assessment Using OHAT Approach for Systematic Review and Evidence Integration*. Research Triangle Park, NC: Office of Health Assessment and Translation, National Toxicology Program. https://ntp.niehs.nih.gov/sites/default/files/ntp/ohat/pubs/handbookmarch2019_508.pdf

Oreskes, N., K. Shrader-Frechette, and K. Belitz. 1994. "Verification, validation, and confirmation of numerical models in the earth sciences." *Science* 263 (5147): 641–646.

Popper, K. R. 1968. *The Logic of Scientific Discovery.* New York: Harper and Row.

Ridley, C. E., S. D. Kaylor, S. S. Lee, J. N. P. Miller, S. Penry, and K. A. Schofield. 2024. *Application of Weight-of-Evidence Methods for Transparent and Defensible Numeric Nutrient Criteria.* Washington, DC: U.S. Environmental Protection Agency. https://www.epa.gov/water-research/application-weight-evidence-methods-transparent-and-defensible-numeric-nutrient.

Rooney, A. A., A. L. Boyles, M. S. Wolfe, J. R. Biucher, and K. A. Thayer. 2014. "Systematic review and evidence integration for literature-based environmental health science assessments." *Environmental Health Perspectives* 122: 711–718.

Rudén, C., J. Adams, M. Ågerstrand, T. C.M. Brock, V. Poulsen, C. E. Schlekat, J. R. Wheeler, and T. R. Henry. 2017. "Assessing the relevance of ecotoxicological studies for regulatory decision making." *Integrated Environmental Assessment and Management* 13 (4): 652–663.

Rudén, C., and S. O. Hansson. 2008. "Evidence-based toxicology: "sound science" in new disguise." *International Journal of Occupational and Environmental Health* 14 (4): 299–306.

vom Saal, F. S., and J. P. Myers. 2010. "Good laboratory practices are not synonymous with good scientific practices, accurate reporting, or valid data." *Environmental Health Perspectives* 118 (2): A60.

Samuel, G. O., S. Hoffmann, R. A. Wright, M. M. Lalu, G. Patlewicz, R. A. Becker, G. L. DeGeorge, D. Fergusson, T. Hartung, and R. J. Lewis. 2016. "Guidance on assessing the methodological and reporting quality of toxicologically relevant studies: A scoping review." *Environment International* 92: 630–646.

Schneider, K., M. Schwarz, I. Burkholder, A. Kopp-Schneider, L. Edler, A. Kinsner-Ovaskainen, T. Hartung, and S. Hoffmann. 2009. ""ToxRTool", a new tool to assess the reliability of toxicological data." *Toxicology Letters* 189 (2): 138–144.

Schünemann, H., J. Brożek, G. Guyatt, and A. Oxman. 2013. *The GRADE Handbook.* London, England: Cochrane Collaboration.

Segal, D., S. L. Makris, A. D. Kraft, A. S. Bale, J. Fox, M. Gilbert, D. R. Bergfelt, K. C. Raffaele, R. B. Blain, and K. M. Fedak. 2015. "Evaluation of the ToxRTool's ability to rate the reliability of toxicological data for human health hazard assessments." *Regulatory Toxicology and Pharmacology* 72 (1): 94–101.

Staveley, J. P., S. A. Law, A. Fairbrother, and C. A. Menzie. 2014. "A causal analysis of observed declines in managed honey bees (*Apis mellifera*)." *Human and Ecological Risk Assessment* 20 (2): 566–591.

Stevenson, R. W., and P. M. Chapman. 2017. "Integrating causation in investigative ecological weight of evidence assessments." *Integrated Environmental Assessment and Management* 13 (4): 702–713.

Susser, M. 1986. "Rules of inference in epidemiology." *Regulatory Toxicoogy and Pharmacology* 6 (2): 116–128.

Suter, G. W., and S. M. Cormier. 2014. "The problem of biased data and potential solutions for health and environmental assessments." *Human and Ecological Risk Assessment* 21 (7): 1736–1752.

Tabuchi, H. 2026. "Study's retraction renews weedkiller concerns: EPA facing pressure to reassess roundup." *New York Times* (National), January 4, 2026, A1, A19.

UK Department of the Environment, Food and Rural Affairs. 2000. *Guidelines for Environmental Risk Assessment and Management.* London: Stationary Office.

USEPA. 2000. *Stressor Identification Guidance Document.* Washington, DC: U.S. Environmental Protection Agency.

USEPA. 2011. *A Field-Based Aquatic Life Benchmark for Conductivity in Central Appalachian Streams*. Washington, DC: Office of Research and Development, National Center for Environmental Assessment, U.S. Environmental Protection Agency.

USEPA. 2014. *An Assessment of Potential Mining Impacts on Salmon Ecosystems of Bristol Bay, Alaska*. Seattle, WA: U.S. Environmental Protection Agency.

USEPA. 2016. *Weight of Evidence in Ecological Assessment*. Washington, DC: Risk Assessment Forum, U.S. Environmental Protection Agency.

USEPA. 2022. *ORD Staff Handbook for Developing IRIS Assessments*. Washington, DC: Office of Research and Development, U.S. Environmental Protection Agency.

Van Der Kraak, G. J., A. J. Hosmer, M. L. Hanson, W. Kloas, and K. R. Solomon. 2014. "Effects of atrazine in fish, amphibians, and reptiles: An analysis based on quantitative weight of evidence." *Critical Reviews in Toxicology* 44 (sup5): 1–66.

Velicer, C., G. St Helen, and S. A. Glantz. 2017. "Tobacco papers and tobacco industry ties in regulatory toxicology and pharmacology." *Journal of Public Health Policy* 39 (1): 34.

Wagner, W., E. Fisher, and P. Pascual. 2018. "Whose science? A new era in regulatory "science wars"." *Science* 362 (6415): 636–639.

Warne, M., G. E. Batley, R. A. van Dam, J. C. Chapman, D. R. Fox, C. W. Hickey, and J. L. Stauber. 2025. *Revised Method for Deriving Australian and New Zealand Water Quality Guideline Values for Toxicants*. Brisbane: Department of Science, Information Technology and Innovation.

Weiss, N. S. 2002. "Can the "specificity" of an association be rehabilitated as a basis for supporting a causal hypothesis?" *Epidemiology* 13 (1): 6–8.

Williams, G. M., R. Kroes, and I. C. Munro. 2000. "Safety evaluation and risk assessment of the herbicide Roundup and its active ingredient, glyphosate, for humans." *Regulatory Toxicology and Pharmacology* 31 (2): 117–165.

Williams, G. M., R. Kroes, and I. C. Munro. 2025. "Retraction notice to "Safety evaluation and risk assessment of the herbicide roundup and its active ingredient, glyphosate, for humans." *Regulatory Toxicology and Pharmacology* 31 (2000): 117–165.

Zheng, L., J. Gerritsen, and S. M. Cormier. 2014. "Clear fork watershed case study: The value of state monitoring programs." In *Ecological Causal Assessment*, edited by S. B. Norton, S. M. Cormier, and G. W. Suter, 353–383. Boca Raton, FL: CRC Press.

8 Weigh the Body of Evidence

We have assembled evidence in Chapter 6 and weighted it in Chapter 7. Now, we have a weighted body of evidence to integrate, weigh, and use to draw conclusions. In the metaphor of weight of evidence, this is the point at which we load the weighted evidence in the pans of the scales and weigh them by noting the relative height of the pans for alternative hypotheses. The weights for the hypotheses are then used to interpret the bodies of evidence to reach a conclusion. If a conclusion is not clear, ambiguities and discrepancies should be described. The steps in this process are presented in Figure 8.1. Evidence synthesis is sometimes a synonym for weighing the evidence. However, it is ambiguous because it does not recognize the importance of the weight of the evidence.

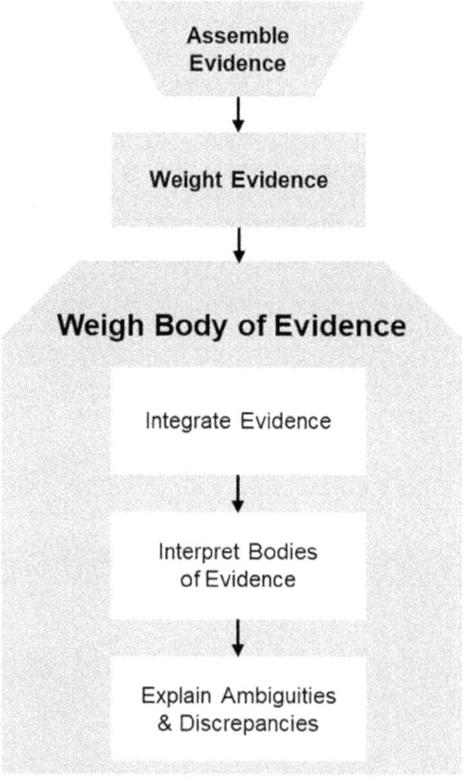

FIGURE 8.1 An elaboration of the process for weighing a body of evidence, the third step in the weight-of-evidence process. (Adapted from USEPA 2016a.)

DOI: 10.1201/9781003626336-10

8.1 INTEGRATE THE BODY OF EVIDENCE

The input to the integration of the body of evidence is the weighted pieces or types of evidence, which should be available in evidence rating tables (Section 7.2.4). Now it is time to evaluate the body of evidence for each hypothesis with respect to the combined weight of evidence for each hypothesis and the collective properties of the bodies of evidence.

8.1.1 Evidence Weighing Tables

The first step is to derive an evidence weighing table. An obvious difference from an evidence weighting table (Section 7.2.4) is that an evidence weighing table includes the collective properties of the body of evidence and includes all hypotheses so as to facilitate comparisons (Table 8.1). The rows may be defined by the same pieces or types of evidence in the evidence weighting table. Alternatively, they may be defined by characteristics or implications.

Defining the rows by the characteristics of the hypothesized quality being assessed (e.g., causation, protection, impairment, teratogenicity, carcinogenicity, or remediation) may be very helpful in weighing the body of evidence (Table 8.1).

TABLE 8.1
A Generic Weight-of-Evidence Table for 3 Alternative Causal Hypotheses (H_1, H_2, H_3), Based on Causal Characteristics (Table 14.1) and Collective Properties of the Bodies of Evidence (Box 8.1)

	Combined Weight		
	H_1	H_2	H_3
Characteristics of Causation			
Co-occurrence			
Sufficiency			
Time order			
Interaction			
Alteration			
Antecedents			
Body of Evidence, Collective Properties			
Number			
Coherence			
Absence of bias			
Diversity			
Integrated Weight of Evidence			
Weight of evidence for the hypothesis			

Adapted from USEPA (2016a).

For example, a causal assessment of honeybee decline weighs evidence for candidate causes with respect to causal characteristics rather than properties (Staveley et al. 2014). Characteristics for some common qualities in environmental or health assessment are presented in Appendix B. For example, time order is a characteristic of causation, and occurrence at concentrations greater than background is a characteristic of contaminants of concern. It can be advantageous to weigh evidence based on the characteristic for which it is evidence rather than its type. You can see what is known about the hypothesized quality based on the evidence for its defining characteristics. Therefore, it is important to clearly associate the characteristics with the pieces of evidence that support or weaken them.

Defining the rows by implications of the evidence (Section 6.5) can also be helpful. Implications might include exposure/response relationship, environmental influence, external exposure, internal exposure, mechanism of action, etc. They do not define the hypothesized quality like the characteristics, but they can organize the evidence in terms of its relevance to the hypotheses. As with the characteristics, it is important to associate the implications with specific pieces of evidence.

8.1.2 Properties of Bodies of Evidence

Bodies of evidence have collective properties that give them more or less weight (Box 8.1). The collective properties should be evaluated and rated like the properties of the pieces or types of evidence. For example, you could evaluate and rate the properties of a body of evidence and combine them to conclude that the weight of the body of evidence collectively is strongly supportive of the hypothesis (++).

BOX 8.1 COLLECTIVE PROPERTIES OF BODIES OF EVIDENCE (MODIFIED FROM NORTON, CORMIER, AND SUTER 2014 AND USEPA 2016A)

The following collective properties of the body of evidence can be considered in addition to considering the relevance, reliability, and strength of the pieces or types of evidence. As with the properties of evidence, any of these properties could be used for weighting, depending on their utility for distinguishing among the hypotheses.

Number: More pieces or types of evidence within a body of evidence increase the reliability, if they are consistent and, ideally, are generated independently.

Coherence: Logical consistency among types of evidence increases the reliability of a body of evidence, particularly when the types of evidence are diverse.

Absence of Bias: Consistent results are more reliable if they are from different types of institutions and funders (e.g., academic, industry, government, nongovernmental organization, government). This is different from the absence of bias in a piece of evidence (Box 7.2).

Diversity: Evidence from more responses, taxa, life stages, or communities under more conditions is more likely to detect important exposures, responses, and relationships.

8.1.3 Weighing the Importance of Effects

In general, the purpose of weight-of-evidence systems is to determine what hypothesis best explains the evidence or whether the evidence supports a hypothesis. However, some systems add a step that weighs the importance of the effects. For example, the GRADE system uses a scale of 1–9 to rate the importance of medical effects ranging from nausea to death (Schünemann et al. 2013). If a standard scale of importance is not available, it may be best to leave value judgments concerning the importance of effects to a designated decision maker. However, some decision makers want help in interpreting effects, particularly unfamiliar ecological effects.

If importance is a property of the outcome, it may be necessary to weigh specific attributes of the assessment endpoints (USEPA 2016b). For ecological assessments, they might include the presence of threatened, endangered, or highly valued species, area of habitat affected, presence of wetlands, aesthetic value, recreational value, loss of species richness, etc.

8.2 INTERPRET THE BODY OF EVIDENCE

Having integrated the evidence in an evidence weighing table, you can proceed to interpret it to reach a conclusion. You will ask, which, if any, hypothesis is well supported, and which is best supported? While you can ask an absolute question such as, are oil field chemicals the cause of kit fox decline (Chapter 1), it is better to ask what caused the kit fox decline and then compare alternative hypotheses. You may have mostly positive evidence for toxicity, but without a comparison of causal hypotheses, you cannot be sure that toxicity is the best-supported hypothesis. A single hypothesis can be judged by comparison to its negation. For example, the hypothesis that a chemical is a teratogen can be compared to the hypothesis that it is not a teratogen. Similarly, tests and field studies that found an association of deformities with a chemical exposure could be weighed and compared to those that did not, based on relevance and reliability, as well as the conventional strength of effects.

If all goes well, your inference by weight of evidence will end when you examine the evidence weighing table, compare the ratings of evidence for the hypotheses, and find a conclusive result (Figure 8.2). However, you may not have a conclusive result. If so, you can proceed to settle in for some critical thinking about what is going on with the evidence relative to the hypotheses. We have found that, in most cases, we have been able to make sense of the evidence and draw a conclusion. Either way, in most cases, you will want to express your conclusions categorically. In the worst case, the conclusion will be that no conclusion can be reached. In such cases, you declare a null result or determine what reinterpretation of the case or what new information would allow a repeat assessment to succeed.

8.2.1 Interpretation Based on the Ratings

The distribution of ratings in an evidence weighing table can often be interpreted directly. That is, you can look at the table and say that one hypothesis is clearly well supported and better supported than the others. It can be helpful to combine the

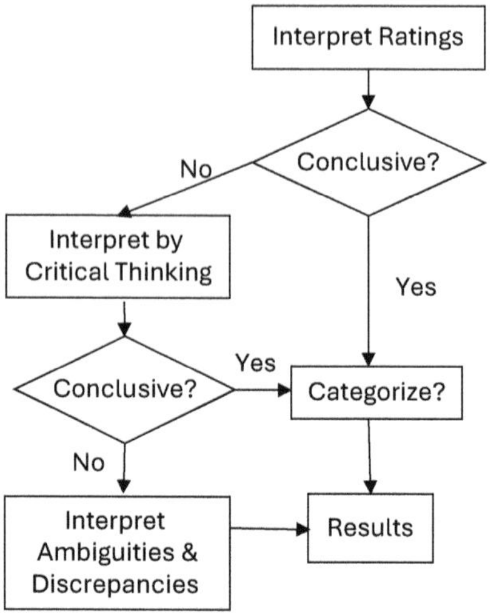

FIGURE 8.2 A diagram of the process of interpreting the body of evidence. *Categorize* has a question mark because it may be skipped.

weights of the evidence for each hypothesis, as indicated by the last row in Table 8.1, to obtain an overall rating for each hypothesis. Table 8.2 is part of a large evidence weighing table that had an obvious interpretation: predation by coyotes was the cause of kit fox decline. The ratings of evidence for predation were all positive, and some were convincing. The evidence for accidents was also positive. However, their contribution was small. Based on evidence from demographic models, the predation effects alone were sufficient to cause the decline (Table 8.2, row 6).

TABLE 8.2
Partial WoE Table for Alternative Possible Causes of the Decline of San Joaquin Kit Foxes.

Categories of Evidence	Predation	Toxics	Accidents	Disease
1. Spatial/temporal co-occurrence	+	+	+	–
2. Temporal sequence	0	0	NE	NE
3. Evidence of exposure or biological mechanism	++	++	++	– –
4. Causal pathway	+	+	+	0
5. Manipulation of exposure	+	NE	NE	NE
6. Stressor-response relationships from demographic models	+++	–	+	– –
7. Coherence	+++	–	+	+

Source: Adapted from (Suter and O'Farrell 2009).
For illustrative purposes, only 4 of 6 potential causes and 7 of 16 types of evidence are shown.

Note that, even when examination of the evidence weighing table does not result in the selection of a hypothesis, it can eliminate hypotheses. Clear negative conclusions tend to be more apparent than clear positive conclusions.

If the results are apparent from the ratings, you have two options. If you want to express the results categorically, proceed to Section 8.2.3. If your results do not have a categorical expression, proceed to Section 8.3 to express the results and your confidence in them. If the results from the ratings are not conclusive, more thoughtful interpretation is needed, based on critical thinking.

8.2.2 Interpretation Based on Critical Thinking

If the conclusion is not evident from examining the evidence weighing table and comparing the evidence ratings for the hypotheses, critical thinking should be applied to the evidence for the remaining hypotheses (Box 8.2). In this interpretation step, it is assumed that the evidence is sufficient, and one of the hypotheses is correct and will be revealed upon further examination. The critical thinking is intended to encourage you to think deeply and carefully. If this process fails to arrive at a conclusion, you may consider whether the hypotheses must be reformulated or more evidence must be obtained (Section 8.3.4).

BOX 8.2 CRITICAL THINKING

Critical thinking is an intellectual approach that relies on habits of thought rather than particular methods. It applies reason to problems in a way that is logical and objective. It differs from deduction and induction in applying the most appropriate reasoning to whatever problems present themselves. The modern usage of the concept is attributed to the pragmatist philosopher John Dewey (1910).

There are many guides to critical thinking and types of critical thinking, such as Socratic questioning. An appropriate approach in this context is to consider whether the relationship of evidence to a hypothesis has "theoretical virtues" that make it more or less fitting as an explanation of the condition to which the hypotheses are applied. An example is the explanatory virtues used to evaluate explanatory hypotheses (Section 3.6). The following list of virtues more completely addresses the evidence, the hypotheses, and the relationships among them than the basic set in Section 3.6. They are based on our experience and on the experience of others (Watson 2024; Lett 1990; Pinkston 2020). They are not criteria or even properties that evidence should have. Rather, they are points to consider that may aid your inference. For example, you would not give more weight to a simpler hypothesis, because Ockham's razor is just a rule of thumb. You might, however, take simplicity into consideration when thinking critically about the evidence and hypotheses (see simplicity below). Then, ask yourself whether a more complex hypothesis is more explanatory, more confusing, or more demanding of evidence.

Exhaustive Evidence: All available and potentially relevant evidence should be considered in the weighing of evidence for each hypothesis. Is there any evidence that you missed or may have eliminated improperly?

Replicated Evidence: Consider whether any of the evidence may be due to coincidence and, if so, whether that evidence has been replicated.

Replicated Weights: You should have carefully weighted the evidence (Chapter 7), but you might enlist one or more independent assessors to reconsider the weights. Differences in weights may indicate the causes of discrepant results.

Falsification: Popperian falsification is strong evidence, and it should be identified when it occurs. A positive body of evidence results from a number of positive pieces or categories of evidence. However, even one clearly negative piece of evidence can falsify a hypothesis. For example, effects on a stream community upstream of a source or a source that began after the effects can eliminate the hypothesis that the effluent caused the stream impairment.

Relative Weights of Properties: Experimental psychological studies have shown that strength is often given too much weight and that reliability is given too little (Griffin and Tversky 1992). Strength is salient and often quantitative, while most evidence of reliability is qualitative and, therefore, is given less credence. However, careful consideration shows that the studies that generate estimates of strength often have weaknesses and biases that are inherent in the study design or are introduced by flawed implementation. Relevance is essential, but it is seldom the deciding factor because the information derivation process seldom results in irrelevant evidence. Consider whether you have underweighted reliability.

Honesty: Consider whether you may have a bias with respect to the assessment, and reexamine the evidence and weights with that potential bias in mind. This is an anti-biasing self-check.

Appropriate Formulation: Glen Fox (1991) inverted the usual question, *is the evidence sufficient to accept the hypothesis?* He asked, "What evidence must I ignore to conclude that a causal relationship does not exist?" (Fox 1991). That is, rather than a priori demanding that certain standards of proof apply, he would start with the assumption that we are assessing a putative cause because we have reason to believe that it is true. He was responding to cases such as acid deposition and forest decline, for which there was considerable evidence, but more was always being demanded. This form of inference is equivalent to distinguishing confirmation (does the evidence confirm the hypothesis?) from explanation (does the hypothesis explain the evidence?) (Pinkston 2020). You could consider whether reformulating the inference can clarify your case.

Simplicity: Occam's law of parsimony indicates that simple explanations should be preferred. This is a scientific rule of thumb that is, despite its name, not a law. However, it is a virtue that gives some weight to simpler hypotheses. That is, if there is equal evidence for two hypotheses, but one is more causally complex, the simpler one has more evidence per causal link.

Sufficiency: There are few criteria for a sufficient body of evidence. Clearly, one positive piece of evidence that comes from a highly relevant and reliable experiment and gives a strong and reasonable result may be sufficient. However, most cases require multiple pieces of evidence to achieve sufficiency. Standards of sufficiency may be available in guidance documents (Section 8.4). If they are not available for your assessment, it is desirable to explicitly define sufficiency for your case. It may change your conclusion.

Conservatism: A well-known rule that dates back to David Hume and Blaise Pascal is, extraordinary claims require extraordinary evidence. A version of this rule that is more commonly applicable is, a good explanation does not contradict well-established knowledge in a field. If it does, the evidence should be extraordinarily weighty.

Explanatory Depth: A good explanation provides details of how an effect occurs or could occur.

Explanatory Breadth: A good explanation explains a range of related instances and phenomena.

Plausibility: A body of evidence should make sense in light of a hypothesis. In your case, are the hypotheses still plausible and not just weighty after weighing the body of evidence?

Precaution: An essential concept when assessments are required to protect human health or the environment is precaution. The evidence is never complete and never absolutely reliable, so some precautionary measures are usually required. The degree of precaution should be a matter of ethics and policy, not science (Weed 2004). The obvious response is to give less weight to evidence that indicates no risk or more weight to evidence that indicates high risk. At the extreme, one can give all weight to the evidence that indicates the greatest risk (Section 2.8). Another approach is to reduce the number of evidential criteria (Section 2.7) that must be met before accepting the existence of a risk or a cause. More generally, you can base a precautionary conclusion on a minimal body of evidence. Your willingness to make precautionary decisions should be informed by the potential consequences of not being protective, but the degree of precaution should be guided by policy.

Appropriate Standard of Evidence: As in civil courts of law, the standard of evidence in most environmental assessments is "the preponderance of evidence," which may be interpreted as, the hypothesis is more likely than not to be true. More demanding standards, such as "beyond a reasonable doubt," are appropriate when protecting humans and endangered species. Consider what standard is appropriate to your case and how it should influence the inference.

If the critical thinking process results in a conclusion, you have two options. If you want to use standard categorical results, proceed to Section 8.2.3. If your results do not have a categorical expression, proceed to Section 8.2.4 to explain the results in a narrative. If you still do not have results, explain the ambiguities and discrepancies that prevent you from reaching a conclusion (Section 8.3).

8.2.3 Categorical Results of Interpretations

The magnitude of the probability of an argument...depends upon a balance between the favorable and unfavorable evidence.

John Maynard Keynes (1921)

You will probably want to express the results of weighing the body of evidence as categorical conclusions concerning the hypotheses. The categories depend on the hypotheses being considered and the decision context. The number of categories depends on the discrimination of conclusions to be achieved.

8.2.3.1 Null

Sometimes there is no evidence to support a conclusion. The information search may have turned up nothing, or the studies that were found were all screened out.

8.2.3.2 Binary

When assessing a single hypothesis, such as the effluent caused the fish kill or atrazine is an anuran teratogen, a dichotomous logic is used. Atrazine either is or is not an anuran teratogen. Binary categories imply that you are confident in your results from weighing the evidence; there is no indeterminate (weight 0) result. In some decision contexts, the binary logic is imposed. Given the evidence, you must reach a thumbs-up or thumbs-down conclusion. However, the degree of confidence may be prescribed. For example, in U.S. courts of law, the decisions are binary and must be made by one of three standards depending on the type of case: preponderance of evidence, clear and convincing evidence, and beyond a reasonable doubt. Most environmental assessments use a preponderance of evidence as their standard for a conclusion. However, higher standards may be applied in some cases, such as registering a pesticide to which an endangered species would be exposed. In such cases, a sufficient demonstration of safety may require proof beyond a reasonable doubt.

Some binary distinctions are quantitative, such as whether a concentration exceeds a regulatory benchmark value. If these decision criteria are created for your case, it is possible to determine the outcome of an assessment by that choice (Grimes and Heathers 2025). Therefore, standard criterion values are preferred, and criteria developed after the data have been examined should be carefully justified by evidence.

8.2.3.3 Trinary

A three-value logic is a dichotomy with a knowledge gap. As with the binary outcomes, a hypothesis may be that a chemical has some property or does not have the property, and the third outcome is no decision. Examples of trinary outcomes include:

- True, False, Unknown,
- Safe, Harmful, Uncertain,
- Cause, Not the Cause, Undetermined,
- Clearly anthropogenic, Clearly natural, Unclear, and
- Teratogen, Non-teratogen, Insufficient evidence.

In some cases, the categories are standardized. For example, at one point the USEPA proposed a trinary carcinogenicity categorization: known/likely, not likely, and cannot be determined (USEPA 1996). Three-value logic can also be used to narrow the options and guide the next assessment step (Box 8.3).

The advantage of a three-value logic is that it is simple without being misleading or failing to acknowledge uncertainty or lack of information. By acknowledging uncertainty, this logic can reduce the skepticism of the public or stakeholders. It also emphasizes that you are confident in some results.

A disadvantage of three-value logic is that it does not provide a bright line and, therefore, if the results fall into the indeterminate category, it places the burden on the decision maker. Unclear results may prompt a decision to gather more information and repeat the assessment, take a precautionary approach, or take a reckless approach.

BOX 8.3 THREE-VALUE LOGIC AND INDETERMINATE INFERENTIAL RESULTS (SHULL AND PULKET 2015; SCHALL ET AL. 2020, 2023)

The CADDIS system for weighing causal evidence was used to assess the cause of a large decline in the abundance of smallmouth bass in the Susquehanna River. The existing information was not sufficient to identify the cause. However, it was sufficient to identify 18 candidate causes and to classify them: 2 are likely, 8 are unlikely, and 8 are uncertain. Guided by these results, several studies were subsequently conducted that suggested that disease in young-of-the-year fish was interacting with other causes. The population has largely recovered, but studies continue.

8.2.3.4 Polynary

More than three categories of conclusions imply more information and more choices. The best-known polynary categories of conclusions in toxicological and environmental assessment are the potential conclusions concerning carcinogens (USEPA 2005).

- Carcinogenic to humans.
- Likely to be carcinogenic to humans.
- Suggestive evidence of carcinogenic potential.
- Inadequate information to assess carcinogenic potential.
- Not likely to be carcinogenic to humans.

These USEPA categories are defined by descriptive text to be used for performing a narrative weight of evidence. However, the equivalent categorization for the International Agency for Research on Cancer provides rules for integrating types of evidence (Table 8.3).

Similar categories can be devised for other specific assessment types, such as conclusions concerning causation in bodies of evidence for assessments of air pollutants

TABLE 8.3
Integration of Types of Evidence in Reaching Overall Cancer Hazard Classifications

Types of Evidence			Classification Based on Weight of Body of Evidence
Evidence of Cancer in Humans	**Evidence of Cancer in Animals**	**Mechanistic Evidence**	
Sufficient Limited or inadequate	Not necessary ***Sufficient***	Not necessary ***Strong (exposed humans)***	Carcinogenic to humans (Group 1)
Limited Inadequate ***Limited*** Limited or Inadequate	***Sufficient*** ***Sufficient*** Less than sufficient Not necessary	***Strong, Limited, or Inadequate*** ***Strong (human cells of tissues)*** ***Strong*** ***Strong (mechanistic class)***	Probably carcinogenic to humans (Group 2A)
Limited Inadequate Inadequate ***Limited***	Less than sufficient ***Sufficient*** Less than sufficient ***Sufficient***	Limited or Inadequate Strong, Limited, or Inadequate ***Strong*** ***Strong (does not operate in humans)***	Possibly carcinogenic to humans (Group 2B)
Inadequate	***Sufficient***	***Strong (does not operate in humans)***	Not classifiable as to its carcinogenicity to humans (Group 3)

All other situations not listed above
Source: Adapted from IARC (2019).
The evidence in bold italic represents the basis of the overall evaluation.

(Table 8.4). Each conclusion should have guidance for when it is applied and implications for regulatory action.

All of these systems for categorical conclusions use terms that are not entirely clear, such as inadequate, limited, or suggestive. They could benefit from more clearly defined rating systems.

8.2.4 Explain Ambiguities and Discrepancies

In some cases, weighing the evidence for the alternative hypotheses will not answer the assessment question, even when supplemented by critical thinking. The results may be discrepant, ambiguous, or nonsensical; examples include the following.

TABLE 8.4
Weight of Evidence for Causal Determination in the USEPA's Integrated Science Assessments for Air Quality Standards (USEPA 2015)

Determination	Description of Weight of Evidence
Causal relationship	Evidence is sufficient to conclude that there is a causal relationship with relevant pollutant exposures (e.g., doses or exposures generally within one to two orders of magnitude of recent concentrations). That is, the pollutant has been shown to result in health effects in studies in which chance, confounding, and other biases could be ruled out with reasonable confidence. For example: (1) controlled human exposure studies that demonstrate consistent effects, or (2) observational studies that cannot be explained by plausible alternatives or that are supported by other lines of evidence (e.g., animal studies or mode-of-action information). Generally, the determination is based on multiple high-quality studies conducted by multiple research groups.
Likely to be a causal relationship	Evidence is sufficient to conclude that a causal relationship is likely to exist with relevant pollutant exposures. That is, the pollutant has been shown to result in health effects in studies where the results are not explained by chance, confounding, and other biases, but uncertainties remain in the evidence overall. For example: (1) observational studies show an association, but copollutant exposures are difficult to address and/or other lines of evidence (controlled human exposure, animal, or mode-of-action information) are limited or inconsistent, or (2) animal toxicological evidence from multiple studies from different laboratories demonstrates effects, but limited or no human data are available. Generally, the determination is based on multiple high-quality studies.
Suggestive of, but not sufficient to infer, a causal relationship	Evidence is suggestive of a causal relationship with relevant pollutant exposures, but is limited, and chance, confounding, and other biases cannot be ruled out. For example: (1) when the body of evidence is relatively small, at least one high-quality epidemiologic study shows an association with a given health outcome and/or at least one high-quality toxicological study shows effects relevant to humans in animal species, or (2) when the body of evidence is relatively large, evidence from studies of varying quality is generally supportive but not entirely consistent, and there may be coherence across lines of evidence (e.g., animal studies or mode-of-action information) to support the determination.
Inadequate to infer a causal relationship	Evidence is inadequate to determine that a causal relationship exists with relevant pollutant exposures. The available studies are of insufficient quantity, quality, consistency, or statistical power to permit a conclusion regarding the presence or absence of an effect.
Not likely to be a causal relationship	Evidence indicates there is no causal relationship with relevant pollutant exposures. Several adequate studies, covering the full range of levels of exposure that human beings are known to encounter and considering at-risk populations and life stages, are mutually consistent in not showing an effect at any level of exposure.

- The bodies of evidence for all assessed hypotheses may be incoherent.
- One hypothesis has sufficient evidence, is supported by abundant positive evidence, and has no negative evidence, but most of the evidence has low reliability. However, some other hypotheses have only a few pieces of positive evidence, but all of it is reliable and relevant.
- No hypothesis has predominantly positive evidence.

In such cases, you may be out of time or funds, in which case you can simply summarize and release whatever you accomplished and move on to your next project. However, if you have the option to continue, you should step back and reframe the problem or obtain new evidence (Section 9).

8.3 RESULTS

The results of weighing evidence include conclusions, confidence in the conclusions, justification of the conclusions, and a narrative explaination of the conclusions, including their bases. When communicating the results of weighing the evidence, you should keep in mind that your purpose is to inform fellow assessors, regulatory decision makers, courts of law, stakeholders, the general public, or other interested nations (e.g., signatories to environmental treaties) (Walker 1996).

8.3.1 Conclusions

The conclusion from interpreting the body of evidence should be a chosen hypothesis and confidence in that hypothesis, or failure to identify a hypothesis with sufficient confidence. In the latter case, the assessment process may end, or the ambiguities and discrepancies that prevented a conclusion should be assessed to reformulate the assessment (Chapter 9).

Conclusions can be expressed in various ways. You may simply state that a hypothesis has been accepted or rejected. For example, the hypothesis that the kit fox decline was due to predation is accepted, based on the weight of evidence. If you use a set of categorical results, you may state which categorical conclusion is accepted. For example, based on sufficient evidence from animal tests, the chemical is judged to be a possible human carcinogen (Table 8.3). Other forms of conclusions may be appropriate for other purposes.

It may also be appropriate to identify the hypotheses that were rejected and why they were rejected. This is particularly important if some of the rejected hypotheses were advocated by stakeholders. For example, California Fish and Game and some environmental advocacy groups believed that pollution from oil development was the cause of the Elk Hills kit fox decline. It was essential for me to explain the weakness of the pollution hypothesis and not just the strength of the predation hypothesis.

8.3.2 Confidence, Uncertainty, and Variability in Conclusions

An advantage of inference by weight of evidence is that the integrated weight of the body of evidence is an expression of confidence in a hypothesis given the evidence (bottom line, Table 8.1). For example, the weight of evidence for predation as the cause of kit fox decline was "strong," so we had strong confidence in that conclusion.

However, if a critical thinking step is added to the interpretation of the evidence ratings (Section 8.2.2), confidence in the conclusion may increase relative to the confidence provided by the weight of evidence alone.

Confidence, rather than uncertainty, is the best expression of the appropriate degree of belief in assessment results (Mastrandrea et al. 2010; SPC 2000; Suter 2023). The Intergovernmental Program on Climate Change (IPCC) defines it in terms of weighing evidence, as follows (Mastrandrea et al. 2010): "Confidence in the validity of a finding, based on the type, amount, quality, and consistency of evidence (e.g., mechanistic understanding, theory, data, models, expert judgment) and the degree of agreement. Confidence is expressed qualitatively."

Uncertainty is a lack of knowledge concerning the state of an organism or system or concerning the true value of a quantity. Uncertainty is a property of the investigator, which may be reduced by measurement or observation. In assessments, uncertainty is often expressed by a list or tabulation of things that are unknown but would be good to know. Those uncertainties may be ranked or rated. Results of an analysis of uncertainty can be presented along with confidence to provide a full understanding of the qualification of the assessment results (ECHA 2024; Mastrandrea et al. 2011).

Variability is a property of the system that cannot be reduced by measurement or observation, such as variance in stream flow and the resulting variance in exposure. It may be expressed as a distribution or range.

8.3.3 Justification of Conclusions

In some cases, the presentation of assessment results and confidence in the results is sufficient to justify a conclusion. However, it is not sufficient for some decision contexts because the conclusion is controversial or unclear. In such cases, an explicit justification of the conclusion is appropriate. The following issues could be considered in the justification:

- Weight of evidence for the conclusion relative to the alternative hypotheses.
- Consistency of the evidence for the conclusion.
- Coherence of the evidence for the conclusion.
- Extent to which the characteristics of causation or other relevant characteristics are satisfied.
- Justification of the conclusion despite the low weight of the body of evidence, if necessary.

REFERENCES

Dewey, J. 1910. *The Way We Think*. Reprinted by Project Gutenberg. Montclair, NJ.

ECHA. 2024. *Weight of Evidence/Uncertainty in Hazard Assessment: Background Document & Examples*. Helsinki: European Chemicals Agency.

Fox, G. A. 1991. "Practical causal inference for ecoepidemiologists." *Journal of Toxicology and Environmental Health* 33: 359–373.

Griffin, D., and A. Tversky. 1992. "The weighing of evidence and the determinants of confidence." *Cognitive Psychology* 24: 411–435.

Grimes, D. R., and J. Heathers. 2025. "Cake causes herpes?"-promiscuous dichotomisation induces false positives." *BMC Medical Research Methodology* 25 (1): 1–10.

IARC. 2019. *IARC Monographs on the Identification of Carcinogenic Hazards to Humans: Preamble*. Lyon: International Agency for Research on Cancer.

Keynes, J. M. 1921. *A Treatis on Probability*. New York: McMillan Press.

Lett, J. 1990. "A field guide to critical thinking." *Skeptical Inquirer* 14 (2): 153–160.

Mastrandrea, M. D., C. B. Field, T. F. Stocker, O. Edenhofer, K. L. Ebi, D. J. Frame, H. Held, E. Kriegler, K. J. Mach, P. R. Matschoss, G.-K. Plattner, G. W. Yohe, and F. W. Zwier. 2010. *Guidance Note for Lead Authors of the IPCC Fifth Assessment Report on Consistent Treatment of Uncertainties*. Geneva, Switzerland. Intergovernmental Panel on Climate Change.

Mastrandrea, M. D., K. J. Mach, G.-K. Plattner, O. Edenhofer, T. F. Stocker, C. B. Field, K. L. Ebi, and P. R. Matschoss. 2011. "The IPCC AR5 guidance note on consistent treatment of uncertainties: A common approach across the working groups." *Climatic Change* 108: 675–691.

Norton, S. B., S. M. Cormier, and G. W. Suter. 2014. *Ecological Causal Assessment*. Boca Raton, FL: CRC Press.

Pinkston, J. A. 2020. *Evidence and Hypothesis in Clinical Medical Science*. Springer.

Schall, M. K., G. D. Smith, V. S. Blazer, H. L. Walsh, Y. Li, and T. Wagner. 2020. "A fishery after the decline: The Susquehanna River smallmouth bass story." *Fisheries* 45 (11): 576–584.

Schall, M. K., G. D. Smith, V. S. Blazer, H. L. Walsh, T. Wertz, D. Shull, and T. Wagner. 2023. "Assembling the right pieces: Developing an interdisciplinary team to study disease, decline, and recovery of a world-class smallmouth bass fishery." *Fisheries* 48 (7): 287–294.

Schünemann, H., J. Brożek, G. Guyatt, and A. Oxman. 2013. *The GRADE Handbook*. London, UK: Cochrane Collaboration.

Shull, D., and M. Pulket. 2015. *Causal Analysis of the Smallmouth Bass Decline in the Susquehanna and Juniata Rivers*. Harrisburg, PA: Pennsylvania Department of Environmental Protection.

SPC. 2000. *Risk Characterization Handbook*. Washington, DC: Science Policy Council, U.S. Environmental Protection Agency.

Staveley, J. P., S. A. Law, A. Fairbrother, and C. A. Menzie. 2014. "A causal analysis of observed declines in managed honey bees (Apis mellifera)." *Human and Ecological Risk Assessment* 20 (2): 566–591.

Suter, G. W. 2023. *Fundamentals of Environmental Assessment*. Boca Raton, FL: CRC Press.

Suter, G. W., and T. P. O'Farrell. 2009. *Analysis of the Causes of a Decline in the San Joaquin Kit Fox Population on the Elk Hills, Naval Petroleum Reserve #1, California*. Cincinnati, OH: U.S. Environmental Protection Agency.

USEPA. 1996. "Proposed guidelines for carcinogen risk assessment." *Federal Register* 61 (79): 1760–1811.

USEPA. 2005. *Guidelines for Carcinogen Risk Assessment*. Washington, DC: U.S. Environmental Protection Agency, Risk Assessment Forum.

USEPA. 2015. *Preamble to the Integrated Science Assessments*. Washington, DC: U.S. Environmental Protection Agency.

USEPA. 2016a. *Weight of Evidence in Ecological Assessment*. Washington, DC: Risk Assessment Forum, U.S. Environmental Protection Agency.

USEPA. 2016b. *Generic Ecological Assessment Endpoints (GEAEs) for Ecological Risk Assessment: Second Edition with Generic Ecosystem Services Endpoints Added*. Washington, DC: U.S. Environmental Protection Agency.

Walker, V. R. 1996. "Risk characterization and the weight of evidence: Adapting gatekeeping concepts from the courts." *Risk Analysis* 16: 793–799.

Watson, J. C., R. Arp, and S. King. 2024. *Critical Thinking*. Bloomsburg, PA: Bloomsberg Academic.

Weed, D. L. 2004. "Precaution, prevention and public health ethics." *Journal of Medicine and Philosophy* 29: 313–332.

9 Reassessment

If at first you don't succeed, try, try again.

Hickson (1836)

As with any assessment process, weight of evidence can result in ambiguous or incomplete conclusions. You may simply explain the inconclusive result (Section 8.2.4), or you can try again. If you have the time and resources, reconsider the evidence and the hypotheses to support a reweighing of the evidence. We have called this process building a case (Suter and Cormier 2011). Consider the courtroom dramas that you have watched. First, the hero attorney is frustrated that the evidence does not make sense, but then he studies the evidence on his cork crime board (think conceptual model) and finally realizes that he had been thinking about the case all wrong and he can make sense of it by changing his assumptions. Another possibility is that a critical piece of evidence is missing that could resolve the case. The attorney's investigators (Paul Drake for Perry Mason or Cisco for Micky Haller) come up with the missing evidence at the last minute. By analogy, you can reconsider the implications and weights of the evidence or generate new evidence by digging deeper into the literature, performing novel analyses of public datasets, or collecting new data or observations. A structured weight of evidence has advantages in this regard because it clearly indicates the weight of the body of evidence, which types of evidence are responsible for the low weight, and which types are missing. In such cases, the weight-of-evidence approach may be extended to determine why the results are ambiguous and how additional evidence or inference can resolve the ambiguities.

The process of building a logical case from a failed inference can take two forms based on logical form and a third based on setting goals.

1. **Change the Evidence**: If weighing the evidence for and against a hypothesis led to an incomplete or inconclusive result, the addition of new evidence, previously unconsidered prior knowledge, or reinterpretation of the evidence may build a consistent case for or against the original hypothesis.
2. **Change the Hypotheses**: If weighing the evidence for and against alternative hypotheses did not reveal an acceptable hypothesis, the a priori hypotheses may be modified and integrated to create new and better hypotheses (i.e., hypotheses that are more consistent with the evidence) or may be more narrowly defined.
3. **Change the Assessment Endpoint**: Relationships between evidence and hypotheses may not be identified because the valued entity and attribute are defined in such a way that they are not related to any agent of concern.

When deciding which of these approaches to use, it is helpful to improve your understanding of the case by listing the inconsistencies and uncertainties and asking, for each, what could explain them.

DOI: 10.1201/9781003626336-11

9.1 CHANGING THE EVIDENCE

During the interpretation, it may become apparent that there are critical gaps in the evidence that could resolve the ambiguities. They may be due to missing evidence, weak evidence, or inconsistent evidence. These gaps may be filled by new experimental or observational studies, by reanalysis of existing data, or by empirical or mechanistic modeling (Chapter 6). For example, copper concentrations in a stream may exceed levels that cause effects in the laboratory, but that evidence is contradicted by the lack of a relationship between copper concentrations and effects in the field. This may be explained by differences in the bioavailability of copper between the laboratory and the field, leading to a new hypothesis that the copper is not bioavailable, but that requires evidence of the form of copper. Another potential hypothesis might be that the inconsistent evidence is explained by poor field data. Assessing that hypothesis would require a review of the field studies to determine whether different sites were sampled by different crews, in different seasons, or using different protocols. Without new independent evidence, neither explanation can be fully supported.

9.2 CHANGING THE HYPOTHESIS

Changing the hypothesis is more conceptually difficult and complex than identifying needed evidence. One approach to revising or generating hypotheses is to consider potential reasons that a hypothesis may fail to explain the evidence. These example prompts may be helpful.

- Combined causes. Co-occurring causal agents may have additive or interactive combined effects.
- Proximate causes converted into intermediate causes. For example, a chemical might not cause the loss of a population through direct toxic effects, but it may reduce food or habitat.
- Hypothetical linking processes. For example, an unknown immunotoxic effect of a chemical may enhance the effect of a pathogen that is not known to be lethal and would link a chemical to a pathogen through a new component—susceptible organisms.
- Hypothetical agents or events. For example, an unknown septic field might be the source of organic matter that lowered dissolved oxygen.
- Extreme events that were thought to be too improbable to include initially.
- Potential sources, exposure pathways, and agents should be reconsidered. If not all potentially causal sources, agents, or actions were considered, then causal hypotheses would be missed, and relevant evidence would not be collected. For example, if the conceptual model for a new biocide does not include all potential exposure pathways, results of field trials may be discrepant.

These prompts for hypothesis revision may require the generation of new evidence.

An approach to hypothesis revision is to revise the conceptual models for each hypothesis. The planning phases of assessments usually include the creation of conceptual models for each hypothesis. Having completed the weighing of evidence without reaching a conclusion, the knowledge gained during that process may allow you to create new conceptual models representing new hypotheses. This may be done by eliminating elements of the model that are not supported by evidence, rearranging the components of the conceptual model, or combining elements of multiple conceptual models. The San Joaquin kit fox assessment provides an example (Suter and O'Farrell 2009). The disease and toxicity hypotheses were rejected, and the predation hypothesis was accepted as the primary cause of the population decline. However, the accidental deaths hypothesis included mortalities that clearly contributed as a secondary cause (Figure 9.1). Only vehicular accidents significantly contributed. Reduced abundance of lagomorph prey very likely resulted in more time spent by the foxes outside their dens, which would increase exposure to accidents and predation. Very weak evidence linked prey and predator (coyote) abundance to habitat. Combining these results and some new assumptions (e.g., coyote control may have limited the predation rate) results in a new complex hypothesis for the direct and indirect causes of the decline in kit fox abundance (Figure 9.1) and can help identify potential remediation options. This new conceptual model accounts for multiple causal hypotheses with some positive evidence and indicates indirect causal links that may be managed.

Reframing the hypotheses and reinterpreting the existing evidence can lead to new insights into the case and new conclusions, as in Figure 9.1. This may be sufficient to convince decision-makers and stakeholders. However, creating hypotheses after the evidence is analyzed constitutes post hoc inference, which is discouraged in science because it can lead to fudging or self-deception. Therefore, reinterpretations may be doubted unless new evidence is found or generated and the new body of evidence is weighed. Ideally, the new evidence would identify a previously undocumented attribute of the system that would occur if the revised hypothesis is true. For example, when a large volume of whisky was spilled into the Kentucky River, the initial hypothesis to explain the fish kill was ethanol toxicity. However, the survival of air-breathing organisms such as frogs and turtles supported the alternative hypothesis that the cause of the fish kill was anoxia due to the biological oxygen demand (Cormier, Suter, and Norton 2010).

Although conceptual models are the best general tool for building a case, it is desirable to supplement them with a prose explanation (Section 10.2.3). If the logic of the case is sound, it can be converted into a narrative that will be compelling to decision makers and stakeholders who are put off by diagrams or tables. In addition, a narrative can clarify the mechanisms that cause the links in the modeled hypothesis. Finally, translating between a graphic and a narrative account can act as a quality check for the assessors' logic.

9.3 REDEFINE THE ASSESSMENT ENDPOINT

Assessments may be unclear because the endpoint effect is unclear or irrelevant. For example, a causal assessment of effects on the biota of a stream may be discrepant because the endpoint was defined as reduced abundance of invertebrates, when only one sensitive taxon was affected. Alternatively, the endpoint for a fish kill might be

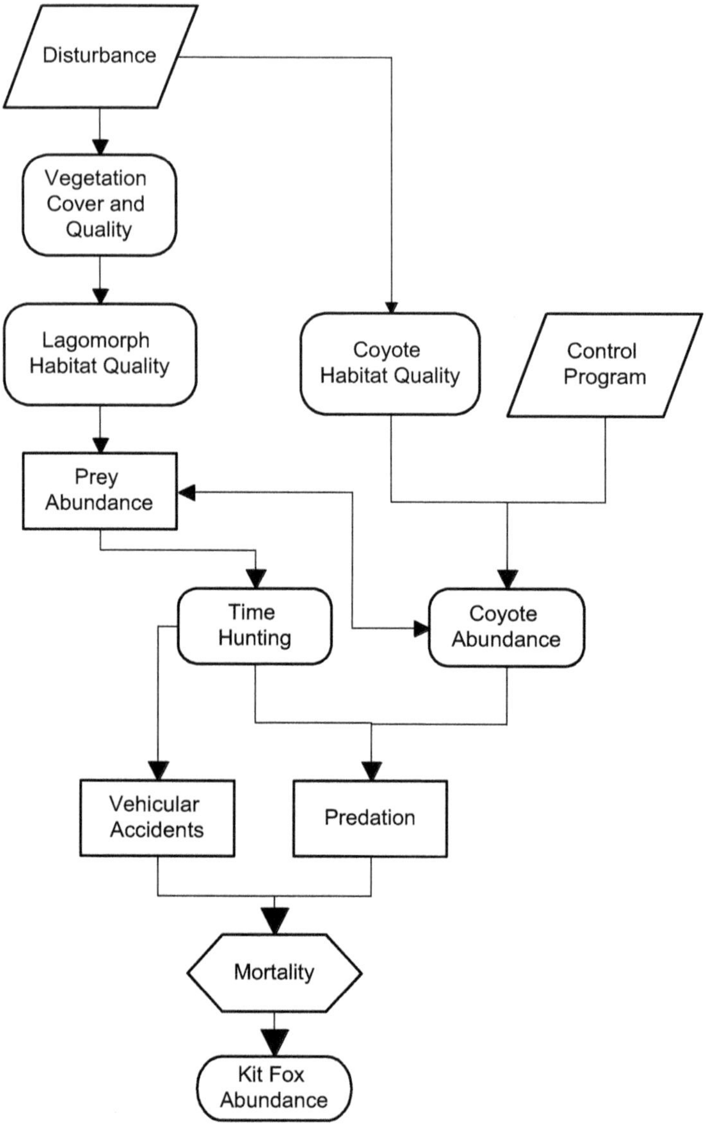

FIGURE 9.1 The final conceptual model for the cause of the kit fox decline on the Elk Hills Petroleum Reserve. It integrates conceptual models for three hypotheses: predation, accidents, and habitat disturbance. The thickness of the arrow lines indicates the weight of evidence for the causal connection (Suter and O'Farrell 2009).

defined as mortality of fish in a multi-kilometer reach of a stream, but the kill may have occurred only in a hundred-meter stretch below an unidentified source. The classic example is the long process of determining the cause of the decline in the abundance of peregrine falcons. The cause was determined to be a toxic effect only after the endpoint was changed from mortality to reproductive failure due to eggshell thinning. Analysis of the eggs identified DDT metabolites, and reproductive tests confirmed that cause. Finally, care should be taken to ensure that the revised assessment endpoint is consistent with the goal of the assessment.

REFERENCES

Cormier, S. M., G. W. Suter, and S. B. Norton. 2010. "Causal characteristics for ecoepidemiology." *Human and Ecological Risk Assessment* 16: 53–73.

Hickson. 1836. *The Singing Master*. Reprint by Boethius Press, the University of Michigan, Ann Arbor, MI.

Suter, G. W., and S. M. Cormier. 2011. "Why and how to combine evidence in environmental assessments: Weighing evidence and building cases." *Science of the Total Environment* 409: 1406–1417.

Suter, G. W., and T. P. O'Farrell. 2009. *Analysis of the Causes of a Decline in the San Joaquin Kit Fox Population on the Elk Hills, Naval Petroleum Reserve #1, California*. Cincinnati, OH: U.S. Environmental Protection Agency.

10 Communication

10.1 COMMUNICATION GUIDELINES

In a regulatory context, it is important to follow the applicable communication guidelines. There are no guidelines that we know of that are specific to assessments based on weight of evidence. However, the abundant guidelines for communication of risk assessments are broadly applicable. Risk communication is defined as communication intended to supply audience members with the information they need to make informed, independent judgments about risks to health, safety, and the environment (USEPA 2025a). Its goal, according to the USEPA, is "to provide meaningful, understandable, and actionable information to our many audiences." Risk communication issues can be difficult because of the common skepticism of audiences (USEPA 2025b).

In some situations, we are asking people to change long-held beliefs, which make them intransigent. When I (Cormier) made the claim that the major ions were more toxic than previously espoused, my own EPA colleagues were among my greatest doubters because they focused on their own work, and the implication was that EPA regulations were not as protective as they had hoped, something they did not want to believe. It slowed down the progress of the work, as I listened to their reservations and concerns and looked for either the evidence or a way to convey the information they needed. It also forced me to assess potential weaknesses in the case and make sure every method, inference, and rationale was clearly communicated. It took longer, but then it wasn't just me explaining the results anymore; it was my colleagues as well.

Weight of evidence has a particular role in communication with respect to transparently explaining the evidence and its use. It serves to inform fellow assessors, regulatory decision-makers, courts reviewing decisions, stakeholders, the general public, and other interested parties, such as signatories to environmental treaties, in a way that encourages deference to the judgment of assessors (Walker 1996).

Some organizations have policies and guidance that relate to communicating evidence quality. In the United States, agencies are required by the Office of Management and Budget to ensure four properties of information that federal agencies disseminate to the public: quality, objectivity, utility, and integrity (OMB 2002). The USEPA meets this requirement by evaluating the quality and relevance of scientific and technical information, taking into account six factors: soundness, applicability and utility, clarity and completeness, uncertainty and variability, and evaluation and review (USEPA 2002). These requirements can be met by weighting the pieces of evidence and weighing the body of evidence. Quality and relevance clearly correspond to the standard properties, reliability and relevance (Chapter 7). The other factors roughly correspond to evidence subproperties. The fundamental difference is that these factors are intended to assure the public and stakeholders of the quality and objectivity

DOI: 10.1201/9781003626336-12

of information that is provided by the Agency, not to support inference. However, weight of evidence for an inference should supply all the information required to communicate evidence quality.

10.2 AUDIENCE CONSIDERATIONS

The first law of communication is to know your audience. In environmental assessment, there are many audiences with various interests and knowledge bases. Communicating to multiple audiences becomes even more complicated when many types of evidence must be weighted, and the body of evidence must be weighed, integrated, interpreted, and, finally, communicated in all its complexity.

10.2.1 Body of the Assessment

The assessment product is a technical document that is aimed at the decision-maker's science adviser, peer reviewers, and stakeholders' technical staffs. It may also be appropriate for decision-makers and stakeholders if it is not too technical or too long. It should include methods and results organized in terms of the assessment framework, from planning to conclusions. For weight-of-evidence-based assessments, it should include the systematic review of the literature and any other processes that went into evidence generation. It should describe the implications of the evidence for the hypotheses, the weighting and weighing of the evidence, integration of the evidence, and interpretation leading to conclusions. The amount of detail to include is a matter of judgment.

10.2.2 Documentation

For the sake of full disclosure, the assessment should include sufficient detail in appendices or data banks to allow replication of the assessment, if anyone should care to. If the assessment problem is contentious, stakeholders may perform alternative assessments or may attempt to find flaws in your assessment. If you do not provide sufficient information, you may be accused of hiding your bias, committing fraud, or covering up critical information. If studies were performed to generate new information, the reports of those studies should be provided. Finally, in the interest of making the body of the assessment readable and short enough to be read, some material that could be appropriate for the body of the assessment may be moved to the appendices and cited where appropriate. A concise executive summary, abstract, or graphical abstract can summarize the main points of the results and make the longer version more accessible.

With respect to weight of evidence, two types of justification are likely to end up in appendices. One is documentation of the literature review, with justification for excluding some studies. Rejection of studies is often a cause of contention, so that justification is important. The second type is justification of weights. The justifications of weights that are included in the body of the assessment are typically brief, often no more than a sentence. Longer justifications, including data, analyses, and cited sources, can be presented in appendices.

Presentation of raw data, metadata, cleanup protocols, and outputs is becoming the norm for regulatory agencies to ensure full transparency. Some information pedigrees are housed in agency data warehouses or associated with publications. For example, the U.S. government currently stores almost 4,000 datasets in data.gov. Zenodo and GitHub provide other options for storing computer code and data outputs. Data sets and methods published in conjunction with papers are also important scientific products.

10.2.3 Narrative Summary

You need narratives to clarify your interpretations of the weight-of-evidence results and to convince the decision-maker that your analysis makes sense. As Daniel Kahnman said to Michael Lewis (2017), "No one ever made a decision because of a number. They need a story."

A narrative also serves as a summary of the weight-of-evidence process. The narrative is the story of how you got to the conclusion and to the associated confidence. It should be credible, so that it is not perceived as arbitrary or incomplete. The narrative summary should be written for non-scientists so that it is appropriate for decision-makers, stakeholders, and the interested public. The narrative summary is often referred to as the executive summary, and it is placed at the front of the assessment document or is a separate document (USEPA 2014).

As discussed in Chapter 1, a weight-of-evidence process may stand alone, but often it is one component of an assessment. In that case, the narrative summary of the weight-of-evidence process may be integrated into an overall summary of the assessment.

As discussed in Section 2.1, the narrative fallacy, accepting the hypothesis with the best story, is an impediment to good inference. It is a particular danger for narrative weight of evidence. However, that danger is greatly reduced in the formally structured weight-of-evidence approach described in Part II. In our approach, the inference is performed by the structured process that includes an analysis of the evidence with respect to the hypotheses. The narrative summary simply explains the process and results.

10.3 COMMUNICATION STRATEGIES

Following the communication guidelines described in Section 10.1 should help you communicate with the public, stakeholders, and decision-makers. However, recent research on risk communication can help you in implementing the guidelines.

When a decision is contentious, communication becomes difficult. During the COVID-19 pandemic, it was difficult to sell the public and politicians on the advisability of wearing masks, isolating, or being vaccinated because of the need to counter misinformation. A review of attempts to communicate public health recommendations found that framing the message in terms of weight of evidence and consensus was more successful than scare tactics (e.g., disease images), appeals to authority, expressions of certainty, humor, misinformation warnings, presumed sympathetic messengers, or debunking (Whitehead et al. 2023). The results of Whitehead's review have a specific context, but they suggest that featuring weight of evidence in risk communication can be advantageous.

More general advice on risk communication is provided by the Winton Center for Risk and Evidence Communication, University of Cambridge, in its summary of research results. They provide five rules of evidence communication (Blastland et al. 2020).

1. **Inform, Not Persuade**: People will not trust you if they believe that you have ulterior motives. Pushing a conclusion makes you seem biased.
2. **Offer Balance, Not False Balance**: A table presenting the weighted body of evidence presents the supporting and weakening evidence for alternative hypotheses (see Table 8.2).
3. **Disclose Uncertainties**: Say what you know and what you do not know. See Section 6.1 for what you should know.
4. **State Evidence Quality**: This is the relevance and reliability of the evidence (Chapter 7).
5. **Inoculate Against Misinformation**: Anticipate people's concerns, misunderstandings, and likely attempts to sow doubt; then engage in prebunking. Communicate evidence in a way that will counter anticipated falsehoods and satisfy concerns.

REFERENCES

Blastland, M., A. L. Freeman, S. van der Linden, T. M. Marteau, and D. Spiegelhalter. 2020. "Five rules for evidence communication." *Nature* 587 (7834): 362–364.

Lewis, M. 2017. *The Undoing Project*. New York: W.W. Norton & Company.

OMB. 2002. *Guidelines for Ensuring and Maximizing the Quality, Objectivity, Utility, and Integrity of Information Disseminated by Federal Agencies*. Washington, DC: US Office of Management and Budget.

USEPA. 2002. *Guidelines for Ensuring and Maximizing the Quality, Objectivity, Utility, and Integrity, of Information Disseminated by the Environmental Protection Agency*. Washington, DC: U.S. Environmental Protection Agency, Office of Environmental Information.

USEPA. 2014. *An Assessment of Potential Mining Impacts on Salmon Ecosystems of Bristol Bay, Alaska: Executive Summary*. Seattle, WA: U.S. Environmental Protection Agency, Region 10.

USEPA. 2025a. *About Risk Communication*. Washington DC: U.S. Environmental Protection Agency. https://www.epa.gov/risk-communication.

USEPA. 2025b. "The SALT framework: A process framework to guide risk communication." https://www.epa.gov/sites/default/files/2021-03/documents/web_salt_framework_v7.pdf.

Walker, V. R. 1996. "Risk characterization and the weight of evidence: Adapting gatekeeping concepts from the courts." *Risk Analysis* 16: 793–799.

Whitehead, H. S., C. E. French, D. M. Caldwell, L. Letley, and S. Mounier-Jack. 2023. "A systematic review of communication interventions for countering vaccine misinformation." *Vaccine* 41 (5): 1018–1034.

11 Verification and Justification for Weight of Evidence

Probably, the most neglected aspect of environmental assessment is verifying the quality of assessment methods and their application. This would be easier if assessments had more standardized methods. If the methods for a type of assessment are ad hoc, you will have difficulty determining which features are working. The difficulty of evaluation is also due to the importance of expert judgment in weight of evidence (Section 1.6). Expert judgment is inevitable in assessments, but it is particularly important when weighing evidence. The options for verifying and justifying weight of evidence and other assessment methods include the following.

- Peer review of assessments,
- Peer review of methodological guidelines,
- Outcome assessments to check the assessment's predictions,
- Comparison of early-tier assessments to later-tier assessments,
- Comparison of assessments of the same agent or action by different organizations,
- Replication of assessments to determine consistency, and
- Self-auditing to minimize bias and to document lessons learned.

All of these require time, personnel, and funds, which may not be available. However, any of these evaluations can occur under the right circumstances. For example, the USEPA's IRIS program was under strong criticism for being slow to complete chemical assessments, for not being transparent, and for being biased against the chemical industry. This all came to a head when a congressman required that the formaldehyde assessment be reviewed by the National Academy of Sciences. The criticisms of the IRIS program in that review led to completely new methods and to rounds of reviews of those methods (NRC 2011, 2014, 2018). Reviews also came from the chemical industry (Hegstad 2015). As a result, a completely transformed IRIS methodology, including formal weighing of evidence (called evidence integration), was developed over 11 years (USEPA 2022). It is too early to tell whether the new method results in more or better assessments, and as of December 2025, it is too late. The Trump-appointed USEPA administrator is eliminating the IRIS program, but there is always a next president.

A simple and practical approach is the comparison of early-tier assessments to later-tier assessments to determine whether the early tiers are sufficiently protective (van Wijngaarden, Maltby, and Brock 2015; Brock et al. 2016; van Wijngaarden and

DOI: 10.1201/9781003626336-13

Arts 2018). The later tiers should weigh a larger body of evidence, including evidence from more realistic studies, so they should serve as standards for verification of the early tiers. However, the later-tier evidence should be weighted to determine its relevance and reliability. In any case, new studies should support early conclusions.

Comparison of assessments by different organizations will inevitably occur because the regulatory agencies are quite aware of what other agencies or industry groups have decided. For example, IARC decided that glyphosate is a likely human carcinogen, and the USEPA did not, and the reasons for the difference have been closely examined. In this case, the difference is in part a matter of differences in methodological requirements and precedents. It is also a matter of differences in weighting and interpretation, which are expected in any weighing of evidence. Recently, fraud in at least one industry study complicates things. In addition to such ad hoc comparisons, some comparisons of assessments are performed deliberately and consistently (Ågerstrand and Rudén 2010; Woodall et al. 2025). If your assessment question has been addressed by prior assessments, the reasons for any differences in the results should be stated.

Part of the variance among assessments is due to differences in expert judgment. Eccentricities in the judgments of experts can be reduced by using the judgment of expert teams. They can also be reduced by using detailed guidance that specifies decisions such as how much weight to give good laboratory practices or replication of reference populations. Such constraints on judgment are most practical when guidance is specific to a limited problem, such as weighing multiple mammalian tests as evidence for human carcinogenicity. An alternative to constraining expert judgment is to make it transparent by requiring that judgments be justified.

Self-auditing should be automatic. Consider whether you have followed the relevant guidance and whether a reasonable expert might make a different judgment. Ensure that you have fairly considered and responded to peer review comments. Bring the assessment team together to compile the results and raise and resolve issues of disagreement.

The ultimate validation is comparison of the assessment results to the real-world consequences of taking action on the basis of the assessment. This is termed an outcome assessment (Figure 4.1). It requires that monitoring be conducted using good study designs. It should monitor the endpoint effects and the exposure processes that were predicted to cause the effects, or not. Because the action taken in response to the risk assessment should leave few or no residual effects, the monitoring methods should be sensitive to small exposures and effects. In any case, the monitoring should include multiple types of evidence, including evidence of exposure and effects on endpoint media and organisms.

In outcome assessments, you should also keep an eye out for collateral damage and unexpected consequences. For example, sediment runoff in the Yakima River was burying salmon spawning areas. A concerted effort drastically reduced the agricultural runoff, and the water ran clear. However, aquatic plants covered the spawning grounds because of a lack of shading by suspended sediments, and legacy nutrients in groundwater continued to flow into the river. These unexpected effects thwarted the goal of improving salmon spawning success.

REFERENCES

Ågerstrand, M., and C. Rudén. 2010. "Evaluation of the accuracy and consistency of the Swedish Environmental Classification and Information System for Pharmaceuticals." *Science of the Total Environment* 408 (11): 2327–2339.

Brock, T. C. M., R. Bhatta, R. P. A. van Wijngaarden, and A. Rico. 2016. "Is the chronic Tier-1 effect assessment approach for insecticides protective for aquatic ecosystems?" *Integrated Environmental Assessment and Management* 12 (4): 747–758.

Hegstad, M. 2015. "Despite IRIS improvements, industry looks for more substantive changes." *Inside EPA's Risk Policy Report* 22 (2): 1–11.

NRC. 2011. *Review of the Environmental Protection Agency's Draft IRIS Assessment of Formaldehyde*. Washington, DC: National Research Council of the National Academies, National Academies Press.

NRC. 2014. *Review of EPA's Integrated Risk Information System (IRIS) Process*. Washington, DC: National Research Council of the National Academies of Science, National Academies Press.

NRC. 2018. *Progress Toward Transforming the Integrated Risk Information System (IRIS) Program: A 2018 Evaluation*. Washington, DC: National Research Council, National Academies of Science, Engineering and Medicine, National Academies Press.

USEPA. 2022. *ORD Staff Handbook for Developing IRIS Assessments*. Washington, DC: Office of Research and Development, U.S. Environmental Protection Agency.

van Wijngaarden, R. P. A., and G. H. P. Arts. 2018. "Is the tier-1 effect assessment for herbicides protective for aquatic algae and vascular plant communities?" *Environmental Toxicology and Chemistry* 37 (1): 175–183.

van Wijngaarden, R. P. A., L. Maltby, and T. C. M. Brock. 2015. "Acute tier-1 and tier-2 effect assessment approaches in the EFSA Aquatic Guidance Document: Are they sufficiently protective for insecticides?" *Pest Management Science* 71 (8): 1059–1067.

Woodall, G. M., S. E. Kobylewski-Saucier, R. M. Shaffer, A. M. Luke, and L. M. Carlson. 2025. "Comparative review of human health reference values of the United States." *Toxicological Sciences* 207 (1): 1–19.

12 Inference for Deriving Quantities

Most of the emphasis in weight of evidence for environmental assessments has been on determining qualities, particularly in determining causation and hazards. However, deriving quantities may also involve weighing multiple pieces of evidence. Examples include benchmark values (e.g., water quality criteria), assessment endpoints (e.g., the meters of stream habitat that will be impaired), and intermediate properties in assessments (e.g., the maximum potential atmospheric concentration). The multiple estimates of these quantities are pieces of evidence for the quantitative result of interest. For example, if the quantitative result of interest is the threshold for aquatic acute lethality, the pieces of evidence could be median lethal concentrations (LC_{50}s) from different laboratories and for different species.

The process of deriving a quantity by weighing evidence has four steps (Figure 12.1). First, you must determine the quality to be quantified. Second, assemble and weight the quantitative evidence and supporting qualitative evidence. Third, derive the quantity by the alternative methods or from alternative datasets. Fourth, determine the weight (confidence) to assign to the results. Two general approaches may be used to derive a quantity from multiple values and multiple studies. The first approach is to choose the best value. The second is to combine the multiple values.

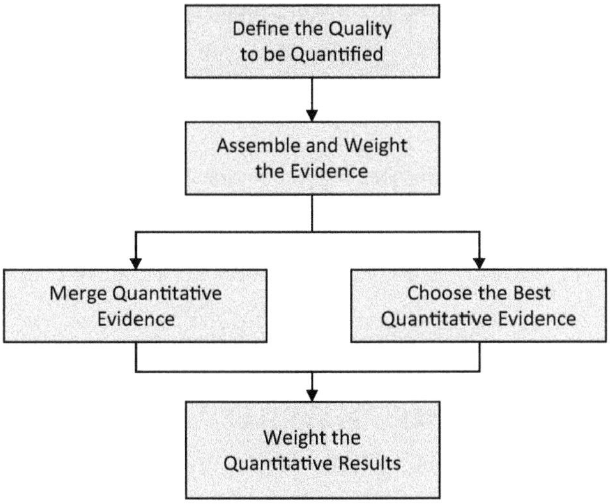

FIGURE 12.1 Steps in a process for using weight of evidence to derive a quantitative result. Quantitative evidence is an estimate of the quantity of interest (i.e., the quantitative results). (Adapted from USEPA 2016; Suter, Cormier, and Barron 2017.)

DOI: 10.1201/9781003626336-14

TABLE 12.1
Qualities That Could Be Identified by Qualitative Weight of Evidence and the Associated Quantities That Could Be Derived by the Quantitative Weight of Evidence Process

Example Quality	Example Corresponding Quantity
Hazard: teratogen	Threshold maternal body burden
Hazard: carcinogen	Slope factor
Specific cause: chloride in effluent as the cause of a fish kill	Estimated maximum one hour average concentration
Specific cause: ammonia in stream as the cause of a biological impairment	Total maximum daily load
Bioaccumulative	Bioaccumulation factor
Exposure metric	Bioavailable concentration at point of contact
Susceptibility	Proportion responding at an exposure level
Alteration	Area of wetland
Remediation required	Soil concentration goal

Source: Adapted from USEPA (2016).

12.1 DETERMINING THE QUALITY TO BE QUANTIFIED

Getting the quantity right depends on first getting the quality right (NRC 2014). In some cases, this is not an issue. The quality of concern is determined by the problem formulation for the assessment. It might be the biodegradation rate of a chemical required by a fate model or the cause of the loss of 28% of aquatic species. The corresponding quantities can then be identified (Table 12.1).

If the quality to be quantified is not specified, you must determine it, which can be accomplished by weighing the evidence for the quality as described in Chapters 6–8. By performing a formal qualitative assessment, you:

- present the alternative qualities and their relevance to the desired endpoint,
- assemble and present evidence for the qualities in a consistent and defensible way,
- express confidence in the body of evidence supporting the endpoint quality, and
- get feedback from interested parties before proceeding.

12.2 CHOOSING THE BEST QUANTITY

Often, multiple estimates of the quantity of concern are derived from studies that have important differences in their weights due to differences in the suitability, relevance, or reliability of their methods and data. If the differences are large, and particularly if one study has much more weight than the others, the quantitative results from that weightiest study may be adopted. The weights of the studies that were used to determine the quality to be quantified are often also useful when quantifying that quality. However, it may be appropriate to reweigh the studies because you may judge some properties to be more or less important to the quantitative result than to the qualitative determination. In particular, it is often inappropriate to use strength as a property when weighing evidence for quantification. For example, when you are choosing the best estimate of an EC_{10}, you would not want to bias the choice by giving greater weight to exposure-response studies that gave low concentrations. However, some sub-properties are more important than others. For example, when choosing the best estimate, you may emphasize study design and execution—evidence generated with a good study design that is well performed and statistically well analyzed is more reliable (Box 7.2).

Using the weight of evidence to choose the best study is not an established practice. However, it is recommended by the European Chemical Agency for choosing among alternative test results even when the tests include the same species, durations, and conditions (ECHA 2008). If the weights are the same, the geometric mean may be used. Choosing a value by weighing the evidence could be particularly useful when one result is an outlier. The weights can indicate the cause of the difference and whether the outlier may be the best value.

It is generally appropriate to choose the best value rather than combine multiple results if the values are from different types of studies. For example, it is inadvisable to combine studies for an aquatic benchmark value if they include laboratory toxicity tests, mesocosm tests, and field biological surveys. They will have inherent differences in what was measured and in the duration, water quality, and habitat conditions. In addition, one type of study is likely to better represent the desired environmental endpoint; that is, it would have greater relevance. For human health assessments, when suitable animal and human studies are available, a human study is used (USEPA 2022). Similarly, the best type of study may be defined by the reliability of the method. When choosing the best evidence for mixtures toxicity, test data for the mixture of concern are better than data for a similar mixture, which are better than test data for component chemicals applied to a model of combined effects (Risk Assessment Forum 2000). However, even the best study types, such as human clinical trials, require weighting of their reliability to determine whether they are acceptable for determining a recommended dosage.

In some assessments, the acceptable study with the most protective result is best. For example, you might choose the study with the lowest acute lethality value (e.g., LC_{50}), the highest bioaccumulation factor, or the highest slope factor for a carcinogen. There are multiple reasons for choosing the most protective value.

- Your choice may be determined by a precautionary policy.
- When deriving screening values to determine whether a chemical requires additional study, you may choose the lowest value to ensure that chemicals that do require more attention are not screened out.
- Because there are so many communities, species, life stages, and responses, it is likely that some effects occur at an even lower toxic value than the lowest value. Therefore, it is likely that the most precautionary value is the closest to the true protective value.

- Finally, the magnitude of the estimate may be a tie breaker. That is, if, after weighing the evidence for each value, multiple estimates have the same relevance and reliability, you may choose the lowest value.

The best value may also come from a case-specific study that is conducted to provide a more relevant or reliable result than any pre-existing study. This approach may be chosen when none of the existing values is adequate, or if the stakeholders do not agree on the best of the existing values, and if it is feasible to perform new studies. For example, when the evidence for a pesticide registration is contentious, the regulator may agree with the registrant on a consensus study design that will resolve the disagreements over existing evidence and provide an agreed-upon best value.

Finally, the best value may be determined by a structured weight-of-evidence process. The simplest way to choose the best is to choose the value with the weightiest body of evidence using the method for weighting properties from Chapters 7 and 8. However, you may devise a weighing method appropriate to your quantification problem. An example is provided in Table 12.2, based on Hope and Clarkson (2014). Their matrix plots effect levels from a risk assessment against evidence weights for each of four evidence groups shown by circled numbers. The risks of effects are probabilities of achieving the endpoint effect, which is <5% reduction in species richness. Evidence group 3 can be considered the best because it has the highest weight and is not an outlier in terms of the level of effect. The result of weighing the evidence might be: group 3 (evidence from an in-stream test) is convincing (+++) and gives an estimate of a 15% probability of the endpoint effect.

Even if a best value is chosen, information from other studies may increase or decrease the overall weight of the body of evidence for the quantitative result. For example, if the results of all or most of the other studies are quantitatively similar, that increases confidence in the best value.

In some cases, you may weigh benchmark values from different agencies and organizations to choose the best benchmark. For example, a group of USEPA assessors compiled human health reference values and showed how to weigh properties to determine the most appropriate value for a particular use (Woodall et al. 2025).

TABLE 12.2
Weight of Evidence Matrix to Summarize Quantitative Risk Estimates for Four Evidence Types or Groups (Numbered Circles) and Their Weights

		Evidence Weight			
		0	+	++	+++
Probability of adverse effect endpoint	>75%				
	51%–75%				
	25%–50%			①	
	5%–24%			②	③
	<5%		④		

An adaptation of the Hope and Clarkson method from (USEPA 2016).

The properties that they weighed are target population; duration and frequency of exposure; health effect severity; confidence in the dataset; use of well-documented and contemporary derivation methods; transparency and documentation; and thoroughness of review.

12.3 COMBINING QUANTITIES

Quantities derived from different datasets and possibly by different methods may be combined if they have certain characteristics.

- They should be concordant. They should have the same endpoints (same effect, species, ecosystem type, and other endpoint attributes), and they should be derived by similar laboratory or field methods.
- They should have the same dimensions or be convertible.
- They should be quantitatively similar. If the range of values is large (e.g., greater than 10×), some property of the studies other than random variance is likely to be responsible. That large difference would lead to choosing the best value.
- They should all be relevant and reliable, which could make each of them a possible best value.

If the raw data are available for all studies, it is possible not only to convert the endpoints to common units, but also to reanalyze them using common statistics.

12.3.1 THE BASICS: MEANS AND DISTRIBUTIONS

The obvious method for combining alternative estimates for a quantity is to determine the mean or median. For example, geometric means are commonly used to combine values for the same species in the same test type (ECHA 2008; USEPA 1985). Meta-analysis and Bayesian methods are more statistically sophisticated. However, they are not always applicable, and in some cases, such as when the sources of the candidate values are heterogeneous, the extra effort is not justified. In addition, decision-makers and stakeholders understand means and probably medians, but may not understand more complex statistics and may not accept what they do not understand.

In some cases, the average value is judged to be not sufficiently conservative or protective. In such cases, a low centile, such as the 10th centile of the distribution of candidate values, may be used to derive a benchmark or other value. This approach is a suggested option for Australia and New Zealand (ANZG 2018).

12.3.2 COMBINING DATASETS

If the results of multiple studies have the same endpoint and exposure metrics, the datasets can be combined, and the desired quantity derived from that larger dataset. This has been termed combined analysis (USEPA 2022), or individual-level analysis because individual observations are retained (NRC 2014). This approach is in contrast to aggregate-level analysis, which combines aggregate data such as means and variances. Advantages of combined analyses are simplicity and retention of all information in the relevant datasets. For example, it is simpler to perform a dose-response regression with a combined dataset than to perform a meta-regression that keeps the multiple studies distinct. It also allows you to select the most appropriate common regression method.

12.3.3 Meta-Analysis

Meta-analysis refers to statistically rigorous approaches for combining quantitative results of multiple studies. It began with a publication by the statistical pioneer Karl Pearson that combined multiple epidemiological studies of enteric fever (Pearson 1904). Meta-analysis is still used to combine results from epidemiological studies of a particular exposure or effect. It is also commonly used to combine results of clinical trials, including assessments by the Cochrane Collaborative (Higgins et al. 2023). Peer-reviewed meta-analyses based on a systematic review, like the Cochrane reviews, are commonly referred to as the strongest scientific evidence. However, even Cochrane reviews can be biased and contentious. Meta-analysis has been used less in ecological assessments because, due to differences in species, communities, and habitats, multiple studies of the same ecological issue are seldom equivalent and cannot be interconverted.

Like regression analysis, meta-analysis is not a method. Rather, it is a collection of methods to solve a statistical problem. Like most statistics, it includes Bayesian and frequentist methods. It even includes meta-regression for combining associations such as exposure-response relationships or quantitative structure-activity relationships. A highly standardized meta-analysis method is in the Cochrane manual (Higgins et al. 2023). There are many books, guides, and software packages for meta-analysis. An easy introduction is provided by Cumming (2012).

Any measure of effects can be combined by meta-analysis as long as it is common to all of the studies to be combined. Examples include absolute or percent difference from the reference, relative risk, odds ratio, and slope. To perform the analyses, the data must be available for each study, not just summary statistics. The studies may be weighted; the most generally accepted weighting factor is the inverse variance. The input studies and output results of a meta-analysis may be represented by a forest plot (Figure 12.2). However, the measure may be common, but the effect is meaningless if the effects are not also common.

Meta-analysis is also used for hypothesis testing (Section 4.2). For example, if the 95% confidence interval on the combined effects estimate does not include 0, the effect is "statistically significant" and therefore assumed to be real. However, when deriving a quantity, it is better to use the confidence interval as an estimate of certainty concerning the effects level (Cumming 2012). This is particularly appropriate when causality is certain because it has been baked into the study by its experimental design. Alternatively, you can avoid the flaws in statistical hypothesis testing by using the weight of evidence to judge whether an apparent cause-effect relationship in a meta-analysis is real (MacGillivray 2019; Suter 1996) (Chapter 7).

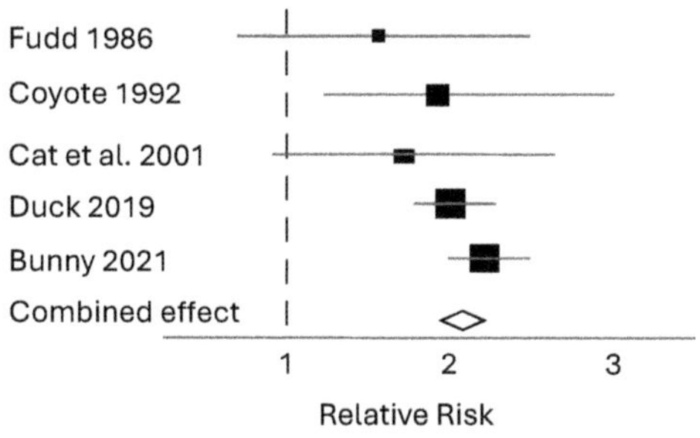

FIGURE 12.2 A hypothetical forest plot combining five studies with effects expressed as relative risk (Suter 2023). The squares indicate the mean effect, and their size indicates weight. The horizontal lines are 95% confidence intervals. The diamond indicates the combined mean, and its width is the combined 95% confidence interval.

The objections to meta-analysis are: (1) the evidence must be homogeneous, (2) differences in relevance and reliability are not incorporated, (3) statistical assumptions are neglected, and (4) the method cannot address causation or mechanisms. Another limitation of meta-analysis as a stand-alone inferential method is that it does not add to the understanding of the phenomenon of interest and therefore does not suggest new hypotheses (Slavin 1995). Further, it gives too much attention to large studies, not enough attention to large effects, and no attention to study quality. Meta-analysis combines multiple studies, but it is not a type of weight of evidence (Section 3.3). However, its results have been used as a piece of evidence.

12.3.4 Composite Values

Composite values may be derived for categories of effects, such as summing the risks of various tumors. They may also be derived for all types of effects on a system to be protected (e.g., hepatic hazards). The IRIS method manual includes the derivation of composite organ/system-specific toxicity values across hazards (USEPA 2022). After applying a set of six "considerations," including "weight of evidence of hazard," to evidence for the non-concordant hazards, two methods may be applied to derive a toxicity value: (1) a best value may be chosen, or (2) the candidate values may be composited by no specified method.

12.3.5 Combining Analytical Methods

In some cases, only one dataset is available, but the best method for analytically deriving a benchmark value is unclear. One solution is to apply all reasonable analytical methods to the data, examine the distribution of results, and choose an appropriate value based on interpretation of the distribution (Wang et al. 2024; Silberzahn et al. 2018). That approach is called specification curve analysis or multiverse analysis.

An example of justified averaging across analytical methods is as follows. A SAB committee was charged with choosing from six different methods used by a USEPA work group to derive candidate sediment benchmarks. The reviewers said that none of the methods was more reliable and the resulting values were similar, so the USEPA should use the mean of the results from the six methods. The mean was judged to be a reasonable benchmark value.

12.3.6 Bayesian Approaches

There are Bayesian approaches for almost any statistical task. A National Research Council committee recommended to the IRIS program a Bayesian hierarchical approach for combining exposure-response relationships from heterogeneous studies of the toxicity of chemicals for human health assessments (NRC 2014). Studies were classified as high-throughput, animal, or human. The high-throughput studies were combined, and the result was used as the Bayesian prior. The prior was updated with the combined animal studies, and that updated result was used as the prior for the combined human studies to obtain the final update. Equivalent hierarchies could be defined for combining ecological studies.

12.4 CHOOSING A METHOD

The fundamental issue when choosing whether to combine values or to choose the best value is the source of the alternative values (i.e., a single type of study or heterogeneous types). Heterogeneous values are not concordant and should not be combined. However, some people have combined heterogeneous values and associated the results with a vague endpoint such as loss of biotic integrity. In general, it is better to weigh heterogeneous values for their relevance and reliability to determine the best value. Values from different studies that are sufficiently similar may be combined by meta-analysis or some other statistical approach. Determining whether results from laboratory and field studies, stream and lake studies, or rat and beagle studies are sufficiently similar is a matter of judgment or policy.

12.5 UNCERTAINTY, VARIABILITY, AND CONFIDENCE

Inconsistency in the use of uncertainty, variability, confidence, and related terms is a significant source of confusion among environmental assessors (Spiegelhalter and Riesch 2011). We have used these terms in different ways over the years and have now settled on the following terminology as the most useful for considering qualities and quantities. See Box 12.1 for a similar usage of terms for the assessment of climate change.

Note that most criteria, standards, or other regulatory benchmarks are not presented with expressions of variability or uncertainty (Chapter 15). Like highway speed limits, to be enforced they must be bright lines.

Variability: Deriving a quantity inevitably uses data that have some degree of variability, which is the subject of statistical analysis. It might, for example, be due to variability among organisms, such as the various times to death of organisms in a toxicity test. Results of its analysis may be expressed as the proportion dead at a particular time or the probability that an organism will have died by a particular time. Variability results from inherent randomness or natural variance and is sometimes termed aleatoric uncertainty. It cannot be reduced, but it can be quantified. The number of applicable techniques from descriptive statistics is large. In general, it is best to use familiar and easily interpreted descriptors such as confidence intervals and ranges. If the best estimate is chosen by weight of evidence, the statistics applied by the authors of the study may be used. If studies are combined, variability may be estimated by meta-analysis or meta-regression.

Variability in Models: If the best estimate is derived from a simulation model rather than statistical analysis of data, a distribution of estimates may be derived by Monte Carlo simulation, bootstrapping, or other methods. Ideally, the model variables will be defined as measured variability, but, unfortunately, in some cases, they are a mixture of variance and uncertainty. Note that these simulations and their variances are pieces of evidence and not weight-of-evidence techniques. They are simply one method for estimating a quantity that may be weighted and weighed along with other types of evidence.

Uncertainty: When we lack knowledge, we are uncertain. For example, we do not know whether a new water treatment plant will fail and spill untreated waste into a stream during a 100-year low flow (scenario uncertainty). Uncertainties due to lack of knowledge are sometimes termed epistemic uncertainties. They can be reduced by obtaining more information, if the relevant information is obtainable. A parameter may have an epistemic component even if it has a variable/aleatory estimate (Beven and Young 2013). For example, perhaps you have measurements of variance in stream flow for a full year (aleatoric), but inter-annual variance is unknown (epistemic). Since you don't have data on the reliability of the plant or the frequency of particular flow rates for the past 100 years or future variability given climate change, you might simply conclude that the 100-year low-flow condition is a possible worst case or subjectively assign a frequency of occurrence (i.e., assume a value). Alternatively, multiple scenarios that cover a reasonable range of assumptions may be analyzed. Another common approach is to simply list the uncertain variables. Epistemic uncertainty also includes current assumptions, judgments, and unknown unknowns that result in "deep uncertainty" (Spiegelhalter and Riesch 2011). That is, they are real but unmanageable causes of a lack of knowledge. However, deep uncertainty might be incorporated into reliability weights if we down-weight evidence that is based on more assumptions than data. Commonly, in risk assessments, uncertain estimates are divided by uncertainty factors. For example, benchmark values estimated from a rodent test are divided by 10 for animal-to-human extrapolation and by 10 again for average-human-to-sensitive-human extrapolation (USEPA 2022).

Confidence: Weight is a metric of confidence in pieces, types, and bodies of evidence (Section 1.2). When deriving a quantity by weight of evidence, confidence in the quantity is the weight of the body of evidence for either the best value or the combined value. When using the best-value approach, only one study is relevant. Your confidence in the quantity may be the weight of the body of evidence for the study from which it is derived. When using a combined-value approach, the weights for the combined studies can be integrated to characterize your confidence in the combined value.

BOX 12.1 CONFIDENCE AND UNCERTAINTY IN THE INTERGOVERNMENTAL PROGRAM ON CLIMATE CHANGE REPORTS (MASTRANDREA ET AL. 2010). NOTE THAT THEY LUMP VARIABILITY AND UNCERTAINTY

"The AR5 [Fifth Assessment Report] will rely on two metrics for communicating the degree of certainty in key findings:

- Confidence in the validity of a finding, based on the type, amount, quality, and consistency of evidence (e.g., mechanistic understanding, theory, data, models, expert judgment) and the degree of agreement. Confidence is expressed qualitatively.
- Quantified measures of uncertainty in a finding expressed probabilistically (based on statistical analysis of observations or model results, or expert judgment)."

12.6 AN EXAMPLE OF THE POTENTIAL COMPLEXITY OF WEIGHT OF EVIDENCE FOR A QUANTITY

The failure of an earth-fill tailings dam was one of the greatest environmental risks associated with the proposed Pebble Mine in the upper Bristol Bay watershed in southwest Alaska. The risk was estimated as failures per dam-year from three types of evidence (USEPA 2014).

- **Evidence Type 1**: Variability among frequencies of failures of tailings dams found in four reviews gave a range of 0.00049–0.0014 failures per mine-year. The combined estimate was 0.0005. The relevance and reliability of these historical data were judged to be low. Dam construction practices have changed and advanced.
- **Evidence Type 2**: The mining company insisted that the past frequencies were irrelevant because modern tailings dams never fail. This expert judgment was judged to be strongly relevant evidence, but the reliability was judged to be low. No structure is immune to failure.
- **Evidence Type 3**: Another estimate was based on the State of Alaska's hazard classification applied to a model relating failure frequencies to engineering safety factors. That gave a range of 0.000001–0.0001 failures per mine-year. The high end of this range is lower than the lowest empirical value (Type 1). However, the modeled low estimates seemed justified by advances in engineering and the safety demands of the state.

Less than a year after the assessment was published, the modern tailings dam at the 17-year-old Mount Polley mine failed catastrophically (0.059 failures per mine-year). That mine was very similar to the proposed Pebble Mine. Its failure belatedly refuted Evidence Type 2 and suggested that the safety factors in dam engineering are not necessarily effective (Evidence Type 3). In sum, the attempt to quantify the risk of dam failure was itself a failure in that the available evidence did not provide a useful estimate. The conclusion from weighing the evidence was simply that dam failure was possible but infrequent and that the results can be catastrophic.

12.7 AN EVEN MORE COMPLEX EXAMPLE OF WEIGHT OF EVIDENCE FOR A QUANTITY

The methods for estimating the response of the earth's climate to input of greenhouse gases are diverse, complex, and important. The basic variable of concern is the "equilibrium climate sensitivity," which is estimated to have a >66% credible interval of 1.5–4.5 K, derived by combining multiple estimates from diverse methods. Because of the importance of extreme values that could result in costly but unnecessary actions or unexpected catastrophic effects, it is important to consider the likelihood of extremes. A dual weight-of-evidence approach to this problem was developed (Sherwood et al. 2020; Stevens et al. 2016). Their approach statistically combines evidence and interprets the evidence in terms of "storylines" that could result in extreme climate sensitivity. The three lines (types) of evidence are (1) modern observations and models of system variability and feedback processes, (2) the rate and trajectory of historical warming, and (3) the paleoclimate record. Each of these types of evidence includes numerous pieces of evidence that vary in their methods. The focus of the stories is on feedback processes that are not well understood but could amplify or suppress climate response. The storylines require judgment, so the authors "attempt to incorporate it in a transparent and consistent way" (Sherwood et al. 2020). Their result was a narrowing of the range of likely equilibrium climate sensitivity.

12.8 PRESENTING QUANTITIES DERIVED BY WEIGHT OF EVIDENCE

Table 12.3 illustrates the presentation of results from a straightforward weighing of evidence for a quantity. For example, you may find four studies of the biodegradation rate in freshwater. You may combine them statistically or choose the best one (most relevant to your ecosystem of interest) to obtain a rate to be used in your assessment. Table 12.4 presents the results of the complex weight-of-evidence evaluation of mine failure risks described above.

Your confidence in a quantitative result is derived in part from the statistical variability in the results. For example, if the range or variance of the data is small, your confidence in a best estimate is increased. We use the term scatter to encompass all of the ways that statistics can be used to express the distribution of values to convey variability or uncertainty.

TABLE 12.3
Hypothetical Example of a Table for Presenting Quantitative Results Including Variability and Confidence

Component	Example
The quality quantified	The biodegradation rate in fresh water
Numerical results and method	0.48/year by combining multiple studies
Variability	95% CI of 0.35–0.62/year
Weights of the body of evidence	Highly relevant and moderately reliable = moderate confidence

TABLE 12.4
Real World Results of Weighing Evidence for the Risk of a Tailings Dam Failure with Heterogeneous Evidence and Low Overall Confidence (USEPA 2014)

Component	Result
The quality quantified	The risk of tailings dam failure
Numerical results and units	1. Combined empirical estimate: 0.0005 failures/mine year 2. Expert judgment: 0 failures/mine year 3. Modeled estimate: 0.0001–0.000001 failures/mine year
Variability	1. Empirical range: 0.00049–0.0014 failures/mine year
Weights of body of evidence	Overall, moderate relevance and low reliability = low confidence, quantitative information, but no suitable quantitative result

The numbers designate the types of evidence.

REFERENCES

ANZG. 2018. *Australian and New Zealand Guidelines for Fresh and Marine Water Quality*. Canberra: Australian and New Zealand Governments.

Beven, K., and P. Young. 2013. "A guide to good practice in modeling semantics for authors and referees." *Water Resources Research* 49 (8): 5092–5098.

Cormier, S. M., J. F. Paul, R. L. Spehar, W. J. Berry, P. Shaw-Allen, and G. W. Suter. 2008. "Using field data and weight of evidence to develop water quality criteria." *Integrated Environmental Assessment and Management* 4: 490–504.

Cumming, G. 2012. *Understanding the New Statistics: Effect Sizes, Confidence Intervals, and Meta-Analysis*. New York: Routledge.

ECHA. 2008. *Guidance on Information Requirements and Chemical Safety Assessment. Chapter R.10: Characterization of Dose [Concentration]-Response for Environment*. Heksinki: European Chemical Agency.

Higgins, J. P. T., J. Thomas, M. Cumpston, J. Chandler, T. Li, M. J. Page, and V. A. Welch. 2023. *Cochrane Handbook for Systematic Reviews of Interventions Version 6.4*. London, UK: Cochrane Collaborative. Accessed 1/6/2023.

Hope, B. K., and J. R. Clarkson. 2014. "A strategy for using weight-of-evidence methods in ecological risk assessments." *Human and Ecological Risk Assessment* 20 (2): 290–315.

MacGillivray, B. H. 2019. "Null hypothesis testing≠ scientific inference: A critique of the shaky premise at the heart of the science and values debate, and a defense of value-neutral risk assessment." *Risk Analysis* 39 (7): 1520–1532.

Mastrandrea, M. D., C. B. Field, T. F. Stocker, O. Edenhofer, K. L. Ebi, D. J. Frame, H. Held, E. Kriegler, K. J. Mach, P. R. Matschoss, G.-K. Plattner, G. W. Yohe, and F. W. Zwier. 2010. *Guidance Note for Lead Authors of the IPCC Fifth Assessment Report on Consistent Treatment of Uncertainties*. Geneva, Switzerland. Intergovernmental Panel on Climate Change.

NRC. 2014. *Review of EPA's Integrated Risk Information System (IRIS) Process (2014)*. Washington, DC: National Research Council of the National Academies of Science, Engineering, and Medicine, National Academies Press.

Pearson, K. 1904. "Report on certain enteric fever inoculation statistics." *British Medical Journal* 2: 1243.

Risk Assessment Forum. 2000. *Supplementary Guidance for Conducting Health Risk Assessment of Chemical Mixtures*. Washington, DC: U.S. Environmental Protection Agency.

Sherwood, S. C., M. J. Webb, J. D. Annan, K. C. Armour, P. M. Forster, J. C. Hargreaves, G. Hegerl, S. A. Klein, K. D. Marvel, and E. J. Rohling. 2020. "An assessment of Earth's climate sensitivity using multiple lines of evidence." *Reviews of Geophysics* 58 (4): e2019RG000678.

Silberzahn, R., E. L. Uhlmann, D. P. Martin, P. Anselmi, F. Aust, E. Awtrey, Š. Bahník, B. Feng, C. Bannard, and E. Bonnier. 2018. "Many analysts, one data set: Making transparent how variations in analytic choices affect results." *Advances in Methods and Practices in Psychological Science* 1 (3): 337–356.

Slavin, R. E. 1995. "Best evidence synthesis: An intelligent alternative to meta-analysis." *Journal of Clinical Epidemiology* 48 (1): 9–18.

Spiegelhalter, D., and H. Riesch. 2011. "Don't know, can't know: Embracing deeper uncertainties when analyzing risks." *Philosophical Transactions of the Royal Society* 369: 4730–4750.

Stevens, B., S. C. Sherwood, S. Bony, and M. J. Webb. 2016. "Prospects for narrowing bounds on Earth's equilibrium climate sensitivity." *Earth's Future* 4 (11): 512–522.

Suter, G. W. 1996. "Abuse of hypothesis testing statistics in ecological risk assessment." *Human and Ecological Risk Assessment* 2: 331–349.

Suter, G. W. 2023. *Fundamentals of Environmental Assessment*. Boca Raton, FL: CRC Press.

Suter, G. W., S. M. Cormier, and M. G. Barron. 2017. "A weight of evidence framework for environmental assessments: Inferring quantities." *Integrated Environmental Assessment and Management* 13: 1045–1051.

USEPA. 1985. *Guidelines for Deriving Numeric National Water Quality Criteria for the Protection of Aquatic Organisms and Their Uses*. Washington, DC: U.S. Environmental Protection Agency.

USEPA. 2014. *An Assessment of Potential Mining Impacts on Salmon Ecosystems of Bristol Bay, Alaska*. Seattle, WA: U.S. Environmental Protection Agency.

USEPA. 2016. *Weight of Evidence in Ecological Assessment*. Washington, DC: Risk Assessment Forum, U.S. Environmental Protection Agency.

USEPA. 2022. *ORD Staff Handbook for Developing IRIS Assessments*. Washington, DC: Office of Research and Development, U.S. Environmental Protection Agency.

Wang, Y., T. Pitre, J. D. Wallach, R. J. de Souza, T. Jassal, D. Bier, C. J. Patel, and D. Zeraatkar. 2024. "Grilling the data: Application of specification curve analysis to red meat and all-cause mortality." *Journal of Clinical Epidemiology* 168: 111278.

Woodall, G. M, Jr., S. E. Kobylewski-Saucier, R. M. Shaffer, A. M. Luke, and L. M. Carlson. 2025. "Comparative review of human health reference values of the United States." *Toxicological Sciences* 207 (1): 1–19.

Part III

Using Weight of Evidence

13 Setting Goals and Endpoints

Clearly defined assessment endpoints provide direction and boundaries for the risk assessment and can minimize miscommunication and reduce uncertainty; where they are poorly defined, inappropriate, or at the incorrect scale, they can be very problematic.

USEPA (1998)

Environmental assessments must have defined management goals that guide the entire process, including the weight-of-evidence processes. In particular, the implementation of goals begins by operationally defining assessment endpoints, potentially by weighing evidence for their relevance and importance. The assessment endpoints are used in deriving the specific hypotheses for which evidence is assembled.

13.1 SETTING MANAGEMENT GOALS

Management goals are statements about the desired environmental conditions for human health or the environment. They are stated in environmental protection laws, such as the U.S. Clean Water Act's goal to "protect and restore the chemical, physical, and biological integrity of the Nation's waters" and the European Union's Water Framework Directive goal "to protect and, where necessary, restore water bodies in order to reach good status, and to prevent deterioration." However, those broad goals must be interpreted by managers for specific contexts. When setting goals, managers have multiple considerations that come from consulting the law, regulatory guidance, stakeholders' values, policies, and precedents. These considerations are implicit in a manager's subjective decision-making. Goal setting is the foundation of any assessment process, including the identification of appropriate hypotheses and evidence. Also, clear management goals can be used to unambiguously identify success.

The water quality guidelines for Australia and New Zealand (2017) provide four considerations for management goals:

- specific problems or threats to established community values,
- key attributes of community values that must be protected (e.g., key ecosystem assemblages and processes, species of importance, or a particular stock or crop), and
- desired levels of protection (for aquatic ecosystems).

Setting the specific goals from such considerations requires some sort of weighing of attributes and values.

DOI: 10.1201/9781003626336-16

Management goals may be site-specific and incorporate the goals of the local stakeholders. For example, "the Mirarr Traditional Owners desire that operations at the Ranger uranium mine should not result in any change to the natural water quality of surface waters outside of the Ranger Project Area" (Supervising Scientist 2021).

13.2 IDENTIFYING ASSESSMENT ENDPOINTS

Ecological assessment endpoints are explicit expressions of the actual environmental value that is to be protected, operationally defined by an ecological entity and its attributes (USEPA 1998). Multiple pieces of evidence and multiple properties of evidence should be weighed when selecting the endpoints for an assessment. The USEPA (2016) has four major properties of evidence for potential assessment endpoints.

- **Ecological Relevance**: Ecologically relevant endpoints are important entities and attributes of the system and are functionally related to other entities.
- **Susceptibility to Known or Potential Adverse Agents**: Ecological entities are considered susceptible when they are sensitive to a stressor to which they are, or may be, exposed.
- **Relevance to Management Goals**: Risk managers are more willing to use the results of an ecological assessment for making decisions when it is based on ecological values that people care about, such as endangered species, commercially or recreationally important species, Indigenous cultural values, functional attributes such as flood control, aesthetic values such as clean air in national parks, or charismatic species such as eagles or whales.
- **Demonstrated Utility**: Endpoints with demonstrated utility are listed as generic assessment endpoints, which means that they have been used by the Agency or are indicated by the language found in statutes, treaties, or regulations.

A formal weighing of evidence can be used to select and justify ecological assessment endpoints (Table 13.1). Clearly, the four salmon species that would have been exposed to the Pebble open-pit copper and gold mine are important endpoint entities. In contrast, evidence that aquatic insects are relevant, useful, and susceptible might be assembled and weighed to demonstrate their importance to decision-makers (Suter and Cormier 2015).

Specific assessment issues require consideration of issue-specific endpoint attributes. For example, economic benefits analysis requires that the assessment endpoints be "economically meaningful" (USEPA 2014b). Similarly, net benefits analysis requires concordant endpoints so that the benefits of a remedial action and the damage done during remediation can be compared (Section 21.6).

TABLE 13.1
Abundance, Production, and Genetic Diversity of Salmon Populations as Potential Assessment Endpoints in the Bristol Bay Watershed Assessment (USEPA 2014a)

Property	Rating	Explanation
Ecological relevance	+++	Salmon are consumed by marine mammals and terrestrial wildlife, and they import marine nutrients to freshwater and terrestrial ecosystems.
Susceptibility	++	Salmon are susceptible to loss of spawning habitat and copper toxicity, which are stressors that arise from open-pit copper mining.
Relevance to human goals	+++	Salmon are essential to subsistence, recreational, and commercial fisheries.
Demonstrated utility	++	Abundance and production are USEPA generic endpoints for vertebrate populations, particularly game and resource species. Genetic diversity is not.

REFERENCES

Australia, & New Zealand. 2017. "Management goals." In *Guidelines for Fresh & Marine Water Quality*. Canberra.

Supervising Scientist. 2021. *Magnesium in Surface Water — Rehabilitation Standard for the Ranger Uranium Mine (Version 2)*. Darwin, NT: Supervising Scientist Branch.

Suter, G. W., and S. M. Cormier. 2015. "Why care about aquatic insects: Uses, benefits, and services." *Integrated Environmental Assessment and Management* 11 (2): 188–194.

USEPA. 1998. *Guidelines for Ecological Risk Assessment*. Washington, DC: Risk Assessment Forum, U.S. Environmental Protection Agency.

USEPA. 2014a. *An Assessment of Potential Mining Impacts on Salmon Ecosystems of Bristol Bay, Alaska*. Seattle, WA: U.S. Environmental Protection Agency.

USEPA. 2014b. *Framework for Human Health Risk Assessment to Inform Decision Making*. Washington, DC: U.S. Environmental Protection Agency.

USEPA. 2016. *Generic Ecological Assessment Endpoints (GEAEs) for Ecological Risk Assessment: Second Edition with Generic Ecosystem Services Endpoints Added*. Washington, DC: U.S. Environmental Protection Agency.

14 Condition, Causation, and Environmental Epidemiology

These opinions respecting the cause of cholera are brought forward, not as matters of certainty, but as containing a greater amount of probability in their favour than any other, in the present state of our knowledge.

John Snow (1849)

14.1 ENVIRONMENTAL EPIDEMIOLOGY

In the conventional sense, environmental epidemiology is the study of the effects of environmental conditions and exposures on human populations. An equivalent term for nonhuman populations and communities is ecological epidemiology. However, environmental epidemiology has also been used to include effects on humans and environmental resources (Box 14.1). We use the broader definition of environmental epidemiology, which may involve characterization of a deleterious environmental condition and/or identification of causes.

Environmental epidemiology includes assessments with two components: (1) determining the condition of human or nonhuman population or community qualities, particularly to determine whether there is an impairment (a condition assessment), and (2) determining the cause of any impairment (a causal assessment) (top row, Figure 4.1). In this chapter, we consider how weight of evidence can be used for condition and causal assessments, the work that must be done before beginning the work of finding solutions. We address weight of evidence for predictive and outcome assessments in Chapters 15–18. Condition assessments and causal assessments may also be performed independently of environmental epidemiology, as described in later sections.

The impetus for environmental epidemiology could be any of the following:

- human activities and socioeconomic developments that exert stress on the natural environment (e.g., tourism, subsidies for biofuels, regulations for commercial fishing, or a warming climate);
- natural drivers and pressures, such as drought, subsidence, and climate change;
- exposures to pollutants or other agents;
- apparent effects, such as disease clusters or mass mortalities;
- ecosystems that may have exceptionally good condition and thereby may be declared exceptional waters, parks, or other protected designations; or
- programs to monitor status and trends.

DOI: 10.1201/9781003626336-17

Typically, adverse effects are noted first, leading to the characterization of the severity, timing, and geographical patterns of the effects; then a cause or source is investigated. Often, one assessment leads to another, and once the cause is known, ways to control the cause are sought through criterion assessments, risk assessments, and management assessments (Cormier and Suter 2013). Example cases include the following:

- a suspected cluster of cholera leading to confirmation of the drinking water source (Snow 1849);
- declining bird populations and the ultimate identification and banning of DDT (Epstein 2014; Carson 1962),
- adult coho salmon deaths leading to the identification of 6PPD-quinone from tires and the release of screening values for 6PPD-quinone and 6PPD (Feist et al. 2017; USEPA 2024),
- litigation of PFAS effects eventually leading to an epidemiological study and PFAS criteria for drinking water and fish tissue (Bilott 2019), and
- a spike in child lead poisoning leading to characterization of the exposure (drinking water), the source (a change in Flint, Michigan's, water supply), and local and even national awareness of child lead poisoning and its long-term adverse health effects (Hanna-Attisha et al. 2016; Ruckart et al. 2019).

In these example cases, the weight-of-evidence process is largely narrative and not as structured as we might like, but it got the investigations started, often in the face of government reluctance or outright opposition. Often, these cases take decades to resolve, and the persistence of the assessors was essential to the ultimate influence of the effort. For all the less renowned cases that don't make it to the box office, I can't help but think how much more accessible epidemiological studies would be if there were tidy weight-of-evidence summaries.

BOX 14.1 EPIDEMIOLOGY FOR HUMANS AND THE ENVIRONMENT

Environmental epidemiology includes human health and ecological epidemiology. Although we emphasize ecological epidemiology, it is appropriate to also consider humans in an integrated assessment or at least coordinated assessments. They both assess conditions and determine the cause of any impairments and their sources. The information available can be similar, such as results of dietary exposure modeling for humans and wildlife, or it can be very different, such as uptake from water by the gills of fish versus drinking water consumption in humans. Effects data for humans come from medical records and epidemiological studies, but data for environmental media and nonhuman organisms often are obtained from environmental monitoring. Some types of information are commonly collected and used in ecological assessments, such as toxicity testing of environmental media, whereas human dose-response associations may be obtained from industrial exposures. Ideally, information would be shared, but human health assessors are, in our experience, reluctant to use ecological data. For example, plant uptake of mercury was measured by ecologists on the Oak Ridge Reservation, but the health assessors insisted on using default uptake factors, which gave unreasonably high dietary exposures. Also, in our experience, human environmental epidemiology is less likely to weigh multiple types of evidence.

14.2 CONDITION

Assessments may determine the condition of the environment for at least four purposes, alone or paired with a causal assessment.

- Condition may be monitored to determine status and trends in the environment. A change may precipitate an epidemiological investigation.
- Condition assessment may be used to determine background condition for comparison in other types of assessments, such as setting goals for remediation (Chapter 17), ascertaining the applicability of criteria to a new area (Section 20.4) (Cormier et al. 2018), or developing benchmark values (Chapter 15).
- Regulatory condition assessments may determine whether environmental quality standards are being violated or whether the system is impaired sufficiently to require regulatory action. Standard sets of evidence may be specified for these assessments.
- Condition assessments may be conducted to determine whether the condition of a system of interest indicates a potential need for management actions. These might include declines in the abundance of a species, elevated concentrations of a substance, altered geophysical regimes such as altered flow or sedimentation, or the degree of recovery of a managed system.

Characterizing condition is tricky. It all turns on the definition of condition and how conditions are determined. Stoddard et al. (2006) described four categories of condition: best attainable condition, least disturbed condition, minimally disturbed condition, and historical condition (e.g., pre-settlement in North America, pre-intensive agriculture (e.g., 1850 in Great Britain, 1700s in Germany)). Condition endpoints include various metrics such as contaminant concentrations, dissolved oxygen, abundance of game fish, species richness, diversity indices, and net primary production. The simplest presentations of status are distributions of monitoring results and summary statistics such as means and centiles. To minimize the volume of data, some types of condition metrics may be combined into indices. However, indices are hard to interpret and may be misleading (Suter 1993, 2023). A more defensible approach is to rate condition metrics from a water body, site, or region into categories like Stoddard's. The endpoint metric is evidence concerning environmental status, and it may be weighted as evidence that the monitored environment is in "best attainable condition" or some other condition. The weights may be based on centiles of the distribution of the metrics, relationships to designated reference sites, records of past conditions, or best professional judgment. The weighted evidence of condition might then be integrated as discussed in Chapter 8.

But if we take a step back, weight of evidence is critical well before the final characterization of condition. Weight of evidence is absolutely essential for choosing what, where, when, and how to measure indicators and how to estimate a reference condition, that is, determining how to characterize the quality or quantity of the condition metrics.

14.2.1 Monitoring Status and Trends

Many nations and other entities conduct environmental monitoring programs to determine whether existing management and regulatory activities have been successful and to inform the public and stakeholders of environmental conditions. For example, U.S. National Aquatic Resource Surveys collect many measurements and report conditions using indicators in several formats, including an interactive dashboard (https://www.epa.gov/national-aquatic-resource-surveys). For example, Figure 14.1

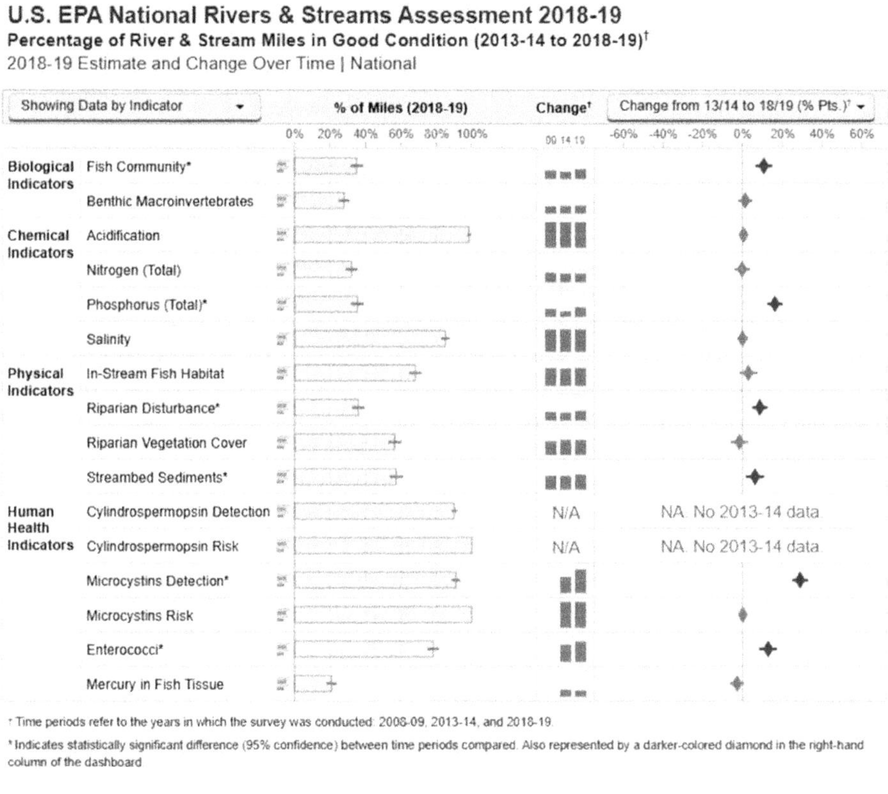

FIGURE 14.1 Example from the National Rivers and Streams Assessment 2018–2019. This screenshot from the data dashboard shows the percentage of river and stream miles in good condition, the direction of change, and the change between the 2013–2014 and the 2018–2019 surveys. Ninety-five percent confidence intervals are shown. https://riverstreamassessment.epa.gov/dashboard/?&view=indicator&studypop=rs&subpop=national&label=none&condition=good&diff=2v3.

shows several individual stream indicators for biological, chemical, physical, and human health monitoring between 2018 and 2019, and changes since 2013–2014. Regional benchmarks, biological presence-absence, or USEPA-recommended water quality criteria are used to characterize good, fair, and poor conditions. You will note that there is often a lag between data collection and reporting. This is typical of large-scale sampling efforts. For national assessments, another challenge is to ensure that samples collected by different organizations are comparable.

You might notice that less than 40% of stream miles were characterized as good for biological indicators and nutrients. Such information may suggest that a causal assessment is needed. You might also notice the absence of indicators for the detection of toxic chemicals except mercury in fish tissue. Early choices in what to measure or assess influence environmental decision-making.

14.2.2 Background Conditions

The distinction between natural background and anthropogenically altered conditions is fundamental. We do not want to remediate, regulate, or isolate a substance in a medium that is, in fact, at background levels and in a naturally occurring form. Condition assessments that distinguish background from contaminated levels of naturally occurring substances are particularly reliant on the availability of many types of relevant and reliable evidence. That is because the divide between natural and altered concentrations is indistinct and spatially and temporally variable.

We define "background" as the range of concentrations naturally occurring in waters, soils, or biotic materials that have not been substantially influenced by human activity. The process of defining that range can be attempted using a variety of approaches. A descriptive approach may define background based on some proportion of samples or a breakpoint in the distribution. For example, one may ask, "At what concentration is there an inflection in a cumulative distribution of measurements?" There are numerous approaches for quantifying background and determining whether an observed condition exhibits altered chemistry (Table 14.1). Alternatively, background condition can be evaluated using a form of causal assessment. For example, one may ask the question, "Is the observed chemistry at a site caused by anthropogenic inputs?" If so, it is not included in the characterization of the background range. If there are no anthropogenic causes of the chemistry, the observed values can be used to characterize the background range.

14.2.3 Impaired or Exceptional Status

Many countries require condition assessments to determine whether water bodies or other environmental media are impaired. Commonly, benchmark values based on single-chemical laboratory toxicity tests are used to define impairment of aquatic life. In the U.S., water quality criteria are estimated from toxicity tests of a minimum of eight taxa from peer-reviewed studies (Chapter 15). These regulatory benchmarks are mandatory for determining the acceptability of ambient waters, but they are not reliable alone. Benchmarks may be supplemented by other evidence, such as results of toxicity tests for ambient waters and sediments and results

TABLE 14.1
Example Methods for Estimating Background

Method	Example
1. Anything that is not an anomaly (e.g., not an ore body, salt spring, or stream reach receiving an effluent) may be background.	This results in background limits, such as the mean plus two standard deviations (Reimann, Filzmoser, and Garrett 2005).
2. When concentrations at background and contaminated sites have distinct distributions, background has been defined by inflections, break points, or deviations from normality (Molinari et al. 2014).	For example, the contaminant lead was distinguished from the natural concentration distribution in soil by deviation from linearity in a probability plot (Zhao, McGrath, and Merrington 2006).
3. Background may be distinguished by differences in isotopic composition of natural and unnatural materials or differences in the "fingerprints" of chemical mixtures (the relative concentrations of constituents) from different sources.	For example, polyaromatic hydrocarbon mixtures may differ among natural oil seeps, bunker fuels, and oil spills (Page et al. 1996). Similarly, stable isotope ratios have been used to determine natural and anthropogenic sources of lead (Luo et al. 2015; Sucharova et al. 2014).
4. Background may be what is found in uncontaminated or undisturbed sites or materials, such as reference streams, deep sediments, ice cores, predevelopment analyses, or museum specimens.	For example, (Hinsby, de Melo, and Dahl 2008) used the 90th or 97.5th centile of uncontaminated ground water concentrations as background, depending on the amount and quality of data. The U.S. EPA commonly uses the 75th centile of reference site surface water concentrations as the limit of background, because some surface water reference sites are the best available (USEPA 2000).
5. Low levels for a chemical may represent background.	U.S. EPA commonly has used the 10th or 25th centile of regional levels as the background level (Cormier et al. 2018; USEPA 2000; Stoddard et al. 2006).
6. Background may be the concentration in the source materials.	For example, background for Rhine River sediments was assumed to be the concentration of elements in upper watershed soils (Van de Meent et al. 1990).
7. Background may be estimated by hydro-bio-geo-chemical models (Runnells, Shepherd, and Angino 1992; Mast et al. 2007)	For example, background nutrient levels are empirically modeled from runoff and other watershed characteristics (Smith, Alexander, and Schwarz 2003).
8. If background may be defined by the absence of input from identified sources, it may be determined from the distribution of concentrations relative to the sources.	For example, background elemental concentrations for soils receiving coal fly ash were defined by the asymptotic concentrations on sampling gradients away from a power plant (Gough and Crock. 1997). Alternatively, source strength, dilution, and losses may be modeled to determine what proportion of downstream concentration is not background (Helgen and Moore 1995).

(Continued)

TABLE 14.1 (*Continued*)
Example Methods for Estimating Background

Method	Example
9. Weight of evidence or background may be the level that best displays the characteristics of background given the body of evidence.	After reviewing and dismissing individual methods to define background, Reimann, Filzmoser, and Garrett (2005) recommended graphical inspection of various maps and plots (particularly box plots) and application of integrative judgment. The USGS similarly applies multiple analyses to background derivation (Mast et al. 2007). Those approaches are limited to analyses of concentration data. Evidence of similar specific conductivity background was weighed in a formal assessment process comparing two areas in an ecoregion (Cormier et al. 2018).

Source: Adapted from Cormier et al. (2018).

of biological surveys that account for effects of mixtures, water chemistry, and population- or whole-community-level chronic effects. These types of evidence may be used to derive benchmark values by weight of evidence for problematic chemicals or mixtures (Chapter 15).

Rather than relying on benchmark values, an impaired condition of a water body may be identified by integrating multiple types of evidence under the U.S. Clean Water Act, in Australia and New Zealand under the Guidelines for Freshwater and Marine Water Quality, and in the European Union under the Water Framework Directive. For example, the EU requires four types of evidence for the final decision on the status class of water bodies (EU Water Directors, 2023).

- **Biological Quality Elements**: Measures of the condition of biological communities.
- **Physico-Chemical Quality Elements**: Measures of the condition of waters including dissolved oxygen, nutrient levels, suspended particles, transparency, temperature, and others.
- **Hydromorphological Quality Elements**: Measures of channel form, continuity, connectivity, variation in flows, and others.
- **Ecotoxicity**: Concentrations of substances relative to harmful levels.

Designations of impairment, based on multiple types of evidence, may be derived in two ways. First, independent applicability requires that if any type of evidence indicates impairment, the site is impaired (Section 2.8). Second, a full weight-of-evidence process may be used. That is, all types of evidence may be weighted, and the body of evidence for impairment may be weighed to determine whether the hypothesis of impairment or nonimpairment is best supported (Chapters 7 and 8).

An example of integrating multiple types of evidence to identify exceptional status is the designation of protected wetlands in the United States. Evidence must be provided for each of three criteria, and the evidence must be based on interpretation of numerous indicators and judgment of sufficient relevance and reliability (Environmental Laboratory 1987). "The weight of evidence for wetland delineation is determined by the total evidence of wetland conditions on site, provided by the indicators present, once subjected to reasonable scientific judgment" (Florida DEP 2026).

- **Hydric Soils**: Wetlands must have soils that are saturated with water at a frequency and duration sufficient to support a prevalence of vegetation typically adapted for life in saturated soil conditions.
- **Hydrophytic Vegetation**: Wetlands must have a prevalence of vegetation that is typically adapted for life in saturated soil conditions.
- **Hydrology**: Wetlands must have ground or surface water that inundates or saturates soils sufficiently to support a prevalence of hydrophytic vegetation.

14.2.4 Need for Management Actions

Sometimes condition assessments are needed to determine whether a particular management action is appropriate. For example, I (Cormier) was asked to ascertain whether the background chloride levels in the Bitter Creek Watershed, Wyoming, USA, exceeded state chloride standards. The issue was that natural levels should not be considered impaired, even if they were greater than national criteria. Thus, it might be appropriate to raise the state's standard for chloride. The Wyoming Department of Environmental Quality (WDEQ) chronic and acute chloride standards are 230 and 860 mg/L. The WDEQ noted that chloride levels exceeded standards in a number of locations, but those levels could be naturally elevated in this arid area with a marine geological history. Because the criteria would apply to the entire watershed, or even the state, it was necessary to characterize the background chloride levels before evaluating what the effect would be if the state were to begin the management action of raising its chloride standards. This is an example of a condition assessment linked to a risk estimation for a management action.

As was often the case, there was no budget for field sampling. Therefore, in addition to planning the estimation approach by selecting the geographical scale, source of the measurements, and statistic of the background condition, we also needed to assess the suitability and quality of publicly available datasets. We weighed evidence for each methodological decision. Once both measured and modeled evidence were estimated throughout the watershed, we evaluated our options for inferring natural background. The heavy lifting was done by weight of evidence. Once the methodological decisions were made, estimating current condition and comparing it to probable background and aquatic life standards were relatively simple calculations.

Tables 14.2 and 14.3 provide a table of types of evidence with their implications, and a weight-of-evidence table from the Wyoming case. They may be adapted to other cases or prompt your own thinking about how to perform and document these activities that, sadly, are often not explicit or are buried in text. Table 14.2 lists potential evidence to consider when using existing field data that had been collected for a variety

TABLE 14.2
Example Types of Evidence for Evaluating Available Water Quality Monitoring Datasets for Reliability

Evidence	Influence on Estimating Background (Implications)
Total # of stations	Larger sample sizes are more likely to include some reference stations.
Number of stations with only one sample	May not be representative of the station.
Percentage of stations with >2 samples	Replicates increase precision and better characterize variability
Number of stations with >100 samples	Biases the centiles and means, but useful for seasonal analysis; can use a data reduction to adjust.
Range (μS/cm) median	Indicative of representativeness of watershed and inclusion of stations near background.
Range (μS/cm) with replicates	Indicative of representativeness of watershed and inclusion of stations near background.
Within year station range	Percentage of stations with variability >1500 μS/cm indicative of effluent.
Percentage of samples with >1500 μS/cm	Suggests whether dataset measures freshwater or altered water.
Skewness	Indicative of representativeness of watershed and inclusion of stations near background. Skew to right is indicative of altered water.
Kurtosis	Could be representative, nonrepresentative of either tail (e.g., background stations), biased by season.
Percentage of stations before 1985	Older data may not have complete documentation.
First and last recorded sample year	Longer records are useful for evaluating trends
Span of sampling in years	Different sample periods may represent different weather conditions, which can influence water chemistry.
Total # of years sampled in the dataset	
Purposive data collection (versus random)	Biased sample toward anthropogenically influenced stations.
Spatial distribution	Area that was sampled dictates area that it represents. Broader spatial coverage is more likely to capture some reference stations.
Bimodality	Indicates that there are at least two populations (e.g., methods, exposures, seasons, etc.) with potentially different background or interpretations.
Percentage of stations in tributaries	Fewer likely inputs and, therefore, some may represent background. However, where there are inputs, small streams provide less dilution.
Absolute change between 10th and 25th centile (μS/cm)	When there are many stations with similar backgrounds, the 10th and 25th are similar. Large changes may be caused by false or nonrepresentative low values or bias toward altered stations.
Percentage of repeated values	May indicate data entry error, detection limit, or repeat samples from same location. (Percentage of identified records in raw dataset from the same station, date, time, and having the same values, and calculated percentage).
Percentage negative values	Specific conductivity cannot be a negative value.

(Continued)

TABLE 14.2 (*Continued*)
Example Types of Evidence for Evaluating Available Water Quality Monitoring Datasets for Reliability

Evidence	Influence on Estimating Background (Implications)
Percentage of zero values	Zero conductivity is not possible and may indicate no measurement or below detection.
Percentage of values <10 μS/cm	Values <10 or even <100 μS/cm are unlikely, especially in Bitter Creek, given the known geology and climate.
Percentage of values >10,000 μS/cm	Does not represent background, indicative of sampling of altered water. May be data entry error, extremely altered system, or salt spring/terminal inland lake.
Percentage of outliers (values larger than mean plus five times the standard deviation)	Suggests nonrepresentative sampling or data entry errors.

This example is from the Bitter Creek case identifying background chloride levels (adapted from Cormier et al. 2022a).

TABLE 14.3
Weight-of-Evidence Table for Selection of Method, Dataset, and Statistic for Surface Flow Background SC and Cl^- in Bitter Creek Watershed (Cormier, Wang, and Wharton 2022a)

A plus sign or signs indicate support (+) or strong support (++) of the consideration based on relevance, strength, and reliability confidence. A minus sign (–) weakens the consideration based on relevance or confidence. Zero (0) indicates ambiguous weighting that neither strengthens nor weakens.

	Relevance		Reliability and Justification	
Scale				
Ecoregion	Regional statistics are only somewhat relevant to a watershed due to variance in geology, ionic composition, and physiography.	+	The Level III Wyoming Basin Ecoregion 25th centile (341 μS/cm) is consistent with the Bitter Creek watershed background from observed values (411 μS/cm) and with the lowest predicted background (384 μS/cm). However, this is a small dataset with different dominant ions compared to the Bitter Creek watershed.	+
Watershed	A watershed is most relevant to itself but may be altered from background.	+	Xeric conditions and the ephemeral and intermittent flows of streams increase the variability likely to occur throughout the watershed and site-specific background may be more accurate.	+

(*Continued*)

TABLE 14.3 (*Continued*)
Weight-of-Evidence Table for Selection of Method, Dataset, and Statistic for Surface Flow Background SC and Cl^- in Bitter Creek Watershed (Cormier, Wang, and Wharton 2022a)

A plus sign or signs indicate support (+) or strong support (++) of the consideration based on relevance, strength, and reliability confidence. A minus sign (–) weakens the consideration based on relevance or confidence. Zero (0) indicates ambiguous weighting that neither strengthens nor weakens.

	Relevance		Reliability and Justification	
		Scale		
Stream segment	Stream segment is most relevant to itself, but the stream may be altered or there may be no observed data.	++	Observed background in a stream segment is dependent on ground truthing stations but are not available for this study.	0
			Predicted background is not dependent on ground truthing but useful for validation.	+
Summary	The smaller the scale, the more relevant the estimate; however, dataset characteristics affect selection. Ecoregional estimate offers no advantage over watershed or stream segment scale.			
		Source of estimate		
Predicted (random forest model)	The model is based on natural factors that affect background and thus is more relevant because it is minimally influenced by human alterations. However, because it is a national model, it may not be optimized for the watershed and require calibration.	+	Model performance was good for the NRSA Xeric Ecoregion and for the few Bitter Creek watershed stations with multiple samples with a geomean within the range of predicted natural background. Observed 10th centile estimates from the USGS and Combined datasets fell within the range of predicted background values (384 μS/cm and 1,004 μS/cm, respectively).	++
Observed ecoregional	Randomly sampled stations reduce bias.	+	Wyoming Basin covers a large geographic area, may not represent an annual average, has a different ionic composition, and is less reliable because of a small sample size.	–
Observed USGS	Measurements are inherently relevant to local conditions, but purpose of site selection is unknown and included altered waters.	++	The USGS dataset was representative of background because it had no apparent errors, was large, was spatially distributed across the watershed, and had a more normal distribution than the other datasets. However, the dataset is older.	++
Observed Combined	Measurements are inherently relevant to local conditions, but site selection was often targeted toward altered waters.	–	Combined observations have less consistent quality assurance. Some are unlikely to have many high-quality stations because they are targeted to areas of concern. Data distribution is more skewed toward greater specific conductivity (SC) and Cl – values. Ionic composition differs at more of the sampling stations.	–

(*Continued*)

TABLE 14.3 (*Continued*)
Weight-of-Evidence Table for Selection of Method, Dataset, and Statistic for Surface Flow Background SC and Cl^- in Bitter Creek Watershed (Cormier, Wang, and Wharton 2022a)

A plus sign or signs indicate support (+) or strong support (++) of the consideration based on relevance, strength, and reliability confidence. A minus sign (–) weakens the consideration based on relevance or confidence. Zero (0) indicates ambiguous weighting that neither strengthens nor weakens.

	Relevance		Reliability and Justification	
	Source of estimate			
Summary	Of the observed datasets, the USGS dataset is suitable, but as noted above, the watershed scale is coarse. Other datasets are not suitable. At the stream segment scale, the predicted estimates are recommended because they are more accurate, scaled to the stream segment, and based on least disturbed stations and thus represent minimal influence by anthropogenic alteration.			
	Background statistic			
Annual mean	Central tendency appropriate for a stream segment for predicted natural background surface water flow.	++	Central tendency is statistically more robust than a low centile. Among stations within the predicted background range, the absolute difference between observed geomean and mean predicted values (MAE) was 210.5 μS/cm ± 150.4 μS/cm. The national model MAE for the Xeric Ecoregion is 62 μS/cm. The difference is attributed to the relative contribution of surface and ground water flow in the Bitter Creek watershed compared to more perennial streams in the Xeric Ecoregion that are likely dominated by surface water flow.	+
	A mean would characterize neither surface flow nor ground water flow background SC from observed data.	–	Annual mean using observed data would be biased toward ground water flow.	–
10th centile	Anthropogenic alterations support the selection of 10th centile of observed data.	++	Low centile stations are more likely to represent background but may be conservative for segments with naturally saline sources.	+
25th centile	Historical anecdotal reports of salinity support the selection of a higher centile of observed data for background but may be a localized situation.	0	The large SC step increase from 10th centile to 25th centile suggests that 25th is not the watershed background. Degree of anthropogenic influence also does not support selecting the 25th centile.	–

(*Continued*)

TABLE 14.3 (*Continued*)
Weight-of-Evidence Table for Selection of Method, Dataset, and Statistic for Surface Flow Background SC and Cl⁻ in Bitter Creek Watershed (Cormier, Wang, and Wharton 2022a)

A plus sign or signs indicate support (+) or strong support (++) of the consideration based on relevance, strength, and reliability confidence.
A minus sign (–) weakens the consideration based on relevance or confidence.
Zero (0) indicates ambiguous weighting that neither strengthens nor weakens.

	Relevance		Reliability and Justification	
	Background statistic			
75th centile	Appropriate with minimally affected samples, but none were identified by dataset metadata.	–	Minimally affected samples were not identified.	–
Summary	Suitable statistics are (1) mean or median predicted background and (2) 10th centile from USGS observed SC.			
Selection of background endpoint	Predicted background appears to be the most reliable estimate of background and may be particularly useful where there are no reference stations or where there are anthropogenic alterations. Of the observed datasets, the 10th centile of the USGS dataset with repeat measurements is also a reasonable choice, but accuracy is affected by the large, predicted range of possible background SC in the watershed. Confidence could be improved with more evidence, including comparison with verified minimally disturbed reference stations, evidence from biological data, and identification of local anthropogenic and natural sources.			
	Because there are no confirmed reference stations in Bitter Creek, a pragmatic process for setting background was described for unknown or nonreference stations using observed and predicted background SC (available at Cormier, Wang, and Wharton 2022a).			

of reasons. It illustrates the complexity of field-derived information that provides opportunities but also challenges for weight of evidence. The evidence could also be weighted for the relevance of sampling locations (e.g., location within the watershed, habitat type, industrial or agricultural development, and physical chemistry).

Utilizing a table like this involves weighting the types of evidence and other considerations and combining them into an expression of confidence in the monitoring data (i.e., weight). The types of evidence may be appropriate to a particular condition assessment like this one, or to an assessment of the confidence in the data from a monitoring program for use by condition assessments in general or environmental epidemiology.

Table 14.3 lists, weights, and weighs the evidence for the selection of the method, dataset, and statistic used to estimate the condition of surface-flow background specific conductivity and chloride in the Bitter Creek Watershed. Table 14.1 lists inferential methods for estimating background to later estimate the risk of relaxing the water quality standards for chloride.

Although these lists and rationales are specific to the case, they can be a good starting point for developing weight-of-evidence tables for other studies to assess datasets and background conditions. Perhaps even more importantly, these detailed tables illustrate that transparency, along with rationales, can help provide a sound scientific basis for decision-making. The Bitter Creek report enabled the Wyoming DEP to holistically assess condition and the implications of potential regulatory changes. The full report, appendices, datasets, and R-code are available at data.gov and can serve as a good example of the full transparency of an assessment effort (Cormier, Wang, and Wharton 2022a, 2022b).

After ambient chloride levels were estimated, these levels were compared to the Wyoming Department of Environmental Quality chloride standard. Many locations had low to moderate levels of chloride below the current WDEQ standards. Based on the background condition assessment, it was concluded that areas are unlikely to be in a natural condition where ionic concentrations are higher than the background predicted from geology, climate, soils, and vegetation factors (Olson and Cormier 2019; Wharton, Olson, and Cormier 2021), and where chloride is the dominant anion and is strongly elevated. Raising the standards across the watershed would risk valued potable water sources for humans and livestock and would reduce the dilution of altered waters that tended to be downstream. Although some of the chloride loadings may be reduced and mitigated with better management practices, the geological alteration of Bitter Creek from legacy mining near Rock Springs and current and abandoned extraction operations may have irrevocably altered the stream's hydrology. This puts additional focus on reducing additional ion inputs and protecting those locations that still provide dilution. As a result of this comprehensive weight-of-evidence background condition assessment and causal assessment, the Wyoming Department of Environmental Quality did not relax its chloride standards.

As you can see, weight of evidence requires flexibility and pragmatic adaptability when applying professional judgment in a structured weight-of-evidence process. Weight of evidence becomes useful when the evidence is accessible and the reasons for the weights and the relevance of the inferences are clearly stated. It is an invitation to explore the evidence and perhaps even to alter or strengthen the conclusions.

14.3 CAUSATION

As we have noted, in environmental assessments, weight of evidence has been used primarily to assess causation. This is the case because causation is never 100% proven, so it's important to consider as much evidence as possible. Weight of evidence helps to organize the abundance of evidence and the inferences we draw from that evidence. In fact, weight of evidence has been equated with determining the degree of credence due to a causal hypothesis (Krimsky 2005; Rhomberg 2013).

Types of evidence are a useful and commonly used way to think about all the evidence that you can generate or find to apply to a causal assessment. But when it comes to organizing the evidence, it doesn't matter so much that you can give a name to a piece of evidence by type. What really matters is how to organize evidence in a way that best facilitates weighting, weighing, and integrating the evidence to identify the probable cause. If you can characterize the information so that it is clear to you, then it is more likely to be clear to others.

When we first began running workshops on causal assessment, we used types of evidence along with an adaptation of Hill's considerations, but participants ran into difficulty keeping the terms straight, dealing with nearly similar evidence, bulking up with one kind of evidence, double-counting the same evidence, and basically burning up energy and time. It was frustrating for them and for us. We changed tactics.

We asked, what are we trying to show? We are trying to show that a relationship is or is not causal while laying out a logical and coherent case. From that frame of reference, we could focus on assembling evidence for characteristics of causal relationships and evaluate whether a relationship is causal based on the presence of evidence of causal characteristics. The emphasis became more about the inferences drawn from the evidence and less about how the evidence was obtained.

After extensively consulting the writings of philosophers and scientists, we developed a list of characteristics common to causal relationships (Table 14.4). To prepare the evidence for weighting and weighing, the pieces of evidence are sorted into sets

TABLE 14.4
Characteristics of Causal Relationships (USEPA 2016; Cormier, Suter, and Norton 2010)

Causal Characteristics	Description	What Evidence of a Characteristic Shows
Time order	The cause precedes the effect.	Change in the entity after exposure to the cause and not before
Antecedence	The causal relationship is a result of a larger web of antecedent cause-and-effect relationships. Evidence includes sources and routes of transport.	Earlier events leading to the particular causal event. It demonstrates that a potential causal event is part of a network of prior causal events, increasing confidence that the potential causal event actually occurred
Co-occurrence	The cause co-occurs with the susceptible entity in space and time. Evidence includes contaminant concentrations in ambient media and habitat alterations.	The presence of both a causal agent and the affected entity in circumstances that create the potential for exposure
Interaction	The cause interacts with the entity in a way that can lead to the effect. Evidence includes injuries, body burdens, and biomarkers.	Signs that the entity has been exposed to the agent in such a way that changes can or have been initiated through a plausible mechanism
Sufficiency	The intensity, frequency and duration of the cause are adequate, and the entity is sufficiently susceptible to produce the type and magnitude of the effect.	Enough of the cause and a sufficiently susceptible entity that can result in the level of the observed effect
Alteration	The entity is changed by physical, chemical, or other mechanisms leading to the defined effect. Evidence includes symptoms and related attributes of the affected entities, or distinctive biological changes to communities associated with the potential cause.	Changes in the entity attributable to or at least appropriate to the cause.

TABLE 14.5
Rules for Determining Causation by Weighing the Body of Evidence for a Causal Relationship

Body of Evidence	Causal Relationship
Evidence refuting[a] 1 or more characteristics	Refuted causation
Evidence discounting[b] 4, 5, or 6 characteristics	Unlikely causation
Evidence discounting 1, 2, or 3 characteristics, others supporting	Unlikely causation but low confidence
Evidence strongly documenting 6 characteristics	Confirmed causation
Evidence documenting 5 or 6 characteristics and none discounting	Very probable causation
Evidence strongly documenting 3 or 4 characteristics and none discounting	Probable causation
Evidence strongly documenting 2 characteristics and none discounting	Probable causation but low confidence
Evidence documenting 1 characteristic	Insufficient evidence to make a determination

[a] Refuting is the logical process of demonstrating the impossibility of a candidate cause, thus allowing it to be eliminated from further consideration.
[b] Discounting is the weighting of evidence that weakens the case for a candidate cause but is insufficient to refute.

that each illustrate a characteristic. If there are multiple similar studies, they can form a subset for weighting, but eventually all the related evidence in a subset is used to infer a characteristic of a causal relationship. Then, the assessor evaluates whether there is enough evidence for enough causal characteristics to identify a cause (Table 14.5). Sorting the evidence by causal characteristics makes it easy for you and your audience to see what evidence is available to judge whether a potential cause is actually causal.

Because the causal characteristics are relatively new and not broadly adopted, we feel the need to support their use. When we published the causal characteristics (Cormier, Suter, and Norton 2010), the editor of *Human and Ecological Risk Assessment*, Barry Johnson, gave them a strong endorsement. Dr. Johnson was a former Director of the Agency for Toxic Substances and Disease Registry, and former U.S. Assistant Surgeon General. He liked the causal characteristics and recognized the validity of our critique of weighting types of evidence or considerations. Legal support for the characteristics is presented in Section 14.4 and Box 14.2. We are comfortable recommending the use of weight of evidence based on causal characteristics because it works and has convincingly conveyed the information to broad audiences.

14.4 EXAMPLE CASES

The U.S. Geological Survey scientists used weight of evidence to show that insecticides adversely affect aquatic invertebrates (Nowell et al. 2024). Multiple types of evidence for each of the six characteristics of causal relationships were strong, diverse, and consistent. Evidence for *antecedence* showed that land-applied pesticides enter streams. Insecticides *co-occurred* with susceptible stream invertebrates, such as mayflies, which led to exposures and *interactions* with aquatic life at *sufficiently* toxic concentrations. These exposures were strong enough to consistently extirpate specific groups of taxa, resulting in an *alteration* of the overall aquatic assemblage of macroinvertebrates. Mesocosm and other *time-order* evidence showed that adverse changes in aquatic organisms and communities occurred after interaction with the pesticides and not before. Thus, a strong case was made that pesticides are probable causes of impairment of aquatic life in agricultural and urban areas.

Margaret Watkins and I (Cormier) collaborated on a criterion assessment and used causal characteristics to show that, throughout most of the Grand Portage Reservation, Minnesota, total phosphorus and total nitrogen levels had natural causes and, therefore, aquatic life criteria could be set at the median of ambient levels (Watkins, 2023). A condition assessment alone was insufficient to determine whether nutrient levels were natural or anthropogenic because ecoregional reference water bodies outside the reservation in existing datasets were not entirely representative of the smaller Grand Portage Reservation water bodies. Therefore, a causal assessment was performed to determine whether current observed nutrient and dissolved organic carbon levels were due to local or remote anthropogenic causes or natural causes. Evidence was compiled for all six causal characteristics. Evidence for natural causes was all strongly supportive, whereas evidence for anthropogenic causes was all nonexistent, weak, or refuted. The smaller and shallower size of the lakes, coupled with abundant beaver activity, probably accounts for the difference between nutrient levels in the Grand Portage Reservation and those outside the reservation.

One of the more interesting pieces of evidence was for time order. Lake total phosphorus levels were inferred from diatoms in lake sediment cores (Swamp Lake: 1781–2005; Speckled Trout Lake: 1825–2006) (Edlund et al., 2007, 2009; Christensen et al., 2010). Sediment core profiles for total nitrogen concentration and stable isotopic ratios (δ15N) related to diatom composition were relatively constant since the 1800s. From that evidence, we inferred that the levels of TP and TN were relatively unchanged over two centuries up to recent times. That indicates that anthropogenic nutrient inputs are minimal. This strongly negative evidence alone refuted anthropogenic causes post-colonization, but evidence for the other causal characteristics strengthened the weight of evidence (Table 14.5, row 2). The causal assessment led to the development of total phosphorus and total nitrogen criteria based on the median total levels as of January 2025, with the intent of maintaining the land and water as close to natural as possible. This differs from most nutrient criteria, which are set at an assimilative threshold while meeting designated uses.

In a coal mining case, Judge Chambers distinguished general and specific causation and required that the plaintiff first establish general causation and then specific causation for the particular streams (Chambers 2015). He explained in a footnote the similarity between the coal case and the Exxon Valdez litigation:

> In the oil spill context, proving general causation would require showing that exposure to oil is capable of causing the alleged injury—e.g., disrupting the reproductive capacity of sea otters. Proving specific causation would require showing that sea otters were exposed to oil in doses capable of causing injury and that the oil came from the defendant's release.
>
> **(Knudsen 2014)**

In the coal mining case, the plaintiffs needed to show that something from valley-fill effluents was generally capable of causing harm and that, in specific streams below valley fills, it had caused biological impairment. The court recognized that environmental causes are probable (i.e., more likely than not) and that there can be more than one contributing cause: "Conductivity levels in the streams at issue need not be the sole cause of observed biological impairment, but must be a substantial contributor. This standard does not require scientific certainty, but rather legal probability" (Chambers 2015).

In that mining case, our assessment of general causation was admitted as evidence in court to establish general causation, and the characteristics were used with additional evidence to establish specific causation. The assessment used the weight of evidence of causal characteristics to establish that dissolved ions can cause aquatic impairments below mountaintop mining and valley fills (i.e., general causation) (USEPA 2011; Cormier et al. 2013). The weight of evidence was understood by the nonscientific members of the court, and the evidence was critical to the determination that a coal company had violated the Clean Water Act (Chambers 2015). Box 14.2 is an excerpt from the closing section of the opinion from the case that highlights key causal characteristics that the court called out. I have added the characteristic in brackets.

BOX 14.2 OHIO VALLEY ENVIRONMENTAL COALITION ET AL. V. FOLA COAL COMPANY, LLC. QUOTATIONS FROM THE JUDGE'S OPINION (CHAMBERS 2015)

Cormier and Suter analyzed six characteristics of causation: co-occurrence, preceding causation, interaction, alteration, sufficiency, and time order, finding all but one strongly supported the causal relationship, with no evidence available for the outstanding characteristic [Time Order]. Though Cormier and Suter were unable to obtain the necessary information to assess the significance of time order in assessing the causal relationship, the record here is replete with such information and evidence... [e.g.,] pre-mining samples taken downstream from Outlet 001 and upstream from the confluence of Road Fork and Leatherwood Creek showed conductivity levels ranging from 40 μS/cm to 73 μS/cm ... Since Defendant's mining activities began, ... conductivity levels ranged from 1,803 to 5,700 μS/cm.

On the basis of the evidence presented, the Court **FINDS** that Plaintiffs have demonstrated the following by a preponderance of the evidence: (1) Road Fork and Cogar Hollow are biologically impaired, as measured by WVSCI scores below the federally approved threshold score of 68; (2) the discharges complained of share the same characteristic composition presently known in the scientific community to cause or materially contribute to impairment in central Appalachian streams [1 & 2 Co-occurrence]; (3) the conductivity levels observed at Defendant's discharges are far in excess of the thresholds identified by the EPA and the available scientific literature known to cause stress to aquatic communities [Sufficiency]; (4) Defendant's mining operations are the only land use that could have caused impairment [Antecedence]; and (5) changes in the biological community particularly show the loss of conductivity-intolerant organisms [Alteration]. Thus, the Court **FINDS** that high-conductivity discharges from Fola Surface Mine No. 2 and No. 6 are causing or materially contributing to the biological impairment of Road Fork and Cogar Hollow in violation of Defendant's current WV/NPDES permits.

14.5 CONCLUSIONS

A real test of an assessment is whether it serves its purpose. In the Bitter Creek Watershed example, the issue was resolved, but the criteria were not relaxed. In the Grand Portage example, the weight of evidence enabled the Grand Portage Band to establish aquatic life standards. We don't know how the USGS insecticide assessment has been used, but as of July 11, 2026, it had been cited 20 times according to ResearchGate. We can hope that its strong weight of evidence will lead to better field and runoff practices for pesticides. In two cases involving coal mine runoff (Section 14.4), the water quality standards were enforced, and the judge cited our weight-of-evidence analysis as a basis for the findings. It is useful to take a peek at how causal assessments are perceived by people and how they consult the weight of evidence so we all can learn how to provide clear and useful scientific facts and reasoning to inform environmental decisions.

REFERENCES

Bilott, R. 2019. *Exposure: Poisoned Water, Corporate Greed, and One Lawyer's Twenty-Year Battle against DuPont.* New York, NY: Simon & Schuster.

Carson, R. 1962. *Silent Spring*. New York: Houghton Mifflin.

Chambers, R. C. 2015. "Memorandum Opinion and Order: Ohio Valley Environmental Coalition, West Virginia Highlands Conservancy, and Sierra Club v. Fola Coal Company, LLC." Civil Action No. 2:2013–21588 (consolidated with 2:2013-16044). Document 119 (S.D.W. Va. 2015). August 12, 2015. Charleston Division, U.S. District court for the Southern District of West Virginia.

Christensen, V. G., P. M. Jones, M. B. Edlund, and J. M. Ramstack. 2010, Water quality (2000–08) and historical phosphorus concentrations from paleolimnological studies of Swamp and Speckled Trout Lakes, Grand Portage Reservation, northeastern Minnesota: archived by the University of Minnesota, Duluth: U.S. Geological Survey Scientific Investigations Report 2010–5192, https://hdl.handle.net/11299/189316.

Cormier, S. M., and G. W. Suter II. 2013. "A method for assessing causation of field exposure-response relationships." *Environmental Toxicology and Chemistry* 32 (2): 272–276.

Cormier, S. M., G. W. Suter II, and S. B. Norton. 2010. "Causal characteristics for ecoepidemiology." *Human and Ecological Risk Assessment* 16: 53–73.

Cormier, S. M., G. W. Suter, L. Zheng, and G. J. Pond. 2013. "Assessing causation of the extirpation of stream macroinvertebrates by a mixture of ions." *Environmental Toxicology and Chemistry* 32 (2): 277–287.

Cormier, S. M., Y.-C. Wang, and C. Wharton. 2022a. *Characterization of Chloride and Conductivity Levels in the Bitter Creek Watershed, Wyoming*. Cincinnati, OH: U.S. Environmental Protection Agency.

Cormier, S. M., Y.-C. Wang, and C. Wharton. 2022b. *Characterization of Chloride and Conductivity Levels in the Bitter Creek Watershed, Wyoming, Appendix: Methods, Figures, Resources, and Weight-of-Evidence Tables*. Cincinnati, OH: U.S. Environmental Protection Agency.

Cormier, S. M., L. Zheng, G. W. Suter, and C. Flaherty. 2018. "Assessing background levels of specific conductivity using weight of evidence." *Science of the Total Environment* 628–629: 1637–1649.

Edlund, M. B., J. Ramstack, and D. Engstrom, 2007. Evaluating Historical Trends in Nutrient Concentrations from Swamp Lake Sediments, Grand Portage Reservation, northeastern Minnesota. Final report Project No. 8607-AZJ00, St. Croix Watershed Research Station, Marine on St. Croix, Minnesota, 19 pp.

Edlund, M. B., J. Ramstack, and D. Engstrom, 2009. Evaluating Historical Trends in Nutrient Concentrations from Trout Lake Sediments, Grand Portage Reservation, northeastern Minnesota. Final report GPIR Environment Dept. Work Order #2007-1, St. Croix Watershed Research Station, Marine on St. Croix, Minnesota, 17 pp.

Environmental Laboratory. 1987. *Wetlands Delineation Manual*. Vicksburg, Mississippi: Army Technical Report Y-87-1. Corps of Engineers Waterways Experiment Station.

Epstein, L. 2014. "Fifty years since silent spring." *Annual Review of Phytopathology* 52 (1): 377–402.

EU Water Directors. 2023. *Water Framework Directive Reporting Guidance 2022*. Copenhagen, Denmark: European Environmental Agency.

Feist, B. E., E. R. Buhle, D. H. Baldwin, J. A. Spromberg, S. E. Damm, J. W. Davis, and N. L. Scholz. 2017. "Roads to ruin: Conservation threats to a sentinel species across an urban gradient." *Ecological Applications* 27 (8): 2382–2396.

Florida DEP. 2026. *Wetland Delineation - Hydrologic Indicators*. Tallahassee, FL: Florida Department of Environmental Protection.

Gough, L., and J. Crock. 1997. *Distinguishing Between Natural Geologic and Anthropogenic Trace Element Sources*. Washington, DC: U.S. Geological Survey.

Hanna-Attisha, M., J. LaChance, R. C. Sadler, and A. C. Schnepp. 2016. "Elevated blood lead levels in children associated with the Flint drinking water crisis: A spatial analysis of risk and public health response." *American Journal of Public Health* 106 (2): 283–290.

Helgen, S., and J. Moore. 1995. "Natural background determination and impact quantification in trace metal-contaminated river sediments." *Environmental Science and Technology* 30: 129–135.

Hinsby, K., M. de Melo, and M. Dahl. 2008. "European case studies supporting the derivation of natural background levels and groundwater threshold values for the protection of dependent ecosystems and human health." *Science of the Total Environment* 401: 1–20.

Knudsen, S. H. 2014. "The long-term tort: In search of a new causation framework for natural resource damages." *Northwestern University Law Review* 108 (2): 475–542.

Krimsky, S. 2005. "The weight of scientific evidence in policy and law." *American Journal of Public Health* 95: S129–S135.

Luo, X.-S., Y. Xue, Y.-L. Wang, L. Cang, B. Xu, and J. Ding. 2015. "Source identification and apportionment of heavy metals in urban soil profiles." *Chemosphere* 127: 152–157.

Mast, M. A., P. L. Verplanck, W. G. Wright, and D. J. Bove. 2007. "Characterization of background water quality." In *Integrated Investigations of Environmental Effects of Historical Mining in the Animas River Watershed, San Juan County, Colorado*, edited by P. von Guerard S. E. Church, and S. E. Finger. 351–386. Washington, DC: U.S. Geological Survey.

Molinari, A., F. Chidichimo, S. Straface, and A. Guadagmini. 2014. "Assessment of natural background levels in potentially contaminated coastal aquifers." *Science of the Total Environment* 476: 38–48.

Nowell, L. H., P. W. Moran, I. R. Waite, T. S. Schmidt, P. M. Bradley, B. J. Mahler, and P. C. Van Metre. 2024. "Multiple lines of evidence point to pesticides as stressors affecting invertebrate communities in small streams in five United States Regions." *Science of the Total Environment* 915: 169634.

Olson, J. R., and S. M. Cormier, 2019. "Modeling spatial and temporal variation in natural background specific conductivity." *Environmental Science and Technology* 53 (8): 4316–4325.

Page, D. S., P. D. Boehm, G. S. Douglas, A. E. Bence, W. A. Burns, and P. J. Mankiewicz. 1996. "The natural petroleum hydrocarbon background in subtidal sediments of Prince William Sound, Alaska, USA." *Environmental Toxicology and Chemistry* 15: 1266–128.

Reimann, C., P. Filzmoser, and R. G. Garrett. 2005. "Background and threshold: Critical comparison of methods and determination." *Science of the Total Environment* 346: 1–16.

Rhomberg, L. R. 2013. *Hypothesis-Based Weight of Evidence: An Approach to Assessing Causation and Its Application to Regulatory Toxicology*. Cambridge, MA: Harvard Center for Risk Analysis.

Ruckart, P. Z., A. S. Ettinger, M. Hanna-Attisha, N. Jones, S. I. Davis, and P. N. Breysse. 2019. "The flint water crisis: A coordinated public health emergency response and recovery initiative." *Journal of Public Health Management and Practice* 25: S84–S90.

Runnells, D., T. Shepherd, and E. Angino. 1992. "Metals in water, determining natural background concentrations in mineralized areas." *Environmental Science and Technology* 26 (12): 2316–2323.

Smith, R., R. Alexander, and G. Schwarz. 2003. "Natural background concentrations of nutrients in streams and rivers of the conterminous United States." *Environmental Science and Technology* 37: 3039–3047.

Snow, J. 1849. *On the Mode of Communication of Cholera*. London: John Churchill, Princes Street, Soho.

Stoddard, J. L., D. P. Larsen, C. P. Hawkins, R. K. Johnson, and R. H. Norris. 2006. "Setting expectations for the ecological condition of streams: The concept of reference condition." *Ecological Applications* 16: 1267–1276.

Sucharova, J., I. Suchara, M. Hola, and C. Reimann. 2014. "Contemporary lead concentration and stable isotope ratio distribution in forest moss across the Czech Republic." *Applied Geochemistry* 40: 51–60.

Suter, G. W. 1993. "A critique of ecosystem health concepts and indices." *Environmental Toxicology and Chemistry* 12: 1533–1539.

Suter, G. W. 2023. *Fundamentals of Environmental Assessment*. Boca Raton, FL: CRC Press.

USEPA. 2000. *Nutrient Criteria Technical Guidance Manual: Rivers and Streams*. Washington, DC: U.S. Environmental Protection Agency.

USEPA. 2016. *Weight of Evidence in Ecological Assessment*. Washington, DC: Risk Assessment Forum, U.S. Environmental Protection Agency.

USEPA. 2024. *Acute Aquatic Life Screening Value for 6PPD-quinone in Freshwater*. Washington, DC: Office of Water, U.S. Environmental Protection Agency.

USEPA. 2011. A Field-Based Aquatic Life Benchmark for Conductivity in Central Appalachian Streams (Final Report). Washington, DC: U.S. Environmental Protection Agency, EPA/600/R-10/023F. https://cfpub.epa.gov/ncea/risk/recordisplay.cfm?deid=233809

Van de Meent, D., T. Aldenberg, J. Canton, C. van Gestel, and W. Sloof. 1990. *Desire for Levels, Background Study for the Policy Document Setting Environmental Quality Standards for Water and Soil*. Bilthoven: RIVM.

Watkins, M. 2023. Technical Support Document for Numeric Site-Specific Criteria for Nutrients, Aluminum, and Specific Conductance within the Reservation of the Grand Portage Band of Lake Superior Chippewa. Grand Portage, MN: Departments of Biology and Environment. Grand Portage Band of Lake Superior Chippewa.

Wharton, C., J. Olson, and S. M. Cormier. 2021. "Data set: Background conductivity data view." In: USEPA Freshwater Explorer 2.0. Washington, DC: U.S. Environmental Protection Agency.

Zhao, F., S. McGrath, and G. Merrington. 2006. "Estimates of ambient background concentrations of trace metals in soils for risk assessment." *Environmental Pollution* 148: 221–229.

15 Benchmark Values

It is the national policy that the discharge of toxic pollutants in toxic amounts be prohibited

U.S. Clean Water Act of 1972

Benchmarks are numerical values (concentrations, doses, etc.) or characteristics (no oil sheen) that determine whether an exposure level is acceptable. Benchmark is a generic term. In specific cases, benchmarks may be called standards, criteria, guidelines, guidance values, or limits. We use benchmarks except when referring to a specific context.

Benchmarks have five uses:

- to set limits on concentrations of contaminants in effluents or ambient media,
- to provide remedial goals for contaminated sites,
- to determine whether a monitored ecosystem is impaired,
- to determine limitations on the use of chemicals in the environment, and
- to serve as screening values to determine which contaminants have negligible risk and which are contaminants of concern that require further assessment.

Weight of evidence has been recommended for deriving benchmarks by multiple agencies and authors (ANZG 2018a; Buchwalter, Clements, and Luoma 2017; Suter, Cormier, and Barron 2017; USEPA 2016; Warne et al. 2018; Leung et al. 2014; Cormier et al. 2008; Merrington et al. 2014; Chapman 2018; Smith and Tran 2010).

15.1 TYPES OF BENCHMARKS

The USEPA (2023b) has three accepted categories of water quality criteria: published numerical criteria, modified published criteria, and narrative criteria. In addition, they accept "other scientifically defensible methods." These approaches for selecting or deriving benchmark values may be appropriate in contexts other than the Clean Water Act. The USEPA's alternate approaches suggest great flexibility that allows for various weight-of-evidence methods.

15.1.1 Published Numeric Criteria

Early U.S. aquatic life criteria were based on informal narrative weight of evidence, with unspecified literature search methods and unstructured expert judgment (USEPA 1976). There were no clear criteria for rejection of studies, and both experimental and field observational studies were included.

DOI: 10.1201/9781003626336-18

In 1980, the USEPA moved away from informal weight of evidence for aquatic life criteria to a formal method based on species sensitivity distributions (USEPA 1986). These are statistical models of the distribution of sensitivity among species to a chemical tested for acute and chronic toxicity (Posthuma, Suter, and Traas 2001). The benchmark values are centiles (e.g., 0.05) of the distribution. This method has been widely adopted for ecological benchmarks. A particularly well-developed example is the Australia and New Zealand water quality guideline values (Warne et al. 2018).

Benchmark values for human health have also been based on informal narrative weight of evidence. However, in recent years, structured methods for assembling and weighing evidence for human health have been adopted (USEPA 2022; EFSA Scientific Committee 2017).

15.1.2 Modified Published Criteria to Reflect Site-Specific Conditions

Published water quality criteria may be modified if (1) the physical/chemical properties of water at a site modify the availability or toxicity of the chemical, or (2) species are added or deleted to avoid non-native species or other sources of bias or insensitivity. The modification for properties of the receiving water may involve corrections for alkalinity, hardness, suspended or dissolved solids, or pH, as described in criteria documents for the chemical or agent. Alternatively, site-specific water-effect ratios based on test results for laboratory and receiving water can be applied to published criteria (USEPA 1994; Welsh, Lipton, and Chapman 2000). Species deletions or additions should be based on knowledge of occurrence and sensitivity. That is, if a tested species does not occur at a site and no similar species occur there, it may be deleted. On the other hand, if a species occurs at a site and neither it nor a sufficiently similar species has been tested, it may be added. These adjustments could be based on multiple pieces or types of evidence that could be weighed. The evidence could include mechanisms of bioavailability or toxicity, empirical relationships, site-specific characteristics and associations, and toxicity tests with site-relevant species and water chemistries.

Site-specific criteria have an arithmetic bias problem. In a species sensitivity distribution for a site-specific criterion, species may be removed or added to be representative of the site. By adding more tolerant species or removing sensitive ones, the lower 5th centile shifts to greater concentrations. The net effect is more permissive, site-specific criteria. Furthermore, a species such as brook trout may be omitted with the claim that it does not occur, but the reason may be that the stream is already impaired and trout are extirpated. I (Cormier) remember sitting in the audience at a SETAC meeting as this process was flaunted. The moral of the story is that not everyone is moral, and, no matter how scientifically sound a method, there are ways to subvert the intended protection. The best defense is transparency and scientific review with diverse representation (Section 7.1.2).

15.1.3 Other Scientifically Defensible Methods

The USEPA (2023b) approves the use of methods that are scientifically defensible to derive criteria, even if they are not standard methods. Examples include methods for applying standard toxicity data to derive water quality-based effluent limits, methods for applying whole-effluent toxicity tests to effluents or ambient waters, and methods for developing and applying biological criteria. Other methods, such as microcosms and mesocosms, are also scientifically defensible.

Nonstandard but scientifically defensible methods are potentially useful for atypical systems such as thermal springs, saline lakes, soils formed from ore-bearing rocks, intermittent streams or pools, and high-conductivity waters. They can also protect species, life stages, life histories, behaviors, and interactions that are not readily tested.

The *Field-based Aquatic Life Benchmark for Conductivity in Central Appalachian Streams* is an example of a defensible novel method (USEPA 2011). Its species sensitivity distributions are derived from biological survey data. It is a designated national criterion method, and it was judged to be appropriate for enforcement of a West Virginia narrative criterion (Chambers 2014).

The weight-of-evidence methods for deriving criteria for aqueous nutrients are also examples of defensibility (Section 15.7) (Ridley et al. 2024). Weight of evidence is particularly appropriate because the adverse effects are due to the complex effects of excess fertility rather than toxicity.

15.1.4 Narrative Criteria

Narrative criteria can serve multiple functions that are not met by numerical criteria.

- They can express qualities that are observable but not readily quantified. Examples of water qualities expressed as narratives include oil sheen; debris, scum, and other matter creating a nuisance; or nuisance or objectionable aquatic life (e.g., algal blooms).
- Even if a chemical is potentially amenable to conventional laboratory toxicity tests and numeric criteria derivation, narrative criteria free you from the constraints of the availability of relatively few published numeric criteria.
- Narrative criteria are particularly useful for wetlands. Those ecosystems may be more susceptible to changes in ecosystem structural and functional properties than to chemical pollutants. These might include water seasonality and depth, biodiversity, or structure provided by vascular plants.

Narrative criteria may provide considerable latitude for method development, including weight-of-evidence methods. Some narratives have no need for numerical benchmarks (e.g., oil sheen), but enforcement of narrative criteria always requires some specificity (e.g., what makes some aquatic life a nuisance?). Various scientifically defensible methods can be used to enforce a narrative, but they must be based on a sound scientific rationale and must protect the designated use of the water body.

15.1.5 Screening Benchmarks

Contaminant screening is a process of categorizing chemicals and other agents to determine their treatment in assessment and management. Screening may be used to identify a subset of chemicals that require further assessment. A common example is the screening stage of Superfund remedial investigations (i.e., assessment of an uncontrolled contaminated site—Section 17.2.2), such as the soil screening levels for human health and ecological assessments (USEPA 1996, 2005). Screening benchmarks are concentrations or doses that bound a safe level of contamination. Because chemicals at levels below screening benchmarks are safe, they are not considered further. Chemicals occurring at levels above lower screening benchmarks are considered contaminants of concern. In many cases, this is the only benchmark that is considered. Because screening benchmarks are meant to assure safety, they should be precautionary. For example, derivation of ecological soil screening values uses only tests of the highest-bioavailability soils (USEPA 2005).

Other bounding benchmarks may be used to screen chemicals as required by the case. Chemicals that exceed levels that clearly require remedial actions require no further consideration because they will be remediated. The remedial-level benchmarks are usually published criteria or standards but may be case-specific. If levels are sufficiently high, they may prompt emergency actions. Levels between safe and remedial bounds may lead to further assessment. Those assessments may include better characterization of the contaminant levels or better characterization of the hazard. Derivation of any of these benchmarks may use weight of evidence.

Screening benchmarks may also be used in the context of water quality criteria. That is, if there is insufficient information to derive a conventional water quality criterion, a simpler benchmark may be used to interpret observed levels of a substance. A recent example from Canada provides an alternative method using assessment factors when the standard Species Sensitivity Distribution method is not possible (Tillmanns and Pearson 2024).

15.1.6 Benchmarks for Hazards

Chemicals and other agents may be the subjects of hazard assessments that determine whether they have a particular hazardous property or determine which hazardous properties they possess (Chapter 16). Such assessments may include the derivation of hazard-specific benchmarks. For example, "the chemical of concern is a teratogen at doses greater than *x*."

15.2 WEIGHT OF EVIDENCE IN BENCHMARK DERIVATION

Weight of evidence is useful for benchmark derivation for the same reason that it is generally useful—it uses all the evidence to achieve a defensible result. It can play six roles.

1. Because of its flexibility, it can be useful for cases in which standard derivation methods are not applicable. For example, the standard method for deriving water quality criteria for aquatic life in the U.S. requires standard laboratory toxicity test data from at least eight families, a requirement that most chemicals do not meet (USEPA 1985). In addition, harmful agents that do not have toxic effects, or whose effects are not displayed in standard laboratory tests (e.g., indirect effects of nutrients), are not applicable.

Neither are tests of large or endangered species, of some life stages (e.g., glochidia of freshwater mussels), of long-term effects (e.g., selenium in freshwaters), or of nonstandard effects (e.g., fish behavior) (Cormier et al. 2008; Bertram et al. 2025). In such cases, nonstandard methods, including weight-of-evidence methods, are needed.
2. It can be used when implementing a standard method. For example, when deriving a species sensitivity distribution, weight of evidence can be used to determine whether the distribution is unimodal or multimodal (see Australia and New Zealand Water Quality Guideline Values for Toxicants, below).
3. Whenever field data are used to determine a benchmark, weight of evidence must be used to determine whether the exposure-response relationship is causal and whether it is confounded by covariates (Chapter 12).
4. It can be used to choose a benchmark value when there are multiple candidate values (see Section 12.1).
5. It can be used to increase the credibility of values derived by standard methods by weighing them along with other evidence (Merrington et al. 2014).
6. It can be used to define the appropriate level of confidence in the benchmark value. That is, the weight of the benchmark can be an expression of confidence.

Weight-of-evidence approaches have been most often used for naturally occurring agents such as nutrients, suspended and bedded sediment, and conductivity. This is because field data are often required to account for complex and indirect causal processes, adaptation, and covariates. However, mesocosm and laboratory data are still useful as mechanistic evidence (Clements et al. 2002).

15.2.1 Weight of Evidence for Evaluating Endpoint Qualities

Default benchmark values are nearly always based on a few generic endpoint effects. For ecological effects, it is desirable to consider whether benchmarks will protect endpoint qualities that are appropriate to the contaminant and biotic community. That is, benchmarks should be (1) protective of valued population and community entities and attributes of those entities that are (2) at risk from ongoing or potential contamination or disturbance. Weight of evidence is used to determine what effects occur in laboratory and field studies and whether they are expressive of biologically or culturally significant endpoint effects that are relevant to the site or region of interest. Information such as biomarkers that is not suitable as an endpoint may be used as supporting evidence. The process of endpoint identification for site-specific endpoint derivations should include field studies to determine condition and causation at the site of concern (Chapter 12).

15.2.2 Approaches for Selection of the Benchmark Value

The selection of a quantity by weight of evidence from among candidate values can be based on two general approaches: choosing the best candidate value, as determined by weight, or statistically combining the candidates. The use of these two methods has been adopted from the USEPA quantitative weight-of-evidence approach (Chapter 12) and is recommended for deriving water quality guideline values in Australia and New Zealand (Water Quality Australia 2020).

- **Best Value**: The best value may be chosen by applying weight of evidence to the candidate values. That is, the best value is the weightiest, based on the relevance and reliability of the study that derived it. Best value is the most flexible approach. The values need not have the same methods, endpoints, or units.
- **Statistics**: The most common statistic for combining numerical benchmark values is simply the arithmetic or geometric mean. Meta-analysis may be used for studies that may be considered replicates, such as exposure-response relationships from multiple surveys of a region. If there are significant differences in methods, endpoints, relevance, or reliability among studies, it is better to choose the best (weightiest) value rather than combine the values.
- **Precedent**: New criteria should be reasonably consistent with preceding benchmarks. For example, when a method for field-based aquatic benchmarks was developed, it was based on protecting 95% of genera, which is consistent with the standard laboratory-based USEPA method for water quality criteria (Cormier, Suter, and Zheng 2013). Similarly, USEPA benchmarks for nutrients and biocriteria that are based on achieving regional background values have used the 75th centile of reference sites as the benchmark because it is a standard limit of background (USEPA 2000, 2002). Part of the reasoning is that there will be fewer points of contention when there are fewer differences between new and already accepted methods.
- **Logic**: Logical arguments can be applied to narrative criteria and evidence. For example, if the goal is to protect exceptional waters, then the numerical properties of designated exceptional waters can be used as benchmarks. For example, if all exceptional waters in a region have at least 12 fish species, that could be a biological benchmark value for new exceptional waters.
- **Precaution**: For example, when deriving benchmarks to protect endangered or highly valued species, use the most protective candidate value or divide a standard value by a safety factor.

15.3 CHARACTERISTICS AND OTHER IMPLICATIONS OF EVIDENCE FOR BENCHMARKS

Benchmark values should have certain characteristics, and evidence that supports a value should imply one or more of those characteristics (Table 15.1). The characteristics for a particular type of benchmark may be defined more specifically. For example, exposure metrics for an aqueous or atmospheric release must specify temporal dimensions (duration and frequency), as well as concentration, but exposure to persistent chemicals in soil, sediment, or food items seldom requires those temporal dimensions. Evidence for each of these characteristics can be weighted for the appropriate properties (relevance, reliability, strength).

TABLE 15.1
Potential Characteristics of a Protective Benchmark

Characteristics of Protective Benchmarks	Description
Causal relationship	The exposure-response relationship used in the derivation should be causal. Evidence from toxicity tests inherently implies causation.
Predictive model of the relationship	The model of the relationship predicts the effects of specified exposures. For example, basing a benchmark on a median acute lethality test endpoint (LC50) predicts that mass mortality will occur above the benchmark exposure.
Represents sensitive endpoints	The benchmark represents a threshold for the occurrence of sensitive responses of sensitive species or other entities. Evidence from field studies may represent more sensitive responses than those from laboratory tests.
Exposure metric is relevant	The exposure metric represents the relevant exposure of the sensitive entities. For example, dissolved concentrations of metals are more relevant for aqueous benchmarks than total concentrations.
Set at an appropriately protective level	Independent data demonstrate, with sufficient confidence, the absence of effects of concern on the aquatic community and constituent taxa at the proposed benchmark.
Discriminatory power	The benchmark discriminates between impaired and unimpaired communities with sufficient confidence. Biotic communities with no exceedances of benchmarks should not be impaired, but where benchmarks are significantly exceeded, impairments should be detectable.

Adapted from USEPA (2016).

15.4 AUSTRALIA AND NEW ZEALAND NUMERIC WATER QUALITY GUIDELINE VALUES

Australia and New Zealand (A&NZ), in our opinion, have the best approach to benchmark derivation. They derive water quality benchmark values (known as guidelines) using multiple types of data and weight of evidence (Warne et al. 2018, 2025). Other guidance addresses benchmarks for site remediation (Box 17.2) (Humphrey and Chandler 2018). Extensive online guidance is provided (ANZG 2018a). They include values based on three types of information:

- reference site data are used to determine whether a candidate guideline value is within the range of natural conditions,
- laboratory effects data and field effects data define an exposure level below which ecologically meaningful effects do not occur, and
- multiple types of evidence—a weight-of-evidence approach.

These methods form a hierarchy—the least reliable is reference, followed by laboratory effects, field effects, and then weighing multiple types of evidence. More reliable methods are used when risks are high or the resource is particularly valued.

TABLE 15.2
Scoring System for Assessing the Quality of Toxicity Test Data for Metals in Freshwater Non-plants to be Used in the Derivation of Guideline Values for Toxicants (Warne et al. 2025)

Question	Score
1. Was the duration of the exposure stated (for example, 48 or 96 h)?	Yes (10), No (0)
2. Was the biological endpoint (for example, immobilization or population growth) stated and defined?	Yes (10), Stated only (5), Neither (0)
3. Was the biological effect stated (for example, LC or NOEC)?	Yes (5), No (0)
4. Was the biological effect quantified (for example, 50% effect, 25% effect)? Note: the effect for NOEC and LOEC data must be quantified.	Yes (5), No (0)
5. Were appropriate controls (for example, a no-toxicant control and/or solvent control) used?	Yes (5), No (0)
6. Was each control and chemical concentration at least duplicated?	Yes (5), No (0)

The first six of 18 questions are presented as examples.

15.4.1 A&NZ Weight of Evidence for Screening Information

After toxicity data have been screened to eliminate clearly unacceptable data (Section 6.2.1), this system scores each piece of evidential data for quality (i.e., reliability) to determine whether it is usable for deriving a guideline value (Warne et al. 2018, 2025). They provide tables of scoring questions for metals and nonmetals, animals and plants, and freshwater and marine tests. A few examples are presented in Table 15.2. Based on the total score, each test result is weighed and designated as high, acceptable, or unacceptable quality.

Field-based, microcosm, and mesocosm data may also be used to derive default guideline values. Because these data are less standardized, their quality is not numerically rated. Rather, nine criteria for acceptable quality are provided, and the pieces of information are rated as meeting each criterion or not.

15.4.2 A&NZ Weight of Evidence for Species Sensitivity Distributions

Although species sensitivity distributions are not a weight-of-evidence method (they are a piece of evidence), their derivation may involve weight of evidence to resolve issues in model implementation. These issues include polymodality, representativeness of the data, and goodness of fit.

One inherent problem when deriving species sensitivity distributions is determining which taxa should be aggregated in a single distribution. This question has a standard answer in some cases. For example, the USEPA combines vertebrate and invertebrate animals but separates plants when deriving water quality criteria (Stephan et al. 1985). However, Warne et al. (2018) weigh evidence to determine which taxa should be aggregated. The statistical models used for species sensitivity distributions are usually assumed to be unimodal, but if taxa with different sensitivity distributions are combined, the distribution is polymodal. Polymodality can cause poor model fits, poor estimates of the centiles of species sensitivity, and poor protection of particularly sensitive taxa.

The polymodality issue is addressed by collecting and weighing evidence to answer four questions (Warne et al. 2018).

- *Is there a specific mode of action that could result in taxon-specific sensitivity?* For example, is the chemical an herbicide, in which case plants should be distributed separately?
- *Does the dataset suggest multimodality?* Does the species distribution show distinct humps, or do statistical tests support bimodality?
- *Do the data show taxon-specific sensitivity (that is, through distinct grouping of taxa)?* This can be answered by inspection, but apparent clusters can be confirmed by box plots or other statistical characterizations of the apparent clusters.
- *Is it likely that indications of bimodality, multimodality, or distinct clustering of taxon groups are not due to artifacts of data selection, small sample size, test procedures, or other reasons unrelated to a specific mode of action?* This last question is answered by weighting the evidence for reliability. Other issues that might be considered include differential effects of water, soil, or sediment chemistry and whether other chemicals with the same mode of action are polymodal.

If the datasets for a chemical are large enough to allow good estimates of taxon-specific distributions, it is good practice to derive them, thereby providing more information (Oginah et al. 2023). However, doing so can result in too few data to create separate, useful sensitivity distributions.

The species toxicity values in SSDs are not weighted in any of the current agency guidance, but weighting has been considered by Australia and New Zealand. Test species have not been randomly selected to represent biotic communities, nor are they selected by any importance criteria. Some taxa, such as fish, are over-represented and others, such as mollusks, are under-represented relative to the number of species in nature. Weighting to compensate has been suggested (Forbes and Calow 2002). Weighting might also be based on economic, cultural, or ecological importance. However, Australia and New Zealand guidance currently does not recommend weighting values in species sensitivity distributions.

A weight-of-evidence process may also be used to evaluate goodness of fit. Conventional statistical tests can quantify goodness of fit, but they are not sufficient. Visual inspection may identify outliers, and their implications may be determined by weight of evidence (Batley et al. 2018). An outlier may be the sole representative of a taxon or effect; it may be from a nonstandard test protocol or conditions; or it may be from a poorly conducted test. It is particularly important to consider the values at the low end of the distribution, which are influential in benchmark derivation.

15.4.3 A&NZ Weight of Evidence for Deriving Numeric Guidelines

Australia and New Zealand follow the USEPA's guidelines for weight of evidence by recommending two options for deriving a benchmark or other quantity (Chapter 7). First, the candidate values may be statistically combined, with or without weighting. If more than one of the categories of evidence provides multiple candidate values (e.g., multiple field toxicity tests), they may be combined if they are concordant. Then, the aggregate value for each category may be statistically combined with the others. These aggregations should be performed only if all the endpoints are concordant. Second, if the values are not statistically combined, the value derived from the best evidence may be chosen. The best candidate value is the one with the greatest weight of evidence. These options are discussed in Chapter 8, as well as in ANZG (2018a), Batley et al. (2018), and Warne et al. (2025). More recent guidance recommends a weight-of-evidence process that integrates default and site-specific guideline values with other types of evidence as preferable to either default or site-specific guideline values alone (ANZG 2023).

15.4.4 A&NZ Derivation of Site-Specific Benchmark Values

Australia and New Zealand place particular emphasis on tailoring their guidelines to local conditions, including biological, physical, and chemical conditions, and community values (ANZG 2018a). When deriving site-specific guideline values, they allow modification of default values (Section 15.1.2), but prefer to derive site-specific values using a weight-of-evidence approach (Batley et al. 2018). Deriving guidelines de novo allows them to consider more factors than are included in the modification of published default values. The following is a list of reasons for site-specific guidelines (adapted from Van Dam et al. 2019).

- The site has high ecological or cultural value.
- The local species are rare or economically or culturally valuable.
- The local species composition is distinct or unusual.
- The water varies in chemistry, temperature, and flow, which affects exposure and sensitivity at different life stages of the biota. The nature of the exposure is different from that in standard laboratory tests (e.g., dietary uptake versus direct uptake from water).
- The timing of exposure events, such as pulsed exposures, is different from standard laboratory tests.
- Natural background concentrations exceed the generic guidelines.
- The activities at the site, or those proposed for the site, are controversial, or the developers are particularly responsible.

The many types of studies that could support a site-specific benchmark require a weight-of-evidence approach. The Australians and New Zealanders employ their own general weight-of-evidence approach, adapted from the USEPA approach (ANZG 2018b; USEPA 2016). However, they also cite other appropriate weight-of-evidence approaches for benchmark derivation, including Cormier et al. (2008), Van Dam et al.

(2014, 2019), and Moore et al. (2017). They reviewed approaches for deriving local water quality benchmarks and provided general guidance (Van Dam et al. 2019).

15.5 USEPA SUSPENDED AND BEDDED SEDIMENTS

The USEPA has developed a weight-of-evidence framework for deriving water quality criteria and applied it to excess or insufficient suspended and bedded sediments (Cormier et al. 2008; USEPA 2006; Paul et al. 2007). It relies heavily on observational field data, but it can include laboratory and mesocosm tests and mechanistic models. The framework includes multiple approaches and methods. One weight-of-evidence approach is to develop candidate criteria from different methods or for different regions or endpoints and choose the best (i.e., most relevant and reliable). Another proposed method is to statistically aggregate candidate criteria by taking a geometric mean, weighted mean, or a proportion of their distribution.

The case study of a benchmark for percent fines used percentiles, conditional probability, quantile regression, linear regression, logistic regression, and species sensitivity distributions (Cormier et al. 2008). The hypothetical criterion is 7% fines, based on the arithmetic mean of results from all methods. Choosing the best evidence by weight of evidence was not useful because no method was judged by the authors or reviewers to be more relevant or reliable. The framework also recommends deriving candidate criteria for multiple endpoints, such as benthic invertebrate species richness or brook trout abundance. The most sensitive endpoint could be chosen to increase the probability of sufficient protection.

This is not a certified criterion method, but it should qualify as a scientifically defensible method. Its approach could be applied to other hazardous agents.

15.6 NEW ZEALAND FINE SEDIMENTS

Independently of the Australia and New Zealand Governments (ANZG 2018a), the National Institute of Water & Atmospheric Research (NIWA) developed a weight-of-evidence-based framework for deriving thresholds for fine sediments in New Zealand waters (Franklin et al. 2019). It is based on the USEPA guidelines (USEPA 2016; Suter, Cormier, and Barron 2017). NIWA identified attributes to be protected from excess sediment and five candidate methods for deriving the relationships between exposure to sediment and the ecological attributes. Each of the methods (boosted regression tree, quantile regression, generalized linear modeling, extirpation analysis, and community deviation) provides evidence of the relationship.

NIWA used the standard properties of evidence, relevance and reliability, and the same sub-properties as the USEPA. They did not use strength as a property because they were deriving a quantitative benchmark. Rather, they created a new property, suitability, which is comprised of sub-properties that deal with the suitability of evidence for derivation of a sediment benchmark (Table 15.3).

The weighting of pieces of evidence was performed using the +, −, and 0 rating symbols, with tables that provide consistent, detailed justification of the ratings. Because of the important differences among the evidence methods and their results,

TABLE 15.3
Description of Suitability Sub-properties for Sediment Benchmarks (Franklin et al. 2019)

Suitability Measure	Description
Bottom-line	Is the method suitable for identifying a bottom-line threshold?
Bands	Is the method suitable for identifying attribute band thresholds?
Global vs. Classes	Can thresholds be derived for different sediment state classes?
Reproducible	Is the method of deriving thresholds easily reproducible? For example, does it rely on expert interpretation or quantitative deviations?
Departure from reference	Can thresholds be derived as a departure from reference state using the method?

the ratings were not statistically combined. Rather, the best method (community deviation) was chosen based on weight of evidence.

15.7 USEPA AND STATE AQUEOUS NUTRIENT CRITERIA

Aqueous nutrients are a difficult problem for benchmark derivation because nutrients are essential, and the boundary between a productive system and a eutrophic system is seldom clear. Also, their mode of action is not the usual direct toxicity. Benchmark derivation is particularly difficult for rivers and streams because eutrophication of running waters is poorly characterized compared with lakes. Guidance for solving this problem by weight of evidence has been provided by the USEPA (2000), the State of New York (Smith and Tran 2010), and the USEPA again (Ridley et al. 2024).

In 2000, the USEPA recommended applying an unspecified weight-of-evidence approach to any or all of three approaches for setting nutrient criteria:

- Reference nutrient levels from identified reference reaches,
- Predictive relationships of nutrients with trophic state classes or biocriteria, or predictive models of nutrient effects, and
- Nutrient concentration thresholds or algal limits from the published literature.

New York followed the USEPA in using weight of evidence, but they specified their own weight-of-evidence method (Smith and Tran 2010). They used three different methods for benchmark derivation: percentile analysis, changepoint analysis, and cluster analysis. They then weighted the values from each method based on its connection with biological responses to nutrients. Their scale was: 1 (no connection to biological assemblages), 1.5 (indirect evaluation of response to nutrients), or 2 (direct or threshold response). This is effectively a relevance scale. Finally, the weighted values were averaged for each nutrient metric: total phosphorus, total nitrogen, nitrate nitrogen, and chlorophyll a.

In 2024, the USEPA developed new guidance for nutrient criteria based on the 2016 weight-of-evidence guidelines (Ridley et al. 2024). Its methods and terminology

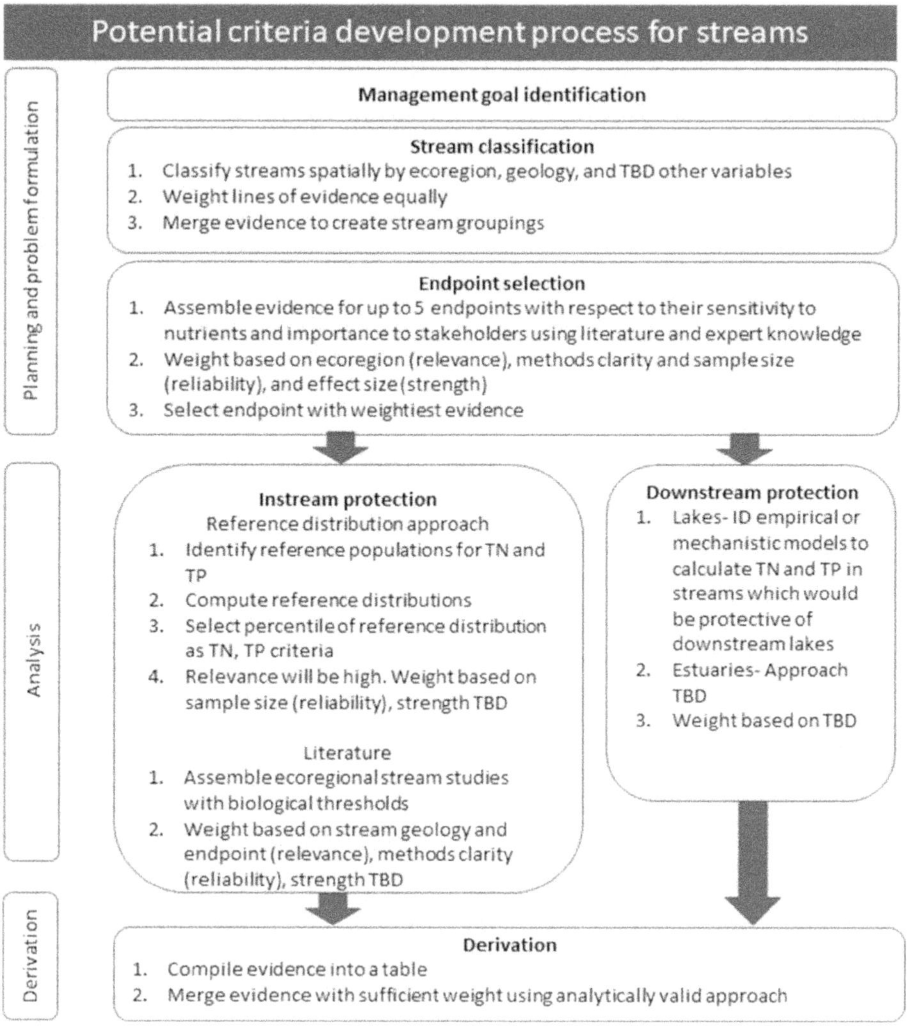

FIGURE 15.1 Process for deriving site-specific aquatic criteria for total phosphorus and total nitrogen by weight of evidence (Ridley et al., 2024).

are intended to provide guidance to the states, including guidance on how to adapt the weight-of-evidence method to a state's particular types of evidence. It includes a full framework for the process and guidance for implementing it (Figure 15.1).

Table 15.4 shows an example of their use of an evidence-rating table to determine the best (weightiest) benchmark value for total phosphorus. The overall weights are derived from the integration of relevance, reliability, and strength of each piece of evidence and for the selected evidence.

The USEPA authors also allowed for taking a mean or otherwise "merge multiple pieces or types of sufficiently weighty evidence if it is technically appropriate and justified to protect the designated use" (Ridley et al. 2024).

TABLE 15.4
Example of Evidence Integration by Selecting Weightiest Evidence (Ridley et al. 2024)

Piece of Evidence	Overall Weight	Candidate Criterion[a]	Explanation
1. TP-Diatom Index S-R curve: Analysis generated with State field data	++	TP = x mg/L with narrow CL	Primary data analysis that resulted in the weightiest evidence.
2. TP-Diatom Index S-R curve: Analysis generated with field data outside of State	0	TP = y mg/L with wide CL	Similar primary data analysis as in (1), but underlying data represent only a single season, so evidence has unacceptably low reliability. Working with neighboring state to include multiple seasons in future analyses.
3. Meta-analysis of TP-Diatom richness relationship: Literature	++	TP = α mg/L	Threshold identified, but endpoint is not sensitive to nutrient change at low TP concentrations, resulting in substantial uncertainty around candidate criterion.
4. Mesocosm phosphorus dosing experiment: Literature	+	TP = β-γ mg/L	TP candidate criterion range identified from low and medium dosage, but not statistically different from high dosage concentration. $\beta < x < \gamma$.
Conclusion	++	TP = x mg/L	(1) selected to inform final criterion value. (4) is not strong, but corroborates the value derived from (1).

[a] TP = total phosphorus, CL = confidence limit

15.8 USEPA SOIL SCREENING LEVELS

Ecological soil screening levels for U.S. contaminated sites are derived using a weight-of-evidence approach (USEPA 2005, 2023a). Individual studies from a systematic literature review are screened using eleven literature exclusion criteria. Evidence from the remaining studies is scored on a 0, 1, 2 scale for each of nine criteria, and the scores are added. A study is included if its total score is at least 10. Studies are then rated for bioavailability, and the benchmark value is derived as the geometric mean of test results for at least three high-bioavailability soils.

15.9 NON-AGENCY PESTICIDE BENCHMARKS

The pesticide industry has a history of funding ecological risk assessments that derive benchmark values for their products as alternatives to benchmarks derived by regulatory agencies. Examples include Giesy et al. (1999), Hall Jr et al. (1999), Hanson and Solomon (2004), and Solomon et al. (1996). They have used multiple pieces and types of evidence, but their approach for integrating evidence has been unstructured narratives (Section 2.1).

In contrast, a recent contractor-derived benchmark value for effects of aqueous atrazine on aquatic plants was derived by a structured weight-of-evidence assessment (Moore et al. 2017). It used four methods to derive candidate benchmarks:

- The Plant Assemblage Toxicity Index is a method for combining the species sensitivity distribution approach with temporal exposure dynamics (Erickson 2012).
- The (Brock, Lahr, and van den Brink 2000) 5-Point Scoring System scored mesocosm studies from 1–5 based on the severity and duration of effects. Scores for each acceptable endpoint were plotted on a graph of time-weighted average exposure concentration versus duration, and a breakpoint was identified by visual inspection.
- The Modified Water Quality Criteria Method uses the USEPA's species sensitivity distribution method but uses distributions of specific growth rate EC_{50} values.
- The Comprehensive Aquatic Systems Model for Atrazine ($CASM_{ATZ2}$) is an aquatic ecosystem simulation model that was modified to incorporate atrazine exposure and effects.

Each of the methods was numerically weighted with respect to seven properties, and the weights were presented and justified in evidence-weighting tables. The $CASM_{ATZ2}$ method was rated most highly, but its benchmark was not chosen. Because of potentially significant limitations in all methods, the authors chose the weighted average of the four values as their recommended benchmark.

15.10 A VENETIAN WEIGHT-OF-EVIDENCE METHOD FOR SSDS

The use of SSDs to derive benchmark values is often limited by the number of available values and their compliance with criteria for inclusion. A group at the University Ca' Foscari in Venice used a weight-of-evidence-based multi-criteria decision analysis tool (PROERA) to increase the number of possible SSDs by accepting heterogeneous data for a chemical (ECETOC 2014; Isigonis et al. 2014). They used a weighted bootstrap approach to derive the distributions. The weights were derived by scoring the toxicological data with respect to 57 subproperties. They fall within three evidential properties: experimental reliability, statistical reliability, and biological relevance.

15.11 BENCHMARKS FOR MIXTURES

Mixtures, such as petroleum, that have reasonably consistent compositions may have assigned benchmark values. For example, the State of Nevada has set a clean-closure benchmark of 100 mg/kg total petroleum hydrocarbons in soil (NDEP 2022). That concentration is said to pose an acceptable level of risk for all exposure scenarios, but other information may also be used to define a clean-enough level. More heterogeneous mixtures may also have benchmark values. For example, the European Union has proposed a standard for the sum of pesticides in surface waters of 0.2 µg/L, covering substances already listed as priority substances (Council of the European Union 2025). Like any other substance, benchmarks for mixtures can be based on the weighing of information from the field and laboratory.

15.12 NATIONAL AMBIENT AIR QUALITY STANDARDS

The U.S. National Ambient Air Quality Standards are peculiar in that they are derived for a small set of criteria chemicals, and they are largely based on epidemiological and eco-epidemiological evidence (USEPA 2015). Because field studies are subject to confounding, the integrated science assessments are primarily concerned with using weight of evidence to determine whether the observed exposure-response relationships are causal. They assess causality using a variation of Hill's considerations (Section 2.2). Once a relationship is found to be causal, another weight-of-evidence process is used to determine the concentration and duration of exposures that best represent the threshold of adverse effects.

15.13 SUMMARY

Using a single standard method like species sensitivity distributions simplifies benchmark derivation. However, any single method has limitations that may be mitigated by a weight-of-evidence approach. Using all available and relevant information in multiple analytical methods may provide greater accuracy, defensibility, flexibility, and confidence (Buchwalter, Clements, and Luoma 2017). Weighing evidence also provides a method for expressing the level of confidence in the benchmark. Weight-of-evidence methods for deriving benchmarks are available, and they are increasingly being advocated and used.

REFERENCES

ANZG. 2018a. *Australian and New Zealand Guidelines for Fresh and Marine Water Quality*. Canberra: Australian and New Zealand Governments.

ANZG. 2018b. *Weight of Evidence*. Canberra: Australian and New Zealand Governments.

ANZG. 2023. "Guideline values for water/sediment quality." Guidelines for Freshwater & Maine Water Quality. Canberra: Australian Government. https://www.waterquality.gov.au/anz-guidelines/guideline-values. Accessed 06/28/2025.

Batley, G. E., R. A. van Dam, M. St. J. Warne, J. C. Chapman, R. Fox, C. W. Hickey, and J. L. Stauber. 2018. *Technical Rational for Changes in the Method for Deriving Australia and New Zealand Water Quality Guideline Values for Toxicants*. Canberra: Australia and New Zealand Governments.

Bertram, M. G., et al. 2025. "EthoCRED: A framework to guide reporting and evaluation of the relevance and reliability of behavioural ecotoxicity studies." *Biological Reviews* 100: 556–585.

Brock, T. C. M., J. Lahr, and P. J. van den Brink. 2000. *Ecological Risks of Pesticides in Freshwater Ecosystems. Part 1: Herbicides*. Wageningen, NL: Alterra, Green World Inc.

Buchwalter, D. B., W. H. Clements, and S. N. Luoma. 2017. "Modernizing water quality criteria in the United States: A need to expand the definition of acceptable data." *Environmental Toxicology and Chemistry* 36 (2): 285–291.

Chambers, R. C. 2014. *Memorandum Opinion and Order, Ohio Valley Environmental Coalition, West Virginia Highlands Conservancy, and Sierra Club, v. Elk Run Coal Company, Inc., and Alex Energy, Inc., In Civil Action No. 3:*12–0785. Huntington, WV: The United States District Court for the Southern District of West Virginia.

Chapman, P. M. 2018. "Environmental quality benchmarks—The good, the bad, and the ugly." *Environmental Science and Pollution Research* 25 (4): 3043–3046.

Clements, W. H., D. M. Carlisle, L. A. Courtney, and E. A. Harrahy. 2002. "Integrating observational and experimental approaches to demonstrate causation in stream biomonitoring studies." *Environmental.Toxicoogy and Chemistry* 21: 1138–1146.

Cormier, S. M., J. F. Paul, R. L. Spehar, W. J. Berry, P. Shaw-Allen, and G. W. Suter. 2008. "Using field data and weight of evidence to develop water quality criteria." *Integrated Environmental Assessment and Management* 4: 490–504.

Cormier, S. M., G. W. Suter, and L. Zheng. 2013. "Derivation of a benchmark for freshwater ionic strength." *Environmental Toxicology and Chemistry* 32 (2): 263–271.

Council of the European Union. 2025. "Water pollution: Council and Parliament reach provisional deal to update priority substances in surface and ground waters." https://www.consilium.europa.eu/en/press/press-releases/2025/09/23/water-pollution-council-and-parliament-reach-provisional-deal-to-update-priority-substances-in-surface-and-ground-waters/

ECETOC. 2014. *Estimating Toxicity Thresholds for Aquatic Ecological Communities from Sensitivity Distributions, 11–13 February 2014, Amsterdam.* Brussels: European Centre for Ecotoxicology and Toxicology of Chemicals.

EFSA Scientific Committee. 2017. "Guidance on the use of the weight of evidence approach in scientific assessments." *EFSA Journal* 15 (8): 1–69.

Erickson, R. J. 2012. *Proposed Methodology for Specifying Atrazine Levels of Concern for Protection of Plant Communities in Freshwater Ecosystems.* Duluth, MN: Office of Research and Development, U.S. Environmental Protection Agency.

Forbes, V. E., and P. Calow. 2002. "Applying weight-of-evidence to retrospective ecological risk assessment when quantitative data are limited." *Human and Ecological Risk Assessment* 8: 1625–1639.

Franklin, P., R. Stoffels, J. Clapcott, D. Booker, A. Wagenhoff, and C. W. Hickey. 2019. *Deriving Potential Fine Sediment Attribute Thresholds for the National Objectives Framework.* Hamilton, NZ: National Institute of Water & Atmospheric Research.

Giesy, J. P., K. R. Solomon, J. R. Coats, K. R. Dixon, J. M. Giddings, and E. E. Kenaga. 1999. Chlorpyrifos*: Ecological Risk Assessment in North American Aquatic Environments.* New York, NY: Springer.

Hall, L. W., J. M. Giddings, K. R. Solomon, and R. Balcomb. 1999. "An ecological risk assessment for the use of Irgarol 1051 as an algaecide for antifoulant paints." *Critical Reviews in Toxicology* 29 (4): 367.

Hanson, M. L., and K. R. Solomon. 2004. "Haloacetic acids in the aquatic environment. Part II: Ecological risk assessment." *Environmental Pollution* 130 (3): 385–401.

Humphrey, C. L., and L. Chandler. 2018. *Use of Field-Effects Information to Derive a Surface Water Guideline Value for Magnesium in Magela Creek.* Darwin: Department of Agriculture, Water, and Environment.

Isigonis, P., A. Zabeo, P. Ciffroy, E. Semenzin, S. Giove, A. Critto, S. Andres, A. R.R. Pery, and A. Marcomini. 2014. "PROERA: A multi-criteria based decision support system for probabilistic ecological risk assessment." *24. SETAC Europe Annual Meeting "Science across Bridges, Borders and Boundaries"*, Bâle, Switzerland, 11 May 2014.

Leung, K. M., G. Merrington, J. Warne, and R. J. Wenning. 2014. "Scientific derivation of environmental quality benchmarks for the protection of aquatic ecosystems: Challenges and opportunities." *Environmental Science and Pollution Research* 21 (1): 1–5.

Merrington, G., A. Youn-Joo, E. P. M. Grist, J. Seung-Woo, C. Rattikansukha, S. Roe, U. Schneider, S. Sthiannopkao, G. W. Suter, and R. Van Dam. 2014. "Water quality guidelines for chemicals: Learning lessons to deliver meaningful environmental metrics." *Environmental Science and Pollution Research* 21: 6–16.

Moore, D. R. J., C. D. Greer, G. Manning, K. Wooding, K. J. Beckett, R. A. Brain, and G. Marshall. 2017. "A weight-of-evidence approach for deriving a level of concern for atrazine that is protective of aquatic plant communities." *Integrated Environmental Assessment and Management* 13 (4): 686–701.

NDEP. 2022. *Petroleum in Soils Closure Checklists July 2022 Update*. Nevada Division of Environmental Protection Bureau of Corrective Actions. Washington, DC: Reprint American Petroleum Institute. https://ndep.nv.gov/uploads/env-sitecleanup-guidance-docs/NDEP_PetroleumInSoils_checklists_Sept14_Version_Petro_Constituents_2022_Update1.pdf

Oginah, S. A., L. Posthuma, M. Hauschild, J. Slootweg, M. Kosnik, and P. Fantke. 2023. "To split or not to split: Characterizing chemical pollution impacts in aquatic ecosystems with species sensitivity distributions for specific taxonomic groups." *Environmental Science and Technology* 57 (39): 14526–14538.

Paul, J. F., S. M. Cormier, W. Berry, P. Kaufmann, R. Spehar, D. Norton, R. Cantilli, R. Stevens, W. Swietlik, and B. Jessup. 2007. "Developing water quality criteria for suspended and bedded sediments–illustrative example application." *TMDLS Conference 2007*. San Antonio, TX.

Posthuma, L., G. W. Suter, and T. P. Traas. (eds.) 2001. *Species Sensitivity Distributions for Ecotoxicology*. Boca Raton, FL: Lewis Publishers.

Ridley, C. E., S. D. Kaylor, S. S. Lee, J. Miller, N. P., Sam, and K. A. Schofield. 2024. *Application of Weight-of-Evidence Methods for Transparent and Defensible Numeric Nutrient Criteria*. Washington, DC: Office of Research and Development, U.S. Environmental Protection Agency.

Smith, A. J., and C. P. Tran. 2010. "A weight-of-evidence approach to define nutrient criteria protective of aquatic life in large rivers." *Journal of the North American Benthological Society* 29 (3): 875–891.

Solomon, K. R., D. B. Baker, R. P. Richards, K. R. Dixon, S. J. Klaine, T. W. La Point, R. J. Kendall, C. P. Weisskopf, J. M. Giddings, and J. P. Giesy. 1996. "Ecological risk assessment of atrazine in North American surface waters." *Environmental Toxicology and Chemistry* 15 (1): 31–76.

Stephan, C. E., D. I. Mount, D. J. Hanson, J. H. Gentile, G. A. Chapman, and W. A. Brungs. 1985. *Guidelines for Deriving Numeric National Water Quality Criteria for the Protection of Aquatic Organisms and Their Uses*. Washington, DC: U.S. Environmental Protection Agency.

Suter, G. W., S. M. Cormier, and M. G. Barron. 2017. "A weight of evidence framework for environmental assessments: Inferring quantities." *Integrated Environmental Assessment and Management* 13: 1045–1051.

Tillmanns, A. R., and A. Pearson. 2024. *wqbench: A Tool for Calculating Aquatic Life Benchmarks for Contaminants of Emerging Concern*. Province of British Columbia, Victoria: Ministry of Water, Land and Resource Stewardship.

USEPA. 1976. *Quality Criteria for Water*. Washington, DC: U.S. Environmental Protection Agency.

USEPA. 1985. *Guidelines for Deriving Numeric National Water Quality Criteria for the Protection of Aquatic Organisms and Their Uses*. Washington, DC: U.S. Environmental Protection Agency.

USEPA. 1986. *Ambient Aquatic Life Water Quality Criteria for Lead.* Duluth, MN: U.S. Environmental Protection Agency.

USEPA. 1994. *Interim Guidance on Determination and Use of Water-Effect Ratios for Metals. Final Report.* Washington, DC: U.S. Envrionmental Protection Agency.

USEPA. 1996. *Soil Screening Guidance: Technical Background Document.* Washington, DC: U.S. Environmental Protection Agency.

USEPA. 2000. *Nutrient Criteria Technical Guidance Manual for Rivers and Streams.* Washington, DC: Office of Water, U.S. Environmental Protection Agency.

USEPA. 2002. *Biological Assessments and Criteria: Crucial Components of Water Quality Programs.* Washington, DC: US Environmental Protection Agency, Office of Water.

USEPA. 2005. *Guidance for Developing Ecological Soil Screening Levels (ECO-SSLs).* Washington, DC: U.S. Environmental Protection Agency, Office of Solid Waste and Emergency Response.

USEPA. 2006. *Framework for Developing Suspended and Bedded Sediments Water Quality Criteria.* Washington, DC. https://assessments.epa.gov/risk/document/&deid%3D164423.

USEPA. 2011. *A Field-Based Aquatic Life Benchmark for Conductivity in Central Appalachian Streams.* Washington, DC: Office of Research and Development, National Center for Environmental Assessment.

USEPA. 2015. *Preamble to the Integrated Science Assessments (ISA).* Washington, DC: U.S. Environmental Protection Agency.

USEPA. 2016. *Weight of Evidence in Ecological Assessment.* Washington, DC: Risk Assessment Forum, U.S. Environmental Protection Agency.

USEPA. 2022. *ORD Staff Handbook for Developing IRIS Assessments.* Washington, DC: Office of Research and Development, U.S. Environmental Protection Agency.

USEPA. 2023a. *Ecological Soils Screening Levels.* U.S. Environmental Protection Agency. https://www.epa.gov/chemical-research/ecological-soil-screening-level.

USEPA. 2023b. *Water Quality Standards Handbook: Chapter 3: Water Quality Criteria.* Washington, DC: U.S. Environmental Protection Agency, Office of Water.

Van Dam, R. A., A. C. Hogan, A. J. Harford, and C. L. Humphrey. 2019. "How specific is site-specific? A review and guidance for selecting and evaluating approaches for deriving local water quality benchmarks." *Integrated Environmental Assessment and Management* 15 (5): 683–702.

Van Dam, R. A., C. L. Humphrey, A. J. Harford, A. Sinclair, D. R. Jones, S. Davies, and A. W. Storey. 2014. "Site-specific water quality guidelines: 1. Derivation approaches based on physicochemical, ecotoxicological and ecological data." *Environmental Science and Pollution Research* 21: 118–130.

Warne, M., G. E. Batley, R. A. van Dam, J. C. Chapman, D. R. Fox, C. W. Hickey, and J. L. Stauber. 2018. *Revised Method for Deriving Australian and New Zealand Water Quality Guideline Values for Toxicants.* Brisbane, Queensland: Department of Science, Information Technology and Innovation.

Warne, M., G. Batley, R. van Dam, J. Chapman, C. Hickey, J. Stauber, and R. Fisher. 2025. *Revised Method for Deriving Australian and New Zealand Water Quality Guideline Values for Toxicants - Update of 2018 Version.* Canberra, ACT: Australian and New Zealand Governments and Australian State and Territory Governments.

Water Quality Australia. 2020. *Deriving Guideline Values Using Multiple Lines of Evidence.* Canberra. Australia/New Zealand Governments.

Welsh, P. G., J. Lipton, and G. A. Chapman. 2000. "Evaluation of water-effect ratio methodology for establishing site-specific water quality criteria." *Environmental Toxicology and Chemistry* 19 (6): 1616–1623.

16 Evaluating the Hazards of Chemicals and Other Substances

Chemicals such as chlorpyrifos, mixtures such as petroleum, products such as plastics, and waste materials such as coal ash may all be evaluated with respect to their environmental properties. For simplicity, we refer to them all as substances. Substance evaluations include hazard identification, hazard assessment, and risk assessment (Box 16.1). Assessments of substances by regulatory bodies are usually based on weight of evidence (e.g., REACH Annex XIII). However, some agencies have needed encouragement from assessors such as Sheldon Krimsky (2005): "When WOE is used consistently and uniformly by a regulatory body, it enables that body to develop a strong comparative approach for assessing the potential health and environmental effects of products."

BOX 16.1 HAZARD AND RISK

Hazard Identification: This is the process of recognizing potential adverse properties of substances. Chemical hazards include carcinogenicity, teratogenicity, reproductive inhibition, asphyxiation, photosynthesis inhibition, etc.

> 2,3,7,8-tetrachlorodibenzo-p-dioxin (TCDD) is a carcinogen.

Hazard Assessment: This process evaluates the severity of hazards and their potential for harm. It can be qualitative, as in assignment to a category, or quantitative, as in the derivation of an exposure-response relationship.

> There is no safe level of TCDD and doubling an individual's background dioxin exposure could increase their lifetime risk of cancer death by 0.1% to 1.0% (Steenland and Deddens 2003).

Risk Assessment: This process estimates the frequency and magnitude of effects given the hazard and exposure. That is, it solves an exposure-response relationship for a particular exposure scenario.

> Risks for any TCDD exposure could be estimated from the models in Steenland and Deddens (2003).

Note that some assessors may merge hazard identification and hazard assessment. Others may merge hazard assessment and risk assessment.

DOI: 10.1201/9781003626336-19

Typically, evaluations of substances are reference documents that summarize information concerning the substance and identify the substance's hazards. For permitting of chemicals, there is considerable controversy about whether assessment of hazards is sufficient or whether a full risk assessment should be required (Josefsson, Gardelli, and Ågerstrand 2025). Assessments that go on to set benchmark values are addressed in Chapter 15.

16.1 PURPOSES OF SUBSTANCE EVALUATIONS

Substance evaluations have multiple potential purposes.

- Regulatory agencies may assess the hazards posed by substances to determine whether they may be marketed for a designated use. In the United States, laws that compel such assessments include the Federal Insecticide, Fungicide, and Rodenticide Act and the Toxic Substances Control Act.
- Components of agencies such as the USEPA's IRIS program, the U.S. Public Health Service's Agency for Toxic Substances and Disease Registry, and the World Health Organization's International Agency for Research on Cancer provide generic evaluations of the hazards posed by substances that are independent of particular regulatory mandates.
- Hazard evaluations may provide input to systems, such as the USEPA's Toxics Release Inventory or the European Union's Classification and Labelling Inventory, that compile hazard classifications and related information for various uses.
- Manufacturers or their contractors may perform hazard assessments to determine whether to proceed with the development and licensing of a product. This may be for in-house purposes or as a preliminary step in the regulatory process. They may also be used to determine the relative reliability of conflicting hazard assessments of a substance (Williams et al. 2016).
- Comparative assessments of hazards may compare alternative substances to determine which is the least hazardous for a designated use. Manufacturers may find this useful when deciding whether to replace an old product with a new one. Advocacy groups may make recommendations concerning the relative hazards of alternative products.
- Evaluations of substances may be developed for legal purposes, such as when manufacturers or other responsible parties sue regulatory agencies or when affected parties sue the manufacturers or marketers.

For all of these purposes, hazard identification is important. However, regulatory actions based on hazard identification may have unintended consequences. For example, the European Union is currently considering whether the carcinogenic and reproductive hazards posed by ethanol should preclude its use as a disinfectant. The hazards are associated with consumption of ethanol, not with disinfecting surfaces. However, ethanol has those hazardous properties. Similarly, California's Proposition 65 requires that the following label be applied to state-listed chemicals: "WARNING: This product can expose you to chemicals including [name of one or more chemicals], which is [are] known to the State of California to cause cancer [and] [or] [birth defects or other reproductive harm]." Currently, a TikTok video is warning viewers not to buy air fryers because they have that warning label. That hazard is associated with chemicals that may form when cooking at high temperatures; the fryers contain no carcinogens, and the warning could equally apply to a wok. A hazard assessment should avoid these problems through thoughtful hazard identification.

The evaluation of chemicals for their hazards is complex enough, but the safe-and-sustainable-by-design approach requires assessment of hazards plus the sustainability of the manufacturing and disposal cycle (OECD 2022). It weighs and integrates heterogeneous evidence for multiple goals (Abbate et al. 2025). This requires consideration of the qualitative nature of much of the evidence and the potential for quantitative normalization to produce misleading integrated weights (Odu 2019; Kalbar et al. 2017).

16.2 FRAMEWORKS FOR SUBSTANCE EVALUATIONS

The basic framework for assessment (Figure 4.2) is applicable to the evaluation of substances, and the USEPA framework for weight of evidence (Figure 5.1) is applicable to weighing evidence in this context. However, some organizations that conduct or guide such assessments have their own frameworks (USEPA 2022; OHAT 2019; EFSA Scientific Committee 2017; ECHA 2024; UNECE 2021).

16.2.1 Choosing Potential Hazards

In hazard assessments, the hazards are equivalent to assessment endpoints in risk assessments. Hence, potential hazards should be identified up front. A hazard is the potential to cause a particular type of effect, such as being carcinogenic, mutagenic, teratogenic, reprotoxic, endocrine disruptive, or affecting certain taxa or life stages. Hazards of substances may also include specific modes of action, such as ion imbalance, or mechanisms of action, such as aryl-hydrocarbon (Ah) receptor binding. Hazards are related to some, usually broadly defined, receptors, such as mammalian carcinogens or anuran teratogens. A common broad hazard, contaminants of concern or emerging concern, is used in the prioritization of substances (Elliott et al. 2021). REACH regulations now include safety and sustainability as hazards with respect to human health, the environment, and socioeconomics (Schwirn et al. 2025). As in selecting endpoints, assessors must determine which potential hazards should be addressed.

Potential hazards of a substance can prompt hazard identification in two ways: case-specific and standard. Case-specific hazards may be hypothesized due to effects that have been associated with the substance. For example, facial deformities occurred in two of three pregnancies of women working with PFOA at the DuPont Teflon plant in Parkersburg, West Virginia (Stein et al. 2014). Hence, it was reasonable to consider fetal deformities as a potential hazard. Potential hazards may also be identified by the similarity of the substance of concern to chemicals with known hazards (Section 20.1).

Sets of standard hazards have been defined by regulatory agencies and associated organizations like the OECD. They guide the selection of a relevant set of hazards for each substance. The most common set is the Globally Harmonized System of Classification and Labeling of Chemicals (GHS), which is the system of symbols in diamonds that you see on railroad cars and truck trailers. The GHS's set of physical, health, and environmental hazards is expanded in specific contexts, such as European Union chemical regulations (Kättström et al. 2025).

16.2.2 Weighing Evidence for Hazard Identification

Hazard identification is a straightforward application of weight of evidence as described in Part II. Your hypotheses are that the substance of concern has one or more of the potential hazards. Generic implications of evidence for hypothesized hazards include: the substance induces the effect in tests, it induces a response that can result in the effect, or it is observed to be associated with the effect. Specific implications of evidence are characteristics of the hazard, such as the characteristics of teratogens (Table B.6 in Appendix B). The properties to be weighed include relevance, reliability, and strength. You weight the evidence, integrate the types of evidence, and weigh the body of evidence to determine your confidence in the hypothesis that the substance has the hazardous property.

An example of weight of evidence for hazard identification is provided by an assessment to determine whether triphenyl phosphate is an endocrine disrupter and, more specifically, a metabolism disrupter (Beausoleil et al. 2024). The assessors used the EFSA weight-of-evidence approach (EFSA Scientific Committee 2017) but increased credibility by expert elicitation from two multidisciplinary panels. The panels weighed evidence from a systematic literature review, and both panels categorized triphenyl phosphate as a suspected (versus known, presumed, or not categorized) metabolism-disrupting chemical.

16.2.3 Weighing Evidence for Hazard Assessment

Hazard assessments are quite variable in their scope and products. The critical distinction is that they add an exposure component to the hazard identification. This includes identifying uses of the substance, the potential routes of environmental contamination, and modes of exposure of human and nonhuman organisms. A second distinction is the derivation of some indication of the strength of the substance as a cause of the identified hazard. That strength may be a qualitative classification, such as "Glyphosate is a weak carcinogen," or it may be a full exposure-response model. Some hazard assessment systems are effectively case-independent risk assessments. They determine whether effects are likely in a particular use based on generic scenarios and default assumptions. The concentrations used in such scenarios should be environmentally relevant to avoid unrealistically high exposures (Weltje and Sumpter 2017).

Hazard assessment may also address generic hazards to determine the relative priority of chemicals in an environment (EU 2025). This is done by first choosing the hazards of concern and then weighting the evidence and weighing the body of evidence (hazard identification). Then, weigh the types of evidence that are indicative of a potentially significant hazard. Examples of evidence of hazard, in fact, include possible exposures that exceed benchmark values, laboratory toxicity tests, or monitored effects endpoints. An example use of hazard assessment is the prioritization of chemicals in Great Lakes tributaries (Maloney et al. 2025).

Hazard assessment may determine whether to proceed with risk assessment or benchmark derivation. For example, the International Agency for Research on Cancer will proceed with a review only if there is evidence of human exposure and evidence suggesting carcinogenicity (Table B.2 in Appendix B) (Samet et al. 2020; IARC 2019).

16.2.4 Weighing Evidence for Risk Assessment

Risk assessments for substances differ from hazard assessments in that they integrate exposure and exposure-response relationships to determine risks to endpoint entities for some defined case or scenario. They are familiar to assessors who deal with contaminated sites, emissions, applications, or uses of substances. Because their results are usually quantitative rather than categorical, alternative risk estimates may be combined or integrated (Chapter 12). Otherwise, weight of evidence is used to support decisions in the assessment, such as determining whether the exposure-response relationship is linear or whether soil ingestion is a significant exposure pathway.

Estimation of exposure is required for risk assessment of substances, but weight of evidence is less commonly associated with the estimation of exposure than with effects. A weight-of-evidence framework for exposure was developed in a Society for Environmental Toxicology and Chemistry (SETAC) workshop (Miglino and Holmes 2023). It uses the relevance, strength, and reliability properties and the + and − rating system, as in Chapter 7 and in USEPA (2016). The framework includes a tiered approach that begins with (1) a national-scale value, then (2) a site-specific estimate, and finally (3) refined estimates based on monitoring data. The assessment stops when the weight of evidence is sufficient.

Weight-of-evidence methods are used to estimate chemical-specific exposure parameters for bioaccumulation, biomagnification, degradation, persistence, and others (Arnot et al. 2023; Armitage et al. 2021; Prosser et al. 2023; Redman et al. 2022; Fremlin, Elliott, and Gobas 2025; Hughes et al. 2026). They have also been used to assess the occurrence of a route of exposure, such as consumption of plastic-associated chemicals (Koelmans, Diepens, and Nor 2022).

16.2.5 Results of Substance Evaluations

Evaluations of substances begin with a review of relevant information concerning the substance. It includes information concerning chemical formulas, molecular weights, measured concentrations in media and organisms, mechanistic information, toxicological information, and human and ecological data concerning exposure and effects in the field.

The primary products of hazard identification may include:

- Identification of a dichotomous hazard category—the chemical is or is not a teratogen in frogs,
- Biological limits—teratogenicity in early life stages,

- Mode or mechanism of action—aryl-hydrocarbon (Ah) receptor binding,
- Mode of exposure—by direct aqueous uptake, and
- Confidence—likely based on the weight of evidence.

Hazards are principally associated with toxicological attributes, but they may also be associated with other attributes. Evaluations of substances may assess source, transport, and fate information. In particular, persistent or bioaccumulative substances are more likely to cause adverse effects than substances with the same toxic effects but less persistence and no bioaccumulation. For new substances, this information should come from the developer and manufacturer of the substance. For existing substances, information may be provided by real-world data.

An example of the need for weight of evidence with respect to environmental chemistry is provided by Burton Jr. et al. (2019). The persistence of metals in water bodies depends on numerous factors, including the intrinsic properties of metals, the rate at which metals bind to suspended particles, the transformation of metals to different forms, and the potential for remobilization from sediment. Although the authors do not present a formal weight-of-evidence method, they do integrate study results to derive a hazard classification for persistence.

More fundamentally, substance evaluations may include basic physicochemical information. That includes formulas and structures, and attributes such as solubility, partitioning coefficients, volatility, and photolysis. However, these properties are unlikely to require weight of evidence because they are based on simple measurements.

16.3 GUIDANCE FOR TYPES OF SUBSTANCE EVALUATION

Different types of substances are evaluated following different guidance. Some of the differences are due to the properties of the substances, some are due to the endpoint entities, and some are due to history and policies.

16.3.1 Substance Evaluations for Human Health

In the U.S., contrary to some other nations, the hazards of substances in the environment are assessed by regulatory program offices in different ways for human health and ecological endpoints. However, regulation-independent human health assessments are performed without ecological components by the Department of Health and Human Services' National Toxicology Program and by the USEPA's Office of Research and Development with two similar systems, OHAT and IRIS (OHAT 2019; USEPA 2022). They differ from the system recommended here in that they do not use the term weight of evidence, even though they clearly are weighing evidence, as acknowledged in a footnote in the IRIS handbook. They emphasize systematic review, and weighting the studies (pieces of evidence) is a step in the systematic review. After the systematic review, the evidence is synthesized and integrated. That is, evidence is integrated within types (animal, human, and mechanistic) and synthesized in the body of evidence. The properties include internal validity, risk of bias, quality, and others. Evidence ratings are presented as in weight-of-evidence tables like those in Chapters 7 and 8. The OHAT results are presented as combined confidence in the conclusions. The IRIS system assigns judgment levels [*evidence demonstrates*, *evidence indicates (likely)*, *evidence suggests*, *evidence is inadequate*, or *strong evidence supports no effect*] and goes on to derive toxicity values.

As we write this, the status of these assessments is uncertain. The IRIS program has been disbanded in response to industry complaints, and some of the staff have been assigned to regulatory program offices. The fate of the National Toxicology Program and its OHAT assessments is unresolved.

16.3.2 Evaluations of Pesticides and Herbicides

Assessments of pesticides and herbicides lend themselves to weight of evidence because they often use many types of ecotoxicological evidence, including laboratory tests, field experiments, and observations. This is particularly the case for incident investigations and product reregistrations, which have access to relevant field data.

The European Food Safety Authority, despite its name, regulates pesticides on the basis of environmental risks as well as human food safety. They have produced numerous guidance documents, including one on the use of weight of evidence (EFSA Scientific Committee 2017). It uses a framework that is similar to the USEPA guidelines. It differs by encouraging assessors to consider which weight-of-evidence method is best for their purpose and justify the choice.

The USEPA's guidance for assessing pesticides acknowledges the availability of multiple types of evidence, but it does not discuss weighing evidence and does not provide a method for integrating results from multiple pieces or types of evidence (USEPA 2004, 2024a). However, because assessments for threatened and endangered species are more spatially and biologically specific, a variety of information and data sources are used and weighted to develop more refined assessments (USEPA 2024a).

For decades, manufacturers of pesticides and herbicides have funded assessments of the health and ecological risks of their products. An important series of such assessments by consultants to industry has emphasized probability or weight of evidence. The two most recent assessments have assigned numerical weights to the "relevance of the observation to adverse effects" and the "strength and quality of the methods" (Giesy et al. 2025; Purdy et al. 2025). The weights were plotted on relative scales for the two properties. However, the conclusions are based on narrative summaries rather than ratings.

16.3.3 Evaluations of Nonagrochemical Substances

Industrial chemicals and other nonpesticides are often assessed differently because they are not designed to be toxic, and their manufacturers are not required to submit as much information. In addition, their evaluations, unlike those of pesticides, seldom include field data. Some prominent examples of guidance for weighing evidence when assessing such chemicals are presented in Table 16.1. A more complete review is provided by Moore et al. (2023).

SETAC has endorsed weighing evidence in assessments of chemicals and conducted a series of workshops to develop weight-of-evidence guidance for developing

TABLE 16.1
Some Prominent Methods for Evaluating Nonpesticide Substances

Institution	Health or Ecological	Properties	Ratings	Features
European Chemicals Agency (ECHA 2024, 2010)	Mostly HH Some Eco	Relevance, reliability, adequacy, quantity	High/medium/low or Klimisch score	Case-specific weights. They provide a template for weight-of-evidence reporting that includes exposure as well as toxicological attributes.
USEPA Office of Pollution Prevention and Toxics (USEPA 2024b)	HH, Eco	Quality and relevance which include study design, fitness for purpose, replicability, peer review, transparency and reliability of data.	Not specified	Weight of scientific evidence is prescribed. It means "an approach to scientific evaluation in which each piece of relevant information is considered based on its quality and relevance, and then transparently integrated with other relevant information to inform the scientific evaluation prior to making a judgment about the scientific evaluation."
Environment & Climate Change Canada (ECCC 2020)	Eco risk for inorganic chemicals	Strength and types of evidence (e.g., modeling vs. monitoring)	High, moderate, low for pieces of evidence and for risk	Exposure and hazard evidence is weighted separately. High-, moderate-, or low-risk categories are based on the number of risk quotients greater than one or greater than ten.
Australian Industrial Chemicals Introduction Scheme (Australian Government 2026)	HH and/or Eco	Persistence, bioaccumulation, toxicity	High, medium, low, or very low	The method applies categorization criteria followed by a weight-of- evidence-based risk assessment using a WoE table of hazard versus exposure.
Organization for Economic Cooperation & Development (OECD 2019)	HH, Eco	Relevance (including sufficiency, quality, and consistency) and reliability	Low, moderate, or high	A conventional method based on a consensus of prior methods.

countries and emerging economies (SETAC 2018; Hall et al. 2017; Lee et al. 2023; Menzie et al. 2024). They emphasize the evaluation and rating of the standard weight-of-evidence properties, relevance and reliability (Rudén et al. 2017; Moermond et al. 2017). They were focused on evidence from laboratory toxicity tests, but they provide a novel matrix for evaluating types of evidence ranging from mechanistic studies to ecoepidemiology. The final paper emphasizes tiered assessment (Menzie et al. 2024). These SETAC publications do not constitute a consistent method, but they provide opinions on aspects of chemical assessment by weight of evidence.

16.3.4 Evaluations of Complex Materials and Mixtures

Assessments of hazards from chemical mixtures and other complex materials may be particularly difficult. For example, the U.S. Save the Seas Act 2.0 required assessment of the hazards from contamination of marine environments by microfibers. Because microfibers are diverse and complex, multiple types of evidence, including multiple analytical chemistry and morphological characterizations, should be assembled and weighted (IMDCC 2024; Zhu et al. 2019). Because of the physical and chemical diversity of fibers, it is important to consider whether the microfibers in different studies are sufficiently similar to be combined in a weight-of-evidence assessment.

The European Union avoids the problem of assessing mixtures by allowing a waste to be declared HP14 ecotoxic if it contains one or more substances that are classified as ozone depleting, aquatic acute, or aquatic chronic (Council Regulation (EU) 2017/997). However, a more reliable approach to assessing mixtures is to integrate types of evidence and substances in a weight-of-evidence assessment (Fulvio et al. 2023).

The literature on assessing mixtures is large, and there are many methods for generating evidence (Suter 2023). Toxicity tests can be performed on mixtures of concern, synthetic mixtures, similar mixtures, or mixture fractions. You can also model the toxicity of mixtures using tests of component chemicals with concentration addition, effects addition, or empirical models. Finally, you can model effects of mixtures, such as the major ions in streams, from field data (Cormier et al. 2013). Evidence from these sources of information may be weighed and integrated.

If effects of mixtures are modeled based on the mode of combined action, you should weigh the evidence relevant to the mode of action. In particular, combined effects of mixtures are often modeled as concentration or dose additive. Evidence for such additivity includes similar molecular structure, similarity of the exposure-response relationship (e.g., similar slopes), and similar toxicokinetics, toxicodynamics, and overt effects.

Aqueous and atmospheric effluents are another type of mixture to be assessed. The assessments may answer the question: Would release of a proposed effluent cause impairment? The types of toxicological evidence may include toxic benchmark values for constituent chemicals or new tests of the constituent chemicals. These values would be compared to modeled concentrations in the potential receiving environment. The individual chemicals may be assessed independently or by a combined-effects model, or a synthetic mixture may be tested. If existing effluents of the same type, such as petroleum refinery effluents, have been assessed, their whole-effluent toxicity and their effects in the receiving waters may be used as surrogates for the proposed effluent. Re-permitting allows for effluent-specific field studies and whole-effluent testing. Permitting may be based on one type of evidence, but it is preferable to weigh the multiple types of evidence (Cailleaud et al. 2019).

16.3.5 Evaluating Mode of Action

The mode of action of a substance can be important information for determining life-stage susceptibility, species relevance, toxicokinetics, or exposure-response relationships. It has been particularly important in assessments of carcinogenesis because different modes of action, such as mutagenesis versus cell death, imply very different exposure-response relationships. In common usage, mode of action is a general way in which a substance causes its effect, such as hepatotoxicity or photosynthesis inhibition. This is in contrast to more specific mechanisms of action, such as cholinesterase inhibition. Some groups have abandoned the mode/mechanism distinction and defined mode of action as a biologically plausible series of key events leading to an effect (Sonich-Mullin et al. 2001).

A well-developed method for analyzing mode of action is provided by Meek et al. (2014a). Their method begins with hypothesized modes of action and then weighs the evidence for key events using modified Hill considerations. As the authors concede, these considerations for determining causality in epidemiological studies have "a number of limitations" in this context. They then evaluate the level of confidence and analyze concordance among and within species. Depending on confidence, assessment-specific data generation may be required. A pseudo-quantitative variant of this method (i.e., qualitative ratings are converted to numerals) was proposed by Becker et al. (2017).

The definition of mode of action as a series of events is equivalent to adverse outcome pathways, but generally not as detailed. Hence, these weight-of-evidence methods for mode of action may be used to develop adverse outcome pathways (Becker et al. 2017).

A weight-of-evidence method for determining mode of action, in a more traditional sense, was developed and applied by the European Center of Ecotoxicology and Toxicology of Chemicals (ECETOC) (Thomas et al. 2008). For a particular mode of action, such as narcosis, they define types of evidence that are indicative of that mode of action. They then determine whether they have each type of evidence for the chemical of interest. If positive evidence is found for 70% of the types of evidence, then the mode of action is accepted.

The concept of mode of action becomes complex when assessing endocrine-disrupting chemicals. Such chemicals may disrupt multiple hormone pathways and cause effects in multiple tissues. This complexity makes it particularly important to weigh evidence and keep track of what the evidence applies to. A weight-of-evidence system for screening endocrine disrupters is provided by the USEPA (2011). It has been influential, but it uses the Agency's standard assessment factors for information quality as properties to be weighted. Most of the approaches to this problem, like Meek et al. (2014b), use some variant of Hill's considerations, but with some endocrine-related additions. A more promising approach, in our opinion, is provided by Mihaich et al. (2017). They assess modes of action in the form of multi-branched, endocrinological adverse outcome pathways, which are weighed in terms of three broad properties: biological plausibility, essentiality, and empirical evidence. An approach to endocrine disrupters from a SETAC workshop consists of a set of five sequential criteria (see Section 2.5) for determining whether there is sufficient evidence for assessing the risks of endocrine disruption (Matthiessen et al. 2017). A review by pesticide industry scientists (Staveley et al. 2023) advocated weight-of-evidence methods for assessing potential endocrine disrupters that are taken largely from USEPA (2016).

16.4 NONTOXICOLOGICAL EFFECTS

Evaluations may determine whether a substance has hazardous nontoxicological attributes, such as being a persistent or bioaccumulative substance as well as being toxic (PBT). It is recommended that they be assessed by weighing evidence from multiple sources (Matthies et al. 2016; ECETOC 2014). Other potential nontoxicological hazards include anthropocentric attributes such as taste and odor in water.

Bioaccumulation and persistence are standard nontoxicological hazards, particularly in the European REACH regulations. They may be determined by weighting and integrating the following types of evidence: chemical structure, steric properties, partitioning properties, ADME properties (Absorption, Distribution, Metabolism, Elimination), laboratory data, and field data (OECD 2019; Arnot et al. 2023; Nichols et al. 2009). A bioaccumulation tool is provided for REACH, and the weight-of-evidence approach is consistent with the OECD guidance. Weight-of-evidence methods are also available for persistence (Hughes et al. 2026). The types of evidence for persistence include standard biodegradation tests, quantitative structure-activity relationships, field tests, monitoring studies, and nonstandard laboratory tests. The evidence properties are the standard relevance and reliability, but they use an unconventional stepwise approach to rating and integration.

Hazard identification and assessment approaches may also be used for the evaluation and potentially for the regulation of claimed positive attributes such as biodegradable, nontoxic, or selectively toxic. These attributes may seem to be the opposite of hazards, but they should have their own assessments. For example, if you do not find that a chemical is persistent, that does not mean that the manufacturer can claim that it is biodegradable. If more than one test has been performed for a positive attribute, the resulting evidence should be weighed.

16.5 THE FUTURE OF SUBSTANCE EVALUATIONS

Currently, assessments of the same substance for multiple purposes are performed by multiple agencies. This can result in the duplication of effort and the waste of resources (Andreassen, Rudén, and Ågerstrand 2025). Coordination and the use of common identifiers could alleviate this problem. However, assessment replication can be an important check on quality, and duplicate assessments are usually staggered in time, so an up-to-date assessment is more likely to be available.

A major change in substance evaluation is the increase in molecular and other sub-organismal data. Currently, those data are not reliable predictors of overt organism-level effects. However, weight of evidence and systems such as adverse outcome pathways are increasingly used along with those data to identify hazards, and they are expected to become more reliable predictors of the nature and magnitude of effects. Artificial intelligence tools should facilitate that trend (Chapter 23).

REFERENCES

Abbate, E., A. M. J. Ragas, C. Caldeira, L. Posthuma, I. G. Aguirre, A. C. Devic, L. G. Soeteman-Hernández, M. A. J. Huijbregts, and S. Sala. 2025. "Operationalization of the safe and sustainable by design framework for chemicals and materials: Challenges and proposed actions." *Integrated Environmental Assessment and Management* 21(2): 245–262.

Andreassen, M., C. Rudén, and M. Ågerstrand. 2025. "Mapping chemicals across EU's legal frameworks towards a one substance, one assessment approach." *Environment International 199*: 109460.

Armitage, J. M., L. Toose, L. Camenzuli, A. D. Redman, T. F. Parkerton, D. Saunders, J. Wheeler, A. Martin, E. Vaiopoulou, and J. A. Arnot. 2021. "A critical review and weight of evidence approach for assessing the bioaccumulation of phenanthrene in aquatic environments." *Integrated Environmental Assessment and Management* 17 (5): 911–925.

Arnot, J. A., L. Toose, J. M. Armitage, M. Embry, A. Sangion, and L. Hughes. 2023. "A weight of evidence approach for bioaccumulation assessment." *Integrated Environmental Assessment and Management* 19 (5): 1235–1253.

Australian Government. 2026. "Australian Industrial Chemicals Evaluation Scheme." https://www.industrialchemicals.gov.au/evaluations-roadmap/outcome-3-knowledge-and-evidence

Beausoleil, C., A. Thébault, P. Andersson, N. J. Cabaton, S. Ermler, B. Fromenty, C. Garoche, J. L. Griffin, S. Hoffmann, and J. H. Kamstra. 2024. "Weight of evidence evaluation of the metabolism disrupting effects of triphenyl phosphate using an expert knowledge elicitation approach." *Toxicology and Applied Pharmacology* 489: 116995.

Becker, R. A., V. Dellarco, J. Seed, J. M. Kronenberg, B. Meek, J. Foreman, C. Palermo, C. Kirman, I. Linkov, and R. Schoeny. 2017. "Quantitative weight of evidence to assess confidence in potential modes of action." *Regulatory Toxicology and Pharmacology* 86: 205–220.

Burton, G. A., M. L. Hudson, P. Huntsman, R. F. Carbonaro, K. J. Rader, H. Waeterschoot, S. Baken, and E. Garman. 2019. "Weight-of-evidence approach for assessing removal of metals from the water column for chronic environmental hazard classification." *Environmental Toxicology and Chemistry* 38 (9): 1839–1849.

Cailleaud, K., A. Bassères, C. Gelber, J. F. Postma, A. T. M. Ter Schure, P. E. G. Leonards, A. D. Redman, G. F. Whale, M. J. Spence, and M. Hjort. 2019. "Investigating predictive tools for refinery effluent hazard assessment using stream mesocosms." *Environmental Toxicology and Chemistry* 38 (3): 650–659.

Cormier, S. M., G. W. Suter, L. Zheng, and G. J. Pond. 2013. "Assessing causation of the extirpation of stream macroinvertebrates by a mixture of ions." *Environmental Toxicology and Chemistry* 32 (2): 277–287.

ECCC. 2020. *Science Approach Document: Ecological Risk Classification of Inorganic Substances*. Gatineau, Quebec: Environment and Climate Change Canada.

ECETOC. 2014. *Information to be Considered in a Weight-of-Evidence-Based PBT/vPvB Assessment of Chemicals (Annex XIII of Reach)*. Brussels: European Centre For Ecotoxicology and Toxicology of Chemicals.

ECHA. 2010. *Practical Guide 2: How to Report Weight of Evidence*. Helsinki: European Chemicals Agency.

ECHA. 2024. *Weight of Evidence/Uncertainty in Hazard Assessment: Background Document & Examples*. Helsinki: European Chemicals Agency.

EFSA Scientific Committee. 2017. "Guidance on the use of the weight of evidence approach in scientific assessments." *EFSA Journal* 15 (8): 1–69.

Elliott, S. M., D. J. Gefell, R. L. Kiesling, S. L. Hummel, C. K. King, C. H. Christen, S. Kohno, and H. L. Schoenfuss. 2021. "Multiple lines of evidence for identifying potential hazards to fish from contaminants of emerging concern in Great Lakes tributaries." *Integrated Environmental Assessment and Management* 18 (5): 1246–1259.

EU. 2025. *Water Pollution: Council and Parliament Reach Provisional Deal to Update Priority Substances in Surface and Ground Waters*. Brussels, Belgium: Press office – General Secretariat of the Council of the EU.

Fremlin, K. M., J. E. Elliott, and F. A. P. C. Gobas. 2025. "Guidance for measuring and evaluating biomagnification factors and trophic magnification factors of difficult substances: Application to decabromodiphenylethane." *Integrated Environmental Assessment and Management* 21 (2): 263–278.

Fulvio, O., B. Micol, T. Andrea, P. Andrea, and M. Chiara. 2023. "An innovative methodological path to attribute the hazard property HP14 "ecotoxic" to waste using a weight of evidence approach." *Journal of Environmental Management* 332: 117208.

Giesy, J. P., K. R. Solomon, J. R. Purdy, and V. J. Kramer. 2025. "Weight of evidence assessment of effects of sulfoxaflor on aquatic invertebrates: Sulfoxaflor environmental science review." *Journal of Toxicology and Environmental Health, Part B*: 28(5): 293–321.

Hall, A. T., S. E. Belanger, P. D. Guiney, M. Galay-Burgos, G. Maack, W. Stubblefield, and O. Martin. 2017. "New approach to weight-of-evidence assessment of ecotoxicological effects in regulatory decision-making." *Integrated Environmental Assessment and Management* 13 (4): 573–579.

Hughes, C. B., M. Griffiths, S. Cook, D. van de Meent, J. Parsons, D. Lyon, and A. Ott. 2026. "Developing a weight-of-evidence methodology for persistence assessment of substances in the environment." *Integrated Environmental Assessment and Management* 22(2): 509–522.

IARC. 2019. *IARC Monographs on the Identification of Carcinogenic Hazards to Humans: Preamble*. Lyon: International Agency for Research on Cancer.

IMDCC. 2024. "Interagency Marine Debris Coordinating Committee Report on Microfiber Pollution." Interagency Marine Debris Coordinating Committee. https://marinedebris.noaa.gov/interagency-marine-debris-coordinating-committee-reports/interagency-marine-debris-coordinating-committee-report-microfiber-pollution

Josefsson, H., C. Gardelli, and M. Ågerstrand. 2025. "Stakeholder perspectives on the REACH revision and the role of the precautionary principle." *Environmental Sciences Europe* 37: 227.

Kalbar, P. P., M. Birkved, S. E. Nygaard, and M. Hauschild. 2017. "Weighting and aggregation in life cycle assessment: Do present aggregated single scores provide correct decision support?" *Journal of Industrial Ecology* 21 (6): 1591–1600.

Kättström, D., A. Beronius, U. B. af Gennäs, C. Rudén, and M. Ågerstrand. 2025. "Impact of the new hazard classes in the CLP regulation on EU chemicals legislation." *Environmental Sciences Europe* 37 (1): 19.

Koelmans, A. A., N. J. Diepens, and N.H.M. Nor. 2022. "Weight of evidence for the microplastic vector effect in the context of chemical risk assessment." In *Microplastic in the Environment: Pattern and Process*, edited by M. S. Bank, 155–191. Cham: Springer.

Krimsky, S. 2005. "The weight of scientific evidence in policy and law." *American Journal of Public Health* 95: S129–S135.

Lee, K.-C., P. D. Guiney, C. A. Menzie, and S. E. Belanger. 2023. "Advancing the weight of evidence approach to enable chemical environmental risk assessment for decision-making and achieving protection goals." *Integrated Environmental Assessment and Management* 19 (5): 1188–1191.

Maloney, E. M., S. R. Corsi, M. A. Pronschinske, L. A. DeCicco, J. R. Frisch, N. Fuller, A. K. Baldwin, K. Kimbrough, M. Edwards, and S. L. Hummel. 2025. "Retrospective stepwise prioritization of chemicals detected in Great Lakes tributaries (2008–2018)." *Environmental Toxicology and Chemistry* 44 (7): 2048–2069.

Matthies, M., K. Solomon, M. Vighi, A. Gilman, and J. V Tarazona. 2016. "The origin and evolution of assessment criteria for persistent, bioaccumulative and toxic (PBT) chemicals and persistent organic pollutants (POPs)." *Environmental Science: Processes and Impacts* 18 (9): 1114–1128.

Matthiessen, P., G. T. Ankley, R. C. Biever, P. Bjerregaard, C. Borgert, K. Brugger, A. Blankinship, J. Chambers, K. K. Coady, and L. Constantine. 2017. "Recommended approaches to the scientific evaluation of ecotoxicological hazards and risks of endocrine-active substances." *Integrated Environmental Assessment and Management* 13 (2): 267–279.

Meek, M. E., A. Boobis, I. Cote, V. Dellarco, G. Fotakis, S. Munn, J. Seed, and C. Vickers. 2014a. "New developments in the evolution and application of the WHO/IPCS framework on mode of action/species concordance analysis." *Journal of Applied Toxicology* 34 (1): 1–18.

Meek, M. E., C. M. Palermo, A. N. Bachman, C. M. North, and R. Jeffrey Lewis. 2014b. "Mode of action human relevance (species concordance) framework: Evolution of the Bradford Hill considerations and comparative analysis of weight of evidence." *Journal of Applied Toxicology* 34 (6): 595–606.

Menzie, C. A, P. D. Guiney, S. E. Belanger, K.-C. Lee, G. Arts, B. O. Opeolu, and H. C. Silva de Assis. 2024. "Incorporating a weight-of-evidence approach into a tiered assessment for chemicals management, with emphasis on program development and applications in developing countries and emerging economies." *Integrated Environmental Assessment and Management* 20 (6): 2312–2325.

Miglino, A. N., and C. M. Holmes. 2023. "Applying weight of evidence methods to assessing exposure in aquatic environments: Comparing lines of evidence." *Integrated Environmental Assessment and Management* 19 (5): 1207–1219.

Mihaich, E. M., C. Schäfers, D. A. Dreier, M. Hecker, L. Ortego, Y. Kawashima, Z.-C. Dang, and K. Solomon. 2017. "Challenges in assigning endocrine-specific modes of action: Recommendations for researchers and regulators." *Integrated Environmental Assessment and Management* 13 (2): 280–292.

Moermond, C., A. Beasley, R. Breton, M. Junghans, R. Laskowski, K. Solomon, and H. Zahner. 2017. "Assessing the reliability of ecotoxicological studies: An overview of current needs and approaches." *Integrated Environmental Assessment and Management* 13 (4): 640–651.

Moore, D. W., B. Ruffle, A. McQueen, S. Thakali, and D. Edwards. 2023. "Frameworks for screening and risk management of chemicals and advanced materials: A critical review." *Integrated Environmental Assessment and Management* 19 (5): 1192–1206.

Nichols, J. W., M. Bonnell, S. D. Dimitrov, B. I. Escher, X. Han, and N. I. Kramer. 2009. "Bioaccumulation assessment using predictive approaches." *Integrated Environmental Assessment and Management* 5 (4): 577–597.

Odu, G. O. 2019. "Weighting methods for multi-criteria decision making techniques." *Journal of Applied Sciences and Environmental Management* 23 (8): 1449–1457.

OECD. 2019. *Guiding Principles and Key Elements for Establishing a Weight of Evidence for Chemical Assessment*. Paris: Organization for Economic Cooperation and Development, Environmental Directorate.

OECD. 2022. *Sustainability and Safe and Sustainable by Design: Working Descriptions for the Safer Innovation Approach (Issue 105)*. Paris: Organization for Economic Cooperation and Development.

OHAT. 2019. *Handbook for Conducting a Literature-Based Health Assessment Using OHAT Approach for Systematic Review and Evidence Integration*. Research Triangle Park, NC: Office of Health Assessment and Translation, National Toxicology Program.

Prosser, C. M., C. W. Davis, G. E. Bragin, and L. Camenzuli. 2023. "Using weight of evidence to assess degradation potential of UVCB hydrocarbon solvents." *Integrated Environmental Assessment and Management* 19 (4): 1120–1130.

Purdy, J. R., K. R. Solomon, V. J. Kramer, and J. P. Giesy. 2025. "Weight of evidence assessment from field studies on effects of the insecticide sulfoxaflor on hymenopteran pollinators: Sulfoxaflor environmental science review part V." *Journal of Toxicology and Environmental Health, Part B*: 28(5): 374–405.

Redman, A. D., J. Bietz, J. W. Davis, D. Lyon, E. Maloney, A. Ott, J. C. Otte, F. Palais, J. R. Parsons, and N. Wang. 2022. "Moving persistence assessments into the 21st century: A role for weight-of-evidence and overall persistence." *Integrated Environmental Assessment and Management* 18 (4): 868–887.

Rudén, C., J. Adams, M. Ågerstrand, T. C. M. Brock, V. Poulsen, C. E. Schlekat, J. R. Wheeler, and T. R. Henry. 2017. "Assessing the relevance of ecotoxicological studies for regulatory decision making." *Integrated Environmental Assessment and Management* 13 (4): 652–663.

Samet, J. M., W. A. Chiu, V. Cogliano, J. Jinot, D. Kriebel, R. M. Lunn, F. A. Beland, L. Bero, P. Browne, and L. Fritschi. 2020. "The IARC monographs: Updated procedures for modern and transparent evidence synthesis in cancer hazard identification." *JNCI: Journal of the National Cancer Institute* 112 (1): 30–37.

Schwirn, K., D. Völker, M. Løfstedt, P. Fantke, C. Bossa, A. Sharma, L. Posthuma, A. Karakoltzidis, F. Nikiforou, and A. Mikołajczyk. 2025. "The European Commission's safe and sustainable by design framework: Bridging innovation and legislation." *Environmental Sciences Europe* 37 (1): 1–18.

SETAC. 2018. *Technical Issue Paper: Weight-of-Evidence in Environmental Risk Assessment of Chemicals*. Pensacola, FL: Society for Environmental Toxicology and Chemistry.

Sonich-Mullin, C., R. Fielder, J. Wiltse, K. Baetcke, J. Dempsey, P. Fenner-Crisp, D. Grant, M. Hartley, A. Knaap, and D. Kroese. 2001. "IPCS conceptual framework for evaluating a mode of action for chemical carcinogenesis." *Regulatory Toxicology and Pharmacology* 34 (2): 146–152.

Staveley, J. P., E. L. Freeman, M. E. McArdle, L. S. Ortego, K. K. Coady, A. Bone, L. Lagadic, L. Weltje, A. Weyers, and J. R. Wheeler. 2023. "Current testing programs for pesticides adequately capture endocrine activity and adversity for protection of vertebrate wildlife." *Integrated Environmental Assessment and Management* 19 (4): 1089–1109.

Steenland, K., and J. Deddens. 2003. "Dioxin: Exposure-response analyses and risk assessment." *Industrial Health* 41 (3): 175–180.

Stein, C. R., D. A. Savitz, B. Elston, P. G. Thorpe, and S. M. Gilboa. 2014. "Perfluorooctanoate exposure and major birth defects." *Reproductive Toxicology* 47: 15–20.

Suter, G. W. 2023. *Fundamentals of Environmental Assessment*. Boca Raton, FL: CRC Press.

Thomas, P., M. Holt, P. Lemaire, I. Malcomber, D. Salvito, and R. Thompson. 2008. "A weight of evidence (WOE) approach for determining mode of action: An ECETOC case study." *Integrated Environmental Assessment and Management* 4 (374): 375.

UNECE. 2021. *Globally Harmonized System of Classification and Labelling of Chemicals (GHS Rev. 9, 2021)*. New York and Geneva: United Nations Economic Commission for Europe.

USEPA. 2004. *Overview of the Ecological Risk Assessment Process in the Office of Pesticide Programs, U.S. Environmental Protection Agency: Endangered and Threatened Species Effects Determinations*. Washington, DC: Office of Prevention, Pesticides and Toxic Substances, Office of Pesticide Programs, U.S. Environmental Protection Agency.

USEPA. 2011. *Endocrine Disruptor Screening Program, Weight-of-Evidence: Evaluating Results of EDSP Tier 1 Screening to Identify the Need for Tier 2 Testing*. Washington, DC: Office of Chemical Safety and Pollution Prevention, U.S. Environmental Protection Agency.

USEPA. 2016. *Weight of Evidence in Ecological Assessment*. Washington, DC: Risk Assessment Forum, U.S. Environmental Protection Agency.

USEPA. 2022. *ORD Staff Handbook for Developing IRIS Assessments*. Washington, DC: Office of Research and Development, U.S. Environmental Protection Agency.

USEPA. 2024a. *Ecological Risk Assessment for Pesticides: Technical Overview*. Washington, DC: U.S. Environmental Protection Agency.

USEPA. 2024b. "Procedures for chemical risk evaluation under the Toxic Substances Control Act (TSCA)." *Federal Register* 89 (87): 37028–37058.

Weltje, L., and J. P. Sumpter. 2017. "What makes a concentration environmentally relevant? Critique and a proposal." *Environmental Science and Technology* 51: 11520–11521.

Williams, G. M., M. Aardema, J. Acquavella, et al. 2016. "A review of the carcinogenic potential of glyphosate by four independent expert panels and comparison to the IARC assessment." *Critical Reviews in Toxicology* 46 (Suppl 1): 3–20.

Zhu, X., B. Nguyen, J. B. You, E. Karakolis, D. Sinton, and C. Rochman. 2019. "Identification of microfibers in the environment using multiple lines of evidence." *Environmental Science and Technology* 53 (20): 11877–11887.

17 Contaminated Sites

Contaminated sites include terrestrial, wetland, marine, groundwater, and freshwater sites that have been contaminated by waste disposal, spills, or emissions to soil, sediment, or water. They are assessed to determine which media and locations are contaminated by what contaminants, what effects are occurring or will occur if there is no remediation, and how much remediation is required. Disturbed sites may be assessed by the same approach as contaminated sites. They include sites that have been dredged, filled, flooded, plowed, paved, or otherwise physically altered. Commonly, sites are both disturbed and contaminated, so site assessments may require assessment of both types of causes. These site assessments are distinguished by the fact that at most sites, the magnitude and nature of contamination, disturbance, and their effects need not be predicted; they are already occurring.

Most contaminated site assessments are environmental risk assessments. They begin by identifying contamination and inferring risks of effects. In some cases, however, effects are observed, and their causes can be determined by causal assessment (Chapter 14). Some cases require both a risk assessment and a causal assessment. That is, you may need to identify and estimate effects of contaminants other than the observed effects. Some site assessment methods are devoted to a single contaminant. For example, the controversies surrounding risks from selenium contamination of surface waters are alleviated by a selenium-specific weight-of-evidence approach (McDonald and Chapman 2007).

In the U.S., contaminated sites are assessed and remediated under the Comprehensive Environmental Response, Compensation, and Liability Act (CERCLA), which is commonly known as Superfund. Under CERCLA, the primary assessment is the remedial investigation, which determines the nature and extent of any current contaminant-induced impairments and the risks of future impairments. CERCLA is most commonly applied to contaminated soils because, unlike the Clean Air Act and Clean Water Act, there are no clean soil or sediment acts. Oil spills are subject to both CERCLA and the Oil Pollution Act. Contaminated sediments are regulated under CERCLA and the Clean Water Act.

Contaminated streams and other surface waters are assessed, and if they are impaired by the contaminant, they are remediated under the Total Maximum Daily Loads provision of the Clean Water Act. Waters may be impaired because a water quality standard is exceeded, because the waters are toxic, or because the biotic community is impaired. Under the independent applicability rule, the detection of impairment by any of these criteria is sufficient weight of evidence to declare the water body in question to be listed as impaired (Section 2.8).

Contaminated groundwater is primarily a drinking water concern which may be assessed and remediated under CERCLA. However, if it intersects with surface waters, it becomes a Clean Water Act concern. Volatile organic compounds released

DOI: 10.1201/9781003626336-20

from soil and groundwater, as well as vapor intrusion into buildings from soil, are also covered by CERCLA.

The Environmental Liability Directive (2004/35/CE) establishes a framework based on the "polluter pays" principle, requiring operators to prevent and remediate environmental damage to biodiversity, water, and land. It also covers management of extractive waste, geological storage of carbon dioxide, and offshore oil/gas operations.

Environmental assessments of contaminated sites have been the subjects of many implemented or proposed assessment systems (Marcomini, Suter, and Critto 2009; Chapman, McDonald, and Lawrence 2002; Swartjes 2011; Suter et al. 2000; Jensen and Mesman 2006; Expert Group 2025; Sprenger and Charters 1997; Bizzotto et al. 2022; Forbes and Calow 2002; Burton et al. 2002; Bates, Massey, and Wood 2018; Ingersoll and Macdonald 2002; Wenning and Havranek 2025). Laws and regulations for contaminated sites from many nations are summarized by Baker McKenzie (2024). Canada has guidance for assessing ecological risks from contaminated sites in its action plan, which takes an approach similar to our weight-of-evidence method (Environment Canada 2012). The Netherlands has particularly well-developed methods for contaminated soils as a result of its 1986 Soil Protection Act (Swartjes 2011). Weight of evidence for contaminated waters and sediments is integrated into Australia and New Zealand's Guidelines for Fresh and Marine Water Quality (ANZG 2018a). On the other hand, the European Union has no framework specifically for risk assessment of contaminated sites (Bizzotto et al. 2022).

17.1 CONTAMINATED SITE RISK ASSESSMENT

Contaminated site risk assessments have some different features from other risk assessments. They begin with a problem formulation including gathering information, development of conceptual models, and identification of assessment endpoints (USEPA 1998). However, they differ in that they must identify the intended use of the land or water body. A nature reserve requires a very different assessment than an abandoned factory site. The problem formulation is typically based on a scoping phase that includes a site characterization. Unlike site-independent assessments, these problem formulations should include spatially explicit conceptual models that include locations of contamination, human and ecological receptors, and routes of transport.

The analysis phase should typically include three stages: scoping, screening, and definitive analyses. Each includes exposure and effects components. Analysis of effects primarily depends on toxicity tests of individual contaminant chemicals. Ideally, the analysis phase would also include realistic information concerning the composition and toxicity of the contaminated media and on-site ecological effects.

The risk characterization could be as simple as deriving hazard indices for each chemical. Preferably, it will be an inference by weight of evidence, as described in this section.

17.2 STAGES OF EVIDENCE GENERATION AND ANALYSIS

Frameworks for contaminated site assessments divide the process into stages, and many include multiple tiers (Box 17.1). Conducting the assessment in tiers is particularly advantageous for contaminated soils because of their complexity, lack of standard benchmark values, and relatively high uncertainty (Rutgers and Jensen 2011). We define three stages: scoping, screening, and definitive, along with an outcomes assessment (Figure 17.1). The assessment stages are defined and named differently by different organizations. For example, the USEPA has eight stages (Sprenger and Charters 1997). However, the information requirements and assessment results are the same. Before each stage, a problem formulation should be derived to guide that stage of the assessment.

BOX 17.1 STAGES VERSUS TIERS

Stages (also called phases) each have a different purpose (e.g., scoping the site, screening contaminants, deriving the remedial goals). A stage may have multiple tiers.

Tiers are successive iterations of an assessment stage that try to accomplish a task such as listing contaminants of concern. Each successive tier uses more types and more complex types of studies and assessment tools. The evidence from each tier is weighed along with evidence from preceding tiers to determine whether the stage is complete or another tier is required. A stage may consist of multiple tiers of evidence triads (Jensen and Mesman 2006; Rutgers and Jensen 2011).

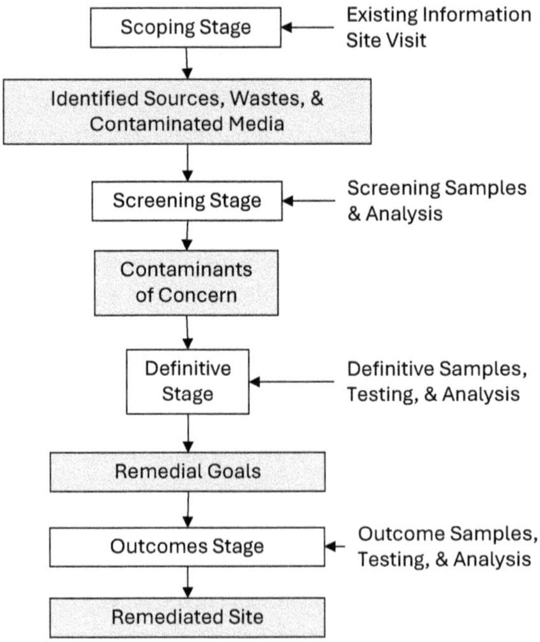

FIGURE 17.1 Stages in the process of generating evidence and assessing contaminated sites. Gray boxes are results of the preceding assessment stage.

17.2.1 Scoping Stage

As the name implies, the scoping stage of an assessment determines its scope, including site history, the location and spatial extent of contamination, the nature of the contaminants, and any signs of effects. It may lead to a conclusion that there is no sign of significant contamination or that the contamination is so limited that remediation should be planned and carried out without further assessment. More often, it leads to a screening assessment.

A scoping stage determines the nature of the contamination (e.g., leaking barrels of oily material) and the areas that are part of the site. Scoping of a large and complex site may define units that may have common characteristics that could lead to a common remediation process. These operable units may include areas where waste was deposited, areas that are contaminated by transport, and land or water bodies that may become contaminated in the future. Information gathering for a scoping phase includes a review of records concerning activities on the site, such as the use of materials that are potential site contaminants. It also includes observational information such as spilled oil, leaking drums or tanks, dead vegetation, discolored soil, odors, floodplains, and flow patterns. It may include preliminary sampling and analysis of possibly contaminated media.

In some cases, scoping may be skipped because the contaminant of concern and its source are defined a priori. For example, an assessment of organotin contamination from ships was performed in specified Dutch harbors (Schipper et al. 2008). Because the contaminant and sites are specified, the evidence was limited. Types of evidence included organotin concentrations and organotin-specific effects: intersex index in gastropods observed in the field and predicted from organotin concentrations, gastropod presence, and a species sensitivity distribution.

Scoping for contaminated waters includes the identification of point sources, groundwater contributions, and overland flows. As with soils, evidence of contaminated waters may be identified by observing signs of contamination and potential sources, and by preliminary sampling and analysis. Observations may include waste deposits near the water body, oil sheens, colors, pipes, ditches, and erosional channels. For example, in a helicopter survey of the upper Potomac River, I observed piles of chicken manure in the floodplain.

Scoping for contaminated sediments provides fewer opportunities for observational evidence. It usually involves sampling and analysis to determine the spatial scope of contaminated sediment.

These multiple pieces of evidence may be weighed, but typically any one of them is sufficient to make a location within scope if it is relevant and reliable. A preliminary conceptual site model should be created showing the locations of contaminants and their sources relative to landscape features and built structures. The apparently contaminated locations are then passed to the screening stage.

The scoping stage may also be appropriate for defining assessment endpoints. Environment Canada (Environment Canada 2012) recommends identifying receptors of concern (equivalent to endpoint entities) based largely on site characteristics, and then adding attributes of those receptors to complete the assessment endpoints. The receptors of concern can be selected based on the weight of evidence for their characteristics (Table 17.1). The attributes for the receptors are generally conventional

TABLE 17.1
Characteristics of Receptors of Concern

Characteristics of Receptors of Concern	Description
Ecological relevance	A species or community that exists on the site or is expected to occur after remediation and which influences other species or ecosystem properties is ecologically relevant.
Relevance to management goals	Appropriate receptors relate to regulatory mandates (e.g., fishable waters) and to proposed land or water uses.
Degree of exposure to contaminants on the site	Behaviors and diet determine the degree to which a receptor is exposed to contaminants.
Sensitivity	Some taxa are more sensitive than others to exposure to the site contaminants.
Availability of information	Taxa or functional groups that are well characterized ecologically and toxicologically are more appropriate.
Likely to bioaccumulate	Chemicals that are persistent and bioaccumulative are of concern.

Source: Adapted from Environment Canada (2012) and USEPA (1998).

(USEPA 2016a). For example, fish in a contaminated water body have strong evidence for the degree of exposure.

17.2.2 Screening Stage

Screening-stage assessments of contaminated sites are primarily devoted to determining whether the contamination identified during scoping contains individual contaminants at hazardous levels. These are known as contaminants of concern. This is conventionally done by analyzing the media and comparing the concentrations or doses to screening benchmark values (Section 15.1.5). Screening benchmarks include soil screening values (USEPA 2005), Canadian Environmental Quality Guidelines (CCME 1999), and the ORNL values (Sample et al. 1998). Screening benchmarks may be developed for the constituents of particular wastes, such as coal ash, and particular sites (Sample and Suter 2002). Regulatory benchmarks may also be used for screening, particularly if they are somewhat precautionary. Precaution is important because you do not want to screen out a contaminant that is actually a significant hazard.

Although contaminants of concern are most commonly determined by comparison of site concentrations to benchmark values, other contaminant characteristics should be considered and weighted to determine which make the list (Table 17.2). These are necessary because risk is not determined by toxicity alone.

17.2.3 Definitive Stage, Assembling Evidence

The definitive stage of an assessment determines whether remediation is required and what remedial goals should be set for contaminants of concern. It may consider both current risks and potential future risks.

TABLE 17.2
Characteristics of a Contaminant of Concern

Characteristics of Contaminants of Concern	Description
Associated with the waste or source	Chemicals that are constituent of the waste or effluent may be contaminants of concern, even if the screening survey does not find them at hazardous levels. The sampling may not be adequate or may not have included all media. Chemicals that are detected but not associated with the waste or effluent of interest could be from an unknown source on the site, an extraneous local source, or regional contamination.
Frequently detected	Chemicals that are infrequently detected at the site are of lesser concern and might be artifacts.
Greater than local background	Chemicals at background levels are not of concern, unless the anthropogenic form is more bioavailable or toxic.
At potentially toxic levels	Concentrations above relevant ecotoxicological benchmarks are of concern.
Analytical detection limits are inadequate	Chemicals that have detection limits at toxic levels are of concern if they are associated with the waste or source.
Belong to the same class as an identified contaminant of concern	A chemical that is not of concern alone might belong to a class, such as polycyclic aromatic hydrocarbons, that could be assessed jointly.
Likely to bioaccumulate	Chemicals that are persistent and bioaccumulative are of concern.

Source: Adapted from Black and Veatch Special Projects Corps (2011), Suter et al. (2000), and USEPA (2016b).

In simple and uncontroversial cases, such as an industrial site, contaminant concentrations are compared to a benchmark value used as a remedial goal. This is similar to the screening stage, but greater confidence may be achieved with more sampling, better characterization of bioavailable concentrations, toxicity testing of site media, or better definition of future land or water body uses. In the U.S. Superfund program, the regulations specify remedial goals (cleanup levels) based on "legally applicable or relevant and appropriate regulatory standards" or on "other peer-reviewed toxicity standards." For ecological endpoints, there are available standards for some chemicals in water (national water quality criteria) but not soil or sediment. This leaves assessors needing site-specific, peer-reviewed toxicity benchmarks to serve as remedial goals or the use of cleanup goals based on weight of evidence for qualities (e.g., degraded vegetation) or quantities (Chapter 12).

Site-specific definitive assessments of contaminated sites determine what effects are being caused or may be caused in the future by contaminants and determine what remedial goals should be met. Weight of evidence for this purpose is a relatively highly developed assessment practice.

Sites that are not simple are typically assessed for ecological risks by weighting multiple types of evidence to achieve confidence in the results. One common basis for assessing contaminated sites is the triad, first presented by Long and Chapman (1985) for assessing contaminated sediments (Section 2.7). The triad consists of three types of evidence that may be used to infer whether contaminants are impairing sediment biota. They are also frequently adapted for contaminated soil or water.

1. **Single Chemicals**: Chemical analysis of the sediment can be compared to sediment toxicity benchmarks to determine whether any chemical is present at levels that can cause effects on benthic biota. Because of the physical-chemical complexity of sediments, the analyses may be performed on whole sediments, sediment pore water, or sieved sediment, and they may be corrected for properties that affect toxicity such as pH, organic matter content, or acid volatile sulfides (Lee and Jones-Lee 2004). If relevant toxicity test data or sediment standards are unavailable, the chemical concentrations that are elevated relative to reference sediments can be identified.
2. **Media Toxicity**: Toxicity tests of site sediments can determine whether the mixture of contaminants in a sediment is toxic to standard test species or site-relevant species. The tests are nearly always performed on sediments brought to the laboratory, but in situ tests are also possible. An example is transplanting bivalve mollusks to contaminated sites (Ellis et al. 2002). These tests are more relevant to site conditions than the generic single-chemical tests used in triad type 1 evidence, and they include combined toxic effects. If contaminants are analyzed, these tests can be used to determine combined exposure-response relationships across a site.
3. **Biological Surveys:** Surveys of species composition and abundances of the benthic communities can be performed on contaminated and uncontaminated sites. These surveys can determine the range of effects of realistic exposures on numerous species. They can be used to develop combined exposure-response relationships for populations or communities.

   More recently, other types of evidence have been proposed to make a tetrad, pentad, or even a hexad (Chapman and Hollert 2006; Simpson et al. 2005; Ingersoll and Macdonald 2002; Wenning et al. 2005; Cecchetto et al. 2025). Any useful evidence may be added to the triad, but we particularly recommend:
4. **Body Burdens**: Chemical analyses of organisms collected on-site can provide evidence of internal exposure, which supports causation, and evidence for assessing food-web effects (effects from consumption by humans or other animals).
5. **Pathologies**: Overt signs of damage to organisms can provide evidence of contaminant effects, particularly if they are characteristic of a contaminant. They may be histopathologies or overt pathologies such as skeletal deformities. A standard set for fish is *DELT abnormalities*—deformities, fin erosion, lesions, and tumors.
6. **Media Properties:** Water, sediment, and soil properties such as pH, texture, or even odor can indicate whether the samples contain waste materials and their potential forms.

All of these types of evidence depend on the availability of good reference sites or a clear gradient. Reference sites should be as similar as possible to the contaminated site, except for the presence of the contaminants and their effects. This can be difficult because contaminated sites often are physically disturbed by the activities that resulted in contamination. It is also desirable to use the same statistical techniques to compare the contaminated and reference sites for each type of evidence. Gradients may be formed by dilution away from a source. They may provide more information concerning the relationship of exposure to effects than a comparison to reference. The generation of useful site-specific environmental information is discussed in Part 2B of Norton, Cormier, and Suter (2014).

The triad and quadrad evidence sets have been weighed using logic tables which determine the outcome based on the assumption that each type of evidence is highly relevant and reliable (Section 2.7). That is, each type of evidence is rated as supporting or not (+ or–), and the conclusions are that contaminants are causing effects or not (Ingersoll and Macdonald 2002; Chapman 1990). We believe that it is better to weight the evidence so as to allow for less than complete relevance and reliability and for the strength of evidence. Ingersoll and Macdonald (2002) provide good guidance on data generation and analysis for triads.

Assessments of contaminated soils sometimes use a weight of evidence based on the sediment triad approach (Jensen and Mesman 2006; Ellis et al. 2002; Swartjes et al. 2012; Expert Group 2025; ISO 2017; Gworek et al. 2018). A soil triad can usually be implemented more easily than a sediment triad. You can scope the site by walking it, and you can often visually determine where to sample and how deeply. You can see soil discoloration and differences in texture, or note where there is no vegetation or depauperate vegetation. You can then sample the contaminated soils for analysis and testing. You can plant native plants in situ and monitor their survival and growth. If the contaminated soils are not apparent, you can sample soils on a grid or gradient like sediments. Assessment of contaminated soils has been more advanced in the Netherlands than in the U.S., and they have led soil protection methods in the EU (Swartjes et al. 2012; Swartjes 2011). In Italy, the ERAMANIA decision support system provides software to implement a soil quality triad (Semenzin et al. 2009). It provides numerical scores for soil toxicity tests, bioavailability metrics, and ecological endpoints, and it combines them into integrated ecological risk indexes.

Assessments of contaminated waters in the U.S. use a triad, even though the three evidence types were not developed for a weight-of-evidence inference. Water analysis and ambient water quality criteria for the U.S. were developed soon after the formation of the USEPA, using a narrative weight of evidence for each chemical (USEPA 1976). Short-term acute and chronic toxicity tests were developed for testing effluents for permitting purposes and were adapted for testing ambient waters to determine whether impairment is occurring (USEPA 2024b). Biological surveys are used to determine whether biocriteria are being met (USEPA 2024a). These triad-equivalent methods differ in that they are not integrated into a weight of evidence under the Clean Water Act, but rather are used as three different ways of identifying impairment (i.e., independent applicability). However, they can be used in a weight of evidence for contaminated waters under Superfund.

Aquatic evidence triads are also used in the European Union. A decision support system called MODELKEY has been developed to rate and integrate evidence to implement the European Union's Water Framework Directive. It has considerable flexibility, including the integration of decision-analysis-based indices or fuzzy logic (Gottardo et al. 2011; Gottardo et al. 2008). Another European system for integrating types of aquatic field evidence with other evidence types is provided by Rico et al. (2025). It is distinguished by an emphasis on inference to explain conflicting results from different evidence. As we write this, new policies from the European Union for implementing the Water Framework Directive call for effects-based monitoring to determine whether contaminant mixtures are causing harm (EU 2025). Where necessary, nations must restore damaged aquatic ecosystems identified by evidence-based monitoring, which requires multiple types of evidence (Neale et al. 2022).

The Canadian framework for aquatic contaminated sites employs a formal weight-of-evidence approach with weighting tables (Environment Canada 2021). However, rather than the usual triad, they rate bulk chemistry, toxicity endpoints, overall toxicity, benthos alteration, and biomagnification as having low, medium, or high weight.

Multiple approaches to integrating the evidence triad have been used to infer that the contaminants are causing effects and that the site should be remediated. They include narratives, triangular plots, logic tables (Section 2.2), histograms, and weight-of-evidence tables (e.g., Table 7.5).

A tradition of using a weight-of-evidence framework for ecological risk assessment of contaminated sites was developed independently of the triad (Suter and Loar 1992; Suter et al. 2000; Menzie et al. 1996). These approaches are based on a risk assessment framework featuring a problem formulation and adapting the weight-of-evidence method to that formulation of the assessment problem (USEPA 1998).

17.2.4 Definitive Stage, Weighing the Evidence

The evidence should be weighted and weighed with respect to the properties of evidence and bodies of evidence (Chapter 7). The three standard properties of evidence (relevance, reliability, and strength—Section 7.1) and their subproperties are appropriate for weighting contaminated site evidence. However, other properties may be included. In particular, when determining whether the site should be remediated, the assessments could consider site-specific properties such as:

Temporal Scope and Variance: A type of evidence may be weightier if it is indicative of long-term contamination and if the data include the relevant range of temporal variance in conditions that raise risks, such as floods and storm events that flush contaminants into a stream or wetland. The risk may change over time if the remediation is incomplete. For example, succession may change the site to a forest and increase biodiversity. A contaminated groundwater plume may spread to intersect surface water. Also, biological surveys are more relevant if they consider the seasonality of the biota.

Spatial Scope: A type of evidence may be weightier if the data encompass the directly and indirectly contaminated areas and the ecosystem types within those areas.

An example of a site-specific approach is provided by an assessment of contamination of the Ottawa River below the Chalk River Laboratories (Bond et al. 2015). They not only developed site-specific properties (called criteria) but also defined high, minor, and negligible ratings and symbols for each of six types of evidence and for the integrated evidence. A similar Canadian case is provided by an assessment of a coal spill in Silver Creek (Trowell et al. 2020). They used seven types of evidence and provided definitions for each evidence type, for high, moderate, or low ratings, and for the integrated evidence. For example, high sediment pore water toxicity is defined as an exposure concentration greater than an LC50 or EC50.

Methods for weighting and then integrating the weighted pieces or types of evidence are presented in Chapters 7 and 8.

17.2.5 Definitive Stage, Cumulative Risks

Contaminated sites are often contaminated by multiple chemicals. Tests of the contaminated media and biological surveys are inherently integrative of multiple contaminants. However, the first leg of the triad includes a comparison of concentrations of the constituents of the mixture to the results of toxicity tests of the constituents. The toxicities of the constituents are combined using an effects-additive model, a concentration- or dose-additive model, or an interactive model. Alternatively, some contaminant mixtures can be treated as a single contaminant. For example, at oil spill sites, total petroleum hydrocarbons can be measured, and that exposure metric can be compared to test results for petroleum hydrocarbons (Kuppusamy et al. 2020). The combination of weighting and weighing the triad of evidence and the various approaches to analyzing, testing, and modeling multiple contaminants requires planning and executing a particularly complex weight of evidence (USEPA 2025a).

17.2.6 Definitive Stage, Inference to Results

A contaminated site assessment based on weighing evidence has three potential types of results, plus uncertainties and confidence in the results.

- Qualitative results that determine what areas and media, if any, require remediation
- Quantitative results that determine what concentration or other remedial goal should be achieved
- Qualitative results that determine the degree of remediation and post-remediation use
- Uncertainties and confidence in the results

The qualitative results come from weighing the results of the evidence triad or some other set of evidence. If the weight of the body of evidence for impairment due to contamination of a medium is sufficient, it will prompt remediation. Some cases are simple; if the contamination is a terrestrial petroleum spill, the area of the spill is remediated. In other cases, you should weigh the evidence to determine which piece or type of evidence is the best evidence for determining what to remediate. These weights should come from the weighing of evidence assembled in the definitive stage. The steps for determining whether remediation is required with respect to ecological risks are as follows:

1. Categorize the pieces of evidence in terms of types, such as the triad types. For example, you may have tests of contaminated soil for multiple plant or invertebrate species or ecosystem processes. The type is contaminated soil tests, and the subtypes are plant, invertebrate, and process.
2. Weight the pieces of evidence in terms of their properties. For example, tests of native plants are more relevant, tests that show a severe effect such as death are strong, and tests that are well designed and conducted are reliable. Some weights are specific to contaminated sites. For example, if one type of evidence identifies significant contamination in a larger area than other evidence, that sensitivity would be an important expression of the strength of the evidence.
3. Integrate the pieces of evidence to determine the weights of evidence for each type of evidence.
4. Weigh the body of evidence to determine whether the site should be remediated. Consider the distribution of positive and negative ratings in the weight-of-evidence tables, and logically interpret the evidence, as in Table 2.2.

An example of the derivation of a qualitative result is provided by an assessment of risks to mink and short-tailed shrews from exposure to PCBs, dioxins, and furans in the floodplain of the Housatonic River (Moore et al. 2015). The assessment used an adaptation of the method of Menzie et al. (1996). The types of evidence were the results of exposure and effects modeling, field surveys, and in situ or whole-medium toxicity tests. Weights were assigned based on three properties:

1. strength of association between exposure and effects endpoints,
2. data and study quality, and
3. study design and execution.

This was followed by an integration of evidence to determine if there is a risk, and if there was, whether effects were low or high. Finally, the concurrence of evidence was determined by examining a weight-of-evidence table. The weight of evidence indicated that the risks to mink were high and to shrews were intermediate.

If remediation is required, quantitative remedial goals may be obtained by either of two approaches: best value or combined values (Chapter 12) (Figure 17.1).

First, the best remedial goal may be chosen based on the relevance and reliability of the evidence for each quantitative result. You may use:

- a regulatory benchmark concentration of a contaminant,
- a concentration at which threshold toxicity occurs in tests of the contaminated medium,
- a threshold concentration at which effects occur in the biotic community.
- a threshold tissue concentration in the biota at which effects occur, or
- concentrations at which overt pathology is observed.

The evidence used to make the qualitative judgment for remediation may also be used to determine the best remedial value. However, the weights for a potential remedial goal may be different.

Second, the effects concentrations for pieces of evidence of the same type can be statistically combined (Section 12.3). For example, a remedial goal (rehabilitation value) for manganese was obtained by averaging values from types of evidence for which there was sufficient confidence (Box 17.2). Averaging different types is appropriate if they are all estimates of the same assessment endpoint. However, combining values for different endpoint effects would result in remedial goals that are benchmarks for no particular endpoint. You are better off if you combine values of the same type, and then choose the best type. However, it is important to interpret the values within a type. For example, plant growth tests of contaminated soil may include multiple plant species. You could average them, but that implies protecting roughly the upper half of the species. It may be more defensible to choose the value from the most sensitive species.

BOX 17.2 REHABILITATION STANDARDS BY WEIGHT-OF-EVIDENCE

Rehabilitation standards for the Ranger uranium mine were derived by the Supervising Scientist to protect aquatic ecosystems exposed to mine leachate and runoff (Humphrey and Chandler 2018; Supervising Scientist 2021). The method was based on the weight-of-evidence approach in the Australian and New Zealand guidelines for fresh and marine water quality (Van Dam et al. 2014; ANZECC 2000; ANZG 2018b). The standards quantify the rehabilitation goals and provide specific values for magnesium.

The types of evidence include laboratory toxicity tests of local aquatic species, local field studies of macroinvertebrates, and mesocosms. Candidate guideline values were derived for each type of evidence, all of which relate to the composition of the macroinvertebrate community (the endpoint). "The magnesium rehabilitation standard was determined as the mean of the lowest available candidate guideline values for which there was deemed to be sufficient confidence (i.e., weight), for each of the laboratory (25 mg/L), mesocosm (2.6 mg/L), and field (3.9 mg/L) studies. This value is 2.9 mg/L" (Supervising Scientist 2021). Other types of evidence, such as toxicity tests of ambient waters, were used as supporting information when weighing the body of evidence. Confidence for field studies was evaluated by weighing evidence for causation and confounding (Cormier et al. 2013; Suter and Cormier 2013).

The degree of remediation is often expressed as a concentration scale (i.e., potential remedial goals). It may be simply a single benchmark value such as: all soil with lead concentrations greater than *X* will be remediated. However, it is often useful to define a multi-value scale, such as the one in Figure 17.2, which defines actions to be taken. Remediation may be determined by a qualitative assessment for alternative technologies such as soil removal, bioremediation, capping, or pump and treat.

In the Netherlands, a different scale relates soil concentrations to land uses such as "always applicable" (background), "applicable to residences" (slightly contaminated), "applicable to industry" (somewhat contaminated), or "never applicable" (severely contaminated) (Swartjes et al. 2012). Depending on the case, other use categories may be defined, such as agriculture or nature reserve. This approach is termed fitness for use. Land use-specific benchmark values are derived in three steps (Swartjes et al. 2012). First, land use-specific requirements are derived for each land use and endpoint (e.g., plant production for agriculture). Second, qualitative endpoints are derived that meet those requirements (e.g., plant growth). Third, quantitative benchmark concentrations are derived for each endpoint (e.g., EC_{10} for growth rate). Standard benchmark values may be used, best values may be derived, or equivalent values may be combined. Interpretations of the evidence and the choice of values should be appropriate for the remedial land use goals.

Another land-use-focused approach from the Netherlands assigns categorical weights to 39 specific types of evidence for each of six land uses (Jensen and Mesman 2006). For example, an earthworm growth and reproduction test gets an applicability rating of high for nature and agriculture; reasonable for parks, garden-food, and garden; and moderate for industry.

In the U.S., a concept similar to land use, designated use, is used to determine site-specific water quality standards given the attainability of the uses. The potential uses include aquatic life, consumption of aquatic life, recreation, public water supply, agriculture, industry, and navigation. Evidence for use attainability of an aquatic site can be gathered and weighed, but the consensus of stakeholders is emphasized by the USEPA.

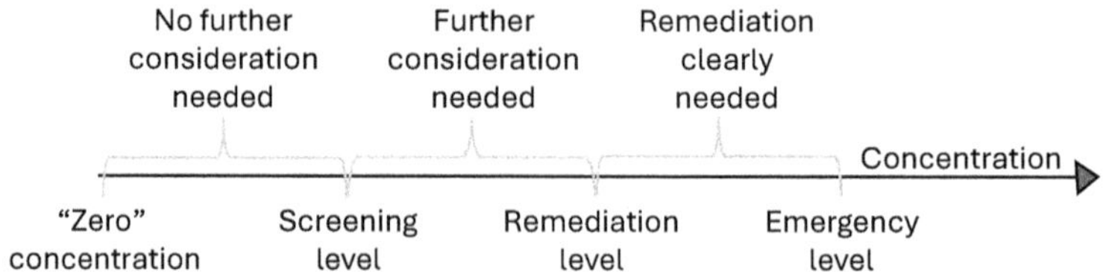

FIGURE 17.2 A scale of assessment and remedial actions based on a trinary logic (Section 8.2.3). If no further consideration is needed, or if remediation is clearly needed, the assessment is complete for the case. If further consideration is needed, more data should be obtained or more analysis should be performed until the contaminant levels no longer fall in the further consideration zone. The categories may be bounded by contaminant concentrations as suggested in the diagram, or may be qualitative categories (e.g., remediation clearly required) based on a qualitative weight of evidence.

17.2.7 Risk-Based Choice of a Remedial Action

Most risk assessments for contaminated sites determine whether remediation is needed and what goals the remedial action should achieve. In Superfund, this is called a remedial investigation. The choice of remedial technology and action plan is left to an engineering-focused assessment called a feasibility study. However, in some cases, these assessments are combined. For example, when planning the remediation of contaminated sediments, the U.S. Army Corps of Engineers weighs evidence related to risk along with issues of practicality to select the remedial action (Bates, Massey, and Wood 2018). The types of evidence are sediment type analysis, contaminant analysis through bioassay, cost, environmental effects, historical management of dredged material, site availability, and socio-political concerns. The properties to be weighted are relevance, quality, and resolution.

17.3 POST-RISK ASSESSMENTS

In some cases, the determination of whether to remediate a contaminated site includes more than the nature and magnitude of risk to human health and the environment. At a contaminated site, multiple issues are raised that can be addressed by assessments based on a single weight-of-evidence approach (Box 17.3).

BOX 17.3 THE CASE OF VENICE'S CONTAMINATED SEDIMENTS

The sediments of Venice's lagoon and canals are contaminated by industry, municipal wastes, runoff, and other sources. Weight of evidence has been used in assessments of the toxicological risks of potentially dredged sediments, limitations on potential beneficial uses of dredged sediments, and hazards to clam farming (Cecchetto et al. 2023; Cecchetto et al. 2025; d'Errico et al. 2021). Each type of assessment weighted multiple types of evidence: sediment chemistry, bioaccumulation, bioassays, biomarkers, and transcriptomics. The evidence of hazard associated with each type was weighted for each location as absent, slight, moderate, major, or severe, with an associated color and 0–100 rating. The weights are based largely on the strength of the difference from reference sites. The numerical ratings along with expert interpretations were used to integrate the evidence.

17.3.1 Net Benefit Analysis

The benefits of remediation should exceed the damage done by the remediation. It is common for remedial actions to involve destroying or damaging valued ecosystems. This issue came to a head when, following the Exxon Valdez oil spill, the rocky shoreline of Prince William Sound was pressure washed to remove the petroleum coating, severely damaging the intertidal community. Similar issues can occur on land. Much of the mature floodplain hardwood forest along the East Fork of Poplar

Creek in Oak Ridge, Tennessee, was bulldozed to remove mercury-contaminated soil despite the absence of toxicity or biological effects from the mercury.

Net benefit analysis could compare the health benefits of remediation to the risks of injury from accidents among workmen and the public. Exposure might even increase if remediation mobilizes contaminants. I remember a public meeting where there were objections to risks from truckloads of contaminated soil passing through a residential community.

Net benefit analysis requires estimating benefits and damages in the same terms so that they are comparable (Efroymson, Nicollette, and Suter 2004). Multiple types of evidence may be available for both costs and benefits, and the evidence may be weighed as it is for a conventional risk assessment. In fact, the body of evidence for the risk can provide the basis for estimating the benefits of remediation.

We describe net benefit analysis in more detail in Section 21.4. That is because there are more reasons to assess the net benefits of resource management than the remediation of contaminants.

17.3.2 Cost-Benefit Analysis

Cost-benefit analysis assesses remedial alternatives in monetary terms by comparing the benefits of remediation to the costs. This analysis is important for many contaminated sites because of the high costs of site remediations and the incompletely monetized benefits (USEPA 2025b).

Estimating the costs of performing the remediation is relatively simple (USEPA 2000). The costs of labor, equipment, waste disposal, and other factors all have monetary market values.

Benefits to human health and the economy are more difficult to monetize than the costs of remediation. Human health and lives are monetized based on standard values. Economic benefits are also well studied and quantified. For example, real estate values go up when nearby contaminated sites are remediated (Timmins 2011). Monetizing environmental conditions is even more difficult, and cost-benefit analysts often default to estimating the human health and economic benefits of enhanced ecological conditions (USEPA 2025b). Analyses of the services of nature can provide a more complete valuation of ecological remediation by weighing evidence (USEPA 2016a; Harwell et al. 2021). Monetization of ecological benefits needs much more research, but non-monetary qualitative and quantitative analysis of benefits can contribute to decision-making. Weight of evidence is important because of the numerous potential benefits and the multiple types of evidence of benefits.

17.3.3 Natural Resource Damage Assessment

Effects of contamination on natural resources are assessed by the resource trustees. If natural resource damages are identified, responsible parties must compensate the trustees monetarily for the cost of resource restoration at the contaminated site or for enhancing resources elsewhere. Damage assessments are a type of condition assessment, and they may include a causal assessment to demonstrate that the damage is caused by the waste or spill. They differ in that they include restoration planning.

Ecological damages result from the loss of ecosystem services relative to the prior or background conditions. Weight-of-evidence methods are used to integrate evidence concerning ecological injury and damage to services.

17.4 ECOLOGICAL VERSUS HUMAN HEALTH SITE ASSESSMENTS

Human health assessments of contaminated sites are less likely to weigh multiple types of evidence than ecological assessments. Contaminated sites seldom expose enough humans to support epidemiology. When a cancer cluster or birth defects cluster is identified, conventional hypothesis-testing statistics are usually applied, and the null hypothesis is not disproven. The consultants who apply this approach for responsible parties are known as cluster busters. Human tests of contaminated site media would be clearly unacceptable, and standard rodent tests are seldom accepted because of the lack of standard protocols for testing media. That leaves one type of evidence: a comparison of estimated site dosages to standard benchmark values, such as IRIS benchmark doses. Hence, the only data gathered at most sites for human health risks are concentrations in soil and possibly water, fish, and wildlife.

In my (Suter's) experience with contaminated site assessments, human health assessors are more methodologically conservative than ecological assessors. On the Oak Ridge Reservation, the USEPA assessors would not accept mercury concentrations from test-garden vegetables for estimating human doses; they used their standard soil-to-plant uptake factors. However, the USEPA ecological assessors accepted plant concentrations from the garden and forest. Similarly, the USEPA would not accept mercury concentrations in human hair as evidence of exposure, but tissue concentrations in fish, birds, mammals, and invertebrates were used in the ecological assessments. Finally, rat chronic tests of mercuric sulfide, the dominant mercury species on the site, showed no effects at far higher doses than maximum human estimated exposures. That test was rejected because it did not have an exposure level that showed effects. Instead, the USEPA used a standard benchmark value for the more toxic mercuric chloride. As a result, one type of low-relevance evidence was used for the human health assessment, while a weight of evidence with multiple types of evidence and multiple taxa was used in the ecological assessment. Because this pattern is common among Superfund assessments in the United States, responsible parties often complain about the higher costs of ecological assessments.

17.5 OUTCOMES OF SITE REMEDIATION

Assessments of the success of remedial actions are common examples of outcome assessments (Figure 4.1). In the U.S., the success of remediation of Superfund sites is reviewed after five years. Remediation is judged in terms of the performance and effectiveness of the intervention (Table 17.3), and evidence for many characteristics constitutes a weighty body of evidence. Because human health risks are nearly always assessed based on contaminant concentrations, outcomes are judged based on the performance of the intervention. Because ecological risks may be based on evidence of exposure and toxicity, success can and should be based on effectiveness in terms of reduced exposure and effects. Site-specific guideline values for

TABLE 17.3
Potential Characteristics of Remediation (USEPA 2016b)

Characteristic of Remediation	Description
	Performance of the Intervention
Functions	The remedial technology functioned as intended.
Reduces the agent of concern	The technology substantially reduces the level and extent of the contaminant or other agent that causes the risk.
Achieves the physical/chemical goal	The levels of the causal agent are reduced sufficiently to achieve the goal.
	Effectiveness of the Intervention
Endpoint entities less exposed	The exposure of the environmental endpoint entity to the agent and their interaction are reduced as predicted.
Endpoint entities improved	The abundance, diversity, function, or other attributes of the endpoint entity improve within the expected time following the remediation.
Endpoint goals achieved	Reference conditions or other ecological endpoint goals are achieved in response to remediation.

chemical, biological, and physical indicators of remediation may be derived from the weight-of-evidence-based remedial goals or may be developed specifically for remediation (Water Quality Australia 2020). To determine whether the remediation actually succeeded, it is important to obtain sufficient evidence to determine whether ecological goals are met due to the remedial action and not just due to natural recovery.

An example of a weight-of-evidence-based outcome assessment is provided by the aquatic effects monitoring and assessment plan for potential aquatic effects of a closed diamond mine at Snap Lake, Alberta (De Beers 2022). The types of evidence from the monitoring program are: water quality (nutrients and chemical contaminants); chronic toxicity at the edge of the effluent mixing zone; sediment quality; fish tissue chemistry; plankton (phytoplankton and zooplankton communities); benthic invertebrates; and fish health. The endpoints from those types were weighted for the strength of support for two hypotheses: nutrient enrichment and toxicological impairment. A rating system represented by a 0 for no response and 1, 2, or 3 up or down arrows representing the magnitude and direction of response (i.e., strength of evidence) was used. In addition to the strength of responses, the pieces of evidence for the endpoints were rated a priori for their inherent strength and relevance. After the evidence was generated, the endpoints were rated for consistency and linkage to effluent releases. Finally, the bodies of evidence for each hypothesis will be integrated qualitatively using a scale of 0 (no support), 1 (weak support), 2 (moderate support), or 3 (strong support).

Another example of an outcomes assessment is provided by a study of spilled mine tailings in Brazil's Rio Doce basin (Abessa et al. 2024). The weight of evidence for the toxicity of metals in whole tailings and pore water was determined six years after the spill. Concentrations were compared to sediment quality and surface water guidelines and literature values. Sediment tests were performed with midges (*Chironomus sancticaroli*) and aqueous tests with water fleas (*Daphnia similis*). Results were compared to tests performed six months after the spill, showing that toxicity was greatly reduced. Although total concentrations of iron and manganese were elevated, no toxicity was reported, apparently due to binding to oxyhydroxides and particles. The method of weighing the pieces of evidence was narrative.

REFERENCES

Abessa, D., G. A. Burton, E. C. Cervi, S. L. Simpson, W. Stubblefield, C. C. Ribeiro, A. C. F. Cruz, G. Kruger, and R. Smith. 2024. "Has the Rio Doce "Time Bomb" been defused? Using a weight-of-evidence approach to determine sediment quality." *Integrated Environmental Assessment and Management* 20 (1): 148–158.

ANZECC. 2000. *Australian and New Zealand Guidelines for Fresh and Marine Water Quality. Vol. 1. The Guidelines*. Canberra: Australia and New Zealand Environment and Conservation Council and Agriculture and Resource Management Council or Austrialia and New Zealand.

ANZG. 2018a. "Australian and New Zealand guidelines for fresh and marine water quality." Australian and New Zealand Governments. www.waterquality.gov.au/anz-guidelines.

ANZG. 2018b. *Weight of Evidence*. Canberra: Australian and New Zealand Governments.

Baker McKenzie. 2024. *Contaminated Land 2024*. London, UK: The Baker McKenzie Global Environmental Group.

Bates, M. E., O. C. Massey, and M. D. Wood. 2018. *Weight-of Evidence Concepts: Introduction and Application to Sediment Management*. Washington, DC: U.S. Army Corps of Engineers.

Bizzotto, E. C., E. Semenzin, E. Giubilato, S. Frisario, L. Zaninetta, G. Bonfedi, F. Villani, and A. Marcomini. 2022. "Ecological risk assessment for contaminated sites in Italy: Guidelines and path forward." *Integrated Environmental Assessment and Management* 19 (4): 913–919.

Black & Veatch Special Projects Corps. 2011. *Ecological Risk Assessment for the Estuary at the LCP Chemical Site in Brunswick, Georgia: Site Investigation/Analysis and Risk Characterization (Revision 4)*. Atlanta, GA: U.S. Environmental Protection Agency, Region 4.

Bond, M. J., R. Silke, M. Stuart, J. Carr, and D. J. Rowan. 2015. "A weight-of-evidence approach to the assessment of ecological risk from historical contamination of Ottawa river sediments near Chalk River Laboratories." *CNL Nuclear Review* 4 (2): 155–170.

Burton, G. A., G. E. Batley, P. M. Chapman, V. E. Forbes, E. P. Smith, T. Reynoldson, C. E. Schlekat, P. J. den Besten, A. J. Bailer, A. S. Green, and R. L. Dwyer. 2002. "A weight-of-evidence framework for assessing sediment (or other) contamination: Improving certainty in the decision-making process." *Human and Ecological Risk Assessment* 8 (7): 1675–1696.

CCME. 1999. *Canadian Environmental Quality Guidelines*. Ottawa: Canadian Council of Ministers of the Environment.

Cecchetto, M., E. Giubilato, I. Bernardini, C. Bettiol, D. Asnicar, C. Bertolini, J. Fabrello, A. Bonetto, L. Peruzza, and M. Ciscato. 2023. "A weight of evidence approach to support the assessment of the quality of Manila clam farming sites in a coastal lagoon." *Marine Pollution Bulletin* 197: 115668.

Cecchetto, M., E. Giubilato, M. Picone, A. V. Ghirardini, C. Bettiol, F. Corami, I. Bernardini, M. Milan, T. Patarnello, and V. Matozzo. 2025. "An extended quantitative weight of evidence with uncertainty evaluation for the risk assessment of dredged sediment." *Environmental Sciences Europe* 37 (1): 169.

Chapman, P. M. 1990. "The sediment quality triad approach to determining pollution-induced degradation." *Science of the Total Environment* 97/98: 815–825.

Chapman, P. M., and H. Hollert. 2006. "Should the sediment quality triad become a tetrad, a pentad, or possibly even a hexad?" *Journal of Soils and Sediments* 6 (1): 4–8.

Chapman, P. M., B. G. McDonald, and G. S. Lawrence. 2002. "Weight-of-evidence issues and frameworks for sediment quality (and other) assessments." *Human and Ecological Risk Assessment* 8 (7): 1489–1515.

Cormier, S. M., G. W. Suter, L. Zheng, and G. J. Pond. 2013. "Assessing causation of the extirpation of stream macroinvertebrates by a mixture of ions." *Environmental Toxicology and Chemistry* 32 (2): 277–287.

d'Errico, G., A. Nardi, M. Benedetti, M. Mezzelani, D. Fattorini, M. Di Carlo, L. Pittura, M. Elisa Giuliani, S. Macchia, and V. Vitiello. 2021. "Application of a multidisciplinary weight of evidence approach as a tool for monitoring the ecological risk of dredging activities." *Frontiers in Marine Science* 8: 765256.

De Beers. 2022. *Snap Lake Mine Aquatic Effects Monitoring Program Design Plan for Closure V.1.2.* Alberta: De Beers Canada Inc.

Efroymson, R. A., J. P. Nicollette, and G. W. Suter. 2004. "A framework for net environmental benefit analysis for remediation or restoration of contaminated sites." *Environmental Management* 34: 315–331.

Ellis, J., V. Cummings, J. Hewitt, S. Thrush, and A. Norkko. 2002. "Determining effects of suspended sediment on condition of a suspension feeding bivalve (*Atrina zelandica*): Results of a survey, a laboratory experiment and a field transplant experiment." *Journal of Experimental Marine Biology and Ecology* 267 (2): 147–174.

Environment Canada. 2012. *Federal Contaminated Sites Action Plan (FCSAP) Ecological Risk Assessment Guidance*. Ottawa: Environment Canada.

Environment Canada. 2021. *Federal Contaminated Sites Action Plan (FCSAP), Framework for Addressing and Managing Aquatic Contaminated Sites under the Federal Contaminated Sites Action Plan (FCSAP), Version 2.1.* Ottawa: Environment Canada.

EU. 2025. *Water Pollution: Council and Parliament Reach Provisional Deal to Update Priority Substances in Surface and Ground Waters*. Brussels, Belgium: Press Office - General Secretariat of the Council of the European Union.

Expert Group. 2025. *Guidance on Best Available Techniques and Best Environmental Practices for Managing Sites Contaminated with Persistent Organic Pollutants (POPs).* Secretariate of the Stockholm Convention, Nairobi, Kenya: United Nations Environmental Program.

Forbes, V. E., and P. Calow. 2002. "Applying weight-of-evidence to retrospective ecological risk assessment when quantitative data are limited." *Human and Ecological Risk Assessment* 8: 1625–1639.

Gottardo, S., E. Semenzin, P. Agostini, A. Critto, and A. Marcomini. 2008. "Supporting tools for decisional process within Water Framework Directive: From EU context to MODELKEY perspective." In *Sustainable Use and Development of Watersheds*, edited by A. Vadineanu, I. E. Gonenc, J. P. Wolflin, and R. C. Russo, 215–225. New York, NY: Springer.

Gottardo, S., E. Semenzin, S. Giove, A. Zabeo, A. Critto, D. De Zwart, A. Ginebreda, and A. Marcomini. 2011. "Integrated risk assessment for WFD ecological status classification applied to Llobregat river basin (Spain). Part I—Fuzzy approach to aggregate biological indicators." *Science of the Total Environment* 409 (22): 4701–4712.

Gworek, B., A. H. Baczewska-Dąbrowska, R. Kalinowski, E. B. Górska, H. Rekosz-Burlaga, D. Gozdowski, I. Olejniczak, M. Graniewska, and W. Dmuchowski. 2018. "Ecological risk assessment for land contaminated by petrochemical industry." *PLoS One* 13 (10): e0204852.

Harwell, M. C., C. Jackson, M. Kravitz, K. Lynch, J. Tomasula, A. Neale, M. Mahoney, C. Pachon, K. Scheuermann, and G. Grissom. 2021. "Ecosystem services consideration in the remediation process for contaminated sites." *Journal of Environmental Management* 285: 112102.

Humphrey, C. L., and L. Chandler. 2018. *Use of Field-Effects Information to Derive a Surface Water Guideline Value for Magnesium in Magela Creek*. Darwin: Department of Agriculture, Water, and Environment.

Ingersoll, C. G., and D. D. Macdonald. 2002. *A Guidance Manual to Support the Assessment of Contaminated Sediments in Freshwater Ecosystems*. Chicago, IL: U. S. Environmental Protection Agency, Great Lakes National Program Office.

ISO. 2017. *Soil Quality--Procedure for Site-Specific Ecologicl Risk Assessment of Soil Contamination (Soil Quality Triad Approach)*. Geneva, Switzerland: International Organizatio for Standardization

Jensen, J., and M. Mesman. 2006. *Ecological Risk Assessment of Contaminated Land*. Bilthoven: National Institute of Public Health and the Environment (RIVM).

Kuppusamy, S., N. R. Maddela, M. Megharaj, and K. Venkateswarlu. 2020. *Total Petroleum Hydrocarbons: Environmental Fate, Toxicity, and Remediation*. New York, NY: Springer.

Lee, G. F., and A. Jones-Lee. 2004. "Appropriate use of chemical information in a best professional judgment triad weight-of-evidence evaluation of sediment quality." *Aquatic Ecosystem Health and Management* 7 (3): 351–356.

Long, E. R., and P. M. Chapman. 1985. "A sediment quality triad: measures of sediment contamination, toxicity and infaunal community composition in Puget Sound." *Marine Pollution Bulletin* 16: 405–415.

Marcomini, A., G. W. Suter, and A. Critto, eds. 2009. *Decision Support Systems for Risk-Based Management of Contaminated Sites*. New York: Springer.

McDonald, B. G., and P. M. Chapman. 2007. "Selenium effects: A weight-of-evidence approach." *Integrated Environmental Assessment and Management* 3: 129–136.

Menzie, C., M. H. Henning, J. Cura, K. Finkelstein, J. Gentile, J. Maughan, D. Mitchell, S. Petron, B. Potocki, S. Svirsky, and P. Tyler. 1996. "A weight-of-evidence approach for evaluating ecological risks: Report of the Massachusetts Weight-of-Evidence Work Group." *Human and Ecological Risk Assessment* 2 (2): 277–304.

Moore, D. R. J., R. L. Breton, T. R. DeLong, S. Ferson, J. P. Lortie, D. B. MacDonald, R. McGrath, A. Pawlisz, S. C. Svirsky, and R. S. Teed. 2015. "Ecological risk assessment for mink and short-tailed shrew exposed to PCBs, dioxins, and furans in the Housatonic River area." *Integrated Environmental Assessment and Management* 12 (1): 174–184.

Neale, P. A., B. I. Escher, M. L. De Baat, M. Dechesne, D. A. Deere, J. Enault, S. A. E. Kools, J.-F. Loret, P. W.M.H. Smeets, and F. D.L. Leusch. 2022. "Effect-based monitoring to integrate the mixture hazards of chemicals into water safety plans." *Journal of Water and Health* 20 (12): 1721–1732.

Norton, S. B., S. M. Cormier, and G. W. Suter. 2014. *Ecological Causal Assessment*. Boca Raton, FL: CRC Press.

Rico, A., U. Hommen, B. I. Escher, A. Koch, A. B.-N., B. González-Gaya, E. Cody, F. Sylvester, G. Treu, and G. Alurralde. 2025. "The use of diagnostic tools to assess the risks of chemicals to freshwater ecosystems: Towards a unified evaluation framework." *Environmental Management*: 75: 3433–3448.

Rutgers, M., and J. Jensen. 2011. "Site-specific ecological risk assessment." In *Dealing With Contaminated Sites*, edited by F. A. Swartjes. 693–720. Dordrecht: Springer.

Sample, B. E., and G. W. Suter. 2002. "Screening evaluation of the ecological risks to terrestrial wildlife associated with a coal ash disposal site." *Human and Ecological Risk Assessment* 8: 637–656.

Sample, B. E., G. W. Suter, R. A. Efroymson, and D. S. Jones. 1998. *A Guide to the ORNL Ecotoxicological Screening Benchmarks: Background, Development, and Application*. Oak Ridge, TN: Environmental Sciences Division, Oak Ridge Natioinal Laboratory.

Schipper, C. A., M. G. D. Smit, N. H. B. M. Kaag, and A. D. Vethaak. 2008. "A weight-of-evidence approach to assessing the ecological impact of organotin pollution in Dutch marine and brackish waters; combining risk prognosis and field monitoring using common periwinkles (*Littorina littorea*)." *Marine Environmental Research* 66 (2): 231–239.

Semenzin, E., A. Critto, M. Rutgers, and A. Marcomini. 2009. "DSS-ERAMANIA: Decision support system for site-specific ecological risk assessment of contaminated sites." In *Decision Support Systems for Risk-Based management of Contaminated Sites*, edited by A. Marcomini, G. W. Suter, and A. Critto. 205–237. New York: Springer.

Simpson, S. L., G. E. Batley, A. A. Chariton, J. L. Stauber, C. K. King, J. C. Chapman, R. V. Hyne, S. A. Gale, A. C. Roach, and W. A. Maher. 2005. *Handbook of Sediment Quality Assessment*. Bangor, NSW: CSIRO.

Sprenger, M. D., and D. W. Charters. 1997. *Ecological Risk Assessment Guidance for Superfund*. Edison, NJ: U.S. Environmental Protection Agency, Environmental Response Team.

Supervising Scientist. 2021. *Magnesium in Surface Water—Rehabilitation Standard for the Ranger Uranium Mine (Version 2)*. Darwin, NT: Supervising Scientist Branch.

Suter, G. W., and S. M. Cormier. 2013. "A method for assessing the potential for confounding applied to ionic strength in central Appalachian streams." *Environmental Toxicology and Chemistry* 32 (2): 288–295.

Suter, G. W., R. A. Efroymson, B. E. Sample, and D. S. Jones. 2000. *Ecological Risk Assessment for Contaminated Sites*. Boca Raton, FL: Lewis Publishers.

Suter, G. W., and J. M. Loar. 1992. "Weighing the ecological risks of hazardous waste sites: The Oak Ridge case." *Environmental Science and Technology* 26 (3): 432–438.

Swartjes, F. A., ed. 2011. *Dealing with Contaminated Sites: From Theory Towards Practial Application*. Dordrecht: Springer Science & Business Media.

Swartjes, F. A., M. Rutgers, J. P. A. Lijzen, P. J. C. M. Janssen, P. F. Otte, A. Wintersen, E. Brand, and L. Posthuma. 2012. "State of the art of contaminated site management in The Netherlands: Policy framework and risk assessment tools." *Science of the Total Environment* 427: 1–10.

Timmins, C. 2011. "The benefits of cleaning superfund sites." Duke University, Nicholas School of the Environment. https://sites.nicholas.duke.edu/loribennear/2011/09/19/the-benefits-of-cleaning-superfund-sites/.

Trowell, J., G. Gilron, K. Graf, L. Patterson, C. Chan, F. Perelló, and S. Bard. 2020. "Potential effects and impacts of a coal spill on sensitive aquatic habitat: A weight-of-evidence sediment quality assessment." *Water Quality Research Journal* 55 (3): 295–309.

USEPA. 1976. *Quality Criteria for Water*. Washington, DC: U.S. Environmental Protection Agency.

USEPA. 1998. *Guidelines for Ecological Risk Assessment*. Washington, DC: Risk Assessment Forum, U.S. Environmental Protection Agency.

USEPA. 2000. *A Guide to Developing and Documenting Cost Estimates during the Feasibility Study*. Washington, DC: U.S. Environmental Protection Agency.

USEPA. 2005. *Guidance for Developing Ecological Soil Screening Levels (ECO-SSLs)*. Washington, DC: U.S. Environmental Protection Agency, Office of Solid Waste and Emergency Response.

USEPA. 2016a. *Generic Ecological Assessment Endpoints (GEAEs) for Ecological Risk Assessment: Second Edition with Generic Ecosystem Services Endpoints Added*. Washington, DC: U.S. Environmental Protection Agency.

USEPA. 2016b. *Weight of Evidence in Ecological Assessment*. Washington, DC: Risk Assessment Forum, U.S. Environmental Protection Agency.

USEPA. 2024a. *Biological Water Quality Criteria*. Washington, DC: U.S. Environmental Protection Agency.

USEPA. 2024b. *CWA Methods Regulatory History*. Washington, DC: U.S. Environmental Protection Agency.

USEPA. 2025a. *Guidelines for Cumulative Risk Assessment Planning and Problem Formulation*. Washington, DC: U.S. Environmental Protection Agency, Risk Assessment Forum.

USEPA. 2025b. *The Superfund Program: Protecting Healthy Communities, Advancing Environmental Protection*. Washington, DC: U.S. Environmental Protection Agency, Office of Solid Waste and Emergency Response.

Van Dam, R. A., C. L. Humphrey, A. J. Harford, A. Sinclair, D. R. Jones, S. Davies, and A. W. Storey. 2014. "Site-specific water quality guidelines: 1. Derivation approaches based on physicochemical, ecotoxicological and ecological data." *Environmental Science and Pollution Research* 21: 118–130.

Water Quality Australia. 2020. "Assessing a remediation study." Accessed May 7, 2024.

Wenning, R. J., G. E. Bartley, C. Ingersoll, and D. R. J. Moore, eds. 2005. *Use of Sediment Quality Guidelines and Related Tools for the Assessment of Contaminated* Sediments. Pensacolla, FL: SETAC Press.

Wenning, R. J., and T. Havranek. 2025. "Decision-support tools for contaminated sites." *JDSUPRA* (blog).

18 Proposed Actions

In many nations, environmental assessments are required for proposed projects. These assessments have many potential effects, which lends them to weighing multiple pieces of evidence.

18.1 ENVIRONMENTAL IMPACT ASSESSMENT

In 1970, the U.S. National Environmental Policy Act (NEPA) included a requirement for an environmental impact assessment (EIA) for all major federal actions. Subsequently, many other nations have mandated EIAs, as have the World Bank and other development banks (UNEP 2018). They are predictive assessments that estimate the effects of projects to aid planning and implementation and to inform the public and stakeholders of expected effects. Strategic Environmental Assessments are a variant of EIAs that assess a multi-project plan or program.

The mandates for EIAs vary. In the U.S., they are not regulatory, and no response to their findings is required. Also, they are environmental and do not require human health, cultural, or economic assessments. Their primary legal implication is that the agency responsible for the project may be sued for providing no EIA or providing one that does not fully disclose expected impacts.

Federal projects to be assessed include obvious examples such as building a dam or highway, or dredging a harbor. They also include permitting a nuclear power plant, issuing a forest management plan, or changing the designation of a place, such as a wilderness designation.

Because EIAs are predictive, direct evidence of effects, such as toxicity tests of site soils, not available. Often, the best evidence for project impacts is an analogy to projects of the same type in similar locations. For example, the best predictors of the effects of a proposed reservoir on aquatic life turned out to be the conditions in existing similar reservoirs, not the mechanistic mathematical model (Goodman 1976). Because models can give mechanistic insight while analogies with other reservoirs give real-world information, weighing and integrating both types of evidence might have given even better results. However, EIAs in the United States do not have a strong practice of formal, structured weight-of-evidence analyses.

Because EIAs put more emphasis on public participation than risk assessments, there are often more sources and types of evidence. For example, in Canada, impact assessors are told to consider science, indigenous knowledge, community knowledge, and gender analysis (IAAC 2018). Integrating evidence of these types may be problematical. Weighing the relevance and reliability of scientific evidence along with indigenous or community knowledge could be perceived as inappropriate and disrespectful (Section 6.2.7).

DOI: 10.1201/9781003626336-21

18.2 PERMITS FOR PROPOSED PROJECTS

Assessments for issuing an air or water quality permit do not require an environmental impact assessment and are typically routine. However, Clean Water Act dredge and fill permits can be difficult.

A proposal to create an open-pit copper and gold mine in the headwaters of the Nushagak and Kvichak rivers in Alaska led indigenous communities, commercial fishermen, and environmental advocates to petition the USEPA to save the salmon fisheries by blocking the mine. An assessment was performed in response to that petition under the auspices of the Clean Water Act (USEPA 2014). A weight-of-evidence approach was used to predict the effects of a mine failure and tailings spill, a pipeline failure and diesel spill, the blockage of salmon migration at road crossings, and altered hydrology. The weight-of-evidence table for diesel spills is shown in Table 7.5. Other endpoints, particularly the area of wetlands and length of streams that would be lost, had only one type of evidence. The critical regulatory finding was that, in the undisturbed wilderness of the proposed mine site, the loss of wetlands would be adverse to spawning salmon and could not be repaired or replaced (USEPA 2022). However, as of this date, the Pebble Limited Partnership is once again suing the USEPA.

Permits may also be issued for projects. The permitting of an offshore wind project in Sweden was supported by a weight-of-evidence-based assessment of risks to cod (Hammar, Wikström, and Molander 2014). The evidence was weighed with respect to two properties: the reliability and the applicability of the evidence to the case. The endpoints were the likelihood of effects and the magnitude of effects on the cod population.

18.3 INCORPORATING THE DECISION

These assessments of proposed actions are devoted to predicting the environmental effects of the actions based on the evidence. Actions often include multiple component actions, each of which can have multiple effects, and multiple pieces of evidence for each effect. That can result in a complex weight of evidence leading to an overall judgment concerning environmental acceptability. It is one input to making a decision. It is also possible to integrate the environmental effects with other aspects of the decision using decision analysis (Section 3.4) or an evidence-to-decision framework (Chapter 22).

REFERENCES

Goodman, D. 1976. "Ecological expertise." In *Boundaries of Analysis, An Enquiry into the Tocks Island Dam Controversy*, edited by H. A. Feiveson, F. W. Sinden, and R. H. Socolow, 317–360. Cambridge, MA: Ballinger.

Hammar, L., A. Wikström, and S. Molander. 2014. "Assessing ecological risks of offshore wind power on Kattegat cod." *Renewable Energy* 66: 414–424.

IAAC. 2018. *Basics of Environmental Assessment under the Canadian Environmental Assessment Act, 2012*. Ottawa: Impact Assessment Agency of Canada.

UNEP. 2018. *Assessing Environmental Impacts: A Global Review of Legislation*. Nairobi: United Nations Environment Programme.

USEPA. 2014. *An Assessment of Potential Mining Impacts on Salmon Ecosystems of Bristol Bay, Alaska*. Washington, DC: U.S. Environmental Protection Agency.

USEPA. 2022. *Recommended Determination of the U.S Environmental Protection Agency Region 10 Pursuant to Section 404(C) of the Clean Water Act, Pebble Deposit Area*. Seattle, WA: U.S. Environmental Protection Agency, Region 10.

19 Deriving and Selecting Models

The relationship between mathematical modeling and weight of evidence is not immediately apparent. However, Clayton Aldern (2024) makes it clear:

> And if the goal of the agent's modeling efforts is to best represent the outside world—to best generate hypotheses that explain surprising events—then the version of the model that is able to do so is also the version of the model with the best underlying evidence.

Weight of evidence can be used to select a model from among alternatives or to determine confidence in a model (i.e., its weight). Four approaches may be used to achieve those purposes for statistical or mathematical models.

The first approach involves statistical methods that are nominally weight of evidence but do not meet our definition (Chapter 1). These methods select a model from a set of alternative models based on the support provided to each model by a data set (i.e., the evidence) (Section 3.2). Good (1950) defined the weight of evidence for members of a set of alternative models as their relative likelihoods, given the data. Rhomberg (2015) described the use of relative likelihoods to choose human toxicological hazards. He admits, however, that it is difficult to calculate likelihoods of models of toxicological hazards, and in practice, he uses qualitative weight of evidence. Models can also be compared using the sum of squares or another simple goodness–of-fit statistic, Bayes' factor, or Akaike's information criterion (Anderson 2008; Hilborn and Mangel 1997). If the models represent alternative causal hypotheses, choice of the model that best explains the data can be said to have chosen the best causal explanation of the modeled effect (Newman and Clements 2008). This use of the term weight of evidence is distinct from the others in this document (Chapter 1). Such statistical approaches to model selection are included here for completeness because they have appeared in the ecological assessment literature as weight of evidence (Linkov et al. 2009; Smith, Lipkovich, and Ye 2002). We recommend using statistical results as pieces of evidence in a more inclusive weighing of evidence.

Second, Bayesian model averaging extends this concept to provide a more recognizable example of weighing evidence to derive a model (USEPA 2009). That is, rather than choosing a single best model, one might combine suitable models after weighting them based on their weights in Good's sense (i.e., their consistency with the data).

Third, a more conventional use of the concept of weight of evidence is the weighing of evidence concerning model assumptions. For example, to determine whether certain phenomena should be included in a model, one might assemble, weight, and weigh evidence concerning their occurrence. Such phenomena might include compensatory processes in a fish population, dietary toxicity in aquatic invertebrates, nonlinearities in exposure–response relationships, or avoidance behavior at subtoxic

DOI: 10.1201/9781003626336-22

exposures. The methods for weighing evidence for these model assumptions would be the same as for other qualitative conclusions.

Finally, model selection can be based on weighing a mixture of statistical and empirical considerations. The benchmark dose guidance recommends choosing a dose–response model based on both biological plausibility and Akaike's information criterion (USEPA 2012). A Science Advisory Board panel has recommended broadening the statistical and nonstatistical considerations to be weighed during model selection and using a more formal and transparent weighing process (SAB 2015). For dose–response modeling, they recommended prioritizing regression models that directly use exposure data for individuals.

Models have properties that may be weighed beyond relevance and reliability: verification and validation (Thacker et al. 2004). Verification is concerned with identifying and removing errors in the model, such as coding errors, incorrect functional forms, or incorrect datasets. Verification may entail comparing the implemented model to the conceptual model. Evidence generated by a model is more reliable if it has been carefully verified. Validation evaluates the accuracy of the model's simulation or estimation of the real-world system of interest. Validation requires an experimental or observational study of the system of interest or components of it. Evidence generated by a model is more reliable if it has been validated, but some validations are more relevant or reliable than others.

Because any causal hypothesis can, in theory, be represented by a mathematical model, uses of weight of evidence for model selection could be equivalent to the approaches for deriving qualitative conclusions presented in Part II.

REFERENCES

Aldern, C. P. 2024. *The Weight of Nature*. New York: Dutton.

Anderson, D. R. 2008. *Model Based Inference in the Life Sciences*. New York: Springer Science and Business Media.

Good, I. J. 1950. *Probability and the Weighing of Evidence*. New York: Hafner Press.

Hilborn, R., and M. Mangel, eds. 1997. *The Ecological Detective: Confronting Models with Data, Monographs in Population Biology*. Princeton, NJ: Princeton University Press.

Linkov, I., D. Loney, S. M. Cormier, F. K. Satterstrom, and T. Bridges. 2009. "Weight-of-evidence evaluation in environmental assessment: Review of qualitative and quantitative approaches." *The Science of the Total Environment* 407 (19): 5199–205.

Newman, M. C., and W. H. Clements. 2008. *Ecotoxicology: A Comprehensive Treatment*. Boca Raton, FL: CRC Press.

Rhomberg, L. 2015. "Hypothesis-based weight of evidence: An approach to assessing causation and its application to regulatory toxicology." *Risk Analysis* 35 (6): 1114–1124.

SAB. 2015. *Science Advisory Board Review of the EPA's Evaluation of the Inhalation Carcinogenicity of Ethylene Oxide (Revised External Review Draft – August 2014). Letter to Administrator Gina McCarthy*. Washington, DC: SAB, U.S. Environmental Protection Agency.

Smith, E. P., I. Lipkovich, and K. Ye. 2002. "Weight-of-Evidence (WOE): Quantitative estimation of probability of impairment for individual and multiple lines of evidence." *Human and Ecological Risk Assessment* 8 (7): 1585–1596.

Thacker, B. H., S. W. Doebling, F. M. Hemez, M. C. Anderson, J. E. Pepin, and E. A. Rodriguez. 2004. *Concepts of Model Verification and Validation*. Los Alamos, NM: Los Alamos National Laboratory.

USEPA. 2009. *Guidance on the Development, Evaluation, and Application of Environmental Models*. Washington, DC: Council for Regulatory Environmental Modeling, Office of the Science Advisor, U.S. Environmental Protection Agency.

USEPA. 2012. *Benchmark Dose Technical Guidance*. Washington, DC: Risk Assessment Forum, U.S. Environmental Protection Agency.

20 Analogies

We may define a cause to be an object, followed by another, and where all the objects similar to the first are followed by objects similar to the second.

David Hume (1739)

A.B. Hill (1965) listed analogy as one of his causal considerations. It has been criticized as unreliable and an invitation to imagination (Rothman and Greenland 1998). However, if it is approached seriously and if a structured weight-of-evidence procedure is used to determine its weight, it can be reliable and enlightening. For example, Darwin conceived evolution by natural selection by an analogy to artificial selection by animal breeders. As indicated by the epigram above, David Hume defined causation in terms of analogy among associations. We can simplify it as: similar causes have similar effects. For example, if the impairment of concern involves a large and rapid decline in the abundance of aquatic insects, and if insecticides have been found to cause similar declines in other cases, then by analogy, that evidence supports an insecticide as the cause. Weight of evidence can provide a means of determining the evidential support for similarity of hypothesized analogues.

Analogies are scattered throughout environmental assessments. This chapter presents some examples.

20.1 READ-ACROSS

An inherent difficulty of environmental assessment is the lack of fate and effects data for potential endpoint effects of many chemicals. Quantitative structure-activity relationships can provide missing data, but they require data for several chemicals of the same class. In the absence of sufficient data to model a relationship, the missing data may be provided by read-across.

Read-across is a process in which you read across from source agents (i.e., well-studied agents) to a target agent (i.e., a poorly studied agent), and you posit that what is true of the first agent is also true of the second agent. In toxicology, it is a technique for filling data gaps by analogy between a poorly tested chemical and an adequately tested chemical that is sufficiently similar (ECHA 2015; ECETOC 2012; OECD 2017). For example, if there are no reproductive toxicity data for a target chemical, its reproductive toxicity may be represented by data from tests of a sufficiently similar analogue. The problem is to determine which, if any, of the potential source chemicals is sufficiently similar. Read-across is inherently a weight-of-evidence process that weights evidence of attributes of the chemicals that may be relevant to the toxicological endpoint. However, it has been criticized for not being sufficiently rigorous. Most read-across procedures do not have an explicit procedure for weighting the evidence and weighing the bodies of evidence for analogy. Rather, a narrative procedure is employed for weighing the evidence (Section 2.1).

DOI: 10.1201/9781003626336-23

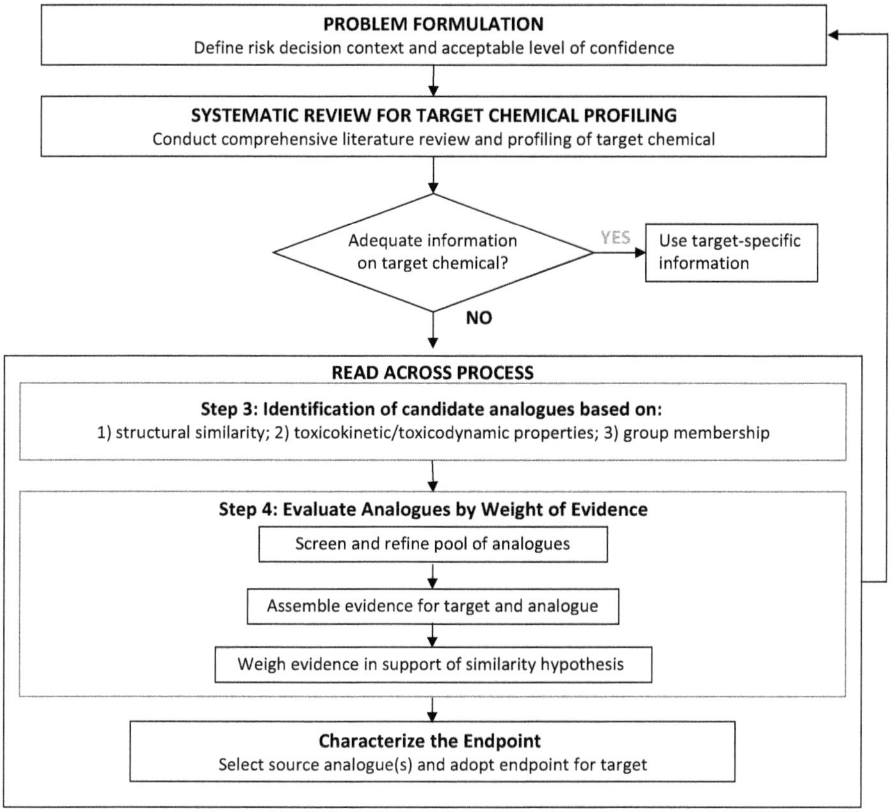

FIGURE 20.1 A read-across framework from Lizarraga et al. (2023).

A read-across process has been developed that formalizes the process and incorporates the USEPA's weight-of-evidence guidelines (Figure 20.1) (Lizarraga et al. 2023; Suter and Lizarraga 2022). An example is the application of WoE to select a source analogue to support screening-level quantitative assessment of human noncancer oral toxicity of the data-poor organochlorine contaminant, *p,p′*-dichlorodiphe nyldichloroethane (DDD) (Suter and Lizarraga 2022). DDD is the target compound, and the potential analogue compounds were *p,p′*-dichlorodiphenyltrichloroethane (DDT), dichlorodiphenyldichloroethylene (DDE), and *p,p′*-dimethoxydiphenyltrich loroethane (methoxychlor). These compounds are structurally similar to DDD and have oral noncancer toxicity data. DDT is the analogue with the greatest weight of evidence for analogy.

Types of evidence for chemical analogy are presented in Table 20.1. Each may have multiple subtypes, resulting in large weight-of-evidence tables (Suter and Lizarraga 2022). A summary weight of evidence (Table 20.2) for analogues of DDD combines the types of evidence, the bodies of evidence, and the integrated weight of evidence.

TABLE 20.1
Types of Evidence for Read-across in Environmental Toxicology

Types of Evidence	Specific Types of Evidence
Structural attributes	Major structural features, moieties, functional groups, structural alerts and membership in an existing chemical category or group.
Physiochemical attributes	Experimental or model predictions related to bioavailability, bioaccumulation, or environmental fate (e.g., water solubility, log Kow, BCF).
Toxicokinetics—absorption	Studies in humans, animals, in vitro assays, or model predictions examining rates and extents of absorption relevant to the endpoint effect.
Toxicokinetics—distribution	Studies in humans, animals, in vitro assays, or model predictions examining patterns and extents of internal distribution relevant to the endpoint effect.
Toxicokinetics—metabolism	Studies in human, animals, in vitro assays, or models examining pathways, metabolites, rates, and extents of metabolism relevant to the endpoint effect.
Toxicokinetics—excretion	Studies in humans, animals, in vitro assays, or models examining routes and rates of excretion relevant to the endpoint effect.
Toxicodynamics—toxicity	Studies in humans, animals, in vitro assays, or model predictions examining target organs, health effects, and dose-response relationships for repeat-dose or acute exposure relevant to the endpoint effect.
Toxicodynamics—mechanisms	Studies in humans, animals, in vitro assays, or model predictions related to the potential mechanisms of toxicity relevant to the endpoint effect.

Source: Adapted from Suter and Lizarraga (2022).

20.2 ADVERSE OUTCOME PATHWAYS

Adverse outcome pathways (AOPs) describe a series of causally linked events that lead from an initiating event to an adverse health effect in an organism following exposure to a chemical or other agent (Ankley et al. 2010). Although they may represent different levels of biological organization, AOPs are primarily used to describe molecular events beginning with an initiating event in which a chemical of concern binds to a biological molecule, and the results of that event lead to other key molecular events until an adverse outcome is reached. Chemicals with the same initiating event should be analogues. Evidence of any key event in common supports the analogy. AOPs may be derived using weight of evidence as in the identification of modes of action (Section 16.3.5).

Established AOPs for chemicals that are sufficiently similar to a target chemical of interest may be used as analogues. They are used in that way in the Integrated Risk Information System (IRIS). "Thus, the outcomes from other exposures with similar molecular initiating events or key events can be predicted from measurable upstream events" (USEPA 2022). The more steps in common, the greater the weight for the analogy.

TABLE 20.2
Summary of the Weight of Evidence for each Potential Analogue of DDD to Determine which is the Best

	Weight		
	DDT	DDE	Methoxychlor
Weight of Evidence for Types of Evidence			
Structure—general structural features	+	+	+
Structure—similarity scores	0	0	0
Structure—functional group	++	++	−
Physiochemical properties—related to bioaccumulation	++	++	−
Toxicokinetics—absorption	0	0	0
Toxicokinetics—distribution	+	+	NE
Toxicokinetics—metabolism	++	−	−−
Toxicokinetics—excretion	+	+	NE
Toxicodynamics—acute lethality studies	+	+	+
Toxicodynamics—repeat-dose studies (cancer)	+	+	0
Toxicodynamics—in vitro bioactivity	+	+	+
Body of Evidence, Collective Properties			
Number	++	++	++
Diversity	+++	+++	++
Coherence	++	+	−
Integrated WoE			
WoE for the hypothesized analogue	++	+	−

The three sections of the table are: weights for types of evidence integrated across properties (relevance, strength, and reliability), weights for collective properties for the bodies of evidence, and the integrated weight of evidence (Suter and Lizarraga 2022). The rating symbols are explained in Chapter 7.

Determining the similarity of a chemical of concern to a chemical that initiates an AOP is similar to read-across, except that the initiating event is a specific target property. Weight of evidence plays three roles when using AOPs:

- Weighing evidence for the reliability of the analogue's AOP.
- Weighing evidence of key events for the chemical of interest that are in the analogue's AOP.
- Weighing evidence that the chemical of interest has the same initiating event as the analogue.

20.3 EXTRAPOLATION BETWEEN SPECIES

Test species are treated as analogues of entire taxa or specific species. For example, fathead minnows are treated as analogues of freshwater fish in general. They are also used as analogues of specific valued species such as endangered dusky darters.

Typically, ecological test species are treated as sufficient analogues without justification. However, multiple types of evidence of similarity can be weighed and integrated.

Taxonomic distance is the most common type of evidence for interspecies analogies. It can be used as evidence to choose the test species that is most closely related to an endpoint species as the best analogue. It can be used in extrapolation models such as those in the *Interspecies Correlation Estimation* system to derive a benchmark value for endpoint species (Raimondo, Jackson, and Barron 2010; Lilavois, Nelson, and Raimondo 2024; Suter and Rosen 1988). Models may also be based on:

- Morphology, particularly size, as evidence for interspecies analogy;
- Diet, which may be important evidence for interspecies analogy because of the potential for dietary exposure and because diet has implications for uptake and metabolism of contaminants; and
- Physiology, which may be used as evidence for species analogies, particularly toxicokinetic and toxicodynamic data (USEPA 2014).

In most human health risk assessments, all test species are treated equally as nonhuman. Their test results are divided by ten to derive a human effects level. However, more types of evidence are being weighed.

If the necessary biochemical information is available, AOPs may be used to extrapolate between tested and untested species (USEPA 2026). For example, if fathead minnows are used in toxicity testing and dusky darters have structurally conserved estrogen receptors (see the SeqAPASS tool), then an AOP linking estrogen receptor activation (a molecular initiating event) to altered sex ratios (an adverse outcome) in the minnows would support the extrapolation that biologically active concentrations of the chemical in minnows will also disrupt sex ratios in the darters. Extrapolation tools employing multiple types of evidence and multiple methods are under development (LaLone et al. 2026).

20.4 SPATIAL EXTRAPOLATION

A common problem when using field data in environmental assessments is determining whether data and relationships among data from one location can be used in a different location. This problem can be resolved by weighing relevant attributes of each location as evidence for the hypothesis of sufficient similarity.

An example is provided by field-based water quality criteria for specific conductivity (Cormier et al. 2018). The criteria derivation required specific conductivity values for background. Sufficient data were available for Ecoregion 70 in West Virginia, but not in Ohio. If the variables that influence background water quality are sufficiently similar, and if the chemical and biological attributes influenced by specific conductivity are similar, the West Virginia criterion could, by analogy, be used in Ohio. The types of evidence were regional attributes, empirically estimated background, spatial distribution of low specific conductivity sites, and biological indicators. Each evidence type had multiple subtypes. The properties were relevance and reliability. Table 20.3 is the weight-of-evidence scoring table for regional properties.

TABLE 20.3
Weight of Evidence Table for Regional Properties of Ecoregion 70 in Ohio and West Virginia

Types of Evidence	Results
Similar climate across the region	Climate is similar across the ecoregion (+)
Similar geology across the region	Geological strata are similar across the ecoregion (++)
Similar physiography across the region	The entire ecoregion has a hilly physiography (+)
Geophysical modeling	Empirically estimated background Base-flow SC estimated by a geophysical model for the new area (195 μS/cm) was 10.8% higher than modeled in the original area (176 μS/cm). The model estimated that 76.2% of the new area has base-flow SC less than the background SC level in the original area (210 μS/cm) (++)

The hypothesis is that the ecoregion is sufficiently similar in the two states (Cormier et al. 2018).

20.5 CASE-BASED REASONING

Humans and artificial intelligence systems can reason by analogy using case-based reasoning. Case-based reasoning by AI uses the following general process:

1. Retrieve the most similar case(s) by weighing attributes of the new case to compare to attributes of past cases.
2. Use the retrieved similar case to try to solve the current problem.
3. Revise and adapt the proposed solution if necessary.
4. Retain the final solution in the library of cases.

Examples include diagnostic systems that retrieve past medical cases with similar symptoms and could include plant, fish, or wildlife pathology assessment systems that identify similarly impaired biota. The evidence is the organismal symptoms or the attributes of the impaired community. The properties of the evidence to be weighted depend on the case of interest. Environmental assessors are less likely than medical diagnosticians to have access to a library of relevant cases, but they can appeal to experience or the literature. A case-based analogy can be used as one piece of evidence in an inference.

20.6 CONCLUSIONS

Many analogies are used in environmental assessment without thought to the appropriate method. For example, reference sites are analogues of contaminated sites but without the contaminants. They are often chosen without weighing the evidence of similarity. Whenever you use an analogy, you should assemble and weigh the evidence.

REFERENCES

Ankley, G. T., R. S. Bennett, R. J. Erickson, D. J. Hoff, M. W. Hornung, R. D. Johnson, D. R. Mount, J. W. Nichols, C. L. Russom, and P. K. Schmieder. 2010. "Adverse outcome pathways: A conceptual framework to support ecotoxicology research and risk assessment." *Environmental Toxicology and Chemistry* 29 (3): 730–741.

Cormier, S. M., L. Zheng, G. W. Suter, and C. M. Flaherty. 2018. "Assessing background levels of specific conductivity using weight of evidence." *Science of The Total Environment* 628: 1637–1649.

ECETOC. 2012. *Categorical Approaches, Read-Across, (Q)SAR*. Brussels: Europen Centre for Ecotoxicology and Toxicology of Chemicals.

ECHA. 2015. *Read-Across Assessment Framework (RAAF)*. Helsinki: European Chemicals Agency.

Hill, A. B. 1965. "The environment and disease: Association or causation?" *Proceedings of the Royal Society of Medicine* 58: 295–300.

Hume, D. 1739. *A Treatice of Human Nature*, edited by D. F. Norton and M. J. Norton. p. 76. Oxford: Oxford University Press.

LaLone, C. A., P. Schumann, C. Rivetti, J. Ewald, S. Edwards, M. Embry, N. Burden, G. Hodges, and N. Basu. 2026. "Best practices for the development and application of bioinformatics approaches for species extrapolation in chemical safety assessment." *Integrated Environmental Assessment and Management*, vjag058.

Lilavois, C. R., Nelson, S. A., and Raimondo, S. 2024. *Web-Based Interspecies Correlation Estimation (Web-ICE) for Acute Toxicity: User Manual. Version 4.0*. Gulf Breeze, FL: U.S. Environmental Protection Agency.

Lizarraga, L. E., G. W. Suter, J. C. Lambert, G. Patlewicz, J. Q. Zhao, J. L. Dean, and P. Kaiser. 2023. "Advancing the science of a read-across framework for evaluation of data-poor chemicals incorporating systematic and new approach methods." *Regulatory Toxicology and Pharmacology* 137: 105293.

OECD. 2017. *Guidance on Grouping of Chemicals*, 2nd edition. Paris: Organisation for Economic Co-operation and Development.

Raimondo, S., C. R. Jackson, and M. G. Barron. 2010. "Influence of taxonomic relatedness and chemical mode of action in acute interspecies estimation models for aquatic species." *Journal of Environmental Science and Technology* 44 (19): 7711–7716.

Rothman, K. J., and S. Greenland. 1998. *Modern Epidemiology*, 2nd edition. Philadelphia, PA: Lippincott, Williams & Wilkins.

Suter, G. W., and L. E. Lizarraga. 2022. "Clearly weighing the evidence in read-across can improve assessments of data-poor chemicals." *Regulatory Toxicology and Pharmacology* 129: 105111.

Suter, G. W., II, and A. E. Rosen. 1988. "Comparative toxicology for risk assessment of marine fishes and crustaceans." *Journal of Environmental Science and Technology* 22: 548–556.

USEPA. 2014. *Guidance for Applying Quantitative Data to Develop Data-Derived Extrapolation Factors for Interspecies and Intraspecies Extrapolation*. Washington, DC: U.S. Environmental Protection Agency.

USEPA. 2022. *ORD Staff Handbook for Developing IRIS Assessments*. Washington, DC: Office of Research and Development, U.S. Environmental Protection Agency. https://cfpub.epa.gov/ncea/iris_drafts/recordisplay.cfm?deid=356370.

USEPA. 2026. *Adverse Outcome Pathway (AOP) Research Brief*. Washington, DC: U.S. Environmental Protection Agency.

21 Resource Management

As presented in the preceding sections, environmental assessments are primarily concerned with determining environmental conditions, causes of those conditions, and risks associated with those causes. However, environmental assessment may extend to the processes of remediation, restoration, and management of resources.

- Resource remediation assessment involves stopping and reversing environmental damage caused by contaminants or other agents. As discussed in Section 17, the U.S. Superfund process includes remediation of contaminated media based on remedial goals and feasibility studies. However, remediation may also involve isolation of wastes, treatment of surface and ground waters, remediation of agricultural land, and other processes.
- Resource restoration assessments are primarily concerned with environmental restoration in disturbed areas such as remediated sites, cleared or burned forests, mines, overgrazed grasslands, lands covered by invasive plants, dredged or filled wetlands, and channelized or dammed streams. It includes replanting, recreating structure, reintroducing native species, etc.
- Resource management assessments may involve the management and protection of fisheries, game species, endangered species, forests, scenery, or valued systems. Such assessments are concerned primarily with managing harvests, recreation, and other resource uses. For example, the European Union has recently proposed a Nature Restoration Law which requires identification of important habitat to be restored (Hering et al. 2023).

Evidence is weighed to determine the need for remediation, restoration, or management and to determine what actions are best supported.

21.1 SYSTEMATIC REVIEW-BASED EVIDENCE SYNTHESIS

Many literature reviews are performed each year to answer questions concerning environmental resources, and most of them are simply narrative descriptions of the results of a literature search. An advance over that practice is the evidence syntheses performed using the Collaboration for Environmental Evidence's protocol (CEE 2018). Examples of the types of questions that they address include:

- How effective are marine protected areas at enhancing commercial fish populations?
- What is the impact of wind farm installations on bird populations?
- What is the optimal grazing regime for maximizing plant diversity in upland meadows?

DOI: 10.1201/9781003626336-24

The components of their evidence syntheses are:

- Assemble evidence: This is information relevant to belief in the truth or validity of a hypothesis. Evidence is assumed to have been generated by primary scientific research. Reviewers are encouraged to predefine their methods to minimize bias in the way the evidence is found and selected.
- Systematic review: This is an evidence synthesis method that aims to answer a specific question as precisely as possible in an unbiased way. The method collates, critically appraises, and synthesizes all available evidence relevant to the question. Reviewers use predefined methods to identify risks of bias in the evidence and to minimize bias. As presented in Section 4.2, the synthesis may use meta-analysis or narratives and may include weighting of the environmental evidence. However, the CEE specifies no procedure for evidence weighting.
- Systematic map: This is an evidence synthesis method that aims to provide an accurate description of the evidence base relating to a particular question. The method collates, codes, and configures all available evidence relevant to the question. It then uses narrative, mapping, and visualization methods for the synthesis. This type of synthesis presents the state of knowledge without answering the assessment question.

CEE's 111-page guidelines contain some familiar components of medical and toxicological systematic reviews. For example they recommend generating PICO-type structures for the questions to be answered (Section 6.1.1). The question, "Aare marine protected areas effective at conserving commercial fish stocks?" could translate into these key elements:

P = Populations of commercially important fish species
I = Establishment of a marine protected area
C = Area with no protection or little protection
O = Relative change in fish populations

They also provide useful advice for synthesizing and presenting evidence. For example, they provide a discussion of seven types of bias in evidence review that go beyond actual alteration of data. The results of evidence syntheses that follow the CEE guidelines are published in a dedicated journal, *Environmental Evidence*.

21.2 SEESAW VERSUS SCALES

A group of conservation biologists developed a structured weight-of-evidence framework for ecosystem management (Christie et al. 2023). They used the standard properties: relevance and reliability. Weights are rated as: refutes, mixed, weakly supports, or strongly supports. Instead of the conventional metaphor of loading evidence into scales, they use a beam pivoting on a center point like a seesaw as weighted pieces of evidence are loaded onto the beam. This beam and the evidence are diagrammed to integrate the body of evidence as support for a hypothesis. Hence, they call their method the Balanced Evidence Assessment Method.

21.3 REMEDIATION AND RESTORATION BY THE CORPS OF ENGINEERS

Conventionally, contaminated media are subject to two assessments: a risk assessment to determine what risks require correction and an engineering assessment to determine how the goal should be achieved (Section 17). The U.S. Army Corps of Engineers takes an integrated approach to sediment management that combines risk, remediation, and restoration into one weight-of-evidence assessment that includes hazards beyond toxicity (Bates, Massey, and Wood 2018). They call for a structured weight-of-evidence process but allow considerable latitude in the design of the process.

The weight of evidence addresses remedial alternatives such as alternative methods and alternative locations for dredge spoil disposal. The Corps' weight-of-evidence approach is similar to the USEPA's but with different terminology. The evidence properties are referred to as metadata, and they are relevance, quality, and resolution. They are rated as high, medium, and low. Because the weight of evidence addresses sediment management, some of the Corps' types of evidence are not found in risk assessments. They include cost, site availability, sociopolitical concerns, and environmental effects of the action. Bates, Massey, and Wood (2018) present a case study that includes the conversion of qualitative ratings to numerical values that are arithmetically integrated. We discourage that pseudo-quantitative practice due to the possibility of illogical results (Section 7.4). However, the Corps' overall weight-of-evidence approach is recommended for management assessments.

21.4 WILDLIFE MANAGEMENT

Hone and Krebs (2023) recommend using weight of evidence for inference in wildlife management and not statistical inference. They identified 12 criteria for causality and selected a limited subset for inference. They rate the strength of the body of evidence for alternative hypotheses using a scaffolding approach in which each level assumes evidence for each of the lower levels. It is assumed that in all cases you have evidence for association (level 0). They provide two alternative sets of criteria. For example, in set 1, if you have a unique mechanism (level 2), it is assumed that you have evidence for levels 0 and 1.

	Criteria Set 1	**Criteria Set 2**
0	Association	Association
1	Experiment and temporality	Unique mechanism and temporality
2	Unique mechanism	Validated prediction
3	Validated prediction	

21.5 PROTECTED AREA MANAGEMENT

Assessment of protected areas such as parks and refuges is analogous to assessment of contaminated sites (Chapter 17), except that they are expected to have good quality. The purpose is to ensure that the quality is as intended. The scoping, screening, and definitive stages are the same, except that both positive attributes (e.g., diverse and mature ecosystems) and negative attributes (contamination and other injuries) are weighted. For example, an assessment of a marine protected area in the Gulf of Trieste determined that the sediment was not contaminated based on weighing the evidence from a sediment quality triad (Piccardo et al. 2024). Protected area assessments may include potential negative effects of harvesting, physical disturbance, overuse, and other agents, and positive effects of restoration and regulation of use. Case-specific evidence should be obtained and weighted using appropriate properties.

21.6 NET ENVIRONMENTAL BENEFIT ANALYSIS

Conventional environmental risk assessments of contaminated or damaged sites address only the hazardous agents or actions, and, if the risks are significant, they imply the need for remedial action. However, remediation can cause environmental damage greater than the adverse effects of the hazard. An assessment is needed to determine whether the cure (remediation) is worse than the disease (effects of toxicity or disturbance). Such assessments are called net environmental benefit analyses. They originated after the *Exxon Valdez* oil spill when the pressure-washing and fertilization of oiled shores and the use of fires and dispersants on marine slicks may have increased the effects of the spilled oil. As a result, the U.S. federal government and the oil industry developed and promoted Natural Resource Damage Assessments for oil spills (IPIECA 2000; NOAA 1990).

Many actions other than remediating contamination are subject to some sort of net benefits analyses. Should pesticides be applied to forests to control gypsy moths? Should controlled burns be applied to old-growth forests? Should wolves be culled to increase the abundance of caribou? In each case, the nominally beneficial action poses environmental hazards.

Efroymson, Nicollette, and Suter (2003, 2004) developed a framework for net environmental benefit analysis that formalized the process and related it to terrestrial sites and to contaminants other than petroleum. It has been further modified to include remediation of structures such as offshore drilling platforms (Nicolette et al. 2023).

Net environmental benefit analysis is an environmental management assessment in that it determines the action that will result in the best environmental condition, rather than just determining the acceptability of the risk posed by the hazardous agents. It is a difficult assessment problem because of the number and diversity of potential effects from hazards and remedial actions (see Box 21.1). Natural remediation (the no-action alternative) may allow contaminants to degrade and dissipate, thereby reducing remedial effects, but potentially causing toxic effects in the interim. It may also allow the contaminants to spread and preclude human use of the site for some time. Active remediation may destroy terrestrial and aquatic ecosystems by digging, dredging, capping, burning, or chemically treating contaminated soils, sediments, and waters. It may also damage uncontaminated ecosystems by building roads, borrow pits, landfills, and other facilities. If the analysis includes human health effects, active remediation may also cause death, injury, and contaminant exposures of workers and the public.

BOX 21.1 BENEFITS AND RISK ANALYSIS JARGON

Some risk jargon is related to net environmental benefits analysis. It might be useful to distinguish component analyses:

- Inherent Risk Analysis addresses the risk due to the contamination prior to remediation (also termed baseline risk analysis).
- Risk of Remedy Analysis addresses risks due to the remedial action such as dredging, excavation, or revegetation with non-native species.
- Residual Risk Analysis addresses the risks due to contaminant residues that are not removed or degraded.

Many other environmental management practices involve balancing the benefits and damages of the management actions. Currently, the most prominent example is "controlled" burns. The deliberately set fires are intended to reduce the risk of large, uncontrolled burns. They are also said to recreate "natural" fire regimes, although they often resemble indigenous fire regimes. There are many examples of nominally controlled or natural burns that cause great damage to forests, streams, human infrastructure, and the lives and health of humans and wildlife. Another example of resource management is the treatment of water bodies with rotenone to remove alien fish species and clear the way to reestablish native game fish, while incidentally killing native nongame fish and invertebrates. Other examples include "regenerative grazing" of prairie by cattle, selective harvesting of forests, and hunting to prevent overconsumption by herbivores.

Weighing evidence for environmental harm and benefits may be performed by a narrative or other method. However, the structured approach to weight of evidence in Part II is adapted to NEBA assessments of contaminants as five weighing steps:

1. **Weight the Evidence for Each Hypothesized Contaminant Effect**: The weighting of evidence for contaminant effects is the same as in the conventional risk assessments for contaminated sites. In the U.S., these include Superfund remedial investigations and Clean Water Act determinations of impairment. The types of evidence include the triad or tetrad (Section 2.7), and the interpretations and ratings are the same as in Chapter 17. The questions are: Are effects occurring, and are they caused by the contaminants?
2. **Weight the Evidence for Occurrence of Each Hypothesized Remedial Effect**: Assessments of remedial options under Superfund focus on the economic and engineering feasibility of achieving the remedial goal (i.e., benchmark contamination level). In net environmental benefit analysis, the

risks from remedial actions are assessed by the same approach as risks from the contaminants. However, some of the effects of remedial actions such as excavation, capping, and dredging are absolute ecosystem destruction, so they require relatively little weighing of evidence.

3. **Weight the Damages of the Contaminant Risks**: To determine net benefits, the effects of contaminants should, as far as is feasible, be expressed in comparable terms. The endpoints may be conventional assessment endpoints such as species richness of invertebrates, integrative endpoints such as habitat value, or services of nature, such as recreational fisheries (USEPA 2016; Munns et al. 2016; Nicolette, Burr, and Rockel 2013; Efroymson, Peterson, Giffen, et al. 2008; Efroymson, Peterson, Welsh, et al. 2008). For the contaminated state and for each remedial action, the effects are rated in terms of properties. We suggest five properties of the damage: intensity, importance, and duration of effects, and the area affected (Box 21.2).
4. **Weigh the Benefits and Damages from Each Remedial Alternative**: The benefits and damages of each remedial alternative should be expressed in comparable terms to allow comparisons among them. They should be based on the same anthropocentric or natural properties as the contaminant damages. Ideally, at least one remedial alternative will provide benefits greater than damages.
5. **Compare the Weighted Benefits and Damages of Each Remedial Alternative Relative to the Unremediated Environment**: The difficult part is deriving a scale of benefits and damages based on either anthropogenic properties or natural properties as suggested in step 3.

BOX 21.2 ATTRIBUTES OF ENVIRONMENTAL EFFECTS

In order to balance benefits and damages, you should rate the effects based on defined attributes. We suggest these four, but others may be more appropriate for your case:

Intensity: Some effects are more or less intense, which may be expressed as the magnitude or severity of beneficial or harmful effects. Examples include chronic reduced survivorship versus acute mass mortality, devegetation versus reduced primary production, etc.

Importance: Some species are more important because they are threatened or endangered, support fisheries, are culturally important, etc. Some have negative importance because they are invasive pests. Some biotic communities are more important because they have recreational, timber, aesthetic, habitat, or other values.

Duration: How long will a contaminant take to naturally degrade to nontoxic levels, for the ecosystem on a site to recover from dredging or excavation, for a population to recover from overharvesting, for groundwater pumping to stop the spread of a pollutant, etc.?

Area: How large is the area that is contaminated, excavated, harvested, dredged, chemically treated, etc.?

Rather than rating effects on conventional assessment endpoints, you may use aggregate endpoints. That approach has been proposed for creating comparable scales of benefit and damages. The obvious scale is monetary, but we do not advise that approach unless the benefits and damages have market values. An alternative is to create a habitat metric that captures all important ecological values at a site. Efroymson, Peterson, Welsh, et al. (2008) reviewed habitat metrics and suggested three broad categories of metrics: site use by groups of organisms, site rarity, and use value. A companion paper presents a case study of a habitat-based assessment of remedial alternatives (Efroymson, Peterson, Giffen, et al. 2008).

If you have found the general weighing process that we presented in Part II to be useful, you can adapt it to weighing benefits and damages. The first two steps are weight of evidence for risk assessments—first for the contaminants and then for the remedial actions. The properties of the evidence can be weighted and the bodies of evidence integrated using weight-of-evidence tables as presented in Chapter 17. The result of these steps is a set of effects of contaminants and of remedial actions that are sufficiently supported by the evidence to be carried forward.

The third and fourth steps are an extension of those risk assessments to quantify the benefits and damages of remedial actions relative to the damages caused by unremediated contamination. Tables 21.1 and 21.2 are for the implementation of steps 3 and 4. The use of ecological endpoints and ecosystem services endpoints in the two tables is for illustrative purposes. The same endpoints should carry through the analysis, if appropriate. Integration of the ratings for each property of effects and each type of effects should be based on judgment and not the addition of the ratings. For example, if an endangered terrestrial plant species is damaged by remediation, the rating for importance and the overall rating should be "highly damaging," no matter what the rating is for the other properties.

Integration of benefits and damages for remedial alternatives relative to no action can use a table like Table 21.3. Because the alternatives have qualitatively different environmental effects, this final step requires careful consideration of what the results of the analysis represent with respect to the contaminated or mismanaged site.v

TABLE 21.1
Potential Weight of Effects Table for Unremediated Contaminant Effects on Ecological Endpoint Entities

Types of Ecological Entities	Properties of Contaminant Effects				
	Intensity	Importance	Duration	Area	Integrated Properties
Terrestrial plants					
Soil biota					
Terrestrial wildlife					
Aquatic invertebrates					
Integrated effects					

TABLE 21.2
Potential Weight of Effects Table for Effects of a Remediation Alternative on Services of Nature

Types of Services of Nature	Properties of Remedial Effects				
	Intensity	Importance	Duration	Area	Integrated Properties
Fisheries					
Recreation					
Water quality					
Critical habitat					
Climate moderation					
Integrated services					

TABLE 21.3
Integrated Weight of Effects Table for No Remediation and Three Potential Remedial Actions

Alternative Remedial Actions	Ecological Benefits and Damages				
	Terrestrial Plants	Soil Biota	Terrestrial Wildlife	Aquatic Invertebrates	Net Benefits and Damages
No action					
Action 1					
Action 2					
Action 3					

The cells, except for the bottom row, are for integrated ratings from a table like 21.1 (none) and from three tables like 21.2 (remedial alternatives 1–3).

The ratings of the benefits and damages can be represented in the same way as ratings of evidence, but with different interpretations.

+++	Highly beneficial
++	Moderately beneficial
+	Slightly beneficial
0	No effect
–	Slightly damaging
– –	Moderately damaging
– – –	Highly damaging

Note that this common scale should be supported by data when available. For example, in most cases, area is quantified, but its interpretation depends on the case.

Loss of a two-hectare habitat may, depending on the nature of the habitat and the species using it, be considered slightly, moderately, or highly damaging.

Net benefits are more readily inferred than risks because they are comparative rather than quantified. That is, a proposed remedial action is acceptable if the benefits exceed the damages.

The assessment of net environmental benefits is more difficult when application of the precautionary principle to the environment is mandated. Precaution is problematic for any comparative assessment because, to avoid bias, precautionary assumptions must somehow have equal influence on each alternative action. Normally, precaution is applied to risks from contamination, but in net environmental benefit analysis, it could equally be argued that risks from remedial actions should be assessed with precautionary assumptions. Choosing just one alternative to be the object of precaution would be an arbitrary bias.

Human health risks may override net environmental benefits. For example, when the remedial action was chosen for the floodplain of the East Fork of Poplar Creek in Oak Ridge, mature floodplain forest was destroyed to protect humans in a highly unlikely floodplain settler scenario with unrealistic exposure assumptions. Because there was no plausible evidence of human health effects and none of the evidence for terrestrial ecological endpoints indicated adverse toxic effects, there were negative net benefits.

21.7 NATURAL RESOURCE DAMAGE ASSESSMENTS

In some nations and circumstances, polluters are required to compensate resource trustees for natural resource damages. In the United States, the resource trustees are NOAA for coastal and marine oil spills, and the Department of the Interior for other spills or releases of hazardous materials (USEPA 2025). The natural resource damages are defined as the loss of services of nature. The damages may be paid in currency or by replacing the services of the lost resources.

The natural resource damage assessment must not only determine the damages but also must demonstrate that the contaminants were the cause. Assessment to determine the damages is the same as other assessments for contaminated sites, including weighing evidence (Chapter 17), except that the effects are expressed as losses of ecosystem services so that they can be converted to monetary damages. The monetary damages are used by the trustees to restore or replace the lost natural resources. Assessing causation is performed as in Chapter 14. If the damages are associated with a Superfund site, the natural resource damage assessment can build upon the ecological risk assessment for the site remediation. However, the damage assessment is performed by the resource trustee, and the damages include those from the remedial action as well as the contamination.

21.8 MANAGING INTRODUCED SPECIES

Species may be deliberately introduced as biocontrol agents, crops, decorative plants, pets, livestock, gamefish, terrestrial game, or for other purposes. Other species are accidentally introduced. Weight of evidence may be used to assess whether a species is present, whether it is likely to spread, whether it is or is likely to become a pest, weed, or pathogen, and whether it has specific undesirable attributes or is already causing specific effects. Assessments can determine the risks of allowing an introduction or of failing to control a species after introduction.

Introduced species may be cryptic or difficult to identify before they disrupt ecosystems. As a result, it may be difficult to classify them for management. A program in South Africa assembled and weighed evidence for the occurrence of three species (Matthys, Wilson, and Geerts 2025). Their properties of evidence are credibility, specificity, sensitivity, and quantity, which were rated as high, moderate, or low. Their sources of evidence included published data and studies, ecological measurements, observations, location records, scientific literature, and expert opinion. They integrated the evidence using argument maps.

21.9 ASSESSMENT OF MANAGEMENT ACTIONS AS CAUSES OF RECOVERY

When management actions have apparently been successful, a causal assessment should be performed to determine which, if any, of the actions is the cause. A good example is a weight-of-evidence application to determine which management action or natural process is responsible for the recovery of Okanagan sockeye salmon (Alexander et al. 2024). Salmon in the Columbia River basin drastically declined, apparently due to harvesting, hydroelectric dams, irrigation diversions, mining, logging, urban development, agricultural use of floodplains, non-native species introductions, and climate change. While two other sockeye stocks in the basin have failed to respond positively to management actions, the Okanagan stock has increased by over 400%.

Six causal hypotheses for recovery were proposed:

1. Changes in escapement concurrent with improving local fish passage;
2. the implementation of fish-friendly flows in the Okanagan River;
3. initiating a hatchery restocking program;
4. potential improvements to Columbia River dam operations to support higher relative survival of out-migrating juvenile fish;
5. possible shifts in survival-favorable conditions in the coastal marine environment for ocean-going life stages; and
6. broader changes to multi-stock harvest regimes in the Columbia River.

The first three causal hypotheses are specific to management actions in the Okanagan River basin. The last three are out-of-basin causes associated with shifts in natural environmental conditions, larger-scale flow management, or regional-scale management of harvests.

The properties that were rated are the standard relevance, reliability, and strength. The two subproperties of relevance were logical plausibility and exposure. Subproperties of reliability were trustworthiness and corroboration. Strength was the strength of a cause-effect association. The properties were rated using the standard +, −, and 0 symbols.

The ratings of the properties for each hypothesis were integrated to give an overall credibility rating of H (highly credible and very likely), M (moderately credible and likely), or L (low credibility and unlikely). Four of the hypotheses were rated H, indicating that multiple causes contributed to the recovery. Both local management actions and natural events contributed.

Another example is provided by the large research program and causal assessment for the decline in sockeye salmon production in the Fraser River (Marmorek et al. 2011). The weighing of evidence was structured as a sequential criteria weight of evidence (Section 2.5), but with a mix of binary and trinary logic modeled on Burkhardt-Holm (2008). For each hypothesized causal factor, the properties of the evidence correspond to stages in the sequence. They are expressed as questions concerning the plausibility of the mechanism, the occurrence of exposure, the correlation of adverse effects with the causal factor, and other evidence. Based on the results of the sequence, the causal hypotheses were rated as unlikely (representing the lowest level of confidence), possible, likely, or very likely (representing the highest level of confidence) to have been a primary causal factor behind the overall pattern of declining productivity in Fraser sockeye.

21.10 WATER QUALITY MANAGEMENT BASED ON CONNECTIVITY

In the United States, protection of water quality extends to all waters of the United States. Defining those waters has been a contentious issue. In 2006, the Supreme Court defined it as having a "continuous surface connection" or a "nexus of connection" to "traditional" waters of the United States. As a result, the USEPA (2015) developed a weight-of-evidence approach to determine whether such connections occur at a site based on the nexus definition. Using a systematic literature search, they identified types of connections (physical, chemical, and biological), categories of water (streams, riparian/floodplain wetlands, and non-floodplain wetlands), and types of evidence (field-based methods, modeling, and remote sensing). Evidence was weighted for five properties: strength, consistency, mechanistic plausibility, and relevance to the endpoints. Since then, the nexus definition has been withdrawn by the Supreme Court and new regulations have been proposed by the Trump administration's USEPA. However, states and tribes may continue to use the methods based on the nexus definitions to fill the gaps resulting from reduced federal oversight.

REFERENCES

Alexander, C. A. D., I. Alameddine, D. Machin, and K. Alex. 2024. "A weight-of-evidence approach for understanding the recovery of Okanagan Sockeye Salmon." *Environmental Management* 74 (6): 1063–1085.

Bates, M. E., O. C. Massey, and M. D. Wood. 2018. *Weight-of Evidence Concepts: Introduction and Application to Sediment Management*. Washington, DC: U.S. Army Corps of Engineers.

Burkhardt-Holm, P. 2008. "The decline of brown trout (*Salmo trutta*) in Switzerland - How to assess potential causes in a multi-factorial cause-effect relationship." *Marine Environmental Research* 66: 181–182.

CEE. 2018. *Guidelines and Standards for Evidence Synthesis in Environmental Management Version 5.0*. Collaboration for Environmental Evidence.

Christie, A. P., W. H. Morgan, N. Salafsky, T. B. White, R. Irvine, N. Boenisch, R. M. Chiaravalloti, K. Kincaid, A. M. Rezaie, and H. Yamashita. 2023. "Assessing diverse evidence to improve conservation decision-making." *Conservation Science and Practice* 5 (10): e13024.

Efroymson, R. A., J. P. Nicollette, and G. W. Suter. 2003. *A Framework for Net Environmental Benefit Analysis for Remediation or Restoration of Contaminated Sites*. Oak Ridge, TN: Oak Ridge national Laboratory.

Efroymson, R. A., J. P. Nicollette, and G. W. Suter 2004. "A framework for net environmental benefit analysis for remediation or restoration of contaminated sites." *Environmental Management* 34: 315–331.

Efroymson, R. A., M. J. Peterson, N. R. Giffen, M. G. Ryon, J. G. Smith, W. W. Hargrove, W. K. Roy, C. J. Welsh, D. L. Druckenbrod, and H. D. Quarles. 2008. "Investigating habitat value to inform contaminant remediation options: Case study." *Journal of Environmental Management* 88 (4): 1452–1470.

Efroymson, R. A., M. J. Peterson, C. J. Welsh, D. L. Druckenbrod, M. G. Ryon, J. G. Smith, W. W. Hargrove, N. R. Giffen, W. K. Roy, and H. D. Quarles. 2008. "Investigating habitat value to inform contaminant remediation options: Approach." *Journal of Environmental Management* 88 (4): 1436–1451.

Hering, D., C. Sürings, F. Wenskus, K. Blackstock, A. Borja, S. Birk, C. Bullock, L. Carvalho, M. B. Dagher-Kharrat, and S. Lakner. 2023. "Securing success for the nature restoration law." *Science* 382 (6676): 1248–1250.

Hone, J., and C. J. Krebs. 2023. "Causality and wildlife management." *The Journal of Wildlife Management* 87 (5): e22412.

IPIECA. 2000. *Choosing Spill Response Options to Minimize Damage: Net Environmental Benefit Analysis*. London: International Petroleum Industry Environmental Conservation Association.

Marmorek, D., D. Pickard, A. Hall, K. Bryan, L. Martell, C. Alexander, K. Wieckowski, L. Greig, and C. Schwarz. 2011. *Fraser River Sockeye Salmon: Data Synthesis and Cumulative Impacts*. Vancouver, BC: ESSA Technologies Ltd.

Matthys, C., J. R. Wilson, and S. Geerts. 2025. "Argument maps can support decisions to declare the presence of alien species: South Africa as a case study." *ARPHA Preprints*. https://doi.org/10.3897/arphapreprints.e165209

Munns, W. R., V. Poulsen, W. R. Gala, S. J. Marshall, A. W. Rea, M. T Sorensen, and K. von Stackelberg. 2016. "Ecosystem services in risk assessment and management." *Integrated Environmental Assessment and Management* 13 (1): 62–73.

Nicolette, J., S. Burr, and M. Rockel. 2013. "A practical approach for demonstrating environmental sustainability and stewardship through a Net Ecosystem Service Analysis." *Sustainability* (5): 2152–2177.

Nicolette, J. P., N. A. Nelson, M. K. Rockel, M. L. Rockel, A. N. Testoff, L. L. Johnson, L. D. Williamson, and V. L. G. Todd. 2023. "A framework for a net environmental benefit analysis based comparative assessment of decommissioning options for anthropogenic subsea structures: A North Sea case study." *Frontiers in Marine Science* 9: 1020334.

NOAA. 1990. *Excavation and Rock Washing Treatment Technology: Net Environmental Benefit Analysis*. Seattle, WA: National Oceanic and Atmospheric Administration.

Piccardo, M., V. Vellani, S. Anselmi, E. Grazioli, M. Renzi, A. Terlizzi, L. Pittura, G. D'Errico, F. Regoli, and S. Bevilacqua. 2024. "The application of the Weight-Of-Evidence approach for an integrated ecological risk assessment of marine protected sites." *Ecological Indicators* 159: 111676.

USEPA. 2015. *Connectivity of Streams and Wetlands to Downstream Waters: A Review and Synthesis of the Scientific Evidence*. Washington, DC: Office of Research and Development, U.S. Environmental Protection Agency.

USEPA. 2016. *Generic Ecological Assessment Endpoints (GEAEs) for Ecological Risk Assessment: Second Edition with Generic Ecosystem Services Endpoints Added.* Washington, DC: U.S. Environmental Protection Agency.

USEPA. 2025. *Natural Resource Damages: ERAs and NRDAs*. Washington, DC: U.S. Environmental Protection Agency.

22 Evidence-to-Decision Frameworks

Environmental assessments can provide information to inform decision-making, but conventionally, they do not determine the decision. However, in public health, evidence-to-decision frameworks have been developed to combine the derivation of hazards or risks with decision-making methods. An early and important evidence-to-decision framework was developed by the Grading of Recommendations, Assessment, Development, and Evaluations (GRADE) Working Group. It was intended to provide defensible and transparent recommendations for clinical practices (GRADE Working Group 2004), including those in the Cochrane Collaborative (Higgins and Thomas 2023). It has been extended to causation in epidemiology and the assessment of hazards in environmental and occupational health (Shunemann et al. 2011; Morgan et al. 2016; Senerth et al. 2025). It uses an evidence-grading system adapted from the Hill "criteria" for the evidence part of the framework and weight-of-evidence-informed judgments for the decision-making part. Note that it emphasizes properties of the intervention and minimizes properties of the evidence.

The basic GRADE framework has three main sections:

1. Formulating the Question
   - Define the health problem and the intervention options.
   - Clarify the population, setting, and outcomes of interest.
2. Making an Evidence-Informed Assessment
   - Evaluate multiple properties of the intervention, including:
     - Benefits and harms of the intervention
     - Certainty of the evidence
     - Values and preferences of stakeholders
     - Resource use and cost-effectiveness
     - Equity, acceptability, and feasibility
3. Drawing Conclusions
   - Weigh the evidence and judgments to reach a recommendation or decision.
   - Document the rationale and make it transparent for others to understand or adapt.

The WHO-INTEGRATE framework is an adaptation of GRADE for the development of public health policies, such as COVID-19 guidelines for schools (Rehfuess et al. 2019, 2024). It emphasizes societal issues in decision-making. In particular, WHO-INTEGRATE includes several decision-specific criteria—balance of health benefits and harms, human rights and sociocultural acceptability, health equity, equality and nondiscrimination, societal implications, financial and economic considerations,

DOI: 10.1201/9781003626336-25

feasibility and health system considerations, as well as evidence quality. It uses a + and – rating system, like that in Chapter 7, to weight features of the potential interventions. The GRADE Working Group has developed a similar evidence-to-decision framework for environmental and occupational health (Senerth et al. 2025).

Some organizations use approaches termed evidence-based decision-making that are equivalent to evidence-to-decision frameworks. The U.S. Centers for Disease Control and Prevention (CDC) has developed an equivalent but less formalized approach for decision-making in public health practice called Evidence-Based Guidelines (Carande-Kulis, Elder, and Koffman 2022). Its process has three steps: gathering evidence, interpreting evidence, and applying what you learn from the evidence. The evidence is weighed based on an adaptation of the GRADE system. The decision concerning best practices is informed by considering the benefits and harms indicated by the evidence in the context of feasibility, values, and preferences. It is also influenced by value judgments and policy considerations (Carande-Kulis, Elder, and Koffman 2022).

Assessments of contaminated sites determine whether the contaminants should be remediated (Chapter 17). However, to choose a method for remediation, criteria like those in WHO-INTEGRATE should be considered. Similarly, assessments of the impacts or risks of proposed actions (Chapter 18) could use a decision framework like WHO-INTEGRATE to determine what design constraints would make the project acceptable.

These integrated risk and decision assessment methods have emphasized human health but can be adapted to decision-making for ecological effects.

REFERENCES

Carande-Kulis, V., R. W. Elder, and D. M. Koffman. 2022. "Standards required for the development of CDC evidence-based guidelines." *Morbidity and Mortality Weekly Report Suppl* 71 (Suppl 1): 1–6.

Grade Working Group. 2004. "Grading quality of evidence and strength of recommendations." *BMJ* 328: 1–8.

Higgins, J., and J. Thomas. 2023. *Cochrane Handbook for Systematic Reviews of Interventions Version 6.4*. London, UK: Cochrane Collaborative.

Morgan, R. L., K. A. Thayer, L. Bero, N. Bruce, Y. Falck-Ytter, D. Ghersi, G. Guyatt, C. Hooijmans, M. Langendam, and D. Mandrioli. 2016. "GRADE: Assessing the quality of evidence in environmental and occupational health." *Environment International* 92: 611–616.

Rehfuess, E. A., L. Pfadenhauer, M. Nothacker, and B. Strahwald. 2024. "Using the WHO-INTEGRATE framework to develop a COVID-19 guideline for schools, Germany." *Bulletin of the World Health Organization* 102 (10): 742.

Rehfuess, E. A., J. M. Stratil, I. B. Scheel, A. Portela, S. L. Norris, and R. Baltussen. 2019. "The WHO-INTEGRATE evidence to decision framework version 1.0: Integrating WHO norms and values and a complexity perspective." *BMJ Global Health* 4 (Suppl 1): e000844.

Senerth, E., P. Whaley, E. Akl, B. Beverly, P. Alonso-Coello, A. Rooney, H. J. Schünemann, K. A. Thayer, K. Tsaioun, and R. L. Morgan. 2025. "GRADE guidance 40: The GRADE evidence-to-decision framework for environmental and occupational health." *Environment International* 197: 109314.

Shunemann, H., S. Hill, G. Guyatt, E. A. Akl, and F. Ahmed. 2011. "The GRADE approach and Bradford Hill's criteria for causation." *Journal of Epidemiology and Community Health* 65 (5): 392–395.

23 Future Weight of Evidence

23.1 EFFICIENT WEIGHT OF EVIDENCE

The practice of weighing evidence in environmental assessments has increased greatly in the decades that we have been environmental assessors. Environmental assessors have developed explicit procedures and practices to make the practice more reliable and acceptable. However, as a result, it has become more time- and labor-intensive. In agencies like the USEPA, which are routinely understaffed, that is problematical. For example, we have encouraged the USEPA Office of Water to use weight of evidence when deriving water quality criteria for aquatic life, but they have stuck with the original method (Stephan et al. 1985). Efficiency is a particular issue now, as the current administration eliminated about a quarter of USEPA positions in 2025 alone and also most of the contracted support that augmented most of the EPA's work.

A potential solution is to automate the process to some extent. There are already decision support tools for environmental assessment (Marcomini, Suter, and Critto 2009; Wenning and Havranek 2025). The USEPA has developed procedures and tools for the rapid assessment of human health risks of chemicals (Lambert et al. 2025). This trend will continue as we begin to incorporate artificial intelligence, which could be the ultimate efficiency generator (Section 23.4).

23.2 NEW TYPES OF EVIDENCE

Standard toxicological evidence in ecological assessments typically has three endpoint attributes: survival, growth, and reproduction, plus species richness in field studies. Sub-organismal attributes have been treated as biomarkers that may be indicative of certain classes of contaminants. Now, with the opposition to whole-animal toxicity tests, particularly in Europe, and the increase in chemical, tissue, and cellular effects information, the utility of sub-organismal attributes as endpoints is an issue. In particular, what are the population or community implications of sub-organismal attributes? Adverse outcome pathways are useful for hazard identification and may be useful as a piece of evidence, but currently they are not useful for quantifying risks or setting benchmark values (https://www.epa.gov/comptox-tools/adverse-outcome-pathway-wiki). Computational toxicology provides some useful tools for interpreting evidence from new approach methods (https://comptox.epa.gov/dashboard/).

A new method may be evaluated by weight of evidence to determine whether it is relevant, reliable, or suitable to derive evidence. If so, its results may be treated as a type of evidence in applications (Box 6.2). A review of methods for toxicological

DOI: 10.1201/9781003626336-26

risk assessment for wildlife argued for a weight-of-evidence approach to evaluating methods (Bean et al. 2024). They made the point that novel methods, particularly sub-organismal new approach methods, seldom have decisive endpoints, but they can provide important mechanistic information to supplement standard methods. This trend will make weight of evidence more complex because of the mixing of evidence from multiple levels of organization, potentially from molecular to population levels.

23.3 FUTURE APPROACHES AND FRAMEWORKS

As discussed in Chapter 2, there is great diversity among approaches to weight of evidence. Ideally, a consensus will develop among organizations and practitioners concerning the best approach for weight of evidence in environmental assessments, a kind of evolutionary survival of the most practical and useful. Ecological risk assessment provides a precedent. Since 1992, when the USEPA developed the framework for ecological risk assessment, other organizations have developed their own frameworks. Most have shown some influence of the USEPA framework. Recently, there has been some acceptance of the basic three-part framework for weight of evidence (Figure 5.1). Currently, any framework more specific than that is context-dependent.

23.4 ARTIFICIAL INTELLIGENCE

Artificial intelligence has the potential to contribute to inference by weight of evidence in environmental assessment (Vinuesa et al. 2020; Al Khourdajie 2025). In particular, software is available for performing literature reviews and data extraction, and it is getting better. With the increasing power of artificial intelligence, it should be possible to add the capacity to identify types of evidence and evidence of characteristics, and determine their relevance, reliability, and strength. We would need to clearly identify the rules to be applied. We may also need to interact with the software to make it reliable. For example, the software might prompt an assessor to examine its judgments and the text from the literature that it used and then to verify or correct them. It is also likely that we would leave the final weighing of the bodies of evidence and identification of the best-supported hypothesis to human assessors.

The foundation of artificial intelligence is artificial neural networks, which numerically weigh evidence associated with nodes of the networks. Neural network analysis has even been analogized to the scales of justice (Kucharski 2025). However, unlike other networks (Section 3.8), these networks learn from models, like using thousands of model images of cats to learn to recognize cats. An example from ecotoxicology is the use of machine learning to estimate thousands of LC_{50}s to create species sensitivity distributions for poorly tested chemicals (Posthuma, Price, and Viljanen 2025).

An immediately practical use of artificial intelligence is in generating evidence. For example, it can identify and count animals in images from camera traps (Norouzzadeh et al. 2018) or identify and count phytoplankton (Embleton, Gibson, and Heaney 2003). Using artificial intelligence to analyze images of potential histopathologies is already a common medical practice. The use of artificial intelligence to generate relevant information will increase the need for weighting and weighing evidence.

Artificial intelligence can make errors, including inventing evidence, known as AI slop. Another risk is the accumulation of junk science, which AI may not be able to recognize. It will be important to determine the validity of your AI-generated evidence by weighing relevant non-AI evidence to create standards and comparing the results.

REFERENCES

Al Khourdajie, A. 2025. "The role of artificial intelligence in climate change scientific assessments." *PLOS Climate* 4 (9): e0000706.

Bean, T. G., V. R. Beasley, P. Berny, K. M. Eisenreich, J. E. Elliott, M. L. Eng, P. C. Fuchsman, M. S. Johnson, M. D. King, and R. Mateo. 2024. "Toxicological effects assessment for wildlife in the 21st century: Review of current methods and recommendations for a path forward." *Integrated Environmental Assessment and Management* 20 (3): 699–724.

Embleton, K. V., C. E. Gibson, and S. I. Heaney. 2003. "Automated counting of phytoplankton by pattern recognition: A comparison with a manual counting method." *Journal of Plankton Research* 25 (6): 669–681.

Kucharski, A. 2025. *Proof, the Art and Science of Certainty*. New York: Basic Books.

Lambert, J. C., J. Brown, H. Gong, C. Kilburn, J. Krysa, B. Kuntzelman, J. Lee, A. Luke, J. Powell, and A. Rashid. 2025. ""RapidTox": A decision-support workflow to inform rapid toxicity and human health assessment." *Computational Toxicology*: 100377.

Marcomini, A., G. Suter, and A. Critto (eds.). 2009. *Decision Support Systems for Risk-Based Management of Contaminated Sites*. New York: Springer.

Norouzzadeh, M. S., A. Nguyen, M. Kosmala, A. Swanson, M. S. Palmer, C. Packer, and J. Clune. 2018. "Automatically identifying, counting, and describing wild animals in camera-trap images with deep learning." *Proceedings of the National Academy of Sciences* 115 (25): E5716–E5725.

Posthuma, L., T. Price, and M. Viljanen. 2025. "Improving the ecotoxicological hazard assessment of chemicals by pairwise learning." *Environmental Science and Technology* 59 (31): 16250–16260.

Stephan, C. E., D. I. Mount, D. J. Hanson, J. H. Gentile, G. A. Chapman, and W. A. Brungs. 1985. *Guidelines for Deriving Numeric National Water Quality Criteria for the Protection of Aquatic Organisms and Their Uses*. Washington, DC: U.S. Environmental Protection Agency.

Vinuesa, R., H. Azizpour, I. Leite, M. Balaam, V. Dignum, S. Domisch, A. Felländer, S. D. Langhans, M. Tegmark, and F. F. Nerini. 2020. "The role of artificial intelligence in achieving the Sustainable Development Goals." *Nature Communications* 11 (1): 233.

Wenning, R. J., and T. Havranek. 2025. *Decision-Support Tools for Contaminated Sites*. Sausalito, CA: JDSUPRA.

Part IV

Supporting Material

24 Weight of Evidence Epilogue

Evidence is precious, do not waste it.

Give evidence proper recognition for its relevance, reliability, and strength.
Consider plausible and meaningful hypotheses.
Evidence and hypotheses are a couple.

Evidence validates hypotheses.
Hypotheses give meaning to evidence.

Do not reject hypotheses prematurely.

Poorly supported hypotheses may have useful information.
New evidence may revive a failing hypothesis.

Be mindfully conscious of what you are assessing and why.

Be objective and beware of appealing biases.
Do not hide your work.

Lay it out for stakeholders, decision makers, and fellow scientists.

DOI: 10.1201/9781003626336-28

Appendix A
Qualitative Questions for Which Evidence Is Weighed in Different Types of Assessments (Adapted from USEPA 2016)

Assessments begin with a question to be answered. Then, hypotheses are proposed that might answer the question, and evidence is weighed to determine which of them is best supported. The following questions appear in different types of assessments.

A.1 CAUSAL ASSESSMENT

Causal assessments identify causes of observed effects, and they take one of two general forms.

- What is the cause of general effect *y*? This is the general causal question. Examples include: What causes colony collapse disorder in honeybees? and What causes coral bleaching?
- What is the cause of specific effect *y*? This is the specific causal question. Examples include: What caused the low invertebrate species richness in the Coal River? and What caused the decline in San Joaquin kit foxes on the Elk Hills in the 1980s?

A.2 HAZARD ASSESSMENT

Hazard assessments have two purposes. Most risk assessments begin with a problem formulation that includes a causal question concerning what hazard is associated with an agent and the risk that the hazard will be expressed. Stand-alone hazard assessments determine whether an agent has an undesired attribute. Note that these hazard identification questions are not the risk question, but they are essential to defining the risk question.

- Does agent *x* cause general effect *y*? This is a general hazard identification question for assessments such as setting benchmarks or permitting the use of a chemical. Examples include: Does smoking cause lung cancer? and Does atrazine cause deformities in frogs?

- Might agent *x* cause effect *y*? This is a hazard identification question asked in risk assessments for specific actions. Examples include: Might a tailings spill at the proposed mine cause a loss of salmon spawning? Might permitting a new use of atrazine result in an increase in the frequency of frog deformities?
- Is agent *x* causing effect *y*? This is a hazard identification question asked in risk assessments for an agent under existing conditions. Examples include: Are the sediment contaminants causing reduced invertebrate species richness or abundance? Is acid deposition reducing forest production?

Without one of these hazard-identifying assessments, there is no basis for defining the endpoint entity and attribute beyond using conventional endpoints such as mortality of fish.

A.3 RISK ASSESSMENT

Other qualitative questions beyond hazard identification must also be answered for risk assessments. Examples that might involve WoE include the following.

- Should dietary exposure be considered?
- Will endangered species *x* occur on the site in the future?
- Is the site large enough to support an assessment population of species *y*?
- Is long-range transport a significant source?
- What are the contaminants of concern?

A.4 CONDITION ASSESSMENT

Qualitative questions concerning conditions, such as the following, are relatively straightforward to assess when a standard of condition has been defined.

- Is the ecosystem impaired?
- Does the ecosystem have exceptional quality?
- Is the species endangered?
- Is the water contaminated?
- Are wildfires more frequent?

However, framing the question may be conceptually difficult, particularly with respect to the standard of comparison: natural conditions, reference conditions, historical conditions, or a defined standard.

A.5 OUTCOME ASSESSMENT

Qualitative outcome questions are like condition questions but include issues concerning the efficacy of remedial or regulatory actions.

- How much has the ecosystem improved relative to predictions?
- Has bioremediation sufficiently reduced exposure of endpoint species?

- Is the cap leaking?
- Is the site acceptable for recreational fishing?

A.6 BENCHMARK ASSESSMENTS

Benchmark assessments refer to the development of quantities for screening or regulatory values for chemical, physical, or biological agents (see Chapter 15). Weight of evidence is used to identify the best value or to determine whether multiple values are sufficiently similar to be combined. After development, benchmarks can be compared to existing levels of the agent to assess condition, acceptable risks, and objectives for resource management and restoration, or to evaluate management performance.

- What level of a contaminant will likely extirpate 5% of genera?
- What duration of dissolved oxygen lower than 5 mg/L will cause salmonid decline?
- What pH level in rainwater will reduce soil buffering capacity?
- How frequently can an exposure be exceeded without an adverse effect?

REFERENCE

USEPA. 2016. *Weight of Evidence in Ecological Assessment.* Washington, DC: Risk Assessment Forum, U.S. Environmental Protection Agency.

Appendix B
Characteristics of Inferred Qualities (Adapted and Expanded from USEPA 2016)

Weight of evidence addresses qualities, each of which is characterized by certain characteristics (Chapters 5 and 8). For example, if the quality of interest is causation, the hypothesized relationship should have the characteristics of causation, such as the causal agent and the affected entity interact. Associating evidence with characteristics of the quality being addressed can help to perform and justify the inference. For example, a weight-of-evidence table organized in terms of causal characteristics may show that there is abundant evidence for co-occurrence but none for interaction. Potential characteristics for five qualities are presented here. They can serve as models for you to develop characteristics of other qualities, as needed.

B.1 CHARACTERISTICS OF CAUSES

Characteristics of causal relationships were developed first because of the manifest need for causal assessments (Cormier, Suter, and Norton 2010). Characteristics are not mandatory criteria. Few bodies of evidence for causal assessments include evidence for all of these characteristics. Even in very well-studied cases, evidence of conditions prior to induction of the effect is seldom available to allow consideration of time order. Even an incomplete WoE table based on characteristics, however, provides more insight into what is known about causation in a case than a table based only on types of evidence.

Although these characteristics were developed for causation in specific cases, they also can be relevant to general causation (i.e., Does agent x cause effect y?), if multiple cases are available. The characteristics can be used to demonstrate that, at least in those actual cases, the agent could cause the effect. They can be particularly useful when the general question is framed in terms of real-world conditions. An example is the use of the characteristics to determine that, where aqueous conductivity is elevated due to major mineral ions in West Virginia and Kentucky, the ion mixture causes the extirpation of invertebrate genera (Cormier and Suter 2013; Cormier et al. 2013; USEPA 2011).

The characteristics apply to, and can be observed in, actual cases, but they are not observable in the same way when the general causal question is hypothetical. That is, hypothetically, can x cause y rather than, in fact, did x cause y or is x causing y? In particular, when the only evidence of effects is from toxicity tests, the co-occurrence and its antecedents are simply the design and setup of the test, and the time order is inevitable. Sufficiency in hypothetical causal assessments is not of interest because no particular effect level is identified, so no particular exposure is sufficient. Instead, we use the test data to establish an exposure-response relationship that can be applied generally. The real causal concerns are with the relevance of the co-occurrence, interaction, and alteration in the test to those in the field. Are the conditions, durations, species, and life stages in the tests relevant to the co-occurrence of interest in the field? Are the interactions of the stressor and test organisms and the resulting alterations plausible in the field? Are other interactions likely to be more sensitive or important? (Table B.1)

The characteristics of causation can be arranged in a logical sequence: *antecedent* causes lead to *co-occurrence* of the proximate cause with susceptible or affected

TABLE B.1
Characteristics of Causal Relationships[a]

Causal Characteristics	Description	What Evidence of a Characteristic Shows
Time order	The cause precedes the effect.	Change in the affected entity after exposure to the cause and not before
Antecedence	The causal relationship is a result of a larger web of antecedent cause-and-effect relationships. Evidence includes sources and routes of transport.	Earlier events that led to the particular causal event. It demonstrates that a potential causal event is part of a network of prior causal events, which increases confidence that the potential causal event actually occurred (e.g., that it was not a result of a measurement error or hoax).
Co-occurrence	The cause co-occurs with the susceptible entity in space and time. Evidence includes contaminant concentrations in ambient media and habitat alterations.	The presence of both a causal agent and the affected entity in circumstances that create the potential for exposure
Interaction	The cause interacts with the entity in a way that can lead to the effect. Evidence includes injuries, body burdens, and biomarkers.	Signs that the entity has been exposed to the agent in such a way that changes can be initiated through a plausible mechanism
Sufficiency	The intensity, frequency, and duration of the cause are adequate, and the entity is sufficiently susceptible to produce the type and magnitude of the effect.	Enough of the cause and a sufficiently susceptible entity that can result in the level of the observed effect
Alteration	The entity is changed by physical, chemical, or other mechanisms leading to the defined effect. Evidence includes symptoms and related attributes of the affected entities.	Changes in the entity attributable to or at least appropriate to the cause, which can indicate that the mechanism is acting

[a] Modified from Cormier, Suter, and Norton (2010); Norton, Cormier, and Suter (2014), and USEPA (2016).

organisms, which leads to *interaction* with the affected organisms and, if the interaction is *sufficient*, leads to *alteration* of the organisms. The characteristics also serve to prompt assessors to make full use of the evidence with respect to all characteristics of causation. For example, it is obvious that laboratory test data provide evidence concerning sufficiency, but the test data also might provide evidence of alterations in organisms affected by the tested agent that might also be observed in the field.

B.2 CHARACTERISTICS OF CARCINOGENS

Although cancer is primarily a concern in human health risk assessments, it may also be a concern in ecological risk assessments when tumors are found in fish or wildlife. It is not uncommon for fishermen or hunters to report tumors and ask about the cause and any risks to consumers. Tumors may also be considered evidence of biological impairment in waters of the United States. The characteristics in Table B.2 are descriptive of mechanisms of action.

B.3 CHARACTERISTICS OF PROTECTION

Criteria, standards, screening levels, and other benchmarks are intended to define a limit on exposure that would be adequately protective (Chapter 15). Protection, like causation, is a difficult concept with multiple characteristics (Table B.3). Weight of evidence might be used to assess the confidence that a benchmark is appropriately protective.

TABLE B.2
Key Characteristics of Carcinogens from IARC (2019) and Smith et al. (2016)

Characteristic	Examples of Relevant Evidence
1. Is electrophilic or can be metabolically activated	Parent compound or metabolite with an electrophilic structure (e.g., epoxide, quinone), formation of DNA and protein adducts
2. Is genotoxic	DNA damage (DNA strand breaks, DNA–protein cross-links, unscheduled DNA synthesis), intercalation, gene mutations, cytogenetic changes (e.g., chromosome aberrations, micronuclei)
3. Alters DNA repair or causes genomic instability	Alterations of DNA replication or repair (e.g., topoisomerase II, base-excision, or double-strand break repair)
4. Induces epigenetic alterations	DNA methylation, histone modification, microRNA expression
5. Induces oxidative stress	Oxygen radicals, oxidative stress, oxidative damage to macromolecules (e.g., DNA, lipids)
6. Induces chronic inflammation	Elevated white blood cells, myeloperoxidase activity, altered cytokine and/or chemokine production
7. Is immunosuppressive	Decreased immunosurveillance, immune system dysfunction
8. Modulates receptor-mediated effects	Receptor in/activation (e.g., ER, PPAR, AhR) or modulation of endogenous ligands (including hormones)
9. Causes immortalization	Inhibition of senescence, cell transformation
10. Alters cell proliferation, cell death or nutrient supply	Increased proliferation, decreased apoptosis, changes in growth factors, energetics and signaling pathways related to cellular replication, cell cycle control, or angiogenesis

TABLE B.3
Potential Characteristics of a Protective Benchmark (USEPA 2016)

Characteristics of Protective Benchmarks	Description
Causal relationship	The exposure-response relationship used in the derivation must be causal.
Predictive model of the relationship	The model of the relationship predicts the effects of specified exposures.
Represents sensitive endpoints	Known sensitive endpoint species or other entities and responses can be represented by the relationship.
Exposure metric is relevant	The exposure metric represents the relevant exposure of the sensitive entities.
Set at an appropriately protective level	Independent data demonstrate, with sufficient confidence, a lack of effects of concern on the biotic community and constituent taxa.
Discriminatory power	The benchmark discriminates between impaired and unimpaired communities with sufficient confidence.

TABLE B.4
Characteristics of Contaminants of Concern at Contaminated Sites

Characteristics of Contaminants of Concern	Description
Associated with the waste or source	Chemicals that are not associated with a waste or effluent could be from an extraneous local source, regional contamination, or natural background.
Frequently detected	Chemicals that are infrequently detected at the site are of lesser concern and might be artifacts.
Greater than local background	Chemicals at background levels are not of concern, unless the anthropogenic form is more bioavailable or toxic.
At potentially toxic levels	Concentrations above relevant toxicological benchmarks are of concern.
Detection limits are inadequate	Chemicals that could not be detected at toxic levels are of concern, particularly if they are associated with the waste or source.
Belongs to the same class as an identified contaminant of concern	A chemical that is not of concern alone might belong to a class, such as polycyclic aromatic hydrocarbons, that could be assessed jointly.
Likely to bioaccumulate	Chemicals that are persistent and bioaccumulative are of concern.

B.4 CHARACTERISTICS OF CONTAMINANTS OF CONCERN

Risk assessments of contaminated sites typically develop lists of contaminants of concern based on chemical analyses of site media. The contaminants of concern are then the subjects of the risk assessment. Concern may be determined by comparison to ecological benchmarks, but more often, multiple characteristics of a contaminant are weighed to determine which make the list. The characteristics in Table B.4 are based on a book and a Superfund ecological risk assessment (Black & Veatch Special Projects Corps 2011; Suter et al. 2000).

B.5 CHARACTERISTICS OF REMEDIATION

Assessments of the success of remedial actions are common and important examples of outcome assessments. Remediation is judged in terms of the performance and effectiveness of the intervention (Table B.5). Obtaining sufficient evidence to determine whether ecological goals are attained because of the remedial action and not just natural recovery is valuable.

B.6 CHARACTERISTICS OF TERATOGENS

Although developmental defects are primarily a concern in human health risk assessments, they may also be a concern in ecological risk assessments when deformities are found by fishermen and hunters in fish or wildlife. Deformities may also be considered evidence of biological impairment in waters of the United States. An example is the cranial deformities in fish caused by selenium (Table B.6).

TABLE B.5
Potential Characteristics of Remediation

Characteristic of Remediation	Description
	Performance of the Intervention
Functions	The remedial technology functions as intended.
Reduces the agent	The technology substantially reduces the level and extent of the contaminant or other agent that causes the risk.
Achieves the physical/ chemical goal	The levels of the causal agent are reduced sufficiently to achieve the goal.
	Effectiveness of the Intervention
Endpoint less exposed	The exposure of the environmental endpoint entity to the agent and their interaction are reduced as predicted.
Endpoint improved	The abundance, diversity, function, or other attributes of the endpoint entity improve within the expected time following the remediation.
Endpoint goal achieved	Reference conditions or other ecological endpoint goal are achieved in response to remediation.

TABLE B.6
Characteristics of Teratogens

Characteristics of Teratogens	Examples of Relevant Evidence
Chemical Structure	Certain chemical structures are more likely to be associated with teratogenic effects (Patlewicz 2019).
DNA Reactivity	Chemicals that can directly interact with DNA, causing mutations, are often teratogenic.
Mechanisms of Action	Chemicals that disrupt folate metabolism, neural crest cell development, endocrine function, or cause oxidative stress are often teratogenic (Van Gelder et al. 2010).
Transfer to Embryo	Chemicals that are teratogenic must be capable of reaching the developing embryo. For mammals, the chemical must cross the placental barrier.

TABLE B.7
Key Characteristics of Endocrine Disrupters

Characteristic	Examples of Relevant Evidence
1. Interacts with or activates hormone receptors	Binding or agonism of hormone receptors
2. Antagonizes hormone receptors	Antagonism of nuclear or cell surface hormone receptors
3. Alters hormone receptor expression	Abundance, distribution, and degradation of hormone receptors
4. Alters signal transduction in hormone- responsive cells	Abundance of post-translational modifications, cofactors, transcription factors and transcripts, and activity of associated enzymes
5. Induces epigenetic modifications in hormone- producing or hormone responsive cells	Abundance of post-translational modifications, cofactors, transcription factors and transcripts, and activity of associated enzymes
6. Alters hormone synthesis	Expression or activity of enzymes or substrates in hormone synthesis
7. Alters hormone transport across cell membranes	Intracellular transport, vesicle dynamics, or cellular secretion
8. Alters hormone distribution or circulating hormone levels	Blood protein expression and binding capacity, blood levels of pro- hormones and hormones
9. Alters hormone metabolism or clearance	Inactivation, breakdown, recycling, clearance, excretion, or elimination of hormones
10. Alters fate of hormone-producing or hormone-responsive cells	Atrophy, hyperplasia, hypertrophy, differentiation, migration, proliferation, or apoptosis

Source: Adapted from La Merrill et al. (2020)

B.7 KEY CHARACTERISTICS OF ENDOCRINE DISRUPTERS

Chemicals that disrupt endocrine systems may interfere with development, reproduction, and other processes, which could result in endpoint impairments. Table B.7 lists key characteristics based on an expert consensus (La Merrill et al. 2020). They are applicable to humans and other vertebrates.

REFERENCES

Black & Veatch Special Projects Corps. 2011. *Ecological Risk Assessment for the Estuary at the LCP Chemical Site in Brunswick, Georgia: Site Investigation/Analysis and Risk Characterization (Revision 4)*. Atlanta, GA: U.S. Environmental Protection Agency, Region 4.

Cormier, S. M., and G. W. Suter II. 2013. "A method for assessing causation of field exposure-response relationships." *Environmental Toxicology and Chemistry* 32 (2): 272–276.

Cormier, S. M., G. W. Suter II, and S. B. Norton. 2010. "Causal characteristics for ecoepidemiology." *Human and Ecological Risk Assessment* 16: 53–73.

Cormier, S. M., G. W. Suter II, L. Zheng, and G. J. Pond. 2013. "Assessing causation of the extirpation of stream macroinvertebrates by a mixture of ions." *Environmental Toxicology and Chemistry* 32 (2): 277–287.

IARC. 2019. *IARC Monographs on the Identification of Carcinogenic Hazards to Humans: Preamble*. Lyon: International Agency for Research on Cancer.

La Merrill, M. A., L. N. Vandenberg, M. T. Smith, W. Goodson, P. Browne, H. B. Patisaul, K. Z. Guyton, A. Kortenkamp, V. J. Cogliano, and T. J. Woodruff. 2020. "Consensus on the key characteristics of endocrine-disrupting chemicals as a basis for hazard identification." *Nature Reviews Endocrinology* 16 (1): 45–57.

Norton, S. B., S. M. Cormier, and G. W. Suter II. 2014. *Ecological Causal Assessment*. Boca Raton, FL: CRC Press.

Patlewicz, G. 2019. "Can teratogenic risk be predicted from chemical structure?" In *Teratology Primer*, 3rd edition, edited by B. Hales, A. Scialli, and M. Tassinari. Reston, VA: Society for Birth Defects Research and Prevention.

Smith, M. T., K. Z. Guyton, C. F. Gibbons, J. M. Fritz, C. J. Portier, I. Rusyn, D. M. DeMarini, J. C. Caldwell, R. J. Kavlock, P. F. Lambert, S. S. Hecht, J. R. Bucher, B. W. Stewart, R. A. Baan, V. J. Cogliano, and K. Straif. 2016. "Key characteristics of carcinogens as a basis for organizing data on mechanisms of carcinogenesis." *Environmental Health Perspectives* 124 (6): 713–721.

Suter, G. W., II, R. A. Efroymson, B. E. Sample, and D. S. Jones. 2000. *Ecological Risk Assessment for Contaminated Sites*. Boca Raton, FL: Lewis Publishers.

USEPA. 2011. *A Field-Based Aquatic Life Benchmark for Conductivity in Central Appalachian Streams*. Washington, DC: Office of Research and Development, National Center for Environmental Assessment, U.S. EPA.

USEPA. 2016. *Weight of Evidence in Ecological Assessment*. Washington, DC: Risk Assessment Forum, U.S. Environmental Protection Agency.

Van Gelder, M. M. H. J., I. A. L. M. Van Rooij, R. K. Miller, G. A. Zielhuis, L. T. W. de Jong-van den Berg, and N. Roeleveld. 2010. "Teratogenic mechanisms of medical drugs." *Human Reproduction Update* 16 (4): 378–394.

Appendix C
Strategic Tips

The examples in this appendix are intended to show how weight of evidence can be applied to real assessments and how it can translate complex scientific information to non-scientists as well as scientists.

C.1 DOCUMENT, DIAGRAM, AND EXPLAIN YOUR PROCESS

The weight-of-evidence process is described and explained in Part II and diagrammed in the figures in Chapters 4–8. This material describes how to weigh evidence within an environmental assessment in general terms. Following the USEPA weight-of-evidence guidelines is not mandatory in the United States, but using them will make your assessment approach neither arbitrary nor capricious. However, you will need to fill in the details and elaborate the process so that reviewers, decision makers, and stakeholders understand it. It is also valuable to explain and justify your process and decisions. This will make your assessment more credible and ensure that you can justify your work.

A simple example of diagramming a weight-of-evidence process is presented in Figure C.1. It comes from a causal assessment to develop a water quality benchmark for specific conductivity. An elaboration of the USEPA weight-of-evidence process to derive nutrient criteria is shown in Figure 15.1 and explained by Ridley et al. (2024).

C.2 TEXT, MAPS, GRAPHICS, AND TABLES

Different people grasp ideas more easily in one format than another. Use them all to reach more people. Graphics are useful for showing how the analysis or weight-of-evidence process is performed (e.g., Figure C.2). Maps show spatial relationships. Tables are succinct ways to summarize evidence and weights, and they make it easy to depict comparisons among hypothesized causes. Using text helps to ensure that details are included, and some people are overwhelmed by flow charts and respond better to written descriptions. When material gets too redundant, rely on appendices for completeness.

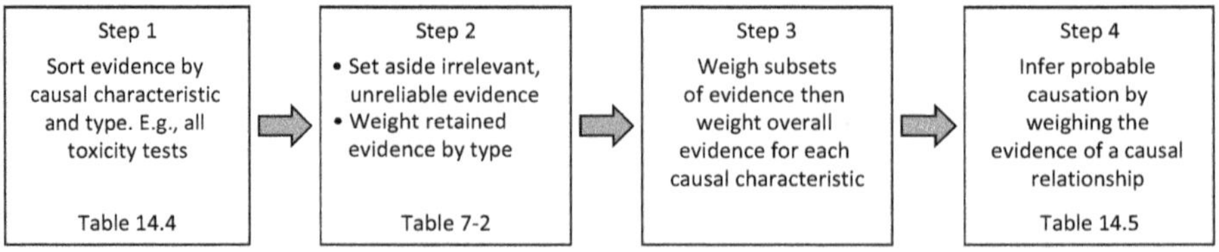

FIGURE C.1 A criteria-guided process to weight (score) and weigh the evidence for or against causation. Tables called out below each box contain the criteria for that step (USEPA 2011).

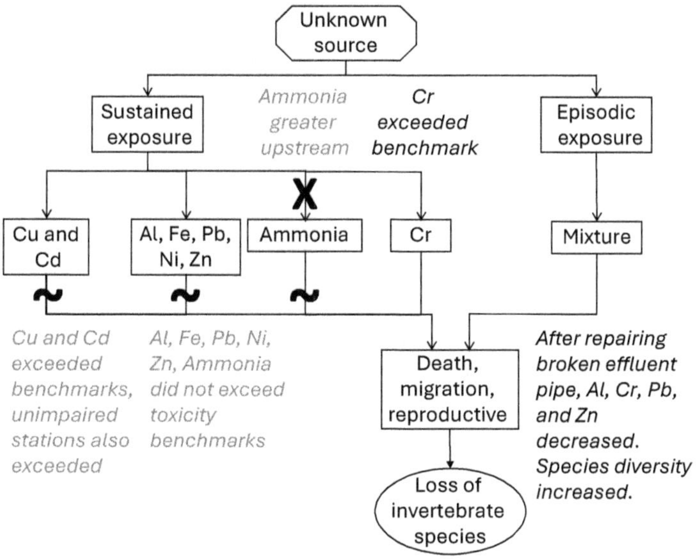

FIGURE C.2 Conceptual model and evidence discounting and supporting the causal relationship between a toxic mixture and reduced biodiversity in the Middle River, CT. A weakened connection in the pathway is illustrated by an arrow bisected by a "~" where the pathway is weakened, with evidence in gray italics. A refuted connection in the pathway is shown as "X," with evidence in gray italics. Evidence supporting a causal pathway is shown in black italics with an uninterrupted pathway. (Adapted from Bellucci, Hoffman, and Cormier 2010.)

Conceptual models can show interactions in complex ecosystems and the type of evidence related to a component or step. For example, in a tributary to the Willimantic River, CT, a causal assessment indicated that there was an unknown source of a toxic mixture and pinpointed where additional sampling was needed. New sampling by Christopher Bellucci and Guy Hoffman of the Connecticut Department of Energy and Environmental Protection revealed a 19th-century mill raceway hidden behind vegetation. Serendipitously, on the day of sampling, a gray discharge was observed from the raceway, which was traced to a broken wastewater line from a nearby mill that held an NPDES permit for the release of organic waste, metals, and ammonia to the publicly owned wastewater treatment facility. The mill operated by batch production with sporadic releases. Figure C.2 shows part of the conceptual model dealing with toxicity as the hypothesized cause of biological impairment. It shows unlikely causal pathways with relevant evidence and pathways with evidence confirming the episodic mixture as the cause.

C.3 MAKE IT EASY FOR YOUR READER

Err on the side of stating the obvious. Too many times, I (Cormier) have skipped explanations of steps because I'm in a rush, only to have colleagues gently remonstrate with comments of "please explain in more detail."

Define your terms. Many words have colloquial meanings or are used differently in different disciplines. Even in scientific papers, a table with key words and definitions has been welcomed by peer reviewers of our causal assessments. We have included a glossary in this book.

C.4 EXPERTISE AND CREDENTIALS

Build up your reputation and earn respect for your trustworthiness, expertise, and problem-solving skills. Publish in peer-reviewed journals. Be active in scientific organizations. Take on editorial responsibility for a respected scientific journal. If you are an attorney, vet your experts.

To establish a scientific consensus, leverage the power of independent peer review and the power of scientific advisory boards, scientific society workshops, and review panels, including review by the National Academy of Sciences. Expert review panels are the closest things we have to a scientific court, and our actual judicial system puts significant weight on them. In the mining case in Chapter 14, Judge (Chambers 2015) noted that: "As previously noted by the Court, in developing the Benchmark, the EPA's findings were subjected to review and comment by the Science Advisory Board. [Fola (Stillhouse), 2015 WL 362643, at *6.] As noted by Dr. Garabrant, not only are there epidemiologists on the Science Advisory Board, 'there are some very fine [epidemiologists]' serving in that capacity. [Tr. 1 at 71, ECF No. 105.] While offering constructive criticisms for further analysis, these epidemiologists evidently accepted the EPA's Benchmark as adequate, to say the least."

C.5 DEFENDING YOUR WORK

There's money to be made casting doubt, so it is important to inspire confidence by being diligent and responding to criticism. Journal editors are generally very willing to publish a rebuttal to an editorial or manuscript critical of your work. A letter explaining your desire to do so and a follow-up rebuttal is a straightforward solution to criticism. It can be annoying, but consider it an opportunity to make your position clearer and to strengthen any vulnerable or incomplete aspects of the work. It's how scientific understanding has advanced through the centuries.

REFERENCES

Bellucci, C., G. Hoffman, and S. Cormier. 2010. *An Iterative Approach for Identifying the Causes of Reduced Benthic Macroinvertebrate Diversity in the Willimantic River, Connecticut*. Cincinnati, OH: U.S. Environmental Protection Agency, Office of Research and Development, National Center for Environmental Assessment. EPA/600/R-08/144.

Chambers, R. C. 2015. *Memorandum Opinion and Order: Ohio Valley Environmental Coalition, West Virginia Highlands Conservancy, and Sierra Club v. Fola Coal Company, LLC*. Civil Action No. 2:2013-21588 (consolidated with 2:2013-16044). Document 119 (S.D.W. Va. 2015). August 12, 2015. Charleston Division, U.S. District Court for the Southern District of West Virginia.

Ridley, C. E., S. D. Kaylor, S. S. Lee, J. Miller, N. P. Sam, and K. A. Schofield. 2024. *Application of Weight-of-Evidence Methods for Transparent and Defensible Numeric Nutrient Criteria*. Washington, DC: Office of Research and Development, U.S. Environmental Protection Agency.

USEPA. 2011. *A Field-Based Aquatic Life Benchmark for Conductivity in Central Appalachian Streams*. Washington, DC: U.S. Environmental Protection Agency. Office of Research and Development, National Center for Environmental Assessment.

Glossary

The terms and definitions in this glossary follow the USEPA (2016), but with some additional terms and edits to improve clarity.

Alteration: (1) A characteristic of causal relationships; it specifies that the affected entity is changed by physical, chemical, or other mechanisms leading to the endpoint effect. (2) A change in an entity that is consistent with an interaction with a cause.

Ambiguous: Evidence that has no clear meaning or more than one possible meaning. Evidence may be ambiguous because it is weak (e.g., shows no clear relationships) or unreliable (e.g., reference sites may have important extraneous differences from exposed sites).

Antecedence: A characteristic of causal relationships; it specifies that the causal interaction is itself connected to processes that precede it, potentially leading back to a source.

Assessment, environmental: (1) A process of generating and presenting scientific information to inform an environmental regulatory or management decision. (2) The product of an environmental assessment process.

Attribute: (1) Along with entity, one of two components of an assessment endpoint's definition. (2) A generic term that encompasses properties, qualities, implications, types, and characteristics.

Benchmark: A criterion, standard, screening value, effect threshold, or other value used to differentiate potential exposure levels that are of concern from levels that are not of concern.

Case: (1) The situation that is the subject of an environmental assessment; for example, the case may be a water body experiencing an algal bloom, a hazardous waste site, or a proposed pesticide use. (2) The weighted body of evidence for an assessment hypothesis; for example, "the lack of co-occurrence weakens the case for fine sediments as the cause."

Categorical Results: Results of weighing evidence expressed as standard categories such as the USEPA's categories of carcinogens: 1. carcinogenic to humans, 2. likely to be carcinogenic to humans, 3. suggestive evidence of carcinogenic potential, and 4. inadequate information to assess carcinogenic potential.

Categories of Evidence: Sets of evidence aggregated to reduce duplication during the weighting of evidence and to identify sets of evidence that are effectively replicates. Evidence is commonly categorized by type but may also be categorized by implication or characteristic.

Characteristics: Properties that define a quality of interest and that could be supported by evidence. For example, when weighing evidence for causation, evidence is sought to support six characteristics of causal relationships (Table B.1).

Coherence: The property of a body of evidence whereby its constituent pieces are logically linked together, thus forming a reasonable explanation.

Combination of evidence: Any method of inference that combines evidence without weighing it.

Confidence: The degree of support for a hypothesis provided by information. The weight of the evidence is a specific expression of confidence derived from the integration of multiple weighted pieces or types of evidence. For quantitative results, confidence is determined by the scatter of estimates as well as the weight of the evidence.

Consideration: A property, type, or implication of evidence used to guide the weighing of the evidence; Hill's considerations are the most common. Although they are often called criteria, the term considerations is more consistent with Hill's intent (Phillips and Goodman 2006).

Consilience: William Whewell originated this term to describe the unification of knowledge among the different branches of learning. For example, toxicology must be consilient with biochemistry.

Co-occurrence: (1) A characteristic of causal relationships; it specifies that the cause and effect are collocated in space and time at a scale appropriate to the cause and effect. (2) An instance of collocation in space and time.

Correspondence: The similarity of a piece or type of evidence to the entity or conditions to which the evidence will be applied. Evidence is relevant if it corresponds well to all aspects of the case (biological, physical/chemical, and environmental conditions).

Corroboration: Supporting evidence for an assessment proposition from one or more independent studies providing similar results.

Data: Unanalyzed results of measurements, counts, or observations used as a basis for reasoning or calculation.

Discrepant: Evidence that is inconsistent with or contrary to established facts or theory.

Endpoint, assessment: An explicit expression of the environmental or health qualities to be protected, operationally defined as an entity and its attributes.

Evaluation: The determination of how much influence should be assigned to a piece or category of evidence. Evaluation plus the rating of its results constitutes the weighting of evidence.

Evidence: Information that informs inferences regarding a condition, cause, prediction, or outcome.

Evidence, body of: All the applicable evidence used to make inferences concerning a proposition.

Evidence, category of: A grouping of evidence to facilitate weighting. In WoE for environmental assessments, evidence is categorized in terms of conventional *types of evidence* or in terms of characteristics that the evidence supports.

Evidence, heterogeneous: A body of evidence obtained from multiple types of studies.

Evidence, line of: (1) A complex piece of evidence including multiple causal or logical steps that establishes a line of reasoning. (2) A general term used to refer to a piece or type of evidence—this use is discouraged to reduce ambiguous terminology.

Evidence, piece of: The basic unit of evidence; examples include the results of a toxicity test or a stream survey.

Evidence, type of: A category of evidence that is based on information from a distinct form of study. Conventional types include biological surveys, biomarkers, ambient media toxicity tests, single-substance toxicity tests, population and ecosystem models, and quantitative structure-activity relationships.

Explanation: The process of translating pieces or types of evidence into a characteristic of causation or some other attribute that is relevant to a hypothesis. It is part of evidence integration in the weighing of the body of evidence step.

Expert judgment: The making of a finding or decision by recognized experts based on their knowledge, training, or experience in a relevant area of interest.

Hazard: The potentially adverse properties of an agent that, with potential exposure of a receptor, impart a risk.

Hypothesis: (1) A proposition proposed to be a potential explanation of a phenomenon (the fish kill was caused by high temperatures). (2) A potential outcome of a phenomenon (building more parking areas will increase maximum flows in streams).

Implication: The meaning of evidence with respect to a hypothesis. For example, evidence may imply that a relationship demonstrates spatial co-occurrence, that a particular process links two states in a conceptual model, or simply that a hypothesis is supported or weakened. Information becomes evidence when it has implications for a hypothesis.

Inference: (1) The act of reasoning from evidence. (2) A result of such reasoning.

Information: Data or other facts used to derive evidence.

Integration: The first step in the weighing of the body of evidence, in which the pieces or categories of evidence are combined to characterize the body of evidence.

Interaction: (1) A characteristic of causal relationships; it specifies that a causal agent impinges upon, enters, binds to, or initiates a response in a susceptible entity in a way that can lead to the effect of concern. (2) An instance of interaction of a causal agent with a susceptible entity that can lead to an effect.

Interpretation: The second step in the weighing of the body of evidence, in which the body of evidence for each hypothesis is compared to other hypotheses or judged against a standard to determine which conclusion is best supported by the weight of evidence.

IRIS (Integrated Risk Information System): A USEPA system for the assessment of the hazards to human health posed by chemicals and the derivation of toxic values.

ISA (Integrated Science Assessment): A USEPA assessment of the hazards posed by realistic exposures to a criteria air pollutant using weight of evidence.

Mechanism: A natural process or series of events by which a cause elicits a particular effect. It may be a sequence of altered states or a network of exchanged matter, energy, or information.

Meta-analysis: A set of statistical methods for combining the numerical results of multiple equivalent studies of the same problem to obtain a common judgment of a hypothesis or a numerical estimate. In WoE, it may be used to generate a piece of evidence by combining statistical results from multiple instances of an individual type of study.

Narrative: A common inferential method for WoE in which a prose summary of the evidence leads to a conclusion by professional judgment.

Property: A feature of evidence that determines how much weight it should be assigned. Examples of high-level properties of evidence include relevance, strength, and reliability.

Property, collective: A feature of bodies of evidence that, along with the properties of the constituent pieces or types of evidence, determines how much weight a body of evidence should be assigned. In this book, the collective properties are number, coherence, diversity, and absence of bias.

Proposition: A condition, cause, prediction, or outcome hypothesized as a possible result of an assessment.

Qualitative weight of evidence: Weight of evidence used to infer a quality such as causation, impairment, or carcinogenicity.

Quality of interest: A distinctive attribute of an entity, relationship, or system that should be determined to exist or not for a hypothesis in an assessment.

Quantitative weight of evidence: (1) Weight of evidence used to estimate a quantity such as a benchmark value or a biodegradation rate. (2) Weight of evidence that numerically weights the evidence and combines the weights arithmetically. We use only the first definition because the second definition is often misleading.

Rating: The process of formalizing the evaluation of evidence by assigning a symbol, word, or numeral. Synonyms include scoring and grading. The term scoring is generally considered inappropriate for weight of evidence because it implies quantification.

Reasonable explanation: (1) A statement or account that coherently explains a body of evidence. (2) Informed reasons for apparent inconsistencies in a body of evidence that provide coherence.

Refutation: The logical process of demonstrating the impossibility of a condition, cause, predicted effect, or outcome.

Relevance: A property of a piece or type of evidence that expresses the degree of correspondence between the evidence and the assessment endpoint to which it is applied.

Reliability: A property of evidence determined by the degree to which it has attributes that inspire confidence.

Risk: The chance of adverse effects associated with exposure to an agent.

Scatter: The distribution of measured or estimated values due to uncertainty, variability, or both.

Strength: A property of evidence determined by the degree of differentiation from randomness or from control, reference, or background conditions.

Sufficiency: (1) A characteristic of a causal relationship; it specifies that the causal agent or event is adequate to induce the effect in susceptible entities. (2) An occurrence of enough of an agent or process to affect a susceptible entity.

Systematic Review: A literature review that uses a defined procedure to search the literature, screen the results, extract information, and document the process. It may be used to create a summary of evidence or to support meta-analysis, weight of evidence, or some other inferential process.

Suitability: A property that determines whether a piece of evidence is appropriate for a particular case or class of cases, such as water quality criteria. The requirements for suitability are often specified in guidance documents.

Synthesis: The process of obtaining multiple pieces of information and drawing inferences from them by any of various processes, including weight of evidence, systematic review, and meta-analysis.

Table, evidence weighing: A table created in the weighing step that summarizes the bodies of evidence and their weights for one or more hypotheses.

Table, evidence weighting: A table created in the weighting step that presents the pieces or categories of evidence and their ratings with respect to relevant properties.

Table, logic: A table that logically derives conclusions from weights of standard sets of types of evidence, particularly the sediment quality triad.

Time order: (1) A characteristic of a causal relationship; it specifies that the cause precedes the effect. (2) The sequence, in time, of the occurrence of a candidate cause and the effect of concern. It is sometimes called temporality or temporal sequence.

Type of evidence: See Evidence, type of.

Type of study: Conventional categories of scientific investigation such as whole-organism toxicity tests, epidemiological studies, biomarker monitoring, in vitro tests, and exposure modeling.

Uncertainty: Lack of knowledge concerning the state of an organism or system or concerning the value of a quantity. Uncertainty is a property of the investigator and, unlike variability, may be reduced by measurement or observation.

Variability: Heterogeneity over time, space, or among members of a population. Variability is a property of nature and may not be reduced by measurement or observation.

Weigh: The process of considering the properties of the body of evidence to assess confidence in a hypothesis.

Weight: (noun) The importance to an inference of a piece or category of evidence. (verb) To assign an importance descriptor or rating to a piece or category of evidence.

Vote counting: Making an inference by counting and comparing the number of pieces of evidence that support each hypothesis. For example, comparing the number of tests that show increased deformities to the number of tests that do not.

Weight of evidence: (1) A process of making inferences from multiple pieces of evidence, adapted from the legal metaphor of the scales of justice. (2) The relative degree of support for a conclusion provided by evidence; the result of weighing the body of evidence.

REFERENCES

Phillips, C. V., and K. J. Goodman. 2006. "Causal criteria and counterfactuals; nothing more (or less) than scientific common sense." *Emerging Themes in Epidemiology* 3 (1): 5.

USEPA. 2016. *Weight of Evidence in Ecological Assessment*. Washington, DC: Risk Assessment Forum, U.S. Environmental Protection Agency.

Index

Note: **Bold** page numbers refer to tables and *italic* page numbers refer to figures.

For Product Safety Concerns and Information please contact our EU
representative GPSR@taylorandfrancis.com
Taylor & Francis Verlag GmbH, Kaufingerstraße 24, 80331 München, Germany

www.ingramcontent.com/pod-product-compliance
Ingram Content Group UK Ltd.
Pitfield, Milton Keynes, MK11 3LW, UK
UKHW051001151026
476466UK00023BA/66
* 9 7 8 1 0 4 1 0 3 9 9 8 3 *